Richard Talman

Geometric Mechanics

1807–2007 Knowledge for Generations

Each generation has its unique needs and aspirations. When Charles Wiley first opened his small printing shop in lower Manhattan in 1807, it was a generation of boundless potential searching for an identity. And we were there, helping to define a new American literary tradition. Over half a century later, in the midst of the Second Industrial Revolution, it was a generation focused on building the future. Once again, we were there, supplying the critical scientific, technical, and engineering knowledge that helped frame the world. Throughout the 20th Century, and into the new millennium, nations began to reach out beyond their own borders and a new international community was born. Wiley was there, expanding its operations around the world to enable a global exchange of ideas, opinions, and know-how.

For 200 years, Wiley has been an integral part of each generation's journey, enabling the flow of information and understanding necessary to meet their needs and fulfill their aspirations. Today, bold new technologies are changing the way we live and learn. Wiley will be there, providing you the must-have knowledge you need to imagine new worlds, new possibilities, and new opportunities.

Generations come and go, but you can always count on Wiley to provide you the knowledge you need, when and where you need it!

William J. Pesce
President and Chief Executive Officer

Peter Booth Wiley
Chairman of the Board

Richard Talman

Geometric Mechanics

Toward a Unification of Classical Physics

Second, Revised and Enlarged Edition

BICENTENNIAL 1807 WILEY 2007 BICENTENNIAL

WILEY-VCH Verlag GmbH & Co. KGaA

The Author

Prof. Richard Talman
Cornell University
Laboratory of Elementary Physics
Ithaca, NY 14853
USA

talman@mail.lepp.cornell.edu

Library of Congress Card No.:
applied for

British Library Cataloguing-in-Publication Data
A catalogue record for this book is available from the British Library.

Bibliographic information published by the Deutsche Nationalbibliothek
The Deutsche Nationalbibliothek lists this publication in the Deutsche Nationalbibliografie; detailed bibliographic data are available in the Internet at <http://dnb.d-nb.de>.

Composition Uwe Krieg, Berlin
Printing Strauss GmbH, Mörlenbach
Binding Litges & Dopf Buchbinderei GmbH, Heppenheim
Wiley Bicentennial Logo Richard J. Pacifico

Printed in the Federal Republic of Germany
Printed on acid-free paper

ISBN: 978-3-527-40683-8

Contents

Geometric Mechanics: Toward a Unification of Classical Physics. 2nd Edition. Richard Talman

ISBN: 978-3-527-40683-8

Preface

This text is designed to accompany a junior/senior or beginning graduate student course in mechanics for students who have already encountered Lagrange's equations. As the title *Geometric Mechanics* indicates, the content is classical mechanics, with emphasis on geometric methods, such as differential geometry, tensor analysis, and group theory. Courses for which the material in the text has been used and is appropriate are discussed in the Introduction. To reflect a substantial new emphasis in this second edition, compared to the first, the subtitle "Toward a Unification of Classical Physics" has been added. Instead of just laying the groundwork for follow-on, geometry-based, physics subjects, especially general relativity and string theory, this edition contains substantial introductions to both of those topics. To support this, introductory material on classical field theory, including electrodynamic theory (also formulated as mechanics) has been included. The purpose of these "other physics" chapters is to show how, based on Hamilton's principle of least action, *all*, or at least most, of classical physics is naturally subsumed into classical mechanics.

Communications pointing out errors, or making comments or suggestions will be appreciated; E-mail address; talman@mail.lepp.cornell.edu. Because of its complete reorganization, there are undoubtedly more minor errors and dangling references than might be expected for a second edition.

The institutions contributing (in equal parts) to this text have been the public schools of London, Ontario, and universities U.W.O., Caltech, and Cornell. I have profited, initially as a student, and later from my students, at these institutions, and from my colleagues there and at accelerator laboratories worldwide. I have also been fortunate of family; parents, brother, children, and, especially my wife, Myrna.

Ithaca, New York
May, 2007

Richard Talman

Geometric Mechanics: Toward a Unification of Classical Physics. 2nd Edition. Richard Talman

ISBN: 978-3-527-40683-8

Introduction

The first edition of this text was envisaged as a kind of *Mathematical Methods of Classical Mechanics for Pedestrians*, with geometry playing a more important role than in the traditional pedagogy of classical mechanics. Part of the rationale was to prepare the student for subsequent geometry-intensive physics subjects, especially general relativity. Subsequently I have found that, as a text for physics courses, this emphasis was somewhat misplaced. (Almost by definition) students of physics want to learn "physics" more than they want to learn "applied mathematics." Consistent with this, there has been a tendency for classical mechanics to be squeezed out of physics curricula in favor of general relativity or, more recently, string theory. This second edition has been revised accordingly. Instead of just laying the groundwork for subjects such as electromagnetic theory, string theory, and general relativity, it subsumes these subjects into classical mechanics. After these changes, the text has become more nearly a *Classical Theory of Fields for Pedestrians*.

Geometric approaches have contributed importantly to the evolution of modern physical theories. The best example is general relativity; the most modern example is string theory. In fact general relativity and string theory are the theories for which the adjective "geometric" is most unambiguously appropriate. There is now a chapter on each of these subjects in this text, along with material on (classical) field theory basic to these subjects. Also, because electromagnetic theory fits the same template, and is familiar to most students, that subject is here also formulated as a "branch" of classical mechanics.

In grandiose terms, the plan of the text is to arrogate to classical mechanics *all* of *classical* physics, where "classical" means nonquantum-mechanical and "all" means old-fashioned classical mechanics plus the three physical theories mentioned previously. Other classical theories, such as elasticity and hydrodynamics, can be regarded as having already been subsumed into classical mechanics, but they lie outside the scope of this text.

In more technical terms, the theme of the text is that *all* of classical physics starts from a Lagrangian, continues with Hamilton's principle (also known

Geometric Mechanics: Toward a Unification of Classical Physics. 2nd Edition. Richard Talman

ISBN: 978-3-527-40683-8

as the principle of least action) and finishes with solving the resultant equations and comparison with experiment. This program provides a *unification* of classical physics. General principles, especially symmetry and special relativity, limit the choices surprisingly when any new term is to be added to the Lagrangian. Once a new term *has* been added the entire theory and its predictions are predetermined. These results can then be checked experimentally. The track record for success in this program has been astonishingly good. As far as classical physics is concerned the greatest triumphs are due to Maxwell and Einstein. The philosophic basis for this approach, apparently espoused by Einstein, is not that we live in the best of all possible worlds, but that we live in the *only* possible world. Even people who find this philosophy silly find that you don't have to subscribe to this philosophy for the approach to work well.

There is an ambitious program in quantum field theory called "grand unification" of the four fundamental forces of physics. The present text can be regarded as preparation for this program in that it describes classical physics in ways consistent with this eventual approach. As far as I know, any imagined grand unification scheme will, when reduced to the classical level, resemble the material presented here. (Of course most of the *essence* of the physics is quantum mechanical and cannot survive the reduction to classical physics.)

Converting the emphasis from applied mathematics to pure physics required fewer changes to the text than might be supposed. Much of the earlier book emphasized specialized mathematics and computational descriptions that could be removed to make room for the "physics" chapters already mentioned. By no means does this mean that the text has been gutted of practical worked examples of classical mechanics. For example, most of the long chapters on perturbation theory and on the application of adiabatic invariants (both of which are better thought of as physics than as mathematics) have been retained. All of the (admittedly unenthusiastic) discussion of canonical transformation methods has also been retained.

Regrettably, some material on the boundary between classical and quantum mechanics has had to be dropped. As well as helping to keep the book length within bounds, this deletion was consistent with religiously restricting the subject matter to *nothing but* classical physics. There was a time when classical Hamiltonian mechanics seemed like the best introduction to quantum mechanics but, like the need to study Latin in school, that no longer seems to be the case. Also, apart from its connections to the Hamilton–Jacobi theory (which every educated physicist has to understand) quantum mechanics is not very geometric in character. It was relatively painless therefore, to remove unitary geometry, Bragg scattering (illustrating the use of covariant tensors), and other material on the margin between classical and quantum mechanics.

In this book's first manifestation the subject of mechanics was usefully, if somewhat artificially, segmented into Lagrangian, Hamiltonian, and Newtonian formulations. Much was made of Poincaré's extension to the Lagrangian approach. Because this approach advances the math more than the physics, it now has had to be de-emphasized (though most of the material remains). On the other hand, as mentioned already, the coverage of Lagrangian field theory, and especially its conservation laws, needed to be expanded. Reduced weight also had to be assigned to Hamiltonian methods (not counting Hamilton's principle.) Those methods provide the most direct connections to quantum mechanics but, with quantum considerations now being ignored, they are less essential to the program. Opposite comments apply to Newtonian methods, which stress fictitious forces (centrifugal and Coriolis), ideas that led naturally to general relativity. Gauge invariant methods, which play such an important role in string theory, are also naturally introduced in the context of direct Newtonian methods. The comments in this paragraph, taken together, repudiate much of the preface to the first edition which has, therefore, been discarded.

Everything contained in this book is explained with more rigor, or more depth, or more detail, or (especially) more sophistication, in at least one of the books listed at the end of this introduction. Were it not for the fact that most of those books are fat, intimidating, abstract, formal, mathematical and (for many) unintelligible, the reader's time would be better spent reading them (in the right order) than studying this book. But if this text renders books like these both accessible and admirable, it will have achieved its main purpose. It has been said that bridge is a simple game; dealt thirteen cards, one has only to play them in the correct order. In the same sense mechanics is easy to learn; one simply has to study readily available books in a sensible order. I have tried to chart such a path, extracting material from various sources in an order that I have found appropriate. At each stage I indicate (at the end of the chapter) the reference my approach most closely resembles. In some cases what I provide is a kind of *Reader's Digest* of a more general treatment and this may amount to my having systematically specialized and made concrete, descriptions that the original author may earlier have systematically labored to generalize and make abstract. The texts to which these statements are most applicable are listed at the end of each chapter, and keyed to the particular section to which they relate. It is not suggested that these texts should be systematically referred to as they tend to be advanced and contain much unrelated material. But if particular material in this text is obscure, or seems to stop short of some desirable goal, these texts should provide authoritative help.

Not very much is original in the text other than the selection and arrangement of the topics and the style of presentation. Equations (though not text) have been "borrowed," in some cases verbatim, from various sources. This is especially true of Chapters 9, on special relativity, 11, on electromagnetic theory, and 13, on general relativity. These chapters follow Landau and Lifschitz quite closely. Similarly, Chapter 12 follows Zwiebach closely. There are also substantial sections following Cartan, or Arnold, or others. As well as occasional reminders in the text, of these sources, the bibliography at the end of each chapter lists the essential sources. Under "General References" are books, like the two just mentioned, that contain one or more chapters discussing much the same material in at least as much, and usually far more, detail, than is included here. These references could be used *instead of* the material in the chapter they are attached to, and *should* be used to go deeper into the subject. Under "References for Further Study" are sources that can be used *as well as* the material of the chapter. In principle, none of these references should actually be necessary, as the present text is supposed to be self-sufficient. In practice, obscurities and the likelihood of errors or misunderstandings, make it appropriate, or even necessary, to refer to other sources to obtain anything resembling a deep understanding of a topic.

The mathematical level strived for is only high enough to support a persuasive (to a nonmathematician) trip through the physics. Still, "it can be shown that" almost never appears, though the standards of what constitutes "proof" may be low, and the range of generality narrow. I believe that much mathematics is made difficult for the less-mathematically-inclined reader by the absence of concrete instances of the abstract objects under discussion. This text tries to provide essentially correct instances of otherwise hard to grasp mathematical abstractions. I hope and believe that this will provide a broad base of general understanding from which deeper, more specialized, more mathematical texts can be approached with a respectable general comprehension. This statement is most applicable to the excellent books by *Arnold*, who tries hard, but not necessarily successfully, to provide physical lines of reasoning. Much of this book was written with the goal of making one or another of his discussions comprehensible.

In the early days of our weekly Laboratory of Nuclear Studies Journal Club, our founding leader, Robert Wilson, imposed a rule – though honored as much in the breach as in the observance, it was not intended to be a joke – that the Dirac γ-matrices never appear. The (largely unsuccessful) purpose of this rule was to force the lectures to be intelligible to us theory-challenged experimentalists. In this text there is a similar rule. It is that hieroglyphics such as

$$\phi : \{x \in R^2 : |x| = 1\} \to R$$

not appear. The justification for this rule is that a "physicist" is likely to skip such a statement altogether or, once having understood it, regard it as obvious. Like the jest that the French "don't care what they say as long as they pronounce it properly" one can joke that mathematicians don't care what their mappings do, as long as the spaces they connect are clear. Physicists, on the other hand, care primarily what their functions represent physically and are not fussy about what spaces they relate. Another "rule" has just been followed; the word *function* will be used in preference to the (synonymous) word *mapping*. Other terrifying mathematical words such as *flow*, *symplectomorphism*, and *manifold* will also be avoided except that, to avoid long-winded phrases such as "configuration space described by generalized coordinates," the word *manifold* will occasionally be used. Of course one cannot alter the essence of a subject by denying the existence of mathematics that is manifestly at its core. In spite of the loss of precision, I hope that sugar-coating the material in this way will make it more easily swallowed by nonmathematicians.

Notation: "Notation isn't everything, it's the only thing." Grammatically speaking, this statement, like the American football slogan it paraphrases, makes no sense. But its clearly intended meaning is only a mild exaggeration. After the need to evaluate some quantity has been expressed, a few straightforward mathematical operations are typically all that is required to obtain the quantity. But specifying quantities is far from simple. The conceptual depth of the subject is substantial and ordinary language is scarcely capable of defining the symbols, much less expressing the relations among them. This makes the introduction of sophisticated symbols essential. Discussion of notation and the motivation behind its introduction is scattered throughout this text – probably to the point of irritation for some readers. Here we limit discussion to the few most important, most likely to be confusing, and most deviant from other sources: the *qualified equality* $\stackrel{q}{=}$, *the vector*, the *preferred reference system*, the *active/passive interpretation* of transformations, and the terminology of *differential forms*.

A fairly common occurrence in this subject is that two quantities A and B are equal or equivalent from one point of view but not from another. This circumstance will be indicated by "qualified equality" $A \stackrel{q}{=} B$. This notation is intentionally vague (the "q" stands for qualified, or questionable, or query? as appropriate) and may have different meanings in different contexts; it only warns the reader to be wary of the risk of jumping to unjustified conclusions. Normally the qualification will be clarified in the subsequent text.

Next vectors. Consider the following three symbols or collections of symbols: $\longrightarrow$, $\mathbf{x}$, and $(x, y, z)^T$. The first, $\longrightarrow$, will be called an arrow (because it is one) and this word will be far more prevalent in this text than any other of which I am aware. This particular arrow happens to be pointing in a horizontal direction (for convenience of typesetting) but in general an arrow can point in any direction, including out of the page. The second, bold face, quantity, $\mathbf{x}$, is an *intrinsic* or *true* vector; this means that it is a symbol that "stands for" an arrow. The word "intrinsic" means "it doesn't depend on choice of coordinate system." The third quantity, $(x, y, z)^T$, is a column matrix (because the T stands for transpose) containing the "components" of $\mathbf{x}$ relative to some pre-established coordinate system. From the point of view of elementary physics these three are equivalent quantities, differing only in the ways they are to be manipulated; "addition" of arrows is by ruler and compass, addition of intrinsic vectors is by vector algebra, and addition of coordinate vectors is component wise. Because of this multiplicity of meanings, the word "vector" is ambiguous in some contexts. For this reason, we will often use the word arrow in situations where independence of choice of coordinates is being emphasized (even in dimensionality higher than 3.) According to its definition above, the phrase *intrinsic vector* could usually replace *arrow*, but some would complain of the redundancy, and the word *arrow* more succinctly conveys the intended geometric sense. Comments similar to these could be made concerning higher order *tensors* but they would be largely repetitive.

A virtue of arrows is that they can be plotted in figures. This goes a long way toward making their meaning unambiguous but the conditions defining the figure must still be made clear. In classical mechanics "inertial frames" have a fundamental significance and we will almost always suppose that there is a "preferred" reference system, its rectangular axes fixed in an inertial system. Unless otherwise stated, figures in this text are to be regarded as "snapshots" taken in that frame. In particular, a plotted arrow connects two points fixed in the inertial frame at the instant illustrated. As mentioned previously, such an arrow is symbolized by a *true* vector such as $\mathbf{x}$.

It is, of course, essential that these vectors satisfy the algebraic properties defining a vector space. In such spaces "transformations" are important; a "linear" transformation can be represented by a matrix symbolized, for example, by $\mathbf{M}$, with elements $M^i{}_j$. The result of applying this transformation to vector $\mathbf{x}$ can be represented symbolically as the "matrix product" $\mathbf{y} \stackrel{q}{=} \mathbf{M}\,\mathbf{x}$ of "intrinsic" quantities, or spelled out explicitly in components $y^i = M^i{}_j x^j$. Frequently both forms will be given. This leads to a notational difficulty in distinguishing between the "active" and "passive" interpretations of the transformation. The new components y^i can belong to a new arrow in the old frame (active interpretation) or to the old arrow in a new frame (passive interpretation). On the other hand, the *intrinsic* form $\mathbf{y} \stackrel{q}{=} \mathbf{M}\,\mathbf{x}$ *seems* to support only an active interpretation according to which $\mathbf{M}$ "operates" on vector $\mathbf{x}$ to yield a different vector $\mathbf{y}$. To avoid this problem, when we wish to express a passive interpretation we will ordinarily use the form $\overline{\mathbf{x}} \stackrel{q}{=} \mathbf{M}\,\mathbf{x}$ and will insist that $\mathbf{x}$ and $\overline{\mathbf{x}}$ stand for *the same* arrow. The significance of the overhead bar then is that $\overline{\mathbf{x}}$ is simply an abbreviation for an array of barred-frame coordinates $\overline{x}^i$. When the active interpretation is intended the notation will usually be expanded to clarify the situation. For example, consider a moving point located initially at $\mathbf{r}(0)$ and at $\mathbf{r}(t)$ at later time t. These vectors can be related by $\mathbf{r}(t) = \mathbf{O}(t)\,\mathbf{r}(0)$ where $\mathbf{O}(t)$ is a time-dependent operator. This is an *active* transformation.

The beauty and power of vector analysis as it is applied to physics is that a bold face symbol such as **V** indicates that the quantity is *intrinsic* and also abbreviates its multiple components V^i into one symbol. Though these are both valuable purposes, they are *not* the same. The abbreviation works in vector analysis only because vectors are the only multiple component objects occurring. That this will no longer be the case in this book will cause considerable notational difficulty because the reader, based on experience with vector analysis, is likely to jump to unjustified conclusions concerning bold face quantities.[1] We will not be able to avoid this problem however since we wish to retain familiar notation. Sometimes we will be using bold face symbols to indicate intrinsically, sometimes as abbreviation, and sometimes both. Sometimes the (redundant) notation $\vec{\mathbf{v}}$ will be used to *emphasize* the intrinsic aspect. Though it may not be obvious at this point, notational insufficiency was the source of the above-mentioned need to differentiate verbally between *active* and *passive* transformations. In stressing this distinction the text differs from a text such as Goldstein that, perhaps wisely, de-emphasizes the issue.

According to Arnold "it is impossible to understand mechanics without the use of differential forms." Accepting the validity of this statement only grudgingly (and trying to corroborate it) but knowing from experience that typical physics students are innocent of any such knowledge, a considerable portion of the text is devoted to this subject. Briefly, the symbol dx will stand for an old-fashioned differential displacement of the sort familiar to every student of physics. But a new quantity $\widetilde{\mathbf{dx}}$ to be known as a differential form, will also be used. This symbol is distinguished from dx both by being bold face and having an overhead tilde. Displacements $dx^1, dx^2, \ldots$ in spaces of higher dimension will have matching forms $\widetilde{\mathbf{dx}}^1, \widetilde{\mathbf{dx}}^2, \ldots$. This notation is mentioned at this point only because it is unconventional. In most treatments one or the other form of differential is used, but not both at the same time. I have found it impossible to cause classical formulations to morph into modern formulations without this distinction (and others to be faced when the time comes.)

It is hard to avoid using terms whose meanings are vague. (See the previous paragraph, for example.) I have attempted to acknowledge such vagueness, at least in extreme cases, by placing such terms in quotation marks when they are first used. Since quotation marks are also used when the term is actually being defined, a certain amount of hunting through the surrounding sentences may be necessary to find if a definition is actually present. (If it is not clear whether or not there is a definition then the term is *without any doubt* vague.) Italics are used to emphasize key phrases, or pairs of phrases in opposition, that are

1) Any computer programmer knows that, when two logically distinct quantities have initially been given the same symbol, because they are expected to remain equal, it is hard to unscramble the code when later on it becomes necessary to distinguish between the two usages.

central to the discussion. Parenthesized sentences or sentence fragments are supposedly clear only if they are included *right there* but they should not be allowed to interrupt the logical flow of the surrounding sentences. Footnotes, though sometimes similar in intent, are likely to be real digressions, or technical qualifications or clarifications.

The text contains at least enough material for a full year course and far more than can be covered in any single term course. At Cornell the material has been the basis for several distinct courses: (a) Junior/senior level classical mechanics (as that subject is traditionally, and narrowly, defined.) (b) First year graduate classical mechanics with geometric emphasis. (c) Perturbative and adiabatic methods of solution and, most recently, (d) "Geometric Concepts in Physics." Course (d) was responsible for the math/physics reformulation of this edition. The text is best matched, therefore, to filling a curricular slot that allows variation term-by-term or year-by-year.

Organization of the book: Chapter 1, containing review/examples, provides appropriate preparation for any of the above courses; it contains a brief overview of elementary methods that the reader may wish (or need) to review. Since the formalism (primarily Lagrangian) is assumed to be familiar, this review consists primarily of examples, many worked out partially or completely. Chapter 2 and the first half of Chapter 3 contain the geometric concepts likely to be both "new" and needed. The rest of Chapter 3 as well as Chapter 4 contain geometry that can be skipped until needed. Chapters 5, 6, 7, and 8 contain, respectively, the Lagrangian, Newtonian, Hamiltonian and Hamilton–Jacobi, backbone of course labeled (a) above. The first half of Chapter 10, on conservation laws, is also appropriate for such a course, and methods of solution should be drawn from Chapters 14, 15, and 16.

The need for relativistic mechanics is what characterizes Chapters 9, 11, 12, and 13. These chapters can provide the "physics" content for a course such as (d) above. The rest of the book does not depend on the material in these chapters. A course should therefore include either none of this material or all of it, though perhaps emphasizing either, but not both, of general relativity and string theory.

Methods of solution are studied in Chapters 14, 15, and 16. These chapters would form an appreciable fraction of a course such as (c) above.

Chapter 17 is concerned mainly with the formal structure of mechanics in Hamiltonian form. As such it is most likely to be of interest to students planning to take a subsequent courses in dynamical systems, chaos, plasma or accelerator physics. Somehow the most important result of classical mechanics – Liouville's theorem – has found its way to the last section of the book.

The total number of problems has been almost doubled compared to the first edition. However, in the chapters covering areas of physics not traditionally regarded as classical mechanics, the problems are intended to require no special knowledge of those subjects beyond what is covered in this text.

Some Abbreviations Used in This Text

E.D.	exterior derivative
B.C.	bilinear covariant
O.P.L.	optical path length
I.I.	integral invariant
H.I.	Hamiltonian variational line integral
L.I.I.	Lagrange invariant integral
R.I.I.	relative integral invariant

Bibliography

General Mechanics Texts

1 V.I. Arnold, *Mathematical Methods of Classical Mechanics*, Springer, New York, 1978.

2 N.G. Chetaev, *Theoretical Mechanics*, Springer, Berlin, 1989.

3 H. Goldstein, *Classical Mechanics*, Addison-Wesley, Reading, MA, 1980.

4 L.D. Landau and E.M. Lifshitz, *Mechanics*, Pergamon, Oxford, 1976.

5 L.A. Pars, *Analytical Dynamics*, Ox Bow Press, Woodbridge, CT, 1979.

6 K.R. Symon, *Mechanics*, Addison-Wesley, Reading, MA, 1971.

7 D. Ter Haar, *Elements of Hamiltonian Mechanics*, 2nd ed., Pergamon, Oxford, 1971.

8 E.T. Whittaker, *Treatise on the Analytical Dynamics of Particles and Rigid Bodies*, Cambridge University Press, Cambridge, UK, 1989

Specialized Mathematical Books on Mechanics

9 V.I. Arnold, V.V. Kozlov, and A.I. Neishtadt, *Dynamical Systems III*, Springer, Berlin, 1980.

10 J.E. Marsden, *Lectures on Mechanics*, Cambridge University Press, Cambridge, UK, 1992.

11 K.R. Meyer and R. Hall, *Introduction to Hamiltonian Dynamical Systems and the N-Body Problem*, Springer, New York, 1992.

Relevant Mathematics

12 E. Cartan, *The Theory of Spinors*, Dover, New York, 1981.

13 E. Cartan, *Leçons sur la géometrie des espaces de Riemann*, Gauthiers-Villars, Paris, 1951. (English translation available.)

14 B.A. Dubrovin, A.T. Fomenko, and S.P. Novikov, *Modern Geometry I*, Springer, Berlin, 1985

15 H. Flanders, *Differential Forms With Applications to the Physical Sciences*, Dover, New York, 1989

16 D.H. Sattinger and O.L. Weaver, *Lie Groups and Algebras, Applications to Physics, Geometry, and Mechanics*, Springer, New York, 1986

17 B.F. Schutz, *Geometrical Methods of Mathematical Physics*, Cambridge University Press, Cambridge, UK, 1980

18 V.A. Yakubovitch and V.M. Starzhinskii, *Linear Differential Equations With Periodic Coefficients*, Wiley, New York, 1975

Physics

19 L.D. Landau and E.M. Lifshitz, *The Classical Theory of Fields*, Pergamon, Oxford, 1975.

20 S. Weinberg, *Gravitation and Cosmology*, Wiley, New York, 1972.

21 B. Zwiebach, *A First Course in String Theory*, Cambridge University Press, Cambridge, UK, 2004.

1
Review of Classical Mechanics and String Field Theory

1.1
Preview and Rationale

This introductory chapter has two main purposes. The first is to review Lagrangian mechanics. Some of this material takes the form of worked examples, chosen both to be appropriate as examples and to serve as bases for topics in later chapters.

The second purpose is to introduce the mechanics of classical strings. This topic is timely, being introductory to the modern subject of (quantum field theoretical) string theory. But, also, the Lagrangian theory of strings is an appropriate area in which to practice using supposedly well-known concepts and methods in a context that is encountered (if at all) toward the end of a traditional course in intermediate mechanics. This introduces the topic of Lagrangian field theory in a well-motivated and elementary way. Classical strings have the happy properties of being the simplest system for which Lagrangian field theory is appropriate.

The motivation for emphasizing strings from the start comes from the daring, and apparently successful, introduction by Barton Zwiebach, of string theory into the M.I.T. undergraduate curriculum. This program is fleshed out in his book *A First Course in String Theory*. The present chapter, and especially Chapter 12 on relativistic strings, borrows extensively from that text. Unlike Zwiebach though, the present text stops well short of quantum field theory.

An eventual aim of this text is to *unify* "all" of classical physics within suitably generalized Lagrangian mechanics. Here "all" will be taken to be adequately represented by the following topics: mechanics of particles, special relativity, electromagnetic theory, classical (and, eventually, relativistic) string theory, and general relativity. This list, which is to be regarded as defining by example what constitutes "classical physics," is indeed ambitious, though it leaves out many other important fields of classical physics, such as elasticity and fluid dynamics.[1] The list also includes enough varieties of geometry

1) By referring to a text such as *Theoretical Mechanics of Particles and Continua*, by Fetter and Walecka, which covers fluids and elastic solids in very much the same spirit as in the present text, it should be clear that these two topics can also be included in the list of fields *unified* by Lagrangian mechanics.

Geometric Mechanics: Toward a Unification of Classical Physics. 2nd Edition. Richard Talman

ISBN: 978-3-527-40683-8

to support another aim of the text, which is to illuminate the important role played by geometry in physics.

An introductory textbook on Lagrangian mechanics (which this is not) might be expected to begin by announcing that the reader is assumed to be familiar with Newtonian mechanics – kinematics, force, momentum and energy and their conservation, simple harmonic motion, moments of inertia, and so on. In all likelihood such a text would then proceed to review these very same topics before advancing to its main topic of Lagrangian mechanics. This would not, of course, contradict the original assumption since, apart from the simple pedagogical value of review, it makes no sense to study Lagrangian mechanics without anchoring it firmly in a Newtonian foundation. The student who had not learned this material previously would be well advised to start by studying a less advanced, purely Newtonian mechanics textbook. So many of the most important problems of physics can be solved cleanly without the power of Lagrangian mechanics; it is uneconomical to begin with an abstract formulation of mechanics before developing intuition better acquired from a concrete treatment. One might say that Newtonian methods give better "value" than Lagrangian mechanics because, though ultimately less powerful, Newtonian methods can solve the most important problems and are easier to learn. Of course this would only be true in the sort of foolish system of accounting that might attempt to rate the relative contributions of Newton and Einstein. One (but not the only) purpose of this textbook, is to go beyond Lagrange's equations. By the same foolish system of accounting just mentioned, these methods could be rated less valuable than Lagrangian methods since, though more powerful, they are more abstract and harder to learn.

It is assumed the reader has had some (not necessarily much) experience with Lagrangian mechanics.[2] Naturally this presupposes familiarity with the above-mentioned elementary concepts of Newtonian mechanics. Nevertheless, for the same reasons as were described in the previous paragraph, we start by reviewing material that is, in principle, already known. It is assumed the reader can define a Lagrangian, can write it down for a simple mechanical system, can write down (or copy knowledgeably) the Euler–Lagrange equations and from them derive the equations of motion of the system, and finally (and most important of all) trust these equations to the same extent that she or he trusts Newton's law itself. A certain (even if grudging) acknowledgement of the method's power to make complicated systems appear simple is also helpful. Any reader unfamiliar with these ideas would be well advised

2) Though "prerequisites" have been mentioned, this text still attempts to be "not too advanced." Though the subject matter deviates greatly from the traditional curriculum at this level (as represented, say, by Goldstein, *Classical Mechanics*) it is my intention that the level of difficulty and the anticipated level of preparation be much the same as is appropriate for Goldstein.

to begin by repairing the defect with the aid of one of the numerous excellent textbooks explaining Lagrangian mechanics.

Since a systematic review of Newtonian and Lagrangian mechanics would be too lengthy, this chapter starts with worked examples that illustrate the important concepts. To the extent possible, examples in later chapters are based on these examples. This is especially appropriate for describing the evolution of systems that are close to solvable systems.

1.2
Review of Lagrangians and Hamiltonians

Recall the formulas of Lagrangian mechanics. For the next few equations, for mnemonic purposes, each equation will be specialized (sometimes in parenthesis) to the simplest prototype, mass and spring. The kinetic and potential energies for this system are given by

$$T = \frac{1}{2} m\dot{x}^2, \quad V = \frac{1}{2} k x^2, \tag{1.1}$$

where $\dot{x} \equiv dx/dt \equiv v$. The Lagrangian, a function of x and $\dot{x}$ (and, in general though not in this special case, t) is given by

$$L(x, \dot{x}, t) = T - V \quad \left(= \frac{1}{2} m\dot{x}^2 - \frac{1}{2} k x^2 \right). \tag{1.2}$$

The Lagrange equation is

$$\frac{d}{dt}\frac{\partial L}{\partial \dot{x}} = \frac{\partial L}{\partial x} \quad \left(\text{or} \quad m\ddot{x} = -kx \right). \tag{1.3}$$

The momentum p, "canonically conjugate to x," is defined by

$$p = \frac{\partial L}{\partial \dot{x}} \quad (= m\dot{x}). \tag{1.4}$$

The Hamiltonian is derived from the Lagrangian by a transformation in which both independent and dependent variables are changed. This transformation is known as a "Legendre transformation." Such a transformation has a geometric interpretation,[3] but there is no harm in thinking of it as purely a formal calculus manipulation. Similar manipulations are common in thermodynamics to define quantities that are constant under special circumstances. For a function $L(x, v, t)$, one defines a new independent variable $p = \partial L/\partial v$ and a new function $H(x, p, t) = vp - L(x, v, t)$, in which v has to be expressed in

3) The geometric interpretation of a Legendre transformation is discussed in Arnold, *Mathematical Methods of Classical Mechanics*, and Lanczos, *The Variational Principles of Mechanics*.

terms of x and p by inverting $p = \partial L/\partial v$. The motivation behind this definition is to produce cancellation of second and fourth terms in the differential

$$\begin{aligned} dH &= v\,dp + p\,dv - \frac{\partial L}{\partial x}\,dx - \frac{\partial L}{\partial v}\,dv \\ &= v\,dp - \frac{\partial L}{\partial x}\,dx. \end{aligned} \tag{1.5}$$

Applying these substitutions to our Lagrangian, with v being $\dot{x}$, one obtains the "Hamiltonian" function,

$$H(x,p,t) = p\,\dot{x}(x,p) - L(x,\dot{x}(x,p),t). \tag{1.6}$$

With (1.5) being the differential of this function, using Eq. (1.4), one obtains Hamilton's equations;

$$\dot{x} = \frac{\partial H}{\partial p}, \quad \dot{p} = -\frac{\partial H}{\partial x}, \quad \frac{\partial H}{\partial t} = -\frac{\partial L}{\partial t}. \tag{1.7}$$

The third equation here, obvious from Eq. (1.6), has been included for convenience, especially in light of the following argument. As well as its formal role, as a function to be differentiated to obtain the equations of motion, the Hamiltonian $H(x,p,t)$ can be evaluated for the actually evolving values of its arguments. This evolution of H is governed by

$$\dot{H} = \frac{\partial H}{\partial x}\dot{x} + \frac{\partial H}{\partial p}\dot{p} + \frac{\partial H}{\partial t} = \frac{\partial H}{\partial t}, \tag{1.8}$$

where Eqs. (1.7) were used in the final step. This equation implies that the absence of explicit dependence on t implies the constancy of H.

To be able to apply Hamiltonian mechanics it is necessary to be able to express $\dot{x}$ as a function of p – trivial in our example;

$$\dot{x} = \frac{p}{m}, \tag{1.9}$$

and to express the combination $\dot{x}p - L(x,\dot{x})$ in terms of x and p, thereby defining the Hamiltonian;

$$H(x,p) = \frac{p^2}{m} - L = \frac{p^2}{2m} + \frac{1}{2}kx^2 = \mathcal{E}. \tag{1.10}$$

Since $H(x,p)$ does not depend explicitly on time (in this example) $H(x,p)$ is a constant of the motion, equal to the "energy" $\mathcal{E}$.

1.2.1 Hamilton's Equations in Multiple Dimensions

Given coordinates $\mathbf{q}$ and Lagrangian L, "canonical momenta" are defined by

$$p_j = \frac{\partial L(\mathbf{q},\dot{\mathbf{q}},t)}{\partial \dot{q}^j}; \tag{1.11}$$

p_j is said to be "conjugate" to q^j. To make partial differentiation like this meaningful it is necessary to specify what variables are being held fixed. We mean implicitly that variables q^i for all i, $\dot{q}^i$ for $i \neq j$, and t are being held fixed. Having established variables $\mathbf{p}$ it is required in all that follows that velocities $\dot{\mathbf{q}}$ be explicitly expressible in terms of the $\mathbf{q}$ and $\mathbf{p}$, as in

$$\dot{q}^i = f^i(\mathbf{q}, \mathbf{p}, t), \quad \text{or} \quad \dot{\mathbf{q}} = \mathbf{f}(\mathbf{q}, \mathbf{p}, t). \tag{1.12}$$

Hamilton's equations can be derived using the properties of differentials. Define the "Hamiltonian" by

$$H(\mathbf{q}, \mathbf{p}, t) = p_i f^i(\mathbf{q}, \mathbf{p}, t) - L(\mathbf{q}, \mathbf{f}(\mathbf{q}, \mathbf{p}, t), t), \tag{1.13}$$

where the functions f^i were defined in Eq. (1.12). If these functions are, for any reason, unavailable, the procedure cannot continue; the velocity variables *must* be eliminated in this way. Furthermore, as indicated on the left-hand side of Eq. (1.13), it is essential for the formal arguments of H to be $\mathbf{q}$, $\mathbf{p}$ and t. Then, when writing partial derivatives of H, it will be implicit that the variables being held constant are all but one of the $\mathbf{q}$, $\mathbf{p}$, and t. If all independent variables of the Lagrangian are varied independently the result is

$$dL = \frac{\partial L}{\partial q^i} dq^i + \frac{\partial L}{\partial \dot{q}^i} d\dot{q}^i + \frac{\partial L}{\partial t} dt. \tag{1.14}$$

(It is important to appreciate that the q^i and the $\dot{q}^i$ are being treated as formally independent at this point. Any temptation toward thinking of $\dot{q}^i$ as some sort of derivative of q^i must be fought off.) The purpose of the additive term $p_i f^i$ in the definition of H is to cancel terms proportional to $d\dot{q}^i$ in the expression for dH;

$$\begin{aligned} dH &= f^i dp_i + p_i df^i - \frac{\partial L}{\partial q^i} dq^i - \frac{\partial L}{\partial \dot{q}^i} d\dot{q}^i - \frac{\partial L}{\partial t} dt \\ &= -\frac{\partial L}{\partial q^i} dq^i + f^i dp_i - \frac{\partial L}{\partial t} dt \\ &= -\dot{p}_i dq^i + \dot{q}^i dp_i - \frac{\partial L}{\partial t} dt, \end{aligned} \tag{1.15}$$

where the Lagrange equations as well as Eq. (1.12) have been used. Hamilton's first-order equations follow from Eq. (1.15);

$$\dot{p}_i = -\frac{\partial H}{\partial q^i}, \quad \dot{q}^i = \frac{\partial H}{\partial p_i}, \quad \frac{\partial H}{\partial t} = -\frac{\partial L}{\partial t}. \tag{1.16}$$

Remember that in the partial derivatives of H the variables $\mathbf{p}$ are held constant but in $\partial L/\partial t$ the variables $\dot{\mathbf{q}}$ are held constant.

Example 1.2.1. Charged Particle in Electromagnetic Field. *To exercise the Hamiltonian formalism consider a nonrelativistic particle in an electromagnetic field. In Chapter 11 it is shown that the Lagrangian is*

$$L = \frac{1}{2}m(\dot{x}^2 + \dot{y}^2 + \dot{z}^2) + e(A_x\dot{x} + A_y\dot{y} + A_z\dot{z}) - e\Phi(x, y, z), \tag{1.17}$$

where $\mathbf{A}(\mathbf{x})$ *is the vector potential and* $\Phi(\mathbf{x})$ *is the electric potential. The middle terms, linear in velocities, cannot be regarded naturally as either kinetic or potential energies. Nevertheless, their presence does not impede the formalism. In fact, consider an even more general situation,*

$$L = \frac{1}{2}A_{rs}(\mathbf{q})\,\dot{q}^r\dot{q}^s + A_r(\mathbf{q})\,\dot{q}^r - V(\mathbf{q}). \tag{1.18}$$

Then

$$p_r = A_{rs}\dot{q}^s + A_r, \quad \textit{and} \quad \dot{q}^r = B_{rs}(p_s - A_r). \tag{1.19}$$

It can be seen in this case that the momentum and velocity components are inhomogeneously, though still linearly, related. The Hamiltonian is

$$H = \frac{1}{2}B_{rs}(p_r - A_r)(p_s - A_s) + V, \tag{1.20}$$

and Hamilton's equations follow easily.

1.3
Derivation of the Lagrange Equation from Hamilton's Principle

The Lagrange equation is derivable from the "principle of least action" (or Hamilton's principle) according to which the actual trajectory taken by a particle as it moves from point P_0 to P between times t_0 and t, is that trajectory that minimizes the "action" function S defined by

$$S = \int_{t_0}^{t} L(x, \dot{x}, t)\,dt. \tag{1.21}$$

As shown in Fig. 1.1, a possible small deviation from the true orbit $x(t)$ is symbolized by $\delta x(t)$. Except for being infinitesimal and vanishing at the end points, the function $\delta x(t)$ is an arbitrary function of time. Note that the expressions $(d/dt)\delta x(t)$, $\delta\dot{x}(t)$, and $\dot{\overbrace{\delta x(t)}}$ all mean the same thing. The second form might be considered ambiguous but, for the sake of brevity, it is the symbol we will use.

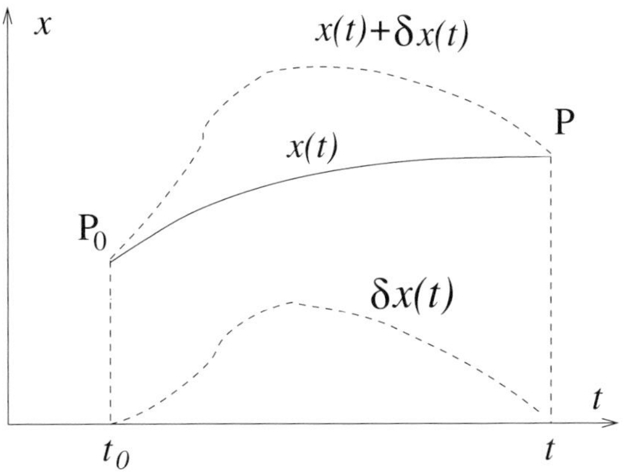

Fig. 1.1 Graph showing the extremal trajectory $x(t)$ and a nearby nontrajectory $x(t) + \delta x(t)$.

Using elementary calculus, the variation in S that accompanies the replacement $x(t) \Rightarrow x(t) + \delta x(t)$ is

$$\delta S = \int_{t_0}^{t} dt \left(\frac{\partial L}{\partial x} \delta x(t) + \frac{\partial L}{\partial \dot{x}} \delta \dot{x}(t) \right). \tag{1.22}$$

Preparing for integration by parts, one substitutes

$$\frac{d}{dt}\left(\frac{\partial L}{\partial \dot{x}} \delta x \right) = \left(\frac{d}{dt} \frac{\partial L}{\partial \dot{x}} \right) \delta x + \frac{\partial L}{\partial \dot{x}} \delta \dot{x}(t), \tag{1.23}$$

to obtain

$$\delta S = \int_{t_0}^{t} dt \left(\frac{d}{dt}\left(\frac{\partial L}{\partial \dot{x}} \delta x \right) - \left(\frac{d}{dt} \frac{\partial L}{\partial \dot{x}} - \frac{\partial L}{\partial x} \right) \delta x \right). \tag{1.24}$$

The first term, being a total derivative, can be integrated directly, and then be expressed in terms of initial and final values. For now we require δx to vanish in both cases. The motivation for performing this manipulation was to make δx be a common factor of the remaining integrand.

Since δx is an arbitrary function, the vanishing of δS implies the vanishing of the other factor in the integrand. The result is the Lagrange equation,

$$\frac{d}{dt} \frac{\partial L}{\partial \dot{x}} = \frac{\partial L}{\partial x}. \tag{1.25}$$

The very meaning of the Lagrange equations requires a clear understanding of the difference between d/dt and $\partial/\partial t$. The former refers to the time rate of change along the actual particle trajectory, while the latter refers to a formal derivative with respect to time with the other independent variables (called out in the argument list of the function being differentiated) held constant. When operating on an arbitrary function $F(\mathbf{x}(t), t)$ these derivatives are related by

$$\frac{d}{dt}F = \frac{\partial}{\partial t}F + (\mathbf{v}\cdot\nabla)F. \tag{1.26}$$

The first term gives the change in F at a fixed point, while the second gives the change due to the particle's motion.

This derivation has been restricted to a single Cartesian coordinate x, and the corresponding velocity $\dot{x}$, but the same derivation also applies to y and z and, for that matter to any generalized coordinates and their corresponding velocities. With this greater generality the Lagrange equations can be written as

$$\frac{d}{dt}\frac{\partial L}{\partial \mathbf{v}} = \frac{\partial L}{\partial \mathbf{r}} \equiv \nabla L. \tag{1.27}$$

Figure 1.1 shows the dependence of just one of the coordinates, x, on time t. Similar figures for other independent variables need not be drawn since we need only one variable, say $x(t)$, to obtain its Lagrange equation.

Problem 1.3.1. *The action S in mechanics is the analog of the optical path length, O.P.L., of physical optics. The basic integral like (1.21) in optics has the form*

$$\frac{1}{c}\text{O.P.L.} = \frac{1}{c}\int_{z_1}^{z_2} L\left(x, y, \frac{dx}{dz}, \frac{dy}{dz}, z\right)dz = \frac{1}{c}\int_{z_1}^{z_2} n(\mathbf{r})\sqrt{x'^2 + y'^2 + 1}\, dz. \tag{1.28}$$

Here x, y, and z are Cartesian coordinates with x and y "transverse" and z defining a longitudinal axis relative to which x and y are measured. The optical path length is the path length weighted by the local index of refraction n. O.P.L./c, is the "time" in "principle of least time." Though it is not entirely valid in physical optics to say that the "speed of light" in a medium is c/n, acting as if it were, the formula gives the time of flight of a particle (photon) following the given trajectory with this velocity.

The calculus of variations can be used to minimize O.P.L. Show that the differential equation (which will reappear as Eq. (7.18)) satisfied by an optical ray is

$$\frac{d}{ds}\left(n\frac{d\mathbf{r}}{ds}\right) = \nabla \mathbf{n}, \tag{1.29}$$

where $n(\mathbf{r})$ is index of refraction, $\mathbf{r}$ is a radius vector from an arbitrary origin to a point on a ray, and s is arc length s along the ray.

1.4
Linear, Multiparticle Systems

The approximate Lagrangian for an n-dimensional system with coordinates $(q_1, q_2, \ldots, q_n)$, valid in the vicinity of a stable equilibrium point (that can be taken to be $(0, 0, \ldots, 0)$), has the form

$$L(\mathbf{q}, \dot{\mathbf{q}}) = T - V, \quad \text{where} \quad T = \frac{1}{2} \sum_{r,s=1}^{n} m_{(rs)} \dot{q}_{(r)} \dot{q}_{(s)}, \qquad V = \frac{1}{2} \sum_{r,s=1}^{n} k_{(rs)} q_{(r)} q_{(s)}. \tag{1.30}$$

It is common to use the summation convention for summations like this, but in this text the summation convention is reserved for tensor summations. When subscripts are placed in parenthesis (as here) it indicates they refer to different variables or parameters (as here) rather than different components of the same vector or tensor. Not to be obsessive about it however, for the rest of this discussion the parentheses will be left off, but the summation signs will be left explicit. It is known from algebra that a linear transformation $q_i \rightarrow y_j$ can be found such that T takes the form

$$T = \frac{1}{2} \sum_{r=1}^{n} m_r \dot{y}_r^2, \tag{1.31}$$

where, in this case each "mass" m_r is necessarily positive because T is positive definite. By judicious choice of the scale of the y_r each "mass" can be adjusted to 1. We will assume this has already been done.

$$T = \frac{1}{2} \sum_{=1}^{n} \dot{y}_r^2. \tag{1.32}$$

For these coordinates y_r the equation

$$\sum_{r=1}^{n} y_r^2 = 1 \tag{1.33}$$

defines a surface (to be called a hypersphere). From now on we will consider only points $\mathbf{y} = (y_1, \ldots, y_n)$ lying on this sphere. Also two points $\mathbf{u}$ and $\mathbf{v}$ will be said to be "orthogonal" if the "quadratic form" $\mathcal{I}(\mathbf{u}, \mathbf{v})$ defined by

$$\mathcal{I}(\mathbf{u}, \mathbf{v}) \equiv \sum_{r=1}^{n} u_r v_r \tag{1.34}$$

vanishes. Being linear in both arguments $\mathcal{I}(\mathbf{u}, \mathbf{v})$ is said to be "bilinear." We also define a bilinear form $\mathcal{V}(\mathbf{u}, \mathbf{v})$ by

$$\mathcal{V}(\mathbf{u}, \mathbf{v}) \equiv \sum_{r,s=1}^{n} k_{rs} u_r v_s, \tag{1.35}$$

where coefficients k_{rs} have been redefined from the values given above to correspond to the new coordinates y_r so that

$$V(\mathbf{y}) = \frac{1}{2}\,\mathcal{V}(\mathbf{y},\mathbf{y}). \tag{1.36}$$

The following series of problems (adapted from Courant and Hilbert, Vol. 1, p. 37) will lead to the conclusion that a further linear transformation $y_i \rightarrow z_j$ can be found that, on the one hand, enables the equation for the sphere in Eq. (1.33) to retain the same form,

$$\sum_{r=1}^{n} z_r^2 = 1, \tag{1.37}$$

and, on the other, enables V to be expressible as a sum of squares with positive coefficients;

$$V = \frac{1}{2}\sum_{r=1}^{n} \kappa_r z_r^2, \quad \text{where} \quad 0 < \kappa_n \le \kappa_{n-1} \le \cdots \le \kappa_1 < \infty. \tag{1.38}$$

Pictorially the strategy is, having deformed the scales so that surfaces of constant T are spherical and surfaces of constant V ellipsoidal, to orient the axes to make these ellipsoids erect. In the jargon of mechanics this process is known as "normal mode" analysis.

The "minimax" properties of the "eigenvalues" to be found have important physical implications, but we will not go into them here.

Problem 1.4.1.

(a) Argue, for small oscillations to be stable, that V must also be positive definite.

(b) Let $\mathbf{z}_1$ be the point on sphere (1.33) for which $V(\stackrel{\text{def.}}{=} \frac{1}{2}\kappa_1)$ is maximum. (If there is more than one such point pick any one arbitrarily.) Then argue that

$$0 < \kappa_1 < \infty. \tag{1.39}$$

(c) Among all the points that are both on sphere (1.33) and orthogonal to $\mathbf{z}_1$, let $\mathbf{z}_2$ be the one for which $V(\stackrel{\text{def.}}{=} \frac{1}{2}\kappa_2)$ is maximum. Continuing in this way, show that a series of points $\mathbf{z}_1, \mathbf{z}_2, \ldots \mathbf{z}_n$, each maximizing V consistent with being orthogonal to its predecessors, is determined, and that the sequence of values, $V(\mathbf{z}_r) = \frac{1}{2}\kappa_r$, $r = 1, 2, \ldots, n$, is monotonically nonincreasing.

(d) Consider a point $\mathbf{z}_1 + \epsilon\boldsymbol{\zeta}$ which is assumed to lie on surface (1.33) but with $\boldsymbol{\zeta}$ otherwise arbitrary. Next assume this point is "close to" $\mathbf{z}_1$ in the sense that ϵ is arbitrarily small (and not necessarily positive). Since $\mathbf{z}_1$ maximizes V it follows that

$$\mathcal{V}(\mathbf{z}_1 + \epsilon\boldsymbol{\zeta}, \mathbf{z}_1 + \epsilon\boldsymbol{\zeta}) \le 0. \tag{1.40}$$

Show therefore that

$$\mathcal{V}(\mathbf{z}_1, \boldsymbol{\zeta}) = 0. \tag{1.41}$$

This implies that

$$\mathcal{V}(\mathbf{z}_1, \mathbf{z}_r) = 0 \quad \textit{for} \quad r > 1, \tag{1.42}$$

because, other than being orthogonal to $\mathbf{z}_1$, $\boldsymbol{\zeta}$ *is arbitrary.*

Finally, extend the argument to show that

$$\mathcal{V}(\mathbf{z}_r, \mathbf{z}_s) = \kappa_r \delta_{rs}, \tag{1.43}$$

where the coefficients κ_r *have been shown to satisfy the monotonic conditions of Eq. (1.38) and* δ_{rs} *is the usual Kronecker-*δ *symbol.*

(e) *Taking these* $\mathbf{z}_r$ *as basis vectors, an arbitrary vector* $\mathbf{z}$ *can be expressed as*

$$\mathbf{z} = \sum_{r=1}^{n} z_r \mathbf{z}_r. \tag{1.44}$$

In these new coordinates, show that Eqs. (1.30) become

$$L(\mathbf{z}, \dot{\mathbf{z}}) = T - V, \quad T = \frac{1}{2} \sum_{r=1}^{n} \dot{z}_r^2, \quad V = \frac{1}{2} \sum_{r=1}^{n} \kappa_r z_r^2. \tag{1.45}$$

Write and solve the Lagrange equations for coordinates z_r.

Problem 1.4.2. *Proceeding from the previous formula, the Lagrange equations resulting from Eq. (1.30) are*

$$\sum_{s=1}^{n} m_{rs} \ddot{q}_s + \sum_{s=1}^{n} k_{rs} q_s = 0. \tag{1.46}$$

These equations can be expressed compactly in matrix form;

$$\mathbf{M}\ddot{\mathbf{q}} + \mathbf{K}\mathbf{q} = 0; \tag{1.47}$$

or, assuming the existence of $\mathbf{M}^{-1}$, *as*

$$\ddot{\mathbf{q}} + \mathbf{M}^{-1}\mathbf{K}\mathbf{q} = 0. \tag{1.48}$$

Seeking a solution of the form

$$q_r = A_r e^{i\omega t} \quad r = 1, 2, \ldots, n, \tag{1.49}$$

the result of substitution into Eq. (1.46) is

$$(\mathbf{M}^{-1}\mathbf{K} - \omega^2 \mathbf{1})\mathbf{A} = 0. \tag{1.50}$$

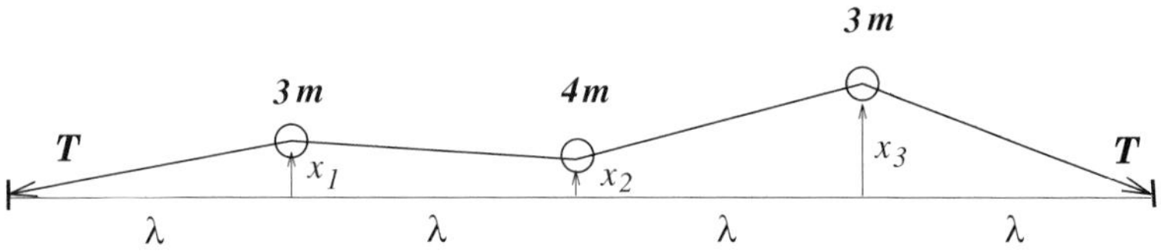

Fig. 1.2 Three beads on a stretched string. The transverse displacements are much exaggerated. Gravity and string mass are negligible.

These equations have nontrivial solutions for values of ω that cause the determinant of the coefficients to vanish;

$$|\mathbf{M}^{-1}\mathbf{K} - \omega^2\mathbf{1}| = 0. \tag{1.51}$$

Correlate these ω "eigenvalues" with the constants κ_r defined in the previous problem.

Problem 1.4.3. *As shown in Fig. 1.2, particles of mass 3m, 4m, and 3m, are spaced at uniform intervals λ along a light string of total length 4λ, stretched with tension $\mathcal{T}$, and rigidly fixed at both ends. To legitimize ignoring gravity, the system is assumed to lie on a smooth horizontal table so the masses can oscillate only horizontally. Let the transverse displacements be x_1, x_2, and x_3. Find the normal modes frequencies and find and sketch the corresponding normal mode oscillation "shapes." Discuss the "symmetry" of the shapes, their "wavelengths," and the (monotonic) relation between mode frequency and number of nodes (axis crossings) in each mode.*

Already with just three degrees of freedom the eigenmode calculations are sufficiently tedious to make some efforts at simplifying the work worthwhile. In this problem, with the system symmetric about its midpoint it is clear that the modes will be either symmetric or antisymmetric and, since the antisymmetric mode vanishes at the center point, it is characterized by a single amplitude, say $y = x_1 = -x_3$. Introducing "effective mass" and "effective strength coefficient" the kinetic energy of the mode, necessarily proportional to $\dot{y}$, can be written as $T_2 = \frac{1}{2}m_{\text{eff}}\dot{y}^2$ and the potential energy can be written as $V_2 = \frac{1}{2}k_{\text{eff}}y^2$. The frequency of this mode is then given by $\omega_2 = \sqrt{k_{\text{eff}}/m_{\text{eff}}}$ which, by dimensional analysis, has to be proportional to $\eta = \sqrt{\mathcal{T}/(m\lambda)}$. (The quantities T_2, V_2, and ω_2 have been given subscript 2 because this mode has the second highest frequency.) Factoring this expression out of Eq. (1.51), the dimensionless eigenvalues are the eigenfrequencies in units of η.

Problem 1.4.4. *Complete the analysis to show that the normal mode frequencies are $(\omega_1, \omega_2, \omega_3) = (1, \sqrt{2/3}, \sqrt{1/6})$, and find the corresponding normal mode "shapes."*

1.4.1
The Laplace Transform Method

Though the eigenmode/eigenvalue solution method employed in solving the previous problem is the traditional method used in classical mechanics, equations of the same form, when they arise in circuit analysis and other engineering fields, are traditionally solved using Laplace transforms – a more robust method, it seems to me. Let us continue the solution of the previous problem using this method. Individuals already familiar with this method or not wishing to become so should skip this section. Here we use the notation

$$\overline{x}(s) = \int_0^\infty e^{-st} x(t)\, dt, \tag{1.52}$$

as the formula giving the Laplace transform $\overline{x}(s)$, of the function of time $x(t)$. $\overline{x}(s)$ is a function of the "transform variable" s (which is a complex number with positive real part.) With this definition the Laplace transform satisfies many formulas but, for present purposes we use only

$$\overline{\frac{dx}{dt}} = s\overline{x} - x(0), \tag{1.53}$$

which is easily demonstrated. Repeated application of this formula converts time derivatives into functions of s and therefore converts (linear) differential equations into (linear) algebraic equations. This will now be applied to the system described in the previous problem.

The Lagrange equations for the beaded string shown in Fig. 1.2 are

$$\begin{aligned} 3\ddot{x}_1 + \eta^2(2x_1 - x_2) &= 0, \\ 4\ddot{x}_2 + \eta^2(2x_2 - x_1 - x_3) &= 0, \\ 3\ddot{x}_3 + \eta^2(2x_3 - x_2) &= 0. \end{aligned} \tag{1.54}$$

Suppose the string is initially at rest but that a transverse impulse I is administered to the first mass at $t = 0$; as a consequence it acquires initial velocity $v_{10} \equiv \dot{x}(0) = I/(3m)$. Transforming all three equations and applying the initial conditions (the only nonvanishing initial quantity, v_{10}, enters via Eq. (1.53))

$$\begin{aligned} (3s^2 + 2\eta^2)\overline{x}_1 - \eta^2\overline{x}_2 &= I/m, \\ -\eta^2\overline{x}_1 + (4s^2 + 2\eta^2)\overline{x}_2 - \eta^2\overline{x}_3 &= 0, \\ -\eta^2\overline{x}_2 + (3s^2 + 2\eta^2)\overline{x}_3 &= 0. \end{aligned} \tag{1.55}$$

Solving these equations yields

$$\overline{x}_1 = \frac{I}{10m}\left(\frac{2/3}{s^2+\eta^2/6} + \frac{1}{s^2+\eta^2} + \frac{5/3}{s^2+2\eta^2/3}\right),$$
$$\overline{x}_2 = \frac{I}{10m}\left(\frac{1}{s^2+\eta^2/6} - \frac{1}{s^2+\eta^2}\right),$$
$$\overline{x}_3 = \frac{I}{10m}\left(\frac{2/3}{s^2+\eta^2/6} + \frac{1}{s^2+\eta^2} - \frac{5/3}{s^2+2\eta^2/3}\right). \quad (1.56)$$

It can be seen, except for factors $\pm i$, that the poles (as a function of s) of the transforms of the variables, are the normal mode frequencies. This is not surprising since the determinant of the coefficients in Eq. (1.55) is the same as the determinant entering the normal mode solution, but with ω^2 replaced with $-s^2$. Remember then, from Cramer's rule for the solution of linear equations, that this determinant appears in the denominators of the solutions. For "inverting" Eq. (1.56) it is sufficient to know just one inverse Laplace transformation,

$$\mathcal{L}^{-1}\frac{1}{s-\alpha} = e^{\alpha t}, \quad (1.57)$$

but it is easier to look in a table of inverse transforms to find that the terms in Eq. (1.56) yield sinusoids that oscillate with the normal mode frequencies. Furthermore, the "shapes" asked for in the previous problem can be read off directly from (1.56) to be (2:3:2), (1:0:1), and (1:-1:1).

When the first mass is struck at $t = 0$, all three modes are excited and they proceed to oscillate at their own natural frequencies, so the motion of each individual particle is a superposition of these frequencies. Since there is no damping the system will continue to oscillate in this superficially complicated way forever. In practice there is always some damping and, in general, it is different for the different modes; commonly damping increases with frequency. In this case, after a while, the motion will be primarily in the lowest frequency mode; if the vibrating string emits audible sound, an increasingly pure, low-pitched tone will be heard as time goes on.

1.4.2 Damped and Driven Simple Harmonic Motion

The equation of motion of mass m, subject to restoring force $-\omega_0^2 m x$, damping force $-2\lambda m\dot{x}$, and external drive force $f\cos\gamma t$ is

$$\ddot{x} + 2\lambda\dot{x} + \omega_0^2 = \frac{f}{m}\cos\gamma t. \quad (1.58)$$

Problem 1.4.5.

(a) *Show that the general solution of this equation when* $f = 0$ *is*

$$x(t) = ae^{-\lambda t}\cos(\omega t + \phi), \tag{1.59}$$

where a *and* ϕ *depend on initial conditions and* $\omega = \sqrt{\omega^2 - \lambda^2}$. *This "solution of the homogeneous equation" is also known as "transient" since when it is superimposed on the "driven" or "steady state" motion caused by* f *it will eventually become negligible.*

(b) *Correlate the stability or instability of the transient solution with the sign of* λ. *Equivalently, after writing the solution (1.59) as the sum of two complex exponential terms, Laplace transform them, and correlate the stability or instability of the transient with the locations in the complex s-plane of the poles of the Laplace transform.*

(c) *Assuming* $x(0) = \dot{x}(0) = 0$, *show that Laplace transforming equation (1.58) yields*

$$\overline{x}(s) = f\frac{s}{s^2 + \gamma^2}\,\frac{1}{s^2 + 2\lambda s + \omega_0^2}. \tag{1.60}$$

This expression has four poles, each of which leads to a complex exponential term in the time response. To neglect transients we need to only drop the terms for which the poles are off the imaginary axis. (By part (b) they must be in the left half-plane for stability.) To "drop" these terms it is necessary first to isolate them by partial fraction decomposition of Eq. (1.60). Performing these operations, show that the steady-state solution of Eq. (1.58), is

$$x(t) = \frac{f}{m}\sqrt{\frac{1}{(\omega_0^2 - \gamma^2)^2 + 4\lambda^2\gamma^2}}\cos(\gamma t + \delta),) \tag{1.61}$$

where

$$\omega_0^2 - \gamma^2 - 2\lambda\gamma i = \sqrt{(\omega_0^2 - \gamma^2)^2 + 4\lambda^2\gamma^2}\, e^{i\delta}. \tag{1.62}$$

(d) *The response is large only for* γ *close to* ω_0. *To exploit this, defining the "small" "frequency deviation from the natural frequency"*

$$\epsilon = \gamma - \omega_0, \tag{1.63}$$

show that $\gamma^2 - \omega^2 \approx 2\epsilon\omega$ *and that the approximate response is*

$$x(t) = \frac{f}{2m\omega_0}\sqrt{\frac{1}{\epsilon^2 + \lambda^2}}\cos(\gamma t + \delta). \tag{1.64}$$

Find the value of ϵ *for which the amplitude of the response is reduced from its maximum value by the factor* $1/\sqrt{2}$.

1.4.3
Conservation of Momentum and Energy

It has been shown previously that the application of energy conservation in one-dimensional problems permits the system evolution to be expressed in terms of a single integral – this is "reduction to quadrature." The following problem exhibits the use of momentum conservation to reduce a two-dimensional problem to quadratures, or rather, because of the simplicity of the configuration in this case, to a closed-form solution.

Problem 1.4.6. *A point mass m with total energy E, starting in the left half-plane, moves in the (x,y) plane subject to potential energy function*

$$U(x,y) = \begin{cases} U_1, & \text{for} \quad x < 0, \\ U_2, & \text{for} \quad 0 < x. \end{cases} \tag{1.65}$$

The "angle of incidence" to the interface at $x = 0$ is θ_i and the outgoing angle is θ. Specify the qualitatively different cases that are possible, depending on the relative values of the energies, and in each case find θ in terms of θ_i. Show that all results can be cast in the form of "Snell's law" of geometric optics if one introduces a factor $\sqrt{E - U(r)}$, analogous to index of refraction.

1.5
Effective Potential and the Kepler Problem

Since one-dimensional motion is subject to such simple and satisfactory analysis, anything that can reduce the dimensionality from two to one has great value. The "effective potential" is one such device. No physics problem has received more attention over the centuries than the problem of planetary orbits. In later chapters of this text the analytic solution of this so-called "Kepler problem" will be the foundation on which perturbative solution of more complicated problems will be based. Though this problem is now regarded as "elementary" one is well-advised to stick to traditional manipulations as the problem can otherwise get seriously messy.

The problem of two masses m_1 and m_2 moving in each other's gravitational field is easily converted into the problem of a single particle of mass m moving in the gravitational field of a mass m_0 assumed very large compared to m; that is $\mathbf{F} = -K\hat{\mathbf{r}}/r^2$, where $K = Gm_0m$ and r is the distance to m from m_0. Anticipating that the orbit lies in a plane (as it must) let χ be the angle of the radius vector from a line through the center of m_0; this line will be later taken as the major axis of the elliptical orbit. The potential energy function is given by

$$U(r) = -\frac{K}{r}, \tag{1.66}$$

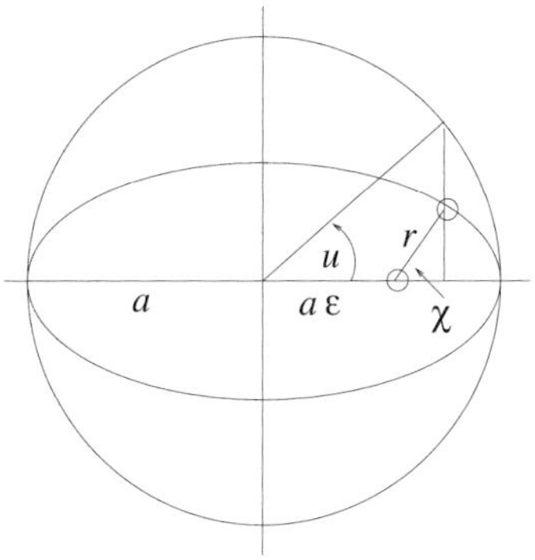

Fig. 1.3 Geometric construction defining the "true anomaly" χ and "eccentric anomaly" u in terms of other orbit parameters.

and the orbit geometry is illustrated in Fig. 1.3. Two conserved quantities can be identified immediately: energy E and angular momentum M. Show that they are given by

$$\begin{aligned} E &= \frac{1}{2}m(\dot{r}^2 + r^2\dot{\chi}^2) - \frac{K}{r}, \\ M &= mr^2\dot{\chi}. \end{aligned} \tag{1.67}$$

One can exploit the constancy of M to eliminate $\dot{\chi}$ from the expression for E,

$$E = \frac{1}{2}m\dot{r}^2 + U_{\text{eff.}}(r), \quad \text{where} \quad U_{\text{eff.}}(r) = \frac{M^2}{2mr^2} - \frac{K}{r}. \tag{1.68}$$

The function $U_{\text{eff.}}(r)$, known as the "effective potential," is plotted in Fig. 1.4. Solving the expressions for E and M individually for differential dt

$$dt = \frac{mr^2}{M}d\chi, \quad dt = \left(\frac{2}{m}\left(E + \frac{K}{r}\right) - \frac{M^2}{m^2r^2}\right)^{-1/2} dr. \tag{1.69}$$

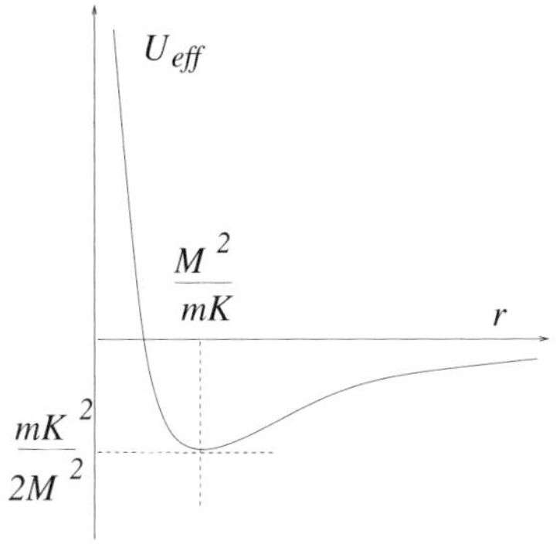

Fig. 1.4 The effective potential $U_{\text{eff.}}$ for the Kepler problem.

Equating the two expressions yields a differential equation that can be solved by "separation of variables." This has permitted the problem to be "reduced to quadratures,"

$$\chi(r) = \int^{r} \frac{M dr'/r'^2}{\sqrt{2m(E + K/r') - M^2/r'^2}}. \tag{1.70}$$

Note that this procedure yields only an "orbit equation," the dependence of χ on r (which is equivalent to, if less convenient than, the dependence of r on χ.) Though *a priori* one should have had the more ambitious goal of finding a solution in the form $r(t)$ and $\chi(t)$, no information whatsoever is given yet about time dependence by Eq. (1.70).

Problem 1.5.1.

(a) *Show that all computations so far can be carried out for any central force – that is radially directed with magnitude dependent only on r. At worst the integral analogous to (1.70) can be performed numerically.*

(b) *Specializing again to the Kepler problem, perform the integral (1.70) and show that the orbit equation can be written as*

$$\epsilon \cos\chi + 1 = \frac{M^2}{mK}\frac{1}{r}. \tag{1.71}$$

where $\epsilon \equiv \sqrt{1 + \frac{2EM^2}{m^2K^2}}$.

(c) *Show that (1.71) is the equation of an ellipse if* $\epsilon < 1$ *and that this condition is equivalent to* $E < 0$.

(d) *It is traditional to write the orbit equation purely in terms of "orbit elements" which can be identified as the "eccentricity"* ϵ, *and the "semimajor axis"* a;

$$a = \frac{r_{max} + r_{min}}{2} = \frac{M^2}{mK}\frac{1}{1-\epsilon^2}. \tag{1.72}$$

The reason a *and* ϵ *are special is that they are* intrinsic *properties of the orbit unlike, for example, the orientations of the semimajor axis and the direction of the perpendicular to the orbit plane, both of which can be altered at will and still leave a "congruent" system. Derive the relations*

$$E = -\frac{K}{2a}, \quad M^2 = (1-\epsilon^2)\,\mathrm{mKa}, \tag{1.73}$$

so the orbit equation is

$$\frac{a}{r} = \frac{1 + \epsilon\cos\chi}{1-\epsilon^2}. \tag{1.74}$$

(e) Finally derive the relation between r and t;

$$t(r) = \sqrt{\frac{ma}{K}} \int^{r} \frac{r'dr'}{\sqrt{a^2\epsilon^2 - (r'-a)^2}}. \tag{1.75}$$

An "intermediate" variable u that leads to worthwhile simplification is defined by

$$r = a(1 - \epsilon \cos u). \tag{1.76}$$

The geometric interpretation of u is indicated in Fig. 1.3. If (x, z) are Cartesian coordinates of the planet along the major and an axis parallel to the minor axis through the central mass, they are given in terms of u by

$$x = a\cos u - a\epsilon, \quad z = a\sqrt{1-\epsilon^2}\,\sin u, \tag{1.77}$$

since the semimajor axis is $a\sqrt{1-\epsilon^2}$ and the circumscribed circle is related to the ellipse by a z-axis scale factor $\sqrt{1-\epsilon^2}$. The coordinate u, known as the "eccentric anomaly" is a kind of distorted angular coordinate of the planet, and is related fairly simply to t;

$$t = \sqrt{\frac{ma^3}{K}}(u - \epsilon \sin u). \tag{1.78}$$

This is especially useful for nearly circular orbits, since then u is nearly proportional to t. Because the second term is periodic, the full secular time accumulation is described by the first term.

Analysis of this Keplerian system is continued using Hamilton–Jacobi theory in Section 8.3, and then again in Section 14.6.3 to illustrate action/angle variables, and then again as a system subject to perturbation and analyzed by "variation of constants" in Section 16.1.1.

Problem 1.5.2. *The effective potential formalism has reduced the dimensionality of the Kepler problem from two to one. In one dimension, the linearization (to simple harmonic motion) procedure, can then be used to describe motion that remains close to the minimum of the effective potential (see Fig. 1.4). The radius $r_0 = M^2/(mK)$ is the radius of the circular orbit with angular momentum M. Consider an initial situation for which M has this same value and $\dot{r}(0) = 0$, but $r(0) \neq r_0$, though $r(0)$ is in the region of good parabolic fit to U_{eff}. Find the frequency of small oscillations and express $r(t)$ by its appropriate simple harmonic motion. Then find the orbit elements a and ϵ, as defined in Problem 1.5.1, that give the matching two-dimensional orbit.*

1.6 Multiparticle Systems

Solving multiparticle problems in mechanics is notoriously difficult; for more than two particles it is usually impossible to get solutions in closed form. But the equations of motion can be made simpler by the appropriate choice of coordinates as the next problem illustrates. Such coordinate choices exploit exact relations such as momentum conservation and thereby simplify subsequent approximate solutions. For example, this is a good pre-quantum starting point for molecular spectroscopy.

Problem 1.6.1. *The position vectors of three point masses, m_1, m_2, and m_3, are $\mathbf{r}_1$, $\mathbf{r}_2$, and $\mathbf{r}_3$. Express these vectors in terms of the alternative configuration vectors $\mathbf{s}_C$, $\mathbf{s}_3{}'$, and $\mathbf{s}_{12}$ shown in the figure. Define "reduced masses" by*

$$m_{12} = m_1 + m_2, \quad M = m_1 + m_2 + m_3, \quad \mu_{12} = \frac{m_1 m_2}{m_{12}}, \quad \mu_3 = \frac{m_3 m_{12}}{M}. \tag{1.79}$$

Calculate the total kinetic energy in terms of $\dot{\mathbf{s}}$, $\dot{\mathbf{s}}_3{}'$, and $\dot{\mathbf{s}}_{12}$ and interpret the result. Defining corresponding partial angular momenta $\mathbf{l}$, $\mathbf{l}_3{}'$, and $\mathbf{l}_{12}$, show that the total angular momentum of the system is the sum of three analogous terms.

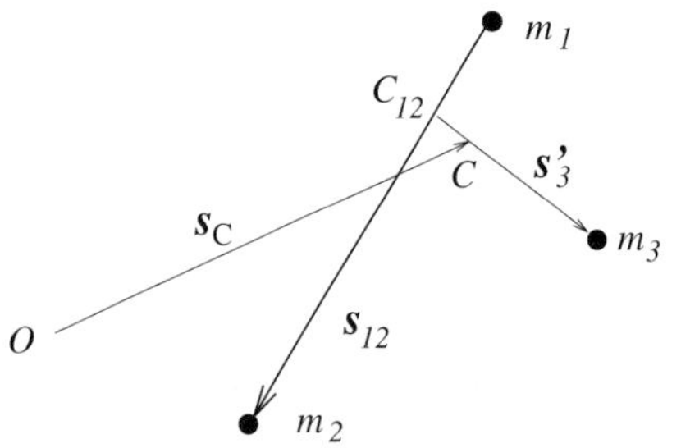

Fig. 1.5 Coordinates describing three particles. C is the center of mass and $\mathbf{s}_C$ its position vector relative to origin O. C_{12} is the center of mass of m_1 and m_2 and $\mathbf{s}_3{}'$ is the position of m_3 relative to C_{12}.

In Fig. 1.5, relative to origin O, the center of mass C is located by radius vector $\mathbf{s}_C$. Relative to particle 1, particle 2 is located by vector $\mathbf{s}_{12}$. Relative to the center of mass at C_{12} mass 3 is located by vector $\mathbf{s}_3{}'$. In terms of these quantities the position vectors of the three masses are

$$\mathbf{r}_1 = \mathbf{s}_C - \frac{m_3}{M}\mathbf{s}_3{}' + \frac{m_2}{m_{12}}\mathbf{s}_{12}, \tag{1.80}$$

$$\mathbf{r}_2 = \mathbf{s}_C - \frac{m_3}{M}\mathbf{s}_3{}' + \frac{m_1}{m_{12}}\mathbf{s}_{12}, \tag{1.81}$$

$$\mathbf{r}_3 = \mathbf{s}_C + \frac{m_{12}}{M}\mathbf{s}_3{}'. \tag{1.82}$$

Substituting these into the kinetic energy of the system

$$T = \frac{1}{2}m_1\dot{\mathbf{r}}_1^2 + \frac{1}{2}m_2\dot{\mathbf{r}}_2^2 + \frac{1}{2}m_3\dot{\mathbf{r}}_3^2, \tag{1.83}$$

the "cross terms" proportional to $\mathbf{s}_C \cdot \mathbf{s}_3{}'$, $\mathbf{s}_C \cdot \mathbf{s}_{12}$, and $\mathbf{s}_3{}' \cdot \mathbf{s}_{12}$ all cancel out, leaving the result

$$T = \frac{1}{2} M\, v_C^2 + \frac{1}{2}\mu_3\, v_3{}'^2 + \frac{1}{2}\mu_{12}\, v_{12}^2, \tag{1.84}$$

where $v_C = |\dot{\mathbf{s}}_C|$, $v_3' = |\dot{\mathbf{s}}_3'|$, and $v_{12} = |\dot{\mathbf{s}}_{12}|$. The angular momentum (about O) is given by

$$\mathbf{L} = \mathbf{r}_1 \times (m_1 \dot{\mathbf{r}}_1) + \mathbf{r}_2 \times (m_2 \dot{\mathbf{r}}_2) + \mathbf{r}_3 \times (m_3 \dot{\mathbf{r}}_3). \tag{1.85}$$

Upon expansion the same simplifications occur, yielding

$$\mathbf{L} = \frac{1}{2}\, M\, \mathbf{r}_C \times \mathbf{v}_C + \frac{1}{2}\, \mu_3\, \mathbf{r}_3{}' \times \mathbf{v}_3{}' + \frac{1}{2}\, \mu_{12}\, \mathbf{r}_{12} \times \mathbf{v}_{12}. \tag{1.86}$$

Problem 1.6.2. *Determine the moment of inertia tensor about center of mass C for the system described in the previous problem. Choose axes to simplify the problem initially and give a formula for transforming from these axes to arbitrary (orthonormal) axes. For the case $m_3 = 0$ find the principal axes and the principal moments of inertia.*

Setting $\mathbf{s}_C = 0$, the particle positions are given by

$$\mathbf{r}_1 = -\frac{m_3}{M}\mathbf{s}_3{}' + \frac{m_2}{m_{12}}\mathbf{s}_{12}, \quad \mathbf{r}_2 = -\frac{m_3}{M}\mathbf{s}_3{}' + \frac{m_1}{m_{12}}\mathbf{s}_{12}, \quad \mathbf{r}_3 = \frac{m_{12}}{M}\mathbf{s}_3{}'. \tag{1.87}$$

Since the masses lie in a single plane it is convenient to take the z-axis normal to that plane. Let us orient the axes such that the unit vectors satisfy

$$\mathbf{s}_3{}' = \hat{\mathbf{x}}, \quad \mathbf{s}_{12} = a\hat{\mathbf{x}} + b\hat{\mathbf{y}}, \tag{1.88}$$

and hence $a = \mathbf{s}_3{}' \cdot \mathbf{s}_{12}$. So the particle coordinates are

$$x_1 = -\frac{m_3}{M} + \frac{m_2}{m_{12}}a, \quad y_1 = \frac{m_2}{m_{12}}b, \tag{1.89}$$

$$x_2 = -\frac{m_3}{M} + \frac{m_1}{m_{12}}a, \quad y_2 = \frac{m_1}{m_{12}}b, \tag{1.90}$$

$$x_3 = \frac{m_{12}}{M}, \quad y_3 = 0. \tag{1.91}$$

In terms of these, the moment of inertia tensor $\mathbf{I}$ is given by

$$\begin{pmatrix} \sum m_i y_i^2 & -\sum m_i x_i y_i & 0 \\ -\sum m_i x_i y_i & \sum m_i x_i^2 & 0 \\ 0 & 0 & \sum m_i (x_i^2 + y_i^2) \end{pmatrix}. \tag{1.92}$$

For the special case $m_3 = 0$ these formulas reduce to

$$\mathbf{I} = \mu_{12} \begin{pmatrix} b^2 & -ab & 0 \\ -ab & a^2 & 0 \\ 0 & 0 & a^2 + b^2 \end{pmatrix}. \tag{1.93}$$

Problem 1.6.3. *A uniform solid cube can be supported by a thread from the center of a face, from the center of an edge, or from a corner. In each of the three cases the system acts as a torsional pendulum, with the thread providing all the restoring torque and the cube providing all the inertia. In which configuration is the oscillation period the longest? [If your answer involves complicated integrals you are not utilizing properties of the inertia tensor in the intended way.]*

1.7 Longitudinal Oscillation of a Beaded String

A short length of a stretched string, modeled as point "beads" joined by light stretched springs, is shown in Fig. 1.6. With a being the distance between beads in the stretched, but undisturbed, condition and, using the fact that the spring constant of a section of a uniform spring is inversely proportional to the length of the section, the parameters of this system are:

$$\begin{aligned}
\text{unstretched string length} &= L_0,\\
\text{stretched string length} &= L_0 + \Delta L,\\
\text{extension, } \Delta L \times \text{string constant of full string, } K &= \text{tension, } \tau_0,\\
\text{number of springs, } N &= \frac{L_0 + \Delta L}{a}\\
\text{spring constant of each spring, } k &= NK,\\
\text{mass per unit length, } \mu_0 &= m/a.
\end{aligned} \tag{1.94}$$

The kinetic energy of this system is

$$T = \frac{m}{2}\left(\cdots + \dot{\eta}_{i-1}^2 + \dot{\eta}_i^2 + \dot{\eta}_{i+1}^2 + \cdots\right), \tag{1.95}$$

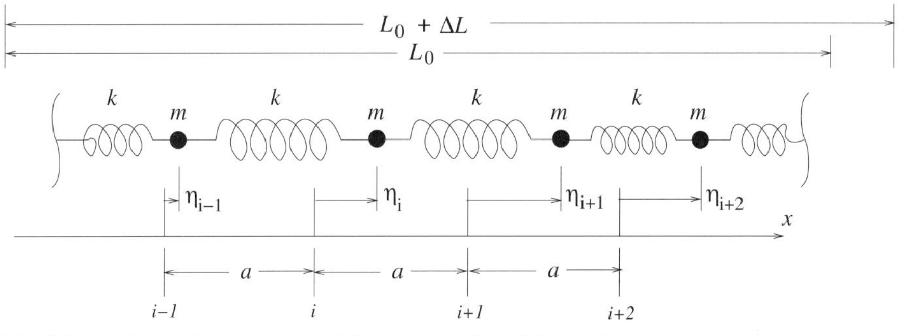

Fig. 1.6 A string under tension modeled as point "beads" joined by light stretched springs.

and the potential energy is

$$V = \frac{k}{2}\left(\cdots + (\eta_i - \eta_{i-1})^2 + (\eta_{i+1} - \eta_i)^2 + \cdots\right). \tag{1.96}$$

The Lagrangian being $L = T - V$, the momentum conjugate to η_i is $p_i = \partial L/\partial\dot{\eta}_i = m\dot{\eta}_i$, and the Lagrange equations are

$$m\ddot{\eta}_i = \frac{\partial L}{\partial \dot{\eta}_i} = k(\eta_{i-1} - 2\eta_i + \eta_{i+1}), \quad i = 1, 2, \ldots, N. \tag{1.97}$$

1.7.1
Monofrequency Excitation

Suppose that the beads of the spring are jiggling in response to sinusoidal excitation at frequency ω. Let us conjecture that the response can be expressed in the form

$$\eta_i(t) = \begin{pmatrix}\sin\\ \cos\end{pmatrix}(\omega t + \Delta\psi\, i), \tag{1.98}$$

where $\Delta\psi$ is a phase advance per section that remains to be determined, and where "in phase" and "out of phase" responses are shown as the two rows of a matrix – their possibly different amplitudes are not yet shown. For substitution into Eq. (1.97) one needs

$$\eta_{i+1} = \begin{pmatrix}\sin(\omega t + \Delta\psi\, i)\, \cos\Delta\psi + \cos(\omega t + \Delta\psi\, i)\, \sin\Delta\psi\\ \cos(\omega t + \Delta\psi\, i)\, \cos\Delta\psi + -\sin(\omega t + \Delta\psi\, i)\, \sin\Delta\psi\end{pmatrix}, \tag{1.99}$$

along with a similar equation for η_{i-1}. Then one obtains

$$\eta_{i-1} - 2\eta_i + \eta_{i+1} = (2\cos\Delta\psi - 2)\eta_i, \tag{1.100}$$

and then, from the Lagrange equation,

$$-m\omega^2 = 2\cos\Delta\psi - 2. \tag{1.101}$$

Solving this, one obtains

$$\Delta\psi(\omega) = \pm\cos^{-1}\left(1 - \frac{m\omega^2}{2k}\right), \tag{1.102}$$

as the phase shift per cell of a wave having frequency ω. The sign ambiguity corresponds to the possibility of waves traveling in either direction, and the absence of real solution $\Delta\psi$ above a "cut-off" frequency $\omega_{co} = \sqrt{4k/m}$ corresponds to the absence of propagating waves above that frequency. At low frequencies,

$$m\omega^2/k \ll 1, \tag{1.103}$$

which we assume from here on, Eq. (1.102) reduces to

$$\Delta\psi \approx \pm\sqrt{\frac{m}{k}}\,\omega. \tag{1.104}$$

Our assumed solution (1.98) also depends sinusoidally on the longitudinal coordinate x, which is related to the index i by $x = ia$. At fixed time, after the phase $i\Delta\psi$ has increases by 2π, the wave returns to its initial value. In other words, the wavelength of a wave on the string is given by

$$\lambda = \frac{2\pi}{\Delta\psi}\,a \approx \frac{2\pi}{\omega}\sqrt{\frac{k}{m}}\,a, \tag{1.105}$$

and the wave speed is given by

$$v = \lambda\,\frac{\omega}{2\pi} = \sqrt{\frac{k}{m}}\,a. \tag{1.106}$$

(In this low frequency approximation) since this speed is independent of ω, low frequency pulses will propagate undistorted on the beaded string. Replacing the index i by a continuous variable $x = ia$, our conjectured solution therefore takes the form

$$\eta(x,t) = \begin{pmatrix}\sin\\ \cos\end{pmatrix}\omega\left(t \pm \frac{x}{v}\right). \tag{1.107}$$

These equations form the basis for the so-called "lumped constant delay line," especially when masses and springs are replaced by inductors and capacitors.

1.7.2
The Continuum Limit

Propagation on a continuous string can be described by appropriately taking a limit $N \to \infty$, $a \to 0$, while holding $Na = L_0 + \Delta L$. Clearly, in this limit, the low frequency approximations just described become progressively more accurate and, eventually, exact. One can approximate the terms in the Lagrangian by the relations

$$\begin{aligned}
\frac{\eta_{i+1} - \eta_i}{a} &\approx \left.\frac{\partial\eta}{\partial x}\right|_{i+1/2}, \\
\frac{\eta_i - \eta_{i-1/2}}{a} &\approx \left.\frac{\partial\eta}{\partial x}\right|_{i-1/2}, \\
\frac{\eta_{i+1} - 2\eta_i + \eta_{i-1}}{a^2} &\approx \left.\frac{\partial^2\eta}{\partial x^2}\right|_{i},
\end{aligned} \tag{1.108}$$

and, substituting from Eqs. (1.94), the Lagrange equation (1.97) becomes

$$\frac{\partial^2 \eta}{\partial t^2} = \frac{k}{m} a^2 \frac{\partial^2 \eta}{\partial x^2} = \frac{N\tau_0/\Delta L}{\mu_0(L+\Delta L)/N} \frac{(L_0+\Delta L)^2}{N^2} \frac{\partial^2 \eta}{\partial x^2}$$
$$= \frac{\tau_0}{\mu_0} \frac{L_0+\Delta L}{\Delta L} \frac{\partial^2 \eta}{\partial x^2}. \tag{1.109}$$

In this form there is no longer any reference to the (artificially introduced) beads and springs, and the wave speed is given by

$$v^2 = \frac{\tau_0}{\mu_0} \frac{L_0+\Delta L}{\Delta L}. \tag{1.110}$$

Though no physically realistic string could behave this way, it is convenient to imagine that the string is *ideal*, in the sense that with zero tension its length vanishes, $L_0 = 0$, in which case the wave equation becomes

$$\frac{\partial^2 \eta}{\partial t^2} = \frac{\tau_0}{\mu_0} \frac{\partial^2 \eta}{\partial x^2}. \tag{1.111}$$

1.7.2.1 **Sound Waves in a Long Solid Rod**

It is a bit of a digression, but a similar setup can be used to describe sound waves in a long solid rod. Superficially, Eq. (1.110) seems to give the troubling result that the wave speed is infinite since, the rod not being stretched at all, $\Delta L = 0$. A reason for this "paradox" is that the dependence of string length on tension is nonlinear at the point where the tension vanishes. (You can't "push on a string.") The formulation in the previous section only makes sense for motions per bead small compared to the extension per bead $\Delta L/N$. Stated differently, the instantaneous tension τ must remain small compared to the standing tension τ_0.

A solid, on the other hand, resists both stretching and compression and, if there is no standing tension, the previously assumed approximations are invalid. To repair the analysis one has to bring in Young's modulus Y, in terms of which the length change ΔL of a rod of length L_0 and cross sectional area A, subject to tension τ, is given by

$$\Delta L = L_0 \frac{\tau/A}{Y}. \tag{1.112}$$

This relation can be used to eliminate ΔL from Eq. (1.109). Also neglecting ΔL relative to L_0, and using the relation $\mu_0 = \rho_0 A$ between mass density and line density, the wave speed is given by

$$v^2 = \frac{\tau}{\mu_0} \frac{L_0}{L_0 \frac{1}{Y}\frac{\tau}{A}} = \frac{Y}{\rho_0}. \tag{1.113}$$

This formula for the speed of sound meets the natural requirement of depending only on intrinsic properties of the solid.

Prescription (1.112) can also be applied to evaluate the coefficient in Eq. (1.109) in the (realistic) stretched string case;

$$ka = YA, \quad \text{and} \quad \frac{a}{m} = \frac{1}{\mu}, \quad \text{give} \quad v^2 = \frac{YA}{\mu}. \tag{1.114}$$

Here Y is the "effective Young's modulus" in the particular stretched condition.

1.8
Field Theoretical Treatment and Lagrangian Density

It was somewhat artificial to treat a continuous string as a limiting case of a beaded string. The fact is that the string configuration can be better described by a continuous function $\eta(x,t)$ rather than by a finite number of discrete generalized coordinates $\eta_i(t)$. It is then natural to express the kinetic and potential energies by the integrals

$$T = \frac{\mu}{2}\int_0^L \left(\frac{\partial\eta}{\partial t}\right)^2 dx, \quad V = \frac{\tau}{2}\int_0^L \left(\frac{\partial\eta}{\partial x}\right)^2 dx. \tag{1.115}$$

In working out V here, the string has been taken to be ideal and Eq. (1.96) was expressed in continuous terms. The Lagrangian $L = T - V$ can therefore be expressed as

$$L = \int_0^L \mathcal{L}\, dx, \tag{1.116}$$

where the "Lagrangian density" $\mathcal{L}$ is given by

$$\mathcal{L} = \frac{\mu}{2}\left(\frac{\partial\eta}{\partial t}\right)^2 - \frac{\tau}{2}\left(\frac{\partial\eta}{\partial x}\right)^2. \tag{1.117}$$

Then the action S is given by

$$S = \int_{t_0}^{t_1}\int_0^L \mathcal{L}(\eta, \eta_{,x}, \eta_{,t}, x, t)\, dx\, dt. \tag{1.118}$$

For $\mathcal{L}$ as given by Eq. (1.117) not all the Lagrangian arguments shown in Eq. (1.118) are, in fact, present. Only the partial derivative of η with respect to x, which is indicated by $\eta_{,x}$ and $\eta_{,t}$, which similarly stands for $\partial\eta/\partial t$, are present. In general, $\mathcal{L}$ could also depend on η, because of nonlinear restoring force, or on x, for example because the string is nonuniform, or on t, for example because the string tension is changing (slowly) with time.

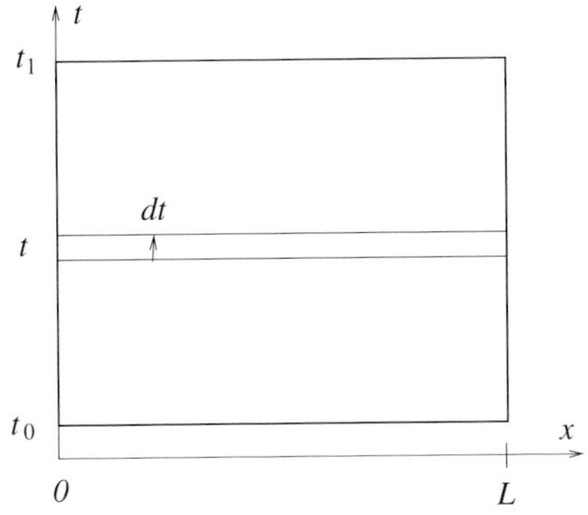

Fig. 1.7 Appropriate slicing of the integration domain for integrating the term $\partial/\partial x\big(\delta\eta\,(\partial\mathcal{L}/\partial\eta_{,x})\big)$ in Eq. (1.119).

The variation of $\mathcal{L}$ needed as the integrand of δS is given by

$$\begin{aligned}
\delta\mathcal{L} &= \mathcal{L}(\eta_{,x}+\delta\eta_{,x},\eta_{,t}+\delta\eta_{,t}) - \mathcal{L}(\eta_{,x}) \\
&\approx \frac{\partial\mathcal{L}}{\partial\eta_{,x}}\delta\eta_{,x} + \frac{\partial\mathcal{L}}{\partial\eta_{,t}}\delta\eta_{,t} \\
&= \frac{\partial}{\partial x}\left(\frac{\partial\mathcal{L}}{\partial\eta_{,x}}\delta\eta\right) + \frac{\partial}{\partial t}\left(\frac{\partial\mathcal{L}}{\partial\eta_{,t}}\delta\eta\right) - \delta\eta\left(\frac{\partial}{\partial x}\left(\frac{\partial\mathcal{L}}{\partial\eta_{,x}}\right) + \frac{\partial}{\partial t}\left(\frac{\partial\mathcal{L}}{\partial\eta_{,t}}\right)\right).
\end{aligned} \tag{1.119}$$

This is expressed as an approximation but, in the limit in which Hamilton's principle applies, the approximation will have become exact. The purpose of the final manipulation, as always in the calculus of variations, has been to re-express the integrand of δS as the sum of two terms, one of which is proportional to $\delta\eta$ and the other of which depends only on values of the functions on the boundaries.

In the present case the boundary is a rectangle bounded by $t = t_0$, $t = t$, $x = 0$, and $x = L$, as shown in Fig. 1.7. The region can, if one wishes, be broken into strips parallel to the x-axis, as shown. When integrated over any one of these strips, the first term on the right-hand side in the final form of Eq. (1.119) can be written immediately as the difference of the function in parenthesis evaluated at the end points of the strip. The integral over the second term can be evaluated similarly, working with strips parallel to the t-axis. In this way the integral over the first two terms can be evaluated as a line integral around the boundary. There is a form of Green's theorem that permits this line integral to be expressed explicitly but, for simplicity, we simply assume that this boundary integral vanishes, for example because $\delta\eta$ vanishes everywhere on the boundary.

Finally δS can be expressed as an integral over the remaining term of Eq. (1.119) and required to be zero. Because $\delta\eta$ is arbitrary, the quantity in parenthesis must therefore vanish;

$$\frac{\partial}{\partial x}\left(\frac{\partial \mathcal{L}}{\partial \eta_{,x}}\right) + \frac{\partial}{\partial t}\left(\frac{\partial \mathcal{L}}{\partial \eta_{,t}}\right) = 0. \tag{1.120}$$

This is the form taken by the Lagrange equations in this (simplest possible) continuum example. When applied to Lagrangian density (1.117), the result is a wave equation identical to Eq. (1.111).

For comparison with relativistic string theory in Chapter 12, one can introduce generalized momentum densities

$$\mathcal{P}^{(x)} = \frac{\partial \mathcal{L}}{\partial \eta_{,x}} = -\tau\frac{\partial \eta}{\partial x}, \tag{1.121}$$

$$\mathcal{P}^{(t)} = \frac{\partial \mathcal{L}}{\partial \eta_{,t}} = \mu\frac{\partial \eta}{\partial t}. \tag{1.122}$$

In terms of these quantities the wave equation is

$$\frac{\partial \mathcal{P}^{(x)}}{\partial x} + \frac{\partial \mathcal{P}^{(t)}}{\partial t} = 0. \tag{1.123}$$

Boundary conditions at the ends of the string are referred to as Dirichlet (fixed ends) or Neumann (free ends). The Dirichlet end condition can be expressed by $\mathcal{P}^{(t)}(t, x = 0, L) = 0$; the Neumann end condition by $\mathcal{P}^{(x)}(t, x = 0, L) = 0$.

A closer analogy with relativistic string theory is produced by generalizing the disturbance $\eta \rightarrow \eta^\mu$ in order to represent the disturbance as just one of the three components of a vector – transverse-horizontal, or transverse-vertical, or longitudinal. Also we introduce the abbreviations of overhead dot for $\partial/\partial t$ and prime for $\partial/\partial x$. With these changes the momentum densities become

$$\mathcal{P}^{(t)}_\mu = \frac{\partial \mathcal{L}}{\partial \dot{\eta}^\mu}, \quad \mathcal{P}^{(x)}_\mu = \frac{\partial \mathcal{L}}{\partial \eta^{\mu\prime}}. \tag{1.124}$$

An unattractive aspect of the dot and prime notation is that the indices on the two sides of these equations seem not to match. The parentheses on (t) and (x) are intended to mask this defect. In this case the Lagrange equation (1.123) also acquires an index μ, one value for each possible component of displacement;

$$\frac{\partial \mathcal{P}^{(x)}_\mu}{\partial x} + \frac{\partial \mathcal{P}^{(t)}_\mu}{\partial t} = 0, \quad \mu = x, y, z. \tag{1.125}$$

If η corresponds to, say, y-displacement, in Eq. (1.123), then that equation is reproduced by Eq. (1.125) by setting μ to y.

1.9
Hamiltonian Density for Transverse String Motion

The generalization from discrete to continuous mass distributions is less straightforward for Hamiltonian analysis than for Lagrangian analysis. In defining the Lagrangian density the spatial and time coordinates were treated symmetrically, but the Hamiltonian density has to single out time for special treatment. Nevertheless, starting from Eq. (1.122), suppressing the (t) superscript, and mimicking the discrete treatment, the Hamiltonian density is defined by

$$\mathcal{H} = \mathcal{P}\dot{\eta} - \mathcal{L}(\dot{\eta}(x,\mathcal{P}),\eta'). \tag{1.126}$$

In terms of this equation the arguments are shown only for $\mathcal{L}$, and only to make the points that $\mathcal{L}$ is independent of η and t and that, as usual, $\dot{\eta}$ has to be eliminated. Exploiting these features, $\partial\mathcal{H}/\partial t$ is given by

$$\frac{\partial\mathcal{H}}{\partial t} = \frac{\partial\mathcal{P}}{\partial t}\dot{\eta} - \frac{\partial L}{\partial\eta'}\dot{\eta}' = -\frac{\partial}{\partial x}\left(\frac{\partial\mathcal{L}}{\partial\eta'}\right)\dot{\eta} - \frac{\partial L}{\partial\eta'}\dot{\eta}'. \tag{1.127}$$

In the first form here, the usual cancellation on which the Hamiltonian formalism is based has been performed and, in the second the Lagrange equation has been used. The equation can be further simplified to

$$\frac{\partial\mathcal{H}}{\partial t} = -\frac{\partial}{\partial x}\left(\frac{\partial\mathcal{L}}{\partial\eta'}\dot{\eta}\right). \tag{1.128}$$

The Hamiltonian for the total system is defined by

$$H = \int_0^L \mathcal{H}\,dx. \tag{1.129}$$

Because energy can "slosh around" internally, one cannot expect $\mathcal{H}$ to be conserved, but one can reasonably evaluate

$$\frac{dH}{dt} = \int_0^L \frac{\partial\mathcal{H}}{\partial t}\,dx = -\int_0^L \frac{\partial}{\partial x}\left(\frac{\partial\mathcal{L}}{\partial\eta'}\dot{\eta}\right)dx = -\left[\frac{\partial\mathcal{L}}{\partial\eta'}\dot{\eta}\right]_0^L. \tag{1.130}$$

where, under the integral, because x is fixed, only the partial derivative of $\mathcal{H}$ is needed. In this form one sees that any change in total energy H is ascribable to external influence exerted on the string at its ends.

Problem 1.9.1. *For a string the Lagrangian density can be expressed in terms of T and V given in Eq. (1.115). Define kinetic energy density $\mathcal{T}$ and potential energy density $\mathcal{V}$ and show that*

$$\mathcal{H} = \mathcal{T} + \mathcal{V}. \tag{1.131}$$

Problem 1.9.2. *Show that, for a string with* passive *(fixed or free) connections at its ends, the total energy is conserved.*

Problem 1.9.3. *For a nonuniform string the mass density $\mu(x)$ depends on position x, though not on time. The tension $\tau(x)$ may also depend on x, perhaps because its own weight causes tension as the string hangs vertically. Repeat the steps in the discussion of Hamiltonian density and show how all equations can be generalized so that the same conclusions can be drawn.*

1.10
String Motion Expressed as Propagating and Reflecting Waves

(Following Zwiebach) the general motion of a string can be represented as a superposition of traveling waves, with reflections at the ends dependent on the boundary conditions there. For simplicity here, let us assume the boundaries are free – so-called Neumann boundary conditions. Such boundary conditions can be achieved, in principle, by attaching the ends of the string to rings that are free to ride frictionlessly on rigid, transverse posts. The slope of the string at a free end has to vanish since there can be no transverse external force capable of balancing a transverse component of the force of tension.

The general solution for transverse displacement of a string stretched on the range $0 \leq x \leq L$, for which the wave speed is $v = \sqrt{T/\mu}$, is

$$y = \frac{1}{2}\Big(f(vt + x) + g(vt - x)\Big). \tag{1.132}$$

Here f and g are arbitrary functions. Because of the free end at $x = 0$, one has

$$0 = \frac{\partial y}{\partial x}\bigg|_{x=0} = \frac{1}{2}\Big(f'(vt) - g'(vt)\Big). \tag{1.133}$$

As time evolves, since the argument vt takes on all possible values, this equation can be expressed as

$$f'(u) = g'(u), \tag{1.134}$$

for arbitrary argument u. One therefore has

$$f(u) = g(u) + \text{constant}. \tag{1.135}$$

Since the "constant" can be suppressed by redefinition of f, this can be expressed, without loss of generality, by $f(u) = g(u)$ and the general solution written as

$$y = \frac{1}{2}\Big(f(vt + x) + f(vt - x)\Big). \tag{1.136}$$

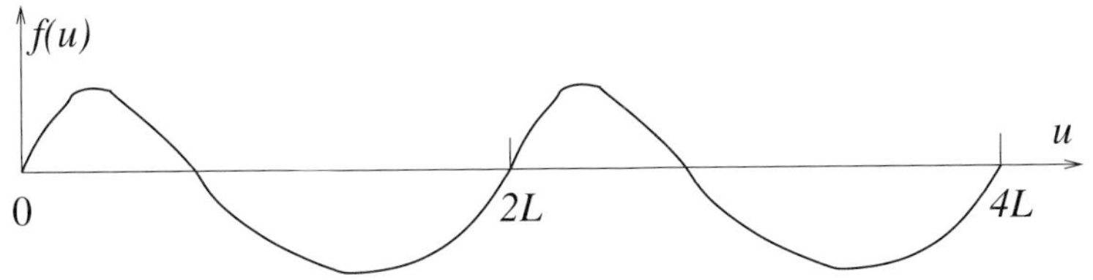

Fig. 1.8 The shape of a $2L$-periodic function $f(u)$ which can produce general string motion as a superposition of the form (1.136).

Because there is also a free end at $x = L$ we have

$$0 = \left.\frac{\partial y}{\partial x}\right|_{x=L} = \frac{1}{2}\Big(f'(vt+L) - f'(vt-L)\Big). \tag{1.137}$$

Again using the variable u to express a general argument, it follows that $f'(u)$ is a periodic function of u with period $2L$;

$$f'(u+2L) = f'(u). \tag{1.138}$$

This relation is consistent with a term in $f(u)$ proportional to u, but if one or both of the string ends are fixed such an inexorable growth would be excluded and $f(u)$ would be a function such as shown in Fig. 1.8. Any function satisfying Eq. (1.138) can be expressed as a Fourier series;

$$f'(u) = f_1 + \sum_{n=1}^{\infty}\Big(a_n \cos\frac{\pi}{L}nu + b_n \sin\frac{\pi}{L}nu\Big). \tag{1.139}$$

This can be integrated and, for simplicity, new coefficients introduced to swallow the multiplicative factors;

$$f(u) = f_0 + f_1 u + \sum_{n=1}^{\infty}\Big(A_n \cos\frac{\pi}{L}nu + B_n \sin\frac{\pi}{L}nu\Big). \tag{1.140}$$

The general solution can then be written by inserting this sum into Eq. (1.136). Restoring the explicit dependences on x and t, and using well-known trigonometric identities yields

$$y = f_0 + f_1 vt + \sum_{n=1}^{\infty}\Big(A_n \cos n\frac{\pi v t}{L} + B_n \sin n\frac{\pi v t}{L}\Big)\cos n\frac{\pi x}{L}. \tag{1.141}$$

Stated as an initial value problem, the initial displacements and velocities would be given functions $y|_0$ and $\partial y/\partial t|_0$, which are necessarily expressible as

$$\begin{aligned} y|_0(x) &= f_0 + \sum_{n=1}^{\infty}\Big(A_n \cos n\frac{\pi x}{L}\Big),\\ \left.\frac{\partial y}{\partial t}\right|_0(x) &= f_1 v + \frac{\pi v}{L}\sum_{n=1}^{\infty}\Big(n\, B_n \cos n\frac{\pi x}{L}\Big). \end{aligned} \tag{1.142}$$

This permits the coefficients to be determined:

$$\begin{aligned}\int_0^L y|_0(x)\,dx &= f_0 L,\\ \int_0^L \cos\left(\frac{m\pi x}{L}\right) y|_0(x)\,dx &= A_m \frac{L}{2},\\ \int_0^L \sin\left(\frac{m\pi x}{L}\right) \frac{\partial y}{\partial t}\bigg|_0(x)\,dx &= \frac{m\pi v}{L} B_m \frac{L}{2}.\end{aligned} \tag{1.143}$$

Motion always "back and forth" between two limits, say a and b, in one dimension, due to a force derivable from a potential energy function $V(x)$, is known as "libration." Conservation of energy then requires the dependence of velocity on position to have the form

$$\dot{x}^2 = (x-a)(b-x)\,\psi(x), \quad \text{or} \quad \dot{x} = \pm\sqrt{(x-a)(b-x)\,\psi(x)}, \tag{1.144}$$

where $\psi(x) > 0$ through the range $a \leq x \leq b$, but is otherwise an arbitrary function of x (derived, of course, from the actual potential function). It is necessary to toggle between the two $\pm$ choices depending on whether the particle is moving to the right or to the left. Consider the change of variable $x \to \theta$ defined by

$$x = \alpha - \beta\cos\theta, \quad \text{where} \quad \alpha - \beta = a, \quad \alpha + \beta = b. \tag{1.145}$$

1.11 Problems

Problem 1.11.1. *Show that* $(x-a)(b-x) = \beta^2\sin^2\theta$ *and that energy conservation is expressed by*

$$\dot{\theta} = \sqrt{\psi(\alpha - \beta\cos\theta)}, \tag{1.146}$$

where there is no longer a sign ambiguity because $\dot{\theta}$ *is always positive. The variable* θ *is known as an "angle variable." One-dimensional libration motion can always be expressed in terms of an angle variable in this way, and then can be "reduced to quadrature" as*

$$t = \int^{\theta} \frac{d\theta'}{\sqrt{\psi(\alpha - \beta\cos\theta')}}. \tag{1.147}$$

This type of motion is especially important in the conditionally periodic motion of multidimensional oscillatory systems. This topic is studied in Section 14.6.

Problem 1.11.2. *The Lagrangian*

$$L = \frac{1}{2}(\dot{x}^2 + \dot{y}^2) - \frac{1}{2}(\omega^2 x^2 + \omega_2^2 y^2) + \alpha x y, \tag{1.148}$$

with $|\alpha| \ll \omega^2$ *and* $|\alpha| \ll \omega_2^2$, *describes two oscillators that are weakly coupled.*

(a) *Find normal coordinates and normal mode frequencies* Ω_1 *and* Ω_2.

(b) *For the case* $\omega = \omega_2$, *describe free motion of the oscillator.*

(c) *Holding* α *and* ω_2 *fixed, make a plot of* Ω *versus* ω *showing a branch for each of* Ω_1 *and* Ω_2. *Do it numerically or with a programming language if you wish. Describe the essential qualitative features exhibited? Note that the branches do not cross each other.*

Problem 1.11.3. *In integral calculus the vanishing of a definite integral does not, in general, imply the vanishing of the integrand; there can be cancellation of negative and positive contributions to the integral. Yet, in deriving Eqs. (1.25) and (1.120), just such an inference was drawn. Without aspiring to mathematical rigor, explain why the presence of an arbitrary multiplicative factor in the integrand makes the inference valid.*

Problem 1.11.4. *Transverse oscillations on a string with just three beads, shown in Fig. 1.2, has been analyzed in Problem 1.4.3. The infinite beaded string shown in Fig. 1.6 is similarly capable of transverse oscillation, with transverse bead locations being* $\dots, y_{i-1}, y_i, y_{i+1}, \dots$. *Using the parameters k and m of the longitudinal model, replicate, for transverse oscillations, all the steps that have been made in analyzing longitudinal oscillations of the beaded string. Start by finding the kinetic and potential energies and the Lagrangian and deriving the Lagrange equations of the discrete system, and finding the propagation speed. Then proceed to the continuum limit, deriving the wave equation and the Lagrangian density.*

Problem 1.11.5. Struck string. *To obtain a short pulse on a stretched string it is realistic to visualize the string being struck with a hammer, as in a piano, rather than being released from rest in a distorted configuration. Consider an infinite string with tension* T_0 *and mass density* μ_0. *An impulse I (which is force times time) is administered at position* $x = x_0$ *to a short length* Δx *of the string. Immediately after being struck, while the string is still purely horizontal, the string's transverse velocity can be expressed (in terms of unit step function U) by a square pulse*

$$\frac{\partial y}{\partial t}(0+, x) = K\Big(U(x - x_0) - U\big(x - (x_0 + \Delta x)\big)\Big). \tag{1.149}$$

(a) *Express the constant K in terms of the impulse I and establish initial traveling waves on the string which match the given initial excitation. Sketch the shape of the string for a few later times.*

(b) *Consider a length of the same string stretched between smooth posts at* $x = 0$ *and* $x = a$. *(i.e., Neumann boundary conditions). Describe (by words or sketches) the subsequent motion of the string. Does the motion satisfy the conservation of momentum?*

Problem 1.11.6. *In the same configuration as in Problem 1.11.5, with the string stretched between smooth posts, let the initial transverse velocity distribution be given by*

$$y(0+, x) = 0, \quad \frac{1}{v_0}\frac{\partial y}{\partial t}(0+, x) = \frac{x - a/2}{a} - 4\left(\frac{x - a/2}{a}\right)^3. \tag{1.150}$$

Find the subsequent motion of the string.

Problem 1.11.7. *On a graph having* x *as abscissa and* vt *as ordinate the physical condition at* $t = 0$ *of a string with wave speed* v *stretched between* 0 *and* a *can be specified by values of the string displacement at all points on the horizontal axis between* 0 *and* a. *An "area of influence" on this graph can be specified by a condition such as "the area containing points at which a positive-traveling wave could have been launched that affects the* $t = 0$ *condition of the string." Other areas of influence can be specified by replacing "positive-traveling" by "negative-traveling" or by reversing cause and effect. From these areas find the region on the plot containing points at which the nature of the end connection* cannot *be inferred from observation of the string.*

Bibliography

General References

1 L.D. Landau and E.M. Lifshitz, *Classical Mechanics*, Pergamon, Oxford, 1976.

References for Further Study

Section 1.4

2 R. Courant and D. Hilbert, *Methods of Mathematical Physics*, Vol. 1, Interscience, New York, 1953, p. 37.

Section 1.10

3 B. Zwiebach, *A First Course in String Theory*, Cambridge University Press, Cambridge, UK, 2004.

2
Geometry of Mechanics, I, Linear

Even before considering geometry as physics, one can try to distinguish between geometry and algebra, starting, for example, with the concept of "vector." The question "What is a vector?" does not receive a unique answer. Rather, two answers are perhaps equally likely: "an arrow," or "a triplet of three numbers (x, y, z)." The former answer could legitimately be called geometric, the latter algebraic. Yet the distinction between algebra and geometry is rarely unambiguous. For example, experience with the triplet (x, y, z) was probably gained in a course with a title such as "Analytic Geometry" or "Coordinate Geometry." For our purposes it will not be necessary to have an iron-clad postulational basis for the mathematics to be employed, but it is important to have a general appreciation of the ideas. That is one purpose of this chapter.

Since the immediate goal is *unlearning* almost as much as learning, the reader should not expect to find a completely self-contained, unambiguous, development from first principles. For progress in physics it is usually sufficient to have only an intuitive grasp of mathematical foundations. For example, the Pythagorean property of right-angle triangles is remembered even if its derivation from Euclidean axioms is not. Still, some mulling over of "well established" ideas is appropriate, as they usually contain implicit understandings and definitions, possibly different for different individuals. Some of the meanings have to be discarded or modified as an "elementary" treatment morphs into a more "abstract" formulation. Faced with this problem a mathematician might prefer to "start from scratch," discard all preconceived notions, define everything unambiguously, and proceed on a firm postulational basis.[1] The physicist, on the other hand, is likely to find the "mathematician's" approach too formal and poorly motivated. Unwilling to discard ideas that have served well, and too impatient or too inexperienced to follow abstract argument, when taking on new baggage, he or she prefers to rearrange the baggage already loaded, in an effort to make it all fit. The purpose of this

1) Perhaps the first treatment from first principles, and surely the most comprehensive text to base mechanics on the formal mathematical theory of smooth manifolds, was Abraham and Marsden, *Foundations of Mechanics*. Other editions with new authors have followed.

Geometric Mechanics: Toward a Unification of Classical Physics. 2nd Edition. Richard Talman

ISBN: 978-3-527-40683-8

chapter is to help with this rearrangement. Elaborating the metaphor, some bags are to be removed from the trunk with the expectation they will fit better later, some fit as is, some have to be reoriented; only at the end does it become clear which fit and which must be left behind. While unloading bags it is not necessary to be fussy, when putting them back one has to be more careful.

The analysis of spatial rotations has played a historically important part in the development of mechanics. In classical (both with the meaning nonrelativistic and the meaning "old fashioned") mechanics courses this has largely manifested itself in the analysis of rigid body motion. Problems in this area are among the most complicated for which the equations can be "integrated" in closed analytic form in spite of being inherently "nonlinear," a fact that gives them a historical importance. But since these calculations are rather complicated, and since most people rapidly lose interest in, say, the eccentric motion of an asymmetric top, it has been fairly common, in the pedagogy of mechanics courses, to skim over this material.

A "modern" presentation of mechanics has a much more qualitative and geometric flavor than the "old fashioned" approach just mentioned. From this point of view, rather than being just a necessary evil encountered in the solution of hard problems, rotations are the easiest-to-understand prototype for the analysis of motion using abstract group theoretical methods. The connection between rotational symmetry and conservation of angular momentum, both because of its importance in quantum mechanics, and again as a prototype, provides another motivation for studying rotations.

It might be said that classical mechanics has been left mainly in the hands of mathematicians – physicists were otherwise occupied with quantum questions – for so long that the language has become nearly unintelligible to a physicist. Possibly unfamiliar words in the mathematician's vocabulary include bivectors, multivectors, differential forms, dual spaces, Lie groups, irreducible representations, pseudo-Euclidean metrics, and so on. Fortunately all physicists are handy with vector analysis, including the algebra of dot and cross products, and the calculus of gradients, divergences, and curls, and in the area of tensor analysis they are familiar with covariant (contravariant) tensors as quantities with lower (upper) indices that (for example) conveniently keep track of the minus sign in the Minkowski metric of special relativity. Tools like these are much to be valued in that they permit a very compact, very satisfactory, formulation of classical and relativistic mechanics, of electricity and magnetism, and of quantum mechanics. But they also leave a physicist's mind unwilling to jettison certain "self-evident" truths that stand in the way of deeper levels of abstraction. Perhaps the simplest example of this is that, having treated vector cross products as ordinary vectors for many years, one's mind has difficulty adopting a mathematician's view of cross products as being quite dissimilar to, and certainly incommensurable with, ordinary vectors.

Considerable effort will be devoted to motivating and explaining ideas like these in ways that are intended to appeal to a physicist's intuition. Much of this material has been drawn from the work of Elie Cartan which, though old, caters to a physicist's intuition.[2] To begin with, *covariant* vectors will be introduced from various points of view, and contrasted with the more familiar, *contravariant* vectors.

2.1
Pairs of Planes as Covariant Vectors

The use of coordinates (x, y, z) – shortly we will switch to (x^1, x^2, x^3) – for locating a point in space is illustrated in Fig. 2.1. Either orthonormal (Euclidean) or skew (Cartesian)[3] axes can be used. It is rarely required to use skew axes rather than the simpler rectangular axes but, in the presence of continuous deformations, skew axes may be unavoidable. Next consider Fig. 2.2 which shows the intersections of a plane with the same axes as in Fig. 2.1. The equation of the plane on the left, in terms of generic point (x, y, z) on the plane, is

$$ax + by + cz = d, \tag{2.1}$$

and, because the coordinates of the intersections with the axes are the same, the equation of the plane on the right in terms of generic point (x', y', z') is also linear, with the same coefficients (a, b, c, d),

$$ax' + by' + cz' = d, \tag{2.2}$$

The figures are "similar," not in the conventional sense of Euclidean geometry, but in a newly defined sense that lines correspond to lines, planes to planes, intersections to intersections, and that the coordinates of the intersections of the plane and the axes are numerically the same. The unit measuring sticks along the Euclidean axes are $\mathbf{e}_x$, $\mathbf{e}_y$, $\mathbf{e}_z$, and along the skew axes $\mathbf{e}_{x'}$, $\mathbf{e}_{y'}$, $\mathbf{e}_{z'}$. The coordinates $(d/a, d/b, d/c)$ of the intersection points are determined by laying out the measuring sticks along the respective axes. Much as (x, y, z) "determines" a point, the values (a, b, c), along with d, "determine" a plane.

2) Cartan is usually credited as being the "father" (though I think not the inventor) of differential forms, as well as the discoverer of spinors (long before and in greater generality than) Pauli or Dirac. That these early chapters draws so much from Cartan simply reflects the lucidity of his approach. Don't be intimidated by the appearance of *spinor's*; only elementary aspect of them will be required.

3) Many authors use the term "Cartesian" to imply orthogonal axes, but we use "Euclidean" in that case and use "Cartesian" to imply (possibly) skew axes.

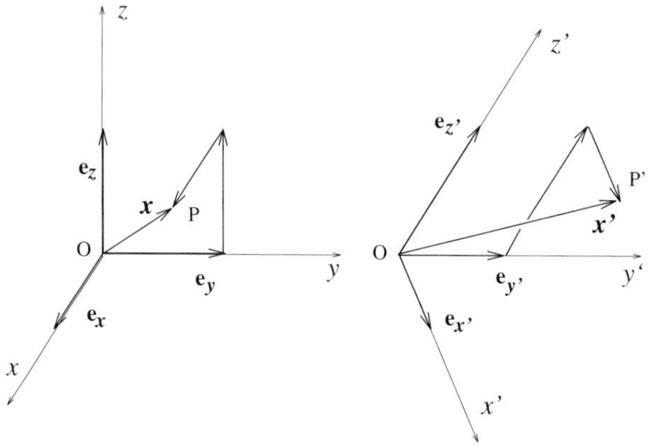

Fig. 2.1 Attaching coordinates to a point with Euclidean (or orthogonal) axes (on the left) and Cartesian (or possibly skew) axes (on the right). One of several possible interpretations of the figure is that the figure on the right has been obtained by elastic deformation of the figure on the left. In that case the primes on the right are superfluous since the coordinates of any particular point (such as the point P) is the same in both figures, namely $(1, 1, 1)$.

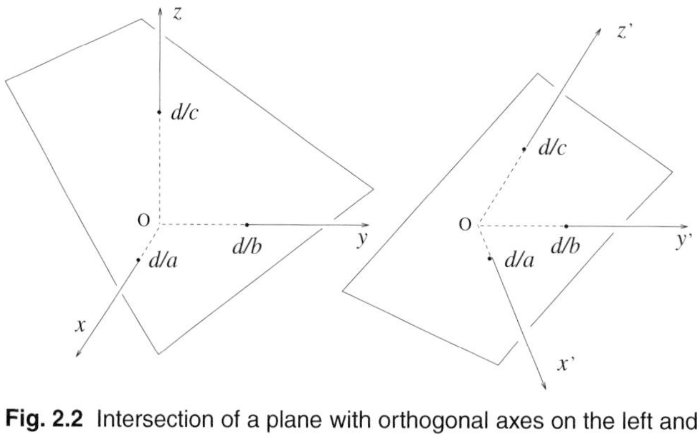

Fig. 2.2 Intersection of a plane with orthogonal axes on the left and a "similar" figure with skew axes on the right. The equations of the planes are "the same," though expressed with unprimed and primed coordinates.

Commonly the values (x, y, z) are regarded as projections onto the axes of an arrow that is allowed to slide around the plane with length and direction held fixed. Similarly, any two planes sharing the same triplet (a, b, c) are parallel. (It would be wrong though to say that such planes have the same normals since, with the notion of *orthogonality* not yet having been introduced, there is no such thing as a vector normal to the plane. Saying that two planes have the same "direction" can only mean that they are parallel – that is, their intercepts are proportional.)

The analogy between plane coordinates (a, b, c) and point coordinates (x, y, z) is not quite perfect. For example, it takes the specification of a definite value d in the equation for the plane to pick out a definite plane, while it takes three values, say the (x_0, y_0, z_0) coordinates of the tail, to pick out a particular vector. Just as one regards (x, y, z) as specifying a sliding vector, it is possible to define a "plane-related" geometric structure specified by (a, b, c) with no reference to d. To suppress the dependence on parameter d, first observe that the shift from d to $d + 1$ corresponds to the shift from the plane $ax + by + cz = d$ to the plane $ax + by + cz = d + 1$. Each member of this pair of unit-separated planes is parallel to any plane with the same (a, b, c) values. The pair of planes is said to have an "orientation,"[4] with positive orientation corresponding to increasing d. This is illustrated in Fig. 2.3. Since it is hard to draw planes, only lines are shown there, but the correspondence with Fig. 2.2 should be clear – and the ideas can be extended to higher dimensionality as well. In this way the triplet (a, b, c) – or (a_1, a_2, a_3), a notation we will switch to shortly – stands for any oriented, unity-spaced, pair of planes, both parallel to the plane through the origin $ax + by + cz = 0$.

Without yet justifying the terminology we will call $\mathbf{x}$ a *contravariant* vector, even though this only makes sense if we regard $\mathbf{x}$ as an abbreviation for the three numbers (x^1, x^2, x^3); it would be more precise to call $\mathbf{x}$ a *true* vector with *contravariant* components $(x, y, z) \equiv (x^1, x^2, x^3)$, so that $\mathbf{x} \equiv x^1\mathbf{e}_1 + x^2\mathbf{e}_2 + x^3\mathbf{e}_3$.

In some cases, if it appears to be helpful, we will use the symbol $\vec{\mathbf{x}}$, instead of just $\mathbf{x}$, to emphasize its *contravariant* vector nature. Also we symbolize by $\widetilde{\mathbf{a}}$, an object with components $(a, b, c) = (a_1, a_2, a_3)$, that will be called *covariant*.[5]

4) The orientation of a pair of planes is said to be "outer." The meaning of *outer* orientation is that two points related to each other by this orientation must be *in separate planes*. This can be contrasted to the *inner* orientation of a vector, by which two points can be related only if they lie *in the same line* parallel to $\mathbf{x}$. An *inner* orientation for a plane would be a clockwise or counter-clockwise orientation of circulation within the plane. An *outer* orientation of a vector is a left or right-handed screw-sense about the vector's arrow.

5) "New" notation is to be discouraged but, where there appears to be no universally agreed-to notation, our policy will be to choose symbols that cause formulas to look like elementary physics, even if their meanings are more general. The most important convention is that multicomponent objects are denoted by bold face symbols, as for example the vector $\mathbf{x}$. This is more compact, though less expressive, than $\vec{x}$. Following Schutz, *Geometrical Methods of Mathematical Physics*, we use an overhead tilde to distinguish a 1-form (or covariant vector, such as $\widetilde{\mathbf{a}}$), from a (contravariant) vector, but we also retain the bold face symbol. The use of tildes to distinguish between covariant and contravariant quantities will break down when *mixed* quantities enter. Many mathematicians use no notational device at all to distinguish these quantities, and we will be forced to that when encountering mixed tensors. When it matters, the array of contravariant components will be regarded as a column vector and the array of covariant components as a row vector, but consistency in this regard is not guaranteed.

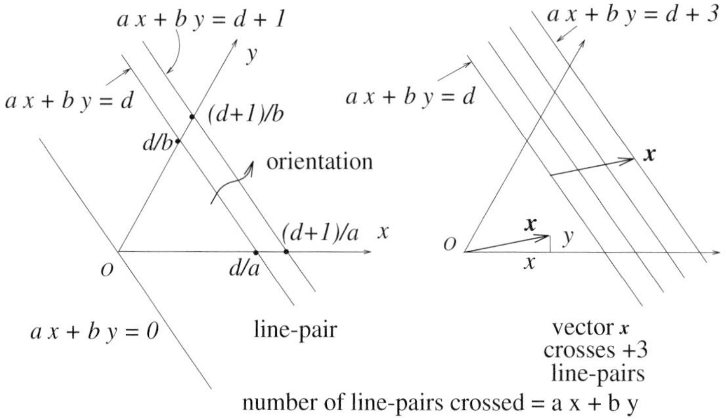

Fig. 2.3 Parallel planes. How many plane-pairs $\widetilde{\mathbf{a}}$ are crossed by vector x?

In Fig. 2.3 the (outer) orientation of two planes is indicated by an arrow (wavy to indicate that no definite vector is implied by it.) It is meaningful to say that the contravariant vector **x** and the covariant vector $\widetilde{\mathbf{a}}$ have the same orientation; it means that the arrow **x** points from the negative side of the plane toward the positive side. Other than being able to compare their orientations, is there any other meaningful geometric question that can be asked of **x** and $\widetilde{\mathbf{a}}$? The answer is yes; the question is "How many plane-pairs $\widetilde{\mathbf{a}}$ does **x** cross?" In Fig. 2.3 the answer is "3." Is there any physics in this question and answer? The answer is yes again. Visualize the right plot of Fig. 2.3 as a topographical map, with the parallel lines being contours of equal elevation. (One is looking on a fine enough scale that the ground is essentially plane and the contours are straight and parallel.) The "trip" **x** entails a change of elevation of three units. This permits us to anticipate/demand that the following expressions (all equivalent):

$$ax + by + cz \equiv a_i x^i \equiv \langle \widetilde{\mathbf{a}}, \mathbf{x} \rangle \equiv \langle \mathbf{x}, \widetilde{\mathbf{a}} \rangle \equiv \widetilde{\mathbf{a}}(\mathbf{x}) \tag{2.3}$$

have an intrinsic, *invariant*, significance, unaltered by deformations and transformations (if and when they are introduced.) This has defined a kind of "invariant product"$\langle \widetilde{\mathbf{a}}, \mathbf{x} \rangle$.[6 7] of a covariant and a contravariant vector. It has also introduced the repeated-index summation convention. This product is

6) The left-hand side of Eq. (2.3), being homogeneous in (x, y, z), is known as a "form" and, being first order in the x^i, as a "1-form." The notations $a_i x^i$, $\langle \widetilde{\mathbf{a}}, \mathbf{x} \rangle$, and $\widetilde{\mathbf{a}}(\mathbf{x})$ are interchangeable.

7) In this elementary context, the invariant significance of $\widetilde{\mathbf{a}}(\mathbf{x})$ is utterly trivial, and yet when the same concept is introduced in the abstract context of tangent spaces and cotangent spaces, it can seem obscure. See, for example Arnold, p. 203, Fig. 166.

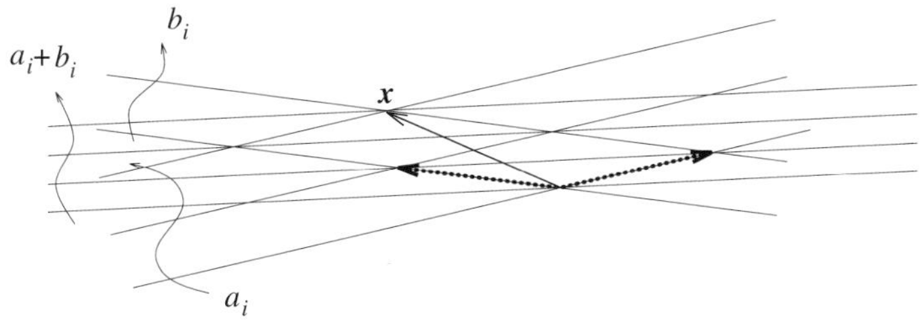

Fig. 2.4 Geometric interpretation of the addition of covariant vectors, $\widetilde{\mathbf{a}} + \widetilde{\mathbf{b}}$. The solid arrow crosses 2 of the $\widetilde{\mathbf{a}}$ plane-pairs, 1 of the $\widetilde{\mathbf{b}}$ plane-pairs, and hence 3 of the $\widetilde{\mathbf{a}} + \widetilde{\mathbf{b}}$ plane-pairs. Tips of the two dotted arrows necessarily lie on the same $\widetilde{\mathbf{a}} + \widetilde{\mathbf{b}}$ plane.

clearly related to the "dot product" of elementary vector analysis $\mathbf{a} \cdot \mathbf{x}$, but that notation would be inappropriate at this point since nothing resembling an angle between vectors, or the cosine thereof, has been introduced. The geometric interpretation of the sum of two vectors as the arrow obtained by attaching the tail of one of the arrows to the tip of the other is well known. The geometric interpretation of the addition of covariant vectors is illustrated in Fig. 2.4. As usual, the natural application of a covariant vector is to determine of the number of plane-pairs crossed by a general contravariant vector. Notice that the lines of $\widetilde{\mathbf{a}} + \widetilde{\mathbf{b}}$ are more closely spaced than the lines of either $\widetilde{\mathbf{a}}$ or $\widetilde{\mathbf{b}}$. The geometry of this figure encapsulates the property that a general vector $\mathbf{x}$ crosses a number of planes belonging to $\widetilde{\mathbf{a}} + \widetilde{\mathbf{b}}$ equal to the number of planes belonging to $\widetilde{\mathbf{a}}$ it crosses plus the number it crosses belonging to $\widetilde{\mathbf{b}}$.

There are various ways of interpreting figures like Fig. 2.2. The way intended so far can be illustrated by an example. Suppose you have two maps of Colorado, one having lines of longitude and latitude plotted on a square grid, the other using some other scheme – say a globe. To get from one of these maps to the other some distortion would be required, but one would not necessarily say there had been a coordinate transformation since the latitude and longitude coordinates of any particular feature would be the same on both maps; call them (x, y). One can consider the right figure of Fig. 2.2 to be the result of a deformation of the figure on the left – both the physical object (the plane or planes) and the reference axes have been deformed – preserving the coordinates of every particular feature. The map analog of the planes in Fig. 2.2 are the equal-elevation contours of the maps. By counting elevation contours one can, say, find the elevation of Pike's Peak relative to Denver. (It would be necessary to break this particular trip into many small segments in order that the ground could be regarded as a perfect plane in each segment.) With local contours represented by $\widetilde{\mathbf{a}}_{(i)}$ and local transverse displacements vectors by $\mathbf{x}_{(i)}$, the overall change in elevation is obtained by

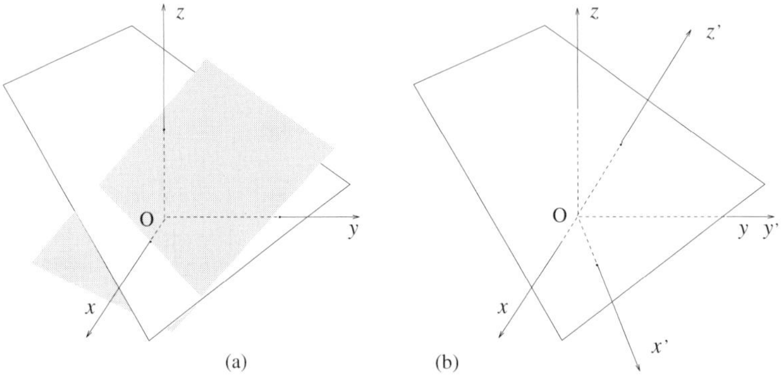

Fig. 2.5 "Active" (a) and "passive" (b) interpretations of the relations between elements of Fig. 2.2 as transformations. In each case, though they were plotted separately in Fig. 2.2 the plots are here superimposed with the common origin O.

summing the contributions $\langle \widetilde{\mathbf{a}}_{(i)}, \mathbf{x}_{(i)} \rangle$ from each segment.[8] Clearly one will obtain the same result from both maps. This is a virtue of the form $\langle \widetilde{\mathbf{a}}, \mathbf{x} \rangle$. As stated previously, no coordinate transformation has yet occurred, but when one does, we will wish to preserve this feature – if $\mathbf{x} \to \mathbf{x}'$ we will insist that $\widetilde{\mathbf{a}} \to \widetilde{\mathbf{a}}'$ such that the value of the form is preserved. That is an *invariant*.

Figure 2.2 can also be interpreted in terms of "transformations" either active or passive. The elements of this figure are redrawn in Fig. 2.5 but with origins superimposed. In part (a) the plane is "actively" shifted. Of course its intersections with the coordinate axes will now be different. The coefficients (covariant components) in the equation of the shifted plane expressed in terms of the original axes are altered from their original values. The new coefficients are said to be the result of an active transformation in this case. Part (b) of Fig. 2.5 presents an alternative view of Fig. 2.2 as a "passive" change of coordinates. The plane is now unshifted but its covariant components are still transformed because of the different axes. Similar comments apply to the transformation properties of contravariant components.[9] From what has been stated previously, we must require the form $\langle \widetilde{\mathbf{a}}, \mathbf{x} \rangle$ to be invariant under transformation. This is true whether the transformation is viewed actively or passively.

8) As always in this text, subscripts (i) are enclosed in parenthesis to protect against their being interpreted as vector indices. There is no implied summation over repeated parenthesized indices.

9) While it is always clear that two possible interpretations exist, it is often difficult to understand which view is intended. A certain fuzziness as to whether an active or a passive view is intended is traditional – a tradition this text will regrettably continue to respect. In many cases the issue is inessential, and in any case it has nothing to do with the contravariant/covariant distinction.

2.2 Differential Forms

2.2.1 Geometric Interpretation

There is a formalism which, though it seems curious at first, is in common use in modern mechanics. These so-called *differential forms* will not be used in this chapter, but they are introduced at this point, after only a minimal amount of geometry has been introduced, in order to emphasize that the concepts involved are very general, independent of any geometry yet to be introduced. In particular there is no dependence on lengths, angles, or orthogonality. Since the new ideas can be adequately illustrated by considering functions of two variables, x and y, we simplify accordingly and *define* elementary differential forms $\widetilde{\mathbf{dx}}$ and $\widetilde{\mathbf{dy}}$ as *functions* (of a vector) satisfying

$$\widetilde{\mathbf{dx}}(\mathbf{\Delta x}) = \Delta x, \quad \widetilde{\mathbf{dy}}(\mathbf{\Delta x}) = \Delta y; \tag{2.4}$$

these *functions* take the displacement vector $\mathbf{\Delta x} = \mathbf{x} - \mathbf{x}_0$ as argument and produce components $\Delta x = x - x_0$ and $\Delta y = y - y_0$ as values.[10] A linear superposition of $\widetilde{\mathbf{dx}}$ and $\widetilde{\mathbf{dy}}$ with coefficients a and b is defined by[11]

$$(a\,\widetilde{\mathbf{dx}} + b\,\widetilde{\mathbf{dy}})(\mathbf{\Delta x}) = a\,\Delta x + b\,\Delta y. \tag{2.5}$$

In practice Δx and Δy will always be infinitesimal quantities and the differentials will be part of a "linearized" or "first term in Taylor series" procedure. Consider a scalar function $h(x,y)$ – for concreteness let us take $h(x,y)$ to be the elevation above sea level at location (x,y). By restricting oneself to a small enough region about some reference location (x_0,y_0), $h(x,y)$ can be *linearized* – i.e., approximated by a linear expansion

$$h(x,y) = h(x_0,y_0) + a\,\Delta x + b\,\Delta y. \tag{2.6}$$

In the language of differential forms this same equation is written as

$$\widetilde{\mathbf{dh}} = a\,\widetilde{\mathbf{dx}} + b\,\widetilde{\mathbf{dy}}, \tag{2.7}$$

10) Though the value of a differential form acting on a vector is a real number, it is not a scalar in general. A possibly helpful mnemonic feature of the notation is that to produce a regular face quantity from a bold face quantity requires the use of another bold face quantity.

11) That the symbols $\widetilde{\mathbf{dx}}$ and $\widetilde{\mathbf{dy}}$ *are not ordinary differentials* is indicated by the bold face type and the overhead tildes. They are being newly defined here. Unfortunately, a more common notation is to represent a differential form simply as dx; with this notation it is necessary to distinguish by context between differential forms and ordinary differentials. A converse ambiguity in our terminology is that it may not be clear whether the term *differential form* means $a\,dx + b\,dy$ or $\widetilde{\mathbf{dh}}$.

where, evidently,

$$a = \left.\frac{\partial h}{\partial x}\right|_{x_0,y_0}, \quad b = \left.\frac{\partial h}{\partial y}\right|_{x_0,y_0}. \tag{2.8}$$

This shows that $\widetilde{\mathbf{dh}}$ is closely connected to the *gradient* of ordinary vector analysis. It is not the same thing, though, since the ordinary gradient is orthogonal to contours of constant h and the concept of orthogonality has not yet been introduced. Note that $\widetilde{\mathbf{dh}}$ is independent of $h(x_0, y_0)$. (Neglecting availability of oxygen and dependence of g on geographic location, the difficulty of climbing a hill is independent of the elevation at its base.)

Returning to the map of Colorado, imagine a trip made up of numerous path intervals $\mathbf{x}_{(i)}$. The change in elevation $h_{(i)}$ during the incremental path interval $\mathbf{x}_{(i)}$ is given by

$$h_{(i)} = a_{(i)}x_{(i)} + b_{(i)}y_{(i)} = (a_{(i)}\widetilde{\mathbf{dx}} + b_{(i)}\widetilde{\mathbf{dy}})(\mathbf{x}_{(i)}) = \widetilde{\mathbf{dh}}_{(i)}(\mathbf{x}_{(i)}). \tag{2.9}$$

Since this equation resembles Eq. (2.3), it can also be written as

$$h_{(i)} = \langle \widetilde{\mathbf{dh}}_{(i)}, \mathbf{x}_{(i)} \rangle. \tag{2.10}$$

The total change of elevation, h, can be obtained by summing over the incremental paths

$$h = \sum_i \langle \widetilde{\mathbf{dh}}_{(i)}, \mathbf{x}_{(i)} \rangle. \tag{2.11}$$

As usual, such a summation becomes an integral in the limit of small steps, with integration limits at beginning B and end E of the trip,

$$h = \int_B^E \langle \widetilde{\mathbf{dh}}, \mathbf{dx} \rangle, \quad \text{or simply,} \quad h = \int_C \widetilde{\mathbf{dh}}, \tag{2.12}$$

where C is the trip curve. The first form has an unambiguous, coordinate-free meaning that makes it clear that the result is invariant, in the sense discussed above. In the second form the integration symbol has been *re-defined* to implicitly include the left out part of the first form. When expanded in components, the formula takes on a more customary appearance.

$$h = \int_B^E (a(\mathbf{x})\,dx + b(\mathbf{x})\,dy). \tag{2.13}$$

Example 2.2.1. *Three points $P_i, i = 1,2,3$, with coordinates $(x_{(i)}, y_{(i)}, z_{(i)})$, are fixed in ordinary space. (i) Dividing Eq. (2.1) by d, find the coefficients in the equation*

$$x\frac{a}{d} + y\frac{b}{d} + z\frac{c}{d} = 1, \tag{2.14}$$

of the plane passing through the points. (ii) Defining $h(x,y)$ as the elevation z at point (x,y), evaluate $\widetilde{\mathbf{dh}}$ at the point P_1. (iii) For a general point P whose horizontal displacements relative to P_1 are given by $(\Delta x = x - x_{(1)}, \Delta y = y - y_{(1)})$, find its elevation $\Delta h = z - z_{(1)}$ relative to P_1.

Solutions

(i) *Ratios of the coefficients (a,b,c,d) are obtained by substituting the known points into Eq. (2.14) and inverting;*

$$\begin{pmatrix} a' \\ b' \\ c' \end{pmatrix} \equiv \begin{pmatrix} a/d \\ b/d \\ c/d \end{pmatrix} = \begin{pmatrix} x_{(1)} & y_{(1)} & z_{(1)} \\ x_{(2)} & y_{(2)} & z_{(2)} \\ x_{(3)} & y_{(3)} & z_{(3)} \end{pmatrix}^{-1} \begin{pmatrix} 1 \\ 1 \\ 1 \end{pmatrix}. \tag{2.15}$$

(ii) *Replacing z by h and defining $h_1 = d/c - (a/c)x_{(1)} - (b/c)y_{(1)}$, Eq. (2.6) becomes*

$$h(x,y) = h_1 - \frac{a}{c}(x - x_{(1)}) - \frac{b}{c}(y - y_{(1)}). \tag{2.16}$$

Since the ratios a'/c' and b'/c' are available from Eq. (2.15) the required differential form is given by Eq. (2.7)

$$\widetilde{\mathbf{dh}} = -\frac{a'}{c'}\widetilde{\mathbf{dx}} - \frac{b'}{c'}\widetilde{\mathbf{dy}}. \tag{2.17}$$

(iii)

$$\Delta h = -\frac{a'}{c'}\Delta x - \frac{b'}{c'}\Delta y. \tag{2.18}$$

Problem 2.2.1.

(i) *For points P_1, P_2, P_3 given by (1,0,0), (0,1,0), (0,0,1), check the formula just derived for Δh by applying it to each of P_1, P_2, and P_3.*

(ii) *The coordinates of three well-known locations in Colorado, Denver, Pike's Peak, and Colorado Springs are, respectively W. longitudes, 105.1°, 105.1°, 104.8°; N. latitudes, 39.7°, 38.8°, 38.8°; and elevation, 5280 feet, 14,100 feet, and 5280 feet. Making the (thoroughly unwarranted) assumption that Golden, situated at 105.2°W, 39.7°N, lies on the plane defined by the previous three locations, find its elevation.*

At this point we can anticipate one implication of these results for mechanics. Recall the connection between elevation h and potential energy $U = mgh$ in the earth's gravitational field. Also recall the connection between work W and potential energy U. To make the equation traditionally expressed as

$\Delta U = \Delta W = \mathbf{F} \cdot \mathbf{\Delta x}$ meaningful, the vectorial character of force $\mathbf{F}$ has to differ from that of the displacement $\mathbf{\Delta x}$. In particular, since $\mathbf{\Delta x}$ is a contravariant vector, the force $\mathbf{F}$ should be a covariant vector, (meaning its symbol should be $\widetilde{\mathbf{F}}$) for the work to be coordinate independent.

In the traditional pedagogy of physics, *covariant* and *contravariant* vectors are usually differentiated on the basis of the behavior of their components under coordinate transformations. Note, though, that in our discussion the quantities $\widetilde{\mathbf{a}}$ and $\mathbf{x}$ have been introduced and distinguished, and meaning was assigned to the form $\langle \widetilde{\mathbf{a}}, \mathbf{x} \rangle$ before any change of coordinates has even been contemplated. This, so far, is the essential relationship between *covariant* and *contravariant* vectors.[12]

Since the ideas expressed so far, though not difficult, may seem unfamiliar, recapitulation in slightly different terms may be helpful. It has been found useful to associate *contravariant* vectors with *independent* variables, like x and y, and *covariant* vectors (or *1-forms*) with *dependent* variables, like h. Knowing $h(x,y)$, one can prepare a series of contours of constant h, separated from each other by one unit of "elevation," and plot them on the (x,y) plane. For a (defined by a contravariant vector) change $(\Delta x, \Delta y)$ in independent variables, by counting the number of contours (defined by a covariant vector) crossed, the change in dependent variable can be determined.

We have been led to a somewhat unconventional and cumbersome notation (with $\widetilde{\mathbf{dx}}$ being the function that picks out the x-component of an arbitrary vector) so that the symbol dx can retain its traditional physics meaning as an infinitesimal deviation of x. In mathematical literature the symbol dx all by itself typically stands for a differential 1-form. Furthermore, we have so far only mentioned *one*-forms. When a 2-form such as $dx\,dy$ is introduced in mathematical literature, it may be taken implicitly (roughly speaking) to pick out the area defined by dx and dy rather than the product of two infinitesimals. We will return to these definitions shortly.

There is an important potential source of ambiguity in traditional discussion of mechanics by physicists and it is one of the reasons mathematicians prefer different terminology for differentials: a symbol such as x is used to stand both for where a particle *is* and where it *could conceivably be*.[13] This is arguably made

12) Because the components of vectors vary in such a way as to preserve scalar invariants, a common though somewhat archaic terminology, is to refer to vectors as *invariants* or as invariant vectors, in spite of the facts that (a) their components vary and (b) the expression is redundant anyway. (Note especially that invariant here does not mean constant.) Nowadays the term *tensor* automatically carries this connotation of invariance. In special relativity the phrase *manifestly covariant* (or simply *covariant*) means the same thing, but this is a different (though related) meaning of our word *covariant*. Our policy, whenever the invariant aspect is to be specially emphasized is to use the term *true* vector, even though it is redundant.

13) If constraints are present x can also stand for a location where the mass could not conceivably be.

clearer by mathematician's notation. Since we wish to maintain physics usage (not to defend it, only to make formulas look familiar) we will use differential forms as much to de-mystify them as to exploit their power.

2.2.2 Calculus of Differential Forms

Even more than the previous section, since the material in this section will not be required for some time, the reader might be well-advised only to skim over it, planning to address it more carefully later – not because the material is difficult but because its motivation may be unclear. Furthermore, the notation here will be far from standard as we attempt to morph from old-fashioned notation to more modern notation. (In any case, since there is no universally accepted notation for this material, it is impossible to use "standard" notation.) For the same reason, there may seem to be inconsistencies even internal to this section. All this is a consequence mainly of our insistence on maintaining a distinction between two types of "differential," dx and $\widetilde{\mathbf{dx}}$. Eventually, once the important points have been made, it will be possible to shed some of the notational complexity.

A notation we will use temporarily for a differential form such as the one defined in Eq. (2.7) is

$$\widetilde{\boldsymbol{\omega}}[d] = f(x,y)\,\widetilde{\mathbf{dx}} + g(x,y)\,\widetilde{\mathbf{dy}}. \tag{2.19}$$

The only purpose of the "argument" d in square brackets here is to correlate $\widetilde{\boldsymbol{\omega}}$ with the particular coordinate differentials $\widetilde{\mathbf{dx}}$ and $\widetilde{\mathbf{dy}}$, as contrasted say with two independent differentials $\widetilde{\boldsymbol{\delta}\mathbf{x}}$ and $\widetilde{\boldsymbol{\delta}\mathbf{y}}$.

$$\widetilde{\boldsymbol{\omega}}[\delta] = f(x,y)\,\widetilde{\boldsymbol{\delta}\mathbf{x}} + g(x,y)\,\widetilde{\boldsymbol{\delta}\mathbf{y}}. \tag{2.20}$$

The δ symbol does not signify some kind of differential operator other than d; it simply allows notationally for the later assignment of independent values to the differently named differentials. Square brackets are used to protect against interpretation of d or δ as an ordinary argument of $\widetilde{\boldsymbol{\omega}}$.

One can develop a calculus of such differential forms. Initially we proceed to do this by treating the differentials as if they were the "old-fashioned" type familiar from physics and freshman calculus. Notationally, we indicate this by leaving off the overhead tildes and not using bold face symbols; hence

$$\omega[d] = f(x,y)\,dx + g(x,y)\,dy. \tag{2.21}$$

With dx and dy being treated as constants, the differential $\delta\omega[d] \equiv \delta(\omega[d])$ is given by

$$\delta\omega[d] = \left(\frac{\partial f}{\partial x}\delta x + \frac{\partial f}{\partial y}\delta y\right)dx + \left(\frac{\partial g}{\partial x}\delta x + \frac{\partial g}{\partial y}\delta y\right)dy. \tag{2.22}$$

Since these are ordinary differentials, if f and g were force components, $\delta\omega[d]$ would be the answer to the question "How much more work is done in displacement (dx, dy) from displaced location $P + \delta P = (x + \delta x, y + \delta y)$ than is done in displacement (dx, dy) from point $P = (x, y)$?" $\delta\omega[d]$ is not the same as $d\omega[\delta]$ but, from the two, the combination

$$\text{B.C.}[\omega] \equiv \delta\omega[d] - d\omega[\delta], \tag{2.23}$$

can be formed; it is to be known as the "bilinear covariant" of ω. After further manipulation it will yield the "exterior derivative" of ω.

Example 2.2.2. *Consider the example*

$$\omega[d] = y\,dx + x\,dy. \tag{2.24}$$

Substituting into Eq. (2.22), we obtain

$$\delta\omega[d] = \delta y\,dx + \delta x\,dy, \quad d\omega[\delta] = dy\,\delta x + dx\,\delta y. \tag{2.25}$$

In this case the bilinear covariant vanishes,

$$\delta\omega[d] - d\omega[\delta] = 0. \tag{2.26}$$

This is not always true however, operating with d and with δ do not "commute" – that is $\delta\omega[d]$ and $d\omega[\delta]$ are, in general, different. But products such as $dx\,\delta y$ and $\delta y\,dx$ are the same; they are simply the products of two (a physicist might say tiny) independently assignable coordinate increments. When its bilinear covariant does, in fact, vanish, ω is said to be "closed."

In the case of Eq. (2.24), $\omega[d]$ is "derivable from" a function of position $h(x, y) = xy$ according to

$$\omega[d] = dh(x, y) = y\,dx + x\,dy. \tag{2.27}$$

In this circumstance (of being derivable from a single-valued function) ω is said to be "an exact differential."

Problem 2.2.2. *Show that the bilinear covariant B.C.$[\omega]$ of the differential 1-form, $\omega[d] = dh(x, y)$ vanishes for arbitrary function $h(x, y)$.*

Example 2.2.3. *For the differential form*

$$\omega[d] = y\,dx, \tag{2.28}$$

one sees that

$$\delta\omega[d] - d\omega[\delta] = \delta y\,dx - dy\,\delta x, \tag{2.29}$$

which does not vanish. But if we differentiate once again (introducing D as yet another symbol to indicate a differential operator), we obtain

$$D(\delta y\,dx - dy\,\delta x) = 0, \tag{2.30}$$

since the coefficients of the differentials being differentiated are now simply constants.

Problem 2.2.3. *For $\omega[d]$ given by Eq. (2.21), with $f(x,y)$ and $g(x,y)$ being general functions, show that its bilinear covariant does not vanish in general.*

We have been proceeding as if our differentials were "ordinary," but to be consistent with our "new" notation Eq. (2.22) would have been written as

$$\widetilde{\boldsymbol{\omega}}[d][\delta] = \left(\frac{\partial f}{\partial x}\widetilde{\boldsymbol{\delta}\mathbf{x}} + \frac{\partial f}{\partial y}\widetilde{\boldsymbol{\delta}\mathbf{y}}\right)\widetilde{\mathbf{dx}} + \left(\frac{\partial g}{\partial x}\widetilde{\boldsymbol{\delta}\mathbf{x}} + \frac{\partial g}{\partial y}\widetilde{\boldsymbol{\delta}\mathbf{y}}\right)\widetilde{\mathbf{dy}}, \tag{2.31}$$

with the result being a 2-form – a function of two vectors, say $\boldsymbol{\Delta}\mathbf{x}_{(1)}$ and $\boldsymbol{\Delta}\mathbf{x}_{(2)}$. Applying Eq. (2.2), this equation leads to

$$\begin{aligned}\widetilde{\boldsymbol{\omega}}[d][\delta](\boldsymbol{\Delta}\mathbf{x}_{(1)}, \boldsymbol{\Delta}\mathbf{x}_{(2)}) &= \left(\frac{\partial f}{\partial x}\Delta x_{(1)} + \frac{\partial f}{\partial y}\Delta y_{(1)}\right)\Delta x_{(2)} \\ &+ \left(\frac{\partial g}{\partial x}\Delta x_{(1)} + \frac{\partial g}{\partial y}\Delta y_{(1)}\right)\Delta y_{(2)}.\end{aligned} \tag{2.32}$$

Except for the re-naming of symbols, $\delta x \to \Delta x_{(1)}$, $dx \to \Delta x_{(2)}$, $\delta y \to \Delta y_{(1)}$, and $dy \to \Delta y_{(2)}$, this is the same as Eq. (2.22). Hence Eqs. (2.22) and (2.31) have equivalent content. For this to be true we have implicitly assumed that the first of the two arguments $\boldsymbol{\Delta}\mathbf{x}_{(1)}$ and $\boldsymbol{\Delta}\mathbf{x}_{(2)}$ is acted on by $\boldsymbol{\delta}$ and the second by d. Since these appear in the same order in every term we could as well say that the first operator acts on $\boldsymbol{\Delta}x_{(1)}$ and the second on $\boldsymbol{\Delta}x_{(2)}$. Furthermore, since Eq. (2.19) made no distinction between δ and d forms, we might as well have written Eq. (2.31) as

$$\widetilde{\boldsymbol{\omega}}[d][d] = \left(\frac{\partial f}{\partial x}\widetilde{\mathbf{dx}} + \frac{\partial f}{\partial y}\widetilde{\mathbf{dy}}\right)\widetilde{\mathbf{dx}} + \left(\frac{\partial g}{\partial x}\widetilde{\mathbf{dx}} + \frac{\partial g}{\partial y}\widetilde{\mathbf{dy}}\right)\widetilde{\mathbf{dy}}, \tag{2.33}$$

as long as it is understood that in a product of two differential forms the first acts on the first argument and the second on the second. Note though that, in spite of the fact that it *is* legitimate to reverse the order in a product of actual displacements like $\Delta x_{(1)}\Delta y_{(2)}$, it is *illegitimate* to reverse the order of the terms in a product like $\widetilde{\mathbf{dx}}\widetilde{\mathbf{dy}}$; that is,

$$\widetilde{\mathbf{dx}}\widetilde{\mathbf{dy}} \neq \widetilde{\mathbf{dy}}\widetilde{\mathbf{dx}}. \tag{2.34}$$

The failure to commute of our quantities, which will play such an important role in the sequel, has entered here as a simple consequence of our notational convention specifying the meaning of the differential of a differential.

How then to express the bilinear covariant, without using the distinction between d and δ? Instead of antisymmetrizing with respect to d and δ we can antisymmetrize with respect to the arguments. A "new notation" version of

Eq. (2.23), with $\widetilde{\boldsymbol{\omega}}$ still given by Eq. (2.19), can be written as

$$\begin{aligned}
\text{B.C.}[\widetilde{\boldsymbol{\omega}}](\boldsymbol{\Delta x}_{(1)},\boldsymbol{\Delta x}_{(2)}) &= \widetilde{\boldsymbol{\omega}}[d][d](\boldsymbol{\Delta x}_{(1)},\boldsymbol{\Delta x}_{(2)}) - \widetilde{\boldsymbol{\omega}}[d][d](\boldsymbol{\Delta x}_{(2)},\boldsymbol{\Delta x}_{(1)}) \\
&= \left(-\frac{\partial f}{\partial y}+\frac{\partial g}{\partial x}\right)(\Delta x_{(1)}\Delta y_{(2)} - \Delta y_{(1)}\Delta x_{(2)}) \\
&= \left(-\frac{\partial f}{\partial y}+\frac{\partial g}{\partial x}\right)\det\begin{vmatrix}\Delta x_{(1)} & \Delta x_{(2)}\\ \Delta y_{(1)} & \Delta x_{(2)}\end{vmatrix}.
\end{aligned} \tag{2.35}$$

This can be re-expressed by defining the "wedge product"

$$\widetilde{\mathbf{dx}}\wedge\widetilde{\mathbf{dy}} \equiv \widetilde{\mathbf{dx}}\widetilde{\mathbf{dy}} - \widetilde{\mathbf{dy}}\widetilde{\mathbf{dx}} \equiv \begin{vmatrix}\widetilde{\mathbf{dx}} & \widetilde{\mathbf{dy}}\\ \widetilde{\mathbf{dx}} & \widetilde{\mathbf{dy}}\end{vmatrix}, \tag{2.36}$$

where element order has to be preserved in expanding the determinant. Note from its definition, that

$$\widetilde{\mathbf{dx}}\wedge\widetilde{\mathbf{dy}} = -\widetilde{\mathbf{dy}}\wedge\widetilde{\mathbf{dx}}, \quad \text{and} \quad \widetilde{\mathbf{dx}}\wedge\widetilde{\mathbf{dx}} = 0. \tag{2.37}$$

We obtain

$$\widetilde{\mathbf{dx}}\wedge\widetilde{\mathbf{dy}}(\boldsymbol{\Delta x}_{(1)},\boldsymbol{\Delta x}_{(2)}) = \det\begin{vmatrix}\Delta x_{(1)} & \Delta x_{(2)}\\ \Delta y_{(1)} & \Delta y_{(2)}\end{vmatrix}, \tag{2.38}$$

which can be substituted into Eq. (2.35) to eliminate the determinant. Since the arbitrary increments $\boldsymbol{\Delta x}_{(1)}$ and $\boldsymbol{\Delta x}_{(2)}$ then appear as common arguments on both sides of Eq. (2.35) they can be suppressed as we define a form E.D.$[\widetilde{\boldsymbol{\omega}}]$, which is B.C.$[\widetilde{\boldsymbol{\omega}}](\boldsymbol{\Delta x}_{(1)},\boldsymbol{\Delta x}_{(2)})$ with its arguments unevaluated;

$$\text{E.D.}[\widetilde{\boldsymbol{\omega}}] = \left(-\frac{\partial f}{\partial y}+\frac{\partial g}{\partial x}\right)\widetilde{\mathbf{dx}}\wedge\widetilde{\mathbf{dy}}. \tag{2.39}$$

When operating on any two vector increments, E.D.$[\widetilde{\boldsymbol{\omega}}]$ generates the bilinear covariant of $\widetilde{\boldsymbol{\omega}}$ evaluated for the two vectors. This newly defined differential 2-form is known as the "exterior derivative" of the differential 1-form $\widetilde{\boldsymbol{\omega}}$. From here on this will be written as[14]

$$\widetilde{\mathbf{d}}\,(f\widetilde{\mathbf{dx}}+g\widetilde{\mathbf{dy}}) = \left(-\frac{\partial f}{\partial y}+\frac{\partial g}{\partial x}\right)\widetilde{\mathbf{dx}}\wedge\widetilde{\mathbf{dy}}. \tag{2.40}$$

Of the four terms in the expansion of the left-hand side, two vanish because they are the self-wedge product of $\widetilde{\mathbf{dx}}$ or $\widetilde{\mathbf{dy}}$ and one acquires a negative sign

14) The notation of Eq. (2.40) is still considerably bulkier than is standard in the literature of differential forms. There, the quantity (exterior derivative) that we have called E.D.$[\widetilde{\boldsymbol{\omega}}]$ is often expressed simply as $d\omega$, and Eq. (2.40) becomes $d\omega = (-\partial f/\partial y + \partial g/\partial x)\,dx\wedge dy$ or even $d\omega = (-\partial f/\partial y + \partial g/\partial x)\,dx\,dy$.

when the order of differentials is reversed. These rules, along with natural extension of wedge products to higher dimensionality, will be sufficient to evaluate all exterior derivatives to appear in the sequel.

The vectors $\Delta \mathbf{x}_{(1)}$ and $\Delta \mathbf{x}_{(2)}$ can be said to have played only a "catalytic" role in the definition of the exterior derivative since they no longer appear in Eq. (2.40). From its appearance, one might guess that the exterior derivative is related to the *curl* operator of vector analysis. This is to be pursued next.

2.2.3
Familiar Physics Equations Expressed Using Differential Forms

Like nails and screws, the calculus of vectors and the calculus of differential forms can be regarded as essentially similar or as essentially different depending on ones point of view. Both can be used to hold physical theories together. A skillful carpenter can hammer together much of a house while the cabinet maker is still drilling the screw holes in the kitchen cabinets. Similarly, the physicist can derive and solve Maxwell's equations using vector analysis while the mathematician is still tooling up the differential form machinery. The fact is though, that, just as some structures cannot be held together with nails, some mechanical systems cannot be analyzed without differential forms.

There is a spectrum of levels of ability in the use of vectors, starting from no knowledge whatsoever, advancing through vector algebra, to an understanding of gradients, curls, and divergences, to a skillful facility with the methods. The corresponding spectrum is even broader for differential forms, which can be used to solve all the problems that vectors can solve plus others as well. In spite of this most physicists remain at the "no knowledge whatsoever" end of the spectrum. This is perhaps partly due to some inherent advantage of simplicity that vectors have for solving the most commonly encountered problem of physics. But the accidents of pedagogical fashion probably also play a role.

According to Arnold, *Mathematical Methods of Classical Mechanics*, p. 163, "Hamiltonian mechanics cannot be understood without differential forms."[15] It behooves us therefore to make a start on this subject. But in this text only a fairly superficial treatment will be included; (the rationale being that the important and hard thing is to get the general idea, but that following specialized texts is not so difficult once one has the general idea.) The whole of advanced calculus can be formulated in terms of differential forms, as can more advanced topics, and there are several texts concentrating narrowly yet accessibly on these subjects. Here we are more interested in giving the general ideas than in either rigorous mathematical proof or practice with the combi-

15) It might be more accurate to say "without differential forms one cannot understand Hamiltonian mechanics as well as Arnold" but this statement would be true with or without differential forms.

natorics that are needed to make the method compete with vector analysis in compactness.

The purpose of this section is to show how formulas that are (assumed to be) already known from vector calculus can be expressed using differential forms. Since these results *are* known, it will not be necessary to prove them in the context of differential forms. This will permit the following discussion to be entirely formal, its only purpose being to show that definitions and relations being introduced are consistent with results already known. We work only with ordinary, three dimensional, Euclidean geometry, using rectangular co-ordinates. It is far from true that the validity of differential forms is restricted to this domain, but our purpose is only to motivate the basic definitions.

One way that Eq. (2.19) can be generalized is to go from two to three dimensions;

$$\widetilde{\omega}^{(1)} = f(x,y,z)\,\widetilde{\mathbf{dx}} + g(x,y,z)\,\widetilde{\mathbf{dy}} + h(x,y,z)\,\widetilde{\mathbf{dz}}, \tag{2.41}$$

where the superscript (1) indicates that $\widetilde{\omega}^{(1)}$ is a 1-form. Calculations like those leading to Eq. (2.40) yield

$$\begin{aligned}\widetilde{d}\widetilde{\omega}^{(1)} &= \Big(-\frac{\partial f}{\partial y} + \frac{\partial g}{\partial x}\Big)\widetilde{\mathbf{dx}}\wedge\widetilde{\mathbf{dy}} + \Big(-\frac{\partial g}{\partial z} + \frac{\partial h}{\partial y}\Big)\widetilde{\mathbf{dy}}\wedge\widetilde{\mathbf{dz}} \\ &\quad + \Big(-\frac{\partial h}{\partial x} + \frac{\partial f}{\partial z}\Big)\widetilde{\mathbf{dz}}\wedge\widetilde{\mathbf{dx}}.\end{aligned} \tag{2.42}$$

Next let us generalize Eq. (2.36) by defining

$$\widetilde{\mathbf{dx}}\wedge\widetilde{\mathbf{dy}}\wedge\widetilde{\mathbf{dz}} \equiv \begin{vmatrix} \widetilde{\mathbf{dx}} & \widetilde{\mathbf{dy}} & \widetilde{\mathbf{dz}} \\ \widetilde{\mathbf{dx}} & \widetilde{\mathbf{dy}} & \widetilde{\mathbf{dz}} \\ \widetilde{\mathbf{dx}} & \widetilde{\mathbf{dy}} & \widetilde{\mathbf{dz}} \end{vmatrix}. \tag{2.43}$$

This definition is motivated by the relation

$$\widetilde{\mathbf{dx}}\wedge\widetilde{\mathbf{dy}}\wedge\widetilde{\mathbf{dz}}(\mathbf{\Delta x}_{(1)}, \mathbf{\Delta x}_{(2)}, \mathbf{\Delta x}_{(3)}) = \det\begin{vmatrix} \Delta x_{(1)} & \Delta x_{(2)} & \Delta x_{(3)} \\ \Delta y_{(1)} & \Delta y_{(2)} & \Delta y_{(3)} \\ \Delta z_{(1)} & \Delta z_{(2)} & \Delta z_{(3)} \end{vmatrix}. \tag{2.44}$$

Consider a differential form

$$\widetilde{\omega}^{(2)} = f(x,y,z)\widetilde{\mathbf{dx}}\wedge\widetilde{\mathbf{dy}} + g(x,y,z)\widetilde{\mathbf{dy}}\wedge\widetilde{\mathbf{dz}} + h(x,y,z)\widetilde{\mathbf{dz}}\wedge\widetilde{\mathbf{dx}}, \tag{2.45}$$

where the superscript (2) indicates that $\widetilde{\omega}^{(2)}$ is a 2-form. At first glance this may seem to be a rather *ad hoc* and special form, but any 2-form that is anti-symmetric in its two arguments can be expressed this way.[16] We then *define*

$$\widetilde{d}\widetilde{\omega}^{(2)} = \widetilde{\mathbf{df}}\wedge\widetilde{\mathbf{dx}}\wedge\widetilde{\mathbf{dy}} + \widetilde{\mathbf{dg}}\wedge\widetilde{\mathbf{dy}}\wedge\widetilde{\mathbf{dz}} + \widetilde{\mathbf{dh}}\wedge\widetilde{\mathbf{dz}}\wedge\widetilde{\mathbf{dx}}. \tag{2.46}$$

16) In most treatments of differential forms the phrase "antisymmetric 2-form" would be considered redundant, since "2-forms" would have been already defined to be antisymmetric.

These definitions are special cases of more general definitions but they are all we require for now. From Eq. (2.46), using Eqs. (2.37), we obtain

$$\widetilde{d}\,\widetilde{\omega}^{(2)} = \left(\frac{\partial f}{\partial z} + \frac{\partial g}{\partial x} + \frac{\partial h}{\partial y}\right)(\widetilde{\mathbf{dx}} \wedge \widetilde{\mathbf{dy}} \wedge \widetilde{\mathbf{dz}}). \tag{2.47}$$

Let us recapitulate the formulas that have been derived, but using notation for the coefficients that is more suggestive than the functions $f(x,y,z)$, $g(x,y,z)$, and $h(x,y,z)$ used so far.

$$\widetilde{\omega}^{(0)} = \phi(x,y,z),$$

$$\widetilde{\omega}^{(1)} = E_x(x,y,z)\,\widetilde{\mathbf{dx}} + E_y(x,y,z)\,\widetilde{\mathbf{dy}} + E_z(x,y,z)\,\widetilde{\mathbf{dz}}, \tag{2.48}$$

$$\widetilde{\omega}^{(2)} = B_z(x,y,z)\widetilde{\mathbf{dx}} \wedge \widetilde{\mathbf{dy}} + B_x(x,y,z)\widetilde{\mathbf{dy}} \wedge \widetilde{\mathbf{dz}} + B_y(x,y,z)\widetilde{\mathbf{dz}} \wedge \widetilde{\mathbf{dx}}. \tag{2.49}$$

Then Eqs. (2.7), (2.42), and (2.47) become

$$\begin{aligned}
\widetilde{d}\,\widetilde{\omega}^{(0)} &= \frac{\partial \phi}{\partial x}\widetilde{\mathbf{dx}} + \frac{\partial \phi}{\partial y}\widetilde{\mathbf{dy}} + \frac{\partial \phi}{\partial z}\widetilde{\mathbf{dz}},\\
\widetilde{d}\,\widetilde{\omega}^{(1)} &= \left(-\frac{\partial E_x}{\partial y} + \frac{\partial E_x}{\partial y}\right)\widetilde{\mathbf{dx}} \wedge \widetilde{\mathbf{dy}} + \left(-\frac{\partial E_y}{\partial z} + \frac{\partial E_z}{\partial y}\right)\widetilde{\mathbf{dy}} \wedge \widetilde{\mathbf{dz}}\\
&\quad + \left(-\frac{\partial E_z}{\partial x} + \frac{\partial E_x}{\partial z}\right)\widetilde{\mathbf{dz}} \wedge \widetilde{\mathbf{dx}},\\
\widetilde{d}\,\widetilde{\omega}^{(2)} &= \left(\frac{\partial B_x}{\partial x} + \frac{\partial B_y}{\partial y} + \frac{\partial B_z}{\partial z}\right)\widetilde{\mathbf{dx}} \wedge \widetilde{\mathbf{dy}} \wedge \widetilde{\mathbf{dz}}.
\end{aligned} \tag{2.50}$$

We can now write certain familiar equations as equations satisfied by differential forms. For example,

$$\widetilde{d}\,\widetilde{\omega}^{(2)} = 0, \quad \text{is equivalent to} \quad \boldsymbol{\nabla} \cdot \mathbf{B} = 0. \tag{2.51}$$

The 3-form $\widetilde{d}\,\widetilde{\omega}^{(2)}$ is "waiting to be evaluated" on coordinate increments as in Eq. (2.44); this includes the "Jacobean factor" in a volume integration of $\boldsymbol{\nabla} \cdot \mathbf{B}$. The equation $\widetilde{d}\,\widetilde{\omega}^{(2)} = 0$ therefore represents the "divergence-free" nature of the vector $\mathbf{B}$. While $\boldsymbol{\nabla} \cdot \mathbf{B}$ is the integrand in the integral form of this law, $\widetilde{d}\,\widetilde{\omega}^{(2)}$ also includes the Jacobean factor in the same integral. When, as here, orthonormal coordinates are used as the variables of integration, this extra factor is trivially equal to 1, but in other coordinates the distinction is more substantial. But, since the Jacobean factor cannot vanish, it cannot influence the vanishing of the integrand. An expanded discussion of integrands is in Section 4.2.

Other examples of familiar equations expressed using differential forms:

$$\widetilde{\omega}^{(1)} = -\widetilde{d}\,\widetilde{\omega}^{(0)}, \quad \text{equivalent to} \quad \mathbf{E} = -\boldsymbol{\nabla}\phi, \tag{2.52}$$

yields the "electric field" as the (negative) gradient of the potential. Also

$$\widetilde{d}\widetilde{\omega}^{(1)} = 0, \quad \text{equivalent to} \quad \nabla \times \mathbf{E} = 0, \tag{2.53}$$

states that **E** is "irrotational;" (that is, the curl of **E** vanishes.)

The examples given so far have been applicable only to time-independent problems such as electrostatics. But let us define

$$\widetilde{\omega}^{(3)} = \Big(J_x(x,y,z,t)\widetilde{\mathbf{dy}} \wedge \widetilde{\mathbf{dz}} + J_y(x,y,z,t)\widetilde{\mathbf{dz}} \wedge \widetilde{\mathbf{dx}} + J_z(x,y,z,t)\widetilde{\mathbf{dx}} \wedge \widetilde{\mathbf{dy}}\Big) \wedge \widetilde{\mathbf{dt}}$$
$$- \rho(x,y,z,t)\widetilde{\mathbf{dx}} \wedge \widetilde{\mathbf{dy}} \wedge \widetilde{\mathbf{dz}}. \tag{2.54}$$

Then

$$\widetilde{d}\widetilde{\omega}^{(3)} = 0 \quad \text{is equivalent to} \quad \nabla \times \mathbf{J} + \frac{\partial \rho}{\partial t} = 0. \tag{2.55}$$

which is known as the "continuity equation." In physics such relations relate "fluxes" to "volume densities." This is developed further in Section 4.3.4. Another familiar equation can be obtained by defining

$$\widetilde{\eta}^{(1)} = A_x\, \widetilde{\mathbf{dx}} + A_y\, \widetilde{\mathbf{dy}} + A_z\, \widetilde{\mathbf{dz}} - \phi\, \widetilde{\mathbf{dt}}, \tag{2.56}$$

Then the equation

$$\widetilde{d}\widetilde{\eta}^{(1)} = \widetilde{\omega}^{(1)} \wedge \widetilde{\mathbf{dt}} + \widetilde{\omega}^{(2)} \tag{2.57}$$

is equivalent to the pair of equations

$$\mathbf{B} = \nabla \times \mathbf{A}, \quad \mathbf{E} = -\frac{\partial \mathbf{A}}{\partial t} - \nabla \phi. \tag{2.58}$$

These examples have shown that familiar vector equations can be re-expressed as equations satisfied by differential forms. All these equations are developed further in Chapter 9.

The full analogy between forms and vectors, in particular including cross products, requires the introduction of "supplementary" multivectors, also known as "the star (*) operation." This theory is developed in Section 4.2.5.

What are the features of these newly introduced differential forms derived by exterior differentiation? We state some of them, without proof for now:

Differential forms inevitably find themselves acting as the "differential elements" of multidimensional integrals. When one recalls two of the important difficulties in formulating multidimensional integrals – introducing the appropriate Jacobians and keeping track of sign reversals – one will be happy to know that exterior derivatives "take care of both problems." They also, automatically, provide the functions which enter when gradients, divergences, and curls are calculated in curvilinear coordinates. Furthermore, the exterior calculus works for spaces of arbitrary dimension, though formidable combinatorial calculations may be necessary. We will return to this subject in Chapter 4.

Differential forms "factor out" the arbitrary incremental displacements, such as $\mathbf{\Delta x}_{(1)}$ and $\mathbf{\Delta x}_{(2)}$ in the above discussion, leaving the arbitrary displacements implicit rather than explicit. This overcomes the inelegant need for distinguishing among different differential symbols such as d and δ. Though this aspect is not particularly hard to grasp – it has been thoroughly expounded here – not being part of the traditional curriculum encountered by scientists, it contributes to the unfamiliar appearance of the equations of physics.

The quantities entering equations of physics such as Maxwell's equations as they are traditionally written are physically measurable vectors, such as electric field **E**, that are naturally visualized as arrows. When the equations are expressed in terms of forms, invariant combinations of forms and vectors, such as $\langle \widetilde{\mathbf{E}}, \overrightarrow{\mathbf{\Delta x}} \rangle$, more naturally occur. Products like this fit very intuitively into Maxwell's equations in integral form. This is the form in which Maxwell's equations are traditionally first encountered in sophomore physics. Only later does one use vector analysis to transform these integral equations into the vector differential equations that fit so compactly on tee shirts. But the integral versions are just as fundamental. Only after these integral equations have been expressed in terms of exterior derivatives do they acquire their unfamiliar appearance.

The most fundamental property of the exterior calculus forms is that it makes the equations *manifestly invariant*; that is, independent of coordinates. Of course this is also the chief merit of the vector operators, gradient, divergence, and curl. Remembering the obscurity surrounding these operators when they were first encountered (some of which perhaps still lingers in the case of curl) one has to anticipate a considerable degree of difficulty in generalizing these concepts – which is what the differential forms do. In this section only a beginning has been made toward establishing this invariance; the operations of vector differentiation, known within vector analysis to have invariant character, have been expressed by differential forms.

Having said all this, it should also be recognized that differential forms really amounts to being just a sophisticated form of advanced calculus.

2.3 Algebraic Tensors

2.3.1 Vectors and Their Duals

In traditional physics (unless one includes graphical design) there is little need for geometry without algebra – synthetic geometry – but algebra without geometry is both possible and important. Though vector and tensor analysis were both motivated initially by geometry, it is useful to isolate their purely algebraic aspects. Everything that has been discussed so far can be distilled into pure algebra. That will be done in this section, though in far less generality than in the references listed at the end of the chapter. Van der Waerden allows numbers more general than the real numbers we need; Arnold pushes further into differential forms.

Most of the algebraic properties of vector spaces are "obvious" to most physicists. Vectors $\mathbf{x}$, $\mathbf{y}$, etc., are quantities for which superposition is valid – for scalars a and b, $a\mathbf{x} + b\mathbf{y}$ is also a vector. The dimensionality n of the vector space containing $\mathbf{x}$, $\mathbf{y}$, etc., is the largest number of independent vectors that can be selected. Any vector can be expanded uniquely in terms of n independent basis vectors $\mathbf{e}_1, \mathbf{e}_2, \dots, \mathbf{e}_n$;

$$\mathbf{x} = \mathbf{e}_i x^i. \tag{2.59}$$

This provides a one-to-one relationship between vectors $\mathbf{x}$ and n-component multiplets $(x^1, x^2, \dots, x^n)$ – for now at least, we will say they are *the same thing*.[17] In particular, the basis vectors $\mathbf{e}_1, \mathbf{e}_2, \dots, \mathbf{e}_n$ correspond to $(1, 0, \dots, 0)$, $(0, 1, \dots, 0), \dots (0, 0, \dots, 1)$. Component-wise addition of vectors and multiplication by a scalar is standard.

Important new content is introduced when one defines a real-valued *linear function* $\widetilde{\mathbf{f}}(\mathbf{x})$ of a vector $\mathbf{x}$; such a function, by definition, satisfies relations

$$\widetilde{\mathbf{f}}(\mathbf{x}+\mathbf{y}) = \widetilde{\mathbf{f}}(\mathbf{x}) + \widetilde{\mathbf{f}}(\mathbf{y}), \quad \widetilde{\mathbf{f}}(a\mathbf{x}) = a\,\widetilde{\mathbf{f}}(\mathbf{x}). \tag{2.60}$$

Expanding $\mathbf{x}$ as $x^i \mathbf{e}_i$, in terms of basis vectors $\mathbf{e}_i$, this yields

$$\widetilde{\mathbf{f}}(\mathbf{x}) = f_i\, x^i \equiv \langle \widetilde{\mathbf{f}}, \mathbf{x} \rangle, \quad \text{where} \quad f_i = \widetilde{\mathbf{f}}(\mathbf{e_i}). \tag{2.61}$$

This exhibits the value of $\widetilde{\mathbf{f}}(\mathbf{x})$ as a *linear form* in the components x^i with coefficients f_i. Now we have a one-to-one correspondence between linear functions $\widetilde{\mathbf{f}}$ and n-component multiplets, $(f_1, f_2, \dots, f_n)$. Using language similarly loose

17) As long as possible we will stick to the colloquial elementary physics usage of refusing to distinguish between a vector and its collection of components, even though the latter depends on the choice of basis vectors while the former does not.

to what was applied to vectors, we can say that a linear function of a vector and a linear form in the vector's components are *the same thing*. But, unlike $\widetilde{\mathbf{f}}$, the f_i depend on the choice of basis vectors. This space of linear functions of vectors-in-the-original-space is called *dual* to the original space. With vectors in the original space called *contravariant*, vectors in the dual space are called *covariant*.

Corresponding to basis vectors $\mathbf{e}_i$ in the original space there is a natural choice of basis vectors $\widetilde{\mathbf{e}}^i$ in the dual space. When acting on $\mathbf{e}_i$, $\widetilde{\mathbf{e}}^i$ yields 1; when acting on any other of the $\mathbf{e}_j$ it yields 0. Just as the components of $\mathbf{e}_1$ are $(1,0,\dots,0)$ the components of $\widetilde{\mathbf{e}}^1$ are $(1,0,\dots,0)$, and so on. More concisely,[18]

$$\widetilde{\mathbf{e}}^i(\mathbf{e}_j) = \delta^i{}_j. \tag{2.62}$$

By taking all linear combinations of a subset of the basis vectors, say the first m of them, where $0 < m < n$, one forms a sub vector space $\mathcal{S}$ of the original space. Any vector $\mathbf{x}$ in the whole space can be decomposed uniquely into a vector $\mathbf{y} = \sum_1^m \mathbf{e}_i x^i$ in this space and a vector $\mathbf{z} = \sum_{n-m}^n \mathbf{e}_i x^i$. A "projection operator" $\mathcal{P}$ onto the subspace can then be defined by $\mathbf{y} = \mathcal{P}\mathbf{x}$. It has the property that $\mathcal{P}^2 = \mathcal{P}$. Since $\mathbf{x} = \mathcal{P}\mathbf{x} + (\mathbf{1} - \mathcal{P})\mathbf{x}$ one has that $\mathbf{z} = (\mathbf{1} - \mathcal{P})\mathbf{x}$ and that $\mathbf{1} - \mathcal{P}$ projects onto the space formed from the last $n - m$ basis vectors.

There is a subspace $\mathcal{S}^0$ in the dual space, known as the "annihilator" of $\mathcal{S}$; it is the vector space made up of all linear combinations of the $n - m$ forms $\widetilde{\mathbf{e}}^{n-m}, \widetilde{\mathbf{e}}^{n-m+1}, \dots, \widetilde{\mathbf{e}}^n$. These are the last m of the natural basis forms in the dual space, as listed in Eq. (2.62). Any form in $\mathcal{S}^0$ "annihilates" any vector in $\mathcal{S}$, which is to say yields zero when acting on the vector. This relationship is reciprocal in that $\mathcal{S}$ annihilates $\mathcal{S}^0$. Certainly there are particular forms not in $\mathcal{S}^0$ that annihilate certain vectors in $\mathcal{S}$ but $\mathcal{S}^0$ contains *all* forms, and *only* those forms, that annihilate *all* vectors in $\mathcal{S}$. This concept of annihilation is reminiscent of the concept of the orthogonality of two vectors in ordinary vector geometry. It is a very different concept however, since annihilation relates a vector in the original space and a form in the dual space. Only if there is a rule associating vectors and forms can annihilation be used to define orthogonality of two vectors in the same space.

By introducing linear functions of more than one vector variable we will shortly proceed to the definition of tensors. But, since all other tensors are introduced in the same way as was the dual space, there is no point in proceeding to this without first having grasped this concept. Toward that end we should eliminate an apparent asymmetry between contravariant vectors and covectors. The asymmetry has resulted from the fact that we *started with* contravariant vectors, and hence might be inclined to think of them as *more basic*. But consider the space of linear-functions of covariant-vectors – that is,

18) There is no immediate significance to the fact that one of the indices of $\delta^i{}_j$ is written as a subscript and one as a superscript. Equal to δ_{ij}, $\delta^i{}_j$ is also a Kronecker-δ.

the space that is dual to the space that is dual to the original space. (As an exercise) it can be seen that the dual of the dual is *the same thing as the original space.* Hence, algebraically at least, which is which between contravariant and covariant vectors is entirely artificial, just like the choice of which is to be designated by superscripts and which by subscripts.

2.3.2
Transformation of Coordinates

When covariant and contravariant vectors are introduced in physics, the distinction between them is usually expressed in terms of the matrices accompanying a change of basis vectors. Suppose a new set of basis vectors $\mathbf{e}'_j$ is related to the original set $\mathbf{e}_j$ by

$$\mathbf{e}'_j = \mathbf{e}_i \Lambda^i{}_j. \tag{2.63}$$

(If one insists on interpreting this relation as a matrix multiplication it is necessary to regard $\mathbf{e}'_j$ and $\mathbf{e}_j$ as being the elements of row vectors, even though the row elements are vectors rather than numbers, and to ignore the distinction between upper and lower indices.[19]) Multiplying on the right by the inverse matrix, the inverse relation is

$$\mathbf{e}'_j (\Lambda^{-1})^j{}_k = \mathbf{e}_i \Lambda^i{}_j (\Lambda^{-1})^j{}_k = \mathbf{e}_k. \tag{2.64}$$

For formal manipulation of formulas the index conventions of tensor analysis are simple and reliable, but for numerical calculations it is sometimes convenient to use matrix notation in which multicomponent *objects* are introduced so that the indices can be suppressed. This is especially useful when using a computer language that can work with matrices as supported *types* that satisfy their own algebra of addition, multiplication, and scalar multiplication.

To begin the attempt to represent the formulas of mechanics in matrix form some recommended usage conventions will now be formulated, and some of the difficulties in maintaining consistency will be addressed. Already in defining the symbols used in Eq. (2.63) a conventional choice was made. The new basis vectors were called $\mathbf{e}'_j$ when they could have been called $\mathbf{e}_{j'}$; that is, the prime was placed on the vector symbol rather than on the index. It is a common, and quite powerful notation, to introduce both of these symbols

19) Since our convention is that the up/down location of indices on matrices is irrelevant, Eq. (2.63) is the same as $\mathbf{e}'_j = \mathbf{e}_i \Lambda_{ij}$. This in turn is the same as $\mathbf{e}'_i = (\Lambda^T)_{ij} \mathbf{e}_j$, which may seem like a more natural ordering. But one sees that whether it is the matrix or its transpose that is said to be the transformation matrix depends on whether it multiplies on the left or on the right and is not otherwise significant.

and to use them to express two distinct meanings (see for example Schutz). In this notation, even as one "instantiates" an index, say replacing i by 1, one must replace i' by $1'$, thereby distinguishing between $\mathbf{e}_1$ and $\mathbf{e}_{1'}$. In this way, at the cost of further abstraction, one can distinguish change of axes with fixed vector from change of vector with fixed axes. At this point this may seem like pedantry but confusion attending this distinction between *active* and *passive* interpretations of transformations will dog us throughout this text and the subject in general. One always attempts to define quantities and operations unambiguously in English, but everyday language is by no means optimal for avoiding ambiguity. Mathematical language, such as the distinction between $\mathbf{e}_1$ and $\mathbf{e}_{1'}$ just mentioned, can be much more precise. But, sophisticated as it is, we *will not* use this notation, because it seems *too* compact, *too* mathematical, and *too* cryptic.

Another limitation of matrix notation is that, though it works well for tensors of one or two indices, it is not easily adapted to tensors with more than two indices. Yet another complication follows from the traditional row and column index-order conventions of matrix formalism. It is hard to maintain these features while preserving other desirable features such as lower and upper indices to distinguish between covariant and contravariant quantities which, with the repeated-index summation convention yield very compact formulas.[20] Often, though, one can restrict calculations to a single frame of reference, or to use only rectangular coordinate systems. In these cases there is no need to distinguish between lower and upper indices.

When the subject of vector fields is introduced an even more serious notational complication arises since a new kind of "multiplication" of one vector by another is noncommutative. As a result the validity of an equation such as $(\mathbf{A}\mathbf{x})^T = \mathbf{x}^T\mathbf{A}^T$ is called into question. One is already accustomed to matrix multiplication being not commutative, but the significance of failure of vector fields to commute compromises the power of matrix notation and the usefulness of distinguishing between row and column vectors.

In spite of all these problems, matrix formulas will still often be used, and when they are, the following conventions will be adhered to:

- As is traditional, contravariant components $x^1, x^2, \ldots, x^n$ are arrayed as a column vector. From this it follows that,

20) The repeated-index convention is itself used fairly loosely. For example, if the summation convention is used as in Eq. (2.63), to express a vector as a superposition of basis vectors, the usage amounts to a simple abbreviation without deeper significance. But when used (as it was by Einstein originally) to form a scalar from a contravariant and a covariant vector, the notation includes deeper implication of *invariance*. In this text both of these conventions will be used but for other summations, such as over particles in a system, the summation symbol will be shown explicitly.

- (Covariant) components f_i of form $\widetilde{f}$ are to be arrayed in a row.
- The basis vectors $\mathbf{e}_i$, though not components of an intrinsic quantity, will be arrayed as a row for purposes of matrix multiplication.
- Basis covectors $\widetilde{\mathbf{e}}^i$ will be arrayed in a column.
- Notations such as $x_{1'}$ will not be used; the indices on a components are necessarily $1, 2, 3, \ldots$.. Symbolic indices with primes, as in $x_{\alpha'}$ are legitimate however.
- The indices on a quantity like $\Lambda^i{}_j$ are spaced apart, and in order, to make it unambiguous which is to be taken as the row, in this case i, and which as the column index. The up/down location is to be ignored when matrix multiplication is intended.

In terms of the new basis vectors introduced by Eq. (2.63), using Eq. (2.64), a general vector $\mathbf{x}$ is re-expressed as

$$\mathbf{e}'_j x'^j \equiv \mathbf{x} \equiv \mathbf{e}_k x^k = \mathbf{e}_j (\Lambda^{-1})^j{}_k x^k, \tag{2.65}$$

from which it follows that

$$x'^j = (\Lambda^{-1})^j{}_k x^k, \quad \text{or} \quad x^i = \Lambda^i{}_j x^k. \tag{2.66}$$

Because the matrix giving $x^i \to x^{i'}$ is inverse to the matrix giving $\mathbf{e}_i \to \mathbf{e}'_i$, this is conventionally known as *contravariant* transformation.

If the column of elements x'^j and x^k are symbolized by $\mathbf{x}'$ and $\mathbf{x}$ and the matrix by $\boldsymbol{\Lambda}^{-1}$ then Eq. (2.66) becomes

$$\mathbf{x}' = \boldsymbol{\Lambda}^{-1}\mathbf{x}. \tag{2.67}$$

When bold face symbols are used to represent vectors in vector analysis the notation implies that the bold face quantities have an invariant geometric character and in this context an equation like (2.67) might by analogy be expected to relate two different "arrows" $\mathbf{x}$ and $\mathbf{x}'$. The present bold face quantities have not been shown to have this geometric character and, in fact, they do not. As they have been introduced, since $\mathbf{x}$ and $\mathbf{x}'$ stand for the same geometric quantity, it is redundant to give them different symbols. This is an instance of the above-mentioned ambiguity in specifying transformations. Our notation is simply not powerful enough to distinguish between active and passive transformations in the same context. For now we ignore this redundancy and regard Eq. (2.67) as simply an abbreviated notation for the algebraic relation between the components. Since this notation is standard in linear algebra, it should be acceptable here once the potential for misinterpretation has been understood.

Transformation of covariant components f_i has to be arranged to secure the invariance of the form $\langle \widetilde{\mathbf{f}}, \mathbf{x} \rangle$ defined in Eq. (2.61). Using Eq. (2.66)

$$f_k x^k = \langle \widetilde{\mathbf{f}}, \mathbf{x} \rangle = f'_j x'^j = f'_j (\Lambda^{-1})^j{}_k x^k, \tag{2.68}$$

and from this

$$f_k = f'_j (\Lambda^{-1})^j{}_k \quad \text{or} \quad f'_k = f_j \Lambda^j{}_k. \tag{2.69}$$

This is known as *covariant* transformation because the matrix is the same as the matrix $\boldsymbol{\Lambda}$ with which basis vectors transform. The only remaining case to be considered is the transformation of basis 1-forms; clearly they transform with $\boldsymbol{\Lambda}^{-1}$.

Consider next the effect of following one transformation by another. The matrix representing this "composition" of two transformations is known as the "concatenation" of the individual matrices. Calling these matrices $\boldsymbol{\Lambda}_1$ and $\boldsymbol{\Lambda}_2$, the concatenated matrix $\boldsymbol{\Lambda}$ can be obtained by successive applications of Eq. (2.67);

$$\mathbf{x}'' = \boldsymbol{\Lambda}_2^{-1} \boldsymbol{\Lambda}_1^{-1} \mathbf{x}, \quad \text{or} \quad \boldsymbol{\Lambda}^{-1} = \boldsymbol{\Lambda}_2^{-1} \boldsymbol{\Lambda}_1^{-1}. \tag{2.70}$$

This result has used the fact that the contravariant components are arrayed as a column vector. On the other hand, with $\widetilde{\mathbf{f}}$ regarded as a row vector of covariant components, Eq. (2.69) yields

$$\widetilde{\mathbf{f}''} = \widetilde{\mathbf{f}} \boldsymbol{\Lambda}_1 \boldsymbol{\Lambda}_2, \quad \text{or} \quad \boldsymbol{\Lambda} = \boldsymbol{\Lambda}_1 \boldsymbol{\Lambda}_2. \tag{2.71}$$

It may seem curious that the order of matrix multiplications can be opposite for "the same" sequence of transformations, but the result simply reflects the distinction between covariant and contravariant quantities. Since general matrices $\mathbf{A}$ and $\mathbf{B}$ satisfy $(\mathbf{AB})^{-1} = \mathbf{B}^{-1}\mathbf{A}^{-1}$, the simultaneous validity of Eq. (2.70) and Eq. (2.71) can be regarded as mere self-consistency of the requirement that $\langle \widetilde{\mathbf{f}}, \mathbf{x} \rangle$ be invariant.

The transformations just considered have been *passive*, in that basis vectors were changed but the physical quantities not. Commonly in mechanics, and even more so in optics, one encounters *active* linear transformations that instead describe honest-to-goodness evolution of a physical system. If the configuration at time t_1 is described by $\mathbf{x}(t_1)$ and at a later time t_2 by $\mathbf{x}(t_2)$ *linear* evolution is described by

$$\mathbf{x}(t_2) = \mathbf{A}(t_1, t_2)\mathbf{x}(t_1), \tag{2.72}$$

and the equations of this section have to be re-interpreted appropriately.

2.3.3 Transformation of Distributions

Often one wishes to evolve not only one particle in the way just mentioned, but rather an entire ensemble or distribution of particles. Suppose that the distribution, call it $\rho(\mathbf{x})$, has the property that all particles lie in the same plane at time t_1. Such a distribution could be expressed as $\hat{\rho}(\mathbf{x})\delta(ax + by + cz - d)$, where δ is the Dirac δ-"function" with argument which, when set to zero, gives the equation of the plane. Let us ignore the distribution within the plane (described by $\hat{\rho}(\mathbf{x})$) and pay attention only to the most noteworthy feature of this ensemble of points, namely the plane itself and how it evolves. If $\mathbf{x}_{(1)}$ is the displacement vector of a generic particle at an initial time $t_{(1)}$, then initially the plane is described by an equation

$$f_{(1)i}\, x^i = 0. \tag{2.73}$$

For each of the particles, setting $x^i = x^i_{(1)}$ in Eq. (2.73) results in an equality. Let us call the coefficients $f_{(1)i}$ "distribution parameters" at time t_1 since they characterize the region containing the particles at that time.

Suppose that the system evolves in such a way that the individual particle coordinates are transformed (linearly) to $x^i_{(2)}$, and then to $x^i_{(3)}$, according to

$$x^i_{(2)} = A^i{}_j\, x^j_{(1)}, \quad x^k_{(3)} = B^k{}_j\, x^j_{(2)}, \quad \text{or} \quad x^k_{(3)} = (\mathbf{BA})^k{}_i\, x^i_{(1)}. \tag{2.74}$$

With each particle having been subjected to this transformation, the question is, what is the final distribution of particles? Since the particles began on the same plane initially and the transformations have been linear it is clear they will lie on the same plane finally. We wish to find that plane, which is to say to find the coefficients $f_{(3)k}$ in the equation

$$f_{(3)k}\, x^k = 0. \tag{2.75}$$

This equation must be satisfied by $x^k_{(3)}$ as given by Eq. (2.74), and this yields

$$f_{(3)k}\, (\mathbf{BA})^k{}_i\, x^i = 0. \tag{2.76}$$

It follows that

$$f_{(3)k} = f_{(1)i}((\mathbf{BA})^{-1})^i{}_k = f_{(1)i}(\mathbf{A}^{-1}\mathbf{B}^{-1})^i{}_k. \tag{2.77}$$

This shows that the coefficients f_i describing a distribution of particles transform covariantly when individual particle coordinates x^i transform contravariantly.

We have seen that the composition of successive linear transformations represented by matrices $\mathbf{A}$ and $\mathbf{B}$ can be either $\mathbf{BA}$ or $\mathbf{A}^{-1}\mathbf{B}^{-1}$ depending on the nature of the quantity being transformed and it is necessary to determine from the context which one is appropriate. If contravariant components compose with matrix **BA** then covariant components compose with matrix $\mathbf{A}^{-1}\mathbf{B}^{-1}$.

Though these concatenation relations have been derived for linear transformations, there is a sense in which they are the only possibilities for (sufficiently smooth) nonlinear transformations as well. If the origin maps to the origin, as we have assumed implicitly, then there is a "linearized transformation" that is approximately valid for "small amplitude" (close to the origin) particles, and the above concatenation properties must apply to that transformation. The same distinction between the transformation properties of particle coordinates and distribution coefficients must therefore also apply to nonlinear transformations, though the equations can be expected to become much more complicated at large amplitudes. It is only *linear* transformations that can be concatenated in closed form using matrix multiplication but the opposite concatenation order of covariant and contravariant quantities also applies in the nonlinear regime.

There is an interesting discussion in Schutz, Section 2.18, expanding on the interpretation of the *Dirac delta function* as a *distribution* in the sense the word is being used here. If the argument of the delta function is said to be transformed contravariantly then the "value" of the delta function transforms covariantly.

2.3.4 Multi-index Tensors and their Contraction

This section is rather abstract. The reader willing to accept that the *contraction* of the upper and lower index of a tensor is invariant can skip it. A footnote on the next page hints how this result can be obtained more quickly.

We now turn to tensors with more than one index. Two-index covariant tensors are defined by considering real-valued bilinear functions of two vectors, say $\mathbf{x}$ and $\mathbf{y}$. Such a function $\widetilde{\mathbf{f}}(\mathbf{x},\mathbf{y})$ is called *bilinear* because it is linear in each of its two arguments separately. When the arguments $\mathbf{x}$ and $\mathbf{y}$ are expanded in terms of the basis introduced in Eq. (2.61) one has,[21]

$$\widetilde{\mathbf{f}}(\mathbf{x},\mathbf{y}) = f_{ij}\, x^i\, y^j, \quad \text{where} \quad f_{ij} = \widetilde{\mathbf{f}}(\mathbf{e_i},\mathbf{e_j}). \tag{2.78}$$

As usual, we will say that the function $\widetilde{\mathbf{f}}$ and the array of coefficients f_{ij} are *the same thing* and that $\widetilde{\mathbf{f}}(\mathbf{x},\mathbf{y})$ is the same thing as the bilinear form $f_{ij}\, x^i\, y^j$. The coefficients f_{ij} are called covariant components of $\widetilde{\mathbf{f}}$. Pedantically it is only $\widetilde{\mathbf{f}}(\mathbf{x},\mathbf{y})$, with arguments inserted, that deserves to be called a *form*, but common usage seems to be to call $\widetilde{\mathbf{f}}$ a *form* all by itself. An expressive notation

21) Equation (2.78) is actually unnecessarily restrictive, since $\mathbf{x}$ and $\mathbf{y}$ could be permitted to come from different spaces.

that will often be used is $\widetilde{\mathbf{f}}(\cdot,\cdot)$, which indicates that $\widetilde{\mathbf{f}}$ is "waiting for" two vector arguments.

Problem 2.3.1. *Show that the transformation $f_{ij} \to f'_{\alpha\beta}$ of the covariant 2-tensor elements f_{ij}, corresponding to transformation (2.59) of the x^1 elements, can be expressed in the following form, ready for immediate evaluation as a triple matrix multiplication:*

$$f'\alpha\beta = (\Lambda^T)_\alpha^{\ i} f_{ij} \Lambda^j_{\ \beta}. \tag{2.79}$$

Especially important are the *antisymmetric* bilinear functions $\widetilde{\mathbf{f}}(\mathbf{x},\mathbf{y})$ that change sign when $\mathbf{x}$ and $\mathbf{y}$ are interchanged

$$\widetilde{\mathbf{f}}(\mathbf{x},\mathbf{y}) = -\widetilde{\mathbf{f}}(\mathbf{y},\mathbf{x}), \quad \text{or} \quad f_{ij} = -f_{ji}. \tag{2.80}$$

These *alternating* or *antisymmetric* tensors are the only multi-index quantities that represent important geometric objects. The theory of determinants can be based on them as well (see Van der Waerden, Section 4.7).

To produce a contravariant two-index tensor requires the definition of a bilinear function of two covariant vectors $\widetilde{\mathbf{u}}$ and $\widetilde{\mathbf{v}}$. One way of constructing such a bilinear function is to start with two fixed contravariant vectors $\mathbf{x}$ and $\mathbf{y}$ and to define

$$\mathbf{f}(\widetilde{\mathbf{u}},\widetilde{\mathbf{v}}) = \langle\widetilde{\mathbf{u}},\mathbf{x}\rangle\langle\widetilde{\mathbf{v}},\mathbf{y}\rangle, \tag{2.81}$$

This tensor is called the *tensor product* $\mathbf{x}\otimes\mathbf{y}$ of vectors $\mathbf{x}$ and $\mathbf{y}$. Its arguments are $\widetilde{\mathbf{u}}$ and $\widetilde{\mathbf{v}}$. (The somewhat old-fashioned physics terminology is to call $\mathbf{f}$ the dyadic product of $\mathbf{x}$ and $\mathbf{y}$.) In more expressive notation,

$$\mathbf{x}\otimes\mathbf{y}(\cdot,\cdot) = \langle\cdot,\mathbf{x}\rangle\langle\cdot,\mathbf{y}\rangle. \tag{2.82}$$

The vectors $\mathbf{x}$ and $\mathbf{y}$ can in general belong to different spaces with different dimensionalities, but for simplicity in the following few paragraphs we assume they belong to the same space having dimension n. The components of $\mathbf{x}\otimes\mathbf{y}$ are

$$f^{ij} = (\mathbf{x}\otimes\mathbf{y})(\widetilde{\mathbf{e}}^i,\, \widetilde{\mathbf{e}}^j) = \langle\widetilde{\mathbf{e}}^i,\mathbf{x}\rangle\langle\widetilde{\mathbf{e}}^j,\mathbf{y}\rangle = x^i\, y^j. \tag{2.83}$$

Though the linear superposition of any two such tensors is certainly a tensor, call it $\mathbf{t} = (t^{ij})$, it does not follow, for general tensor $\mathbf{t}$, that two vectors $\mathbf{x}$ and $\mathbf{y}$ can be found for which $\mathbf{t}$ is their tensor product. However, all such superpositions can be expanded in terms of the tensor products $\mathbf{e}_i\otimes\mathbf{e}_j$ of the basis vectors. These products form a natural basis for such tensors $\mathbf{t}$. In the next paragraph the n^2-dimensional vector space of two-contravariant-index tensors $\mathbf{t}$ will be called $\mathcal{T}$.

At the cost of greater abstraction, we next prove a result needed to relate a function of two vectors to a function of their vector product. The motivation is less than obvious, but the result will prove to be useful straightaway – what a mathematician might call a lemma;[22]

Theorem 2.3.1. *For any function $B(\mathbf{x},\mathbf{y})$ linear in each of its two arguments* $\mathbf{x}$ *and* $\mathbf{y}$ *there exists an intrinsic linear function of the single argument $\mathbf{x}\otimes\mathbf{y}$, call it $S(\mathbf{x}\otimes\mathbf{y})$, such that*

$$B(\mathbf{x},\mathbf{y}) = S(\mathbf{x}\otimes\mathbf{y}). \tag{2.84}$$

The vectors $\mathbf{x}$ *and* $\mathbf{y}$ *can come from different vector spaces.*

Proof. In terms of contravariant components x^i and y^j the given bilinear function has the form

$$B(\mathbf{x},\mathbf{y}) = s_{ij}\,x^i\,y^j. \tag{2.85}$$

This makes it natural, for arbitrary tensor $\mathbf{t}$ drawn from space $\mathcal{T}$, to define a corresponding function $S(\mathbf{t})$ that is *linear* in the components t^{ij} of $\mathbf{t}$;

$$S(\mathbf{t}) = s_{ij}\,t^{ij}. \tag{2.86}$$

When this function is applied to $\mathbf{x}\otimes\mathbf{y}$, the result is

$$S(\mathbf{x}\otimes\mathbf{y}) = s_{ij}\,x^i\,y^j = B(\mathbf{x},\mathbf{y}), \tag{2.87}$$

which is the required result. Since components were used only in an intermediate stage the theorem relates *intrinsic* (coordinate-free) quantities. (As an aside one can note that the values of the functions S and B could have been allowed to have other (matching) vector or tensor indices themselves without affecting the proof. This increased generality is required to validate contraction of tensors with more than two indices.) □

Other tensor products can be made from contravariant and covariant vectors. Holding $\widetilde{\mathbf{u}}$ and $\widetilde{\mathbf{v}}$ fixed while $\mathbf{x}$ and $\mathbf{y}$ vary, an equation like (2.81) can also be regarded as defining a covector product $\widetilde{\mathbf{u}}\otimes\widetilde{\mathbf{v}}$. A *mixed* vector product $\mathbf{f} = \widetilde{\mathbf{u}}\otimes\mathbf{y}$ can be similarly defined by holding $\widetilde{\mathbf{u}}$ and $\mathbf{y}$ constant.[23]

$$\mathbf{f}(\mathbf{x},\widetilde{\mathbf{v}}) = \langle\widetilde{\mathbf{u}},\mathbf{x}\rangle\,\langle\widetilde{\mathbf{v}},\mathbf{y}\rangle. \tag{2.88}$$

22) The reader impatient with abstract argumentation may consider it adequate to base the invariance of the trace of a mixed tensor on the inverse transformation properties of covariant and contravariant indices.

23) A deficiency of our notation appears at this point since it is ambiguous whether or not the symbol $\mathbf{f}$ in $\mathbf{f} = \widetilde{\mathbf{u}}\otimes\mathbf{y}$ should carry a tilde.

The components of this tensor are

$$f_i^{\;j} = (\widetilde{\mathbf{u}} \otimes \mathbf{y})(\mathbf{e}_i,\, \widetilde{\mathbf{e}}^j) = \langle \widetilde{\mathbf{u}}, \mathbf{e}_i \rangle \langle \widetilde{\mathbf{e}}^j, \mathbf{y} \rangle = u_i\, y^j. \tag{2.89}$$

It is also useful to define antisymmetrized tensor products, or "wedge products" by

$$\begin{aligned} \mathbf{x} \wedge \mathbf{y}(\widetilde{\mathbf{u}}, \widetilde{\mathbf{v}}) &= \langle \mathbf{x}, \widetilde{\mathbf{u}} \rangle \langle \mathbf{y}, \widetilde{\mathbf{v}} \rangle - \langle \mathbf{x}, \widetilde{\mathbf{v}} \rangle \langle \mathbf{y}, \widetilde{\mathbf{u}} \rangle, \\ \widetilde{\mathbf{u}} \wedge \widetilde{\mathbf{v}}(\mathbf{x}, \mathbf{y}) &= \langle \widetilde{\mathbf{u}}, \mathbf{x} \rangle \langle \widetilde{\mathbf{v}}, \mathbf{y} \rangle - \langle \widetilde{\mathbf{u}}, \mathbf{y} \rangle \langle \widetilde{\mathbf{v}}, \mathbf{x} \rangle. \end{aligned} \tag{2.90}$$

The generation of a new tensor by "index contraction" can now be considered. Consider the tensor product $\mathbf{t} = \widetilde{\mathbf{u}} \otimes \mathbf{x}$, where $\widetilde{\mathbf{u}}$ and $\mathbf{x}$ belong to dual vector spaces. The theorem proved above can be applied to the function

$$B(\widetilde{\mathbf{u}}, \mathbf{x}) = \langle \widetilde{\mathbf{u}}, \mathbf{x} \rangle = u_i\, x^i, \tag{2.91}$$

bilinear in $\widetilde{\mathbf{u}}$ and $\mathbf{x}$, to prove the existence of intrinsic linear function S such that

$$S(\widetilde{\mathbf{u}} \otimes \mathbf{x}) = u_i\, x^i = \mathrm{trace}(\widetilde{\mathbf{u}} \otimes \mathbf{x}), \tag{2.92}$$

where $\mathrm{trace}(\mathbf{t})$ is the sum of the *diagonal* elements of tensor $\widetilde{\mathbf{u}} \otimes \mathbf{x}$ in the particular coordinate system shown (or any other, since $\langle \widetilde{\mathbf{u}}, \mathbf{x} \rangle$ is invariant). Since any mixed two-component tensor can be written as a superposition of such covector/contravector products, and since the trace operation is distributive over such superpositions, and since $S(\widetilde{\mathbf{u}} \otimes \mathbf{x})$ is an intrinsic function, it follows that $\mathrm{trace}(\mathbf{t}) = t_i^{\;i}$ is an invariant function for any mixed tensor. Here $\mathrm{trace}(\mathbf{t})$ is called the *contraction* of $\mathbf{t}$.

2.3.5
Representation of a Vector as a Differential Operator

Before leaving the topic of tensor algebra, we review the differential form $\widetilde{\mathbf{dh}}$ obtained from a function of position $\mathbf{x}$ called $h(\mathbf{x})$. We saw a close connection between this quantity and the familiar gradient of vector calculus, $\boldsymbol{\nabla} h$. There is little to add now except to call attention to a potentially confusing issue of terminology.

A physicist thinking of vector calculus, thinks of gradients, divergences, and curls (the operators needed for electromagnetism) to be on the same footing in some sense – they are all "vector derivatives." On the other hand, in mathematics books discussing tensors, gradients are normally considered to be "tensor algebra" and only the divergence and curl are the subject matter of "tensor calculus."

It is probably adequate for a physicist to file this away as yet another curiosity not to be distracted by, but contemplation of the source of the terminology may be instructive. One obvious distinction among the operators in question is that gradients act on scalars whereas divergences and curls operate on vectors, but this is too formal to account satisfactorily for the difference of terminology.

Recall from the earlier discussion of differential forms, in particular Eqs. (2.6) and (2.7) that, for a linear function $h = ax + by$, the coefficients of $\widetilde{\mathbf{dh}}$ are a and b. In this case selecting the coefficient a or b, an algebraic operation, and differentiating h with respect to x or y, a calculus operation, amount to the same thing. Even for nonlinear functions, the gradient operator can be regarded as extracting the coefficients of the linear terms in a Taylor expansion about the point under study. In this linear "tangent space" the coefficients in question are the components of a covariant vector, as has been discussed. What is calculus in the original space is algebra in the tangent space. Such conundrums are not unknown in "unambiguously physics" contexts. For example, both in Hamilton–Jacobi theory and in quantum mechanics there is a close connection between the x-component of a momentum vector and a partial-with-respect-to-x derivative.

Yet one more notational variant will be mentioned before leaving this topic. There is a convention for vectors that is popular with mathematicians but not commonly used by physicists (though it should be since it is both clear and powerful.) We introduce it now, only in a highly specialized sense, intending to expand the discussion later. Consider a standard plot having x as abscissa and y as ordinate, with axes rectangular and having the same scales – in other words ordinary analytic geometry. A function $h(x, y)$ can be expressed by equal h-value contours on such a plot. For describing arrows on this plot it is customary to introduce "unit vectors," usually denoted by $(\mathbf{i}, \mathbf{j})$ or $(\hat{\mathbf{x}}, \hat{\mathbf{y}})$. Let us now introduce the recommended new notation as

$$\frac{\boldsymbol{\partial}}{\boldsymbol{\partial} x} \equiv \mathbf{i}, \quad \frac{\boldsymbol{\partial}}{\boldsymbol{\partial} y} \equiv \mathbf{j}. \tag{2.93}$$

Being equal to $\mathbf{i}$ and $\mathbf{j}$ these quantities are represented by bold face symbols.[24] *$\mathbf{i}$ is that arrow that points along the axis on which x varies and y does not, and if the tail of $\mathbf{i}$ is at $x = x_0$, its tip is at $x = x_0 + 1$.* The same italicized sentence serves just as well to define $\boldsymbol{\partial}/\boldsymbol{\partial} x$ – the symbol in the denominator signifies the coordinate being varied (with the other coordinates held fixed.). This same definition will be defined also to hold if the axes are skew, or if their scales are different, and even if the coordinate grid is curvilinear. (Discontinuous scales

24) Whether or not they are *true* vectors depends on whether or not $\mathbf{i}$ and $\mathbf{j}$ are defined to be *true* vectors. The answer to this question can be regarded as a matter of convention; if the axes are regarded as fixed once and for all then $\mathbf{i}$ and $\mathbf{j}$ are true vectors; if the axes are transformed, they are not.

will not be allowed, however.) Note that, though the notation does not exhibit it, the basis vector $\boldsymbol{\partial}/\boldsymbol{\partial}x$ depends on the coordinates other than x because it points in the direction in which the other coordinates are constant.

One still wonders why this notation for unit vectors deserves a partial derivative symbol. What is to be differentiated? – the answer is $h(x, y)$ (or any other function of x and y). The result, $\partial h/\partial x$ yields the answer to what question? – the answer is, how much does h change when x varies by one unit with y held fixed? Though stretching or twisting the axes would change the appearance of equal-h contours, it would not affect these questions and answers, since they relate only dependence of the function $h(x, y)$ on its arguments and do not depend on how it is plotted. One might say that the notation has *removed the geometry* from the description. One consequence of this is that the application of vector operations such as divergence and curl will have to be re-thought, since they make implicit assumptions about the geometry of the space in which arguments x and y are coordinates. But the gradient requires no further analysis.

From a 1-form $\widetilde{\mathbf{a}}$ and a vector $\mathbf{x}$ one can form the scalar $\langle \widetilde{\mathbf{a}}, \mathbf{x} \rangle$. What is the quantity formed when 1-form $\widetilde{\mathbf{dh}}$, defined in Eq. (2.7), operates on the vector $\boldsymbol{\partial}/\boldsymbol{\partial}x$ just defined? By Eq. (2.2) and the defined meaning of $\boldsymbol{\partial}/\boldsymbol{\partial}x$ we have $\widetilde{\mathbf{dx}}(\boldsymbol{\partial}/\boldsymbol{\partial}x) = 1$. Combining this with Eqs. (2.7) and (2.8) yields

$$\widetilde{\mathbf{dh}}\left(\frac{\boldsymbol{\partial}}{\boldsymbol{\partial}x}\right) = \frac{\partial h}{\partial x} = (\nabla h)_x, \tag{2.94}$$

where the final term is the traditional notation for the x-component of the gradient of h. In this case the new notation can be thought of simply as a roundabout way of expressing the gradient. Some modern authors, Schutz for example, (confusingly in my opinion) simply call $\widetilde{\mathbf{d}}h$ "the gradient of h." This raises another question: should the symbol be $\widetilde{\mathbf{dh}}$ as we have been using or should it be $\widetilde{\mathbf{d}}h$? The symbol $\widetilde{\mathbf{d}}$ was introduced earlier and used to indicate "exterior differentiation." *A priori* the independently defined quantities $\widetilde{\mathbf{dh}}$ are $\widetilde{\mathbf{d}}h$ are distinct. It will be shown that these quantities are in fact equal, so it is immaterial which notation is used.

From these considerations one infers that for contravariant basis vectors $\mathbf{e}_x \equiv \boldsymbol{\partial}/\boldsymbol{\partial}x$ and $\mathbf{e}_y \equiv \boldsymbol{\partial}/\boldsymbol{\partial}y$ the corresponding covariant basis vectors are $\widetilde{\mathbf{e}}^1 \equiv \widetilde{\mathbf{dx}}$ and $\widetilde{\mathbf{e}}^2 \equiv \widetilde{\mathbf{dy}}$. Why is this so? For example, because $\widetilde{\mathbf{dx}}(\boldsymbol{\partial}/\boldsymbol{\partial}x) = 1$. To recapitulate:

$$\mathbf{e}_1 = \frac{\boldsymbol{\partial}}{\boldsymbol{\partial}x^1},\ \mathbf{e}_2 = \frac{\boldsymbol{\partial}}{\boldsymbol{\partial}x^2},\ \dots,\ \mathbf{e}_n = \frac{\boldsymbol{\partial}}{\boldsymbol{\partial}x^n}, \tag{2.95}$$

are the natural contravariant basis vectors and the corresponding covariant basis vectors are

$$\widetilde{\mathbf{e}}^1 = \widetilde{\mathbf{dx}}^1,\ \widetilde{\mathbf{e}}^2 = \widetilde{\mathbf{dx}}^2,\ \dots,\ \widetilde{\mathbf{e}}^n = \widetilde{\mathbf{dx}}^n. \tag{2.96}$$

The association of $\partial/\partial x^1, \partial/\partial x^2, \ldots, \partial/\partial x^n$, with vectors will be shown to be of far more than formal significance in Section 3.6.1 where vectors are associated with directional derivatives.

2.4
(Possibly Complex) Cartesian Vectors in Metric Geometry

2.4.1
Euclidean Vectors

Now, for the first time, we hypothesize the presence of a "metric" (whose existence can, from a physicists point of view, be taken to be a "physical law," for example the Pythagorean "law" or the Einstein–Minkowski "law"). We will use this metric to "associate" covariant and contravariant vectors. Such associations are being made constantly and without a second thought by physicists. Here we spell the process out explicitly. The current task can also be expressed as one of assigning covariant components to a *true* vector that is defined initially by its contravariant components.

A point in three-dimensional Euclidean space can be located by a vector

$$\mathbf{x} = \mathbf{e}_1 x^1 + \mathbf{e}_2 x^2 + \mathbf{e}_3 x^3 \equiv \mathbf{e}_i x^i, \tag{2.97}$$

where $\mathbf{e}_1$, $\mathbf{e}_2$, and $\mathbf{e}_3$ form an orthonormal (defined below) triplet of basis vectors. Such a basis will be called "Euclidean." The final form again employs the repeated-index summation convention even though the two factors have different tensor character in this case. In this expansion the components have upper indices and are called "contravariant" though, as it happens, because the basis is Euclidean, the covariant components x_i to be introduced shortly will have the same values. For skew bases (axes not necessarily orthogonal and to be called "Cartesian") the contravariant and covariant components will be distinct. Unless stated otherwise, x^1, x^2, and x^3 are allowed to be complex numbers – we defer concerning ourselves with the geometric implications of this. We are restricting the discussion to $n = 3$ here only to avoid inessential abstraction; in Cartan's book, mentioned above, most of the results are derived for general n, using arguments like the ones to be used here.

The reader may be beginning to fear a certain repetition of discussion of concepts already understood, such as covariant and contravariant vectors. This can best be defended by observing that, even though these concepts are essentially the same in different contexts, they can also differ in subtle ways, depending upon the implicit assumptions that accompany them.

All vectors start at the origin in this discussion. According to the Pythagorean relation, the distance from the origin to the tip of the arrow can be expressed by a "fundamental form" or "scalar square"

$$\Phi(\mathbf{x}) \equiv \mathbf{x} \cdot \mathbf{x} = (x^1)^2 + (x^2)^2 + (x^3)^2. \tag{2.98}$$

Three distinct cases will be of special importance:

1. The components x^1, x^2, and x^3 are required to be real. In this case $\Phi(\mathbf{x})$, conventionally denoted also by $|\mathbf{x}|^2$, is necessarily positive, and it is natural to divide any vector by $|\mathbf{x}|$ to convert it into a "unit vector." This describes ordinary geometry in three dimensions, and constitutes the Pythagorean law referred to above.

2. The components x^1, x^2, and x^3 are complex. Note that the fundamental form $\Phi(\mathbf{x})$ is *not* defined to be $\overline{x}^1\, x^1 + \overline{x}^2\, x^2 + \overline{x}^3\, x^3$ and that it has the possibility of being complex or of vanishing even though $\mathbf{x}$ does not. If $\Phi(\mathbf{x})$ vanishes, $\mathbf{x}$ is said to be "isotropic." If $\Phi(\mathbf{x}) \neq 0$ then $\mathbf{x}$ can be normalized, by dividing by $\Phi(\mathbf{x})$, thereby converting it into a "unit vector."

3. In the "pseudo-Euclidean" case the components x^1, x^2, and x^3 are required to be real, but the fundamental form is given not by Eq. (2.98) but by

$$\Phi(\mathbf{x}) = (x^1)^2 + (x^2)^2 - (x^3)^2. \tag{2.99}$$

Since this has the possibility of vanishing, a vector can be "isotropic," or "on the light cone" in this case also. For $\Phi > 0$ the vector is "space-like;" for $\Phi < 0$ it is "time-like." In these cases a "unit vector" can be defined as having fundamental form of magnitude 1. In this pseudo-Euclidean case, "ordinary" space–time requires $n = 1 + 3$. This metric could legitimately be called "Einstein's metric," but it is usually called "Minkowski's." In any case, its existence can be regarded as a physical law, not just a mathematical construct.

To the extent possible these cases will be treated "in parallel," in a unified fashion, with most theorems and proofs applicable in all cases. Special properties of one or the other of the cases will be interjected as required.

The "scalar" or "invariant product" of vectors $\mathbf{x}$ and $\mathbf{y}$ is defined in terms of their *Euclidean* components by

$$\mathbf{x} \cdot \mathbf{y} \equiv x^1 y^1 + x^2 y^2 + x^3 y^3. \tag{2.100}$$

Though similar looking expressions have appeared previously, this is the first one deserving of the name "dot product." If $\mathbf{x} \cdot \mathbf{y}$ vanishes, $\mathbf{x}$ and $\mathbf{y}$ are said to be orthogonal. An "isotropic" vector is orthogonal to itself. The vectors orthogonal to a given vector span a plane. (In n-dimensional space this is called a "hyperplane" of $n-1$ dimensions.) In the pseudo-Euclidean case there is one minus sign in the definition of scalar product as in Eq. (2.99).

Problem 2.4.1. *Show that definition (2.100) follows from definition (2.99) if one assumes "natural" algebraic properties for "lengths" in the evaluation of* $(\mathbf{x}+\lambda\mathbf{y})\cdot(\mathbf{x}+\lambda\mathbf{y})$*, where* $\mathbf{x}$ *and* $\mathbf{y}$ *are two different vectors and* λ *is an arbitrary scalar.*

2.4.2
Skew Coordinate Frames

The basis vectors $\boldsymbol{\eta}_1$, $\boldsymbol{\eta}_2$, and $\boldsymbol{\eta}_3$, in a skew, or "Cartesian," frame are not orthonormal in general. They must however be "independent;" geometrically this requires that they not lie in a single plane; algebraically it requires that no vanishing linear combination can be formed from them. As a result, a general vector $\mathbf{x}$ can be expanded in terms of $\boldsymbol{\eta}_1$, $\boldsymbol{\eta}_2$, and $\boldsymbol{\eta}_3$,

$$\mathbf{x} = \boldsymbol{\eta}_i x^i. \tag{2.101}$$

and its scalar square is then given by

$$\Phi(\mathbf{x}) = \boldsymbol{\eta}_i \cdot \boldsymbol{\eta}_j x^i x^j \equiv g_{ij} x^i x^j. \tag{2.102}$$

Here "metric coefficients," and the matrix $\mathbf{G}$ they form, have been defined by

$$g_{ij} = g_{ji} = \boldsymbol{\eta}_i \cdot \boldsymbol{\eta}_j, \quad \mathbf{G} = \begin{pmatrix} \boldsymbol{\eta}_1\cdot\boldsymbol{\eta}_1 & \boldsymbol{\eta}_1\cdot\boldsymbol{\eta}_2 & \boldsymbol{\eta}_1\cdot\boldsymbol{\eta}_3 \\ \boldsymbol{\eta}_2\cdot\boldsymbol{\eta}_1 & \boldsymbol{\eta}_2\cdot\boldsymbol{\eta}_2 & \boldsymbol{\eta}_2\cdot\boldsymbol{\eta}_3 \\ \boldsymbol{\eta}_3\cdot\boldsymbol{\eta}_1 & \boldsymbol{\eta}_3\cdot\boldsymbol{\eta}_2 & \boldsymbol{\eta}_3\cdot\boldsymbol{\eta}_3 \end{pmatrix}. \tag{2.103}$$

As in Section 2.3 the coefficients x^i are known as "contravariant components" of $\mathbf{x}$. When expressed in terms of them the formula for length is more complicated than the Pythagorean formula because the basis vectors are skew. Nevertheless it has been straightforward, starting from a Euclidean basis, to find the components of the metric tensor. It is less straightforward, and not even necessarily possible in general, given a metric tensor, to find a basis in which the length formula is Pythagorean.

2.4.3
Reduction of a Quadratic Form to a Sum or Difference of Squares

The material in this and the next section is reasonably standard in courses in algebra. It is nevertheless spelled out here in some detail since, like some of the other material in this chapter, analogous procedures will be used when "symplectic geometry" is discussed.

For describing scalar products, defined in the first place in terms of orthonormal axes, but now using skew coordinates, a quadratic form has been introduced. Conversely, given an arbitrary quadratic form $\Phi = g_{ij}u^i u^j$, with $g_{ij} = g_{ji}$, can we find a coordinate transformation $x_i = a_{ij}u^j$ to variables for which Φ takes the form of Eq. (2.97)?[25]

In general, the components can be complex. If the components are required to be real then the coefficients a_{ij} will also be required to be real; otherwise they can also be complex. The reader has no doubt been subjected to such an analysis before, though perhaps not with complex variables allowed.

Theorem 2.4.1. *Every quadratic form can be reduced to a sum of (positive or negative) squares by a linear transformation of the variables.*

Proof. (a) Suppose one of the diagonal elements is nonzero. Re-labeling as necessary, let it be g_{11}. With a view toward eliminating all terms linear in u^1, define

$$\Phi_1 \equiv g_{ij}u^i u^j - \frac{1}{g_{11}}(g_{1i}u^i)^2, \tag{2.104}$$

which no longer contains u^1. Hence, defining

$$y_1 \equiv g_{1i}u^i \tag{2.105}$$

the fundamental form can be written as

$$\Phi = \frac{1}{g_{11}}y_1^2 + \Phi_1; \tag{2.106}$$

the second term has one fewer variable than previously.

(b) If all diagonal elements vanish, one of the off-diagonal elements, say g_{12} does not. In this case define

$$\Phi_2 \equiv g_{ij}u^i u^j - \frac{2}{g_{12}}(g_{21}u^1 + g_{23}u^3)(g_{12}u^2 + g_{13}u^3), \tag{2.107}$$

which contains neither u^1 nor u^2. Defining

$$y_1 + y_2 = g_{21}u^1 + g_{23}u^3, \quad y_1 - y_2 = g_{12}u^2 + g_{13}u^3, \tag{2.108}$$

we obtain

$$\Phi = \frac{2}{g_{12}}(y_1^2 - y_2^2) + \Phi_2, \tag{2.109}$$

again reducing the dimensionality.

25) For purposes of this proof, which is entirely algebraic, we ignore the traditional connection between upper/lower index location, and contravariant/covariant nature. Hence the components x_i given by $x_i = a_{ij}u^j$ are *not* to be regarded as covariant, or contravariant either, for that matter.

The form can be reduced to a sum of squares step-by-step in this way. In the real domain, no complex coefficients are introduced, but some of the coefficients may be negative. In all cases, normalizations can be chosen to make all coefficients be 1 or -1. □

Problem 2.4.2. Sylvester's law of inertia. *The preceding substitutions are not unique but, in the domain of reals, the relative number of negative and positive coefficients in the final form* is *unique. Prove this, for example by showing a contradiction resulting from assuming a relation*

$$\Phi = y^2 - z_1^2 - z_2^2 = v_1^2 + v_2^2 - w^2, \tag{2.110}$$

between variables y, z_1, and z_2 with two negative signs on the one hand, and v, v_2, and w_1 with only one negative sign on the other. In "nondegenerate" (that is det $|g_{ij}| \neq 0$*) ordinary geometry of real numbers the number of positive square terms is necessarily 3.*

2.4.4 Introduction of Covariant Components

The contravariant components x^i are seen in Eq. (2.101) to be the coefficients in the expansion of $\mathbf{x}$ in terms of the $\boldsymbol{\eta}_i$. In ordinary vector analysis one is accustomed to identifying each of these coefficients as the "component of $\mathbf{x}$" along a particular coordinate axes and being able to evaluate it as $|\mathbf{x}|$ multiplied by the cosine of the corresponding angle. Here we define lowered-index components x_i, to be called "covariant," (terminology to be justified as an exercise) as the "invariant products" of $\mathbf{x}$ with the $\boldsymbol{\eta}_i$;

$$x_i = \mathbf{x} \cdot \boldsymbol{\eta}_i = g_{ij} x^j, \quad \text{or as a matrix equation} \quad \widetilde{\mathbf{x}} = (\mathbf{G}\mathbf{x})^T = \mathbf{x}^T \mathbf{G}^T, \tag{2.111}$$

where $\widetilde{\mathbf{x}}$ stands for the array $(x_1, \ldots, x_n)$. Now the scalar product defined in Eq. (2.100) can be written as

$$\mathbf{x} \cdot \mathbf{y} = x_i y^i = y_i x^i. \tag{2.112}$$

By inverting Eq. (2.111) contravariant components can be obtained from covariant ones

$$\mathbf{x}^T = \widetilde{\mathbf{x}} (\mathbf{G}^T)^{-1}, \text{ or as components, } x^i = g^{ij} x_j \text{ where } g^{ij} = (\mathbf{G}^{-1})^{ij}. \tag{2.113}$$

For orthonormal bases, $\mathbf{G} = \mathbf{1}$ and, as mentioned previously, covariant and contravariant components are identical. Introduction of covariant components can be regarded as a simple algebraic convenience with no geometric significance. However, if the angle θ between vectors $\mathbf{x}$ and $\mathbf{y}$ is defined by

$$\cos\theta = \frac{\mathbf{x} \cdot \mathbf{y}}{\sqrt{\Phi(\mathbf{x})\Phi(\mathbf{y})}}, \tag{2.114}$$

then a general vector $\mathbf{x}$ is related to the basis vectors $\boldsymbol{\eta}_1$, $\boldsymbol{\eta}_2$, and $\boldsymbol{\eta}_3$ by direction cosines $\cos\theta_1$, $\cos\theta_2$, $\cos\theta_3$, and its covariant components are

$$x_i = \mathbf{x}\cdot\boldsymbol{\eta}_i = \sqrt{\Phi(\mathbf{x})\Phi(\boldsymbol{\eta}_i)}\,\cos\theta_i. \tag{2.115}$$

This definition is illustrated in Fig. 2.6.

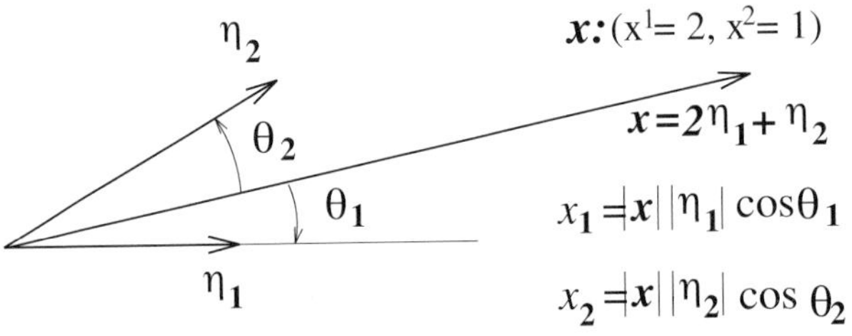

Fig. 2.6 The *true* vector $2\boldsymbol{\eta}_1 + \boldsymbol{\eta}_2$ expressed in terms of contravariant components and, by Eq. (2.115), using covariant components and direction cosines. For Euclidean geometry $\sqrt{\Phi(\mathbf{x})}$ is normally symbolized by $|\mathbf{x}|$.

2.4.5
The Reciprocal Basis

Even in Euclidean geometry there are situations in which skew axes yield simplified descriptions, which makes the introduction of covariant components especially useful. The most important example is in the description of a crystal for which displacements by integer multiples of "unit cell" vectors $\boldsymbol{\eta}_1$, $\boldsymbol{\eta}_2$, and $\boldsymbol{\eta}_3$ leave the lattice invariant.

Let these unit cell vectors form the basis of a skew frame as in Section 2.4.2. For any vector $\mathbf{x}$ in the original space we can associate a particular form $\widetilde{\mathbf{x}}$ in the dual space by the following rule giving the result of evaluating $\widetilde{\mathbf{x}}$ for arbitrary argument $\mathbf{y}$: $\langle\widetilde{\mathbf{x}},\mathbf{y}\rangle = \mathbf{x}\cdot\mathbf{y}$. In particular, "reciprocal basis vectors" $\boldsymbol{\eta}^i$ and basis forms $\widetilde{\boldsymbol{\eta}}^i$ are defined to satisfy

$$\langle\widetilde{\boldsymbol{\eta}}^i,\boldsymbol{\eta}_j\rangle \equiv \boldsymbol{\eta}^i\cdot\boldsymbol{\eta}_j = \delta^i{}_j, \tag{2.116}$$

$\widetilde{\boldsymbol{\eta}}^1$, $\widetilde{\boldsymbol{\eta}}^2$, and $\widetilde{\boldsymbol{\eta}}^3$ are the basis dual to $\boldsymbol{\eta}_1$, $\boldsymbol{\eta}_2$, and $\boldsymbol{\eta}_3$ as in Eq. (2.62). The vectors $\boldsymbol{\eta}^i$ in this equation need to be determined to satisfy the final equality. This can be accomplished mentally;

$$\boldsymbol{\eta}^1 = \frac{\boldsymbol{\eta}_2\times\boldsymbol{\eta}_3}{\sqrt{g}},\quad \boldsymbol{\eta}^2 = \frac{\boldsymbol{\eta}_3\times\boldsymbol{\eta}_1}{\sqrt{g}},\quad \boldsymbol{\eta}^3 = \frac{\boldsymbol{\eta}_1\times\boldsymbol{\eta}_2}{\sqrt{g}},\tag{2.117}$$
$$\text{where}\quad \sqrt{g} = (\boldsymbol{\eta}_1\times\boldsymbol{\eta}_2)\cdot\boldsymbol{\eta}_3$$

where the orientation of the basis vectors is assumed to be such that $\sqrt{g}$ is real and nonzero. (From vector analysis one recognizes $\sqrt{g}$ to be the volume of the unit cell.) One can confirm that Eqs. (2.116) are then satisfied. The vectors $\boldsymbol{\eta}^1$, $\boldsymbol{\eta}^2$, and $\boldsymbol{\eta}^3$ are said to form the "reciprocal basis."

Problem 2.4.3. *In terms of skew basis vectors $\boldsymbol{\eta}_1$, $\boldsymbol{\eta}_2$, and $\boldsymbol{\eta}_3$ in three-dimensional Euclidean space a vector $\mathbf{x} = \boldsymbol{\eta}_i x^i$ has covariant components $x_i = g_{ij}x^j$. Show that*

$$\mathbf{x} = x_1\boldsymbol{\eta}^1 + x_2\boldsymbol{\eta}^2 + x_3\boldsymbol{\eta}^3. \tag{2.118}$$

where the $\boldsymbol{\eta}^i$ are given by Eq. (2.117).

By inspection one sees that the reciprocal base vector $\boldsymbol{\eta}^1$ is normal to the plane containing $\boldsymbol{\eta}_3$ and $\boldsymbol{\eta}_2$. This is illustrated in Fig. 2.7 which shows the unit cell vectors superimposed on a crystal lattice. ($\boldsymbol{\eta}_3$ points normally out of the paper.) Similarly, $\boldsymbol{\eta}^2$ is normal to the plane containing $\boldsymbol{\eta}_3$ and $\boldsymbol{\eta}_1$.

Consider the plane passing through the origin and containing both $\boldsymbol{\eta}_3$ and the vector $\boldsymbol{\eta}_1 + N\boldsymbol{\eta}_2$, where N is a fixed integer. Since there is an atom situated at the tip of this vector, this plane contains this atom as well as the atom at the origin, the atom at $2(\boldsymbol{\eta}_1 + N\boldsymbol{\eta}_2)$ and at $3(\boldsymbol{\eta}_1 + N\boldsymbol{\eta}_2)$ and so on. For the case $N = 1$, these atoms are joined by a line in the figure and several other lines, all parallel and passing through other atoms are shown as well. The vector

$$\frac{(\boldsymbol{\eta}_1 + N\boldsymbol{\eta}_2) \times \boldsymbol{\eta}_3}{\sqrt{g}} = -\boldsymbol{\eta}^2 + N\boldsymbol{\eta}^1 \tag{2.119}$$

is perpendicular to this set of planes. Again for $N = 1$ the figure confirms that $\boldsymbol{\eta}^1 - \boldsymbol{\eta}^2$ is normal to the crystal planes shown.

Problem 2.4.4. *Show (a) for any two atoms in the crystal, that the plane containing them and the origin is normal to a vector expressible as a superposition of reciprocal basis vectors with integer coefficients, and (b) that any superposition of reciprocal basis vectors with integer coefficients is normal to a set of planes containing atoms. [Hint: for practice at the sort of calculation that is useful, evaluate $(\boldsymbol{\eta}^1 + \boldsymbol{\eta}^2) \cdot (\boldsymbol{\eta}_1 + \boldsymbol{\eta}_2)$.]*

It was only because the dot product is meaningful, that Eq. (2.116) results in the association of an ordinary vector $\boldsymbol{\eta}^i$ with the form $\widetilde{\boldsymbol{\eta}}^i$. But once that identification is made all computations can be made using straightforward vector analysis. A general vector $\mathbf{x}$ can be expanded either in terms of the original or the reciprocal basis

$$\mathbf{x} = \boldsymbol{\eta}_i x^i = x_i\boldsymbol{\eta}^i. \tag{2.120}$$

(The components x_i can be thought of either as covariant components of $\mathbf{x}$ or as components of $\widetilde{\mathbf{x}}$ such that $\widetilde{\mathbf{x}}(\mathbf{y}) = x_i y^i$.) In conjunction with Eqs. (2.113)

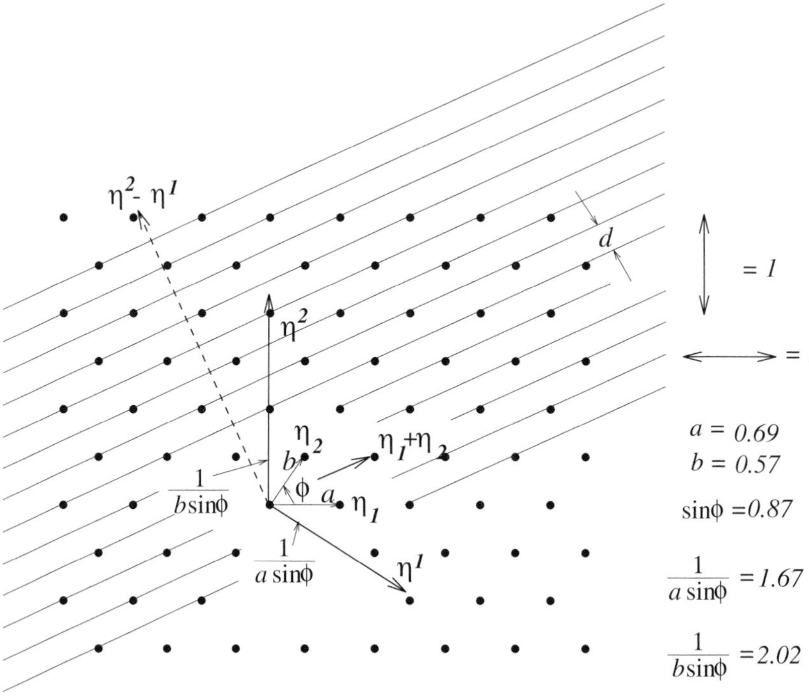

Fig. 2.7 The crystal lattice shown has unit cell vectors $\boldsymbol{\eta}_1$ and $\boldsymbol{\eta}_2$ as shown, as well as $\boldsymbol{\eta}_3$ pointing normally out of the paper. Reciprocal basis vectors $\boldsymbol{\eta}^1$ and $\boldsymbol{\eta}^2$ are shown. The particular lattice planes indicated by parallel lines correspond to the reciprocal lattice vector $\boldsymbol{\eta}^2 - \boldsymbol{\eta}^1$. It is coincidental that $\boldsymbol{\eta}^2$ appears to lie in a crystal plane.

and (2.116) this yields

$$(g^{ij}) = \mathbf{G}^{-1} = \begin{pmatrix} \boldsymbol{\eta}^1 \cdot \boldsymbol{\eta}^1 & \boldsymbol{\eta}^1 \cdot \boldsymbol{\eta}^2 & \boldsymbol{\eta}^1 \cdot \boldsymbol{\eta}^3 \\ \boldsymbol{\eta}^2 \cdot \boldsymbol{\eta}^1 & \boldsymbol{\eta}^2 \cdot \boldsymbol{\eta}^2 & \boldsymbol{\eta}^2 \cdot \boldsymbol{\eta}^3 \\ \boldsymbol{\eta}^3 \cdot \boldsymbol{\eta}^1 & \boldsymbol{\eta}^3 \cdot \boldsymbol{\eta}^2 & \boldsymbol{\eta}^3 \cdot \boldsymbol{\eta}^3 \end{pmatrix}. \tag{2.121}$$

Problem 2.4.5. *Confirm the* Lagrange identity *of vector analysis*

$$(\mathbf{A} \times \mathbf{B}) \cdot (\mathbf{C} \times \mathbf{D}) = \det \begin{vmatrix} \mathbf{A} \cdot \mathbf{C} & \mathbf{A} \cdot \mathbf{D} \\ \mathbf{B} \cdot \mathbf{C} & \mathbf{B} \cdot \mathbf{D} \end{vmatrix}. \tag{2.122}$$

This is most simply done by expressing the cross products with the three index anti-symmetric symbol ϵ_{ijk}. With the vectors **A**, **B**, **C**, *and* **D** *drawn from $\boldsymbol{\eta}_1$, $\boldsymbol{\eta}_2$, and $\boldsymbol{\eta}_3$, each of these determinants can be identified as a co-factor in Eq. (2.103). From this show that*

$$g = \det |\mathbf{G}|. \tag{2.123}$$

Problem 2.4.6. *Show that original basis vectors $\boldsymbol{\eta}_i$ are themselves reciprocal to the reciprocal basis vectors $\boldsymbol{\eta}^i$.*

Bibliography

General References

1 B.F. Schutz, *Geometrical Methods of Mathematical Physics*, Cambridge University Press, Cambridge, UK, 1995.

2 V.I. Arnold, *Mathematical Methods of Classical Mechanics*, 2nd ed., Springer, New York, 1989.

References for Further Study

Section 2.1

3 E. Cartan, *Leçons sur la géométrie des espaces de Riemann*, Gauthiers-Villars, Paris, 1951. (English translation available.)

4 J.A. Schouten, *Tensor Analysis for Physicists*, 2nd ed. Oxford University Press, Oxford, 1954.

Section 2.3

5 B.L. Van der Waerden, *Algebra*, Vol. 1, Springer, New York, 1991.

Section 2.4

6 E. Cartan, *The Theory of Spinors*, Dover, New York, 1981.

…cs, II, Curvilinear

…of orbits in n-dimensional Euclidean space is …ngular coordinates. The case $n = 3$ will be …neralizing to cases with $n > 3$ is unneces- … ordinary space, but it begins to approach …e realistic problems require the introduc- …lized coordinates. Unfortunately the Eu- …rean theorem) is typically not satisfied in …, analysis of curvilinear coordinates in or- …y requires the introduction of mathematical methods …ued in more general situations. It seems sensible to digest this …thematics in this intuitively familiar setting rather than in the more abstract mathematical setting of differentiable manifolds.

In the $n = 3$ case much of the analysis to be performed may be already familiar, for example from courses in electricity and magnetism. For calculating fields from symmetric charge distributions, for example radially symmetric, it is obviously convenient to use spherical coordinates rather than rectangular. This is even more true for solving boundary value problems with curved boundaries. For solving such problems, curvilinear coordinate systems that conform with the boundary must be used. It is therefore necessary to be able to express the vector operations of gradient, divergence, and curl in terms of these "curvilinear" coordinates. Vector theorems such as Gauss's and Stokes' need to be similarly generalized.

In electricity and magnetism one tends to restrict oneself to geometrically simple coordinate systems such as spherical or cylindrical and in those cases some of the formulas to be obtained can be derived by reasonably elementary methods. Here we consider general curvilinear coordinates where local axes are not only not parallel at different points in space (as is true already for spherical and cylindrical coordinates) but may be skew, not orthonormal. Even the description of force-free particle motion in terms of such curvilinear coordinates is not trivial – you could confirm this, for example, by describing force-free motion using cylindrical coordinates. Commonly one is interested in particle motion in the presence of forces that are most easily described using

Geometric Mechanics: Toward a Unification of Classical Physics. 2nd Edition. Richard Talman

ISBN: 978-3-527-40683-8

particular curvilinear coordinates. Consider, for example, a beam of particles traveling inside an elliptical vacuum tube which also serves as a waveguide for an electromagnetic wave. Since solution of the wave problem requires the use of elliptical coordinates, one is forced to analyze the particle motion using the same coordinates. To face this problem seriously would probably entail mainly numerical procedures, but the use of coordinates conforming to the boundaries would be essential. The very setting up of the problem for numerical solution requires a formulation such as the present one.

The problem just mentioned is too specialized for detailed analysis in a text such as this; these comments have been intended to show that the geometry to be studied has more than academic interest. But, as stated before, our primary purpose is to assimilate the necessary geometry as another step on the way to the geometric formulation of mechanics. Even such a conceptually simple task as describing straight line motion using curvilinear coordinates will be instructive.

3.1 (Real) Curvilinear Coordinates in n-Dimensions

3.1.1 The Metric Tensor

An n-dimensional "Euclidean" space is defined to consist of vectors $\mathbf{x}$ whose components along rectangular axes are $x^1, x^2, \ldots, x^n$, now assumed to be real. The "length" of this vector is

$$\mathbf{x} \cdot \mathbf{x} = x^{1^2} + x^{2^2} + \cdots + x^{n^2}. \tag{3.1}$$

The "scalar product" of vectors $\mathbf{x}$ and $\mathbf{y}$ is

$$\mathbf{x} \cdot \mathbf{y} = x^1 y^1 + x^2 y^2 + \cdots + x^n y^n. \tag{3.2}$$

The angle θ between $\mathbf{x}$ and $\mathbf{y}$ is defined by

$$\cos\theta = \frac{\mathbf{x} \cdot \mathbf{y}}{\sqrt{(\mathbf{x} \cdot \mathbf{x})(\mathbf{y} \cdot \mathbf{y})}}, \tag{3.3}$$

repeating the earlier result (2.114). That this angle is certain to be real follows from a well-known inequality (Schwarz). A fundamental "orthonormal" set of "basis vectors" can be defined as the vectors having rectangular components $\mathbf{e}_1 = (1, 0, \ldots, 0)$, $\mathbf{e}_2 = (0, 1, \ldots, 0)$, etc.

More general "Cartesian," or "skew" components x'^i are related to the Euclidean components by linear transformations

$$x'^i = A^i{}_j x^j, \quad x^i = (A^{-1})^i{}_j x'^j. \tag{3.4}$$

Such a *homogeneous* linear transformation between Cartesian frames is known as a "centered-affine" transformation. If the equations are augmented by additive constants, shifting the origin, the transformation is given the more general name "affine." In terms of the new components the scalar product in (3.2) is given by

$$\mathbf{x}\cdot\mathbf{y} = \sum_i (A^{-1})^i{}_j (A^{-1})^i{}_k x'^j y'^k \equiv g'_{jk} x'^j y'^k, \tag{3.5}$$

where the coefficients g'_{jk} are the primed-system components of the metric tensor. Clearly they are symmetric under the interchange of indices and the quadratic form with $\mathbf{x} = \mathbf{y}$ has to be positive definite. In the original rectangular coordinates $g_{jk} = \delta_{jk}$, where δ_{jk} is the Kronecker symbol with value 1 for equal indices and 0 for unequal indices. In the new frame the basis vectors $\mathbf{e}'_1 = (1, 0, \dots, 0)$, $\mathbf{e}'_2 = (0, 1, \dots, 0)$, etc., are not orthonormal in general, in spite of the fact that their given contravariant components superficially suggest it; rather

$$\mathbf{e}'_i \cdot \mathbf{e}'_j = g'_{ij}. \tag{3.6}$$

As defined so far, the coefficients g_{jk} are constant, independent of position in space. Here, by "position in space" we mean "in the original Euclidean space." For many purposes the original rectangular coordinates $x^1, x^2, \dots, x^n$ would be adequate to locate objects in this space and, though they will be kept in the background throughout most of the following discussion, they will remain available for periodically "getting our feet back on the ground." These coordinates will also be said to define the "base frame" or, when mechanics intrudes, as an "inertial" frame.[1] As mentioned previously, "curvilinear," systems such as radial, cylindrical, elliptical, etc., are sometimes required. Letting $u^1, u^2, \dots, u^n$ be such coordinates, space is filled with corresponding coordinate curves; on each of the "u^1 curves" u^1 varies while $u^2, \dots, u^n$ are fixed, and so on. Sufficiently close to any particular point P, the coordinate curves are approximately linear. In this neighborhood the curvilinear infinitesimal deviations $\Delta u^1, \Delta u^2, \dots, \Delta u^n$ can be used to define the scalar product of deviations $\mathbf{\Delta x}$ and $\mathbf{\Delta y}$;

$$\mathbf{\Delta x}\cdot\mathbf{\Delta y} = g_{jk}(P)\Delta u^j \Delta u^k. \tag{3.7}$$

1) There is no geometric significance whatsoever to a coordinate frame's being inertial, but the base frame will occasionally be called inertial as a mnemonic aid to physicists, who are accustomed to the presence of a preferred frame such as this. The curvilinear frame under study may or may not be rotating or accelerating.

This equation differs from Eq. (3.5) only in that the coefficients $g_{jk}(P)$ are now permitted to be functions of position P.[2]

Problem 3.1.1. *Indices can be raised or lowered as in Eq. (2.113). Use the fact that the matrices of covariant and contravariant components are inverse to show that*

$$g_{il}\frac{\partial g^{lk}}{\partial u^m} = -g^{ik}\frac{\partial g_{il}}{\partial u^m}. \tag{3.8}$$

3.1.2
Relating Coordinate Systems at Different Points in Space

One effect of the coordinates' being curvilinear is to complicate the comparison of objects at disjoint locations. The quantities that will now enter to discipline such comparisons are called "Christoffel coefficients." Deriving them is the purpose of this section.

Consider the coordinate system illustrated in Fig. 3.1, with $\mathbf{M} + \mathbf{dM}(u^1 + du^1, u^2 + du^2, \dots, u^n + du^n)$ being a point close to the point $\mathbf{M}(u^1, u^2, \dots, u^n)$. (The figure is planar but the discussion will be n-dimensional.) For example, the curvilinear coordinates $(u^1, u^2, \dots, u^n)$ might be spherical coordinates (r, θ, ϕ). The vectors $\mathbf{M}$ and $\mathbf{M} + \mathbf{dM}$ can be regarded as vectors locating the points relative to an origin not shown; their base frame coordinates $(x^1, x^2, \dots, x^n)$ refer to a rectangular basis in the base frame centered there; one assumes the base frame coordinates are known in terms of the curvilinear coordinates and *vice versa*. At every point, "natural" basis vectors[3] $(\mathbf{e}_1, \mathbf{e}_2, \dots \mathbf{e}_n)$ can be defined having the following properties:

- $\mathbf{e}_i$ is tangent to the coordinate curve on which u^i varies while the other coordinates are held constant. Without loss of generality i can be taken to be 1 in subsequent discussion.
- With the tail of $\mathbf{e}_1$ at $\mathbf{M}$, its tip is at the point where the first component has increased from u^1 to $u^1 + 1$.
- However, the previous definition has to be qualified since unit increment of a coordinate may be great enough to cause the coordinate curve

2) In this way, a known coordinate transformation has determined a corresponding metric tensor $g_{jk}(P)$. Conversely, one can contemplate a space described by components $u^1, u^2, \dots, u^n$ and metric tensor $g_{jk}(P)$, with given dependence on P, and inquire whether a transformation to components for which the scalar product is Euclidean can be found. The answer, in general, is no. A condition that is needed to be satisfied to assure that the answer to this question be yes is given in the Cartan reference listed at the end of the chapter.

3) The basis vectors being introduced at this point are none other than the basis vectors called $\partial/\partial u_1$, $\partial/\partial u_2$, etc., in Section 2.3.5, but we refrain from using that notation here.

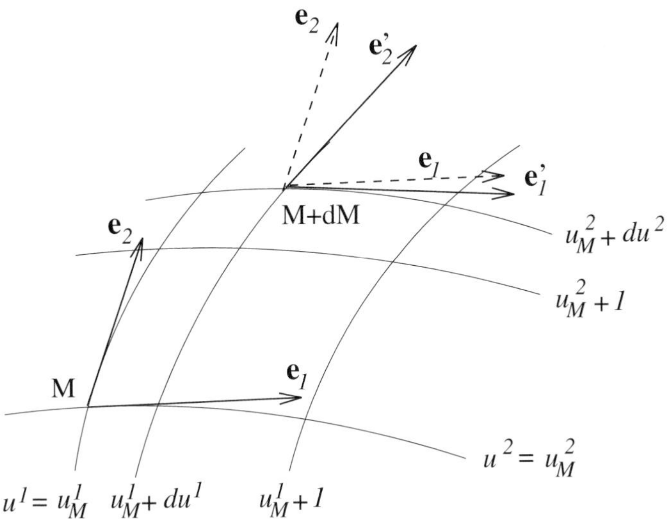

Fig. 3.1 Relating the "natural" local coordinate axes at two different points in ordinary space described by curvilinear coordinates. Because this is a Euclidean plane, the unit vectors $\mathbf{e}_1$ and $\mathbf{e}_2$ at $\mathbf{M}$ can be "parallel slid" to point $\mathbf{M} + \mathbf{dM}$ without changing their lengths or directions; they are shown there as dashed arrows. The curve labeled $u^1_M + 1$ is the curve on which u^1 has increased by 1 and so on.

to veer noticeably away from the straight basis vector – think, for example, of a change in polar angle $\phi \to \phi + 1$ radian. Clearly the rigorous definition of the "length" of a particular basis vector, say $\mathbf{e}_1$, requires a careful limiting process. Instead, forsaking any pretense of rigor, let us assume the scale along the u^1 coordinate curve has been expanded sufficiently by "choosing the units" of u^1 to make the unit vector coincide with the coordinate curve to whatever accuracy is considered adequate.

- One is tempted to use the term "unit vector" to describe a basis vector $\mathbf{e}_i$, but doing so is likely to be misleading since, at least in science, the term "unit vector" usually connotes a vector of unit length. A common notation for a vector parallel to $\mathbf{e}_i$ and having unit length, is $\widehat{\mathbf{u}}_\mathbf{i}$. Here we have $\mathbf{e}_i \parallel \widehat{\mathbf{u}}_i$ but $\widehat{\mathbf{u}}_i = \mathbf{e}_i / |\mathbf{e}_i|$.

- If one insists on ascribing physical dimensions to the $\mathbf{e}_i$ one must allow the dimensions to be different for different i. For example, if $(\mathbf{e}_1, \mathbf{e}_2, \mathbf{e}_3)$ correspond to (r, θ, ϕ), then the first basis vector has units of length while the other two are dimensionless. Though this may seem unattractive, it is not unprecedented in physics – one is accustomed to a relativistic 4-vector having time as one coordinate and distances as the others. On the other hand, the vectors $(\hat{\mathbf{r}}, \hat{\boldsymbol{\theta}}, \hat{\boldsymbol{\phi}})$ all have units of meters – but this is not much of an advantage since, as mentioned already, the lengths of these vectors are somewhat artificial in any case.

Hence, deviating from traditional usage in elementary physics, we will use the basis vectors $\mathbf{e}_i$ almost exclusively, possibly even calling them unit vectors in spite of their not having unit length. Dimensional consistency will be enforced separately.

Dropping quadratic (and higher) terms, the displacement vector $\mathbf{dM}$ can be expanded in terms of basis vectors at point $\mathbf{M}$ as[4]

$$\mathbf{dM} = du^1\mathbf{e}_1 + du^2\mathbf{e}_2 + \cdots du^n\mathbf{e}_n \equiv du^i\mathbf{e}_i. \tag{3.9}$$

At each point other than $\mathbf{M}$ the coordinate curves define a similar "natural" n-plet of unit vectors. The reason that "natural" is placed in quotation marks here and above is that what is natural in one context may be unnatural in another. Once the particular coordinate curves $(u^1, u^2, \ldots, u^n)$ have been selected the corresponding n-plet $(\mathbf{e}_1, \mathbf{e}_2, \ldots, \mathbf{e}_n)$ is natural, but that does not imply that the coordinates $(u^1, u^2, \ldots, u^n)$ themselves were in any way fundamental.

Our present task is to express the frame $(\mathbf{e'}_1, \mathbf{e'}_2, \ldots, \mathbf{e'}_n)$ at $\mathbf{M} + \mathbf{dM}$ in terms of the frame $(\mathbf{e}_1, \mathbf{e}_2, \ldots, \mathbf{e}_n)$ at $\mathbf{M}$. Working with just two components for simplicity, the first basis vector can be approximated as

$$\mathbf{e'}_1 = \mathbf{e}_1 + \mathbf{de}_1 = \mathbf{e}_1 + \omega_1{}^1\mathbf{e}_1 + \omega_1{}^2\mathbf{e}_2 \tag{3.10}$$

$$\equiv \mathbf{e}_1 + (\Gamma^1{}_{11}du^1 + \Gamma^1{}_{12}du^2)\,\mathbf{e}_1 + (\Gamma^2{}_{11}du^1 + \Gamma^2{}_{12}du^2)\,\mathbf{e}_2. \tag{3.11}$$

The (yet to be determined) coefficients $\omega_i{}^j$ can be said to be "affine-connecting" as they connect quantities in affinely related frames; the coefficients $\Gamma^i{}_{jk}$ are known as Christoffel symbols or as an "affine connection." Both equations (3.10) and (3.11), will occur frequently in the sequel, with the (b) form being required when the detailed dependence on coordinates u^i has to be exhibited, and the simpler (a) form being adequate when all that is needed is an expansion of new basis vectors in terms of old. Here, for the first of many times, we employ a standard, but bothersome notational practice; the incremental expansion coefficients have been written as $\omega_i{}^j$ rather than as $d\omega_i{}^j$ – a notation that would be harmless for the time being but would clash later on when the notation $d\omega$ is conscripted for another purpose. To a physicist it seems wrong for a differential quantity $\mathbf{de}_1$ to be a superposition of quantities like $\omega_1{}^1\mathbf{e}_1$ that appear, notationally, to be nondifferential. But, having already accepted the artificial nature of the units of the basis vectors, we can adopt this notation, promising to sort out the units and differentials later.

4) A physicist might interpret Eq. (3.9) as an approximate equation in which quadratic terms have been neglected, a mathematician might regard it as an exact expansion in the "tangent space" at $\mathbf{M}$.

The terminology "affine connection" anticipates more general situations in which such connections do not necessarily exist. This will be the case for general "manifolds" (spaces describable, for example, by "generalized coordinates" and hence essentially more general than the present Euclidean space). For general manifolds there is no "intrinsic" way to relate coordinate frames at different points in the space. Here "intrinsic" means "independent of a particular choice of coordinates." This can be augmented to include the following prohibition against illegitimate vector superposition:

A vector at one point cannot be expanded in basis vectors belonging to a different point.[5]

After this digression we return to the Euclidean context and Eq. (3.11). This equation appears to be doing the very thing that is not allowed, namely expanding $\mathbf{e}'_1$ in terms of the $\mathbf{e}_i$. The reason it is legitimate in this case is that there *is* an intrinsic way of relating frames at $\mathbf{M}$ and $\mathbf{M}+\mathbf{dM}$ – it is the traditional parallelism of ordinary geometry, as shown in Fig. 3.1. One is really expanding $\mathbf{e}'_1$ in terms of the vectors $\mathbf{e}_i$ slid parallel from $\mathbf{M}$ to $\mathbf{M}+\mathbf{dM}$. All too soon, the concept of "parallelism" will have to be scrutinized more carefully but, for now, since we are considering ordinary space, the parallelism of a vector at $\mathbf{M}$ and a vector at $\mathbf{M}+\mathbf{dM}$ has its usual, intuitively natural, meaning – for example basis vectors $\mathbf{e}_1$ and $\mathbf{e}'_1$ in the figure are almost parallel while $\mathbf{e}_2$ and $\mathbf{e}'_2$ are not.

With this interpretation, Eq. (3.11) is a relation entirely among vectors at $\mathbf{M}+\mathbf{dM}$. The coefficients $\omega_i^{\ j}$ and $\Gamma^j_{\ ik}$ being well defined, we proceed to determine them, starting by re-writing Eq. (3.11) in compressed notation;

$$\mathbf{e}'_i = \mathbf{e}_i + \mathbf{de}_i = \mathbf{e}_i + \omega_i^{\ j}\,\mathbf{e}_j \tag{3.12}$$

$$= \mathbf{e}_i + \Gamma^j_{\ ik} du^k\,\mathbf{e}_j. \tag{3.13}$$

The quantities

$$\omega_i^{\ j} = \Gamma^j_{\ ik} du^k \tag{3.14}$$

are 1-forms, linear in the differentials du^k.[6] The new basis vectors must satisfy Eq. (3.6);

$$\mathbf{e}'_i \cdot \mathbf{e}'_r = g_{ir} + dg_{ir} = (\mathbf{e}_i + \omega_i^{\ j}\,\mathbf{e}_j)\cdot(\mathbf{e}_r + \omega_r^{\ s}\,\mathbf{e}_s). \tag{3.15}$$

Dropping quadratic terms, this can be written succinctly as

$$dg_{ir} = \omega_i^{\ j} g_{jr} + \omega_r^{\ s} g_{is} \stackrel{\text{def.}}{=} \omega_{ir} + \omega_{ri}. \tag{3.16}$$

5) This may seem counter intuitive; if you prefer, for now replace "cannot" by "must not" and regard it as a matter of dictatorial edict.
6) In old references they were known as Pfaffian forms.

(Because the quantities $\omega_i^{\ j}$ are not the components of a *true* tensor, the final step is not a manifestly covariant, index lowering tensor operation, but it can nonetheless serve to define the quantities ω_{ij}, having two lower indices.) Because $dg_{ir} \stackrel{also}{=} (\partial g_{ir}/\partial u^j)\, du^j$, one obtains

$$
\begin{aligned}
-\frac{\partial g_{ij}}{\partial u^k} &= -g_{ir}\Gamma^r_{\ jk} - g_{jr}\Gamma^r_{\ ik},\\
\frac{\partial g_{jk}}{\partial u^i} &= \ g_{jr}\Gamma^r_{\ ki} + g_{kr}\Gamma^r_{\ ji},\\
\frac{\partial g_{ki}}{\partial u^j} &= \ g_{kr}\Gamma^r_{\ ij} + g_{ir}\Gamma^r_{\ kj}.
\end{aligned} \tag{3.17}
$$

For reasons to be clear shortly, we have written the identical equation three times, but with indices permuted, substitutions like $g_{ri} = g_{ir}$ having been made in some terms, and the first equation having been multiplied through by -1.

Problem 3.1.2. *Show that Eqs. (3.17) yield $n^2(n+1)/2$ equations that can be applied toward determining the n^3 coefficients $\Gamma^j_{\ ik}$. Relate this to the number of parameters needed to fix the scales and relative angles of a skew basis set. For the $n = 3$, ordinary geometry case, how many more parameters are needed to fix the absolute orientation of a skew frame? How many more conditions on the $\Gamma^j_{\ ik}$ does this imply? Both for $n = 3$ and general n, how many more conditions will have to be found to make it possible to determine all of the Christoffel coefficients?*

Digression concerning "flawed coordinate systems:" *Two city dwellers part company intending to meet after taking different routes. The first goes east for N_E street numbers, then north for N_N street numbers. The second goes north for N_N street numbers, then east for N_E street numbers. Clearly they will not meet up in most cases because the street numbers have not been established carefully enough. Will their paths necessarily cross if they keep going long enough? Because cities are predominantly two dimensional, they usually will. But it is not hard to visualize the presence of a tunnel on one of the two routes that leads one of the routes below the other without crossing it. In dimensions higher than two that is the generic situation.*

Though it was not stated before, we now require our curvilinear coordinate system to be free of the two flaws just mentioned. At least sufficiently locally for higher-than-quadratic products of the Δu^i factors, this can be assured by requiring

$$
\frac{\partial}{\partial u^i}\frac{\partial \mathbf{M}}{\partial u^j} = \frac{\partial}{\partial u^j}\frac{\partial \mathbf{M}}{\partial u^i}. \tag{3.18}
$$

When expressed in vector terms using Eq. (3.9), the quantities being differentiated here can be expressed as

$$
\frac{\partial \mathbf{M}}{\partial u^i} = \mathbf{e}_i. \tag{3.19}
$$

Hence, using Eq. (3.13), we require

$$\frac{\partial \mathbf{e}_i}{\partial u^j} = \Gamma^k{}_{ij}\,\mathbf{e}_k = \Gamma^k{}_{ji}\,\mathbf{e}_k, \quad \text{or} \quad \Gamma^k{}_{ij} = \Gamma^k{}_{ji}. \tag{3.20}$$

This requirement that $\Gamma^k{}_{ji}$ be symmetric in its lower indices yields $n^2(n-1)/2$ further conditions which, along with the $n^2(n+1)/2$ conditions of Eq. (3.17), should permit us to determine all n^3 of the Christoffel coefficients.

It can now be seen why Eq. (3.17) was written three times. Adding the three equations and taking advantage of Eq. (3.20) yields

$$g_{kr}\Gamma^r{}_{ij} = \frac{1}{2}\left(\frac{\partial g_{jk}}{\partial u^i} + \frac{\partial g_{ki}}{\partial u^j} - \frac{\partial g_{ij}}{\partial u^k}\right). \tag{3.21}$$

For any particular values of "free indices" i and j (and suppressing them to make the equation appear less formidable) this can be regarded as a matrix equation of the form

$$g_{kr}\Gamma^r = R_k \quad \text{or} \quad \mathbf{G}\boldsymbol{\Gamma} = \mathbf{R}. \tag{3.22}$$

Here $\mathbf{G}$ is the matrix (g_{kr}) introduced previously, $\boldsymbol{\Gamma} = (\Gamma^r)$ is the set of Christoffel symbols for the particular values of i and j, $\mathbf{R} = (R_k)$ is the corresponding right-hand side of Eq. (3.21). The distinction between upper and lower indices is unimportant here.[7] Being a matrix equation, this can be solved without difficulty to complete the determination of the Christoffel symbols;

$$\boldsymbol{\Gamma} = \mathbf{G}^{-1}\mathbf{R}. \tag{3.23}$$

Though these manipulations may appear overly formal at this point, an example given below will show that they are quite manageable.

3.1.3
The Covariant (or Absolute) Differential

There is considerable difference between mathematical and physical intuition in the area of differentiation. Compounding this, there is a plethora of distinct types of derivative, going by names such as total, invariant, absolute, covariant, variational, gradient, divergence, curl, Lie, exterior, Frechét, Lagrange,

7) Failing to distinguish between upper and lower indices ruins the invariance of equations as far as transformation between different frames is concerned, but it is valid in any particular frame. In any case, since the quantities on the two sides of Eq. (3.22) are not tensors, distinction between upper and lower indices would be unjustified.

etc. Each of these – some are just different names for the same thing – combines the common concepts of differential calculus with other concepts. In this chapter some of these terms are explained, and eventually nearly all will be.

The differential in the denominator of a derivative is normally a scalar, or at least a one component object, often dt, while the numerator is often a multicomponent object. The replacement of t by a monotonically related variable, say $s = f(t)$, makes a relatively insignificant change in the multicomponent derivatives – all components of the derivative are multiplied by the same factor dt/ds. This makes it adequate to work with differentials rather than derivatives in most cases, and that is what we will do. We will disregard as inessential the distinction between the physicist's view of a differential as an approximation to a small but finite change and the mathematician's view of a differential as a finite yet exact displacement along a tangent vector.

We start with a type of derivative that may be familiar to physicists in one guise, yet mysterious in another; the familiar form is that of coriolis or centrifugal acceleration. Physicists know that Newton's first law – free objects do not accelerate – applies only in inertial frames of reference. If one insists on using an accelerating frame of reference – say fixed to earth, such as latitude, longitude, and altitude – the correct description of projectile motion requires augmenting the true forces, gravity, and air resistance, by "fictitious" coriolis and centrifugal forces. These extra forces compensate for the fact that the reference frame is not inertial. Many physicists, perhaps finding the introduction of fictitious forces artificial and hence distasteful, or perhaps having been too-well taught in introductory physics that "there is no such thing as centrifugal force," resist this approach and prefer a strict inertial frame description. Here we instead develop a noninertial description using the curvilinear coordinates introduced in the previous section.

A particle trajectory can be described by $u^1(t), u^2(t), \ldots, u^n(t)$ curvilinear coordinates that give its location as a function of time t. For example, uniform motion on a circle of radius R is described by $r = R, \phi = \omega t$. The velocity $\mathbf{v}$ has curvilinear velocity components that are *defined* by

$$v^i \equiv \frac{du^i}{dt} \equiv \dot{u}^i. \tag{3.24}$$

In the circular motion example $\dot{r} = 0$. Should one then define curvilinear acceleration components by

$$a^i \overset{?}{\equiv} \frac{dv^i}{dt} = \frac{d^2u^i}{dt^2}. \quad \text{No!} \tag{3.25}$$

One *could* define acceleration this way, but it would lead, for example, to the result that the radial acceleration in uniform circular motion is zero – certainly not consistent with conventional terminology.

Here is what has gone wrong: while **v** is a perfectly good arrow, and hence a true vector, its components v^i are projections onto axes parallel to the local coordinate axes. Though these local axes are not themselves rotating, a frame moving so that its origin coincides with the particle and having its axes always parallel to local axes has to be rotating relative to the inertial frame. One is violating the rules of Newtonian mechanics. Here is what can be done about it: calculate acceleration components relative to the base frame.

Before doing this we establish a somewhat more general framework by introducing the concept of *vector field*. A *vector field* $\mathbf{V}(P)$ is a vector function of position that assigns an arrow **V** to each point P in space. An example with $\mathbf{V} = \mathbf{r}$, the radius vector from a fixed origin, is illustrated in Fig. 3.2. (Check the two bold face arrows with a ruler to confirm $\mathbf{V} = \mathbf{r}$.) In the figure the same curvilinear coordinate system as appeared in Fig. 3.1 is assumed to be in use. At each point the curvilinear components V^i of the vector **V** are defined to be the coefficients in the expansion of **V** in terms of local basis vectors;

$$\mathbf{V} = V^i \mathbf{e}_i. \tag{3.26}$$

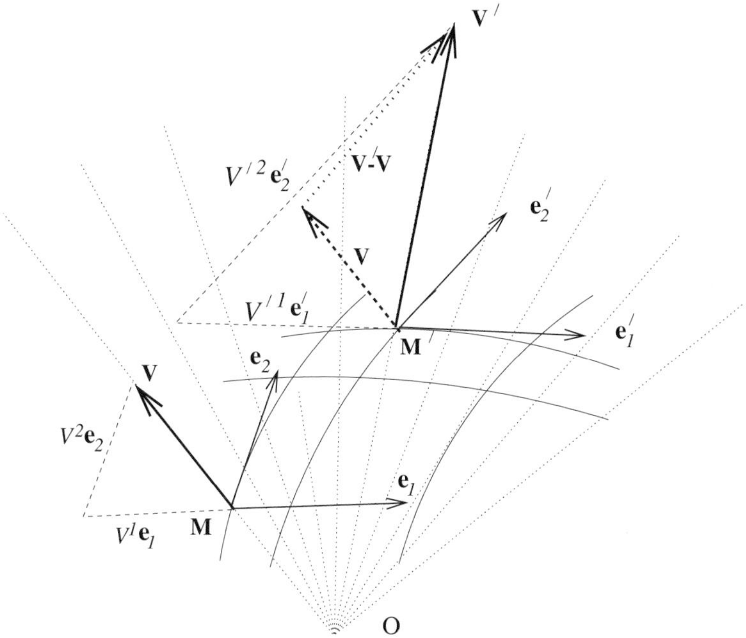

Fig. 3.2 The vector field $\mathbf{V}(P) = \mathbf{r}(P)$, where $\mathbf{r}(P)$ is a radius vector from point O to point P, expressed in terms of the local curvilinear coordinates shown in Fig. 3.1. The change $\mathbf{V}' - \mathbf{V}$ in going from point $\mathbf{M}$ to point $\mathbf{M}'$ is shown.

The *absolute* differential $D\mathbf{V}$ of a vector function $\mathbf{V}(P)$, like any differential, is the change in **V** that accompanies a change in its argument, in the small change limit. For this to be meaningful it is, of course, necessary to specify what is meant by the changes. In Fig. 3.2, in the (finite) change of position from point M to point M', the change in $\mathbf{V}$ is indicated by the arrow labeled $\mathbf{V}' - \mathbf{V}$; being an arrow, it is manifestly a *true* vector. In terms of local coordinates, the vectors at M and M' are given, respectively, by

$$\mathbf{V} = \mathbf{V}(M) = V^1\mathbf{e}_1 + V^2\mathbf{e}_2, \quad \mathbf{V}' = \mathbf{V}(M') = V^{1'}\mathbf{e'}_1 + V^{2'}\mathbf{e'}_2. \tag{3.27}$$

In the limit of small changes, using Eq. (3.12), one has

$$D\mathbf{V} \equiv d(V^i)\mathbf{e}_i + V^j d(\mathbf{e}_j) = d(V^i)\mathbf{e}_i + V^j \omega_j^{\ i}\,\mathbf{e}_i = \left(d(V^i) + V^j\omega_j^{\ i}\right)\mathbf{e}_i. \tag{3.28}$$

This differential (a true vector by construction) can be seen to have contravariant components given by

$$DV^{\ i} \equiv (D\mathbf{V})^i = dV^i + V^j\,\omega_j^{\ i} \equiv dV^i + V^j\Gamma^i_{\ jk}du^k, \tag{3.29}$$

where the du^k are the curvilinear components of M' relative to M. (Just this time) a certain amount of care has been taken with the placement of parentheses and indices in these equations. The main thing to notice is the definition $DV^{\ i} \equiv (D\mathbf{V})^i$. Note that, since the components u^k and V^i are known functions of position, their differentials du^k and dV^i are unambiguous; there is no need to introduce symbols $D(u^k)$ and $D(V^i)$ since, if one did, their meanings would just be du^k and dV^i. On the other hand the quantity $D\mathbf{V}$ is a newly defined true vector whose components are being first evaluated in Eq. (3.29). (It might be pedagogically more helpful if these components were always symbolized by $(DV)^i$ rather than by $DV^{\ i}$; but since that is never done it is necessary to remember the meaning some other way. For the moment the superscript i has been moved slightly away to suggest that it "binds somewhat less tightly" to V than does the D.) Note then that the $DV^{\ i}$ are the components of a true vector, while dV^i, differential changes in local coordinates, are not.

$D\mathbf{V}$ is commonly called the *covariant* differential; this causes $DV^{\ i}$ to be the "contravariant components of the covariant differential." Since this is unwieldy, we will use the term *absolute* differential rather than *covariant* differential. If the vector being differentiated is a constant vector $\mathbf{A}$, it follows that

$$DA^{\ i} = 0, \quad \text{and hence,} \quad dA^i = -A^j\omega_j^{\ i}. \tag{3.30}$$

How to obtain the *covariant components* of the absolute differential of a variable vector $\mathbf{V}$ is the subject of the following problem.

Problem 3.1.3. *Consider the scalar product,* $\mathbf{V}\cdot\mathbf{A}$, *of a variable vector* $\mathbf{V}$ *and an arbitrary constant vector* $\mathbf{A}$. *Its differential, as calculated in the base frame, could be designated* $D\,(\mathbf{V}\cdot\mathbf{A})$, *while its differential in the local frame could be designated* $d\,(\mathbf{V}\cdot\mathbf{A})$. *Since the change of a scalar should be independent of frame, these two* *differentials must be equal. Use this, and Eq. (3.30) and the fact that* **A** *is arbitrary, to show that the covariant components of the absolute differentials of vector* **V** *are given by*

$$DV_i = dV_i - V_k\,\omega_i^{\ k}. \tag{3.31}$$

Problem 3.1.4. *The line of reasoning of the previous problem can be generalized to derive the absolute differential of more complicated tensors. Consider, for example, a mixed tensor $a_i^{\ j}$ having one upper and one lower index. Show that the absolute differential of this tensor is given by*

$$Da_i^{\ j} = da_i^{\ j} - a_k^{\ j}\,\omega_i^{\ k} + a_i^{\ k}\,\omega_k^{\ j}. \tag{3.32}$$

Problem 3.1.5. Ricci's theorem. *Derived as in the previous two problems,*

$$Da_{ij} = da_{ij} - a_{kj}\,\omega_i^{\ k} - a_{ik}\,\omega_j^{\ k}. \tag{3.33}$$

Using this formula and Eq. (3.16), show that $Dg_{ij} = 0$; i.e. the absolute differential of the metric tensor g_{ij}, vanishes. Use this result to show that the absolute differential $D(\mathbf{A}\cdot\mathbf{B})$ of the scalar product of two constant vectors **A** *and* **B** *vanishes (as it must).*

Problem 3.1.6. *Use the result of the previous problem (in the form $Dg_{ij}/du^l = 0$) to show that*

$$\frac{\partial g_{ij}}{\partial u^l} = g_{kj}\Gamma^k_{\ il} + g_{ik}\Gamma^k_{\ jl}. \tag{3.34}$$

Problem 3.1.7. *Show that*

$$\frac{\partial g^{ij}}{\partial u^l} = -g^{mj}\Gamma^i_{\ ml} - g^{im}\Gamma^j_{\ ml}. \tag{3.35}$$

and, as a check, using also the previous problem, confirm Eq. (3.8).

Equation (3.29) can be used to define an *absolute* derivative

$$V^i_{\ ;k} = \frac{\partial v^i}{\partial u^k} + V^j\Gamma^i_{\ jk}. \tag{3.36}$$

The ";k" index, as well as abbreviating the absolute differentiation, is placed as a subscript to indicate that the result is a *covariant* component. The symbol $V^{i;l}$ would stand for $g^{lk}V^i_{\ ;k}$.

Problem 3.1.8. *A result that will be required while applying the calculus of variations within metric geometry is to find the variation δg^{ij} accompanying an infinitesimal change of coordinates $u^i \to u^i + \xi^i \equiv u'^i$. In this change*

$$g'^{ij}(u'^n) = g^{kl}\left(\delta^i_k + \frac{\partial \xi^i}{\partial u^k}\right)\left(\delta^j_l + \frac{\partial \xi^j}{\partial u^l}\right) \approx g^{ij}(x^n) + g^{il}\frac{\partial \xi^j}{\partial u^l} + g^{kj}\frac{\partial \xi^i}{\partial u^k}. \tag{3.37}$$

This gives the new metric coefficients at displaced location u'^n. Correcting back to the original position, and subtracting from the original

$$\delta g^{ij} \approx -g^{il}\frac{\partial \xi^j}{\partial u^l} - g^{jl}\frac{\partial \xi^i}{\partial u^l} + \frac{\partial g^{ij}}{\partial u^l}\xi^l. \tag{3.38}$$

Use this result in the limit that the approximation is exact, and results from previous problems, to obtain the results

$$\delta g^{ij} = \xi^{i;j} + \xi^{j;i}, \quad \delta g_{ij} = -\xi_{i;j} - \xi_{j;i}. \tag{3.39}$$

Setting one or the other of these to zero gives a condition that needs to be satisfied for the metric coefficients to be unchanged by the change of coordinates.

Because the concept of *absolute differentiation* is both extremely important and quite confusing, some recapitulation may be in order. Since the difference of two arrows is an arrow, the rate of change of an arrow is an arrow. Stated more conventionally, the rate of change of a *true* vector is a *true* vector. Confusion enters only when a vector is represented by its components. It is therefore worth emphasizing:

The components of the rate of change of vector **V** *are not, in general, the rates of change of the components of* **V**.

This applies to all true tensors. Unfortunately, since practical calculations almost always require the introduction of components, it is necessary to develop careful formulas, expressed in component form, for differentiating vectors (and all other tensors.) The derivation of a few of these formulas is the subject of the set of problems just above.

3.2 Derivation of the Lagrange Equations from the Absolute Differential

In mechanics one frequently has the need for coordinate systems that depend on position (curvilinear) or time (rotating or accelerating). Here we analyze the former case while continuing to exclude the latter. That is, the coefficients of the metric tensor can depend on position but are assumed to be independent of t. On the other hand, the positions of the particle or particles being described certaeinly vary with time.

In this section we symbolize coordinates by q^i rather than the u^i used to this point. This could be regarded as a pedantic distinction between the position u^i where the particle *could be* and the position q^i where the particle *is*. But physicists are rarely this fussy so there is no content to this change other than the fact that generalized coordinates in mechanics are usually assigned the symbol q.

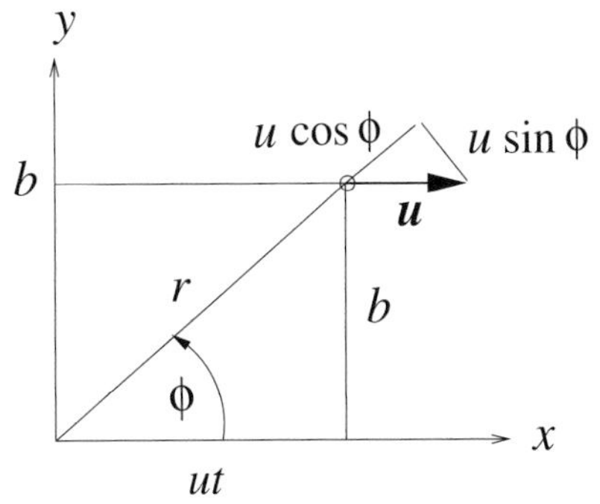

Fig. 3.3 A particle moves parallel to the x-axis at constant speed u.

Example 3.2.1. *Motion that is simple when described in one set of coordinates may be quite complicated in another. For example, consider a particle moving parallel to the x-axis at $y = b$ with constant speed u. That is $(x = ut, y = b)$. In spherical coordinates, with $\theta = \pi/2$, the particle displacement is given by*

$$r = \frac{b}{\sin\phi}, \quad \phi = \tan^{-1}\frac{b}{ut}. \tag{3.40}$$

The first time derivatives are

$$v^r \equiv \dot{r} = u\cos\phi, \quad v^\phi \equiv \dot{\phi} = -\frac{u}{b}\sin^2\phi; \tag{3.41}$$

following our standard terminology for velocities, we have defined $v^r = \dot{r}$ and $v^\phi = \dot{\phi}$. (This terminology is by no means universal, however. It has the disagreeable feature that the components are not the projections of the same arrow onto mutually orthonormal axes, as Fig. 3.3 shows. Also they have different units. They are, however, the contravariant components of a true *vector along well-defined local axes.) Taking another time derivative yields*

$$\ddot{r} = \frac{u^2}{b}\sin^3\phi, \quad \ddot{\phi} = \frac{2u^2}{b^2}\cos\phi\,\sin^3\phi. \tag{3.42}$$

Defining by the term "absolute acceleration" the acceleration in an inertial coordinate frame, the absolute acceleration obviously should vanish in this motion. And yet the quantities $\ddot{r}$ and $\ddot{\phi}$ are nonvanishing. We will continue this example below in Example 3.2.4.

Example 3.2.2. *In cylindrical (r, ϕ, z) coordinates the nonvanishing Christoffel elements are*

$$\Gamma^2{}_{12} = \Gamma^2{}_{21} = \frac{1}{r}, \quad \Gamma^1{}_{22} = -r, \tag{3.43}$$

(as will be shown shortly.) The vector expansion of $\mathbf{r}$ *is* $\mathbf{r} = q^1\hat{\mathbf{e}}_r + q^2\hat{\mathbf{e}}_\phi = r\hat{\mathbf{e}}_r$. *and the components of the covariant derivative with respect to t are*

$$\frac{Dq^1}{dt} = \dot{q}^1 + q^j \Gamma^1{}_{jk}\dot{q}^k = \dot{r}, \quad \frac{Dq^2}{dt} = \dot{q}^2 + q^j \Gamma^1{}_{jk}\dot{q}^k = \dot{\phi}. \tag{3.44}$$

In this example the components of $\mathrm{D}\mathbf{r}/dt$ *are seen to be* $(\dot{r}, \dot{\phi})$, *or*

$$\frac{\mathrm{D}\mathbf{r}}{dt} = \frac{d\mathbf{r}}{dt}. \tag{3.45}$$

This is misleadingly simple however. It follows from the fact that, with r being the first and only nonvanishing component, $Dq^i/dt = \dot{q}^i + r\Lambda_1{}^i{}_1\dot{r}$. *For this particular choice of coordinates, the second term vanishes for all i.*

Because they are macroscopic quantities, q^1 *and* q^2 *cannot themselves be expected to be related by linear relations derived by linearization. Individually* q^1 *and* q^2 *are scalar functions for which* $\dot{q}^1$ *and* $\dot{q}^2$ *can be evaluated. Since the triplet* $(\dot{q}^1, \dot{q}^2, \dot{q}^3)$ does *consist of the components of a true vector, relations (3.44), though suspect, are therefore not obviously wrong. Since* $q^1\hat{\mathbf{e}}_r + q^2\hat{\mathbf{e}}_\phi$ *is, in fact, a true vector, Eqs. (3.44) do generate a true vector.*

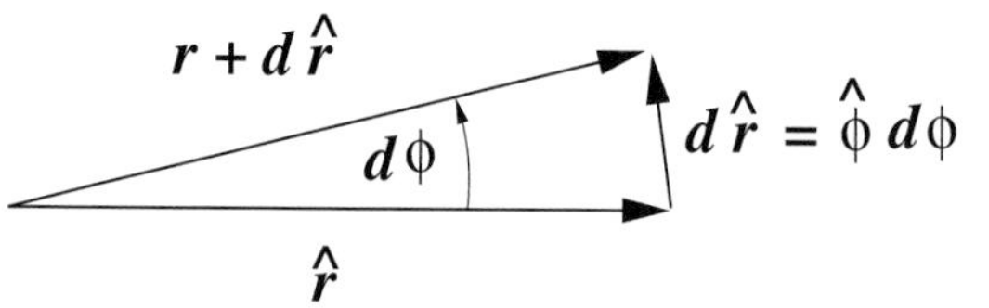

Fig. 3.4 Time rate of change of a unit vector.

Example 3.2.3. *The result of the previous example can be obtained simply using traditional vector analysis on the vectors shown in Fig. 3.4;*

$$\frac{d}{dt}(r\hat{\mathbf{r}}) = \dot{r}\hat{\mathbf{r}} + r\frac{d\hat{\mathbf{r}}}{dt} = \dot{r}\hat{\mathbf{r}} + \dot{\phi}(r\hat{\boldsymbol{\phi}}). \tag{3.46}$$

The factor r in the final term reflects the difference between our basis vector $\mathbf{e}_\phi$ *and the unit length vector* $\hat{\boldsymbol{\phi}}$ *of ordinary vector analysis.*

If the changes discussed in the previous section occur during time dt, perhaps because a particle that is at M at time t moves in such a way as to be at M' at time $t + dt$, the differentials DV^i of vector $\mathbf{V}$ can be converted to time derivatives;[8]

$$\frac{DV^i}{dt} = \frac{dV^i}{dt} + V^j \Gamma^i{}_{jk} \frac{dq^k}{dt} \equiv \dot{V}^i + V^j \Gamma^i{}_{jk} \dot{q}^k. \tag{3.47}$$

The quantity **V** being differentiated here is any vector field. One such possible vector field, defined at every point on the trajectory of a moving particle, is its instantaneous velocity $\mathbf{v} = d\mathbf{x}/dt$. Being an arrow (tangent to the trajectory) **v** is a *true* vector; its local components are $v^i = dq^i/dt \equiv \dot{q}^i$. The *absolute* acceleration **a** is defined by

$$\mathbf{a} = \frac{D\mathbf{v}}{dt}, \quad \text{or} \quad a^i = \frac{Dv^i}{dt}. \tag{3.48}$$

Substituting $V^i = v^i$ in Eq. (3.47) yields

$$a^i = \dot{v}^i + v^j \Gamma^i{}_{jk} \frac{dq^k}{dt} = \ddot{q}^i + \Gamma^i{}_{jk} \dot{q}^j \dot{q}^k. \tag{3.49}$$

As the simplest possible problem of mechanics let us now suppose that the particle being described is subject to no force and, as a result, to no acceleration. Setting $a^i = 0$ yields

$$\ddot{q}^i = -\Gamma^i{}_{jk} \dot{q}^j \dot{q}^k. \tag{3.50}$$

This is the equation of motion of a *free* particle. In rectangular coordinates, since $\Gamma^i{}_{jk} = 0$, this degenerates to the simple result that **v** is constant; the motion in question is along a straight line with constant speed. This implies that the solution of Eq. (3.50) is the equation of a straight line in our curvilinear coordinates.

Since a line is a purely geometric object, it seems preferable to express its equation in terms of arc length s along the line rather than time t. As observed previously, such a transformation is easy – especially so in this case since the speed is constant. The equation of a straight line is then

$$\frac{d^2 q^i}{ds^2} = -\Gamma^i{}_{jk} \frac{dq^j}{ds} \frac{dq^k}{ds}. \tag{3.51}$$

Suppose next that the particle is not free, but rather is subject to a force **F**. Substituting $\mathbf{a} = \mathbf{F}/m$, where m is the particle mass, into Eq. (3.49) can be expected

8) In expression (3.47) the common shorthand indication of *total* time derivative by an overhead dot has been used. One can inquire why $\dot{V}^i$ has been defined to mean dV^i/dt, rather than DV^i/dt. It *is* just convention (due originally to Newton), but the convention is well established, and it *must* be respected if nonsense is to be avoided. The vector field **V**, though dependent on position, has been assumed to be constant in time; if **V** has an explicit time dependence, the term $\dot{V}^i$ would have to include also a contribution $\partial V^i/\partial t$.

to yield Newton's law expressed in these coordinates, but a certain amount of care is required before components are assigned to **F**. Before pursuing this line of inquiry, we look at free motion from another point of view.

There are two connections with mechanics that deserve consideration – variational principles and Lagrange's equations. The first can be addressed by the following problems. The first of which is reasonably straightforward. The second is less so and could perhaps be deferred or looked up.[9]

Problem 3.2.1. *Consider the integral* $S = \int_{t_1}^{t_2} L(q^i, \dot{q}^i, t)\, dt$*, evaluated along a candidate path of a particle from starting position at initial time* t_1 *to final position at time* t_2*, where* $L = \frac{m}{2} g_{ij}\dot{q}^i\dot{q}^j$*. (Since this "Lagrangian" L depends only on velocities, the apparent functional dependences on position and time in the expression for S are superfluous in this case.) Using the calculus of variations, show that Eq. (3.51) is the equation of the path for which S is extreme. In other words, show that Eq. (3.51) is the same as the Euler–Lagrange equation for this Lagrangian L.*

Problem 3.2.2. *It is "obvious" also, since free particles travel in straight lines and straight lines have minimal lengths, that the Euler–Lagrange equation for the trajectory yielding extreme value to integral* $I = \int ds$ *where* $ds^2 = g_{ij}dq^i dq^j$ *should also lead to Eq. (3.51). Demonstrate this.*

These two problems suggest a close connection between Eq. (3.49) and the Lagrange equations that will now be considered. The key dynamic variable that needs to be defined is the *kinetic energy T*. In the present context, using Eq. (3.7),

$$T = \frac{m}{2}\sum_{i=1}^{n} \dot{x}^{i\,2} = \frac{m}{2} g_{jk}\dot{q}^j\dot{q}^k. \tag{3.52}$$

(Though not exhibited explicitly, in general the metric coefficients depend on the coordinates q^i.)

When one thinks of "force" one thinks either of what its source is, for example an electric charge distribution, or what it must be to account for an observed acceleration. Here we take the latter tack and (on speculation) introduce a quantity Q_i (to be interpreted later as the "generalized force" corresponding to q^i) by

$$Q_l = \frac{d}{dt}\frac{\partial T}{\partial \dot{q}^l} - \frac{\partial T}{\partial u^l}. \tag{3.53}$$

9) Both problems are solved, starting on page 317, in Dubrovin, Fomenko, and Novikov.

We proceed to evaluate Q_l by substituting for T from Eq. (3.52), and using Eq. (3.21);

$$\begin{aligned}\frac{Q_l}{m} &= \frac{d}{dt}(g_{lk}\dot{q}^k) - \frac{1}{2}\frac{\partial g_{jk}}{\partial q^l}\dot{q}^j\dot{q}^k \\ &= g_{lk}\ddot{q}^k + \frac{1}{2}\left(\frac{\partial g_{lk}}{\partial q^h} + \frac{\partial g_{lh}}{\partial q^k} - \frac{\partial g_{hk}}{\partial q^l}\right)\dot{q}^h\dot{q}^k \\ &= g_{lr}\left(\ddot{q}^r + \Gamma^r{}_{hk}\dot{q}^h\dot{q}^k\right). \qquad (3.54)\end{aligned}$$

This formula resembles the right-hand side of Eq. (3.49); comparing with Eq. (2.113) it can be seen that they are covariant and contravariant components of the same vector. Expressed as an *intrinsic* equation, this yields

$$\frac{Q^i}{m} = \ddot{q}^i + \Gamma^i{}_{jk}\dot{q}^j\dot{q}^k. \qquad (3.55)$$

This confirms that the Lagrange equations are equivalent to Newton's equations since the right-hand side is the acceleration a^i. For this equation to predict the motion it is of course necessary for the force Q^i to be given.

Recapitulation: From a given particle trajectory it is a *kinematical* job to infer the acceleration, and the absolute derivative is what is needed for this task. The result is written on the right-hand side of Eq. (3.55), in the form of contravariant components of a vector. It was shown in Eq. (3.54) that this same quantity could be obtained by calculating the "Lagrange derivatives" $d/dt(\partial T/\partial \dot{q}) - \partial T/\partial q$, where $T = (m/2)g_{jk}\,\dot{q}^j\dot{q}^k$. (The occurrence of mass m in the definition of T suggests that it is a dynamical quantity, but inclusion of the multiplier m is rather artificial; T and the metric tensor are essentially equivalent quantities.) It is only a minor complication that the Lagrange derivative of T yields covariant components which need to "have their indices raised" before yielding the contravariant components of acceleration. *Dynamics* only enters when the acceleration is ascribed to a force according to Newton's law, $a = F/m$. When a in this equation is evaluated by the invariant derivative as in Eq. (3.55), the result is called "Newton's equation." When a is evaluated by the Lagrange derivative of T the result is called "Lagrange's equation."

Commonly force is introduced into the Lagrange equation by introducing $L = T - V$, where V is "potential energy." This is an artificial abbreviation, however, since it mixes a kinematic quantity T and a dynamic quantity V. From the present point of view, since it is not difficult to introduce forces directly, it is a logically clearer procedure than introducing them indirectly in the form of potential energy.

The prominent role played in mechanics by the kinetic energy T is due, on the one hand, to its close connection with ds^2 and, on the other hand to the fact that virtual force components Q_l can be derived from T using Eq. (3.53).

3.2.1
Practical Evaluation of the Christoffel Symbols

The "direct" method of obtaining Christoffel symbols for a given coordinate system is by substituting the metric coefficients into Eq. (3.21) and solving Eqs. (3.22). But this involves much differentiation and is rather complicated. A practical alternative is to use of the equations just derived. Suppose, for example, that spherical coordinates are in use; $(q^1, q^2, q^3) \equiv (r, \theta, \phi)$. In terms of these coordinates and the formula for distance ds, one obtains metric coefficients from Eqs. (3.5) and (2.113);

$$
\begin{aligned}
&ds^2 = dr^2 + r^2\, d\theta^2 + r^2 \sin^2\theta\, d\phi^2, \\
&g_{11} = 1, \quad g_{22} = r^2, \quad g_{33} = r^2 \sin^2\theta, \\
&g^{11} = 1, \quad g^{22} = \frac{1}{r^2}, \quad g^{33} = \frac{1}{r^2 \sin^2\theta},
\end{aligned}
\tag{3.56}
$$

and all off-diagonal coefficients vanish. (g^{11}, g^{22}, g^{33} are defined by the usual index raising.) Acceleration components a^i can then be obtained using Eq. (3.55), though it is necessary first to raise the index of Q_l using the metric tensor. From Eq. (3.55) one notes that the Christoffel symbols are the coefficients of terms quadratic in velocity components, $\dot{r}$, $\dot{\theta}$, and $\dot{\phi}$ in Eq. (3.54), and this result can be used to obtain them.

Carrying out these calculations (for spherical coordinates with $m = 1$) the kinetic energy and the contravariant components of virtual force are given by

$$
2T = \dot{r}^2 + r^2\dot{\theta}^2 + r^2 \sin^2\theta\, \dot{\phi}^2, \tag{3.57}
$$

$$
\begin{aligned}
Q^1 &= \left(\frac{d}{dt}\frac{\partial T}{\partial \dot{r}} - \frac{\partial T}{\partial r}\right) = \ddot{r} - r\dot{\theta}^2 - r\sin^2\theta\, \dot{\phi}^2, \\
Q^2 &= \frac{1}{r^2}\left(\frac{d}{dt}\frac{\partial T}{\partial \dot{\theta}} - \frac{\partial T}{\partial \theta}\right) = \ddot{\theta} + \frac{2}{r}\dot{r}\dot{\theta} - \sin\theta\, \cos\theta\, \dot{\phi}^2, \\
Q^3 &= \frac{1}{r^2 \sin^2\theta}\left(\frac{d}{dt}\frac{\partial T}{\partial \dot{\phi}} - \frac{\partial T}{\partial \phi}\right) = \ddot{\phi} + \frac{2}{r}\dot{r}\dot{\phi} + 2\frac{\cos\theta}{\sin\theta}\dot{\theta}\dot{\phi}.
\end{aligned}
\tag{3.58}
$$

Matching coefficients, noting that the coefficients with factors of 2 are the terms that are duplicated in the (symmetric) off-diagonal terms of (3.54), the nonvanishing Christoffel symbols are

$$
\begin{aligned}
&\Gamma^1{}_{22} = -r, \quad \Gamma^1{}_{33} = -r\sin^2\theta, \\
&\Gamma^2{}_{12} = \frac{1}{r}, \quad \Gamma^2{}_{33} = -\sin\theta\, \cos\theta, \\
&\Gamma^3{}_{13} = \frac{1}{r}, \quad \Gamma^3{}_{23} = \frac{\cos\theta}{\sin\theta}.
\end{aligned}
\tag{3.59}
$$

Example 3.2.4. *We test these formulas for at least one example by revisiting Example 3.2.1. Using Eq. (3.58) the force components acting on the particle in that example are*

$$Q^1 = \frac{u^2}{b}\sin^3\phi - r\dot{\phi}^2 = 0,$$
$$Q^3 = \frac{2u^2}{b^2}\cos\phi\,\sin^3\phi + \frac{2}{r}\dot{r}\dot{\phi} = 0. \tag{3.60}$$

This shows that the particle moving in a straight line at constant speed is subject to no force. This confirms statements made previously.

Problem 3.2.3. *For cylindrical (ρ, ϕ, z) coordinates calculate the Christoffel symbols both directly from their defining Eqs. (3.21) and indirectly using the Lagrange equation. To check your results you can use a program such as MATHEMATICA or MAPLE, which can make the Christoffel symbols readily available for arbitrary coordinate systems. This is most useful for less symmetric coordinates, defined by more complicated formulas.*

3.3 Intrinsic Derivatives and the Bilinear Covariant

Absolute differentials have been defined for contravariant vectors in Eq. (3.29), for covariant vectors in Problem 2.4.2, and for two-index tensors in Problem 3.1.4. The generalization to tensors of arbitrary order is obvious, and the following discussion also transposes easily for tensors of arbitrary order. For simplicity consider the case of Eq. (3.33);

$$Da_{il} = da_{il} - a_{jl}\,\omega_i^{\ j} - a_{ij}\,\omega_l^{\ j} = \left(\frac{\partial a_{il}}{\partial u^k} - a_{jl}\,\Gamma^j{}_{ik} - a_{ij}\,\Gamma^j{}_{lk}\right)du^k. \tag{3.61}$$

Since du^k and Da_{il} are *true* tensors, the coefficients

$$a_{il;k} \equiv \frac{Da_{il}}{du^k} \equiv \frac{\partial a_{il}}{\partial u^k} - a_{jl}\,\Gamma^j{}_{ik} - a_{ij}\,\Gamma^j{}_{lk}, \tag{3.62}$$

also constitute a *true* tensor. As another example, if X^i are the covariant components of a vector field, then

$$X_{i;j} \equiv \frac{DX_i}{du^j} \equiv \frac{\partial X_i}{\partial u^j} - X_k\,\Gamma^k{}_{ij} \tag{3.63}$$

is also a *true* tensor. The tensors of Eqs. (3.62) and (3.63) are called *covariant* (or invariant) derivatives of a_{il} and X_i, respectively.

We now perform an important, though for now somewhat poorly motivated, manipulation. What makes this apology necessary is that our entire discussion up to this point has been more special than may eventually be required. A (subliminal) warning of this was issued as the Christoffel symbols were introduced and described as a "connection." Their unique definition relied on the fact that the curvilinear coordinates being analyzed were embedded in a Euclidean space, with distances and angles having their standard meanings inherited from that space. In more general situations an affine connection exists but is not calculable by Eq. (3.21).

Once one accepts that the $\Gamma^k{}_{ij}$ coefficients are special one must also accept that covariant derivatives like $a_{il;k}$ or $X_{i;j}$, rather than being universal, are specific to the particular connection that enters their definition. But, relying on the fact that the Christoffel symbols are symmetric in their lower indices, it is clear (in Eq. (3.63) for example) that a more universal (independent of connection) derivative can be formed by antisymmetrizing these tensors to *eliminate the Christoffel symbols.* Subtracting Eq. (3.63) and the same equation with indices interchanged yields

$$\frac{DX_i}{du^j} - \frac{DX_j}{du^i} = \frac{\partial X_i}{\partial u^j} - \frac{\partial X_j}{\partial u^i}. \tag{3.64}$$

Being a sum of tensors this is a tensor and it has *intrinsic* significance for any system described by smoothly defined coordinates. It generalizes the "curl" of vector field **X**, familiar from vector analysis. For tensors having more indices similar, antisymmetrized, intrinsic, derivative can also be defined.

That this combination is a true tensor can be used to prove the invariance of the so-called "bilinear covariant[10]" formed from a differential form $\omega[d]$. Here $\omega[d]$ is the (introduced in Section 2.2.3) abbreviation

$$\omega[d] = X_1 du^1 + X_2 du^2 + \cdots + X_n du^n. \tag{3.65}$$

The same differential, but expressed with a different argument δ is

$$\omega[\delta] = X_1 \delta u^1 + X_2 \delta u^2 + \cdots + X_n \delta u^n. \tag{3.66}$$

(More symmetric notation, such as $d_{(1)}$ and $d_{(2)}$ instead of d and δ could have been used but would have caused a clutter of indices.) Interchanging d and δ, then forming another level of differential in Eqs. (3.65) and (3.66), and then subtracting, yields

$$d\omega[\delta] - \delta\omega[d] = \frac{1}{2}\left(\frac{\partial X_k}{\partial u^j} - \frac{\partial X_j}{\partial u^k}\right)\left(\delta u^k\, du^j - \delta u^j\, du^k\right). \tag{3.67}$$

10) For the time being we continue to use the somewhat archaic "bilinear covariant" as it was introduced in Section 2.2.2 rather than rely on the "exterior derivative" formalism because the only result that will be used is explicitly derived in this section. The exterior derivative formalism streamlines the algebra and obviates the need for introducing the distinguished symbols d and δ, but the present terminology is (arguably) better motivated in the present context.

(The factor 1/2 takes advantage of the fact that, when this is expanded, the terms are equal in pairs.) The right-hand side is the tensor contraction of a product of the first factor (shown to be a *true* tensor in Eq. (3.64)) and the bivector (also an invariant[11]) formed from **du** and $\boldsymbol{\delta}\mathbf{u}$. The exterior derivative of a form will also be discussed in Chapter 4. As a result, this combination $d\omega[\delta] - \delta\omega[d]$, called the "bilinear covariant" of the form ω, has been also shown to be independent of choice of coordinates.

The combination $X_{i,j} - X_{j,i}$ has been shown to be *intrinsic* or *a true tensor.* This is true for fields defined on any smooth manifold. Strictly speaking this result has only been proved here for manifolds with a *connection* $\Gamma^k_{\ ij}$ defined, but the only requirement on the connection is contained in Eq. (3.13) that links basis frames at nearby locations. For manifolds encountered in mechanics this weak requirement is typically satisfied.

3.4 The Lie Derivative – Coordinate Approach

The Lie derivative is a tool for analyzing rates of change in one coordinate system from the point of view of another. It is discussed here because it is based on concepts like those required to analyze curvilinear coordinates. This operation will only be used in this chapter, and only when the properties of vector fields are used to derive the Poincaré equations. It is, in fact, possible to avoid the Lie derivative altogether in analyzing the algebra of noncommuting vector fields. Therefore, the concept need never enter mechanics. But vector fields and Lie derivatives are so thoroughly woven together in the literature, that one is eventually forced to understand this material. There are also striking similarities between the Lie derivative and the covariant derivative derived earlier in this chapter.

3.4.1 Lie-Dragged Coordinate Systems

Prototypical coordinate systems discussed so far have been spherical, cylindrical, elliptical, etc., the fixed nonrectangular coordinate systems familiar from electricity and magnetism and other fields of physics. We now consider coordinate systems that are more abstractly defined in terms of a general vector field **V** defined at every point in some manifold of points **M**. The situation will be considerably more general than that of the previous section in that the coordinates of point **M** are allowed to be any generalized coordinates and no metric is assumed to be present.

11) Bivectors will be discussed at length in Chapter 4. For now their tensor character can be inferred from their transformation properties under the transformations defined by Eq. (2.63).

From studying special relativity, one has become accustomed to (seemingly) paradoxical phenomena such as "moving clocks run slow." Closer to one's actual experience, at one time or another everyone has been sitting in a train watching the adjacent train pull out slowly, shortly to be surprised that the watched train is stationary and it is actually one's own train that is moving. The "Lie derivative" is a mathematical device for analyzing phenomena like this. To appreciate the name "the fisherman's derivative" that Arnold gives the Lie derivative, one has to visualize the fisherman sitting on the bank and watching what he sees on the river or (better, because there are more concrete objects to view) sitting in a boat that is coasting with the flow and watching what he sees on the shore.

For a concrete example, visualize yourself by the side of the route of a marathon race, 10 miles from the starting point, 1 h after the start, well after the leaders have passed. As the runners straggle by you say "the runners are aging rapidly" when, in fact, it is just that the older runners have taken longer getting there. If the 30 year old runs at 11 miles per hour and the 40 year old runs at 10 miles an hour, one hour into the race the 40 year old will be at your location and the 30 year old will have passed 0.1 h previously. The aging rate you observe is therefore $(40-30)/0.1 \approx 100\,\text{years/h}$. The same result could have been obtained via the spatial rate of change of age at fixed time which is $-10/1 = -10$ years/mile. To get (the negative of) the observed aging rate from this you have to multiply by the 10 miles/h velocity. The 100 years/h aging rate you observe can be said to be the negative of the "Lie derivative" of runner's age.

From the point of view of physics, age and rate of aging are fundamentally, dimensionally, different, but from the point of view of geometry, apart from the limiting process, they differ only by a scalar multiple dt and hence have the same geometric character. A similar relationship is that position vectors and instantaneous velocity vectors have the same geometric character. In the same way the Lie derivative of any quantity has the same geometric (i.e., tensor) character as the quantity being differentiated.

When one recalls the mental strain that accompanied first understanding the time dilation phenomenon of special relativity mentioned above, one will anticipate serious conceptual abstraction and difficult ambiguity-avoidance in defining the Lie derivative. Here it will be defined in two steps. First, starting from one (henceforth to be called preferred) coordinate system, one defines another (actually a family of other) "Lie-dragged" coordinate system. Then, mimicking procedures from Sections 3.1.2 and 3.1.3, components in the dragged frame will be "corrected" to account for frame rotation or distortion, relative the preferred frame, to form the Lie differential.

For visual accompaniment to this discussion consider the nine rows of birds flying across the sky, and shown at time $t = 0$ in Fig. 3.5. Assume the veloci-

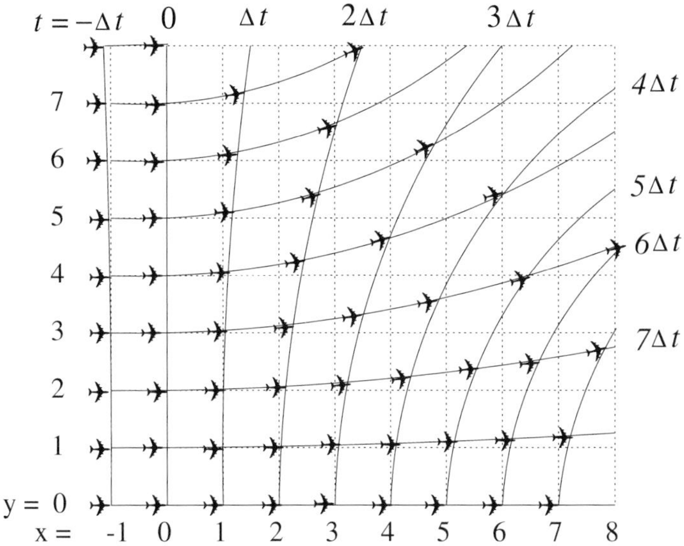

Fig. 3.5 A multiple exposure photograph of a single "line" of birds at successive times. Or, if lines of birds are taking off at regular intervals, the figure is a single, fixed time, exposure. Defining longitudinal and transverse bird indices, a point can be located by coordinates relative to the ground or by (interpolating) between the indices of its nearby birds.

ties $\mathbf{v}$ of the birds depend[12] on location but not on their time of arrival at that location. In other words we assume that $\mathbf{v}(\mathbf{x})$ depends only on $\mathbf{x}$. Though the birds are in a straight line at location $x = 0$ their lines are curved elsewhere. At the instant shown in the figure, the bird locations can serve as a two-dimensional coordinate system, with the longitudinal coordinate being the row number and the transverse coordinate determined by counting birds along the row. However, this is just a moving coordinate system and is not the "Lie-dragged" coordinate system that is to be defined next.

Consider Fig. 3.6 which is derived from Fig. 3.5. Let us suppose that at $t = 0$ there is a bird at each square of the initial rectangular (x, y) grid, and that a single snapshot is taken at a slightly later time $t = \Delta t$. (Though the time interval Δt is arbitrary in principle, it will be useful to think of it as "small" since, for the most important considerations to follow, Δt will approach this limit.) To construct this figure it is necessary to plot displacement-in-time-Δt vectors in Fig. 3.5 and interpolate from them on Fig. 3.6. The rows and lines of birds at $t = \Delta t$ provide the new "Lie-dragged" coordinate system (X, Y).

Since any point in the plane can be located by "old" coordinates (x, y) or "new" coordinates (X, Y), there have to be well-defined functional relations

12) This and all similar dependences are assumed to be smooth.

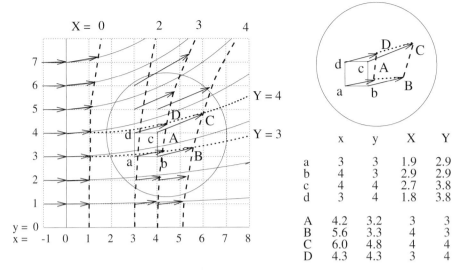

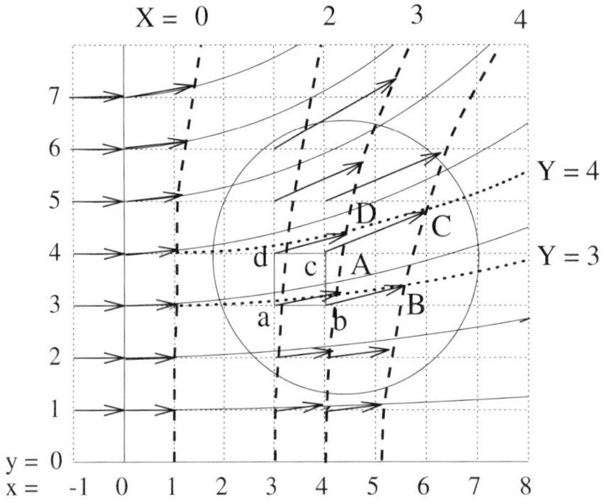

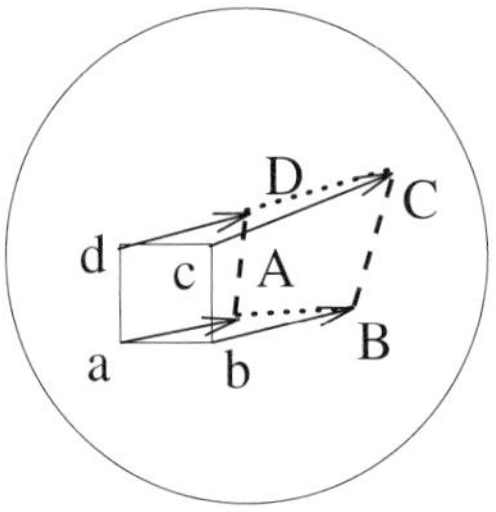

	x	y	X	Y
a	3	3	1.9	2.9
b	4	3	2.9	2.9
c	4	4	2.7	3.8
d	3	4	1.8	3.8
A	4.2	3.2	3	3
B	5.6	3.3	4	3
C	6.0	4.8	4	4
D	4.3	4.3	3	4

Fig. 3.6 Corresponding to Fig. 3.5, the (new) (X, Y) coordinate system derived by "Lie dragging" the (old) (x, y) coordinate system for time Δt with velocity vector field $\mathbf{v}$. The bold face arrows are (almost) parallel to the trajectories of the previous figure but, located at points on the rectangular grid, they represent the velocity a bird would have had, had there been one at that location. To restrict clutter only enough interpolated arrows are shown to illustrate where curves of constant X (shown dashed) and Y (shown dotted) come from.

$x = x(X, Y), y = y(X, Y)$ as well as inverse relations $X = X(x, y), Y = Y(x, y)$. Of course this presupposes that all relevant functions are sufficiently smooth and invertible, and that interpolation between "bird locations" is arbitrarily accurate. The new (X, Y) system is said to be "Lie dragged" from the old system.

Four points a, b, c, and d, defining a unit rectangle in the old coordinates and the corresponding, at-time-Δt, unit "rectangle" with corners A, B, C, and D are emphasized in Fig. 3.6 and broken out on the right where their coordinates are also shown. Note that none of these points lie on the bird paths in Fig. 3.5 but their velocities are parallel (after interpolation) to the bird velocities shown in that figure. In the table the coordinates of all these points are given in both old and new coordinate systems. By the way they have been defined, the new coordinates (X_A, Y_A) of point A are the same as the old coordinates (x_a, y_a) of point a and the same is true for every similarly corresponding pair of points.

We now seek explicit transformation relations between old and new coordinates, though only as Taylor series in powers of Δt. Regarding point a as typical, one has

$$\begin{aligned} x_A &= x_a + v^x(a)\Delta t + \cdots \\ y_A &= y_a + v^y(a)\Delta t + \cdots, \end{aligned} \tag{3.68}$$

where v^x and v^y are known functions of position; they are the the "old" components of $\mathbf{v}$, an arbitrary vector field. (It is not necessary to make the arguments of v^x and v^x more explicit since, to the relevant order in Δt it is not necessary to distinguish among $\mathbf{x}_a$, $\mathbf{x}_A$, $\mathbf{X}_a$, and $\mathbf{X}_A$.) Equation (3.68) can be regarded as the description in the old system of reference of an *active* transformation of point $a \longrightarrow A$. But, since one has *defined* the numerical equalities, $X_A = x_a$ and $Y_A = y_a$, Eq. (3.68) can be re-written as

$$\begin{aligned} x_A &= X_A + v^x \Delta t + \cdots \\ y_A &= Y_A + v^y \Delta t + \cdots . \end{aligned} \tag{3.69}$$

These equations can be regarded as a *passive* $(X, Y) \longrightarrow (x, y)$ coordinate transformation. They can be checked mentally using the entries in the table in Fig. 3.6, making allowance for the nonlinearity (exaggerated for pictorial clarity) of that figure. Since the subscripts A in this equation now refer to the same point, they have become superfluous and can be dropped;

$$\begin{aligned} x &= X + v^x \Delta t + \cdots \\ y &= Y + v^y \Delta t + \cdots . \end{aligned} \tag{3.70}$$

The inverse relations are

$$\begin{aligned} X &= x - v^x \Delta t + \cdots \\ Y &= y - v^y \Delta t + \cdots . \end{aligned} \tag{3.71}$$

Since these equations describe transformation between curvilinear systems, they can be treated by methods described earlier in the chapter. But, as mentioned before, we are not now assuming the existence of any metric. Hence, though the old coordinate system in Fig. 3.6 is drawn as rectangular, it would not be meaningful to say, for example, that the line bc is parallel to the line ad, even though that appears to be the case in the figure. (The pair (x, y) might be polar coordinates (ρ, ϕ) for example.)

3.4.2
Lie Derivatives of Scalars and Vectors

Having Lie dragged the coordinate system, we next define the Lie dragging of a general scalar function $f(\mathbf{x})$, or of a vector function $\mathbf{w}(\mathbf{x})$, or for that matter of tensors of higher rank. For vectors (and tensors of higher rank) this calculation is complicated by the fact that the Jacobean matrix relating coordinate systems depends on position. We defer addressing that problem by starting with scalar functions which, having only one component, transform without reference to the Jacobean matrix.

We define a "Lie-dragged function" f^* (whose domain of definition is more or less the same as that of f) by asserting the relation

$$f^*(a) = f(A) \tag{3.72}$$

to typical point a and its Lie-dragged image point A. This function describes a new physical quantity whose value at a is the same as that of the old quantity f at A. (If A is thought of as having been "dragged forward" from a, then f^* might better be called a "dragged-back" function. This ambiguity is one likely source of sign error.) It could happen that f^* and f have the same value at a point such as a but, because of f's dependence on position, this will not ordinarily be the case. Though it is not shown explicitly in Eq. (3.72), the definition of f^* depends on Δt; rather, this dependence has been incorporated by introducing a new function f^* rather than by giving f another argument. For small Δt, $f(A)$ can be approximated by the leading term of a Taylor series.

$$f(A) = f(a) + \frac{\partial f}{\partial x} v^x \Delta t + \frac{\partial f}{\partial y} v^y \Delta t + \cdots . \tag{3.73}$$

The "Lie derivative" of function f, *relative to vector field* $\mathbf{v}$, evaluated at a, is defined, and then evaluated, by

$$\underset{\mathbf{v}}{\mathcal{L}} f = \lim_{\Delta t \to 0} \frac{f^*(a) - f(a)}{\Delta t} = \frac{\partial f}{\partial x} v^x + \frac{\partial f}{\partial y} v^y . \tag{3.74}$$

Problem 3.4.1. *In the text at the beginning of this section, an observation of runners passing a stationary spectator was described. For the example numerical values given there, assign numerical values to all quantities appearing in Eq. (3.74) and confirm the equality approximately.*

Before evaluating the Lie derivative of a vector field $\mathbf{w}(\mathbf{x})$ we must assign meaning to the Lie dragging of a vector. For this consider Fig. 3.7 which is a blown up version of the circular insert in Fig. 3.6. Consider in particular the arrow $\vec{ac}$, and suppose it to be $\mathbf{w}(a)$, the value of vector field $\mathbf{w}$ evaluated at a. (It happens to connect two intersections of the original grid, but that is just to simplify the picture.) Further suppose that $\mathbf{w}(A)$ is the arrow $\vec{AC'}$. The vectors $\vec{ac}$ and $\vec{AC'}$, because they are defined at different points in the manifold M, cannot be directly compared (or, more to the point, subtracted.) But $\vec{AC'}$ is the result of Lie dragging some vector $\vec{ac^*}$ along the vector $\mathbf{v}$ for the time Δt being considered. In the small Δt limit, the arrow Δ is the Lie differential of $\mathbf{w}$ with respect to $\mathbf{v}$ and the Lie derivative of $\mathbf{w}$ with respect to $\mathbf{v}$ at the point a is defined by

$$\underset{\mathbf{v}}{\mathcal{L}} \mathbf{w} = \lim_{\Delta t \to 0} \frac{\vec{ac^*} - \vec{ac}}{\Delta t} = \lim_{\Delta t \to 0} \frac{\Delta}{\Delta t} . \tag{3.75}$$

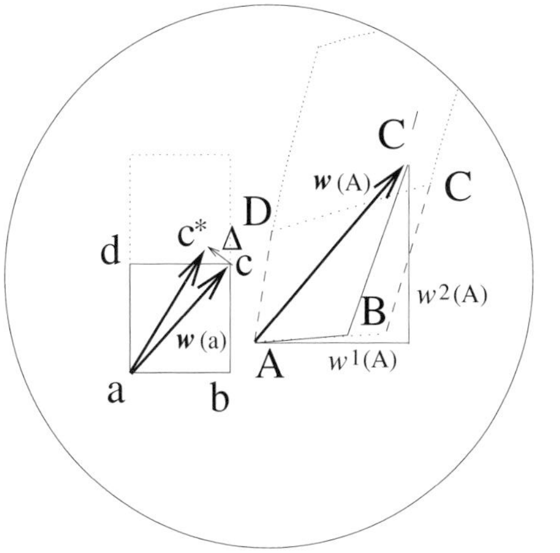

	x	y
a	0.0	0.0
c	1.0	1.0
c*	0.8	1.2
Δ	-0.2	0.2

Fig. 3.7 The circular inset of Fig. 3.6 is blown up to illustrate the Lie dragging and Lie derivative of a true vector field $\mathbf{w}(\mathbf{x})$ whose value at a is the arrow $\vec{ac}$ and whose value at A is the arrow $\vec{AC'}$.

The construction exhibited in Fig. 3.7 is similar to that in Fig. 3.2. In that figure the vector $\mathbf{V}$ was slid from the point $\mathbf{M}$ to $\mathbf{M}'$ without changing its direction. This was possible because "parallelism" was meaningful in the metric geometry valid for that figure. The purpose in sliding $\mathbf{V}$ was to evaluate $\mathbf{V}' - \mathbf{V}$ and from that the derivative of $\mathbf{V}$. The corresponding construction in Fig. 3.7 is to obtain the vector $\vec{ac^*}$ from the vector $\vec{AC'}$ in order to evaluate $\Delta = \vec{cc^*}$ and from that a derivative. The vector $\vec{ac^*}$ is said to be "pseudoparallel" to $\vec{AC'}$. By the same token "unit vectors" $\vec{ab}$ and $\vec{AB}$ can be said to be pseudoparallel, as can $\vec{ad}$ and $\vec{AD}$. Here the "pseudo" means that, their components being proportional, they would be truly parallel except for the fact that the basis vectors do not define parallelism.

Because $\mathbf{w}$ is a true vector, this construction assures that $\underset{\mathbf{v}}{\mathcal{L}}\mathbf{w}$ is also a true vector – the fact that we are able to draw the Lie differential Δ as an unambiguous arrow shows this. This implies that to transform to coordinates other than (x, y) the transformation matrices for $\underset{\mathbf{w}}{\mathcal{L}}$ and $\mathbf{w}$ must be the same. This requirement will now be applied to obtain a formula in component form for $\underset{\mathbf{v}}{\mathcal{L}}\mathbf{w}$. It will be much like the formulas for absolute differential in Section 3.1.3.

Before doing this it is useful to introduce a more expressive notation than (X, Y) – namely $(x^+_{\Delta t}, y^+_{\Delta t})$ – for the new, Lie-dragged, coordinates. As well as making explicit the previously implicit dependence on Δt, this makes manifest the smoothness requirement for the limit $\Delta t \to 0$. The $+$ indicates "new." (Often the notation $*$ is used to indicate "new" but $*$ is already in use.) After this replacement Eq. (3.71) becomes

$$
\begin{aligned}
x^{+}_{\Delta t} &= x - v^x(x,y)\Delta t + \cdots \\
y^{+}_{\Delta t} &= y - v^y(x,y)\Delta t + \cdots .
\end{aligned}
\tag{3.76}
$$

For small variations about the origin these equations yield

$$
\begin{pmatrix} \Delta x^{+}_{\Delta t} \\ \Delta y^{+}_{\Delta t} \end{pmatrix} = \begin{pmatrix} 1 - \frac{\partial v^x}{\partial x}\Delta t + \cdots & -\frac{\partial v^x}{\partial y}\Delta t + \cdots \\ -\frac{\partial v^y}{\partial x}\Delta t + \cdots & 1 - \frac{\partial v^y}{\partial y}\Delta t + \cdots \end{pmatrix} \begin{pmatrix} \Delta x \\ \Delta y \end{pmatrix} .
\tag{3.77}
$$

Example 3.4.1. *For the situation illustrated in Fig. 3.6, rough numerical values are*

$$
\begin{pmatrix} \frac{\partial v^x}{\partial x} & \frac{\partial v^x}{\partial y} \\ \frac{\partial v^y}{\partial x} & \frac{\partial v^y}{\partial y} \end{pmatrix} \approx \begin{pmatrix} 0.4 & 0.1 \\ 0.1 & 0.2 \end{pmatrix} .
\tag{3.78}
$$

Contravariant components of a vector must transform with the same matrix as in Eq. (3.77). Applying this to the contravariant components of the vectors shown in Fig. 3.7 yields

$$
\begin{pmatrix} \vec{AC'}^{+,1} \\ \vec{AC'}^{+,2} \end{pmatrix} = \begin{pmatrix} 1 - \frac{\partial v^x}{\partial x}\Delta t + \cdots & -\frac{\partial v^x}{\partial y}\Delta t + \cdots \\ -\frac{\partial v^y}{\partial x}\Delta t + \cdots & 1 - \frac{\partial v^y}{\partial y}\Delta t + \cdots \end{pmatrix} \begin{pmatrix} \vec{AC'}^{1} \\ \vec{AC'}^{2} \end{pmatrix} ;
\tag{3.79}
$$

the notation on the left indicates that the components are being reckoned in the "new" system. But the prescription for Lie dragging is that these components are numerically equal to the components of $\vec{ac}^*$ in the old system, which are $(\vec{ac}^{*1}, \vec{ac}^{*2})$. (This is illustrated in Fig. 3.7 where the location of point c^* is proportionally the same within the square above dc as the point C' is within the parallelogram above DC.) Remembering that the arrows came originally from the vector **w** whose Lie derivative is being evaluated, in a step analogous to Eq. (3.73), the vector appearing on the right-hand side of Eq. (3.79) can be obtained as

$$
\begin{pmatrix} \vec{AC'}^{1} \\ \vec{AC'}^{2} \end{pmatrix} = \begin{pmatrix} \vec{ac}^1 + \frac{\partial w^1}{\partial x} v^x \Delta t + \frac{\partial w^1}{\partial y} v^y \Delta t \\ \vec{ac}^2 + \frac{\partial w^2}{\partial x} v^x \Delta t + \frac{\partial w^2}{\partial y} v^y \Delta t \end{pmatrix} .
\tag{3.80}
$$

After substituting this in Eq. (3.79) and completing the multiplication, $\vec{ac'}^1$ and $\vec{ac'}^2$ can, to adequate accuracy, be replaced by w^1 and w^2 in the terms proportional to Δt. Combining formulas we obtain

$$
\underset{\mathbf{v}}{\mathcal{L}}\mathbf{w} = \lim_{\Delta t \to 0} \frac{1}{\Delta t} \begin{pmatrix} \vec{ac}^{*1} - \vec{ac}^1 \\ \vec{ac}^{*2} - \vec{ac}^2 \end{pmatrix} = \begin{pmatrix} \frac{\partial w^1}{\partial x} v^x + \frac{\partial w^1}{\partial y} v^y - \frac{\partial v^x}{\partial x} w^1 - \frac{\partial v^x}{\partial y} w^2 \\ \frac{\partial w^2}{\partial x} v^x + \frac{\partial w^2}{\partial y} v^y - \frac{\partial v^y}{\partial x} w^1 - \frac{\partial v^y}{\partial y} w^2 \end{pmatrix} ,
\tag{3.81}
$$

As required $\underset{\mathbf{v}}{\mathcal{L}}\mathbf{w}$ is a tensor of the same order as **w** and its contravariant components are displayed in this equation. For the sake of concreteness this result has been derived in the 2D case, but extending the result to higher dimensions is straightforward (as will be sketched shortly).

It is also possible to define the Lie derivative of covariant vectors and of tensors of higher dimensionality. Toward this end, in order to take advantage of formulas derived previously, we recast result (3.81) in terms of the reminiscent "absolute derivative" derived in Section 3.1.3.

Linear transformation equations like Eq. (3.77) were introduced and analyzed in Section 2.3.2. To make Eq. (3.77) conform with the notation of Eq. (2.66) we define the transformation matrix $\mathbf{\Lambda}$ (truncating higher order terms for simplicity)

$$(\Lambda^{-1})^i{}_j = \begin{pmatrix} 1 - \frac{\partial v^x}{\partial x}\Delta t & -\frac{\partial v^x}{\partial y}\Delta t \\ -\frac{\partial v^y}{\partial x}\Delta t & 1 - \frac{\partial v^y}{\partial y}\Delta t \end{pmatrix}, \quad \Lambda^i{}_j = \begin{pmatrix} 1 + \frac{\partial v^x}{\partial x}\Delta t & \frac{\partial v^x}{\partial y}\Delta t \\ \frac{\partial v^y}{\partial x}\Delta t & 1 + \frac{\partial v^y}{\partial y}\Delta t \end{pmatrix}, \tag{3.82}$$

By Eq. (2.63) these same matrix elements relate unit vectors along the axes according to

$$\begin{aligned} \mathbf{e}^+{}_{\Delta t,1} &= \mathbf{e}_1\Lambda^1{}_1 + \mathbf{e}_2\Lambda^2{}_1 = \mathbf{e}_1 + \frac{\partial v^x}{\partial x}\Delta t\,\mathbf{e}_1 + \frac{\partial v^y}{\partial x}\Delta t\,\mathbf{e}_2, \\ \mathbf{e}^+{}_{\Delta t,2} &= \mathbf{e}_1\Lambda^1{}_2 + \mathbf{e}_2\Lambda^2{}_2 = \mathbf{e}_2 + \frac{\partial v^x}{\partial y}\Delta t\,\mathbf{e}_1 + \frac{\partial v^y}{\partial y}\Delta t\,\mathbf{e}_2. \end{aligned} \tag{3.83}$$

These equations can be related in turn to Eqs. (3.11) in which the connecting quantities $\omega_i{}^j$ were introduced;

$$\begin{aligned} \mathbf{e}^+{}_{\Delta t,1} &= \mathbf{e}_1 + \omega_1{}^1\mathbf{e}_1 + \omega_1{}^2\mathbf{e}_2, \\ \mathbf{e}^+{}_{\Delta t,2} &= \mathbf{e}_2 + \omega_2{}^1\mathbf{e}_1 + \omega_2{}^2\mathbf{e}_2. \end{aligned} \tag{3.84}$$

Here, unlike Section 3.1.2, and as has been stated repeatedly, no metric is being assumed. Still, even though coordinate systems at different points have been connected using the vector field $\mathbf{v}$ rather than a metric, we may as well use the same symbols $\omega_i{}^j$ for the connecting coefficients now;

$$\begin{pmatrix} \omega_1{}^1 & \omega_1{}^2 \\ \omega_2{}^1 & \omega_2{}^2 \end{pmatrix} = \begin{pmatrix} \frac{\partial v^x}{\partial x}\Delta t & \frac{\partial v^x}{\partial y}\Delta t \\ \frac{\partial v^y}{\partial x}\Delta t & \frac{\partial v^y}{\partial y}\Delta t \end{pmatrix}. \tag{3.85}$$

According to Eq. (3.29), the components of the "absolute differential" Dw^i of a vector field $\mathbf{w}$ subject to "connection" $\omega_j{}^i$ are given by

$$Dw^i = dw^i + w^j\,\omega_j{}^i, \tag{3.86}$$

where the second term "adds" the contribution from frame rotation to the "observed" change dw^i. In our present context, wishing to evaluate $\vec{AC}' - \vec{AC}$, (because dragging $\vec{ac}$ forward is as good as dragging $\vec{AC'}$ back) we have to "subtract" the contribution from frame rotation. Hence we obtain

$$(\underset{\mathbf{v}}{\mathcal{L}}\mathbf{w})^i = \lim_{\Delta t \to 0} \frac{\vec{AC'} - \vec{AC}}{\Delta t} - w^j\, \omega_j^{\ i} = v^j \frac{\partial w^i}{\partial x^j} - w^j \frac{\partial v^i}{\partial x^j}, \tag{3.87}$$

in agreement with Eq. (3.81). As mentioned before, this same line of reasoning makes it possible to evaluate the Lie derivative of arbitrary tensors, contravariant, covariant, or mixed, by simple alteration of the formulas for absolute derivatives derived in Section 3.1.2.

Problem 3.4.2. *Evaluate $\underset{\mathbf{v}}{\mathcal{L}}\mathbf{v}$, the Lie derivative of a vector with respect to itself, two ways – one from the formula derived in the text, the other, more intuitively, based on construction of a vector diagram like Fig. 3.7.*

3.5 The Lie Derivative – Lie Algebraic Approach

Here the preceding material concerning the Lie derivative is formulated in *intrinsic*, coordinate-free terms.

3.5.1 Exponential Representation of Parameterized Curves

A family of nonintersecting, space-filling curves such as those encountered in the previous section is known as a congruence. At each point on every curve of the congruence there is a unique tangent vector, call it $\mathbf{v}$. The curves are then known as "flowlines" of $\mathbf{v}$. Two of them are illustrated in Fig. 3.8. The lower curve passes through point A with parameter value $\lambda_{\mathbf{v},0}$ and other points on this curve are given by coordinates $x^i(A, \lambda_{\mathbf{v},0}, \lambda_{\mathbf{v}})$. If another vector field, say $\mathbf{u}$, is to be discussed it will be necessary to introduce another parameter, such as $\lambda_{\mathbf{u}}$ or μ, but for now, since only one vector field is under discussion we can suppress the $\mathbf{v}$ subscript from $\lambda_{\mathbf{v}}$. It will lead to compact formulas to represent an individual curve of the congruence by Taylor series in powers of $\epsilon = \lambda - \lambda_0$; that is, relative to point A;

$$\begin{aligned} x^i(\lambda_0 + \epsilon) &= \left[x^i + \epsilon \frac{dx^i}{d\lambda} + \cdots\right]_{\lambda_0} = \left(1 + \epsilon \frac{d}{d\lambda} + \frac{1}{2}\epsilon^2 \frac{d^2}{d\lambda^2} + \cdots\right) x^i \Big|_{\lambda_0} \\ &\equiv \left[e^{\epsilon \frac{d}{d\lambda}} x^i\right]_{\lambda_0}. \end{aligned} \tag{3.88}$$

The "exponential operator" appearing in the second line can be regarded simply as an abbreviation for the expansion in the previous line. In the sequel,

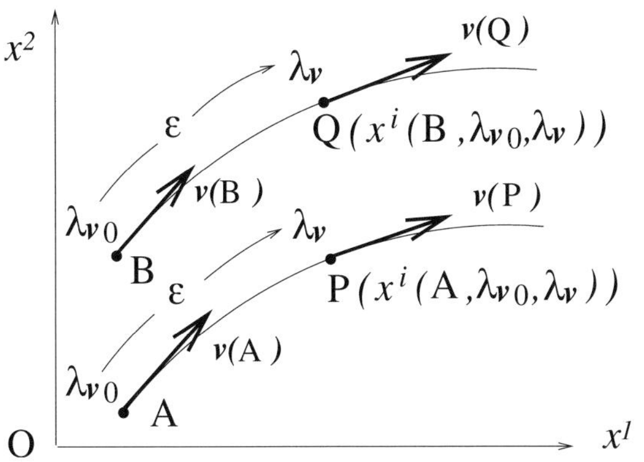

Fig. 3.8 Two out of a *congruence of curves* belonging to vector field **v**, both parameterized by $\lambda_{\mathbf{v}}$.

standard formulas satisfied by the exponential function will be applied to it. Such manipulations will be regarded as formal, subject to later verification, but they could be verified on the spot.

3.6 Identification of Vector Fields with Differential Operators

At this point we take what might be said to be the most important step (at least notationally) on the route to the algebra of vector fields. It is to assert the curious identity

$$\frac{\mathbf{d}}{\mathbf{d}\lambda_{\mathbf{v}}} \equiv \mathbf{v}, \tag{3.89}$$

where **v** is one of the vector fields discussed in the previous section and $\lambda_{\mathbf{v}}$ is the parameter of the **v**-congruence of curves. Like **v**, the arrow corresponding to $\mathbf{d}/\mathbf{d}\lambda_{\mathbf{v}}$ depends on where it is located. By its definition (3.89), both $\mathbf{d}/\mathbf{d}\lambda_{\mathbf{v}}$ and **v** are tangent to the curve passing through that location and, it is assumed that $\lambda_{\mathbf{v}}$ is adjusted so their ratio is equal to 1 everywhere. In short **v** and $\mathbf{d}/\mathbf{d}\lambda_{\mathbf{v}}$ are, by definition, two symbols for the same quantity.

For any usefulness to accrue to this definition it is necessary to ascribe more properties to $\mathbf{d}/\mathbf{d}\lambda_{\mathbf{v}}$. First of all, in a linearized approximation, an increase by 1 unit of the parameter $\lambda_{\mathbf{v}}$ corresponds to the same advance along the curve as does **v**. This is like the relation of ordinary vector analysis in which, if arc length s along a curve is taken as the curve's parameter $\lambda_{\mathbf{v}}$ and **x** is a radius vector to a point on the curve, then $d\mathbf{x}/d\lambda_{\mathbf{v}}$ is a unit-length tangent vector. If time t is taken as the curve's parameter $\lambda_{\mathbf{v}}$ then $d\mathbf{x}/d\lambda_{\mathbf{v}}$ is instantaneous velocity. (It is just a coincidence that, in this case, the symbol **v** is appropriate for "velocity.") These formulas can be interpreted as the result of "operating" on the coordinates of $\mathbf{x}$ (which are functions of position and hence of λ) with the operator $\mathbf{d}/\mathbf{d}\lambda_{\mathbf{v}}$. More generally, if f is any smooth function of position, then[13]

$$\frac{\mathbf{d}}{\mathbf{d}\lambda_{\mathbf{v}}} f \text{ is the (linearized) change in } f \text{ for } \lambda_{\mathbf{v}} \rightarrow \lambda_{\mathbf{v}} + 1. \tag{3.90}$$

With the new notation, to recover velocity components from a trajectory parameterized as $x^i(t)$, one applies the operator $\mathbf{d}/\mathbf{d}t$ to $x^i(t)$. Further justification for the derivative notation will be supplied in the next section.

3.6.1
Loop Defect

Consider next the possibility that two vector fields, the previous one $\mathbf{v}$ and another one $\mathbf{u}$ are defined on the space under study. Since the quantity $g = (\mathbf{d}/\mathbf{d}\lambda_{\mathbf{v}})f$ just introduced is a smooth function of position, it is necessarily possible to evaluate

$$h = \left(\frac{\mathbf{d}}{\mathbf{d}\lambda_{\mathbf{u}}}\right)\left(\frac{\mathbf{d}}{\mathbf{d}\lambda_{\mathbf{v}}}\right) f. \tag{3.91}$$

Then, for consistency with (3.89), the quantity $(\mathbf{d}/\mathbf{d}\lambda_{\mathbf{u}})(\mathbf{d}/\mathbf{d}\lambda_{\mathbf{v}})$, the "composition" of two operators, has to be regarded as being associated with a vector that is some new kind of "product" of two vectors $\mathbf{u}$ and $\mathbf{v}$. With multiplication being the primary operation that is traditionally referred to as "algebraic" one can say then that there is a new algebra of vector fields based on this product. It is *not* yet the *Lie algebra* of vector fields however – the product in that algebra is the "commutator."

An attempt to understand the new "product" of two vectors is illustrated in Fig. 3.9 which brings us the first complication. Though, according to (3.91), the "multiplication" of two vectors is necessarily defined, according to the figure the result of the multiplication depends on the order of the factors. To quantify this we will use Eq. (3.88) to calculate (approximately, for small ϵ) the difference of the coordinates of the two points $B_{(\mathbf{uv})}$ and $B_{(\mathbf{vu})}$ shown in Fig. 3.9;

$$x^i_{(\mathbf{vu})} - x^i_{(\mathbf{uv})} = \left(e^{\epsilon\frac{\mathbf{d}}{\mathbf{d}\lambda_{\mathbf{v}}}} e^{\epsilon\frac{\mathbf{d}}{\mathbf{d}\lambda_{\mathbf{u}}}} - e^{\epsilon\frac{\mathbf{d}}{\mathbf{d}\lambda_{\mathbf{u}}}} e^{\epsilon\frac{\mathbf{d}}{\mathbf{d}\lambda_{\mathbf{v}}}}\right) x^i\Big|_O. \tag{3.92}$$

To abbreviate formulas like this we introduce square brackets to define the "commutator" of two vectors $\mathbf{d}/\mathbf{d}$¯ and $\mathbf{d}/\mathbf{d}$˘ by

$$\left[\frac{\mathbf{d}}{\mathbf{d}\mu}, \frac{\mathbf{d}}{\mathbf{d}\lambda}\right] \equiv \frac{\mathbf{d}}{\mathbf{d}\mu}\frac{\mathbf{d}}{\mathbf{d}\lambda} - \frac{\mathbf{d}}{\mathbf{d}\lambda}\frac{\mathbf{d}}{\mathbf{d}\mu}. \tag{3.93}$$

13) The result (3.90) is also what would result from the replacement $\mathbf{v} \rightarrow \mathbf{v} \cdot \nabla$ in ordinary vector analysis.

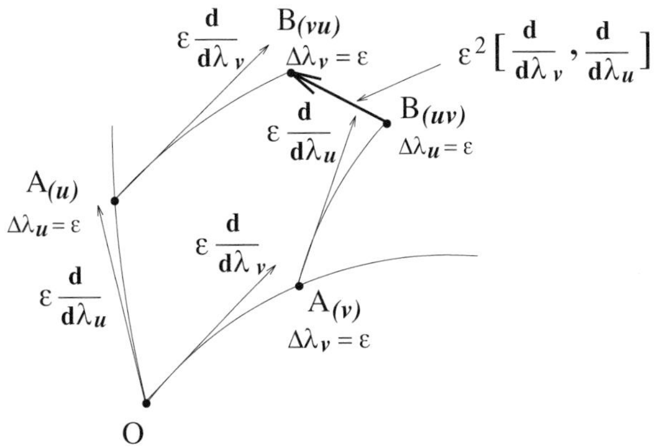

Fig. 3.9 Two routes to (potentially) the same destination. One route starts out from O along the **v**-congruence curve through that point; advancing the parameter by ϵ yields point $A_{(\mathbf{v})}$. The route continues from there to point $B_{(\mathbf{uv})}$ along the **u**-congruence curve as its parameter advances by the same amount ϵ. For the other route the congruences are traversed in reversed order. The deviations between tangent vectors and smooth curves are vastly exaggerated, especially since proceeding to the small ϵ limit is anticipated.

With this notation, dropping terms cubic and higher in ϵ, Eq. (3.92) becomes

$$\begin{aligned} x^i_{(\mathbf{vu})} - x^i_{(\mathbf{uv})} &= \left[e^{\epsilon \frac{\mathbf{d}}{\mathbf{d}\lambda_\mathbf{v}}}, e^{\epsilon \frac{\mathbf{d}}{\mathbf{d}\lambda_\mathbf{u}}} \right] x^i \Big|_O \\ &\approx \left[1 + \epsilon \frac{\mathbf{d}}{\mathbf{d}\lambda_\mathbf{v}} + \frac{1}{2}\epsilon^2 \frac{\mathbf{d}^2}{\mathbf{d}\lambda_\mathbf{v}^2},\ 1 + \epsilon \frac{\mathbf{d}}{\mathbf{d}\lambda_\mathbf{u}} + \frac{1}{2}\epsilon^2 \frac{\mathbf{d}^2}{\mathbf{d}\lambda_\mathbf{u}^2} \right] x^i \Big|_O \\ &\approx \epsilon^2 \left[\frac{\mathbf{d}}{\mathbf{d}\lambda_\mathbf{v}}, \frac{\mathbf{d}}{\mathbf{d}\lambda_\mathbf{u}} \right] x^i \Big|_O . \end{aligned} \tag{3.94}$$

This shows that the commutator, a new vector field, when applied to the position coordinates x^i, provides (to leading, quadratic, order) the coordinate deviation between the two destinations. This justifies representing the closing vector by $\epsilon^2[\mathbf{d}/\mathbf{d}\lambda_\mathbf{v}, \mathbf{d}/\mathbf{d}\lambda_\mathbf{u}]$, as shown in the figure. This has provided us with a geometric interpretation for the commutator of two vector fields.

3.7 Coordinate Congruences

The congruences just analyzed, corresponding to general vector fields **u** and **v**, have much in common with the coordinate curves of ordinary coordinate systems such as the curves on which x^1 varies while x^2 (and, in higher dimensions, all other coordinates) remain constant. We anticipated this connection in Section 2.3.5, Eq. (2.93), where notation

$$\frac{\partial}{\partial x^1} \equiv \mathbf{e}_1, \quad \frac{\partial}{\partial x^2} \equiv \mathbf{e}_2, \quad \cdots \tag{3.95}$$

for unit vectors along coordinate axes was introduced formally. The main difference of the present notation from that of Eq. (3.89), is that partial derivative symbols are used here and total derivative symbols there. This distinction is intentional, for reasons we now investigate. One thing important to recognize is that the quantity x^i plays different roles, usually distinguishable by context:

- The set $(x^1, x^2, \ldots, x^n)$ serves as coordinates of a manifold **M**.
- Any one coordinate, such as x^1, is a one component function of position in **M**. This is the role played by x^i in Eq. (3.88).
- But x^1 can equally well be regarded as the parameter establishing location on the curve resulting from variation of the first coordinate as the remaining coordinates $(x^2, \ldots, x^n)$ are held fixed; this is one of the curves of one of the coordinate congruences. This is the sense of x^1 as it appears in Eq. (3.95). In this context x^1 could just as well be symbolized by $\lambda_{\mathbf{e}_1}$, where $\mathbf{e}_1$ is the vector field yielding this congruence.

The vector $\mathbf{d}/\mathbf{d}\lambda_{\mathbf{v}}$ can presumably be expanded in terms of the basis unit vectors

$$\frac{\mathbf{d}}{\mathbf{d}\lambda_{\mathbf{v}}} = v^1 \frac{\partial}{\partial x^1} + v^2 \frac{\partial}{\partial x^2} + \cdots = v^i \frac{\partial}{\partial x^i}, \tag{3.96}$$

where the v^i are the ordinary components of **v**; they are, themselves, also functions of position. When this expansion is used to evaluate the commutator defined in Eq. (3.93) the result is

$$\left[\frac{\mathbf{d}}{\mathbf{d}\lambda_{\mathbf{v}}}, \frac{\mathbf{d}}{\mathbf{d}\lambda_{\mathbf{u}}}\right] = v^i \frac{\partial}{\partial x^i} u^j \frac{\partial}{\partial x^j} - u^i \frac{\partial}{\partial x^i} v^j \frac{\partial}{\partial x^j} = \left(v^i \frac{\partial u^j}{\partial x^i} - u^i \frac{\partial v^j}{\partial x^i}\right) \frac{\partial}{\partial x^j}, \tag{3.97}$$

where the fact has been used that the order of partial differentiation makes no difference. In this form it can be seen that the failure to commute of **u** and **v** is due to the possibility that their components are nonconstant functions of position – otherwise the partial derivatives on the right-hand side of Eq. (3.97) would vanish. When this observation is applied to the coordinate basis vectors themselves it can be seen that

$$\left[\frac{\partial}{\partial x^i}, \frac{\partial}{\partial x^j}\right] = 0, \tag{3.98}$$

since the expansion coefficients of basis vectors are all constant, either zero or one. In other words, coordinate basis vectors belonging to the same basis commute.

Up to this point one might have been harboring the impression that a "grid" made up of curves belonging to the **u** and **v**-congruences was essentially equivalent to a coordinate grid made of, say, curves of constant x^1 and x^2. It is now clear that this is not necessarily the case, since the latter set "commutes" while the former may not. (Actually it is a special property of two dimensions that the curves of two congruences necessarily form a grid at all. In higher dimensionality the points $B_{(\mathbf{uv})}$ and $B_{(\mathbf{vu})}$ in Fig. 3.9 can be displaced out of the plane of the paper and the curves can pass without intersecting.) We have shown then, that for **u** and **v** to serve as basis vectors it is necessary that they commute. Proof that this condition is also sufficient is not difficult; it can be found in Schutz, p. 49.

Example 3.7.1. Expressing a vector in other coordinates. *Consider the vector* $\mathbf{v} = -y\boldsymbol{\partial}/\boldsymbol{\partial}x + x\boldsymbol{\partial}/\boldsymbol{\partial}y$, *with x and y being rectangular coordinates. How can this vector be expressed in terms of the unit vectors* $\boldsymbol{\partial}/\boldsymbol{\partial}r$ *and* $\boldsymbol{\partial}/\boldsymbol{\partial}\phi$ *where polar coordinates r and* ϕ *are defined by* $r(x,y) = \sqrt{x^2+y^2}$ *and* $\phi(x,y) = \tan^{-1}\frac{y}{x}$? *Evaluating* $\mathbf{v}r$ *and* $\mathbf{v}\phi$ *we find*

$$\mathbf{v}r = 0, \quad \textit{and} \quad \mathbf{v}\phi = 1, \tag{3.99}$$

and from this,

$$\mathbf{v} = (\mathbf{v}r)\frac{\boldsymbol{\partial}}{\boldsymbol{\partial}r} + (\mathbf{v}\phi)\frac{\boldsymbol{\partial}}{\boldsymbol{\partial}\phi} = \frac{\boldsymbol{\partial}}{\boldsymbol{\partial}\phi}. \tag{3.100}$$

This example makes the act of changing coordinates simpler than is the case in general. The simplifying feature here is that both $\mathbf{v}r$ *and* $\mathbf{v}\phi$ *are independent of x and y. In general, to express the coefficients of* **v** *in terms of the new coordinates requires substitution for the old variables in the new coefficients. It is still a useful and straightforward exercise to generalize this procedure to arbitrary coordinate transformations, leaving this substitution implicit.*

3.8 Lie-Dragged Congruences and the Lie Derivative

A gratifying inference can be drawn by combining Eqs. (3.86), (3.96), and (3.97);

$$\underset{\mathbf{v}}{\mathcal{L}}\mathbf{w} = \left(v^j\frac{\partial w^i}{\partial x^j} - w^j\frac{\partial v^i}{\partial x^j}\right)\frac{\boldsymbol{\partial}}{\boldsymbol{\partial}x^i} = \left[\frac{\mathbf{d}}{\mathbf{d}\lambda_{\mathbf{v}}}, \frac{\mathbf{d}}{\mathbf{d}\lambda_{\mathbf{w}}}\right] \equiv [\mathbf{v},\mathbf{w}], \tag{3.101}$$

which can be written succinctly as

$$\underset{\mathbf{v}}{\mathcal{L}} = [\mathbf{v}, \cdot\,], \tag{3.102}$$

where the $\cdot$ appearing as the second argument is a place-holder for an arbitrary vector field, such as $\mathbf{w}(\mathbf{x})$. In words, Lie differentiation with respect to **v** and commutation with respect to **v** are identical operations.

This is such an important result that it is worth re-deriving it in the language of the modern language of vector fields, not because the derivation of Eq. (3.102) has been deficient in any way, but to exercise the methods of reasoning. We return then, to the discussion of Lie-dragged coordinate systems, first encountered in Section 3.4.1, using the new vector field notation.

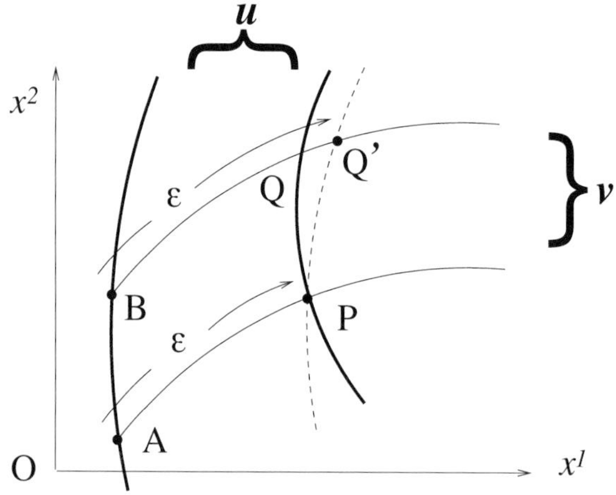

Fig. 3.10 The (heavy) curve through points A and B is Lie dragged by amount $\Delta\lambda_{\mathbf{v}} = \epsilon$ along the **v**-congruence, preserving its parameter values, to yield the dashed curve. If both heavy curves belong to the **u**-congruence, and the curves PQ and PQ' coincide, and their parameters match, the **u** congruence is said to be "Lie dragged along **v**."

Consider Fig. 3.10; it shows the same two curves of the **v**-congruence as are shown in Fig. 3.8. (These curves happen to lie in the (x^1, x^2)-plane but there is the possibility, not shown, of other, out-of-plane coordinates.) Temporarily supposing that some other vector field **u** is also defined, consider the curve of the **u**-congruence that passes through points A and B. From A and B, advancing the **v**-parameter by ϵ results in motions to points P and Q$'$, and advancing other points on the curve through A and B results in the dashed curve PQ$'$. Heavy curve PQ is the member of the **u**-congruence passing through P. As drawn, the point Q lies on the curve BQ$'$, but in more than two dimensions the curves PQ and BQ$'$ might miss completely. In any case the points Q and Q$'$ do not necessarily coincide. On the other hand, if points Q and Q$'$ *do* coincide and the $\lambda_{\mathbf{u}}$ parameter values at P and Q match those at A and B, the **u**-congruence is said to be "Lie-dragged along **v**."

As an alternative, let us drop the assumption that a vector field **u** has been predefined, and proceed to define **u**, retaining only the curve AB to get started. (In higher dimensions it would be a hypersurface.) We assume that $\lambda_{\mathbf{v}}(\mathrm{A}) = \lambda_{\mathbf{v}}(\mathrm{B})$ – this can be achieved easily by "sliding" parameter values by the addition of a constant – and we assume the same is done for all points on the curve AB. Performing the dragging operation shown in Fig. 3.10 for a continuous range of parameter values ϵ yields a "Lie dragged" **u**-congruence. In this dragging points Q and Q′ coincide by definition, and the parameter values along curve AB are dragged (unchanged) to the curve PQ. By construction then, the parameters $\lambda_{\mathbf{v}}$ and $\lambda_{\mathbf{u}}$ can serve as coordinates over the region shown. (In higher dimensionality, if AB is a hypersurface with coordinates $\lambda_{\mathbf{u}_1}, \lambda_{\mathbf{u}_2}, \ldots$, then a similar dragging operation yields hypersurface PQ, and $\lambda_{\mathbf{v}}, \lambda_{\mathbf{u}_1}, \lambda_{\mathbf{u}_2}, \ldots$ form a satisfactory set of coordinates.)

The basis vectors of this newly defined coordinate system are $\mathbf{d}/\mathbf{d}\lambda_{\mathbf{v}}$ and $\mathbf{d}/\mathbf{d}\lambda_{\mathbf{u}}$ since these vectors point along the coordinate curves and (in linearized approximation) match unit advance of their parameter values. Furthermore, since $\lambda_{\mathbf{u}}$ is constant on a curve of the **v**-congruence, the replacement $\mathbf{d}/\mathbf{d}\lambda_{\mathbf{v}} \rightarrow \boldsymbol{\partial}/\boldsymbol{\partial}\lambda_{\mathbf{v}}$ is valid. Similarly, since $\lambda_{\mathbf{v}}$ is constant on a curve of the **u**-congruence, the replacement $\mathbf{d}/\mathbf{d}\lambda_{\mathbf{u}} \rightarrow \boldsymbol{\partial}/\boldsymbol{\partial}\lambda_{\mathbf{u}}$ is also valid. Applying Eq. (3.98) we conclude that the **u**-congruence generated by Lie-dragging along **v** satisfies

$$[\mathbf{u}, \mathbf{v}] = 0. \tag{3.103}$$

We are now prepared to re-visit the Lie derivative concept, to define and then evaluate $\underset{\mathbf{v}}{\mathcal{L}}\mathbf{w}$, the Lie derivative of **w** relative to **v**. The vector **w** will not, in general, satisfy the requirement of having been Lie dragged by **v**. But we can define an auxiliary $\mathbf{w}^*$ congruence that matches a curve of the **w**-congruence, such as the curve AB in Fig. 3.11, and is Lie dragged by **v**. For $\mathbf{w}^*$ constructed in this way, on the AB curve, where $\lambda_v = 0$,

$$\mathbf{w}^*(\lambda_{\mathbf{v}}) = \mathbf{w}(\lambda_{\mathbf{v}}) \quad \text{and} \quad \frac{\mathbf{d}}{\mathbf{d}\lambda_{\mathbf{v}}}\frac{\mathbf{d}}{\mathbf{d}\lambda_{\mathbf{w}^*}} = \frac{\mathbf{d}}{\mathbf{d}\lambda_{\mathbf{w}*}}\frac{\mathbf{d}}{\mathbf{d}\lambda_{\mathbf{v}}}. \tag{3.104}$$

The notation here is not quite consistent with that used in Fig. 3.7 because here the function $\mathbf{w}^*$ is dragged *forward* whereas there it was dragged *back*. This difference will be accounted for by a sign reversal below.

For the following discussion, to avoid having to display vector functions, we introduce an arbitrary scalar function of position f; it could be called "catalytic" since it will appear in intermediate formulas but not in the final result. Using Taylor series expansion to propagate $\mathbf{w}^*$ forward we obtain

$$\left[\frac{\mathbf{d}}{\mathbf{d}\lambda_{\mathbf{w}^*}} f\right]_{\lambda_{\mathbf{v}}+\epsilon} \approx \left[\frac{\mathbf{d}}{\mathbf{d}\lambda_{\mathbf{w}}} f\right]_{\lambda_{\mathbf{v}}} + \epsilon \left[\frac{\mathbf{d}}{\mathbf{d}\lambda_{\mathbf{w}}}\frac{\mathbf{d}}{\mathbf{d}\lambda_{\mathbf{v}}} f\right]_{\lambda_{\mathbf{v}}}, \tag{3.105}$$

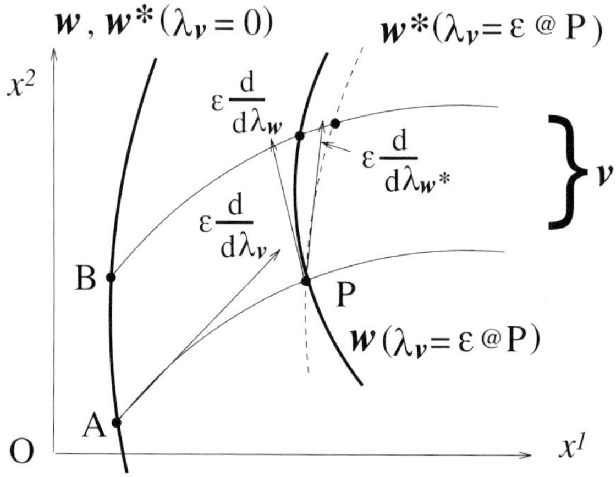

Fig. 3.11 Construction illustrating the vector field derivation of the Lie derivative.

where both of the relations (3.104), have been used. We can also propagate $\mathbf{w}$ backward;

$$\left[\frac{\mathbf{d}}{\mathbf{d}\lambda_{\mathbf{w}}}f\right]_{\lambda_{\mathbf{v}}} \approx \left[\frac{\mathbf{d}}{\mathbf{d}\lambda_{\mathbf{w}}}f\right]_{\lambda_{\mathbf{v}}+\epsilon} - \epsilon\left[\frac{\mathbf{d}}{\mathbf{d}\lambda_{\mathbf{v}}}\frac{\mathbf{d}}{\mathbf{d}\lambda_{\mathbf{w}}}f\right]_{\lambda_{\mathbf{v}}}. \tag{3.106}$$

With adequate accuracy the second coefficient has been evaluated at $\lambda_{\mathbf{v}}$. The two quantities just evaluated both being evaluated at the same place, they can be directly subtracted to (once again) *define* the Lie derivative by

$$\underset{\mathbf{v}}{\mathcal{L}}\mathbf{w}\, f = \lim_{\epsilon\to 0}\frac{[(\mathbf{w}-\mathbf{w}^*)f]_{\lambda_{\mathbf{v}}+\epsilon}}{\epsilon}. \tag{3.107}$$

Combining formulas, ignoring the distinction between $\mathbf{w}$ and $\mathbf{w}^*$ in the double derivatives, and suppressing the subsidiary function f, we obtain

$$\underset{\mathbf{v}}{\mathcal{L}}\mathbf{w} = [\mathbf{v},\mathbf{w}], \tag{3.108}$$

in agreement with Eq. (3.101).

It is possible now (if one is so inclined) to abstract all "geometry" out of the concept of the commutator of two vector fields (or equivalently the Lie derivative.) One can think of a curve not as something that can be drawn with pencil and paper (or by a skywriter in 3D) but as a one-dimensional (smoothly connected, etc.) set of points parameterized by $\lambda_{\mathbf{v}}$ in a space with coordinates $(x^1, x^2, \dots)$, and think of $\mathbf{d}/\mathbf{d}\lambda_{\mathbf{v}}$ as a directional derivative operator (where "directional" means along the set). Then determination of the discrepancy resulting from changing the order of two directional differentiations is a problem of pure calculus. This observation will be put to good use when the Poincaré equation is derived in Chapter 5.

Numerous properties of the Lie algebra of vector fields are investigated in the following series of problems (mainly copied from Schutz). The most important Lie algebraic applications apply to a set of vector fields that is "closed under commutation" (meaning that the commutator of two vectors in the set is also in the set) in spite of the fact that it is a "proper" (not the whole set) subset.

Problem 3.8.1. *Show that*

$$[\underset{\mathbf{v}}{\mathcal{L}}, \underset{\mathbf{w}}{\mathcal{L}}]\mathbf{u} = [[\mathbf{v}, \mathbf{w}], \mathbf{u}], \tag{3.109}$$

and from this, removing the catalytic function $\mathbf{u}$, *show that, when operating on a vector*

$$[\underset{\mathbf{v}}{\mathcal{L}}, \underset{\mathbf{w}}{\mathcal{L}}] = \underset{[\mathbf{v},\mathbf{w}]}{\mathcal{L}}. \tag{3.110}$$

Problem 3.8.2. *Confirm Eq. (3.110) when the terms operate on a scalar function of position.*

Problem 3.8.3. *Using* $\underset{\mathbf{v}}{\mathcal{L}}\,\mathbf{u} \equiv [\mathbf{v}, \mathbf{u}]$, *show that, when acting on a vector* $\mathbf{u}$,

$$[[\underset{\mathbf{x}}{\mathcal{L}}, \underset{\mathbf{y}}{\mathcal{L}}], \underset{\mathbf{z}}{\mathcal{L}}] + [[\underset{\mathbf{y}}{\mathcal{L}}, \underset{\mathbf{z}}{\mathcal{L}}], \underset{\mathbf{x}}{\mathcal{L}}] + [[\underset{\mathbf{z}}{\mathcal{L}}, \underset{\mathbf{x}}{\mathcal{L}}], \underset{\mathbf{y}}{\mathcal{L}}] = 0, \tag{3.111}$$

which is known as the "Jacobi identity."

Problem 3.8.4. *For scalar function* f *and vector function* $\mathbf{w}$ *show that*

$$\underset{\mathbf{v}}{\mathcal{L}}(f\mathbf{w}) = (\underset{\mathbf{v}}{\mathcal{L}}\, f)\mathbf{w} + f\underset{\mathbf{v}}{\mathcal{L}}\,\mathbf{w}, \tag{3.112}$$

which is known as the "Leibniz rule."

Problem 3.8.5. *If* $\mathbf{v} = \partial/\partial x^j$, *which is to say* $\mathbf{v}$ *is one of the coordinate basis vectors, use Eq. (3.86), and the properties of a coordinate basis set to show that*

$$(\underset{\mathbf{v}}{\mathcal{L}}\,\mathbf{w})^i = \frac{\partial w^i}{\partial x^j}. \tag{3.113}$$

Problem 3.8.6. *Consider any two vector "superpositions" of the form*

$$\mathbf{x} = a\mathbf{u} + b\mathbf{v}, \quad \mathbf{y} = c\mathbf{u} + d\mathbf{v}, \quad \textit{where} \quad [\mathbf{u}, \mathbf{v}] = 0, \tag{3.114}$$

and where a, b, c, *and* d *are functions of position with arguments not shown. Show that* $[\mathbf{x}, \mathbf{y}]$ *can be written as a similar superposition of* $\mathbf{u}$ *and* $\mathbf{v}$.

Problem 3.8.7. *Consider any two vector "superpositions" of the form*

$$\mathbf{x} = a\mathbf{u} + b\mathbf{v}, \quad \mathbf{y} = c\mathbf{u} + d\mathbf{v}, \quad \textit{where} \quad [\mathbf{u}, \mathbf{v}] = e\mathbf{u} + f\mathbf{v}, \tag{3.115}$$

and where a, b, c, d, e, *and* f *are functions of position with arguments not shown. Show that* $[\mathbf{x}, \mathbf{y}]$ *can be written as a superposition of* $\mathbf{u}$ *and* $\mathbf{v}$.

Before proceeding to further curvilinear properties it is appropriate, in the next chapter, to specialize again to linear spaces, introducing multivectors and studying their geometric properties. Curvilinear analysis resumes in Section 4.3.

3.9
Commutators of Quasi-Basis-Vectors

A circumstance common in mechanics is that one wishes to use an independent set of vector fields $\boldsymbol{\eta}_1, \boldsymbol{\eta}_2, \ldots, \boldsymbol{\eta}_r$ as "local" bases even though they do not commute. In this case they are called "quasi-basis-vectors." Let their expansions in terms of a true basis set be

$$\boldsymbol{\eta}_\alpha = u^i_\alpha(\mathbf{x}) \frac{\boldsymbol{\partial}}{\boldsymbol{\partial} q^i}, \quad r = 1, 2, \ldots, r, \tag{3.116}$$

where the coefficients are functions of position and, as customary of mechanics, the coordinates of $\mathbf{x}$ are denoted q^i. Sometimes the number r of these vectors is less than the dimensionality n of the space but, for now we assume $r = n$ and that Eqs. (3.116) can be inverted, with the inverse relations being

$$\frac{\boldsymbol{\partial}}{\boldsymbol{\partial} q^k} = (u^{-1})^\gamma_k(\mathbf{x})\, \boldsymbol{\eta}_\gamma. \tag{3.117}$$

Using Eq. (3.97) the commutator of two such quasi-basis-vectors is given by

$$\begin{aligned}
[\boldsymbol{\eta}_\alpha, \boldsymbol{\eta}_\beta] &= \left[u^i_\alpha \frac{\boldsymbol{\partial}}{\boldsymbol{\partial} q^i}, u^j_\beta \frac{\boldsymbol{\partial}}{\boldsymbol{\partial} q^j} \right] = \left(u^i_\alpha \frac{\partial u^k_\beta}{\partial q^i} - u^j_\beta \frac{\partial u^k_\alpha}{\partial q^j} \right) \frac{\boldsymbol{\partial}}{\boldsymbol{\partial} q^k} \\
&= (u^{-1})^\gamma_k \left(u^i_\alpha \frac{\partial u^k_\beta}{\partial q^i} - u^j_\beta \frac{\partial u^k_\alpha}{\partial q^j} \right) \boldsymbol{\eta}_\gamma.
\end{aligned} \tag{3.118}$$

This can be abbreviated as

$$[\boldsymbol{\eta}_\alpha, \boldsymbol{\eta}_\beta] = c^\gamma_{\alpha\beta}\, \boldsymbol{\eta}_\gamma, \quad \text{where} \quad c^\gamma_{\alpha\beta} = (u^{-1})^\gamma_k \left(u^i_\alpha \frac{\partial u^k_\beta}{\partial q^i} - u^j_\beta \frac{\partial u^k_\alpha}{\partial q^j} \right). \tag{3.119}$$

A result based on this formula, to be derived in the next example, will be used for a variational derivation of the Poincaré equation in Chapter 5.

Example 3.9.1. *The trajectory of a "central" particle in a beam of particles is given by a function $\mathbf{x}^*(t)$ and the trajectory of a nearby particle, identified by continuous parameter u, is, expanded in basis vectors $\mathbf{e}_i$,*

$$\mathbf{x}(u, t) = \mathbf{x}^*(t) + u\, \mathbf{w}(t) = q^i(u, t)\, \mathbf{e}_i. \tag{3.120}$$

The velocity of this particle is given by

$$\mathbf{v} = v^i\,\mathbf{e}_i = \frac{\partial}{\partial t}\,\mathbf{x}(u,t) = \mathbf{e}_i\frac{\partial}{\partial t}q^i(u,t) = s^i\,\boldsymbol{\eta}_i. \tag{3.121}$$

In the final step the velocity has been re-expanded in terms of quasi-basis-vectors $\boldsymbol{\eta}_i$ *defined by Eq. (3.116) and the* s^i *are the quasi-velocity components. A "ribbon" beam consists of all particles specified by* u *values in a narrow band centered on* $u = 0$*. Taken together these trajectories span a two-dimensional ribbon-shaped surface; it can be parameterized by* u *and* t*. A function* $f(u,t)$ *defined on this surface, being a single-valued function of* u *and* t*, must satisfy*

$$\frac{\partial^2}{\partial u \partial t} f = \frac{\partial^2}{\partial t \partial u} f. \tag{3.122}$$

Working with coordinates in the quasi-basis, one derivative can be obtained from the final form of Eq. (3.121),

$$\frac{\partial}{\partial t} f\big(\mathbf{x}(u,t)\big) = s^i\boldsymbol{\eta}_i(f). \tag{3.123}$$

Here the interpretation of $\boldsymbol{\eta}_i$ *as a directional derivative operator in the corresponding quasi-basis direction has been used. With this notation the (arbitrary) function can be suppressed;*

$$\frac{\partial}{\partial t} = s^i\boldsymbol{\eta}_i, \quad \textit{and} \quad \frac{\partial}{\partial u} = w^i\boldsymbol{\eta}_i; \tag{3.124}$$

the second of these equations is the result of differentiating Eq. (3.120). Using these same relations, further differentiation leads to

$$\frac{\partial^2}{\partial u \partial t} = \frac{\partial s^i}{\partial u}\boldsymbol{\eta}_i + s^j w^k \boldsymbol{\eta}_k\boldsymbol{\eta}_j, \quad \textit{and} \quad \frac{\partial^2}{\partial t \partial u} = \frac{\partial w^i}{\partial t}\boldsymbol{\eta}_i + s^k w^j \boldsymbol{\eta}_j\boldsymbol{\eta}_k. \tag{3.125}$$

Substituting these results into Eq. (3.122) and rearranging terms and using Eq.(3.119) yields the relation

$$\frac{\partial s^i}{\partial u} = \frac{\partial w^i}{\partial t} + c^i{}_{jk} s^j w^k. \tag{3.126}$$

Bibliography

General References

1 E. Cartan, *Leçons sur la géométrie des espaces de Riemann*, Gauthiers-Villars, Paris, 1951. (English translation available.)

2 B.A. Dubrovin, A.T. Fomenko, and S.P. Novikov, *Modern Geometry: Methods and Applications*, Part. 1, Springer, New York, 1984.

References for Further Study

Section 3.5

3 B.F. Schutz, *Geometrical Methods of Mathematical Physics*, Cambridge University Press, Cambridge, UK, 1995.

4
Geometry of Mechanics, III, Multilinear

4.1
Generalized Euclidean Rotations and Reflections

"Generalized Euclidean" rotations are to be considered next. As in Section 2.4, Euclidean geometry is characterized by the existence of metric form $\Phi(\mathbf{x})$ that assigns a length to vector $\mathbf{x}$. The discussion will be mainly specialized to *three* dimensions, even though, referring to Cartan's book, one finds that most results can be easily generalized to n-dimensions. Certainly relativity requires at least four dimensions and we will need more than three dimensions later on. Though arguments are given mainly in 3D, only methods that generalize easily to higher dimensionality are used. This may make some arguments seem clumsy, but the hope is that maintaining contact with ordinary geometry will better motivate the discussion. In a second pass, one will presumably be better qualified to construct more general relations.

The word "generalized" is intended to convey two ways in which something more general than Euclidean geometry is being studied. One of these, already rather familiar from special relativity, is the "pseudo-Euclidean" case in which one of the signs in the Pythagorean formula is negative. One knows of course, that including time, nature makes use of four coordinates. Without essential loss of generality, to save words without essential loss of generality, we will restrict the discussion to three, say x, y, and t. The more important "generalization" is that the "components" x^1, x^2, and x^3 will be allowed to be complex numbers. In spite of the extra level of abstraction, the theorems and proofs are quite straightforward, and physical meanings can be attached to the results.

In ordinary geometry spatial rotations are described by "orthogonal" matrices. They are sometimes called "proper" to distinguish them from "improper rotations" that combine a reflection and a rotation. But to avoid confusion later on, since the term "proper" will be used differently in connection with the pseudo-Euclidean metric of special relativity, we will use the terms "rotations or reversals" for transformations that preserve the scalar product of any two vectors. Such a transformation has the form

$$x'^i = a^i{}_k x^k, \text{ or as a matrix equation } \mathbf{x}' = \mathbf{A}\mathbf{x}. \tag{4.1}$$

Geometric Mechanics: Toward a Unification of Classical Physics. 2nd Edition. Richard Talman

ISBN: 978-3-527-40683-8

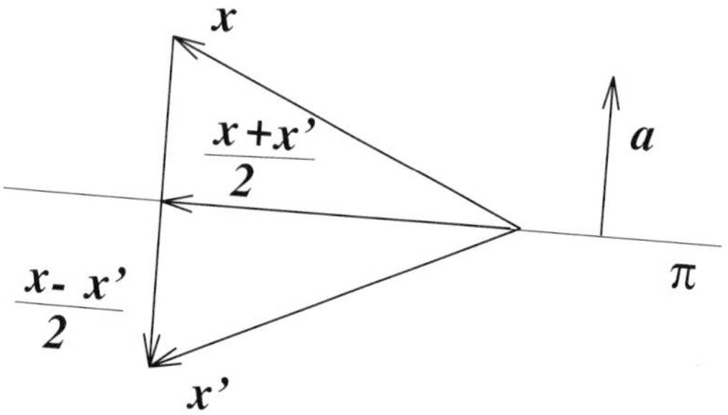

Fig. 4.1 Reflection of vector **x** in plane π associated with vector **a**.

If both frames of reference related by this transformation are "orthonormal," then the orthogonality requirement is

$$\sum_{i=1}^{n} a'^{i}{}_{j} a'^{i}{}_{k} = \delta_{jk}, \tag{4.2}$$

where the usual "Kronecker-δ" symbol satisfies $\delta_{jk} = 1$ for $j = k$ and zero otherwise. These conditions simply express the assumed orthonormality of the new basis vectors. Noting that these terms determine the elements of the matrix product $\mathbf{AA}^T$, they also imply that

$$\det|\mathbf{AA}^T| = 1 \quad \text{and} \quad \det|\mathbf{A}| = \pm 1. \tag{4.3}$$

The same transformation, when expressed in terms of skew axes, related to the orthonormal basis by matrix $\mathcal{T}$, will be described by a matrix equation $\mathbf{x}' = \mathcal{T}\mathbf{A}\mathcal{T}^{-1}\mathbf{x} = \mathbf{Bx}$. As a result, because of the multiplicative property of determinants, $\det|A| = \pm 1$ for any basis vectors, orthonormal or not. Operations for which $\det|\mathbf{A}| = 1$ are to be called "rotations," those for which $\det|\mathbf{A}| = -1$ are "reversals" or "reflection plus rotations,"

4.1.1
Reflections

The equation of a plane (or, in general, hyperplane) $\boldsymbol{\pi}$ containing the origin is

$$\mathbf{a} \cdot \mathbf{x} = a_i x^i = 0. \tag{4.4}$$

This implies that $\boldsymbol{\pi}$ is *associated with* a vector **a** having covariant components a_i and that any vector **x** lying in $\boldsymbol{\pi}$ is orthogonal to **a**; whenever the statement "a hyperplane is associated with (or corresponds to) a vector" appears, this will be its meaning. If **a** has nonvanishing scalar square, as will be required shortly, then **a** can be taken to be a unit vector without loss of generality. A vector $\mathbf{x}'$ resulting from "reflection" of vector x in plane $\boldsymbol{\pi}$ is defined by two conditions:

(i) The vector $\mathbf{x}' - \mathbf{x}$ is orthogonal to hyperplane π;

(ii) The point $\frac{1}{2}(\mathbf{x}' + \mathbf{x})$ lies in π.

The first condition implies that $\mathbf{x}' - \mathbf{x}$ is parallel to $\mathbf{a}$;

$$x'^i - x^i = \lambda a^i, \quad \text{or} \quad x'^i = x^i + \lambda a^i. \tag{4.5}$$

The second condition then implies

$$a_i(2x^i + \lambda a^i) = 0, \quad \text{or} \quad \lambda = -2\frac{a_i x^i}{a_i a^i}. \tag{4.6}$$

Since this formula fails if $a_i a^i = 0$, we insist that $\mathbf{a}$ be *nonisotropic*, and in that case we may as well assume $\mathbf{a}$ is a unit vector. Then the reflection vector $\mathbf{x}'$ is given by

$$x'^i = x^i - 2a_k x^k\, a^i, \quad \text{or} \quad \mathbf{x}' = \mathbf{x} - 2(\mathbf{a}\cdot\mathbf{x})\mathbf{a}. \tag{4.7}$$

For real vectors, using standard vector analysis, this formula is obvious. This transformation can also be expressed in matrix form;

$$\mathbf{x}' = \begin{pmatrix} 1-2a^1a_1 & -2a^1a_2 & -2a^1a_3 \\ -2a^2a_1 & 1-2a^2a_2 & -2a^2a_3 \\ -2a^3a_1 & -2a^3a_2 & 1-2a^3a_3 \end{pmatrix} \begin{pmatrix} x^1 \\ x^2 \\ x^3 \end{pmatrix} \equiv \mathbf{A}\mathbf{x}. \tag{4.8}$$

Just as the reflection plane π can be said to be "associated" with the vector $\mathbf{a}$, the 3×3 matrix $\mathbf{A}$ can also be said to be "associated" with $\mathbf{a}$. Since $\mathbf{a}$ can be any nonisotropic vector, it follows that any such vector can be associated with a reflection plane and a reflection matrix.

Transformation equation (4.8) associated with plane π, or equivalently with unit vector $\mathbf{a}$, is called a "reflection," Reflection preserves the scalar-square, as can be checked. In the real, pseudo-Euclidean case, reflections for which $\mathbf{a}$ is space(time)-like are called "space(time)-like,"

4.1.2
Expressing a Rotation as a Product of Reflections

In this section some properties of rotations are obtained by representing a rotation as the product of two reflections. Though it may seem inelegant and unpromising to represent a continuous object as the product of two discontinuous objects, the arguments are both brief and elementary, and encompass real and complex vectors, as the following theorem shows. The theorem is expressed for general n because the proof proceeds by induction on n; to simplify it, mentally replace n by 3. It applies to all transformations that leave $\mathbf{\Phi}$ (the form introduced in Eq. (2.98)) invariant, but differentiates between the two possibilities, rotations and reversals. For rotations of ordinary geometry the theorem is illustrated in Fig. 4.2.

Theorem 4.1.1. *Any rotation(reversal) in n-dimensional space is a product of an even(odd) number $\leq n$ of reflections.*

Proof. For $n = 1$ the theorem is trivially satisfied since rotation(reversal) amounts to multiplication by $1(-1)$. Assume it holds for $n-1$.

As a special case, suppose that the transformation leaves invariant a nonisotropic vector $\boldsymbol{\eta}_1$. (In ordinary 2D geometry in a plane this could be true for a reflection, but not for a rotation through nonzero angle.) Taking $\boldsymbol{\eta}_1$ as one basis vector, and augmenting it with $n-1$ independent vectors all orthogonal to $\boldsymbol{\eta}_1$ to form a complete set of basis vectors, the fundamental form becomes

$$\Phi = g_{11}(u^1)^2 + \Psi, \tag{4.9}$$

where in $\Psi = g_{ij}u^iu^j$ the summations run $2, 3, \ldots, n$. Since the transformation leaves u^1 invariant, applying the theorem to the $n-1$-dimensional hyperplane orthogonal to $\boldsymbol{\eta}_1$, the domain of applicability of the theorem is increased from $n-1$ to n, and hence to all n in this special case.

Advancing from the special case discussed to the general case, suppose the transformation is such as to transform some nonisotropic vector $\mathbf{a}$ into $\mathbf{a}'$. Consider then the reflection associated with the vector $\mathbf{a} - \mathbf{a}'$. For this reflection to make sense the vector $\mathbf{a} - \mathbf{a}'$ must itself be nonisotropic; we assume that to be the case and defer discussion of the exception for the moment. (In Fig. 4.2 the lines with arrows on both ends, such as the one joining point 1 to point 1′, are difference vectors like this; they are orthogonal to the planes of reflection.) Applying conditions (i) and (ii) above, it can be seen that the vector $\mathbf{a}$ transforms to $\mathbf{a}'$ under this reflection. The original transformation can then be thought of as being composed of this reflection plus another $n-1$-dimensional rotation or reversal that leaves $\mathbf{a}'$ invariant. Having manipulated the problem into the form of the special case of the previous paragraph, the theorem is proved in this case also.

There still remains the possibility that the vector $\mathbf{a} - \mathbf{a}'$ is isotropic for all $\mathbf{a}$. The theorem is true even in this case, but the proof is more difficult; see Cartan. We will cross this particular bridge if and when we come to it. □

4.1.3 The Lie Group of Rotations

That rotations form a group follows from the fact that the concatenation of two reflections conserves the fundamental form and that each reflection has an inverse (which follows because the determinant of the transformation is nonvanishing.) To say that the group is continuous is to say that any rotation can be parameterized by a parameter that can be varied continuously to include the identity transformation. Continuous groups are also called Lie

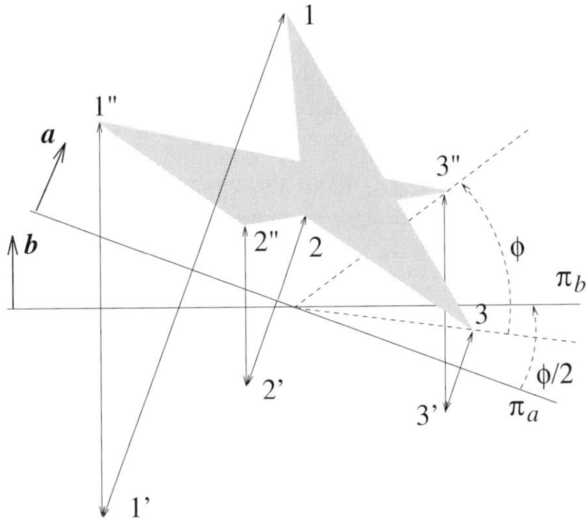

Fig. 4.2 Composition of a pure rotation from two reflections.

groups. One scarcely expects a proof depending on closeness to the identity to be based on transformations that are clearly not close to the identity, such as reflections. But that is what will be done; clearly the product of two reflections can be close to the identity if the successive planes of reflection are almost coincident. Referring again to Fig. 4.2, let **a** and **b** be the vectors associated with those reflections.

Theorem 4.1.2. *In complex Euclidean space, and in real Euclidean space with positive definite fundamental form, the set of rotations (real in the latter case), forms a continuous group.*

Proof. For any two unit vectors **a** and **b**, and for $n \geq 3$ there is at least one unit vector **c** orthogonal to both **a** and **b**. From these vectors one can construct two continuous series of reflections depending on a parameter t; they are the reflections defined by unit vectors

$$\mathbf{a}' = \mathbf{a}\cos t + \mathbf{c}\sin t, \quad \mathbf{b}' = \mathbf{b}\cos t + \mathbf{c}\sin t \quad 0 \leq t \leq \pi/2. \tag{4.10}$$

The planes of reflection are associated, as defined above, with these unit vectors. The product of these reflections is a rotation. Let us suppose that a particular rotation under study results from reflection corresponding to **a** followed by reflection corresponding to **b**. That case is included in Eq. (4.9) as the $t = 0$ limit. In the $t = \pi/2$ limit the transformation is the identity rotation – it is the product of two reflections in the same plane, the one corresponding to **c**. This exhibits the claimed continuity for rotations constructable from two reflections. For dimensions higher than 3, a rotation may need (an even number of) reflections greater than two. Proof that the continuity requirements hold in this case can be be based on pairs of these reflections. □

Every rotation in real 3D space is represented by a 3×3 orthogonal matrix with determinant equal to +1. The group formed from these matrices and their products is called SO(3), where "S" stands for "special" and implies determinant equal to +1, "O" stands for "orthogonal," and "3" is the dimensionality.

4.2 Multivectors

"Multivectors" in n-dimensional space are multicomponent objects, with components that are linear functions of $p \leq n$ vectors, **x**, **y**, **z**, ... Because of this they are also known as "p-vectors," They can be regarded as the generalization of the well-known vector cross product to more than two vectors and/or to dimensionality higher than 3. The number p must have one of the values $1, 2, \ldots, n$. Of these, $p = 1$ corresponds to ordinary vectors, and the case $p = n$, is somewhat degenerate in that, except for sign, all components are equal. In the $n = 3$ case they all are equal the "triple product" $\mathbf{x} \cdot (\mathbf{y} \times \mathbf{z})$. For the case $n = 3$ then, the only nontrivial case is $p = 2$. That is the case that is "equivalent to" the vector cross product of standard physics analysis. Here this geometric object will be represented by a 2-vector, also known as a "bivector," This will permit generalization to spaces of higher dimension. Multivectors are also essentially equivalent to "antisymmetric tensors,"

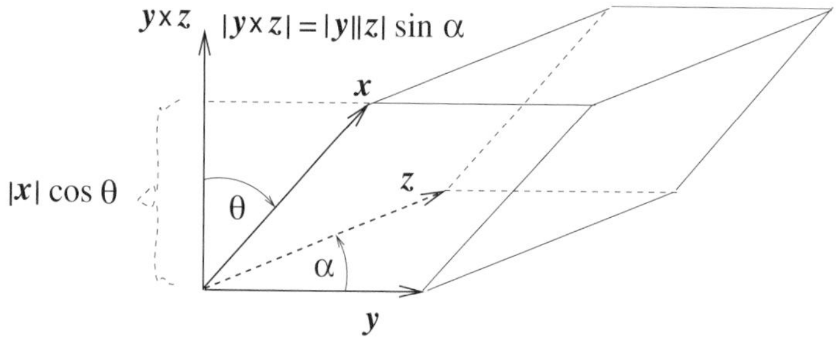

Fig. 4.3 For $n = 3$-dimensional space, the $p = 3$-multivector formed from vectors **x**, **y**, and **z** is essentially equivalent to the triple product $\mathbf{x} \cdot (\mathbf{y} \times \mathbf{z})$. Its magnitude is the volume of the parallelepiped defined by the three vectors and its sign depends on their orientation; this makes it an "oriented volume."

4.2.1 Volume Determined by 3- and by n-Vectors

The (oriented) volume of the parallelepiped defined by vectors **x**, **y**, and **z** is $V = \mathbf{x} \cdot (\mathbf{y} \times \mathbf{z})$. The sign of this product depends on the order of the 3 vectors – that is the essential content of the "oriented" qualifier. The interpretation as "volume" can be inferred from the well-known geometric properties of cross products. By this interpretation it is clear that the volume is invariant, except possibly in sign, if all vectors are subject to the same rotation or reversal. The same result can be derived algebraically from the known properties of determinants. For this the volume is related to the determinant of the array of components

$$\Delta = \det \begin{vmatrix} x^1 & x^2 & x^3 \\ y^1 & y^2 & y^3 \\ z^1 & z^2 & z^3 \end{vmatrix} . \tag{4.11}$$

Assume temporarily that these components are Euclidean, i.e., the basis is orthonormal. If the three vectors are all transformed by the same rotation or reversal (defined previously) the determinant formed the same way from the new components is unchanged, except for being multiplied by the determinant of the transformation. This is known as "the multiplication property of determinants," (This result is regularly used as the "Jacobean" factor in evaluation of integrals.) For rotations or reversals the determinant of the transformation is ± 1 and Δ is at most changed in sign.

Now retract the assumption that the basis in Eq. (4.11) is Euclidean. A determinant can also be formed from the *covariant* components:

$$\Delta' = \det \begin{vmatrix} x_1 & x_2 & x_3 \\ y_1 & y_2 & y_3 \\ z_1 & z_2 & z_3 \end{vmatrix} . \tag{4.12}$$

Its value can be determined from the definition of covariant components (Eq. (2.111)) and the multiplication property of determinants;

$$\Delta' = g\Delta, \tag{4.13}$$

where g is the determinant of the metric coefficients g_{ij}. Taking advantage of the fact that transposing a matrix does not change its determinant, the product $\Delta\Delta'$ is given by

$$\Delta\Delta' = \det \begin{vmatrix} x^1 & x^2 & x^3 \\ y^1 & y^2 & y^3 \\ z^1 & z^2 & z^3 \end{vmatrix} \det \begin{vmatrix} x_1 & x_2 & x_3 \\ y_1 & y_2 & y_3 \\ z_1 & z_2 & z_3 \end{vmatrix} = \det \begin{vmatrix} \mathbf{x.x} & \mathbf{x.y} & \mathbf{x.z} \\ \mathbf{y.x} & \mathbf{y.y} & \mathbf{y.z} \\ \mathbf{z.x} & \mathbf{z.y} & \mathbf{z.z} \end{vmatrix} \equiv V^2, \tag{4.14}$$

where the product has been called V^2; its value is independent of the choice of axes because the final determinant form is expressed entirely in terms of scalars. (They can be evaluated in the Euclidean frame.) From Eqs. (4.13) and (4.14) it follows that

$$V = \sqrt{g}\Delta = \frac{1}{\sqrt{g}}\Delta'. \tag{4.15}$$

Here the sign of V and Δ have been taken to be the same.

All the determinants in this section generalize naturally to higher dimension n. It is natural to define the volume V of the hyper-volume defined by n vectors in n-dimensions by Eq. (4.15) with Δ being the $n \times n$ determinant of the contravariant components of the n vectors.

4.2.2
Bivectors

In 3D, consider the matrix of components of two independent vectors **x** and **y**,

$$\begin{pmatrix} x^1 & x^2 & x^3 \\ y^1 & y^2 & y^3 \end{pmatrix}. \tag{4.16}$$

From this array, $\binom{3}{2} =$ three independent 2×2 determinants can be formed;

$$\begin{aligned} x^{12} &= -x^{21} \equiv \begin{vmatrix} x^1 & x^2 \\ y^1 & y^2 \end{vmatrix}, \\ x^{13} &= -x^{31} \equiv \begin{vmatrix} x^1 & x^3 \\ y^1 & y^3 \end{vmatrix}, \\ x^{23} &= -x^{32} \equiv \begin{vmatrix} x^2 & x^3 \\ y^2 & y^3 \end{vmatrix}, \end{aligned} \tag{4.17}$$

as well as the three others that differ only in sign. (It might be thought to be more natural to define x^{13} with the opposite sign, but it is just a matter of convention and the present definition preserves the order of the columns of the sub-blocks and orders the indices correspondingly with no sign change.) The pair of vectors **x** and **y** can be said to constitute a "bivector" with components given by Eq. (4.17). Note that the components are the "areas" of projections onto the coordinate planes (except for a constant factor which is 1 if the axes are rectangular). This is illustrated in Fig. 4.4. A common (intrinsic) notation for this bivector is $\mathbf{x} \wedge \mathbf{y}$, which is also known as the "wedge product" or "exterior product" of **x** and **y**. Also the components are said to belong to an antisymmetric tensor.[1]

The bivector $\mathbf{x} \wedge \mathbf{y}$ "spans" a two-dimensional space, the space consisting of all linear superpositions of **x** and **y**. The condition that a vector t belong to this

1) Normally the "anti-symmetrization" of tensor x^{ij} is defined to yield $x^{[ij]} \equiv (1/2!)(x^{ij} - x^{ji})$. This means there is a factorial factor $(1/2!)$ by which the wedge product differs from the antisymmetrized product.

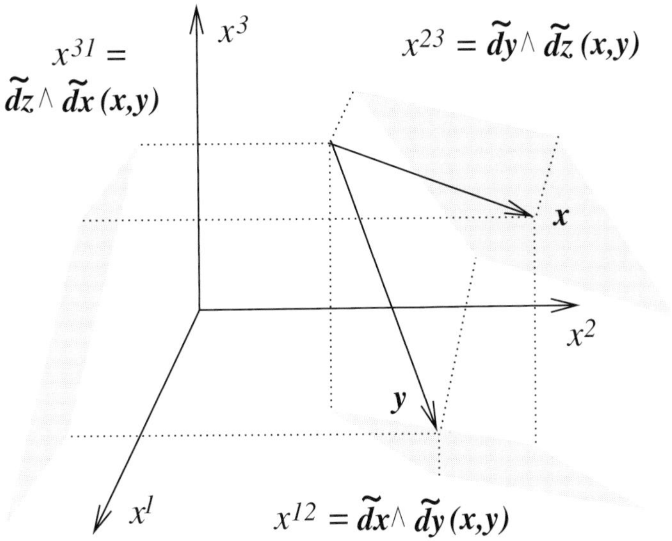

Fig. 4.4 The components x^{ij} of the bivector $\mathbf{x} \wedge \mathbf{y}$, as defined in Eqs. (4.17), are areas of projections onto the coordinate planes. Using Eq.(2.38), their magnitudes (with orientation) can also be expressed as wedge products of coordinate 1-forms evaluated on $\mathbf{x}$ and $\mathbf{y}$.

space is

$$\det \begin{vmatrix} x^1 & x^2 & x^3 \\ y^1 & y^2 & y^3 \\ t^1 & t^2 & t^3 \end{vmatrix} = t^1 x^{23} - t^2 x^{13} + t^3 x^{12} = 0; \tag{4.18}$$

(the volume defined by $\mathbf{x}$, $\mathbf{y}$, and $\mathbf{t}$ is zero.) This means that the necessary and sufficient condition for the two bivectors x^{ij} and y^{ij} to span the same space is that their components be proportional.

One can also define covariant components of a p-vector. They are the same determinants, but with contravariant components replaced by covariant components. From two 1-forms $\widetilde{\mathbf{a}}$ and $\widetilde{\mathbf{b}}$ one can similarly form a 2-form called their *wedge product* $\widetilde{\mathbf{a}} \wedge \widetilde{\mathbf{b}}$. An example has already been exhibited in Fig. 4.4. Also mixed 2-forms, having one factor a vector and the other a form, can be defined. *Symplectic geometry* of Hamiltonian systems is based on such a 2-form. (See, for example, Arnold p. 177, as well as chapter 17 of this text.)

4.2.3
Multivectors and Generalization to Higher Dimensionality

In 3-dimensions one can define a 3-vector from the vectors $\mathbf{x}, \mathbf{y}$, and $\mathbf{z}$. It consists of the $\binom{3}{3} = 1$ independent determinants that can be formed by picking three columns from the 3×3 matrix whose rows are $\mathbf{x}^T$, $\mathbf{y}^T$, and $\mathbf{z}^T$. From Eq. (4.14) it is clear that the value of this component is the oriented volume defined by the three vectors.

In an n-dimensional space, a p-vector can be defined similarly, for $p \leq n$. Its elements are the $\binom{n}{p}$ determinants that can be formed by picking p columns from the matrix with p rows $\mathbf{x}_1^T, \mathbf{x}_2^T, \ldots, \mathbf{x}_p^T$. An (invariant) "measure" or "area" or "volume" (as the case may be) V of a p-vector can be defined by

$$V^2 = \det\begin{vmatrix} \mathbf{x}\cdot\mathbf{x} & \mathbf{x}\cdot\mathbf{y} & \cdots & \mathbf{x}\cdot\mathbf{z} \\ \mathbf{y}\cdot\mathbf{x} & \mathbf{y}\cdot\mathbf{y} & \cdots & \mathbf{y}\cdot\mathbf{z} \\ \cdots & \cdots & \cdots & \cdots \\ \mathbf{z}\cdot\mathbf{x} & \mathbf{z}\cdot\mathbf{y} & \cdots & \mathbf{z}\cdot\mathbf{z} \end{vmatrix} = \det\begin{vmatrix} x^i x_i & x^j y_j & \cdots & x^k z_k \\ y^i x_i & y^j y_j & \cdots & y^k z_k \\ \cdots & \cdots & \cdots & \cdots \\ z^i x_i & z^j y_j & \cdots & z^k z_k \end{vmatrix}$$

$$= x_i y_j \ldots z_k \det\begin{vmatrix} x^i & x^j & \cdots & x^k \\ y^i & y^j & \cdots & y^k \\ \cdots & \cdots & \cdots & \cdots \\ z^i & z^j & \cdots & z^k \end{vmatrix} = \frac{1}{p!}\, x_{ij\ldots k} x^{ij\ldots k}. \tag{4.19}$$

A factor common to all elements in a column of the determinant has been factored out (repeatedly) in going from the first to the second line; also proportional columns have been suppressed. That V^2 is invariant is made manifest by the first expression. In the final summation the only surviving terms have all combinations of indices that are all different. For any such combination there are $p!$ equal terms, which accounts for the $1/p!$ factor.

For example, consider a bivector whose covariant components P_{ij} are given by

$$P_{ij} = \det\begin{vmatrix} x_i & x_j \\ y_i & y_j \end{vmatrix} = \det\begin{vmatrix} g_{ih}x^h & g_{jk}x^k \\ g_{ih}y^h & g_{jk}y^k \end{vmatrix} = g_{ih}g_{jk}P^{hk}. \tag{4.20}$$

The square of the measure of this bivector is

$$V^2 = \frac{1}{2}P^{ij}P_{ij} = \frac{1}{4}(g_{ih}g_{jk} - g_{ik}g_{jh})P^{ij}P^{hk}. \tag{4.21}$$

This has the dimensions of area-squared. For $n = 3$ it is equal to $P^{12}P_{12} + P^{23}P_{23} + P^{31}P_{31}$. If axes 1 and 2 are rectangular, P^{12} and P_{12} are each equal to the projected area on the 1, 2 plane. Since the product $P^{12}P_{12}$ is invariant for other, not necessarily skew, axes $1'$ and $2'$, provided they define the same plane, its value is the squared area of the projections onto that plane. As a result, the square of the measure of the bivector is the sum of the squared areas of the projections onto all coordinate planes. In particular, if $\mathbf{x}$ and $\mathbf{y}$ both lie in one of the basis planes – a thing that can always be arranged – the measure is the area of the parallelogram they define. These relationships can be thought of as a "Pythagorean relation for areas," they are basic to invariant integration over surfaces. Clearly the bivector P_{ij} and the conventional "cross product" $\mathbf{x}\times\mathbf{y}$ are essentially equivalent and the measure of the bivector is equal to $|\mathbf{x}\times\mathbf{y}|$, except possibly for sign. The virtues (and burdens) relative to $\mathbf{x}\times\mathbf{y}$ of P_{ij} are that it is expressible in possibly skew, possibly complex, coordinates and is applicable to arbitrary dimensions.

The invariant measure of the trivector formed from three vectors $\mathbf{x}$, $\mathbf{y}$, and $\mathbf{z}$, in the $n = 3$ case, is

$$V = \sqrt{g}\,\det\begin{vmatrix} x^1 & x^2 & x^3 \\ y^1 & y^2 & y^3 \\ z^1 & z^2 & z^3 \end{vmatrix}. \tag{4.22}$$

An important instance where this combination arises is when $\mathbf{x}$ and $\mathbf{y}$ together represent a bivector (geometrically, an incremental area on the plane defined by $\mathbf{x}$ and $\mathbf{y}$) and $\mathbf{z} = \mathbf{F}$ is a general vector field. In this case V measures the flux of $\mathbf{F}$ through the area. In the next section V will be equivalently regarded as the invariant formed from vector $\mathbf{F}$ and the vector "supplementary" to the bivector formed from $\mathbf{x}$ and $\mathbf{y}$. As mentioned elsewhere, V as defined by Eq. (4.22) and regarded as a function of vectors $\mathbf{x}$, $\mathbf{y}$, and $\mathbf{z}$, is a 3-form, because it is a linear, antisymmetric, function of its three vector arguments.

The measure defined by Eq. (4.19) will be important in generalizing Liouville's theorem in Section 17.6.1.

4.2.4
Local Radius of Curvature of a Particle Orbit

Recall the analysis of a particle trajectory in Section 3.2. As in Eq. (3.52), with local coordinates being u^i, the particle speed v is given in terms of its velocity components $\dot{u}^i$ by

$$v^2 = g_{jk}\dot{u}^j\dot{u}^k, \tag{4.23}$$

where g_{jk} is the metric tensor evaluated at the instantaneous particle location. Since the particle acceleration $a^i = \ddot{u}^i + \Gamma^i{}_{jk}\dot{u}^j\dot{u}^k$ was shown to be a *true* vector in Section 3.2.1, it can be used along with the velocity to form a true bivector

$$P_{ij} = \det\begin{vmatrix} \dot{u}_i & \dot{u}_j \\ a_i & a_j \end{vmatrix}. \tag{4.24}$$

The square of the measure of this bivector (as defined by Eq. (4.21) it is equal to a sum of squared-projected-areas on the separate coordinate planes) is known to be an invariant. In particular, if the particle orbit lies instantaneously in one such plane – a thing that can always be arranged – the measure of the bivector is the area defined by the instantaneous velocity and acceleration vectors, that is, by v^3/ρ, where ρ is the local radius of curvature of the particle trajectory.

Problem 4.2.1. *Write a manifestly invariant expression for local radius of curvature ρ in terms of $\dot{u}^i$, a^i, and g_{jk}. Check it for uniform circular motion on a circle of radius R in the x, y plane.*

4.2.5
"Supplementary" Multivectors

There is a way of associating an $(n-p)$-vector Q to a nonisotropic p-vector P. Cartan calls Q the "supplement" of P, but it is more common to call it the "the Hodge-star of P" or $*P$. It is the mathematician's more-sophisticated-than-cross-product, but in-simple-cases-equivalent, way of obtaining a vector from two other vectors, when a fundamental form exists.
The conditions to be met by Q are:

(i) The $(n-p)$-dimensional manifold spanned by Q consists of vectors orthogonal to the p-dimensional manifold spanned by P.

(ii) The "volume" or "measure" of Q is equal to the "volume" of P.

(iii) The signs of the volumes are the same.

For the case $n = 3$, $p = 2$ the identification proceeds as follows. Let $x^{ij} = (\mathbf{x} \wedge \mathbf{y})^{ij}$ be the 2-vector P and $\mathbf{t}$ be the sought-for 1-vector Q. The conditions for $\mathbf{t}$ to be orthogonal to both $\mathbf{x}$ and $\mathbf{y}$ are

$$t_1 x^1 + t_2 x^2 + t_3 x^3 = 0, \quad t_1 y^1 + t_2 y^2 + t_3 y^3 = 0. \tag{4.25}$$

Eliminating alternately t_2 and t_1 yields

$$t_1 x^{12} + t_3 x^{32} = 0, \quad t_2 x^{21} + t_3 x^{31} = 0. \tag{4.26}$$

On the other hand, as in Eq. (4.18), if the Q_i are the covariant components of Q, the condition for $\mathbf{t}$ to belong to the space spanned by Q is that all 2×2 determinants in the matrix $\begin{pmatrix} t_1 & t_2 & t_3 \\ Q_1 & Q_2 & Q_3 \end{pmatrix}$ must vanish;

$$t_1 Q_3 - t_3 Q_1 = 0, \quad t_2 Q_3 - t_3 Q_2 = 0. \tag{4.27}$$

Comparing Eqs. (4.26) and (4.27), it then follows that the Q_i and the x^{ij} are proportional when the indices (i, j, k) are an even permutation of $(1, 2, 3)$;

$$(Q_1, Q_2, Q_3) = \text{constant} \times (x^{23}, x^{31}, x^{12}). \tag{4.28}$$

Condition (ii) determines the constant of proportionality, and further manipulation yields

$$\begin{aligned} Q_1 &= \sqrt{g}x^{23}, \quad Q_2 = \sqrt{g}x^{31}, \quad Q_3 = \sqrt{g}x^{12}, \\ Q^1 &= \frac{1}{\sqrt{g}}x_{23}, \quad Q^2 = \frac{1}{\sqrt{g}}x_{31}, \quad Q^3 = \frac{1}{\sqrt{g}}x_{12}. \end{aligned} \tag{4.29}$$

As an example, suppose that x^{ij} derives from $\mathbf{x} = (\Delta x, 0, 0)$ and $\mathbf{y} = (0, \Delta y, 0)$, so that its nonvanishing components are $x^{12} = -x^{21} = \Delta x\,\Delta y$. Then the supplementary covector is $(Q_1, Q_2, Q_3) = \sqrt{g}(0, 0, \Delta x\,\Delta y)$. This combination will be used later on in Section 4.3.5, in deriving a generalized version of Gauss's theorem.

This derivation could also have been carried out within traditional vector analysis. All that is being required is that the 3-component "vector" with components $x^2y^3 - x^3y^2, -x^1y^3 + x^3y^2, x^1y^2 - x^2y^3$ is orthogonal to both $\mathbf{x}$ and $\mathbf{y}$. The point of the present derivation is that it works for complex components and for skew axes and furthermore it can be generalized to arbitrary n and p (though it involves relatively difficult combinatorics.)

4.2.6 Sums of p-Vectors

An algebra of p-vectors can be defined according to which two p-vectors can be "added" component wise. All components of a p-vector can also be multiplied by a common factor. Dual $(n - p)$-vectors are obtained using the same formulas as above. After addition of two p-vectors, each derived from p 1-vectors as above, one can inquire whether p 1-vectors can be found that would yield the same p-vector. The answer in general is no. Hence one introduces new terminology. A "simple" p-vector is one obtainable from p 1-vectors as above and the term p-vector is redefined to include sums of simple p-vectors. However, for $n = 3$ all bivectors are simple.

4.2.7 Bivectors and Infinitesimal Rotations

We finally make contact with mechanics by identifying an infinitesimal rotation, such as a physical system might be subject to, with a bivector. It is appropriate to mention that the potential ambiguity between the active and passive interpretations of transformations is nowhere more troublesome than in this area. This difficulty arises mainly in maintaining notational consistency between bold face index-free symbols that stand for geometric objects and regular face symbols with indices that stand for their components. However, the difficulty will mainly come up in later chapters, when the results of this section are applied to mechanics.

Consider a rigid object with a single point, perhaps its center of mass, fixed in space. Taking this point as origin, let $\mathbf{x}(t)$ be a vector from the origin to a point P fixed in the body – this vector depends on time t because the object is rotating. The components $x^i(t)$ of $\mathbf{x}(t)$ are taken with respect to Cartesian axes $(\mathbf{e}_1, \mathbf{e}_2, \mathbf{e}_3)$, not necessarily orthonormal, but fixed in space. The most general possible relation between $\mathbf{x}(t)$ and an initial state $\mathbf{x}(0)$ is a rotation

$$x^i(t) = O^i{}_k(t)x^k(0), \quad \text{with inverse} \quad x^i(0) = (\mathbf{O}^{-1})^i{}_l(t)x^l(t). \tag{4.30}$$

(With skew axes, the (time-dependent) matrix O is not necessarily orthogonal but it will facilitate use of this formula in a later chapter if it is given a symbol that suggests orthogonality.) The velocity of point P is given by

$$v^i(t) = \frac{dO^i{}_k(t)}{dt}x^k(0). \tag{4.31}$$

This shows that velocity components $v^i(t)$ are linear combinations of the $x^k(0)$ which, by the second of Eqs. (4.30), are in turn linear combinations of the instantaneous particle coordinates $x^l(t)$. This implies that the $v^i(t)$ are linear functions of the $x^l(t)$;

$$v^i(t) = \frac{dO^i{}_k}{dt}(\mathbf{O}^{-1})^k{}_l x^l(t) \equiv \Omega^i{}_l(t)x^l(t), \tag{4.32}$$

which serves to define the matrix $\mathbf{\Omega}$;

$$\Omega^i{}_l(t) = \frac{dO^i{}_k(t)}{dt}\,(\mathbf{O}^{-1})^k{}_l(t). \tag{4.33}$$

Since the body is rigid, the velocity is necessarily orthogonal to the position vector;

$$0 = x_i(t)v^i(t) = \Omega^i{}_k x_i x^k = g^{il}\Omega_{lk}x_i x^k = \Omega_{lk}x^l x^k. \tag{4.34}$$

Since this is true for all x^i it follows that $\Omega_{lk} = -\Omega_{kl}$. This means that the components $\Omega^i{}_k$ are the *mixed* components (one index up, one down) of a bivector.

During an "infinitesimal rotation" occurring in the time interval from t to $t + dt$ the displacement of point P is given by $\mathbf{dx} = \mathbf{v}dt$. Then from Eq. (4.32), the infinitesimal rotation can be expressed by

$$x'^i = x^i + \Omega^i{}_k x^k, \tag{4.35}$$

where the $\Omega^i{}_k$ are the mixed components of a bivector. (In an attempt to maintain dimensional consistency, since Ω has inverse time units, a physicist might call $\Omega^i{}_k x^k$ the incremental displacement of P per unit time.) Referring to Fig. 4.5, one can compare this result with rotational formulas from vector analysis;

$$\mathbf{dx} = \mathbf{d}\boldsymbol{\phi} \times \mathbf{x}, \quad \boldsymbol{\omega} = \frac{\mathbf{d}\boldsymbol{\phi}}{dt}, \quad \mathbf{v} = \boldsymbol{\omega} \times \mathbf{x}, \tag{4.36}$$

where $\mathbf{d}\boldsymbol{\phi}$ is an infinitesimal vector, directed along "the instantaneous rotation axis," with magnitude equal to the rotation angle. Our bivector $\Omega^i{}_k$ clearly

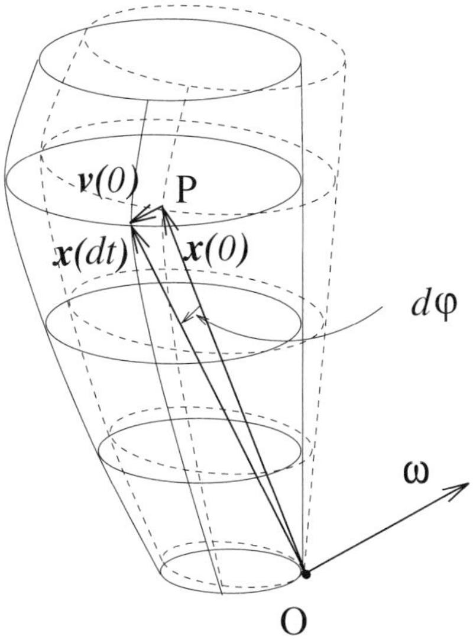

Fig. 4.5 A rigid object with point O fixed rotates by angle $d\phi$ about an axis along vector $\boldsymbol{\omega}$ during time dt. The velocity of point P is given by $\mathbf{v} = \boldsymbol{\omega} \times \mathbf{x}$.

corresponds to the "angular velocity" vector $\boldsymbol{\omega}$; in fact, transcribing the last of Eqs. (4.36) into matrix notation,

$$\mathbf{v} \overset{q}{=} \boldsymbol{\Omega}\mathbf{x} \overset{q}{=} \begin{pmatrix} 0 & -\omega^3 & \omega^2 \\ \omega^3 & 0 & -\omega^1 \\ -\omega^2 & \omega^1 & 0 \end{pmatrix} \begin{pmatrix} x^1 \\ x^2 \\ x^3 \end{pmatrix},$$

$$\text{or} \quad \boldsymbol{\Omega} \overset{q}{=} \begin{pmatrix} 0 & -\omega^3 & \omega^2 \\ \omega^3 & 0 & -\omega^1 \\ -\omega^2 & \omega^1 & 0 \end{pmatrix}. \tag{4.37}$$

(The "qualified equality" symbols are intended to acknowledge the nonintrinsic, i.e., coordinate dependent, nature of the relationships.) Infinitesimal rotations around individual Euclidean base axes can be expressed in terms of the antisymmetric matrices

$$J_1 = \begin{pmatrix} 0 & 0 & 0 \\ 0 & 0 & -1 \\ 0 & 1 & 0 \end{pmatrix}, \quad J_2 = \begin{pmatrix} 0 & 0 & 1 \\ 0 & 0 & 0 \\ -1 & 0 & 0 \end{pmatrix}, \quad J_3 = \begin{pmatrix} 0 & -1 & 0 \\ 1 & 0 & 0 \\ 0 & 0 & 0 \end{pmatrix}. \tag{4.38}$$

For example, an infinitesimal rotation through angle $d\phi_1$ around $\mathbf{e_1}$ is described by

$$\mathbf{x}' = \begin{pmatrix} x'^1 \\ x'^2 \\ x'^3 \end{pmatrix} = \begin{pmatrix} x^1 \\ x^2 - x^3 d\phi_1 \\ x^3 + x^2 d\phi_1 \end{pmatrix} = (1 + J_1 d\phi_1)\mathbf{x}. \tag{4.39}$$

(The sign of the second term depends on whether the transformation is regarded as active – **x** rotates – or passive – the coordinate system rotates. As written, the frame is assumed fixed and the vector **x** is actively rotated – a positive value for $d\phi_1$ corresponds to a vector aligned with the positive x^2-axis being rotated toward the positive x^3-axis.)

Rotation through angle $d\phi$ around unit vector **a** is given by

$$\mathbf{x}' = \left(1 + (a_1 J_1 + a_2 J_2 + a_3 J_3) d\phi\right) \mathbf{x} = (\mathbf{1} + \mathbf{a} \cdot \mathbf{J}\, d\phi)\, \mathbf{x}, \tag{4.40}$$

where the triplet (J_1, J_2, J_3) is symbolized by **J**, as if it were a vector. Notice though, since an infinitesimal relation was seen above to be associated with a bivector, not a vector, this formula may require further clarification later on.

Equation (4.40) also strains our notation in another way. The appearance of a matrix like J_1 as one element of the triplet (J_1, J_2, J_3) suggests it should have a light-face symbol, while the fact that it, itself, has multiple elements suggests that a bold-face symbol is appropriate. However, the latter is somewhat unpersuasive since the elements of J_1 are not different in different coordinate systems.

4.3 Curvilinear Coordinates in Euclidean Geometry (Continued)

At this point several of the threads encountered so far can be woven together: bivectors along with their measures, curvilinear coordinates, invariant differentiation, differential forms, and mechanics.

4.3.1 Repeated Exterior Derivatives

Basis wedge product forms of second and third degree have been defined as determinants of noncommuting basis forms $\widetilde{\mathbf{dx}}$, $\widetilde{\mathbf{dy}}$, ... in Eqs. (2.36) and (2.43). These definitions are to be carried over, but now with the understanding that the affine basis vectors **x**, **y**, and **z**, have arbitrary length and are not necessarily orthogonal. (It would emphasize this re-interpretation to refer to these basis vectors as $\mathbf{x}_1$, $\mathbf{x}_2$, and $\mathbf{x}_3$, but that would require a further, unwelcome, index in all of the quantities in this section.) Evaluated on arbitrary vectors, **u**, **v**, ..., these forms produce results such as

$$\widetilde{\mathbf{dx}} \wedge \widetilde{\mathbf{dy}} \wedge \widetilde{\mathbf{dz}}(\Delta\mathbf{u}, \Delta\mathbf{v}, \Delta\mathbf{w}) = \det \begin{vmatrix} \Delta u^1 & \Delta v^1 & \Delta w^1 \\ \Delta u^2 & \Delta v^2 & \Delta w^2 \\ \Delta u^3 & \Delta v^3 & \Delta w^3 \end{vmatrix}. \tag{4.41}$$

For finite displacements $\Delta\mathbf{u}$, $\Delta\mathbf{v}$, ..., these formulas assume affine geometry, which makes them applicable to differential displacements in more general metric geometries. Extension of this definition to define wedge products of arbitrary degree is obvious.

Defining $\widetilde{\boldsymbol{\omega}}^{(0)} = h(x,y)$ as a "0-form" and $\widetilde{\boldsymbol{\omega}}^{(1)} = f(x,y)\widetilde{\mathbf{dx}} + g(x,y)\widetilde{\mathbf{dy}}$ as a "1-form," a few examples of exterior differentiation such as

$$\widetilde{d}\widetilde{\boldsymbol{\omega}}^{(0)} = \frac{\partial h}{\partial x}\,\widetilde{\mathbf{dx}} + \frac{\partial h}{\partial y}\,\widetilde{\mathbf{dy}},$$

$$\widetilde{\mathbf{d}}\,\widetilde{\boldsymbol{\omega}}^{(1)} = \left(-\frac{\partial f}{\partial y} + \frac{\partial g}{\partial x} \right) \widetilde{\mathbf{dx}} \wedge \widetilde{\mathbf{dy}}. \tag{4.42}$$

have also been given. In deriving results like these the $\widetilde{\mathbf{d}}$ operator treats the basis forms $\widetilde{\mathbf{dx}}, \widetilde{\mathbf{dy}}, \ldots$, as constants (because they will eventually be evaluated on constant displacements) and acts only on the coefficient functions such as $f(\mathbf{x})$ and $g(\mathbf{x})$. The anticommutation formula (2.37) are then used to eliminate and collect terms. Extending formulas like these to functions of more than two or three variables and to forms of higher degree is automatic.

Problem 4.3.1. *Defining a 2-form and a 3-form by*

$$\widetilde{\boldsymbol{\omega}}^{(2)} = f\,\widetilde{\mathbf{dx}} \wedge \widetilde{\mathbf{dy}} + g\,\widetilde{\mathbf{dy}} \wedge \widetilde{\mathbf{dz}} + h\,\widetilde{\mathbf{dz}} \wedge \widetilde{\mathbf{dx}}, \quad \widetilde{\boldsymbol{\omega}}^{(3)} = k\,\widetilde{\mathbf{dx}} \wedge \widetilde{\mathbf{dy}} \wedge \widetilde{\mathbf{dz}}, \tag{4.43}$$

show that

$$\widetilde{\mathbf{d}}\,\widetilde{\boldsymbol{\omega}}^{(2)} = \left(\frac{\partial f}{\partial x} + \frac{\partial g}{\partial y} + \frac{\partial h}{\partial z} \right) \widetilde{\boldsymbol{\omega}}^{(3)}. \tag{4.44}$$

Problem 4.3.2. *For the forms $\widetilde{\boldsymbol{\omega}}^{(m)}$, $m = 0, 1, 2, 3$ just defined, and similarly defined forms of arbitrary degree having an arbitrary number n of independent variables, $x, y, z, \ldots$, show that*

$$(a)\; \widetilde{\mathbf{d}}\widetilde{\mathbf{d}}\widetilde{\boldsymbol{\omega}}^{(0)} = 0, \quad (b)\; \widetilde{\mathbf{d}}\widetilde{\mathbf{d}}\widetilde{\boldsymbol{\omega}}^{(1)} = 0, \quad (c)\; \widetilde{\mathbf{d}}\widetilde{\mathbf{d}}\widetilde{\boldsymbol{\omega}}^{(2)} = 0, \quad (d)\; \widetilde{\mathbf{d}}\widetilde{\mathbf{d}}\widetilde{\boldsymbol{\omega}}^{(m)} = 0. \tag{4.45}$$

Of course case (d) contains (a), (b), and (c) as special cases. They can be regarded as practice toward approaching (d), or possibly even as initial steps of proof by induction.

4.3.2
The Gradient Formula of Vector Analysis

For calculating gradients, divergences, or curls while using nonrectangular coordinate systems one is accustomed to looking for formulas inside the cover of the physics text (usually for electromagnetic theory) in use. Following Arnold, *Mathematical Methods of Classical Mechanics*, Chapter 7, this section uses the method of differential forms to derive these formulas.

Most of classical physics is based on orthonormal coordinate systems, such as rectangular, (x, y, z), cylindrical, (ρ, ϕ, z), and spherical, (r, ϕ, θ). In all such cases the Pythagorean formula of Euclidean geometry is distilled into the metric formulas,

$$\begin{aligned} ds^2 &= dx^2 + dy^2 + dz^2, \\ &= d\rho^2 + \rho^2 d\phi^2 + dz^2, \\ &= dr^2 + r^2 \sin^2\theta dr^2 + r^2 d\theta^2, \\ &= E_1 {dx^1}^2 + E_2 {dx^2}^2 + E_3 {dx^3}^2. \end{aligned} \tag{4.46}$$

In vector analysis a displacement $\Delta\mathbf{x}$ or a field vector $\mathbf{A}$ is expanded in terms of *unit* vectors $\hat{\mathbf{e}}_1$, $\hat{\mathbf{e}}_2$, $\hat{\mathbf{e}}_3$, each having unit length as determined using this metric,

$$\begin{aligned} \Delta\mathbf{x} &= \Delta x^1 \hat{\mathbf{e}}_1 + \Delta x^1 \hat{\mathbf{e}}_1 + \Delta x^1 \hat{\mathbf{e}}_1, \\ \mathbf{A} &= A^1 \hat{\mathbf{e}}_1 + A^1 \hat{\mathbf{e}}_1 + A^1 \hat{\mathbf{e}}_1. \end{aligned} \tag{4.47}$$

An example of a form acting on a basis unit vector is

$$\widetilde{d\mathbf{x}}^i (\hat{\mathbf{e}}_j) = 0, \quad i \neq j, \tag{4.48}$$

which vanishes because $\hat{\mathbf{e}}_j$ has no component along the curve for which x^i varies with all other variables held constant. Another important relation is (with no summation over i)

$$E_i \left(\widetilde{d\mathbf{x}}^i (\hat{\mathbf{e}}_i)\right)^2 = 1, \quad \text{or} \quad \widetilde{d\mathbf{x}}^i (\hat{\mathbf{e}}_i) = \frac{1}{\sqrt{E_i}}, \tag{4.49}$$

which follows from the very definition of $\hat{\mathbf{e}}_i$ as a unit vector.

The crucial step in applying differential forms to vector analysis is to establish an *association* between a vector $\mathbf{A}$ and a 1-form $\widetilde{\omega}_{\mathbf{A}}$. The association is established by defining the result of evaluating $\widetilde{\omega}_{\mathbf{A}}$ on an arbitrary vector

$$\mathbf{v} = v^1 \hat{\mathbf{e}}_1 + v^2 \hat{\mathbf{e}}_2 + v^3 \hat{\mathbf{e}}_3 \tag{4.50}$$

by the relation

$$\widetilde{\omega}_{\mathbf{A}}(\mathbf{v}) = \mathbf{A} \cdot \mathbf{v} = A^1 v^1 + A^2 v^2 + A^3 v^3. \tag{4.51}$$

This definition can be expressed by expanding $\widetilde{\omega}_{\mathbf{A}}$ in basis forms,

$$\widetilde{\omega}_{\mathbf{A}} = a_1 \widetilde{d\mathbf{x}}^1 + a_2 \widetilde{d\mathbf{x}}^2 + a_3 \widetilde{d\mathbf{x}}^3, \quad \text{where} \quad a_i = \sqrt{E_i}\, A^i, \tag{4.52}$$

with no summation implied. This result is obtained by substituting Eq. (4.50) into Eq. (4.51) and using Eqs. (4.48) and (4.49). For convenience in using it, $\widetilde{\omega}_{\mathbf{A}}$

can therefore be expressed in a *hybrid* form, with its coefficients expressed in terms of the vector components,

$$\widetilde{\omega}_{\mathbf{A}} = A^1\sqrt{E_1}\,\widetilde{\mathbf{dx}}^1 + A^2\sqrt{E_2}\,\widetilde{\mathbf{dx}}^2 + A^3\sqrt{E_3}\,\widetilde{\mathbf{dx}}^3. \tag{4.53}$$

Of the forms associated with vectors, the most important is $\widetilde{\omega}_{\boldsymbol{\nabla} f}$, which is the form associated with the gradient $\boldsymbol{\nabla} f$ of an arbitrary function $f(\mathbf{x})$. The defining relation for $\boldsymbol{\nabla} f$,

$$df = \boldsymbol{\nabla} f \cdot \mathbf{dx}, \tag{4.54}$$

expressed in terms of forms, is

$$\widetilde{df} = \frac{\partial f}{\partial x^1}\widetilde{\mathbf{dx}}^1 + \frac{\partial f}{\partial x^2}\widetilde{\mathbf{dx}}^2 + \frac{\partial f}{\partial x^3}\widetilde{\mathbf{dx}}^3. \tag{4.55}$$

Using definition (4.51), and applying $\widetilde{\omega}_{\boldsymbol{\nabla} f}$ to the vector $\mathbf{dx}$, produces

$$\widetilde{\omega}_{\boldsymbol{\nabla} f}(\mathbf{dx}) = \boldsymbol{\nabla} f \cdot \mathbf{dx}. \tag{4.56}$$

For this equation to agree with Eq. (4.54) requires

$$\widetilde{\omega}_{\boldsymbol{\nabla} f} = \widetilde{\mathbf{df}}. \tag{4.57}$$

Combining formulas, the expansion of $\boldsymbol{\nabla} f$ in arbitrary orthonormal coordinates is

$$\boldsymbol{\nabla} f = \frac{1}{\sqrt{E_1}}\frac{\partial f}{\partial x^1}\hat{\mathbf{e}}_1 + \frac{1}{\sqrt{E_2}}\frac{\partial f}{\partial x^2}\hat{\mathbf{e}}_2 + \frac{1}{\sqrt{E_3}}\frac{\partial f}{\partial x^3}\hat{\mathbf{e}}_3. \tag{4.58}$$

For consistency with notation in the next section the form $\widetilde{\omega}_{\mathbf{A}}$ introduced in this section will be denoted by $\widetilde{\omega}^{(1)}_{\mathbf{A}}$, to signify that it is a 1-form.

Problem 4.3.3. *Derive vector analysis formulas giving the gradient of a scalar function in cylindrical and in spherical coordinates and check your results with formulas given, for example, in a text on electromagnetic theory.*

4.3.3 Vector Calculus Expressed by Differential Forms

A vector $\mathbf{A}$ can also be associated with a "flux" 2-form $\widetilde{\omega}^{(2)}_{\mathbf{A}}$ which is defined by its value when evaluated on arbitrary pairs of vectors $\mathbf{u}$ and $\mathbf{v}$;

$$\widetilde{\omega}^{(2)}_{\mathbf{A}}(\mathbf{u}, \mathbf{v}) = \mathbf{A} \cdot (\mathbf{u} \times \mathbf{v}). \tag{4.59}$$

Defined this way, $\widetilde{\omega}^{(2)}_{\mathbf{A}}(\mathbf{u}, \mathbf{v})$ measures the "flux" of vector field $\mathbf{A}$ through the area defined by vectors $\mathbf{u}$ and $\mathbf{v}$.

Problem 4.3.4. *Use arguments like those in the previous section, to show that the form* $\widetilde{\boldsymbol{\omega}}^{(2)}_{\mathbf{A}}$ *just defined can be expanded in the form*

$$\widetilde{\boldsymbol{\omega}}^{(2)}_{\mathbf{A}} = A_1\sqrt{E_2E_3}\,\widetilde{\mathbf{dx}_2}\wedge\widetilde{\mathbf{dx}_3} + A_2\sqrt{E_3E_1}\,\widetilde{\mathbf{dx}_3}\wedge\widetilde{\mathbf{dx}_1} + A_3\sqrt{E_1E_2}\,\widetilde{\mathbf{dx}_1}\wedge\widetilde{\mathbf{dx}_2}. \tag{4.60}$$

Yet another form applicable to vector analysis can be defined naturally. It is the 3-form which, when evaluated on arbitrary vectors **u**, **v**, and **w**, gives the "triple product" of these three vectors,

$$\widetilde{\boldsymbol{\omega}}^{(3)}(\mathbf{u},\mathbf{v},\mathbf{w}) = \mathbf{u}\cdot(\mathbf{v}\times\mathbf{w}). \tag{4.61}$$

As Fig. 4.4 shows, this is the (oriented) volume of the parallelepiped formed from these three vectors, independent of their lengths and orientations.

Problem 4.3.5. *The length of a vector from, say,* x^i *to* $x^i+\epsilon$ *(with the other coordinates fixed) is equal to* $\epsilon\sqrt{E_i}$*. The volume of the parallelepiped defined by such vectors along the three basis directions is therefore* $\epsilon^3\sqrt{E_1E_2E_3}$*. Show therefore, that the form* $\widetilde{\boldsymbol{\omega}}^{(3)}$ *just defined can be expressed as*

$$\widetilde{\boldsymbol{\omega}}^{(3)} = \sqrt{E_1E_2E_3}\,\widetilde{\mathbf{dx}_1}\wedge\widetilde{\mathbf{dx}_2}\wedge\widetilde{\mathbf{dx}_3}. \tag{4.62}$$

This form $\widetilde{\boldsymbol{\omega}}^{(3)}$ can be referred to the form associated with the local volume element defined by the triplet of basis vectors.

Problem 4.3.6. *Consider the basis volume element defined by arbitrary basis vectors* $\epsilon\mathbf{e}_1$, $\epsilon\mathbf{e}_2$, *and* $\epsilon\mathbf{e}_3$*. Referring, for example, to Fig. 4.3, the volume they define, in the intended* $\epsilon\to 0$ *limit, is* ϵ^3*. Show, therefore, that Eq. (4.62) can be generalized to define a 3-form that gives the volume defined by its three infinitesimal vector arguments by*

$$\widetilde{\boldsymbol{\omega}}^{(3)} = \sqrt{g}\,\widetilde{\mathbf{dx}_1}\wedge\widetilde{\mathbf{dx}_2}\wedge\widetilde{\mathbf{dx}_3}, \tag{4.63}$$

where g *is the determinant of the matrix formed from the metric coefficients* g_{ij}*. This form is valid for arbitrary, not necessarily orthonormal, curvilinear coordinates.*

We now wish to show that *all* of vector analysis can be distilled into the following three relations:

$$\widetilde{\mathbf{d}}f = \widetilde{\boldsymbol{\omega}}^{(1)}_{\boldsymbol{\nabla} f}, \qquad \widetilde{\mathbf{d}}\widetilde{\boldsymbol{\omega}}^{(1)}_{\mathbf{A}} = \widetilde{\boldsymbol{\omega}}^{(2)}_{\boldsymbol{\nabla}\times\mathbf{A}}, \qquad \widetilde{\mathbf{d}}\widetilde{\boldsymbol{\omega}}^{(2)}_{\mathbf{A}} = \boldsymbol{\nabla}\cdot\mathbf{A}\,\widetilde{\boldsymbol{\omega}}^{(3)}. \tag{4.64}$$

The first of these equations was already given as Eq. (4.57). The other two can be regarded as *defining* the curl operation, curl $\mathbf{A}\equiv\boldsymbol{\nabla}\times\mathbf{A}$, and the divergence operation, div $\mathbf{A}\equiv\boldsymbol{\nabla}\cdot\mathbf{A}$, both acting on an arbitrary vector **A**. From the manifest invariance of the equations it is clear that the associations implied by these equations determine a vector $\boldsymbol{\nabla} \times \mathbf{A}$ and a scalar $\boldsymbol{\nabla} \cdot \mathbf{A}$ unambiguously.

It remains necessary to correlate these definitions of vector operations with the definitions one is accustomed to in vector analysis. Once this has been done, the integral formulas of vector calculus, namely Gauss's theorem and Green's theorem, can be re-expressed in terms of forms. Though proved initially using rectangular coordinates, these integral formulas will then be applicable using arbitrary orthonormal coordinates.

Let us start by *guessing* a formula for $\boldsymbol{\nabla} \times \mathbf{A}$, with the intention of later showing its invariant character;

$$\begin{aligned}\boldsymbol{\nabla} \times \mathbf{A} &= \frac{1}{\sqrt{E_1E_2E_3}} \begin{vmatrix} \sqrt{E_1}\,\hat{\mathbf{e}}_1 & \sqrt{E_2}\,\hat{\mathbf{e}}_2 & \sqrt{E_3}\,\hat{\mathbf{e}}_3 \\ \partial/\partial x^1 & \partial/\partial x^2 & \partial/\partial x^3 \\ \sqrt{E_1}A_1 & \sqrt{E_2}A_2 & \sqrt{E_3}A_3 \end{vmatrix} \\ &= \frac{1}{\sqrt{E_2E_3}}\left(\frac{\partial(A_3\sqrt{E_3})}{\partial x^2} - \frac{\partial(A_2\sqrt{E_2})}{\partial x^3}\right) + \cdots . \end{aligned} \tag{4.65}$$

This formula obviously reduces to the standard definition of the curl in rectangular coordinates. Formula (4.60) can then be used to write the 2-form associated with curl $\mathbf{A}$;

$$\widetilde{\omega}^{(2)}_{\boldsymbol{\nabla}\times\mathbf{A}} = \left(\frac{\partial(A_3\sqrt{E_3})}{\partial x^2} - \frac{\partial(A_2\sqrt{E_2})}{\partial x^3}\right)\widetilde{\mathbf{dx}}_2 \wedge \widetilde{\mathbf{dx}}_3 + \cdots . \tag{4.66}$$

Alternatively, one can form the exterior derivative of $\boldsymbol{\omega}^{(1)}_{\mathbf{A}}$, working from its expanded form (4.53),

$$\begin{aligned}\widetilde{d}\,\widetilde{\boldsymbol{\omega}}_{\mathbf{A}} &= \widetilde{d}\,(A^1\sqrt{E_1}\,\widetilde{\mathbf{dx}}^1) + \cdots \\ &= \frac{\partial A^1\sqrt{E_1}}{\partial x^2}\,\widetilde{\mathbf{dx}}^2 \wedge \widetilde{\mathbf{dx}}^1 + \frac{\partial A^1\sqrt{E_1}}{\partial x^3}\,\widetilde{\mathbf{dx}}^3 \wedge \widetilde{\mathbf{dx}}^1 + \cdots \\ &= \left(\frac{\partial(A_3\sqrt{E_3})}{\partial x^2} - \frac{\partial(A_2\sqrt{E_2})}{\partial x^3}\right)\widetilde{\mathbf{dx}}_2 \wedge \widetilde{\mathbf{dx}}_3 + \cdots . \end{aligned} \tag{4.67}$$

Since this expansion is identical to the expansion in Eq. (4.66) we have simultaneously proved the consistency of the second of Eqs. (4.64) and proved the validity of Eq. (4.65) as the expansion of the curl operator in arbitrary orthonormal coordinates.

All that remains in the program of expressing vector differential operators in terms of differential forms is to derive the formula for divergence $\boldsymbol{\nabla} \cdot \mathbf{A}$.

Problem 4.3.7. *Evaluate* $\widetilde{\mathbf{d}}\widetilde{\omega}^{(2)}_{\mathbf{A}}$ *by applying* $\widetilde{\mathbf{d}}$ *to Eq. (4.60). Then, using Eq. (4.62), show that the divergence operation defined by*

$$\boldsymbol{\nabla} \cdot \mathbf{A} = \frac{1}{\sqrt{E_1E_2E_3}}\left(\frac{\partial(A_1\sqrt{E_2E_3})}{\partial x^1} + \frac{\partial(A_2\sqrt{E_3E_1})}{\partial x^2} + \frac{\partial(A_3\sqrt{E_1E_2})}{\partial x^3}\right), \tag{4.68}$$

validates the third of Eqs. (4.64).

Finally we are in a position to express the divergence and curl formulas of vector integral calculus in terms of differential forms, starting with Gauss's law. Consider a volume Ω subdivided into differential volume elements $d\Omega$ defined by vectors $\mathbf{du}$, $\mathbf{dv}$, and $\mathbf{dw}$. The boundary of Ω is a closed surface Γ that is subdivided into oriented differential areas $\mathbf{d\Gamma}$, defined by vectors $\mathbf{dr}$ and $\mathbf{ds}$. Consider the two equations,

$$\int_\Omega \nabla\cdot\mathbf{A}\,d\Omega = \int_\Omega \nabla\cdot\mathbf{A}\,\widetilde{\omega}^{(3)}(\mathbf{du},\mathbf{dv},\mathbf{dw}) = \int_\Omega \widetilde{\mathbf{d}}\,\widetilde{\omega}^{(2)}_{\mathbf{A}}(\mathbf{du},\mathbf{dv},\mathbf{dw})$$
$$\stackrel{q}{=} \int_\Omega \widetilde{\mathbf{d}}\,\widetilde{\omega}^{(2)}_{\mathbf{A}}$$
$$\int_\Gamma \mathbf{A}\cdot\mathbf{d\Gamma} = \int_\Gamma \mathbf{A}\cdot(\mathbf{dr}\times\mathbf{ds}) = \int_\Gamma \widetilde{\omega}^{(2)}(\mathbf{dr},\mathbf{ds}) \stackrel{q}{=} \int_\Gamma \widetilde{\omega}^{(2)}_{\mathbf{A}}. \tag{4.69}$$

In both cases the last equation has been expressed as questionable. In these cases the meaning of the integration sign is defined differently on the two sides of the equations. This notation was first encountered in Eq. (2.12). The vector arguments of the final forms are simply dropped, or rather, left implicit. With this interpretation the questionable equalities are validated. The other equalities have all been justified previously. Finally, we note that, according to Gauss's theorem, the first elements in the two equations are equal to each other. We conclude therefore that

$$\int_\Omega \widetilde{\mathbf{d}}\,\widetilde{\omega}^{(2)}_{\mathbf{A}} = \int_\Gamma \widetilde{\omega}^{(2)}_{\mathbf{A}}. \tag{4.70}$$

This is Gauss's theorem (or is it the Newton–Leibniz–Gauss–Green–Ostrogradskii–Stokes–Poincaré formula, which is what Arnold calls it?) expressed in terms of differential forms.

Problem 4.3.8. *Stokes' theorem (or is it ...?) relates a line integral around a closed curve γ to a surface integral over a surface Γ bounded by curve γ. Show that, expressed in terms of differential forms, Stokes' theorem states that*

$$\int_\Gamma \widetilde{\mathbf{d}}\,\widetilde{\omega}^{(1)}_{\mathbf{A}} = \int_\gamma \widetilde{\omega}^{(1)}_{\mathbf{A}}. \tag{4.71}$$

Even though orthonormal coordinates were used in deriving these formulas, from the way they are finally written, they are manifestly valid for all curvilinear coordinates.

4.3.4
Derivation of Vector Integral Formulas

We have just finished *proving* the integral formulas (4.70) and (4.71). The "proofs" started from vector theorems well known from courses in advanced calculus. They were probably proved there using rectangular coordinates.

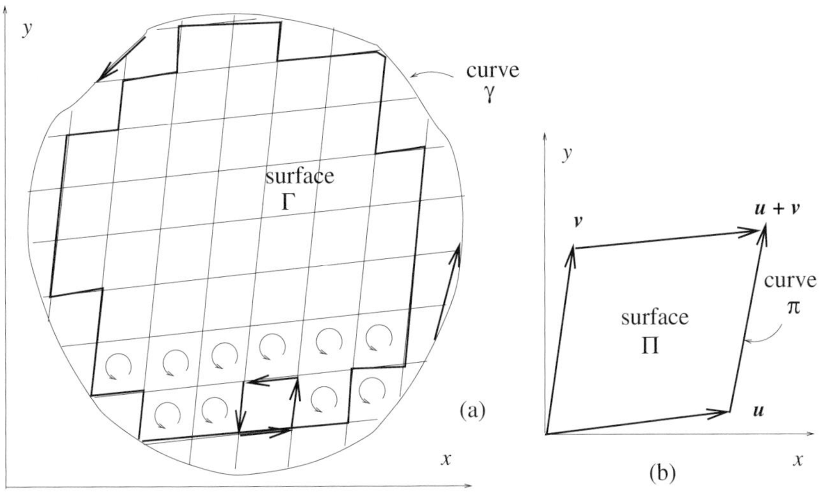

Fig. 4.6 (a) Approximation of the circulation of a 1-form around a contour γ by summing the circulations around the elements of a interpolated grid. For simplicity the surface Γ bounded by γ is here assumed to be a plane. (b) Evaluation of the circulation around one (differential) element of the grid.

One of the beauties of coordinate-free methods is that a result known to be true with one set of coordinates, once appropriately re-expressed, becomes valid in all coordinate systems. There is therefore nothing logically incomplete in the derivations given so far.

Nevertheless, a brief review of one such proof should be helpful in improving intuitive aspects of differential forms. Except for the degree of their forms Eqs. (4.70) and (4.71) are identical and it seems plausible that the same formulas should apply to forms of arbitrary degree. This is another reason for considering such proofs.

In this spirit let us integrate the differential form $\widetilde{\omega} = f(x,y)\widetilde{\mathbf{dx}} + g(x,y)\widetilde{\mathbf{dy}}$ over the curve γ shown in Fig. 4.6(a). A regular grid based on repetition of basis vectors **u** and **v** has been superimposed on the figure. As drawn, since the curve lies in a single plane, a vector normal to the surface is everywhere directed along the z-axis and a surface Γ with γ as boundary can be taken to be the (x,y) plane. (For a general nonplanar curve, establishing the grid and the surface would be somewhat more complicated but the following argument would be largely unchanged.)

Though the curve γ is macroscopic the required circulation can be obtained by summing the circulations around individual microscopic "parallelograms" as shown. Instead of traversing γ directly, one is instead traversing every parallelogram once and summing the results. The interior contributions cancel in pairs and the stair-step path around the periphery can be made arbitrarily close to γ by reducing the grid size – though the number of steps is *inversely proportional* to the grid size, the fractional error made in each step is proportional to the *square* of the step size.

In this way the problem has been reduced to one of calculating the circulation around a microscopic parallelogram as shown in Fig. 4.6(b). The vectors **u** and **v** forming the sides of the parallelogram will be treated as "differentially small" so that their higher powers can be neglected relative to lower powers.

We now wish to introduce for 1-forms the analog of the curl of a vector, so that the integral can be expressed as an integral over the surface Γ rather than along the curve γ. The line integrals along the counter-clockwise vectors in Fig. 4.6(b) coming from the first term $f(x,y)\widetilde{\mathbf{dx}}$ of the form $\widetilde{\boldsymbol{\omega}}$ being integrated are given approximately by

$$f\left(\frac{\mathbf{u}}{2}\right)\langle\widetilde{\mathbf{dx}},\mathbf{u}\rangle \approx \left(f(0)+\frac{\partial f}{\partial x}\frac{u^x}{2}+\frac{\partial f}{\partial y}\frac{u^y}{2}\right)u^x,$$
$$f\left(\mathbf{u}+\frac{\mathbf{v}}{2}\right)\langle\widetilde{\mathbf{dx}},\mathbf{v}\rangle \approx \left(f(0)+\frac{\partial f}{\partial x}\left(u^x+\frac{v^x}{2}\right)+\frac{\partial f}{\partial y}\left(u^y+\frac{v^y}{2}\right)\right)v^x, \tag{4.72}$$

where all partial derivatives are evaluated at the lower left corner. Approximating the clockwise contributions similarly and and summing the four contributions yields

$$\frac{\partial f}{\partial y}(u^y v^x - v^y u^x). \tag{4.73}$$

Performing the same calculations on $g(x,y)\widetilde{\mathbf{dy}}$ and summing all contributions yields

$$\oint_{\partial\Pi}\widetilde{\omega} \approx \left(-\frac{\partial f}{\partial y}+\frac{\partial g}{\partial x}\right)(u^x v^y - u^y v^x). \tag{4.74}$$

Here the notation Π has been introduced for the parallelogram under discussion, described as an area, and $\partial\Pi = \pi$ is the the curve circumscribing it in a counter-clockwise sense. Using Eqs. (4.42) and (2.38), we have

$$\widetilde{\mathbf{d}}\,\widetilde{\omega}(\mathbf{u},\mathbf{v}) = \left(-\frac{\partial f}{\partial y}+\frac{\partial g}{\partial x}\right)\widetilde{\mathbf{dx}}\wedge\widetilde{\mathbf{dy}}\,(\mathbf{u},\mathbf{v}) = u^x v^y - v^x u^y. \tag{4.75}$$

All results can then be combined and abbreviated into the equations

$$\int_{\Pi}\widetilde{\mathbf{d}}\,\widetilde{\omega} = \oint_{\partial\Pi}\widetilde{\omega}, \quad \text{and} \quad \int_{\Gamma}\widetilde{\mathbf{d}}\,\widetilde{\omega} = \oint_{\partial\Gamma}\widetilde{\omega}, \tag{4.76}$$

where the second equation completes the argument implied by Fig. 4.6. Γ is the complete surface and $\partial\Gamma = \gamma$ is the curve circumscribing it in a counter-clockwise sense. This is known as Stokes' theorem for forms.

Strictly speaking we have still not defined such a thing as an area integral since only line integrals have appeared. It is implied however that we intend $\int_\Gamma \widetilde{\mathbf{d}}\,\widetilde{\boldsymbol{\omega}}$ to be regarded as an integral $\int_\Gamma \widetilde{\boldsymbol{\omega}}^{(2)}$ of an antisymmetric 2-form $\widetilde{\boldsymbol{\omega}}^{(2)}$ over the surface Γ. Any surface Γ (planar or otherwise) can be spanned by a grid of infinitesimal parallelograms for which a typical one has sides $\mathbf{u}$ and $\mathbf{v}$. The integral can then be regarded as the sum of infinitesimal contributions $\widetilde{\boldsymbol{\omega}}^{(2)}(\mathbf{u},\mathbf{v})$. Using this result the derivation can be generalized to surfaces Γ and curves $\partial\Gamma$ that are not necessarily planar.

In ordinary vector analysis an integral over a two-dimensional surface can legitimately be called an "area integral" since areas are defined by the usual Pythagorean metric and, if the integrand is 1, the integral over surface Γ yields the total area. Another sort of integral over a two-dimensional surface in ordinary physics is to calculate the "flux" of a vector, say $\mathbf{E}$, through the surface. Not only does the definition of the meaning of such an integral rely on a metric within the surface, it implies the introduction of the concept of "normal to the surface" and the scalar product of $\mathbf{E}$ with that vector. In contrast, the integral

$$\int_\Gamma \widetilde{\boldsymbol{\omega}}^{(2)} \tag{4.77}$$

does not require the existence of a metric on the surface and does not require anything involving "going out of the surface," An important reason for having introduced 2-forms (and forms of other order) is illustrated by Eq. (4.77), where the 2-form serves as the integrand over a two-dimensional surface.

4.3.5
Generalized Divergence and Gauss's Theorem

In this section quantities previously expressed as vectors or as differential forms are represented by tensors and invariant differentials. Absolute differentials have been defined for contravariant vectors in Eq. (3.29), for covariant vectors in Problem 2.4.2, and for two-index tensors in Problem 3.1.4. Tensor definitions and formulas at the start of Section 3.3 should also be reviewed.

In preparation for the formulation of volume integration one must understand the curvilinear description of volume itself. It was shown in Section 4.2.1 that the volume of the parallelepiped defined by n independent-of-position vectors $\mathbf{x}, \mathbf{y}, \dots, \mathbf{z}$ in a Euclidean space of n dimensions, is given by $V = \Delta\sqrt{g}$, where Δ is the determinant of the $n \times n$ matrix of their contravariant coordinates and $g = |\det g_{ij}|$. The vectors $\mathbf{x}, \mathbf{y}, \dots, \mathbf{z}$ also define an n-vector which is an n-index antisymmetric tensor, all of whose components are equal, except for sign, to $a^{12\cdots n}$ (see Section 4.2.3). For $n = 3$,

$$a^{12\cdots n} = \det \begin{vmatrix} x^1 & x^2 & x^3 \\ y^1 & y^2 & y^3 \\ z^1 & z^2 & z^3 \end{vmatrix} \equiv \Delta. \tag{4.78}$$

The covariant differential of this tensor is

$$\begin{aligned} 0 = Da^{12\cdots n} &= da_{12\cdots n} + a^{i2\cdots n}\,\omega_i{}^1 + a^{1i\cdots n}\,\omega_i{}^2 + \cdots + a^{12\cdots i}\,\omega_i{}^n \\ &= da_{12\cdots n} + a^{12\cdots n}(\omega_1{}^1 + \omega_2{}^2 + \cdots + \omega_n{}^n), \end{aligned} \tag{4.79}$$

which vanishes because the vectors of the multivector are assumed to be constant. Expressing this in terms of the volume V yields

$$\frac{d(V/\sqrt{g})}{V/\sqrt{g}} = -\omega_i{}^i. \tag{4.80}$$

Being defined by constant vectors, V itself is constant, which implies

$$\frac{d\sqrt{g}}{\sqrt{g}} = \omega_i{}^i = \Gamma^i{}_{ij}\,du^j. \tag{4.81}$$

This can be rearranged as

$$\Gamma^i{}_{ij} = \frac{1}{\sqrt{g}}\frac{\partial\sqrt{g}}{\partial u^j} = \frac{1}{2g}\frac{\partial g}{\partial u^j}. \tag{4.82}$$

This relation is fundamental to the definition of the divergence in metric geometry and, later, to the introduction of the Lagrangian density of general relativity. In the latter case, because the determinant formed from the metric coefficients is negative, $\sqrt{g}$ is replaced by $\sqrt{-g}$.

Problem 4.3.9. *Confirm Eq. (4.82) by direct differentiation of g.*

Consider next the covariant derivative of a contravariant vector X^i; it is given by Eq. (3.29). It was shown in Section 2.3.4 that *contraction* on the indices of a mixed tensor such as this yields a *true* scalar invariant. In this case it yields what is to be known as the *divergence* of $\mathbf{X}$;

$$\operatorname{div}\mathbf{X} \equiv \frac{DX^i}{du^i} = \frac{\partial X^i}{\partial u^i} + X^k\,\Gamma^i{}_{ki} = \frac{\partial X^i}{\partial u^i} + X^k\frac{1}{\sqrt{g}}\frac{\partial\sqrt{g}}{\partial u^k} = \frac{1}{\sqrt{g}}\frac{\partial(\sqrt{g}\,X^k)}{\partial u^k}, \tag{4.83}$$

where Eq. (4.82) has been used. This quantity does not depend on the Christoffel symbols.

As in ordinary vector analysis, the primary application of the divergence operation is in Gauss's theorem. Cross multiplying the $\sqrt{g}$ factor in Eq. (4.83) and integrating over volume $\mathcal{V}$ (again specializing to $n = 3$ for convenience) yields

$$\iiint_{\mathcal{V}} \sqrt{g}\,\operatorname{div}\mathbf{X}\,du^1du^2du^3 = \iiint_{\mathcal{V}} \frac{\partial(\sqrt{g}\,X^k)}{\partial u^k}\,du^1du^2du^3. \tag{4.84}$$

At this point one should recall the derivation of Gauss's theorem for ordinary rectangular coordinates – the volume is broken up into little parallelepipeds with faces on which one of the coordinates is fixed and the others vary. Applying Taylor's theorem to approximate the integrand's variation, and recognizing that contributions from interior surfaces cancel in pairs, the right-hand side of Eq. (4.84) can be replaced by a surface integral over the closed surface $\mathcal{S}$ bounding the volume $\mathcal{V}$. The result is

$$\iiint_{\mathcal{V}} \operatorname{div}\mathbf{X}\,\sqrt{g}\,du^1du^2du^3 = \iint_{\mathcal{S}} \sqrt{g}(X^1du^2du^3 + X^2du^3du^1 + X^3du^1du^2). \tag{4.85}$$

Comparing with the example below Eq. (4.28), it can be seen that the integrand of the surface integral is the scalar product with $\mathbf{X}$ of the vector supplementary to the bivector formed from vectors along the coordinate axes. This permits Eq. (4.85) to be written more generally as

$$\iiint_{\mathcal{V}} \operatorname{div}\mathbf{X}\,\sqrt{g}\,du^1du^2du^3 = \iint_{\mathcal{S}} \sqrt{g}\det\begin{vmatrix} du^1 & du^2 & du^3 \\ dv^1 & dv^2 & dv^3 \\ X^1 & X^2 & X^3 \end{vmatrix}, \tag{4.86}$$

where **du** and **dv** lie in and define the surface $\mathcal{S}$ locally. This equation can be regarded as a prototypical *modern* version of an ancient formula. For mundane purposes it is simply the elementary Gauss's theorem but, written in this invariant, coordinate-free[2] way, it can be considered more fundamental. If choice of coordinate system were called "choice of gauge" then Eq. (4.85) would be called the *gauge invariant version* of Gauss's theorem.

Calling this result the "generalized Gauss's theorem," clearly the proof given extends easily to arbitrary n, to equate integrals over an n-dimensional "volume" and an $n-1$-dimensional "surface," This result can also be interpreted as a generalization of Stokes' theorem.

4.3.6
Metric-Free Definition of the "Divergence" of a Vector

The theorem expressed by Eq. (4.76), applies to the integration of a differential form $\widetilde{\omega}$ of arbitrary order $n-1$ over the "surface" bounding a closed

2) Though expressed in coordinates, the determinant is invariant.

n-dimensional "volume," This generalization subsumes Gauss's theorem, as in Eq. (4.70), once the divergence has been expressed as a covariant differentiation. While a metric was assumed to exist the definition of div $\mathbf{X} \equiv \nabla \cdot \mathbf{X}$ amounted to requiring Eq. (4.86) to be valid in the limit of vanishingly small ranges of integration. Since this definition depends on the existence of a metric, it needs to be replaced if a divergence theorem for metric-free space is to be established.

It is therefore necessary to define differently the divergence of the given vector $\mathbf{X}(\mathbf{x})$. We have to assume that an n-form $\widetilde{\boldsymbol{\omega}}^{(n)}$ is also defined on the space. The number $\widetilde{\boldsymbol{\omega}}^{(n)}(\mathbf{dp}, \mathbf{dq}, \ldots, \mathbf{dr})$ obtained by supplying the arguments $\mathbf{dp}$, $\mathbf{dq}$, ..., $\mathbf{dr}$ to this form is the *measure* of the hyperparallelogram they delineate. Performing an n-volume integration amounts to filling the interior of the n-volume by such parallelograms and adding the measures. It is possible to choose coordinates such that

$$\widetilde{\boldsymbol{\omega}}^{(n)}(\overbrace{\cdot, \cdot, \cdot, \cdot}^{n \text{ dots}}) = \widetilde{\mathbf{dx}}^1 \wedge \widetilde{\mathbf{dx}}^2 \wedge \cdots \widetilde{\mathbf{dx}}^n(\overbrace{\cdot, \cdot, \cdot, \cdot}^{n \text{ dots}}). \tag{4.87}$$

Expanding in terms of the corresponding basis vectors, $\mathbf{X}$ is given by

$$\mathbf{X} = X^1\mathbf{e}_1 + X^2\mathbf{e}_2 + \cdots + X^n\mathbf{e}_n. \tag{4.88}$$

From $\widetilde{\boldsymbol{\omega}}^{(n)}$ and $\mathbf{X}$ one can define an $(n-1)$-form

$$\widetilde{\boldsymbol{\omega}}^{(n-1)}(\overbrace{\cdot, \cdot, \cdot, \cdot}^{n-1 \text{ dots}}) = \widetilde{\boldsymbol{\omega}}^{(n)}(\mathbf{X}, \overbrace{\cdot, \cdot, \cdot, \cdot}^{n-1 \text{ dots}}). \tag{4.89}$$

Substituting from Eq. (4.88) into Eq. (4.87) yields,

$$\widetilde{\boldsymbol{\omega}}^{(n)}(\mathbf{X}, \overbrace{\cdot, \cdot, \cdot, \cdot}^{n-1 \text{ dots}}) = X^1\widetilde{\mathbf{dx}}^2 \wedge \widetilde{\mathbf{dx}}^3 \wedge \cdots \widetilde{\mathbf{dx}}^n - X^2\widetilde{\mathbf{dx}}^1 \wedge \widetilde{\mathbf{dx}}^3 \wedge \cdots \widetilde{\mathbf{dx}}^n \pm \cdots. \tag{4.90}$$

To obtain this result, for each term in Eq. (4.88), one can rearrange $\widetilde{\boldsymbol{\omega}}^{(n)}$ by transposing the matching differential to be in the first position. Putting the differentials back in order after the next step will restore the original sign because it will require the same number of transpositions. Forming the exterior differential of this expression yields

$$\widetilde{\mathbf{d}}[\widetilde{\boldsymbol{\omega}}^{(n)}(\mathbf{X}, \overbrace{\cdot, \cdot, \cdot, \cdot}^{n-1 \text{ dots}})] = X^i_{\ ,i}\,\widetilde{\boldsymbol{\omega}}^{(n)}(\overbrace{\cdot, \cdot, \cdot, \cdot}^{n \text{ dots}}). \tag{4.91}$$

This has validated the following definition of divergence:

$$\widetilde{\boldsymbol{\omega}}^{(n)}\text{div}_{\,\boldsymbol{\omega}}\mathbf{X} = \widetilde{\mathbf{d}}[\widetilde{\boldsymbol{\omega}}^{(n)}(\mathbf{X}, \overbrace{\cdot, \cdot, \cdot, \cdot}^{n-1 \text{ dots}})]. \tag{4.92}$$

This definition of divergence depends on $\widetilde{\boldsymbol{\omega}}^{(n)}$, a fact that is indicated by the subscript on $\mathrm{div}_{\boldsymbol{\omega}}$. Finally, Eq. (4.70) can be generalized to

$$\int_{\Gamma} \widetilde{\boldsymbol{\omega}}^{(n)} \, \mathrm{div}_{\boldsymbol{\omega}} \mathbf{X} = \oint_{\partial\Gamma} \widetilde{\boldsymbol{\omega}}^{(n)} (\mathbf{X}, \overbrace{\cdot, \cdot, \cdot, \cdot}^{n-1 \text{ dots}}). \tag{4.93}$$

Here the form $\widetilde{\boldsymbol{\omega}}^{(n)}$ is playing the role of relating "areas" on the bounding surface to "volumes" in the interior. This role was played by the metric in the previous form of Gauss's law. The factor $\sqrt{g}$ det| | in Eq. (4.86) constituted the definition of the "volume measure" $\widetilde{\boldsymbol{\omega}}^{(3)}(\mathbf{dp}, \mathbf{dq}, \mathbf{dr})$.

Finally, let us contemplate the extent to which the operations of vector calculus have been carried over to an *intrinsic* calculus of geometric objects defined on a general manifold. A true (contravariant) vector field is, in isolation, subject to no curl-like operation, but a 1-form (or covariant vector) is subject to exterior differentiation, which can be thought of as a generalized curl operation. Furthermore, there is no divergence-like operation by which a true scalar can be extracted from a true (contravariant) vector field $\mathbf{X}$, in the absence of other structure. But we have seen that "divergence" $\mathrm{div}_{\boldsymbol{\omega}}\mathbf{X}$ *can* be formed if a subsidiary n-form $\widetilde{\boldsymbol{\omega}}$ has been given.

4.4 Spinors in Three-Dimensional Space

Some parts of this section should perhaps only be skimmed initially. Apart from its obvious importance in atomic physics, this formalism is necessary for analyzing the propagation of spin directions of moving particles, and is helpful for analyzing rigid body motion.

The treatment here resembles the discussion of "Cayley–Klein" parameters in Goldstein. Basically it is the close connection between groups SO(3) and SU(2) that is to be explored. The treatment follows naturally what has gone before and has the virtue of introducing Pauli matrices in a purely geometric context, independent of quantum mechanics. Our initial purpose is to exploit the representation of a rotation as the product of two reflections, Fig. 4.2, in order to find the transformation matrix for a finite rotation around an arbitrary axis. The three components of certain vectors will, on the one hand, be associated with an object having two complex components (a spinor). On the other hand, and of greater interest to us because it applies to real vectors, a 3D vector will be associated with one 2×2 complex matrix describing rotation about the vector and another describing reflections in the plane orthogonal to the vector.

4.4.1 Definition of Spinors

The Euclidean, complex, components $(x_1 x_2, x_3)$ of an "isotropic" vector **x** satisfy

$$x_1^2 + x_2^2 + x_3^2 = 0. \tag{4.94}$$

Because the coordinates are Euclidean it is unnecessary to distinguish between lower and upper indices. To the vector **x** can be associated an object called a "spinor" with two complex components (ξ_0, ξ_1), defined so that

$$x_1 = \xi_0^2 - \xi_1^2, \quad x_2 = i(\xi_0^2 + \xi_1^2), \quad x_3 = -2\xi_0\xi_1. \tag{4.95}$$

Inverting these equations yields

$$\xi_0 = \pm\sqrt{\frac{x_1 - ix_2}{2}}, \quad \xi_1 = \pm\sqrt{\frac{-x_1 - ix_2}{2}}. \tag{4.96}$$

It is not possible to choose the sign consistently and continuously for all vectors **x**. To see this start, say, with some particular isotropic vector **x** and the positive sign for ξ_0. Rotating by angle α around the $\mathbf{e}_3$-axis causes $x_1 - ix_2$ to be multiplied by $e^{-i\alpha}$, and ξ_0 by $e^{-i\alpha/2}$. Taking $\alpha = 2\pi$ causes **x** to return to its starting value, but the sign of ξ_0 to be reversed. Rotation through 2π around any axis reverses the signs of ξ_0 and ξ_1. Another full rotation restores the signs to their original values.

4.4.2 Demonstration that a Spinor is a Euclidean Tensor

For (ξ_0, ξ_1) to be the components of a tensor, they must undergo a linear transformation when **x** is subjected to an orthogonal transformation

$$\begin{aligned} x_1' &= a_{11}x_1 + a_{12}x_2 + a_{13}x_3, \\ x_2' &= a_{21}x_1 + a_{22}x_2 + a_{23}x_3, \\ x_3' &= a_{31}x_1 + a_{32}x_2 + a_{33}x_3. \end{aligned} \tag{4.97}$$

The corresponding new value ξ_0' satisfies

$$\begin{aligned} \xi_0'^2 &= \frac{1}{2}((a_{11} - ia_{21})x_1 + (a_{12} - ia_{22})x_2 + (a_{13} - ia_{23})x_3) \\ &= \frac{1}{2}(a_{11} - ia_{21})(\xi_0^2 - \xi_1^2) + \frac{1}{2}(a_{12} - ia_{22})i(\xi_0^2 + \xi_1^2) - (a_{13} - ia_{23})\xi_0\xi_1. \end{aligned} \tag{4.98}$$

To see that the right-hand side is a perfect square, write the discriminant

$$\begin{aligned} &(a_{13} - ia_{23})^2 - (a_{11} - ia_{21} + ia_{12} + a_{22})(-a_{11} + ia_{21} + ia_{12} + a_{22}) \\ &\qquad = (a_{11} - ia_{21})^2 + (a_{12} - ia_{22})^2 + (a_{13} - ia_{23})^2 = 0, \end{aligned} \tag{4.99}$$

where the vanishing results because the rows of an orthogonal matrix are orthonormal. As mentioned before, ξ'_0, with square-only determined by Eq. (4.98), can be given either sign. The second spinor component ξ'_1 is given by a similar perfect square, but its sign ξ'_1 follows from the third of Eqs. (4.95) and (4.97);

$$-2\xi'_0\xi'_1 = a_{31}(\xi_0^2 - \xi_1^2) + ia_{32}(\xi_0^2 + \xi_1^2) - 2a_{33}\xi_0\xi_1. \tag{4.100}$$

4.4.3
Associating a 2 × 2 Reflection (Rotation) Matrix with a Vector (Bivector)

It has been seen above in Eq. (4.8) that there is a natural association between a vector $\mathbf{a}$, a plane of reflection $\boldsymbol{\pi}$ orthogonal to $\mathbf{a}$, and a 3×3 transformation matrix describing the reflection of a vector $\mathbf{x}$ in that plane. There is a corresponding 2×2 matrix describing reflection of a spinor (ξ_0, ξ_1) in the plane. It is given by

$$X = \begin{pmatrix} x_3 & x_1 - ix_2 \\ x_1 + ix_2 & -x_3 \end{pmatrix} = x_1\sigma_1 + x_2\sigma_2 + x_3\sigma_3 \equiv \mathbf{x}\cdot\boldsymbol{\sigma}, \tag{4.101}$$

where (known as Pauli spin matrices in quantum mechanics)

$$\sigma_1 = \begin{pmatrix} 0 & 1 \\ 1 & 0 \end{pmatrix}, \quad \sigma_2 = \begin{pmatrix} 0 & -i \\ i & 0 \end{pmatrix}, \quad \sigma_3 = \begin{pmatrix} 1 & 0 \\ 0 & -1 \end{pmatrix}. \tag{4.102}$$

Some useful results follow easily from this definition:

$$\begin{aligned} \det|X| &= -\mathbf{x}\cdot\mathbf{x}, \\ XX &= (\mathbf{x}\cdot\mathbf{x})\mathbf{1}, \\ XY + YX &= 2(\mathbf{x}\cdot\mathbf{y})\mathbf{1}. \end{aligned} \tag{4.103}$$

The latter two equations are especially noteworthy in that they yield matrices proportional to the identity matrix $\mathbf{1}$. In particular, if $\mathbf{x}$ is a unit vector, $X^2 = \mathbf{1}$. Also, if (x_1, x_2, x_3) are real, X is Hermitian;

$$X^* = X^T. \tag{4.104}$$

All these relations can be checked for σ_1, σ_2, and σ_3. For example,

$$\sigma_i\sigma_j = -\sigma_j\sigma_i \quad \text{for} \quad i \neq j. \tag{4.105}$$

Next consider the bivector $\mathbf{x}\times\mathbf{y} = (x_2y_3 - x_3y_2, x_3y_1 - x_1y_3, x_1y_2 - x_2y_1)$. Using Eq. (4.101), the bivector/matrix association is

$$2i(\mathbf{x}\times\mathbf{y}) \rightarrow XY - YX \equiv [X, Y], \tag{4.106}$$

where the matrix "commutator" $[X,Y] \equiv XY - YX$ has made its first appearance. If $\mathbf{x} \cdot \mathbf{y} = 0$ then $XY = -YX$ and

$$i(\mathbf{x} \times \mathbf{y}) \rightarrow XY. \tag{4.107}$$

Problem 4.4.1. *Suppose spinor (ξ_0, ξ_1) is derived from vector $\mathbf{x}$. Show that matrices σ_1, σ_2, and σ_3, when acting on (ξ_0, ξ_1) have the effect of reflecting in the y,z, the x,z and the x,y planes, respectively – that is, of generating the spinor derived from the corresponding reflection of $\mathbf{x}$.*

4.4.4
Associating a Matrix with a Trivector (Triple Product)

Consider a trivector corresponding to three orthogonal vectors $\mathbf{x}$, $\mathbf{y}$, and $\mathbf{z}$. It has six components, one for each permutation of the indices $(1,2,3)$, all equal, except for sign depending on the evenness or oddness of the permutation, to the same determinant, which is $(\mathbf{x} \times \mathbf{y}) \cdot \mathbf{z} = \mathbf{u} \cdot \mathbf{z}$ where $\mathbf{u} = \mathbf{x} \times \mathbf{y}$, a vector necessarily parallel to $\mathbf{z}$. The matrices associated to these vectors by Eq. (4.101) are to be called X, Y, Z, and U. By Eq. (4.103), the scalar product $i\mathbf{u} \cdot \mathbf{z}$ is equal to iUZ. By Eq. (4.107), the matrix iU associated with $i\mathbf{u}$ is XY. Hence

$$XYZ = iUZ\mathbf{1} = i(\mathbf{x} \times \mathbf{y}) \cdot \mathbf{z}\mathbf{1} = iv\mathbf{1}, \tag{4.108}$$

where v is the volume of the trivector. In particular $\sigma_1\sigma_2\sigma_3 = i\mathbf{1}$. The following associations have by now been established:

$$\mathbf{x} \rightarrow X,\ \mathbf{y} \rightarrow Y,\ \mathbf{x} \cdot \mathbf{y} \rightarrow \frac{1}{2}(XY + YX),\ 2i(\mathbf{x} \times \mathbf{y}) \rightarrow [X,Y],\ iv \rightarrow XYZ. \tag{4.109}$$

4.4.5
Representations of Reflections

Reflections in a plane orthogonal to unit vector $\mathbf{a}$ have been described previously, Eq. (4.7);

$$\mathbf{x}' = \mathbf{x} - 2(\mathbf{a} \cdot \mathbf{x})\mathbf{a}. \tag{4.110}$$

Rearranging this into a matrix equation using Eq. (4.103) and $A^2 = 1$ yields

$$X' = X - A(XA + AX) = -AXA. \tag{4.111}$$

By Eq. (4.107) the matrix associated with the bivector corresponding to orthogonal vectors $\mathbf{x}$ and $\mathbf{y}$ is proportional to XY, and reflecting this in the plane defined by A yields

$$X'Y' = (-AXA)(-AYA) = AXYA. \tag{4.112}$$

In terms of the matrix U, defined in the previous section as associated with the bivector,

$$U' = AUA. \tag{4.113}$$

Comparing Eqs. (4.111) and (4.113) one can say that, except for sign, vectors and bivectors transform identically under reflection.

4.4.6
Representations of Rotations

It was demonstrated in Section 4.1.2 that any rotation can be expressed as the product of two reflections. Let the matrices for these reflections be A and B. When subjected to these, the vector-matrix X and the bivector-matrix U of the previous section transform according to

$$X' = BAXAB = (BA)X(BA)^{-1}, \quad U' = BAUAB, \tag{4.114}$$

which is to say identically. Note that $AB = (BA)^{-1}$, since it reverses the two reflections. Defining the matrix $S \equiv BA$ to represent the rotation, the rotations can be written as

$$X' = SXS^{-1}, \quad U' = SUS^{-1}, \tag{4.115}$$

These show that vectors and bivectors transform identically under rotation.

These formulas can be expressed more concretely: let $\mathbf{l}$ be a unit vector along the desired axis of rotation – it is unfortunate that the symbol $\mathbf{l}$, for axis vector, and $\mathbf{1}$, for unit matrix, are so easily confused – L its associated matrix, and θ the desired angle of rotation. By Eq. (4.103) and Eq. (4.106), since the angle between unit vectors $\mathbf{a}$ and $\mathbf{b}$ is $\theta/2$, suppressing the identity matrix for brevity,

$$AB + BA = 2\mathbf{a}\cdot\mathbf{b} = 2\cos\frac{\theta}{2}, \quad AB - BA = 2iL\sin\frac{\theta}{2}. \tag{4.116}$$

Subtracting and adding, these yield

$$\begin{aligned} S = BA &= \cos\frac{\theta}{2} - iL\sin\frac{\theta}{2} = \exp\left(-i\frac{\theta}{2}\mathbf{l}\cdot\boldsymbol{\sigma}\right), \\ S^{-1} = AB &= \cos\frac{\theta}{2} + iL\sin\frac{\theta}{2} = \exp\left(i\frac{\theta}{2}\mathbf{l}\cdot\boldsymbol{\sigma}\right), \end{aligned} \tag{4.117}$$

which, with Eqs. (4.114) and (4.101) yields

$$X' = \left(\cos\frac{\theta}{2} - i\mathbf{l}\cdot\boldsymbol{\sigma}\sin\frac{\theta}{2}\right) X \left(\cos\frac{\theta}{2} + i\mathbf{l}\cdot\boldsymbol{\sigma}\sin\frac{\theta}{2}\right). \tag{4.118}$$

This is a very old formula, derived initially by Hamilton. Stated more compactly

$$\mathbf{x}' \cdot \boldsymbol{\sigma} = \exp\Big(-i\frac{\theta}{2}\mathbf{l} \cdot \boldsymbol{\sigma}\Big)(\mathbf{x} \cdot \boldsymbol{\sigma}) \exp\Big(i\frac{\theta}{2}\mathbf{l} \cdot \boldsymbol{\sigma}\Big). \tag{4.119}$$

A general, real, orthogonal, 3×3 rotation matrix has nine parameters, of which all but three are redundant. This representation of the same rotation has only one constraint – the components of $\mathbf{l}$ must make it a unit vector. The four elements of S are known as Cayley–Klein parameters. Equation (4.119), is somewhat coupled but, by Eq. (4.101), the third component x_3' is not, and it is not difficult to separate x_1' and x_2'. When that is done, if $\mathbf{x}$ is a reference location of a point in a rigid body, then the new location $\mathbf{x}'$ is expressed in terms of the Cayley–Klein parameters. Also formula Eq. (4.119) lends itself naturally to the "concatenation" of successive rotations. Note that the rotated vector $\mathbf{x}'$ can also be obtained from initial vector $\mathbf{x}$ and axis of rotation vector $\mathbf{l}$ using normal vector analysis; see Fig. 4.7;

$$\mathbf{x}' = (\mathbf{l} \cdot \mathbf{x})\mathbf{l} + \cos\theta\,\big((\mathbf{l} \times \mathbf{x}) \times \mathbf{l}\big) + \sin\theta\,(\mathbf{l} \times \mathbf{x}). \tag{4.120}$$

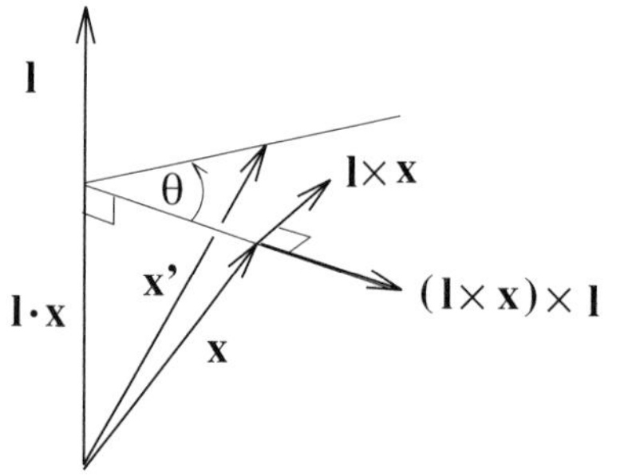

Fig. 4.7 Vector diagram illustrating Eq. (4.120) and giving the result of rotating vector $\mathbf{x}$ by angle θ around the axis $\mathbf{l}$. Except for the factor $|\mathbf{l} \times \mathbf{x}|$, which is the magnitude of the component of $\mathbf{x}$ orthogonal to $\mathbf{l}$, the vectors $\mathbf{l} \times \mathbf{x}$ and $(\mathbf{l} \times \mathbf{x}) \times \mathbf{l}$ serve as orthonormal basis vectors in the plane orthogonal to $\mathbf{l}$.

4.4.7 Operations on Spinors

The defining use of matrix A associated with vector $\mathbf{a}$ is to transform spinor ξ into ξ', its "reflection" in the plane defined by $\mathbf{a}$;

$$\xi' = A\xi. \tag{4.121}$$

It is necessary to show that this definition is consistent with our understanding of the geometry, including the association with the reflected isotropic vector associated with ξ. For special cases this was demonstrated in Problem 4.4.1. Also the spinor rotation is given by

$$\xi' = BA\xi = \exp\Big(-i\frac{\theta}{2}\mathbf{l}\cdot\boldsymbol{\sigma}\Big)\xi. \tag{4.122}$$

4.4.8
Real Euclidean Space

All of the results obtained so far apply to real or complex vectors, either as the components of points in space, or as the vectors associated with reflections or rotations. Now we restrict ourselves to real rotations and reflections in Euclidean (ordinary) geometry.

It has been seen previously that a real vector $\mathbf{x}$ is associated with a Hermitian reflection matrix X; $X^* = X^T$. The matrix U associated with a real bivector satisfies $U^* = -U^T$. Since a rotation is the product of two reflections, $S = BA$, it follows that

$$(S^*)^T = (A^*)^T(B^*)^T = AB = S^{-1}; \tag{4.123}$$

this is the condition for S to be *unitary*. Hence a 2×2 spinor-rotation matrix is unitary. This is the basis of the designation SU(2) for the 2×2 representation of spatial rotations.

Since a spinor is necessarily associated with an *isotropic* vector, and there is no such thing as a *real* isotropic vector, it is not possible to associate a spinor with a real vector. It is however possible to associate "tensor products" of spinors with real vectors. The mathematics required is equivalent to the "addition of angular momenta" mathematics of quantum mechanics.

4.4.9
Real Pseudo-Euclidean Space

In special relativity, taking axis 2 as the time axis to simplify the use of preceding formulas, the position of a (necessarily massless) particle traveling at the speed of light can satisfy

$$x_1^2 + x_3^2 = c^2t^2, \quad x_2 = ct, \quad x_1^2 - x_2^2 + x_3^2 = 0. \tag{4.124}$$

Replacing ix_2 by x_2 in the preceding formalism, and now requiring (x_1, x_2, x_3) to be real, the associated matrix

$$X = \begin{pmatrix} x_3 & x_1 - x_2 \\ x_1 + x_2 & -x_3 \end{pmatrix} \tag{4.125}$$

is real, and there is a spinor (ξ_0, ξ_1), also real, associated with $\mathbf{x}$;

$$x_1 = \xi_0^2 - \xi_1^2, \quad x_2 = \xi_0^2 + \xi_1^2, \quad x_3 = -2\xi_0\xi_1. \tag{4.126}$$

Bibliography

General References

1 E. Cartan, *The Theory of Spinors*, Dover, New York, 1981, p. 10.

2 V.I. Arnold, *Mathematical Methods of Classical Mechanics*, 2nd ed., Springer, New York, 1989.

5
Lagrange–Poincaré Description of Mechanics

5.1
The Poincaré Equation

Before starting on this new topic it is appropriate to review the Lagrangian approach to mechanics, for example as outlined in Chapter 1. We are then ready to apply our geometric ideas to mechanics proper. The plan is to introduce the "Poincaré equation" as an "improvement" upon the Lagrange equation. One aspect of this improvement is its close connection with symmetries and conservation laws. But these features will not be studied until Chapter 10.

A certain amount of circular reasoning creeps into physics naturally (and not necessarily unproductively) as follows. Suppose that by making a special assumption a certain difficult issue can be *finessed*. Then, by the simple expedient of defining "physics," or "fundamental physics," as being limited to systems satisfying the special assumption, one is relieved *by definition* of worrying further about the difficult issue. In the present context here is how it goes. Once one has found the generalized coordinates of a system, the Lagrangian method proceeds reliably and in a purely mechanical way, with no need to be troubled by annoying mathematical concepts such as tangent spaces. The stratagem then is to *define* mechanics to be the theory of systems for which generalized coordinates can be found and *presto*, one has a tidy, self-contained, and powerful tool – Lagrange's equations – for studying it. (It must be acknowledged that even if this approach is judged cowardly as applied mathematics, it may be "high principle" as physics – the principle being that to be Lagrangian is to be fundamental.)

Being unwilling to employ this stratagem, we will face up to the Poincaré equation, which is the tool of choice for studying systems that are Lagrangian except for not being describable by generalized coordinates. This, in turn, requires studying the geometric structure of mechanics. At that point it becomes almost an advantage that the Poincaré equation is novel, since it does not carry with it the baggage of less-than-general truth that already assimilated physics necessarily carries. It is also pedagogically attractive to investigate a brand new subject rather than simply to rehash Lagrangian mechanics.

Geometric Mechanics: Toward a Unification of Classical Physics. 2nd Edition. Richard Talman

ISBN: 978-3-527-40683-8

The greatest virtue of the Lagrange method is that it provides a foolproof scheme for obtaining correct equations of motion. With computers available to solve these equations, the mere writing down of a correct Lagrangian can almost be regarded as the solution of the problem. The Poincaré equation, valid in far less restrictive circumstances (commonly involving constraints or rotational motion) has the same virtue. Its extra terms compared to the Lagrange equation's, though formidable to evaluate by hand unless simplified by symmetry, can be calculated in a computer using symbolic algebra. The resulting differential equations of motion can then be solved numerically, as in the Lagrangian procedure. One might then say that the Poincaré equation is *better than* the Lagrange equation.

The Poincaré equation will be derived in two different ways, first using traditional elementary calculus and then using geometric methods. This is not really an extravagance since it is important to correlate old and new methods. Also the discussion can serve as further review of Lagrangian mechanics, since much of the derivation amounts to studying properties of the Lagrange equations.

There are (at least) two different ways of introducing the Lagrange equations themselves. Both methods start by assuming the configurations of the mechanical system under study are describable *uniquely* by generalized coordinates q^i. From there the quickest route is to postulate Hamilton's principle of least action and then apply the calculus of variations. Though this approach seems to be "black magic" or at least poorly motivated when first encountered, it has become so well established as now to be considered fundamental. This variational approach has the further advantage of exhibiting a remarkable "invariance" to choice of coordinates. The second method amounts to applying "brute force" to the equations given by Newton's second law to transform them into Lagrange's equations using nothing but calculus. This method is also weakly motivated since it is not *a priori* clear what one is looking for. Furthermore, once one has derived the Lagrange equations, one still has to derive their coordinate-invariance property. Having taken the variational approach in Chapter 1 we now take the brute force approach.

Before beginning, we call attention to two limitations of Lagrangian mechanics:

- Fundamental to the Lagrangian formalism are its generalized coordinates q^i and their corresponding velocities $\dot{q}^i$.[1] There are, however, cases in which naturally occurring velocities *cannot* be expressed as the time derivatives of generalized coordinates. Angular velocities are the most familiar example. Because of the noncommutativity of rotations, one

1) In this text, the quantity $\dot{q}^i \equiv dq^i/dt$ is always called "the velocity corresponding to q^i," even though, in some cases, this causes the physical dimensions of different velocity components to be different.

cannot define generalized coordinates whose derivatives are equal to angular velocities around fixed axes.

- In describing constrained motion it is always difficult, and usually impossible, to express the constraints analytically without the use of velocity coordinates. Such constraints are normally not "integrable". The coordinates are said to be nonholonomic and the Lagrange procedure is not directly applicable.

One tends to be not much concerned about the holonomic restriction. This may be partly due to the belief that the most fundamental forces are holonomic or, more likely, because traditional physics courses skip over the problem. It is not because nonholonomic systems are rare in nature. Trains, which are holonomic (at least on cog railways) are far less prevalent than automobiles, which are not.

By permitting "quasicoordinates" the Poincaré equation admits noncommuting coordinates and can be used to describe nonholonomic systems.

It is assumed the reader has already mastered the Lagrange equations, especially concerning the definition of generalized coordinates and generalized forces and the application of d'Alembert's principle to introduce force terms into the equations. We initially assign ourselves the task of changing variables in the Lagrange equations. This will illustrate some of the essential complications that Lagrange *finessed* when he invented his equations. To follow these calculations it is useful to "understand" the *tangent space* of possible instantaneous velocity vectors of the system or (more likely) to accept without protest some steps in the calculus that may seem a bit shady.

A remarkable feature of the Lagrange equations that has already been pointed out is that they maintain the same form when the generalized coordinates q^i are transformed. Since the Lagrangian also depends on velocity components $\dot{q}^i$, one is tempted to consider transformations that mix displacements and velocities. (One is accustomed to mixing displacements and momenta in Hamiltonian mechanics.) This is especially promising in cases where the kinetic energy can be expressed more simply in terms of "new" velocities, call them s^i, rather than in terms of the $\dot{q}^i$. In some cases, the simplest of which are Cartesian velocity components, such velocities can be "integrated". But this is not possible in general.

If two angular velocities vanish then the motion can be "integrated" allowing the system orientation to be specified by the remaining angle. One can attempt to define three global angles, one at a time, in this way. But if rotation occurs around more than one axis, since the order of application of rotations affects the final orientation, these angles would not satisfy the requirement that there be a on-to-one correspondence between generalized coordinates and system configurations. (By carefully specifying their order of application, the so-called "Euler angles" can circumvent this problem in Newtonian mechanics.)

The most familiar example exhibiting noncommutating variables is the rotational motion of an extended object. Let s^1, s^2, and s^3, be the three instantaneous angular velocities around three orthogonal axes Our purpose in this section is derive Lagrange-like equations that are expressed in terms of these "quasivelocities" while being consistent with the noncommutativity of their parent quasicoordinates.

Consider then a mechanical system described by generalized coordinates $q^1, q^2, \ldots, q^n$. To describe the system differently we introduce new quasivelocities $s^1, s^2, \ldots, s^n$, some or all of which differ from $\dot{q}^1, \dot{q}^2, \ldots, \dot{q}^n$. By definition the new velocities are to be invertable superpositions of the original generalized velocities. The transformations therefore have the form[2]

$$s^r = A^r{}_i(\mathbf{q})\dot{q}^i, \quad r = 1, 2, \ldots, n. \tag{5.1}$$

Typically the coefficients $A^r{}_i$ in this equation are functions of the coordinates, but they must not depend on velocities. If the number of quasivelocities is small it is convenient to give them individual symbols such as $(s^1, s^2, \ldots) \rightarrow (s, g, l, \ldots)$. In this case the transformation looks like

$$\begin{pmatrix} s \\ g \\ l \\ . \end{pmatrix} = \begin{pmatrix} \Sigma_1 & \Sigma_2 & \Sigma_3 & \Sigma_4 \\ \Gamma_1 & \Gamma_2 & \Gamma_3 & \Gamma_4 \\ \Lambda_1 & \Lambda_2 & \Lambda_3 & \Lambda_4 \\ . & . & . & . \end{pmatrix} \begin{pmatrix} \dot{q}^1 \\ \dot{q}^2 \\ \dot{q}^3 \\ \dot{q}^4 \end{pmatrix}. \tag{5.2}$$

This form has the advantage of mnemonically emphasizing the close connection between any particular quasivelocity, say g, with its corresponding row $(\Gamma_1, \Gamma_2, \Gamma_3, \Gamma_4)$.[3] As an example, for a single particle with coordinates x, y

2) The slight displacement to the right of the lower index is to facilitate mental matrix multiplication but otherwise has no significance. Also the order of the factors $A^r{}_i$ and $\dot{q}^i$ could be reversed without changing anything except the conventional representation of the equation by matrix multiplication.

3) There are too few letters in the English alphabet. It is conventional to give the coordinates of a mechanical system symbols that are Roman characters such as r, x, y, etc., and similarly, for velocities, v. To emphasize their ephemeral character we will use Greek symbols $\sigma, \gamma, \lambda, \ldots$, to stand for the "quasicoordinates" that are about to be introduced. Corresponding to these will be "quasivelocities" and to emphasize their real existence while preserving their ancestry, the quasivelocities will be symbolized by matching Roman letters $s, g, l, \ldots$. A further (temporary and self-imposed) "requirement" on these characters is that there be upper case Greek characters $\Sigma, \Gamma, \Lambda, \ldots$, available to "match" the quasivelocities (to serve as a mnemonic aid shortly.) The quantities s^1, s^2, s^3 being introduced here are quasivelocities *not* momenta. Probably because the most common quasivelocities are angular velocities, the symbol ω is commonly used in this context, but that symbol is already overworked, especially in the previous chapter. In any case, once general arguments have been made, a less restrictive notational scheme will have to be tolerated.

and z, one could try the definition, $\mathbf{s} = \mathbf{x} \times \dot{\mathbf{x}}$, which can be written

$$\begin{pmatrix} s^x \\ s^y \\ s^z \end{pmatrix} \overset{\text{no good}}{=} \begin{pmatrix} 0 & -z & y \\ z & 0 & -x \\ -y & x & 0 \end{pmatrix} \begin{pmatrix} \dot{x} \\ \dot{y} \\ \dot{z} \end{pmatrix}. \tag{5.3}$$

This choice, though linear in the velocities as required by (5.1), is illegal because the determinant vanishes identically, meaning the relation cannot be inverted. Note that the vanishing does not occur at just a single point in configuration space (which would be tolerable) but rather is identically true for all (x, y, z). (See Problems 5.1.2 and 5.1.3). This failure is unfortunate since the transformation (5.3) seems to be otherwise promising. Its purpose would have been to write the equations of motion in terms of the *angular momentum* variables rather than the linear velocities $(\dot{x}, \dot{y}, \dot{z})$. The characteristics of this transformation will be discussed later to illustrate the concept of *foliation*. To have a sample to visualize, we could try instead

$$\begin{pmatrix} v^x \\ s^x \\ s^y \end{pmatrix} = \begin{pmatrix} 1 & 0 & 0 \\ 0 & -z & y \\ z & 0 & -x \end{pmatrix} \begin{pmatrix} \dot{x} \\ \dot{y} \\ \dot{z} \end{pmatrix}, \tag{5.4}$$

which is invertable.

For s^r defined as in Eq. (5.1) it may happen that coordinates σ^r can be found such that[4]

$$\frac{d\sigma^r}{dt} \overset{q}{=} s^r, \quad r = 1, 2, \ldots, n, \tag{5.5}$$

but this is the exception rather than the rule. (The σ^r would have to be found by "integrating" Eqs. (5.1), which may not be possible, even in principle. The concept of a function being well defined as the derivative of an undefined function is not unusual – an indefinite integral of a definite function is undefined to the extent of admitting an arbitrary constant of integration.) Nevertheless, for the time being, we will pretend that "quasicoordinates" exist, intending to later undo any damage that this incurs. In any case Eq. (5.1) can be written in differential form

$$d\sigma^r = A^r{}_i(\mathbf{q})dq^i \quad r = 1, 2, \ldots, n. \tag{5.6}$$

That this is a differential of the "old fashioned" calculus variety is indicated by the absence of boldface type and overhead tildes.

The concept of "tangent space" – a linear vector space containing sums of and scalar multiples of tangent vectors – is central to the present derivation. The tangent space is a mathematical device for making it legitimate to regard a

4) Recall that the symbol $\overset{q}{=}$ means "qualified" equality.

quantity like dx/dt as not just a formal symbol but as a ratio of two quantities dx and dt that are not even necessarily "small". A physicist is satisfied with the concept of "instantaneous velocity" and does not insist on distinguishing it from an approximation to it that is obtained by taking dt small enough that the ratio dx/dt is a good approximation. For now the tangent space will be considered to constitute a "linearized" approximation for expressing small deviations of the system from an instantaneous configuration.

The following partial derivatives can be derived from Eq. (5.6):

$$\frac{\partial \sigma^r}{\partial q^k} = A^r{}_k, \quad \frac{\partial q^r}{\partial \sigma^k} = (\mathbf{A}^{-1})^r{}_k \equiv B^r{}_k \tag{5.7}$$

(In the final step, the definition $\mathbf{B} \equiv \mathbf{A}^{-1}$ has been introduced solely to reduce clutter in subsequent formulas.) These are the "Jacobean matrices" for the coordinate transformations of Eq. (5.6). The invertability requirement can be stated as a nonvanishing requirement on the determinant of the matrix $A^r{}_k$. We assume then Eq. (5.6) can be inverted;

$$\dot{q}^k = B^k{}_i s^i. \tag{5.8}$$

(It is important to remember that the matrices $\mathbf{A}$ and $\mathbf{B}$ depend on $\mathbf{q}$.)

In mechanics, the most important differential is $dW = Fdx$, the work done by force F acting through displacement dx. For a system described by generalized coordinates q^i this generalizes to

$$dW = Q_i dq^i, \tag{5.9}$$

where the Q_i are said to be "generalized forces". The discussion of contravariant and covariant vectors in Section 3.2, suggests strongly that the Q_i may be expected to be covariant and the notation of (5.9) anticipates that this will turn out to be the case.[5] When expressed in terms of the quasicoordinates, dW is therefore given by

$$dW = S_i d\sigma^i, \quad \text{where} \quad S_i = Q_k B^k{}_i. \tag{5.10}$$

Here the generalized forces S_i have been obtained from the Q_i the same way that generalized forces are always obtained in mechanics. Geometrically (as discussed in Chapter 2) Eq. (5.10) provides a mechanism for counting contours (represented by covariant vector Q_i or S_i) that are crossed by the arrow represented by the dq^i or the $d\sigma^i$.

5) The summation in Eq. (5.9) may be more complicated than it appears. It may include sums over particles or independent systems. Such sums should not be implicitly included in the summation convention but, for brevity, we let it pass.

At this point the "shady" steps mentioned above have to be faced. Regarding the velocities $s^r(\mathbf{q}, \dot{\mathbf{q}})$ as depending on $\mathbf{q}$ and *independently* and *linearly* on $\dot{\mathbf{q}}$ one has

$$\frac{\partial s^r}{\partial q^k} = \frac{\partial A^r{}_k}{\partial q^i}\dot{q}^i; \tag{5.11}$$

the assumed linearity in the $\dot{q}^i$ has made this step simple. Recall that the meaning of partial differentiation is only unambiguous if the precise functional dependence is specified. In this case the implication of $\partial s^r/\partial q^k$ is that all the $\dot{q}^k$ and all the q^k except q^i are being held constant. (Many people have difficulty seeing why it makes sense for q^i to vary and $\dot{q}^i$ to *not* vary. This is at least partly due to the notational ambiguity between the interpretation of q^i as where the system *is* and where it *could be*. Here the latter interpretation is intended, and the same interpretation applies to the velocity components.) Anyway, Lagrange thought it made sense, and everyone since then has either come to their own terms with it or taken their teacher's word for it. As mentioned before, the accepted mathematical procedure for legitimizing this procedure is to introduce "tangent planes" at every location. Displacement in any tangent plane is independent both of displacements in the original space and displacements in any other tangent space.

Once this concept has been accepted, it follows immediately that

$$\frac{\partial s^r}{\partial \dot{q}^i} = A^r{}_i. \tag{5.12}$$

This maneuver will be referred to as "pure tangent plane algebra". Because s^r stands for $d\sigma^r/dt$ and $\dot{q}^i$ stands for dq^i/dt and because $d\sigma^r$ and dq^i reside in the tangent space it is legitimate (in spite of one's possible recollections from introductory calculus) to divide both numerator and denominator in the first half of Eq. (5.7) by dt. This yields Eq. (5.12).

In deriving the Lagrange equations from Newton's equations the only tricky part is more or less equivalent to deriving Eq. (5.12). Since the Lagrange equations have already been derived (from purely geometric considerations) in Section 3.2, we skip this derivation and directly express the Lagrange equations as the equality of two way of evaluating the work during arbitrary displacement δq^k;

$$\left(\frac{d}{dt}\left(\frac{\partial T}{\partial \dot{q}^k}\right) - \frac{\partial T}{\partial q^k}\right)\delta q^k = Q_k \delta q^k. \tag{5.13}$$

(There is no significance to the fact that δq^k is used instead of, say, dq^k. As in Chapter 2 we wish only to preserve flexibility for later assignment of deviations. Later the δq^k will be specialized for our own convenience.) We now wish to transform these equations into the new quasicoordinates.

Since the Lagrange equations are based on the expression for the kinetic energy of the system, the first thing to do is to re-express T. The function expressing kinetic energy in terms of the new coordinates will be called $\overline{T}$;

$$T(\mathbf{q},\dot{\mathbf{q}},t)=\overline{T}(\mathbf{q},\mathbf{s},t). \tag{5.14}$$

If this was a true coordinate transformation then it would be possible for the first argument of $\overline{T}$ to be $\boldsymbol{\sigma}$. But since the very existence of coordinates $\boldsymbol{\sigma}$ cannot be assumed, a hybrid functional dependence on new velocities and old coordinates is all we can count on. What will make this ultimately tolerable is that only *derivatives* of $\overline{T}$ will survive to the final formula.

The terms in Eq. (5.13) can be worked on one at a time, using Eqs. (5.7) and (5.11), to obtain

$$\begin{aligned}
\frac{\partial T}{\partial \dot{q}^k}&=\frac{\partial \overline{T}}{\partial s^r}\frac{\partial \dot{\sigma}^r}{\partial \dot{q}^k}=\frac{\partial \overline{T}}{\partial s^r}A^r{}_k,\\
\frac{d}{dt}\left(\frac{\partial T}{\partial \dot{q}^k}\right)&=\frac{d}{dt}\left(\frac{\partial \overline{T}}{\partial s^r}\right)A^r{}_k+\frac{\partial \overline{T}}{\partial s^r}\frac{d}{dt}\left(A^r{}_k\right)=\frac{d}{dt}\left(\frac{\partial \overline{T}}{\partial s^r}\right)A^r{}_k+\frac{\partial \overline{T}}{\partial s^r}\frac{\partial A^r{}_k}{\partial q^i}\dot{q}^i,
\end{aligned} \tag{5.15}$$

$$\frac{\partial T}{\partial q^k}=\frac{\partial \overline{T}}{\partial \sigma^i}\frac{\partial \sigma^i}{\partial q^k}+\frac{\partial \overline{T}}{\partial s^i}\frac{\partial s^i}{\partial q^k}=\frac{\partial \overline{T}}{\partial \sigma^i}\frac{\partial \sigma^i}{\partial q^k}+\frac{\partial \overline{T}}{\partial s^r}\frac{\partial A^r{}_i}{\partial q^k}\dot{q}^i.$$

The strategy so far, as well as trying to eliminate the q and $\dot{q}$ variables, has been to replace $\dot{\sigma}^i$ by s^i wherever possible. The $\dot{q}^i$ factors remaining can be eliminated using Eq. (5.8). Collecting terms, the left-hand side of Eq. (5.13) contains the following three terms:

$$\begin{aligned}
&\frac{d}{dt}\left(\frac{\partial \overline{T}}{\partial s^i}\right)\delta\sigma^i,\\
&\frac{\partial \overline{T}}{\partial s^r}\left(\frac{\partial A^r{}_k}{\partial q^j}-\frac{\partial A^r{}_j}{\partial q^k}\right)B^j{}_l s^l B^k{}_i\delta\sigma^i\equiv -c^r{}_{li}(\mathbf{q})s^l\frac{\partial \overline{T}}{\partial s^r}\delta\sigma^i,\\
&-\frac{\partial \overline{T}}{\partial \sigma^r}A^r{}_kB^k{}_i\delta\sigma^i=-\frac{\partial \overline{T}}{\partial \sigma^i}\delta\sigma^i.
\end{aligned} \tag{5.16}$$

To abbreviate the writing of the second of these equations, the following coefficients have been introduced:

$$c^r{}_{il}\equiv B^k{}_iB^j{}_l\left(\frac{\partial A^r{}_k}{\partial q^j}-\frac{\partial A^r{}_j}{\partial q^k}\right). \tag{5.17}$$

In anticipation of later developments under special circumstances, these coefficient will be referred to as "structure constants" but, for the time being they are just the abbreviations shown. Since the differentials $\delta\sigma^i$ are arbitrary, they can be replaced by Kronecker δ's. As a result, the Lagrange equations have been transformed into

$$\frac{d}{dt}\left(\frac{\partial \overline{T}}{\partial s^i}\right) - c^r{}_{li}(\mathbf{q})s^l\frac{\partial \overline{T}}{\partial s^r} - \frac{\partial \overline{T}}{\partial \sigma^i} = S_i. \tag{5.18}$$

These are the Poincaré equations. Only the central term on the left-hand side makes the equations look significantly different from the Lagrange equations. It is also true though that, unlike the n Lagrange equations which are second order in time derivatives, these n equations are first order in time derivatives. The defining equations (5.1) provide n more equations, making $2n$ in all. In this regard the Poincaré procedure resembles the transition from Lagrange equations to Hamilton's equations. In spite of this, I consider it appropriate to regard the Poincaré equations as only a modest generalization of the Lagrange equations. No momentum variables have been defined and no phase space introduced.

Apart from the fact that these are complicated equations, an essential complication is that $\overline{T}$ and the coefficients $c^r{}_{li}$ are *a priori* explicitly known only as functions of the original q variables. If the equations were being solved numerically then, at each time step, once the s variables have been updated, the corresponding q's have to be calculated, and from them the $c^r{}_{li}$ coefficients calculated. To avoid this complication it would be desirable to have the $c^r{}_{li}$ coefficients expressed in terms of the σ variables. But this may be impossible, which brings us back to the issue that has been put off so far – what to do when the σ variables do not exist.

Though the quasicoordinates σ^i do not necessarily exist, the quasivelocities certainly do – they are given by Eq. (5.1). Our task then is to evaluate terms in the Poincaré equation that appear to depend on the σ^i in terms of only the s^i. Actually we have already done this once without mentioning it in Eq. (5.10) when we calculated the quasiforces S_i. Because this calculation was "local," it depended only on differentials $d\sigma^i$ which, we have said before, are superpositions of the s^i in linearized approximation. That was enough to relate the S_i's to the Q_i's. Essentially equivalent reasoning allows us to calculate the derivatives $\partial\overline{T}/\partial\sigma^i$ that appear in the Poincaré equation;

$$\frac{\partial \overline{T}}{\partial \sigma^j} = \frac{\partial T}{\partial q^i}\frac{\partial q^i}{\partial \sigma^j} = \frac{\partial T}{\partial q^i}\frac{\partial \dot{q}^i}{\partial s^j} = \frac{\partial T}{\partial q^i}B^i{}_j. \tag{5.19}$$

Earlier we deemphasized the importance of obtaining the generalized forces from a potential energy function U. But if this *is* done, and Lagrangian $L = T - V$ is introduced to take advantage of the simplification then, to complete the transformation to quasicoordinates, we have to introduce a new Lagrangian $\overline{L}$ appropriate to the new coordinates. The formula is

$$\overline{L}(\mathbf{q}, \mathbf{s}, t) = \overline{T}(\mathbf{q}, \mathbf{s}, t) - U(\mathbf{q}), \tag{5.20}$$

where $\overline{T}(\mathbf{q}, \mathbf{s}, t)$ is given by Eq. (5.14). The "force" terms of the Poincaré equations then follow as in Eq. (5.19).

Cartan's witticism that tensor formulas suffer from a *"débauche d'indices"* is certainly born out by formula (5.17) for $c^r{}_{li}$. Expressing it in terms of matrix multiplications can make this formula appear less formidable in preparation for its practical evaluation. After doing Problem 5.1.1 you will have shown that $c^r{}_{li}$ is antisymmetric in its lower indices. Furthermore the upper index r is "free" on both sides of the defining equation for $c^r{}_{li}$. To illustrate this last point suppose that, as in Eqs. (5.2), quasivelocities are symbolized by $s^1 = s$, $s^2 = g, \ldots$. Then definitions $A^1{}_i \equiv \Sigma_i$ and $A^2{}_i \equiv \Gamma_i$ correlate the "rows" of "matrix" $\mathbf{A}$ with the "output variables" s and g in a mnemonically useful way (because upper case Greek (Σ and Γ) and lower case Roman (s and g) symbols form natural pairs). An index has been suppressed in the bargain. With this notation, the defining equations (5.17) become

$$c^{(s)}_{\ il} = (B^T)_l{}^j \left(\frac{\partial \Sigma_k}{\partial q^j} - \frac{\partial \Sigma_j}{\partial q^k} \right) B^k{}_i, \quad c^{(g)}_{\ il} = (B^T)_l{}^j \left(\frac{\partial \Gamma_k}{\partial q^j} - \frac{\partial \Gamma_j}{\partial q^k} \right) B^k{}_i, \quad etc. \quad (5.21)$$

The indices have been manipulated to allow these equations to be processed by matrix multiplication. The order of factors has been changed and one matrix has been transposed in order to switch the order of indices. Also the superscripts (s) and (g) are no longer running indices – they identify the particular quasivelocities previously known as s^1 and s^2. [6] These equations are intended to be reminiscent of Eq. (2.79). They show, for example, that the $c^{(s)}_{\ il}$ elements and the $\partial\Sigma_k/\partial q^j$-$\partial\Sigma_j/\partial q^k$ elements are coordinates of the same two-index tensor-like object in the velocity and the quasi-velocity bases.

Example 5.1.1. *A simple pendulum, bob mass $m = 1$, hanging from the origin at the end of a light rod of length $\ell = 1$ swings in the x, z plane, in the presence of gravity $g = 1$. Let $q^1 \equiv \theta$ define the pendulum angle relative to the z-axis (which is vertical, positive up.) Define quasivelocity $s = \cos\theta\ \dot{\theta}$. Write the Poincaré equation for s. Is there a coordinate σ for which $s = d\sigma/dt$ in this case? Will this result always be true in the $n = 1$ case?*

The equation $d\sigma/dt = \cos\theta\ d\theta/dt$ by which σ is to be found, transforms immediately into $d\sigma/d\theta = \cos\theta$ which, neglecting a constant of integration, yields $\sigma = \sin\theta$. Since σ exists, all the coefficients (5.17) vanish. (In the $n = 1$ case this will clearly always be true.) The potential energy function can be chosen as $U(\theta) = \cos\theta$; here a possible constant contribution to U has been dropped since it would not contribute to the Poincaré equation anyway. The kinetic energy is $T(\dot{\theta}) = \dot{\theta}^2/2$ and

6) Though they have lower indices, the elements Σ_i or Γ_i are not automatically the covariant components of a tensor – recall they are only the row elements of an arbitrary, position dependent matrix. But in their role of relating two tangent plane coordinate systems they will be subject to important invariance considerations.

transformation to quasivelocities yields

$$\overline{T}(\theta, s) = \frac{1}{2}\frac{s^2}{\cos^2\theta}, \quad \overline{L}(\theta, s) = \frac{1}{2}\frac{s^2}{\cos^2\theta} - \cos\theta. \tag{5.22}$$

The derivatives needed are

$$\frac{d}{dt}\frac{\partial \overline{L}}{\partial s} = \frac{\dot{s}}{\cos^2\theta} + \frac{2s^2\sin\theta}{\cos^4\theta}, \quad \frac{\partial \overline{L}}{\partial \sigma} = \left(\frac{s^2\sin\theta}{\cos^3\theta} + \sin\theta\right)\frac{1}{\cos\theta}. \tag{5.23}$$

These are already complicated enough to make it clear that the transformation was ill-advised but, since this is just an example, we persevere. The Poincaré equation (along with the quasivelocity defining equation) is

$$\dot{s} = -\frac{s^2\sin\theta}{\cos^2} - \sin\theta\cos\theta, \quad \dot{\theta} = \frac{s}{\cos\theta}. \tag{5.24}$$

These are easily shown to be equivalent to the well-known equation $\ddot{\theta} = -\sin\theta$.

Example 5.1.2. *Suppose the pendulum in the previous example is a "spherical pendulum," free to swing out of the plane assumed so far. Letting* ϕ *be an azimuthal angle around the z-axis, define quasivelocities by*

$$\begin{pmatrix} \dot{\phi}^x \\ \dot{\phi}^y \\ \dot{\phi}^z \end{pmatrix} \equiv \begin{pmatrix} s^x \\ s^y \\ s^z \end{pmatrix} = \begin{pmatrix} -\sin\phi\,\dot{\theta} \\ \cos\phi\,\dot{\theta} \\ \dot{\phi} \end{pmatrix}. \tag{5.25}$$

As mentioned before, only two of these (any two) are independent. Let us choose s^x *and* s^z. *The matrix A and its inverse are then*

$$\mathbf{A} = \begin{pmatrix} -\sin\phi & 0 \\ 0 & 1 \end{pmatrix}, \quad \mathbf{B} \equiv \mathbf{A}^{-1} = \begin{pmatrix} -1/\sin\phi & 0 \\ 0 & 1 \end{pmatrix}. \tag{5.26}$$

To correlate with the numbering system used above let $\theta \to 1, \phi \to 2$. *For* $r = 1$, *the Poincaré coefficients are obtained from the upper row of* $\mathbf{A}$;

$$\begin{aligned} \mathbf{B}^T\left(\begin{pmatrix} \frac{\partial A_{11}}{\partial\theta} & \frac{\partial A_{12}}{\partial\theta} \\ \frac{\partial A_{11}}{\partial\phi} & \frac{\partial A_{12}}{\partial\phi} \end{pmatrix} - \begin{pmatrix} \frac{\partial A_{11}}{\partial\theta} & \frac{\partial A_{12}}{\partial\theta} \\ \frac{\partial A_{11}}{\partial\phi} & \frac{\partial A_{12}}{\partial\phi} \end{pmatrix}^T\right)\mathbf{B} &= \begin{pmatrix} 0 & -\cot\phi \\ \cot\phi & 0 \end{pmatrix} \\ &= \begin{pmatrix} c^1_{11} & c^1_{12} \\ c^1_{21} & c^1_{22} \end{pmatrix}. \end{aligned} \tag{5.27}$$

5.1.1
Some Features of the Poincaré Equations

In general the factors $c^r_{\;li}(\mathbf{q})$ are functions of position $\mathbf{q}$ and Poincaré equations (5.18) seem likely to be more complicated than the Lagrange equations in cases that are compatible with the existence of Lagrange equations. But there is immediate simplification in important special cases to be considered now.

Suppose that generalized coordinates $\sigma^i(\mathbf{q})$ do, in fact, exist such that $s^i = d\sigma^i/dt$ as in Eq. (5.5). Because the order of taking partial derivatives does not matter, differentiating Eq. (5.7) yields

$$\frac{\partial^2 \sigma^r}{\partial q^k \partial q^i} = \frac{\partial A^r{}_k}{\partial q^i} = \frac{\partial A^r{}_i}{\partial q^k}. \tag{5.28}$$

From Eq. (5.16) it then follows that the array of factors $c^r{}_{li}$ all vanish. In this case the Poincaré equations become simply the Lagrange equations in the new generalized coordinates σ^i. This means that the analysis up to this point amounts to having been an explicit exhibition of the form invariance of the Lagrange equations under a coordinate transformation. Also derived has been a necessary condition for the integrability of a conjectured set of quasivelocities: namely the vanishing of the $c^r{}_{li}$ elements.

Another important simplification occurs when the partial derivatives $\partial\overline{T}/\partial\sigma^i$ vanish. This possibility is closely connected with symmetry. If the coordinate σ^i can be regarded as fixing the gross configuration or orientation or location of the system and the kinetic energy is independent of configuration or orientation of location respectively, then $\partial\overline{T}/\partial\sigma^i = 0$. If the external forces are similarly independent of σ^i, the generalized force factors S_i also vanish. The best known case in which both of these simple features occur is in the force-free rotation of a rigid body; if s is an angular velocity then the kinetic energy depends on s but not on the corresponding angle σ (which would specify spatial orientation of the system) and there is no torque about the axis so the corresponding generalized forces S also vanishes. This example will be pursued later. In the traditional Lagrangian vocabulary the coordinate σ^i is then be said to be "ignorable".

In general the factors $c^r{}_{li}(\mathbf{q})$, defined in Eq. (5.17) depend on position $\mathbf{q}$, but it is the case where these factors are constant, independent of $\mathbf{q}$, that Poincaré had particularly in mind when he first wrote these equations. In this case, it will be shown shortly that the transformations resulting from changing the quasicoordinates form a "Lie group" for which the $c^r{}_{li}$ are "structure constants" that fix the commutation of infinitesimal group transformations. Examples illustrating this case are the subject of a series of problems to appear below.

5.1.2
Invariance of the Poincaré Equation

The newly introduced coefficients $c^r{}_{li}$ have a certain "intrinsic" coordinate independence. To show this they will be expressed in terms of the bilinear covariant introduced in Chapter 2. To accomplish this start by noting that quasicoordinate differentials can be written in the form

$$\sigma[d] = \Sigma_i \, dq^i, \quad \gamma[d] = \Gamma_i \, dq^i, \quad \lambda[d] = \Lambda_i \, dq^i, \quad \text{etc.,} \tag{5.29}$$

where, as in Section 2.2.2, the "argument" d indicates that the coordinate deviation is dq^i (rather than, say, δq^i). We wish to work on the σ^1=σ, σ^2=γ, σ^3=λ, ... quasivelocities one at a time. Let us pick the γ case (rather than, say, the σ case which is too confusing because of the clash of symbols.) The (position dependent) coefficients Γ_i are, on the one hand, elements in the row corresponding to g of matrix $A^r{}_i(\mathbf{q})$ and, on the other hand, coefficients of the form $\gamma[d]$. Introducing a second coordinate deviation δq^i, by Eq. (3.67) the bilinear covariant is

$$d\gamma[\delta] - \delta\gamma[d] = \frac{1}{2}\left(\frac{\partial \Gamma_k}{\partial q^j} - \frac{\partial \Gamma_j}{\partial q^k}\right)(\delta q^k \, dq^j - \delta q^j \, dq^k). \tag{5.30}$$

In Section 3.3 the quantity on the left-hand side was shown to have an *invariant* significance and in Chapter 2 the coefficients of an invariant form linear in the contravariant components of a vector were identified as the covariant components of a two-index tensor. The coefficients $\partial\Gamma_k/\partial q^j$-$\partial\Gamma_j/\partial q^k$ can therefore be regarded as the *covariant* components of an antisymmetric tensor. We also recognize the coefficients $\partial\Gamma_k/\partial q^j$-$\partial\Gamma_j/\partial q^k$ as having appeared in Eqs. (5.21). They were shown there to be related by coordinate transformation to the $c^{(g)}_{\;li}$ elements. These structure elements are therefore covariant tensors, at least in their two lower indices.

The Γ_k elements were chosen arbitrarily in the first place, but elements of the antisymmetric tensor formed from it, after transformation from the velocity basis to the quasivelocity basis produce the $c^{(g)}_{\;li}$ structure elements. The same argument can be applied sequentially to each of the σ^1=σ, σ^2=γ, σ^3=λ, ... cases.

We have seen therefore that the extra terms in the Poincaré equation (over and above those that match terms in the Lagrange equation) exhibit an invariance to coordinate transformation which is much like the invariance to coordinate transformation of the Lagrange equation itself. The invariance is more difficult to express, but that should perhaps have been expected since the class of applicable transformations has been vastly expanded.

5.1.3 Translation into the Language of Forms and Vector Fields

Using notation introduced first in Section 2.1, the defining equations (5.1) or (5.2) can be expressed as

$$\begin{aligned}\widetilde{d\sigma} &= \Sigma_1 \widetilde{\mathbf{dq}}^1 + \Sigma_2 \widetilde{\mathbf{dq}}^2 + \Sigma_3 \widetilde{\mathbf{dq}}^3 + \Sigma_4 \widetilde{\mathbf{dq}}^4 \\ \widetilde{d\gamma} &= \Gamma_1 \widetilde{\mathbf{dq}}^1 + \Gamma_2 \widetilde{\mathbf{dq}}^2 + \Gamma_3 \widetilde{\mathbf{dq}}^3 + \Gamma_4 \widetilde{\mathbf{dq}}^4 \\ & \quad etc.\end{aligned} \tag{5.31}$$

A differential $\widetilde{\mathbf{dq}}^i$ on the right-hand side is a form that, when operating on a displacement vector, projects out the change in coordinate q^i. Similarly $\widetilde{d\sigma}$ projects out the change in quasicoordinate σ. Note that at this stage the equations relate local coordinate systems and have no content whatsoever that depends on or describes actual motion of the system. Proceeding, as in Section 2.3.5, by labeling dual basis vectors as $\widetilde{\mathbf{e}}^i = \widetilde{\mathbf{dq}}^i$ in the original system, and $\widetilde{\mathbf{e}}^{(s)} = \widetilde{d\sigma}$, $\widetilde{\mathbf{e}}^{(g)} = \widetilde{d\gamma}$, etc., in the new system, Eq. (5.31) is equivalently written as

$$\begin{pmatrix} \widetilde{\mathbf{e}}^{(s)} \\ \widetilde{\mathbf{e}}^{(g)} \\ \cdot \\ \cdot \end{pmatrix} = \begin{pmatrix} \Sigma_1 & \Sigma_2 & \Sigma_3 & \Sigma_4 \\ \Gamma_1 & \Gamma_2 & \Gamma_3 & \Gamma_4 \\ \cdot & \cdot & \cdot & \cdot \\ \cdot & \cdot & \cdot & \cdot \end{pmatrix} \begin{pmatrix} \widetilde{\mathbf{e}}^1 \\ \widetilde{\mathbf{e}}^2 \\ \widetilde{\mathbf{e}}^3 \\ \widetilde{\mathbf{e}}^4 \end{pmatrix} \equiv \mathbf{\Sigma} \begin{pmatrix} \widetilde{\mathbf{e}}^1 \\ \widetilde{\mathbf{e}}^2 \\ \widetilde{\mathbf{e}}^3 \\ \widetilde{\mathbf{e}}^4 \end{pmatrix}. \tag{5.32}$$

As defined in Section 2.3, the $\widetilde{\mathbf{e}}^i$ are natural basis forms in the space dual to the space with basis vectors $\mathbf{e}_1$ along the coordinate directions q^i. The $\widetilde{\mathbf{e}}^{(s)}$, $\widetilde{\mathbf{e}}^{(g)}$, etc., are similar basis forms for the quasicoordinates. It was shown in that section that the basis vectors themselves are then related by

$$\begin{aligned}(\mathbf{e}_s \quad \mathbf{e}_g \quad . \quad .) &\equiv (\boldsymbol{\partial}/\boldsymbol{\partial}\sigma \quad \boldsymbol{\partial}/\boldsymbol{\partial}\gamma \quad . \quad .) \\ &= (\mathbf{e}_1 \quad \mathbf{e}_2 \quad \mathbf{e}_3 \quad \mathbf{e}_4) \begin{pmatrix} S_1 & S_2 & S_3 & S_4 \\ G_1 & G_2 & G_3 & G_4 \\ \cdot & \cdot & \cdot & \cdot \\ \cdot & \cdot & \cdot & \cdot \end{pmatrix} \\ &\equiv (\boldsymbol{\partial}/\boldsymbol{\partial}q^1 \quad \boldsymbol{\partial}/\boldsymbol{\partial}q^2 \quad . \quad .)\,\mathbf{S},\end{aligned} \tag{5.33}$$

where the matrix has been expressed as $\mathbf{S}$. The matrix elements are implicitly dependent on $\mathbf{q}$. The velocity vector is expressed in the two bases by

$$\mathbf{v} = \dot{\mathbf{q}} = s\mathbf{e}_s + g\mathbf{e}_g + \cdots = \dot{q}^1\mathbf{e}_1 + \dot{q}^2\mathbf{e}_2 + \cdots. \tag{5.34}$$

Repeating Eqs. (5.2) with the notation of Eq. (5.32) the velocity coordinates in the two frames are related by

$$s^j = \Sigma^j{}_i \dot{q}^i, \quad \text{and} \quad \dot{q}^i = S^i{}_j s^j. \tag{5.35}$$

The transformation formulas for the components of forms have been given in Eq. (2.68). As in that section, the matrices that have been introduced here are related by $\mathbf{\Sigma} = \mathbf{S}^{-1}$.

5.1.4
Example: Free Motion of a Rigid Body with One Point Fixed

The orientation of a rigid body with one point fixed can be described by Euler angles (ϕ, θ, ψ). (See Fig. 5.1.) The problem of noncommuting rotations mentioned above is overcome in this definition by specifying the order, first by angle ϕ about the z-axis, next by θ about the new x-axis, then by ψ about the new z-axis. The unprimed axes in Fig. 5.1 can be regarded as fixed in space, the triply primed axes are fixed in the rigid body, with origin also at the fixed point. The fact that the origin remains fixed in space can be either due to the fact it is held there by a frictionless bearing or because the body is free, the origin is its center of mass, and the origin of space coordinates is taken at that point.

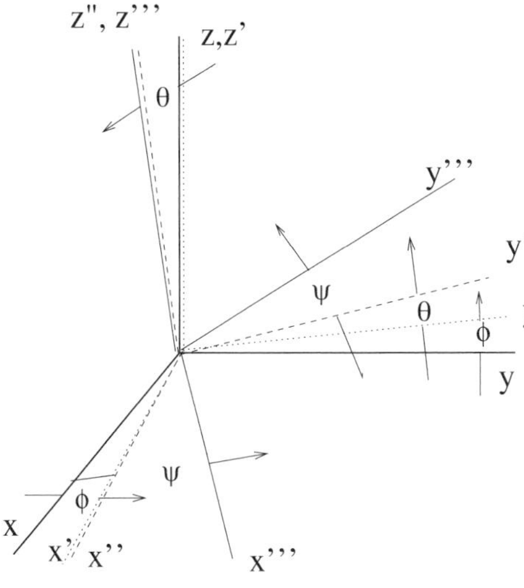

Fig. 5.1 Definition of Euler angles (ϕ, θ, ψ). Initial coordinate axes are (x, y, z), final axes are (x''', y''', z'''), and the order of intermediate frames is given by the number of primes. The initial axes are usually regarded as fixed in space, the final ones as fixed in the rigid body with one point fixed at the origin.

At any instant the rigid body is rotating around some axis, and the angular velocity vector points along that axis, with length equal to the speed of angular rotation around the axis. This vector can be described by coordinates referred to the fixed-in-space ("laboratory") (x, y, z) axes or to the fixed-in-body (x''', y''', z''') axes. For the former choice evaluating the kinetic energy is

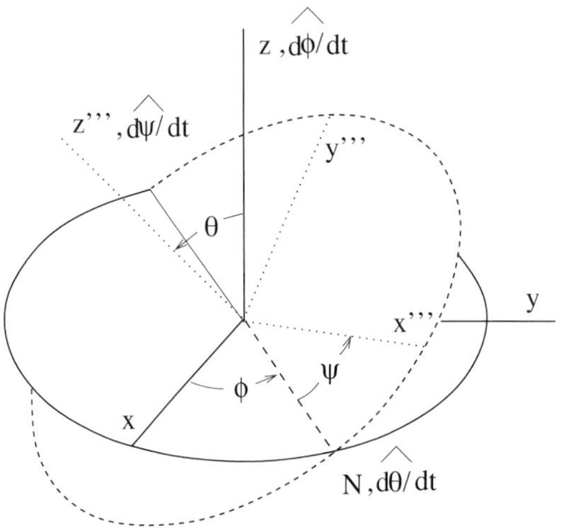

Fig. 5.2 A less cluttered than Fig. 5.1 illustration of Euler angles, showing the rotation axes $\widehat{d\phi/dt}$, $\widehat{d\theta/dt}$, $\widehat{d\psi/dt}$, for angular rotations with one Euler angle varying and the other two held fixed.

complicated because the spatial mass distribution varies with time and with it the moment of inertia tensor. Hence we will take the body angular velocities as quasivelocities, calling them $(\omega^1, \omega^2, \omega^3)$. To calculate them one can treat the Euler angular velocities $\dot\phi$, $\dot\theta$, and $\dot\psi$ one by one. Taking advantage of the fact that they are vectors directed along known axes their components along the body axes can be determined. Finally the components can be superimposed. Figure 5.2 shows the axes that correspond to varying Euler angles one at a time. The transformation to quasivelocities is illustrated in the following series of problems:

Problem 5.1.1. *Show that the factors* $c^r{}_{li}(\mathbf{q}\ )$*, defined in Eq. (5.17) are antisymmetric in their lower indices. For* $n = 3$ *how many independent components* $c^r{}_{li}$ *are there?*

Problem 5.1.2. *Define three vector fields (or operators)*

$$\begin{aligned}
\mathbf{R}_x &= \frac{\partial}{\partial\phi^x} = y\frac{\partial}{\partial z} - z\frac{\partial}{\partial x},\\
\mathbf{R}_y &= \frac{\partial}{\partial\phi^y} = z\frac{\partial}{\partial x} - x\frac{\partial}{\partial z},\\
\mathbf{R}_z &= \frac{\partial}{\partial\phi^z} = x\frac{\partial}{\partial y} - y\frac{\partial}{\partial x},
\end{aligned} \tag{5.36}$$

where ϕ^x*,* ϕ^y*, and* ϕ^z *are azimuthal angles around the respective coordinate axes. In spherical coordinates* ϕ^z *is traditionally called simply* ϕ*. Show that these operators* *satisfy the commutation relations*

$$[\mathbf{R}_x, \mathbf{R}_y] = -\mathbf{R}_z, \quad [\mathbf{R}_y, \mathbf{R}_z] = -\mathbf{R}_x, \quad [\mathbf{R}_z, \mathbf{R}_x] = -\mathbf{R}_y. \tag{5.37}$$

Using (r, θ, ϕ) *spherical coordinates, derive the relations*

$$\begin{aligned} \mathbf{R}_x &= -\sin\phi \frac{\partial}{\partial\theta} - \cos\phi \cot\theta \frac{\partial}{\partial\phi}, \\ \mathbf{R}_y &= -\cos\phi \frac{\partial}{\partial\theta} + \sin\phi \cot\theta \frac{\partial}{\partial\phi}. \end{aligned} \tag{5.38}$$

Problem 5.1.3. *For a single particle, with the three components of quasivelocity* $\mathbf{s}$ *defined as linear functions of Cartesian velocities* $(\dot{x}, \dot{y}, \dot{z})$ *in Eq. (5.3) evaluate the elements* $A^r{}_i$ *according to Eq. (5.1). State why this is an unsatisfactory and interpret the failure geometrically. Show how the result of the previous problem implies the same thing.*

Problem 5.1.4. *The pendulum defined in the previous problem is now allowed to swing freely out of the* x, z *plane, making it a "spherical pendulum". Natural coordinates describing the bob location are the polar angle* θ *relative to the vertical z-axis and azimuthal angle around the z-axis* ϕ*, measured from the* x, z *plane. Instantaneous angular velocities around the x, y, and z-axes are given by*

$$\begin{aligned} \dot{\theta}^x &\equiv s^x = -\sin\phi\, \dot{\theta}, \\ \dot{\theta}^y &\equiv s^y = \cos\phi\, \dot{\theta}, \\ \dot{\theta}^z &\equiv s^z = \dot{\phi}. \end{aligned} \tag{5.39}$$

Choose s^x *and* s^z *as quasivelocities and write the Poincaré equations for these variables along with* θ *and* ϕ*. (This is just an exercise in organizing the work; there is no real merit to the choice of variables.)*

Problem 5.1.5. *With Euler angles* (ϕ, θ, ψ) *as defined in Fig. 5.1 playing the role of generalized coordinates* q^i*, define quasivelocities* $(v^1, v^2, v^3) \equiv (\omega^1, \omega^2, \omega^3)$ *as angular velocities of rotation of a rigid body around body-axes* x'''*,* y'''*, and* z'''*. Evaluate* $(\omega^1, \omega^2, \omega^3)$ *in terms of "Euler angular velocities,"* $(\dot{\phi}, \dot{\theta}, \dot{\psi})$*. Express the transformations in the form* $\omega^r = A^r{}_i(\mathbf{q})\dot{q}^i$ *as in Eq. (5.1). [Since the angular velocity is a true vector (in so far as rotations and not reflections are at issue) it is valid to start with an angular velocity with only one Euler angle changing, say corresponding to* $\dot{\phi} \neq 0$*, and work out its body components, do the same for the other two, and apply superposition.]*

Problem 5.1.6. *For the velocity transformation of the previous problem, evaluate the coefficients* $c^r{}_{li}(\mathbf{q})$ *and show that they are independent of* (ϕ, θ, ψ)*. (Note that* $\mathbf{q}$ *stands for* (ϕ, θ, ψ) *in this case.) If highly satisfactory cancellations do not occur in the calculations of* $c^r{}_{li}$ *you have made some mistake.*

Problem 5.1.7. *The kinetic energy of a rigid body, when expressed relative to "body-axes" (necessarily orthogonal), is a sum of squares:*

$$\overline{T}(\boldsymbol{\omega}) = \frac{1}{2} I_1 \omega^{1^2} + \frac{1}{2} I_2 \omega^{2^2} + \frac{1}{2} I_3 \omega^{3^2}. \tag{5.40}$$

Using this expression for the kinetic energy, write the Poincaré differential equations, Eq. (5.18), for the angular velocities $(\omega^1, \omega^2, \omega^3)$.

Problem 5.1.8. *Specialize the solution of the previous problem to the case of the spherical pendulum.*

5.2 Variational Derivation of the Poincaré Equation

In this section (and only this section) we use notation $\partial/\partial t$ *instead of* d/dt *for the "total time derivative". The reason for this is that a new subsidiary variable* u *will be introduced and the main arguments have to do with functions* $f(u,t)$ *and derivatives holding one or the other of* u *and* t *constant.*

Consider again Fig. 1.1. The particular (dashed) curve $\boldsymbol{\delta}\mathbf{x}(t)$ can be called the "shape" of a variation from the looked-for "true trajectory" $\mathbf{x}(t)$ shown as a solid curve. For the present discussion this function will be renamed as $\mathbf{x}^*(t)$ to free up the symbol $\mathbf{x}$ for a slightly varied function. We now restrict the freedom of variation by replacing $\boldsymbol{\delta}\mathbf{x}(t)$ by $u\,\boldsymbol{\delta}\mathbf{x}(t)$ where u is an artificially introduced multiplicative variable whose range runs from negative to positive and hence certainly includes $u = 0$; the range will not matter but it may as well be thought of as $-1 < u < 1$. With $\boldsymbol{\delta}\mathbf{x}(t)$ being called the shape of the variation, u can be called its "amplitude". Differentiation with respect to amplitude at fixed time and with fixed shape will be expressed as $\partial/\partial u$. Differentiation with respect to time along a varied trajectory whose shape and amplitude are both fixed will be expressed as $\partial/\partial t$.

The function $u\,\boldsymbol{\delta}\mathbf{x}(t)$ is still an arbitrary function of time, but at intermediate points in the analysis, we will insist that only u vary so that the *shape* $\boldsymbol{\delta}\mathbf{x}(t)$ can be held fixed and the action variable S treated as a function only of u. The varied curve joining P_1 and P_2 is given then, in parametric form, as

$$\begin{aligned}
q^1(u,t) &= q^{*1}(t) + u\,\delta q^1(t),\\
q^2(u,t) &= q^{*2}(t) + u\,\delta q^2(t),\\
&\dots,\\
q^n(u,t) &= q^{*n}(t) + u\,\delta q^n(t).
\end{aligned} \tag{5.41}$$

Though being "restricted" in this one sense, variations will be "generalized" in another sense. In the formalism developed so far the variation $\boldsymbol{\delta}\mathbf{x}(t)$ has been described only by deviations δq^i of the generalized coordinates describing the system at time t. But the variation at time t can be regarded as belonging to the tangent space at the point $\mathbf{x}(t)$ on the candidate true trajectory at that time. As a result, the *shape* of a variation can be described by a general vector field $\mathbf{w}(t)$. The vector $u\mathbf{w}(t)$ will be treated as "differentially small" so that it makes sense to add it to a point in the space. Also it is assumed to vanish at the end points. Then Eq. (5.41) is generalized to become

$$\mathbf{x}(u,t) = \mathbf{x}^*(t) + u\,\mathbf{w}(t). \tag{5.42}$$

With this notation the action for Lagrangian $L(\mathbf{x}, \dot{\mathbf{x}}, t)$ is given by

$$S(u) = \int_{t_1}^{t_2} L\Big(\mathbf{x}(u,t), \frac{\partial \mathbf{x}(u,t)}{\partial t}, t\Big)\, dt. \tag{5.43}$$

This is the first instance of our unconventional use of the symbol $\partial/\partial t$ mentioned in the introductory paragraph; its meaning here is that integrand L is being evaluated along a varied curve in which u and the shape of the variation are both held constant. Again $S(u)$ depends on all aspects of the curve along which it is evaluated, but only the dependence on u is exhibited explicitly. The *extremal* condition is

$$0 = \frac{dS(u)}{du} = \int_{t_1}^{t_2} \left(\frac{\partial L}{\partial q^i}\frac{\partial q^i}{\partial u} + \frac{\partial L}{\partial \dot{q}^i}\frac{\partial \dot{q}^i}{\partial u} \right) dt = \int_{t_1}^{t_2} \left(\frac{\partial L}{\partial \mathbf{x}} \cdot \mathbf{w} + \frac{\partial L}{\partial \dot{q}^i}\frac{\partial \dot{q}^i}{\partial u} \right) dt. \tag{5.44}$$

From this integral condition we wish to take advantage of the arbitrariness of the function $\mathbf{w}$ to obtain the differential equation of motion. As usual we must manipulate the integrand in such a way as to leave $\mathbf{w}$ (or one of its components) as a common multiplier. This has already been done with the first term, which has furthermore been written as *manifestly* an invariant – which means it can be evaluated in any convenient coordinate system.

If we proceeded from this point using the original coordinates we would reconstruct the earlier derivation of the Lagrange equations (see Problem 5.2.1). Instead we proceed to obtain the equations of motion satisfied by *quasivelocities*. Basis vectors directed along the original coordinate curves in the tangent space at any particular point are $\boldsymbol{\partial}/\boldsymbol{\partial} q^1, \boldsymbol{\partial}/\boldsymbol{\partial} q^2, \dots, \boldsymbol{\partial}/\boldsymbol{\partial} q^n$. Symbolize the components of the velocity vector in this basis by $v^i = \dot{q}^i$. At every point in configuration space arbitrarily different other basis vectors $\boldsymbol{\eta}_1, \boldsymbol{\eta}_2, \dots \boldsymbol{\eta}_n$ can be introduced and a tangent space vector such as $\mathbf{v}$ can be expanded in terms of them. Such coordinates of the velocity are known as "quasivelocities" s^i. (A typical s^i is an angular velocity around a coordinate axes.) The velocity can then be expressed in either of the forms

$$\mathbf{v} = v^i\, \mathbf{e}_i = s^i\, \boldsymbol{\eta}_i. \tag{5.45}$$

As in Eq. (5.1), the quasivelocity components can be expressed as linear combinations of the v^i components;

$$s^r = A^r{}_j v^j. \tag{5.46}$$

Suppressing the Lagrangian time argument for simplicity, after substitution of Eq. (5.46), the Lagrangian is expressible as a new function

$$\overline{L}(\mathbf{x}, \mathbf{s}) = L(\mathbf{x}, \dot{\mathbf{x}}). \tag{5.47}$$

Proceeding as in Eq. (5.44) we obtain

$$0 = \int_{t_1}^{t_2} \left(\frac{\partial \overline{L}}{\partial \mathbf{x}} \cdot \mathbf{w} + \frac{\partial \overline{L}}{\partial s^i} \frac{\partial s^i}{\partial u} \right) dt. \tag{5.48}$$

As noted before, the first term is automatically expressible in invariant form, but we are still left with the problem of expressing the second term in a form that is proportional to the (arbitrary) vector **w**.

At this point we need to use the result (3.126), a relation based on the commutation relations satisfied by quasi-basis-vectors. Though the physical interpretations of the functions contained in Eqs. (5.42) and (3.120) are different, these equations are identical, and can be subjected to identical manipulations. The required result (3.126) is

$$\frac{\partial s^i}{\partial u} = \frac{\partial w^i}{\partial t} + c^i_{jk}\, s^j\, w^k. \tag{5.49}$$

This formula is quite remarkable in that the s^i and w^i are utterly independent quantities – the last term entirely compensates for the differences between their derivatives. Substituting Eq. (5.49) into Eq. (5.48) yields

$$0 = \int_{t_1}^{t_2} \left(\frac{\partial \overline{L}}{\partial \mathbf{x}} \cdot \mathbf{w} + \frac{\partial \overline{L}}{\partial s^i} \frac{\partial w^i}{\partial t} + \frac{\partial \overline{L}}{\partial s^i} c^i_{jk}\, s^j\, w^k \right) dt. \tag{5.50}$$

The first and third terms are proportional to **w**, but we should express the first term in terms of components w^k to match the representations of the other two terms. Also the second term can be manipulated using integration by parts. Since the function **w** vanishes at both end points the result is

$$0 = \int_{t_1}^{t_2} \left(\boldsymbol{\eta}_k(\overline{L}) - \frac{\partial}{\partial t} \frac{\partial \overline{L}}{\partial s^k} + c^i_{jk} \frac{\partial \overline{L}}{\partial s^i}\, s^j \right) w^k dt. \tag{5.51}$$

With the factor w^k being arbitrary, the integrand must vanish, or

$$\frac{d}{dt} \frac{\partial \overline{L}}{\partial s^k} - c^i_{jk} \frac{\partial \overline{L}}{\partial s^i}\, s^j = \boldsymbol{\eta}_k(\overline{L}). \tag{5.52}$$

Since the subsidiary variable u can now be discarded, the traditional d/dt notation for total time derivative has been restored. *Voila*! the Poincaré equation has re-emerged. As before the n first-order differential equations in Eq. (5.52) have to be augmented by the n defining equations (5.46) in order to solve for the $2n$ unknowns q^i and s^i.

As well as being much briefer than the previous derivation of the Poincaré equation, this derivation makes the interpretation of each term clearer. The derivation has also illustrated the power of the vector field formalism.

Problem 5.2.1. *The discussion in this section has been made neater than earlier treatments by the introduction of the artificial multiplicative amplitude u. Mimicking the treatment in this section, derive the Lagrange equation when there is no need for quasivelocities since the q^i are valid Lagrangian generalized coordinates.*

5.3 Restricting the Poincaré Equation With Group Theory

5.3.1 Continuous Transformation Groups

It seems fair to say that the Poincaré approach as presented so far generalizes the Lagrange approach *too much*. Though the range of coordinate transformations that *could* provide simple and valid descriptions has been greatly expanded, no guidance has been provided toward choosing promising transformations. Operationally, the terms with $c^i_{jk}(\mathbf{x})$ coefficients usually complicate the Poincaré equation seriously. The complication is greatly reduced if the c^i_{jk} coefficients are, in fact, independent of $\mathbf{x}$. Such constancy can only reflect symmetries of the system and these can be analyzed using group theory. For a continuously evolving mechanical system it is continuous, or Lie, groups that enter. In developing his equation Poincaré had this restricted situation primarily in mind. The condition characterizing a Lagrangian system is that the c^i_{jk} all vanish. The condition characterizing a Poincaré system is that the c^i_{jk} are all constant.

It is obvious that the symmetries of a mechanical systems have a significant impact on the possible motions of the system. The mathematical treatment describing this can be formulaic or geometric. The former approach is familiar from the concept of "cyclic" or "ignorable" coordinates in Lagrangian mechanics – if the Lagrangian is independent of a coordinate then its conjugate momentum is conserved. In Newtonian mechanics more purely geometric description is also familiar, for example in the treatment of the angular velocity as a vector subject to the normal rules of vector analysis. In Chapter 6 the use of purely geometric, "Lie algebraic" methods in Newtonian mechanics will be studied. Here we apply group theory within the Lagrange–Poincaré formalism.

The power of the Lagrange procedure is that it becomes entirely mechanical once the coordinates and Lagrangian have been established. But it can be regarded as a weakness that symmetries of the system have an analytical but not a geometric interpretation. We wish to rectify this lack by incorporating group theoretic methods into Lagrangian mechanics, or rather into the Poincaré equation since, as has been mentioned repeatedly, the Lagrange procedure is insufficiently general to handle many systems. Of course we also wish to retain the "turn the crank" potency of the Lagrangian approach.

Though a supposedly "advanced" subject – continuous groups – is to be used, only its simpler properties will be needed and those that *are* will be derived explicitly. Furthermore, only calculus and linear algebra is required. This is consistent with the general policy of the book of expecting as preparation only material with which most physics students are comfortable, and developing theory on a "just in time" basis. It is not possible to claim that a deep understanding of the subject can be obtained this way, but starting from "the particular" – manipulating a Lagrange-like equation – provides a well-motivated introduction to "the general." As mathematics therefore, the treatment in in the rest of this chapter will be "old fashioned" (being due to Lie it is certainly *old*) and perhaps even clumsy.

A change of variables (such as $x' = (1+a^1)x + a^2$)[7] depending on continuously variable parameters (such as a^1 and a^2) with the property that a small change in parameters causes a small change in the transformed variable, is called an r-parameter continuous transformation. ($r = 2$) Such a transformation acts on a space of n-component variables $\mathbf{x}$. ($n = 1$) If there is a parameter choice (such as $a^1 = 0$, $a^2 = 0$) for which the transformation is the identity, and the inverse transformation is included ($x = (1 - a^1/(1+a^1))x' + (-a^2/(1+a^1))$), and parameters can necessarily be found that give the same transformation as two transformations performed sequentially (also known as their concatenation or composition) ($x'' = (1 + a^1 + b^1 + b^1a^1)x + (a^2 + b^2 + b^1a^2)$), the transformation is called a continuous transformation group, or a Lie group.

Let $R(\mathbf{a}) = R(a^1, a^2, \ldots, a^r)$ symbolize the element of the transformation group corresponding to parameters $\mathbf{a}$. (The main transformations of this sort that have been studied up to this point in the text are orthogonal transformations, with the orthogonal matrix $\mathbf{O}$ parameterized by three independent parameters, for example Euler angles. In this case $R(a^1, a^2, a^3) \equiv \mathbf{O}(\psi, \theta, \varphi)$.)

For notational convenience, zero values for the parameters are assumed to

7) In this paragraph, as new quantities are introduced they are illustrated, in parentheses, by an ongoing example, starting here with $x' = (1+a^1)x + a^2$.

correspond to the identity transformation – that is $R(0) \equiv I$. (If this is not true the parameters should be redefined to make it true as it simplifies the algebra.) In general R need not be a matrix but it is the simplest case, as concatenation is then represented by matrix multiplication. In any case, the concatenation of $R(\mathbf{a})$ followed by $R(\mathbf{b})$ is indicated by $R(\mathbf{b})R(\mathbf{a})$.

The existence of transformation inverse to $R(\mathbf{a})$ requires the existence of parameters $\bar{\mathbf{a}}$ such that

$$R(\bar{\mathbf{a}})R(\mathbf{a}) = R(0). \tag{5.53}$$

($\bar{a}^1 = -a^1/(1+a^1), \bar{a}^2 = -a^2/(1+a^1)$.) For transformation $R(\mathbf{a})$ followed by $R(\mathbf{b})$ the group multiplication property is expressed as the requirement that parameters $\mathbf{c}$ exist such that

$$R(\mathbf{c}) = R(\mathbf{b})R(\mathbf{a}). \tag{5.54}$$

($c^1 = a^1 + b^1 + b^1 a^1,\ c^2 = a^2 + b^2 + b^1 a^2$.) It is primarily this concatenation feature that causes these transformations to have useful properties. Expressed functionally, the required existence of parameters $\mathbf{c}$ requires the existence[8] of functions $\phi^\kappa(\mathbf{a};\mathbf{b})$ such that

$$c^\kappa = \phi^\kappa(a^1,\dots,a^r;b^1,\dots,b^r), \quad \kappa = 1,2,\dots,r, \quad \text{or} \quad \mathbf{c} = \boldsymbol{\phi}(\mathbf{a};\mathbf{b}). \tag{5.55}$$

($\phi^1(\mathbf{a};\mathbf{b}) = a^1 + b^1 + b^1 a^1,\ \phi^2(\mathbf{a};\mathbf{b}) = a^2 + b^2 + b^1 a^2$.) For our purposes Eq. (5.55) will be employed primarily in situations where $\mathbf{b}$ is infinitesimal, meaning that it corresponds to a transformation close to the identity; to signify this we change symbols $\mathbf{b} \to \boldsymbol{\delta}\mathbf{a}$ and identify $\mathbf{c}$ as $\mathbf{a} + \mathbf{da}$. Then Eq. (5.55) yields

$$\mathbf{a} + \mathbf{da} = \boldsymbol{\phi}(\mathbf{a};\boldsymbol{\delta}\mathbf{a}). \tag{5.56}$$

(Throughout this chapter, the symbol δ will always be associated with near-identity group transformations.) Differentiating Eq. (5.56) yields a linear relation between the increments $(\delta a^1,\dots,\delta a^r)$ and the increments $(da^1,\dots,da^r)$,

$$da^\lambda = B^\lambda_{\ \mu}(\mathbf{a})\delta a^\mu, \quad \text{where} \quad B^\lambda_{\ \mu}(\mathbf{a}) = \left.\frac{\partial\phi^\lambda(a^1,\dots,a^r;b^1,\dots,b^r)}{\partial b^\mu}\right|_{\mathbf{b}=0}. \tag{5.57}$$

The matrix $\mathbf{B}$ is $r \times r$. Multiplying the vector of parameters of an (arbitrary) infinitesimal transformation by $\mathbf{B}$ gives the parameters of the transformation resulting from the sequential application of the finite transformation followed

8) Though the existence of functions $\phi^\kappa(\mathbf{a};\mathbf{b})$ is analytically assured, their definition is implicit and they are not necessarily available in closed form. Examples for which they *are* explicitly available will be given shortly.

by the infinitesimal transformation. Introducing its inverse, $A = B^{-1}$, and inverting Eq. (5.57) yields

$$\delta a^{\lambda} = A^{\lambda}_{\ \mu}(\mathbf{a}) da^{\mu}. \tag{5.58}$$

In our example,

$$\begin{pmatrix} B^1_{\ 1} & B^1_{\ 2} \\ B^2_{\ 1} & B^2_{\ 2} \end{pmatrix} = \begin{pmatrix} 1+a^1 & 0 \\ a^2 & 1 \end{pmatrix}, \quad \begin{pmatrix} A^1_{\ 1} & A^1_{\ 2} \\ A^2_{\ 1} & A^2_{\ 2} \end{pmatrix} = \begin{pmatrix} 1/(1+a^1) & 0 \\ -a^2/(1+a^1) & 1 \end{pmatrix}. \tag{5.59}$$

Note that **A** and **B**, matrices properties of the group, are independent of **x**.

Continuous transformations, when acting on a configuration space with generalized coordinates q^i (in the simplest case the original coordinates q^i are Cartesian coordinates (x, y, z)) are expressed by functions f^i (nonlinear, in general, and independent in ways to be clarified later) such that

$$q^{i'} = f^i(q^1, \ldots, q^n; a^1, \ldots, a^r), \quad i = 1, 2, \ldots, n \quad \text{or} \quad \mathbf{q}' = \mathbf{f}(\mathbf{q}; \mathbf{a}). \tag{5.60}$$

$(f^1(\mathbf{q}; \mathbf{a}) = (1 + a^1)x + a^2)$ Derivatives of these transformations will be symbolized by

$$u^i_{\ \kappa}(\mathbf{q}) = \left. \frac{\partial f^i(q^1, \ldots, q^n; a^1, \ldots, a^r)}{\partial a^{\kappa}} \right|_{\mathbf{a}=0}. \tag{5.61}$$

$(u^1_{\ 1} = x,\ u^1_{\ 2} = 1)$ Because these derivatives are evaluated at $\mathbf{a} = 0$ the functions $u^i_{\ \kappa}(\mathbf{q})$, though they depend on the form of the transformation equations, are independent of the parameters **a**. If the (finite) parameters **a** are regarded as dependent on a single evolving parameter t (presumably time) there is a corresponding velocity of evolution of the system coordinates,

$$\frac{dq^i}{dt} = u^i_{\ \kappa}(\mathbf{q}) \left. \frac{da^{\kappa}}{dt} \right|_{\mathbf{a}=0}; \tag{5.62}$$

or, for arbitrary differential changes δa^{κ} of the group parameters,

$$dq^i = u^i_{\ \kappa}(\mathbf{q})\, \delta a^{\kappa}. \tag{5.63}$$

A requirement to be used below is that the functions $u^i_{\ k}$ be independent – none of them is allowed to be identically expandable in terms of the others. One says that all of the parameters have to be *essential* – identities like $a^2 = 2a^1$, $a^3 = a^1 a^2$, or $a^1 =$ any-function of $a^2, a^3, \ldots$ are to be excluded.

The concatenation requirement (5.55) for the transformation functions **f** defined in Eq. (5.60) can be expressed as

$$\mathbf{f}(\mathbf{f}(\mathbf{q}; \mathbf{a}); \mathbf{b}) = \mathbf{f}(\mathbf{q}; \boldsymbol{\phi}(\mathbf{a}; \mathbf{b})). \tag{5.64}$$

In our example $r > n$, but commonly, the number of parameters r is less than the number of independent variables n. The arguments in the next section are followed most easily in the case $r = n$.

5.3.2 Use of Infinitesimal Group Parameters as Quasicoordinates

Sometimes the time evolution of a mechanical system can be described using a continuous group (for example the rotation group when rigid body motion is being described) with the coordinates $\mathbf{q}(t)$ being expressed in terms of the transformation functions $\mathbf{f}$;

$$\mathbf{q}(t) = \mathbf{f}\big(\mathbf{q}(0); \mathbf{a}(t)\big). \tag{5.65}$$

Here it is assumed that the configuration can necessarily[9] be described as the result of operating on an initial configuration $\mathbf{q}(0)$ with $R(\mathbf{a}(t))$. At a slightly later time the same equation reads

$$\mathbf{q} + \mathbf{dq} = \mathbf{q}(t + dt) = \mathbf{f}\big(\mathbf{q}(0); \mathbf{a}(t + dt)\big). \tag{5.66}$$

The occurrence of time variable t suggests that Eq. (5.66) describes the actual motion of a particular system but we wish also to describe "virtual" configurations that are close together, but not necessarily realized in an actual motion. For such configurations

$$\mathbf{q} + \mathbf{dq} = \mathbf{f}\big(\mathbf{q}(0); \mathbf{a} + \mathbf{da}\big). \tag{5.67}$$

Equation (5.65) shows that the quantities $\mathbf{a}$, called parameters so far, can be satisfactory generalized coordinates and can serve as independent variables in Lagrange equations for the system.[10] Before writing those equations we pursue some consequences of Eq. (5.64), applying it to a case where $\mathbf{a}$ is macroscopic and $\mathbf{b} \equiv \boldsymbol{\delta}\mathbf{a}$ is differential;

$$\begin{aligned}\mathbf{q} + \mathbf{dq} &= \mathbf{f}\big(\mathbf{f}(\mathbf{q}(0); \mathbf{a}); \boldsymbol{\delta}\mathbf{a}\big) = \mathbf{f}\big(\mathbf{q}(0); \boldsymbol{\phi}(\mathbf{a}; \boldsymbol{\delta}\mathbf{a})\big)\\ &= \mathbf{f}\big(\mathbf{q}(0); \mathbf{a} + \mathbf{da}\big) = \mathbf{f}(\mathbf{q}; \boldsymbol{\delta}\mathbf{a}).\end{aligned} \tag{5.68}$$

As illustrated in Fig. 5.3, parameters $\mathbf{a} + \mathbf{da}$ describe the direct system reconfiguration $\mathbf{q}(0) \xrightarrow{\mathbf{a}+\mathbf{da}} \mathbf{q} + \mathbf{dq}$. But the final configuration can also be produced by the sequence $\mathbf{q}(0) \xrightarrow{\mathbf{a}} \mathbf{q} \xrightarrow{\boldsymbol{\delta}\mathbf{a}} \mathbf{q} + \mathbf{dq}$. In the latter case the

9) A continuous group is said to be "transitive" if it necessarily contains a transformation carrying any configuration into any other. This requires $r \geq n$.

10) There are situations with $r \leq n$ in which reconfigurations are usefully described by variations of $\mathbf{a}$, but they can form a complete set of generalized coordinates only if $r = n$. The parameters $\mathbf{a}$ will in fact be interpreted as generalized coordinates below.

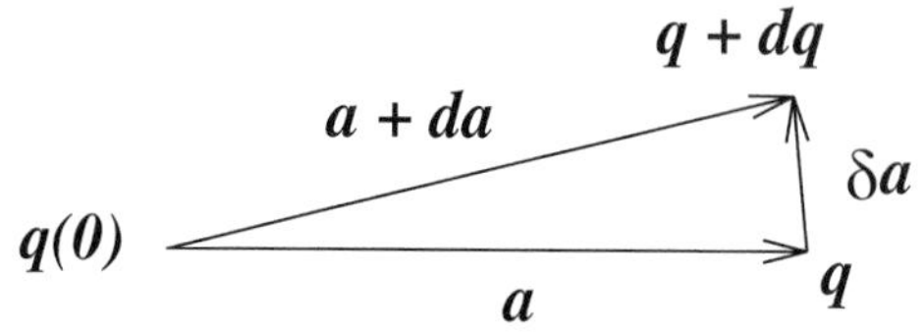

Fig. 5.3 Pictorial representation of alternate sequential transformation leading from an initial configuration to the same later configuration.

final transformation is infinitesimal, close to the identity, and its parameters are $\boldsymbol{\delta}\mathbf{a} = (\delta a^1, \ldots, \delta a^r)$.

We encountered equations like Eqs. (5.62) in Eq. (5.1), while discussing quasicoordinates σ^i with their corresponding quasivelocities s^i. As in that case, though small changes δa^κ can be meaningfully discussed, it is not valid to assume that the quasicoordinates a^κ can be found globally for which these are the resulting differentials. On the other hand, Eqs. (5.65) define the parameters a^κ globally. We will shortly employ the δa^λ's as differentials of quasicoordinates.

We now have three sets of independent differentials – displacements can be expressed in terms of dq^i (not independent if $r < n$), da^κ, or δa^κ. From the outermost members of Eq. (5.68), substituting from Eq. (5.61), variations dq^i can be related to variations da^λ indirectly via variations δa^κ;

$$dq^i = u^i{}_\kappa(\mathbf{q})\delta a^\kappa = u^i{}_\kappa(\mathbf{q})A^\kappa{}_\lambda(\mathbf{a})da^\lambda, \tag{5.69}$$

where δa^κ has been replaced using Eq. (5.57). This leads to

$$\frac{\partial q^i}{\partial a^\lambda} = u^i{}_\kappa(\mathbf{q})A^\kappa{}_\lambda(\mathbf{a}). \tag{5.70}$$

The first factor depends on the configurational coordinates $\mathbf{q}$ (and implicitly on the definition of, but not the values of, the parameters $\mathbf{a}$); the second factor is a property of the transformation group only. The elements of (5.70) can be regarded as the elements of the Jacobian matrix of the transformation $\mathbf{q} \to \mathbf{a}$ only if $r = n$, but they are well-defined even if $r < n$, in which case the variations generated by variation of the group parameters span less than the full tangent space. Expressions (5.70) are fundamental to the proof of Lie's theorem which governs the commutation relations of infinitesimal transformations of the Lie group. Before turning to that, it is useful to associate directional derivative operators with tangent space displacements.

5.3.3 Infinitesimal Group Operators

It has been remarked that different variables can be used to specify displacements in the tangent space at $\mathbf{q}$. Here we concentrate on Lie transformations close to the identity, as parameterized by δa^ρ. Consider any general (presumably real) scalar function $F(\mathbf{q})$ defined on the configuration space. Using the left portion of (5.69), its change dF corresponding to parameter variation δa^ρ is

$$dF = \frac{\partial F}{\partial q^i} dq^i = \frac{\partial F}{\partial q^i} u^i{}_\rho(\mathbf{q})\, \delta a^\rho. \tag{5.71}$$

The purpose of this expansion is to express dF as a superposition of changes in which one parameter a^λ varies while the rest remain constant. To emphasize this, Eq. (5.71) can be rearranged and re-expressed in terms of operators $\mathbf{X}_\rho$ defined by[11]

$$dF = (\delta a^\rho\, \mathbf{X}_\rho)F, \quad \text{where} \quad \mathbf{X}_\rho = u^i{}_\rho(\mathbf{q}) \frac{\boldsymbol{\partial}}{\boldsymbol{\partial} q^i}. \tag{5.72}$$

The operators $\mathbf{X}_\rho$ (which operate on functions $F(\mathbf{q})$) are called "infinitesimal operators of the group." There are r of these operators, as many as there are independent parameters in the group. Each one extracts from function F the rate of change of F per unit change in the corresponding parameter with the other parameters fixed.

Though the motivation for introducing these operators $\mathbf{X}_\rho$ comes entirely from our analysis of continuous groups, they are not different from the *vector fields* discussed in Section 3.5. To conform with notation introduced there they have been assigned bold face symbols, and the $\boldsymbol{\partial}/\boldsymbol{\partial} q^i$ are given similar treatment.

Instead of the differentials δa^ρ one can introduce "quasivelocities"

$$s^\rho = \frac{\delta a^\rho}{\delta t} \equiv \left.\frac{da^\rho}{dt}\right|_{\mathbf{a}=0}. \tag{5.73}$$

Then the system evolution described by Eq. (5.62) results in evolution of the function $F(\mathbf{q})$ according to

$$\frac{dF}{dt} = s^\rho \mathbf{X}_\rho F. \tag{5.74}$$

11) In Eq. (5.72) (and all subsequent equations) the order of the factors $\boldsymbol{\partial}/\boldsymbol{\partial} q^i$ and $u^i{}_\rho(\mathbf{q})$ has been reversed to avoid the nuisance of having to state that $\boldsymbol{\partial}/\boldsymbol{\partial} q^i$ does not act on $u^i{}_\rho$. As a result the indices no longer appear in their conventional, matrix multiplication, order. But since $u^i{}_\rho$ has one upper and one lower index their order doesn't really matter.

Results obtained previously can be re-expressed in terms of the $\mathbf{X}_\rho$. For example, choosing F to be the variable q^i yields a result equivalent to Eq. (5.62);

$$q^i + dq^i = (1 + \delta a^\rho \mathbf{X}_\rho) q^i = q^i + \delta a^\rho u^i{}_\rho(\mathbf{q}), \quad \text{or} \quad \frac{dq^i}{dt} = u^i{}_\kappa \frac{da^\kappa}{dt}. \tag{5.75}$$

Example 5.3.1. *For the infinitesimal transformation* $x' = (1 + a^1)x + a^2$, *it was shown above that* $u^1{}_1 = x$, $u^1{}_2 = 1$ *and the infinitesimal operators are therefore*

$$\mathbf{X}_1 = x\frac{\partial}{\partial x}, \quad \mathbf{X}_2 = \frac{\partial}{\partial x}. \tag{5.76}$$

Example 5.3.2. *For 2D rotations given by*

$$\mathbf{x}' = R(a^1)\mathbf{x}, \quad \begin{pmatrix} x^{1'} \\ x^{2'} \end{pmatrix} = \begin{pmatrix} \cos a^1 & -\sin a^1 \\ \sin a^1 & \cos a^1 \end{pmatrix} \begin{pmatrix} x^1 \\ x^2 \end{pmatrix} \tag{5.77}$$

$$\begin{aligned} \phi^1 &= a^1 + b^1 \\ \left(B^1{}_1\right)^T &= \left(\tfrac{\partial}{\partial b^1}\right)\left(a^1 + b^1\right)\Big|_{\mathbf{b}=0} = (1)\,. \end{aligned}$$

Note that this result followed from the fact that rotation angles about the same axis are simply additive. The transformation formulas and infinitesimal operators are

$$\begin{aligned} f^1 &= \cos a^1\, x^1 - \sin a^1\, x^2, \quad f^2 = \sin a^1\, x^1 + \cos a^1\, x^2, \\ u^1{}_1 &= -\,x^2, \quad u^2{}_1 = x^1, \\ \mathbf{X}_1 &= -\,x^2 \frac{\partial}{\partial x^1} + x^1 \frac{\partial}{\partial x^2}. \end{aligned} \tag{5.78}$$

Anticipating later formulas, the same result could have been obtained using the matrix J_3 *defined in Eq. (6.105). After suppressing the third row and the third column, it satisfies* $J_3^2 = -\mathbf{1}$. $\mathbf{x}' = e^{\phi J_3}\mathbf{x}$, *where a more conventional notation results from setting* $a^1 = \phi$. *Differentiating with respect to* ϕ *yields*

$$\begin{pmatrix} u^1{}_1 \\ u^2{}_1 \end{pmatrix} = \frac{\partial \mathbf{x}'}{\partial \phi}\bigg|_{\phi=0} \mathbf{x} = \begin{pmatrix} 0 & -1 \\ 1 & 0 \end{pmatrix} \mathbf{x} = \begin{pmatrix} -x^2 \\ x^1 \end{pmatrix}. \tag{5.79}$$

Example 5.3.3. *In 3D, consider the transformation*

$$\mathbf{x}' = e^{\mathbf{a}\cdot\mathbf{J}}\mathbf{x}, \quad \textit{where} \quad \mathbf{a}\cdot\mathbf{J} = \begin{pmatrix} 0 & -a^3 & a^2 \\ a^3 & 0 & -a^1 \\ -a^2 & a^1 & 0 \end{pmatrix}, \tag{5.80}$$

and the triplet of vectors $\mathbf{J}$ *was defined in Eq. (4.38). This expresses the matrix for angular rotation around* $\hat{\mathbf{a}}$ *by macroscopic angle* a *in terms of the matrix describing microscopic rotation around the same axis. Expanding it as* $e^{\mathbf{a}\cdot\mathbf{J}} = \mathbf{1} + \mathbf{a}\cdot\mathbf{J}/1! + (\mathbf{a}\cdot\mathbf{J})^2/2! + \cdots$*, differentiating with respect to* a^1*, then setting* $\mathbf{a} = 0$*, yields*

$$\begin{pmatrix} u^1_{\ 1} \\ u^2_{\ 1} \\ u^3_{\ 1} \end{pmatrix} = \left.\frac{\partial \mathbf{x}'}{\partial a^1}\right|_{\mathbf{a}=0} = \begin{pmatrix} 0 & 0 & 0 \\ 0 & 0 & -1 \\ 0 & 1 & 0 \end{pmatrix} \begin{pmatrix} x^1 \\ x^2 \\ x^3 \end{pmatrix} = \begin{pmatrix} 0 \\ -x^3 \\ x^2 \end{pmatrix}. \tag{5.81}$$

Combining this with corresponding results for a^2 *and* a^3 *yields*

$$\begin{pmatrix} u^1_{\ 1} & u^1_{\ 2} & u^1_{\ 3} \\ u^2_{\ 1} & u^2_{\ 2} & u^2_{\ 3} \\ u^3_{\ 1} & u^3_{\ 2} & u^3_{\ 3} \end{pmatrix} = \begin{pmatrix} 0 & x^3 & -x^2 \\ -x^3 & 0 & x^1 \\ x^2 & -x^1 & 0 \end{pmatrix}. \tag{5.82}$$

A standard notation is to change the name of the differential operators from $\mathbf{X}_i$ *to* $\mathbf{R}_i$*;*

$$\mathbf{R}_1 = -x^3\frac{\partial}{\partial x^2} + x^2\frac{\partial}{\partial x^3},\ \mathbf{R}_2 = -x^1\frac{\partial}{\partial x^3} + x^3\frac{\partial}{\partial x^1},\ \mathbf{R}_3 = -x^2\frac{\partial}{\partial x^1} + x^1\frac{\partial}{\partial x^2}. \tag{5.83}$$

These can be written compactly as $\mathbf{R}_i = \epsilon_{ijk}x^j\partial/\partial x^k$*. In the next section it will be shown that* $\mathbf{R}_i \equiv \partial/\partial\phi^i$*, where* ϕ^i *is a rotation angle about axis i.*

Example 5.3.4. The 3D Rotation Group. *To calculate the matrix* $B^i_{\ j}$*, defined in Eq. (5.57), for the 3D rotation group (in a special case) it is sufficient to consider a finite rotation* $R(\mathbf{a})$ *like that of Example 2 followed by an infinitesimal rotation* $R(\mathbf{b})$ *like that in Example 5.3.2;*[12]

$$R(\mathbf{c}) = R(\mathbf{b})\cdot R(\mathbf{a}) = \begin{pmatrix} 1 & -b^3 & b^2 \\ b^3 & 1 & -b^1 \\ -b^2 & b^1 & 1 \end{pmatrix} \begin{pmatrix} 1 & 0 & 0 \\ 0 & \cos a^1 & -\sin a^1 \\ 0 & \sin a^1 & \cos a^1 \end{pmatrix}$$

$$= \begin{pmatrix} 1 & -b^3\cos a^1 + b^2\sin a^1 & b^3\sin a^1 + b^2\cos a^1 \\ b^3 & \cos a^1 - b^1\sin a^1 & -\sin a^1 - b^1\cos a^1 \\ -b^2 & \sin a^1 + b^1\cos a^1 & \cos a^1 - b^1\sin a^1 \end{pmatrix}. \tag{5.84}$$

Ideally, this result would be expressible in the form $R(\mathbf{c}) = e^{\mathbf{c}\cdot\mathbf{J}}$ *since that would mean the coefficients* $\mathbf{c}$ *were known. Not knowing how to do this we have to proceed less directly. The following computations can usefully be performed using, for example, MAPLE.*

The eigenvalues of a 3D orthogonal matrix are, in general, given by $1, e^{\pm i\phi}$ *where* ϕ *is the rotation angle. The trace of an orthogonal matrix is preserved under similarity rotation transformations and is equal to the sum of its eigenvalues. It can be seen from*

12) The notation used here concerning matrices and operators is poorly chosen and confusing. A regular face quantity $R(\mathbf{a})$ is a matrix, a bold face quantity $\mathbf{R}_i$ is an operator. They are not at all commensurable quantities.

the matrix $R(\mathbf{a})$, which represents pure rotation around the x-axis by angle a^1, that the sum of eigenvalues for an orthogonal matrix is $1 + e^{ia^1} + e^{-ia^1} = 1 + 2\cos a^1$. Letting ϕ_C be the rotation angle due to matrix $R(\mathbf{b}) \cdot R(\mathbf{a})$, it follows that

$$\cos\phi_C = \cos a^1 - b^1 \sin a^1. \tag{5.85}$$

To lowest order in b^1, using a standard trigonometric formula, $\phi_C \approx a^1 + b^1$. For the matrix $R(\mathbf{c})$, let $v^{(1)}$ be the eigenvalue corresponding to eigenvalue 1. That is, $Rv^{(1)} = v^{(1)}$, and hence also $R^T v^{(1)} = v^{(1)}$, since $R^T = R^{-1}$. It follows that

$$(R - R^T)\, v^{(1)} = 0, \tag{5.86}$$

and from this follows the proportionality

$$\begin{pmatrix} v_1^{(1)} \\ v_2^{(1)} \\ v_3^{(1)} \end{pmatrix} \sim \begin{pmatrix} R_{23} - R_{32} \\ R_{31} - R_{13} \\ R_{12} - R_{21} \end{pmatrix}. \tag{5.87}$$

This can be converted to a unit vector and then multiplied by $a^1 + b^1$ to produce a vector with both the correct direction and correct magnitude. These calculations can also be carried out in detail using MAPLE. For the argument value $\mathbf{a} = (a^1, 0, 0)$, the functions defined in Eq. (5.55) are given by

$$\phi^1 = a^1 + b^1, \quad \phi^2 = a^1\left(\frac{b^3}{2} + \frac{b^2}{2}\frac{1+\cos a^1}{\sin a^1}\right),$$
$$\phi^3 = a^1\left(-\frac{b^2}{2} + \frac{b^3}{2}\frac{1+\cos a^1}{\sin a^1}\right). \tag{5.88}$$

Then we have

$$B^{\lambda}_{\ \mu}(\mathbf{a}) = \left.\frac{\partial\phi^{\lambda}(\mathbf{a};\mathbf{b})}{\partial b^{\mu}}\right|_{\mathbf{b}=0} = \begin{pmatrix} 1 & 0 & 0 \\ 0 & \frac{a^1}{2}\frac{1+\cos a^1}{\sin a^1} & \frac{a^1}{2} \\ 0 & -\frac{a^1}{2} & \frac{a^1}{2}\frac{1+\cos a^1}{\sin a^1} \end{pmatrix}. \tag{5.89}$$

To de-emphasize the special significance of the first axis let us replace a^1 by a variable ϕ, a rotation angle around arbitrary axis. The Jacobean of transformation (5.89),

$$J(\phi) = |\mathbf{B}| = \frac{\phi^2}{1 - \cos\phi}. \tag{5.90}$$

depends on ϕ, but (obviously) not on the axis of rotation.

Problem 5.3.1. *Consider a matrix of the form e^X where $X = \mathbf{a} \cdot \mathbf{J}$ as given in Eq. (5.80); that is, X is a skew-symmetric 3×3 matrix. Show that*

$$e^X = \mathbf{1} + \frac{\sin\phi}{\phi} X + \frac{1 - \cos\phi}{\phi^2} X^2, \quad \text{where} \quad \phi^2 = -\frac{1}{2}\,\mathrm{tr}(X^2). \tag{5.91}$$

First do it for $a^1 = \phi$, $a^2 = a^3 = 0$, which was the special case appearing in Example 5.3.4.

5.3.4
Commutation Relations and Structure Constants of the Group

The operators $\mathbf{X}_\rho$ operate on functions to describe the effect of infinitesimal group transformations; the functions ${u^i}_\rho$ are expansion coefficients; (see Eq. (5.72)). The property of the infinitesimal operators making them specific to some continuous group is that their commutators can be expressed in terms of so-called group "structure constants." Lie proved that for a continuous group these structure constants are, in fact, constant. This will now be demonstrated. Substituting from Eq. (5.72), the required commutators can be expressed in terms of the operators $\partial/\partial q^i$;

$$[\mathbf{X}_\tau, \mathbf{X}_\sigma] = \left({u^i}_\tau \frac{\partial {u^j}_\sigma}{\partial q^i} - {u^i}_\sigma \frac{\partial {u^j}_\tau}{\partial q^i} \right) \frac{\partial}{\partial q^j}. \tag{5.92}$$

Though quadratic in the functions $\mathbf{X}_\sigma$ (and hence in the functions ${u^i}_\sigma$) these will now be shown to be expressible as a linear superposition (with constant coefficients) of the operators $\mathbf{X}_\sigma$ themselves.

The expression for $[\mathbf{X}_\tau, \mathbf{X}_\sigma]$ in Eq. (5.92) can be simplified using results from the previous section. The quantities δa^κ, being differentials of quasicoordinates, are not necessarily the differentials of globally defined coordinates, but they occur only as intermediate variables. The variables $\mathbf{q}$ and $\mathbf{a}$ are globally related by Eqs. (5.65). By *equality of mixed partials* it follows that

$$\frac{\partial^2 q^i}{\partial a^\lambda \partial a^\mu} = \frac{\partial^2 q^i}{\partial a^\mu \partial a^\lambda}. \tag{5.93}$$

To exploit this relation it is important to exploit the structure of the transformation exhibited in Eq. (5.70), which is repeated here for convenience;

$$\frac{\partial q^i}{\partial a^\mu} = {u^i}_\kappa(\mathbf{q}) {A^\kappa}_\mu(\mathbf{a}). \tag{5.94}$$

In differentiating this equation with respect to a^μ it is necessary to allow for the functional dependence $\mathbf{q} = \mathbf{q}(\mathbf{a})$. In the summations, Roman indices range from 1 to n, and Greek indices run from 1 to r. Until now it has not been required that $r = n$. But, for concreteness, and to make the transformation one-to-one, let us assume $r = n$, Applying Eq. (5.93) yields

$$0 = {u^i}_\kappa \left(\frac{\partial {A^\kappa}_\mu}{\partial a^\lambda} - \frac{\partial {A^\kappa}_\lambda}{\partial a^\mu} \right) + {A^\kappa}_\mu \frac{\partial {u^i}_\kappa}{\partial a^\lambda} - {A^\kappa}_\lambda \frac{\partial {u^i}_\kappa}{\partial a^\mu}. \tag{5.95}$$

Needed terms such as $\partial u^i{}_\kappa / \partial a^\lambda$ can be obtained by differentiating with respect q^j's and again using Eq. (5.94);

$$\frac{\partial u^i{}_\kappa}{\partial a^\lambda} = \frac{\partial u^i{}_\kappa}{\partial q^j}\frac{\partial q^j}{\partial a^\lambda} = \frac{\partial u^i{}_\kappa}{\partial q^j} u^j{}_\nu A^\nu{}_\lambda. \tag{5.96}$$

Substitution into Eq. (5.95) yields a relation satisfied by the coefficient of $\boldsymbol{\partial}/\boldsymbol{\partial} q^j$ on the right-hand side of Eq. (5.92)

$$\left(\frac{\partial A^\kappa{}_\mu}{\partial a^\lambda} - \frac{\partial A^\kappa{}_\lambda}{\partial a^\mu}\right) u^i{}_\kappa = \left(u^j{}_\kappa \frac{\partial u^i{}_\nu}{\partial q^j} - u^j{}_\nu \frac{\partial u^i{}_\kappa}{\partial q^j}\right) A^\kappa{}_\mu A^\nu{}_\lambda. \tag{5.97}$$

To produce a formula for the coefficient in commutator (5.92) in terms of a function linear in the u-functions, multiply by $(A^{-1})^\mu_{\ \tau}$ and by $(A^{-1})^\lambda_{\ \sigma}$ and completing the summations produces

$$\begin{aligned} u^j{}_\sigma \frac{\partial u^i{}_\tau}{\partial q^j} - u^j{}_\tau \frac{\partial u^i{}_\sigma}{\partial q^j} &= (A^{-1})^\mu_{\ \tau}\,(A^{-1})^\lambda_{\ \sigma}\left(\frac{\partial A^\kappa{}_\mu}{\partial a^\lambda} - \frac{\partial A^\kappa{}_\lambda}{\partial a^\mu}\right) u^i{}_\kappa \\ &\equiv c^\kappa{}_{\tau\sigma}\, u^i{}_\kappa(\mathbf{q}). \end{aligned} \tag{5.98}$$

In the last step the coefficients $c^\kappa{}_{\tau\sigma}$ of the u-functions have been assign symbols

$$c^\kappa{}_{\sigma\tau} = (A^{-1})^\mu_{\ \tau}(A^{-1})^\lambda_{\ \sigma}\left(\frac{\partial A^\kappa{}_\lambda}{\partial a^\mu} - \frac{\partial A^\kappa{}_\mu}{\partial a^\lambda}\right). \tag{5.99}$$

Since the left-hand side of Eq. (5.98) is independent of $\mathbf{a}$ the right-hand side must also be independent of $\mathbf{a}$. It could be argued that, in the summation over κ, $\mathbf{A}$-dependent terms could cancel. But such a relation would imply a functional dependency among the u^i_κ, which is excluded by hypothesis. This completes the proof that the $c^\kappa{}_{\tau\sigma}$ coefficients are constant.

Substituting from Eq. (5.97) into Eq. (5.92), the master commutator rule is

$$[\mathbf{X}_\tau, \mathbf{X}_\sigma] = c^\kappa{}_{\tau\sigma}\,\mathbf{X}_\kappa. \tag{5.100}$$

Lie also proved a converse theorem: constancy of the $c^\kappa{}_{\tau\sigma}$ implies that the procedure described here can be reversed. The functions in Eq. (5.94) can be found and those equations integrated to produce a continuous group of transformations. We will not need that result.

Definition (5.99) can be compared with Eq. (5.17), which defined coefficients in the Poincaré equation. The only difference has to do with the choice of independent variables. Instead of using the q^κ variables in the earlier derivation of the Poincaré equation we *could* have treated the a^κ parameters as independent variables, related to the q^i coordinates by Eq. (5.94). In that case formula (5.57) would have been identical, except possibly for sign, to formula (5.1), where quasicoordinates were first introduced. We have seen earlier that the constancy of these coefficients simplifies the Poincaré equation markedly.

5.3.5 Qualitative Aspects of Infinitesimal Generators

The infinitesimal operators $\mathbf{X}_i$ are examples of the *vector fields* introduced in earlier chapters, and are related therefore to *directional derivatives*. Recall, for example, Eqs. (2.93)

$$\frac{\partial}{\partial x} \equiv \mathbf{i} \equiv \mathbf{e}_x, \quad \frac{\partial}{\partial y} \equiv \mathbf{j} \equiv \mathbf{e}_y. \tag{5.101}$$

When first introduced these were justified only for Euclidean axes. These definitions provided a natural association of a basis vector such as $\mathbf{e}_y$ with its corresponding operator $\partial/\partial y$. When applied to function $F(\mathbf{x})$ this operator extracts the (linearized) variation of $F(\mathbf{x})$ as y varies by 1 unit while x and z are held fixed. This is also known as the directional derivative in the y direction.

Before developing this further we make a few qualitative comments. An abstraction that most physicists first grasp in high school is that of *instantaneous* velocity, as well as its distinction from average velocity. Whereas average velocity relates to actual displacements occurring over finite times, instantaneous velocity yields an imaginary displacement that would occur in unit time if conditions remained unchanged. Another abstract concept is that of "virtual displacement," say δq^i, when the generalized coordinates are q^i. This is a possible displacement, consistent with constraints if there are any; it is a conceivable but not necessarily an actual displacement of the system. The actual motion picks one out of the infinity of possible virtual displacements.

Returning to continuous transformations, the operators $\mathbf{X}_\rho$ is a "directional derivative operator" acting in the direction for which only parameter a_ρ varies. Hence the rotation operator $\mathbf{R}_1$ operating on $F(\mathbf{q})$ extracts a quantity proportional to the differential dF in an infinitesimal rotation around the x-axis. (If F depended, say, only on r, then dF would vanish.) Consider the "infinitesimal rotation" of Example 5.3.4.

$$R(\mathbf{b}) = \begin{pmatrix} 1 & -b^3 & b^2 \\ b^3 & 1 & -b^1 \\ -b^2 & b^1 & 1 \end{pmatrix}. \tag{5.102}$$

This equation requires $\mathbf{b}$ to be dimensionless. Substituting $\mathbf{b} = \hat{\boldsymbol{\phi}}\, \delta\phi$, where $\hat{\boldsymbol{\phi}}$ is a unit vector along the rotation axis, is at least dimensionally consistent;

$$R(\hat{\boldsymbol{\phi}}\, \delta\phi) = \begin{pmatrix} 1 & -\delta\phi^3 & \delta\phi^2 \\ \delta\phi^3 & 1 & -\delta\phi^1 \\ -\delta\phi^2 & \delta\phi^1 & 1 \end{pmatrix}. \tag{5.103}$$

In terms of Cartesian coordinates pure rotation around, for example, the x-axis, through an angle $\delta\phi^1$ is described by

$$x' = x, \quad y' = y - \delta\phi^1 z, \quad z' = z + \delta\phi^1 y. \tag{5.104}$$

Though valid for small rotation angles, these equations clearly break down well before the angle becomes comparable with one radian. They have to be regarded as a "linearized" extrapolation of the instantaneous angular motion as follows.

Consider the vector $PT = (\Delta x, \Delta y, \Delta z) = (0, -z, y)$ shown in Fig. 5.4. Being tangential, it can be said to be directed "in the direction of instantaneously increasing ϕ^1." Also, its length being equal to the radius R, it is the tangential motion corresponding to unit increase in coordinate ϕ^1. An angular change of one radian can scarcely be called an infinitesimal rotation but one may proceed indirectly, starting with the vector $(0, -z/\eta, y/\eta)$, where the numerical factor η is large enough that arc and vector are indistinguishable. Scaling this up by the factor η produces the segment PT, which is declared to be the geometric representation of instantaneous rotation of 1 radian. Such a "tangent vector" PT can also be associated with a "directional derivative" symbolized by $\boldsymbol{\partial}/\boldsymbol{\partial}\phi^1$, with operational meaning the same as previously introduced tangent vectors. The awkwardness of this discussion suggests that, generally speaking, it is futile, if not wrong, to plot velocities and displacements on the same graph.

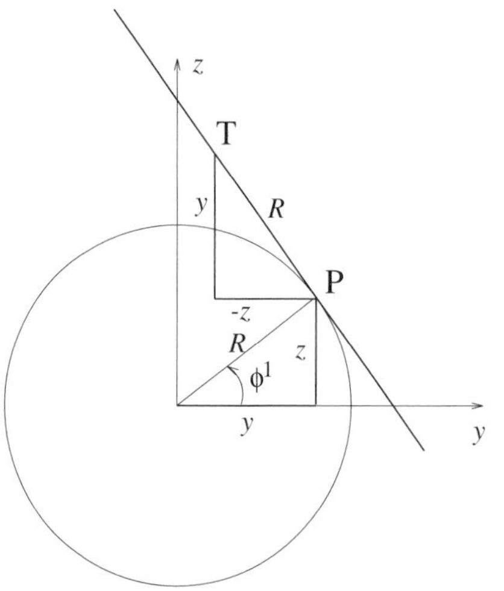

Fig. 5.4 Pictorial illustration of the "vector" $\partial/\partial\phi^1$. The point T is reached from the point P by motion along the tangent vector corresponding to unit increment of ϕ^1.

Referring again to Fig. 5.4, the *linearized* change of arbitrary function $F(x,y,z)$ when (x,y,z) changes as in Eq. (5.104) is

$$\begin{aligned} dF =& F(x, y - \delta\phi^1 z, z + \delta\phi^1 y) - F(x,y,z) \\ =& \delta\phi^1 \Big(-z\frac{\partial}{\partial y} + y\frac{\partial}{\partial z}\Big) F. \end{aligned} \tag{5.105}$$

To be consistent with previous terminology, this change dF should be given by $\delta\phi^1 \mathbf{X}_{\phi^1} F$, where $\mathbf{X}_{\phi^1}$ is the infinitesimal operator corresponding to angle ϕ^1. Guided by this, we *define* the infinitesimal operator $\mathbf{X}_{\phi^1}$ and the symbol $\partial/\partial\phi^1$, by

$$\mathbf{X}_{\phi^1} \equiv \frac{\partial}{\partial\phi^1} \equiv -z\frac{\partial}{\partial y} + y\frac{\partial}{\partial z}. \tag{5.106}$$

$\mathbf{X}_{\phi^1}$ is therefore identical to the operator $\mathbf{R}_1 \equiv \mathbf{R}_x$ defined in Eq. (5.83). This is dimensionally consistent since ϕ^1 is an angle. Again, even though 1 is not a "small" change in ϕ^1, $\mathbf{X}_1 F$ yields the change in F that results from simply "scaling up" the first-order Taylor series approximation in the ratio $1/\delta\phi^1$. This yields the *linearized* change in F for unit change of the parameter ϕ^1 in the direction in which only ϕ^1 changes. Also, even if the existence of variable ϕ^1 as a globally defined variable is problematical, the meaning of the partial derivative $\partial/\partial\phi^1$ is not.

The structure of Eq. (5.99) makes it possible to consistently combine infinitesimal and finite transformations. Infinitesimal transformation (5.104) was previously encountered as Eq. (4.39), which was re-expressed as a transformation $\mathbf{x} \to \mathbf{x}'$,

$$\mathbf{x}' = (1 + \mathbf{a}\cdot\mathbf{J}\, d\phi)\, \mathbf{x}, \tag{5.107}$$

where $\mathbf{a}$ is a unit vector defining an axis of rotation and $\mathbf{J}$ is the triplet of matrices defined in Eq. (4.38). In the present case $\mathbf{a} = \hat{\mathbf{e}}_x$ and Eq. (5.107) becomes

$$\begin{pmatrix} x' \\ y' \\ z' \end{pmatrix} = \left(\begin{pmatrix} 1 & 0 & 0 \\ 0 & 1 & 0 \\ 0 & 0 & 1 \end{pmatrix} + \begin{pmatrix} 0 & 0 & 0 \\ 0 & 0 & -1 \\ 0 & 1 & 0 \end{pmatrix} \delta\phi^1 \right) \begin{pmatrix} x \\ y \\ z \end{pmatrix}. \tag{5.108}$$

The matrices J_1, J_2, J_3 can be said to "represent" the rotation operators $\mathbf{R}_1, \mathbf{R}_2, \mathbf{R}_3$ in that they lead to the same equations. From an arbitrary finite rotation, described by a rotation matrix R, and an arbitrary infinitesimal rotation, defined by matrix J, one can define a similarity transformation $J' = R^{-1}JR$. If the commutation relations of the J operators can be said to characterize the geometry at a location $\mathbf{x}$, by determining the $c^{\kappa}{}_{\tau\sigma}(\mathbf{x})$ coefficients, then the J' operators can be said to similarly characterize the geometry at a location $\mathbf{x}' = R\mathbf{x}$. Substitution into Eq. (5.99) shows that it is the relation $c^{\kappa}{}_{\tau\sigma}(\mathbf{x}) = c^{\kappa}{}_{\tau\sigma}(R\mathbf{x})$ that shows the geometry near $\mathbf{x}$ and near $\mathbf{x}'$ to be identical.

For pure rotation at angular velocity ω^1 around the x-axis the positional changes occurring in time δt are given by

$$x' = x, \quad y' = y - \omega^1 \delta t\, z, \quad z' = z + \omega^1 \delta t\, y. \tag{5.109}$$

Except for the factor δt (needed if nothing else to mollify physicists concerned with dimensional consistency) this is the same transformation as just discussed. The factor δt, strongly suggestive of true particle motion, is misleading in Eq. (5.109) if virtual displacements are intended. We will be willing to simply suppress the factor δt (that is, set it to 1) while continuing to call $\boldsymbol{\omega}$ the angular velocity. Then the changes in Eq. (5.109) can be regarded as changes per unit time. The angular velocity ω^1 is *not* to be identified as a quasicoordinate. Rather the Lie transformation to be used is that of Eq. (5.103) and ϕ^i is the quasicoordinate. Since ω^1 is a velocity, one is perhaps interested in dF/dt, which is given by

$$\begin{aligned} \frac{dF}{dt} &= \frac{1}{dt}\Big(F(x, y - \omega^1 dt\, z, z + \omega^1 dt\, y) - F(x,y,z) \Big) \\ &= \omega^1 \left(-z\frac{\partial}{\partial y} + y\frac{\partial}{\partial z} \right) F = \omega^1 \mathbf{X}_{\phi^1}\, F. \end{aligned} \tag{5.110}$$

Problem 5.3.2. *In defining the operator $\mathbf{X}_{\phi^1}$ in Eq. (5.106) it was assumed implicitly, as always in science, that angles are measured in radians. Modify the definition of $\mathbf{X}_{\phi^1}$ to correspond to measuring ϕ^1 in degrees.*

5.3.6 The Poincaré Equation in Terms of Group Generators

The comments in the previous section have been made in preparation for applying the terminology of continuous groups to the Poincaré equation. A quantity such as da^1, the differential of a continuous group parameter, will be identified with $d\sigma$, the differential of quasicoordinate σ which is related to quasivelocity s by $s = d\sigma/dt$. Suppose, for example, that the role of s is to be played by ω^1, the angular velocity about the x-axis. For continuous group transformation relevant to this variable, we use Eq. (5.104), so that ϕ^1 is the quasicoordinate corresponding to σ and $\dot{\phi}^1 = \omega^1$. The coefficients defined in Eq. (5.61) become $u^i_1 = \partial f^i/\partial \phi^1$, or $u^1_1 = 0$, $u^2_2 = -z$, $u^3_1 = y$.

Finally, we can even express the remaining terms in the Poincaré equation (or for that matter the Lagrange equation) using the infinitesimal operators $\mathbf{X}_\rho$. Derivatives, with respect to position, of functions such as T or V, can be expressed in terms of derivatives with respect to quasicoordinates such as ρ (which may be either a quasicoordinate or a true coordinate) using the relations

$$\frac{\partial T}{\partial \rho} = \mathbf{X}_\rho T, \quad \text{and} \quad \frac{\partial V}{\partial \rho} = \mathbf{X}_\rho V. \tag{5.111}$$

In Chapter 5 the Poincaré equation was written in what we now appreciate was only a preliminary form. At this point we have developed machinery that permits it to be written in terms of group generators;

$$\frac{d}{dt}\frac{\partial T}{\partial s^\rho} - c^\lambda{}_{\mu\rho}\, s^\mu \frac{\partial T}{\partial s^\lambda} - \mathbf{X}_\sigma T = -\mathbf{X}_\sigma V, \quad \rho = 1, \dots, n. \tag{5.112}$$

The left side contains "inertial terms," the right side "force terms." A few points of explanation can be made about these equations, recapitulating the definitions of the symbols.

The quantities s^ρ are quasivelocities. They are related to quasicoordinates σ^ρ by $s^\rho = \dot{\sigma}^\rho$. As we know it is in general not possible to "integrate" these to define σ^ρ globally, but neither will it be necessary. It is assumed that the kinetic energy has been re-expressed in terms of the quasivelocities $T = T(\mathbf{q}, \mathbf{s})$, but we are no longer indicating that with an overhead bar.

The quasivelocities s^ρ are defined in terms of the regular velocities as in Eq. (5.1);

$$s^\rho = A^\rho{}_i(\mathbf{q})\, \dot{q}^i, \quad r = 1, 2, \dots, n. \tag{5.113}$$

In general the coefficients depend on position, as shown. By inverting Eq. (5.113) one obtains expressions for $\dot{q}^i$ which, when substituted in T, provide the kinetic energy in the functional form $T(\mathbf{q}, \mathbf{s})$. The coefficients $c^\lambda{}_{\mu\rho}$, now assumed to be independent of $\mathbf{q}$, were defined in Eq. (5.17). It is usually easy to determine them from the commutation relations

$$[\mathbf{X}_\rho, \mathbf{X}_\sigma] = c^\kappa{}_{\rho\sigma} \mathbf{X}_\kappa. \tag{5.114}$$

As with the Lagrange equations, by defining $L = T - V$, and using the fact that V is independent of velocities (if it is true, that is), these equations can be simplified somewhat.

Unlike Lagrangian analysis, where defining appropriate generalized coordinates is the initial task, it is the choice of velocity variables that is central to the use of the Poincaré equations. Because rotational symmetry is so common, the quasiangles ϕ^x, ϕ^y, and ϕ^z, rotation angles around rectangular axes, are the prototypical quasicoordinates. As we know, these angles do not constitute valid generalized coordinates because of the noncommutativity of rotations, but they are satisfactory as quasicoordinates.

Though attention has been focused on the quasivelocities, once they have been found it remains to "integrate" them to find actual displacements.

5.3.7 The Rigid Body Subject to Force and Torque

5.3.7.1 Infinitesimal Operators

Consider the group of translations and rotations in three dimensions,

$$x^i = b^i + O^i{}_k \overline{x}^k. \tag{5.115}$$

The coordinates x^k belong to a particular particle, say of mass m. A further index distinguishing among particles is not indicated explicitly. If there are N particles this will have introduced $3N$ coordinates. But if the system is a rigid body there must be enough constraints to reduce these to six independent generalized coordinates. The parameters of a transformation group will serve this purpose as quasicoordinates.

Clearly the vector $\mathbf{b}$ is to be the quasicoordinate corresponding to translation. Its corresponding quasivelocity is $\mathbf{v} = \dot{\mathbf{b}}$. The matrix elements $O^i{}_k$ will parameterize rotational motion. The group is transitive – there is a choice of parameters giving every configuration, and *vice versa*. As written, this transformation still has too many parameters, however. There are three parameters b^i and nine parameters of the orthogonal matrix $O^i{}_k$. (Geometrically the elements $O^i{}_k$ are direction cosines of the axes in one frame relative to the axes in the other frame; this can be seen by assuming $(x^j - b^j)\hat{\mathbf{e}}_j = x^i \hat{\overline{\mathbf{e}}}_i$ and using Eq. (5.115) to evaluate $\hat{\mathbf{e}}_i \cdot \hat{\overline{\mathbf{e}}}_j$.) These matrix elements satisfy the orthogonality conditions,

$$O^i{}_k O^j{}_k = \delta_{ij}, \quad O^k{}_i O^k{}_j = \delta_{ij}, \tag{5.116}$$

where summation on k is implied even though both are upper or both lower indices.

The reduction to a minimal set of independent parameters can proceed as follows. Since the transformation to quasicoordinates is actually a velocity transformation we differentiate Eq. (5.115) with respect to time, yielding

$$\dot{x}^i = \dot{b}^i + \dot{O}^i{}_k \overline{x}^k + O^i{}_k \dot{\overline{x}}^k. \tag{5.117}$$

As in Section 4.2.7, introduce the matrix $\mathbf{\Omega} = \mathbf{O}^T \dot{\mathbf{O}}$

$$\Omega_{ij} = O^k{}_i \dot{O}^k{}_j; \tag{5.118}$$

that was shown there to be antisymmetric. The components of $\dot{\mathbf{O}} = \mathbf{O}\mathbf{\Omega}$ are

$$\dot{O}^i{}_k = O^i{}_j \Omega_{jk} = \begin{pmatrix} O^1{}_1 & O^1{}_2 & O^1{}_3 \\ O^2{}_1 & O^2{}_2 & O^2{}_3 \\ O^3{}_1 & O^3{}_2 & O^3{}_3 \end{pmatrix} \begin{pmatrix} 0 & -\omega^3 & \omega^2 \\ \omega^3 & 0 & -\omega^1 \\ -\omega^2 & \omega^1 & 0 \end{pmatrix}, \tag{5.119}$$

which can be written as

$$\dot{O}^n_{\ i} = \epsilon_{ijk} O^n_{\ j} \omega^k. \tag{5.120}$$

This exhibits each of the redundant $\dot{O}$ velocities (i.e., matrix elements) as a linear superposition of the three independent quasivelocities ω^k.

We now interpret Eq. (5.115) in the spirit of Eq. (5.65), with $\overline{x}^k \equiv x^k(0)$. Consider an arbitrary function of position $F(\mathbf{x}(t))$ and evaluate its derivative dF/dt which is its rate of change as observed at a point with moving frame coordinates $\overline{x}^k$;

$$\frac{dF}{dt} = \left(v^k \frac{\partial}{\partial x^k} + \omega^k \epsilon_{ijk} O^n_{\ j} \frac{\partial}{\partial O^n_{\ i}} \right) F \equiv (v^k \mathbf{X}_k + \omega^k \mathbf{R}_k) F. \tag{5.121}$$

As an aside, comparing with Eq. (3.74), it can be seen that the derivative dF/dt in Eq. (5.121) has been written as the sum of two Lie derivatives, with respect to the vectors v^k and $\omega^k \epsilon_{ijk} O^n_{\ j}$.

Because the variation of F can be expressed in terms of the nonredundant variables ω^k along with the v^i, together they comprise a complete set of velocities. The infinitesimal translation operators are

$$\mathbf{X}_1 = \frac{\partial}{\partial b^1}, \quad \mathbf{X}_2 = \frac{\partial}{\partial b^2}, \quad \mathbf{X}_3 = \frac{\partial}{\partial b^3}, \tag{5.122}$$

and the infinitesimal rotation operators are

$$\mathbf{R}_k = -\epsilon_{ijk} O^n_{\ j} \frac{\partial}{\partial O^n_{\ i}}. \tag{5.123}$$

In a problem below this definition will be shown to be equivalent to our earlier definition of $\mathbf{R}_k$.

The infinitesimal translations commute with each other. The structure constants of rotation generators are given by

$$c^1_{\ 32} = -c^1_{\ 23} = 1, \tag{5.124}$$

with cyclic permutation, as the following problems show.

Problem 5.3.3. *From these equations derive the commutation relations*

$$[\mathbf{R}_1, \mathbf{R}_2] F = (\mathbf{R}_1 \mathbf{R}_2 - \mathbf{R}_2 \mathbf{R}_1) F = -\mathbf{R}_3 F, \tag{5.125}$$

and similarly for cyclic permutations. This result could have been obtained differently; for example after solving the next problem.

Problem 5.3.4. *Show that the infinitesimal rotation generators can be written as*

$$\mathbf{R}_x = \frac{\partial}{\partial \phi^x}, \quad \mathbf{R}_y = \frac{\partial}{\partial \phi^y}, \quad \mathbf{R}_z = \frac{\partial}{\partial \phi^z}, \tag{5.126}$$

where ϕ^x, ϕ^y, and ϕ^z, are quasiangles corresponding to ω^x, ω^y, and ω^z, respectively. [Evaluate $d\phi^i$ in terms of variations in $O^j_{\ k}$ and use the fact that $\partial O/\partial\phi$ and $\partial\dot{O}/\partial\dot{\phi}$ mean the same thing.]

Problem 5.3.5. *Evaluate all commutators of the form $[\mathbf{X}_i, \mathbf{R}_j]$. In the next section, to calculate the kinetic energy, v^k will be specialized as being the centroid velocity, and in the following section these commutators $[\mathbf{X}_i, \mathbf{R}_j]$ will be neglected.*

5.3.7.2 Description Using Body Axes

Calculation of the kinetic energy proceeds exactly as in Lagrangian mechanics. Consider a moving body with total mass M, with centroid at $\mathbf{x}_C$ moving with speed v, and rotating with angular velocity $\boldsymbol{\omega}$ about the centroid. With a general point in the body located at $\overline{\mathbf{x}}$ relative to C, one has

$$\sum m\dot{\overline{\mathbf{x}}} = 0, \tag{5.127}$$

by definition of "centroid." Then the kinetic energy is given by

$$\begin{aligned} T &= \frac{1}{2}\sum m(\mathbf{x}_C + \overline{\boldsymbol{\omega}} \times \overline{\mathbf{x}})^2 \\ &= \frac{1}{2}Mv^2 + \frac{1}{2}\sum m(\overline{\boldsymbol{\omega}} \times \overline{\mathbf{x}})^2 = \frac{1}{2}Mv^2 + \frac{1}{2}\overline{\omega}^{\mu}\overline{I}_{\mu\nu}\overline{\omega}^{\nu}, \end{aligned} \tag{5.128}$$

where

$$\overline{\mathbf{I}} = \frac{1}{2}\sum m \begin{pmatrix} \overline{y}^2 + \overline{z}^2 & -\overline{x}\,\overline{y} & -\overline{x}\,\overline{z} \\ -\overline{y}\,\overline{x} & \overline{z}^2 + \overline{x}^2 & -\overline{y}\,\overline{z} \\ -\overline{z}\,\overline{x} & -\overline{z}\,\overline{y} & \overline{x}^2 + \overline{y}^2 \end{pmatrix}. \tag{5.129}$$

A notational clumsiness has appeared in (5.128) that will recur frequently in the text. It has to do with the symbols $\overline{\mathbf{x}}$ and $\overline{\boldsymbol{\omega}}$. Since $\overline{\mathbf{x}}$ is a *true* vector it is not meaningful to associate it with a particular frame as the overhead bar notation seems to imply. In contrast to this, its components $\overline{x}^i$, have well-defined meanings as moving frame coordinates, different from the fixed frame components x^i; as it happens the $\overline{x}^i$ components are constant. The only meaning the overhead bar on $\overline{\mathbf{x}}$ can have is to suggest that these constant components will be the ones to be employed in subsequent calculations. The same comments apply to $\overline{\boldsymbol{\omega}}$. Once these vectors appear in the form $(\overline{\boldsymbol{\omega}} \times \overline{\mathbf{x}})^2$ (which stands for the scalar $(\overline{\boldsymbol{\omega}} \times \overline{\mathbf{x}}) \cdot (\overline{\boldsymbol{\omega}} \times \overline{\mathbf{x}})$) it is clear this quantity could equally well be written $(\boldsymbol{\omega} \times \mathbf{x})^2$. Even a hybrid expression like $\boldsymbol{\omega} \times \overline{\mathbf{x}}$ could enter without error, provided components of $\mathbf{x}$ and $\boldsymbol{\omega}$ and their cross product are all taken in the same frame.[13]

13) It is especially important not to make the mistake of assuming $\overline{\boldsymbol{\omega}} = -\boldsymbol{\omega}$ even though, given that $\boldsymbol{\omega}$ describes the motion of the moving frame relative to the fixed frame, this might seem to be natural meaning of $\overline{\boldsymbol{\omega}}$.

As usual, to simplify the T, one can choose body-fixed axes and orient them along the principal axes, in which case the kinetic energy is given by

$$T = \frac{1}{2}Mv^2 + \frac{1}{2}\left(\overline{I}_1\overline{\omega}^{1^2} + \overline{I}_2\overline{\omega}^{2^2} + \overline{I}_3\overline{\omega}^{3^2}\right). \tag{5.130}$$

For substitution into the Poincaré equation we calculate partial derivatives of Eq. (5.130), assuming the elements $\overline{I}_\mu$ are constant because (in this case) the axes are fixed in the body;

$$\frac{\partial T}{\partial v^\mu} = Mv^\mu, \quad \frac{\partial T}{\partial \overline{\omega}^\lambda} = \overline{I}_{(\lambda)}\overline{\omega}^{(\lambda)}, \tag{5.131}$$

where parentheses indicate absence of summation. Before including external forces we consider the force-free case. Substitution into Eq. (5.112), with $\mathbf{s} \rightarrow \overline{\boldsymbol{\omega}}$, and using structure constants $c^{\kappa}{}_{\rho\sigma}$ from Eq. (5.124), yields

$$M\dot{v}^1 = 0, \quad \overline{I}_1\dot{\overline{\omega}}^1 + \overline{\omega}^2\overline{\omega}^3(\overline{I}_3 - \overline{I}_2) = 0, \tag{5.132}$$

plus four more equations with cyclic permutations of the indices.

Clearly the first equation(s) describe translational motion, the second rotational. It is pleasing that the Euler rotational equations and the centroid translation equations emerge side-by-side without having been subjected to individualized treatment. Furthermore, though developing the machinery has been painful, once developed, the equations of rotational motion have been written down almost by inspection.

Forced motion is described by including the right-hand sides of Eqs. (5.112);

$$M\dot{v}^1 = -\mathbf{X}_1 V, \quad I_1\dot{\overline{\omega}}^1 + \overline{\omega}^2\overline{\omega}^3(\overline{I}_3 - \overline{I}_2) = -\mathbf{R}_1 V, \tag{5.133}$$

and cyclic permutations. The right-hand sides are externally applied force and torque respectively. The three quantities $(\mathbf{X}_1, \mathbf{X}_2, \mathbf{X}_3)V$ are the components of a manifestly *true* vector $\mathbf{X}V$, and $(\mathbf{R}_1, \mathbf{R}_2, \mathbf{R}_3)V$ is a *true* (pseudo) vector $\mathbf{R}V$.[14] Later, this will provide freedom of choice of frame in which, as vectors, they are determined. For substitution into Eq. (5.133) it will be necessary to use their body-frame components though, since the left-hand sides are only valid in that frame. For both the translational and rotational cases, whatever complication results from the spatial dependence of potential energy V has been deferred to this point and is hidden implicitly in the right-hand sides of Eqs. (5.133). The effects of external forces are contained in the variations of V, which, it should be remembered, is a sum over the mass distribution.

$$V = \sum_{(i)} e_{(i)} V'(\mathbf{x}_{(i)}) \equiv \sum eV'(\mathbf{x}). \tag{5.134}$$

14) There is a clash between the use of bold face to indicate that $\mathbf{X}_1$, $\mathbf{X}_2$ and $\mathbf{X}_3$, as well as being vector fields are also the components of a three component object $\mathbf{X}$. In the examples in the next section the bold face notation will be dropped.

Here *potential energy* V has been written as an explicit sum over particles having "gravitational charges" $e_{(i)}$ that make up the rigid body. The (probably ill-advised) step has been taken of using symbol e as the mass of a particle. To support this notation the particle mass $e_{(i)}$ has been referred to as "gravitational charge." Then (by analogy with electromagnetic theory) one has defined "gravitational potential" $V'(\mathbf{x})$, which is the gravitational energy per unit mass, such that the gravitational energy of a particle, as a result of its location being $\mathbf{x}$), is $eV'(\mathbf{x})$. $V'(\mathbf{x})$ will be referred to as an externally imposed field. In this way gravitational forces are described in terms like those used in electromagnetism. For the same equations to be interpreted in electromagnetism V will continue to be the potential energy. Since this preempts the symbol V, the potential has been symbolized as V'. We assume V' is time-independent in the (inertial) space frame. In the final form of Eq. (5.134) the subscripts (i) have been suppressed, as they will be in most of the subsequent equations; they will have to be restored as appropriate. Corresponding to V' we define a "force intensity" field (analogous to electric field)

$$\mathbf{F}'_{(i)} = -\mathbf{X}\, V'\Big|_{\mathbf{x}=\mathbf{x}_{(i)}} = -\frac{\partial V'}{\partial \mathbf{x}}\Big|_{\mathbf{x}_{(i)}}. \tag{5.135}$$

As a result of the (labored) definitions, the equations can now be applied to either gravitational or electrical forces. For simplicity, assume $\mathbf{F}'$ is approximately constant over the body. The first Poincaré equation becomes Newton's law for the motion of a point mass,

$$M\dot{\mathbf{v}} = \sum e\mathbf{F}' \equiv \mathbf{F}_{\text{tot}}. \tag{5.136}$$

Rotational motion is influenced by applied torque. For unconstrained motion in a uniform force field there is no torque about the centroid. In practical cases of unconstrained motion in a nonuniform force field the resulting translational motion has the effect of making the force at the position of the body change with time. Since this would make it impossible to decouple the rotational and the translational motion in general, we exclude that possibility and consider rotational motion with one point, not necessarily the centroid, fixed.

When using body-frame coordinates, tumbling of the body causes the force components $\overline{F}_{(i)\alpha} \equiv e_{(i)}\overline{F}'_{(i)\alpha}$ acting on particle (i) at location $\overline{\mathbf{x}}_{(i)}$ to have seemingly erratic variation with time. To get around this complexity let us work out the right-hand side of the Poincaré equation in the space frame, and later use the fact that force is a vector to obtain its body-frame coordinates. Though it has been necessary to introduce body-frame components $\overline{F}'_{(i)\alpha}$, it is not necessary to introduce a symbol $\overline{\mathbf{F}}'$ since the force intensity $\mathbf{F}'$ is a *true* vector.

Assume $\mathbf{F}'$ is constant in space and time, with its value being $\mathbf{F}'_0$, a vector pointing along a fixed-in-space direction $\hat{\boldsymbol{\eta}}$, whose (space-frame) components are η^l. (For example, in a uniform gravitational force field, $\mathbf{F}' = g\hat{\boldsymbol{\eta}}$, $F_0' \equiv g$, and $\hat{\boldsymbol{\eta}}$ would usually be taken to be $-\hat{\mathbf{k}}$, pointing vertically downward along the z-axis.) Hence we have

$$\frac{\partial V'}{\partial \mathbf{x}} = -F_0'\hat{\boldsymbol{\eta}} \quad \text{or} \quad V'(\mathbf{x}) = -F_0' x^l \eta_l \equiv -F_0' \mathbf{x} \cdot \hat{\boldsymbol{\eta}}. \tag{5.137}$$

According to Eq. (5.83) the rotation generators are $\mathbf{R}_i = \epsilon_{inm} x^n \,\partial/\partial x^m$, and the right-hand side of rotational equation (5.133) becomes

$$-\mathbf{R}_j V = \sum e F_0' \epsilon_{jnm} x^n \eta_l \frac{\partial x^l}{\partial x^m} = \sum e F_0' \,(\mathbf{x} \times \hat{\boldsymbol{\eta}})_j = \sum (\mathbf{x} \times \mathbf{F})_j. \tag{5.138}$$

This can be recognized to be the external torque, which is a *true* vector. Taking advantage of its invariant property, as anticipated above, its body-frame components can be substituted directly into Eq. (5.133);

$$\bar{I}_1 \dot{\bar{\omega}}^1 + \bar{\omega}^2 \bar{\omega}^3 (\bar{I}_3 - \bar{I}_2) = \sum (\overline{\mathbf{x} \times \mathbf{F}})_1, \tag{5.139}$$

and cyclic permutations. The cross product on the right-hand side is a true vector, but its components have to be worked out in the barred frame.

Problem 5.3.6. *To an observer stationed on the rigid body the gravitational field, though spatially uniform, has a varying direction* $\hat{\bar{\boldsymbol{\eta}}}(t)$.

(a) *Show that its time derivative is*

$$\dot{\bar{\boldsymbol{\eta}}} = \bar{\boldsymbol{\eta}} \times \bar{\boldsymbol{\omega}}. \tag{5.140}$$

Justify the sign, and (since the bars on the symbols for the vectors are either ambiguous or redundant) break out the same equation into equations for the separate body-frame coordinates.

(b) *The potential energy V of the body (not to be confused with potential V′) acquires time dependence because it depends on the body's orientation relative to the gravitational axis or, if you prefer, on the orientation of the gravitational axis in the body frame. This can be expressed functionally as* $V = V(\hat{\bar{\boldsymbol{\eta}}}(t))$. *The (time varying) "gradient" of this function is* $\nabla_{\bar{\boldsymbol{\eta}}} V$. *Show that the Poincaré equation can be written as*

$$\bar{I}_1 \dot{\bar{\omega}}^1 + \bar{\omega}^2 \bar{\omega}^3 (\bar{I}_3 - \bar{I}_2) = (\bar{\boldsymbol{\eta}} \times \nabla_{\bar{\boldsymbol{\eta}}} V)_1, \tag{5.141}$$

with cyclic permutations. Paired with Eq. (5.140), this is known as the Euler–Poisson equation. Its virtue is that the vector $\bar{\boldsymbol{\eta}}$ *is independent of position in the body, facilitating the calculation of* $V(\bar{\boldsymbol{\eta}})$.

5.3.7.3 Commutation Relations for Simultaneous Translation and Rotation

It has been rather pedantic to continue to employ bold face symbols for operators and partial derivatives, solely to emphasize their interpretation as vector fields. From here on regular face symbols will be used. We therefore review and simplify somewhat.

The continuous transformation equations (Eq. (5.115))

$$x^i = b^i + O^i{}_k \overline{x}^k, \tag{5.142}$$

form the "Euclidean Lie group in three dimensions" because they preserve lengths and angles. The infinitesimal displacement generators are

$$\mathcal{X}_x = \frac{\partial}{\partial x}, \quad \mathcal{X}_y = \frac{\partial}{\partial y}, \quad \mathcal{X}_z = \frac{\partial}{\partial z}, \tag{5.143}$$

and (as shown in Fig. 5.5), the infinitesimal rotation operators are $\mathcal{R}_i = \epsilon_{ijk} x^j \partial/\partial x^k$;

$$\mathcal{R}_x = y\frac{\partial}{\partial z} - z\frac{\partial}{\partial y}, \quad \mathcal{R}_y = z\frac{\partial}{\partial x} - x\frac{\partial}{\partial z}, \quad \mathcal{R}_z = x\frac{\partial}{\partial y} - y\frac{\partial}{\partial x}, \tag{5.144}$$

which are equivalent to

$$\mathcal{R}_x = \frac{\partial}{\partial \phi^x}, \quad \mathcal{R}_y = \frac{\partial}{\partial \phi^y}, \quad \mathcal{R}_z = \frac{\partial}{\partial \phi^z}. \tag{5.145}$$

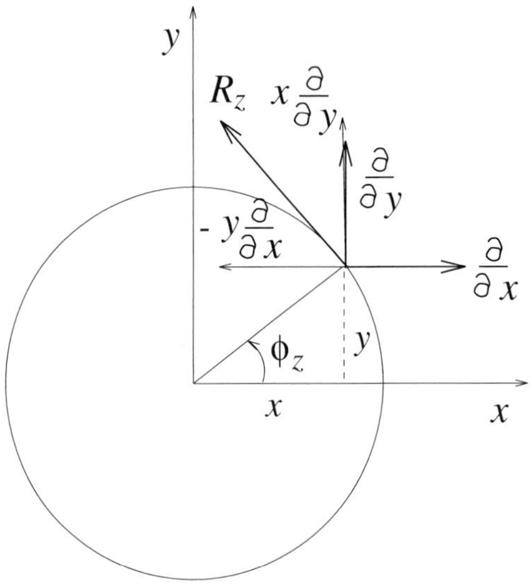

Fig. 5.5 Vector diagram illustrating the R_z, the generator of infinitesimal rotation about the z-axis.

These operators satisfy commutation relations

$$[\mathcal{X}_i, \mathcal{X}_j] = 0, \quad [\mathcal{X}_i, \mathcal{R}_j] = -\epsilon_{ijk}\mathcal{X}_k, \quad [\mathcal{R}_i, \mathcal{R}_j] = -\epsilon_{ijk}\mathcal{R}_k, \tag{5.146}$$

which constitute the "Lie algebra" of the Euclidean group. The Poincaré equations, with potential energy U are

$$\frac{d}{dt}\frac{\partial T}{\partial \omega^\rho} - c^\lambda_{\ \mu\rho}\,\omega^\mu \frac{\partial T}{\partial \omega^\lambda} - X_\rho T = -X_\rho U. \tag{5.147}$$

This is the equation for quasicoordinate π^ρ, whose corresponding quasivelocity is $\omega^\rho \equiv \dot{\pi}^\rho$, and infinitesimal generator is $X_\rho \equiv \partial/\partial \pi^\rho$. We interpret (x, y, z) as the laboratory coordinates of the center of mass of a moving rigid body, (v^x, v^y, v^z) as the corresponding velocities, and $(\omega^x, \omega^y, \omega^z)$ as the instantaneous angular velocity of the body, as measured in the laboratory. The structure coefficients $c^\lambda_{\ \mu\rho}$ were defined in Eq. (5.114);

$$[X_\rho, X_\sigma] = c^\kappa_{\ \rho\sigma} X_\kappa. \tag{5.148}$$

They can be obtained simply by identifying coefficients in Eq. (5.146). An easy mistake to make (for example in the first edition, as pointed out by Chris Gray) is to apply commutation relations (5.146) when rectangular and rotational displacements are not referred to the same origin. This issue is illustrated in Fig. 5.6 and in the following problem.

Problem 5.3.7. *The masses shown in Fig. 5.6 can be located by either (x^1, y^1, x^2, y^2) coordinates or by $(\bar{x}, \bar{y}, r, \theta^z)$ coordinates. Both choices are valid Lagrangian generalized coordinates. But, if the dumb-bell were free to rotate out of the x, y plane, then two other angular coordinates would be required and the set $(\theta^x, \theta^y, \theta^z)$ could only be quasicoordinates, corresponding to quasivelocities $(\omega^x, \omega^y, \omega^z)$. Sticking with the simpler planar case, show that all $c^\kappa_{\ \rho\sigma}$ commutation coefficients for the transformation $(\dot{x}, \dot{y}) \to (\dot{r}, \omega^z)$ vanish. Finally, based on Fig. 5.6, argue that all commutators with one element from $(\dot{x}, \dot{y}, \dot{z})$ and one element from $(\omega^x, \omega^y, \omega^z)$ vanish.*

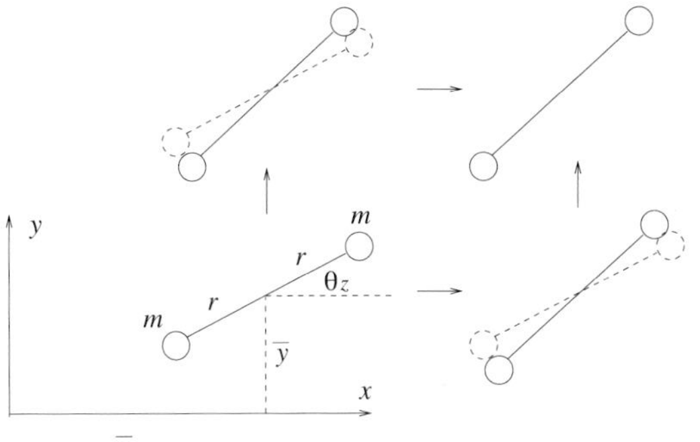

Fig. 5.6 Pictorial demonstration that centroid translation and rotation about centroid commute. The configuration in the upper right is the same, independent of the order of the translation and rotation.

5.3.7.4 **Bowling Ball Rolling Without Slipping**

A sphere of unit mass and unit radius rolls without slipping on a horizontal plane (Fig. 5.7). The moment of inertia of such a sphere about a diameter is $I = 0.4$. A spherical body has the simplifying feature that the elements of its moment of inertia tensor are constant even in the space frame. This permits us to use space-frame velocities as the quasivelocities in the Poincaré equation.

For specifying rotational motion use axes parallel to the fixed frame axes, but with origin at the center of the sphere. Since the Poincaré equation is to be used there is no need for concern that these would not be legitimate as Lagrangian generalized coordinates. There are two conditions for rolling without sliding,

$$\omega^y = v^x, \quad \omega^x = -v^y, \tag{5.149}$$

and these imply

$$\dot{\omega}^y = \dot{v}^x, \quad \dot{\omega}^x = -\dot{v}^y. \tag{5.150}$$

For recording the structure constants in an orderly way let us assign indices according to $(\omega^x, \omega^y, \omega^z, \dot{x}, \dot{y}) \rightarrow (1,2,3,4,5)$. There are five nontrivial Poincaré equations, even though at any instant there are only three degrees of freedom. The excess is accounted for by the two conditions for rolling. The Lagrangian expressed in quasivelocities is

$$L = \frac{1}{2}\left(I\omega^{1^2} + I\omega^{2^2} + I\omega^{3^2} + \omega^{4^2} + \omega^{5^2}\right). \tag{5.151}$$

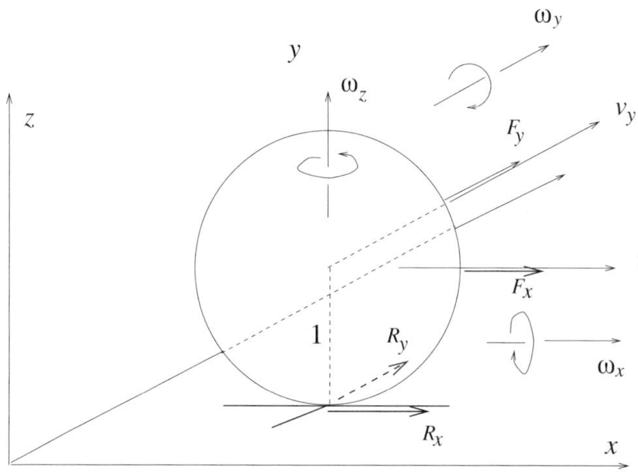

Fig. 5.7 Bowling ball rolling without slipping on a horizontal alley. As well as constraint force components R_x and R_y and external forces F_x and F_y, which are shown, there are also possible external torques about the center of the sphere K_x, K_y, and K_z. All vertical force components cancel.

The nonvanishing derivatives of L are

$$\frac{\partial L}{\partial \omega^1} = I\omega^1, \quad \frac{\partial L}{\partial \omega^2} = I\omega^2, \quad \frac{\partial L}{\partial \omega^3} = I\omega^3, \quad \frac{\partial L}{\partial \omega^4} = \omega^4, \quad \frac{\partial L}{\partial \omega^5} = \omega^5. \quad (5.152)$$

The nonvanishing commutators are

$$\begin{aligned}
[X_1, X_2] &= [\mathcal{R}_1, \mathcal{R}_2] = -\mathcal{R}_3 = -X_3, \\
[X_1, X_3] &= [\mathcal{R}_1, \mathcal{R}_3] = \mathcal{R}_2 = X_2, \\
[X_2, X_3] &= [\mathcal{R}_2, \mathcal{R}_3] = -\mathcal{R}_1 = -X_1.
\end{aligned} \quad (5.153)$$

It is not necessary to keep $X_6 \equiv \partial/\partial z$ since the ball stays in the same plane. The nonvanishing operators appear on the right side of Eq. (5.153) and the corresponding structure constants are their coefficients;

$$\begin{aligned}
c^1{}_{23} &= -1, \quad c^1{}_{32} = 1, \\
c^2{}_{13} &= 1, \quad c^2{}_{31} = -1, \\
c^3{}_{12} &= -1, \quad c^3{}_{21} = 1.
\end{aligned} \quad (5.154)$$

Let the transverse components of the force of constraint be R_4 and R_5, and allow for the possibility of external transverse force components F_4 and F_5 (for example because the plane is tilted) as well as external torques (K_1, K_2, K_3) about the center of the sphere. The constraint force itself provides torque $(R_5, -R_4, 0)$. The vertical components of **F** and **R** need not be introduced as they will always cancel. The Poincaré equations are

$$\begin{aligned}
I\dot{\omega}^1 &= R_5 + K_1, \\
I\dot{\omega}^2 &= -R_4 + K_2, \\
I\dot{\omega}^3 &= K_3, \\
\dot{\omega}^4 &= R_4 + F_4, \\
\dot{\omega}^5 &= R_5 + F_5.
\end{aligned} \quad (5.155)$$

Re-expressed in more intuitive symbols these become

$$\begin{aligned}
I\dot{\omega}^x &= R_y + K_x, \\
I\dot{\omega}^y &= -R_x + K_y, \\
I\dot{\omega}^z &= K_z, \\
\dot{v}^x &= R_x + F_x, \\
\dot{v}^y &= R_y + F_y.
\end{aligned} \quad (5.156)$$

Substituting from Eq. (5.150), the equations become

$$\begin{aligned} I\dot{v}^y &= -R_y - K_x, \\ I\dot{v}^x &= -R_x + K_y, \\ \dot{v}^x &= R_x + F_x, \\ \dot{v}^y &= R_y + F_y. \end{aligned} \tag{5.157}$$

These equations permit the constraint forces to be calculated from the external forces;

$$R_x = \frac{1}{1+I}K_y - \frac{I}{1+I}F_x, \quad R_y = -\frac{1}{1+I}K_x - \frac{I}{1+I}F_y. \tag{5.158}$$

These equations imply that the absence of external forces implies the absence of constraint forces, as would be true if the ball was in free space. But this is fortuitous; in general the forces of constraint have to be allowed for, and then eliminated using the rolling conditions. Substituting Eq. (5.158) into Eq. (5.156) yields

$$\dot{v}^x = \frac{1}{1+I}K_y + \frac{1}{1+I}F_x, \quad \dot{v}^y = -\frac{1}{1+I}K_x + \frac{1}{1+I}F_y. \tag{5.159}$$

These equations show that, in the absence of external torque, the ball responds to external forces like a point mass, but with its apparent mass being increased by the factor $1 + I$.

This result is derived, for example, in Landau and Lifshitz, p. 124, as well as (painfully) in Whittaker. A ten-pins bowling ball appears to violate this result in that it appears to curve as it travels down the alley. To account for this one has to assume that the alley is insufficiently rough to prevent the ball from skidding. So the ball must skid and roll for the first part of its trip and mainly roll for the last part. Whatever curving takes place has to occur during the skidding phase.

Problem 5.3.8. *A (rider-less) "skateboard" is a point-like object supported by a plane surface which has a line defined such that the skateboard slides or rolls without friction along that line, but not at all in the transverse direction. It can also rotate about the axis normal to the surface and passing through the single point of contact. Let the plane be inclined by a fixed angle Θ relative to the horizontal and let (x, y), with y-axis horizontal, be the coordinates of the skateboard in that plane. Let $\phi(t)$ be the instantaneous angle between the skateboard axis and the y-axis. The skateboard mass has m and rotational inertia such that its rotational kinetic energy is $I\dot{\phi}^2/2$. Its potential energy is $V = mg\sin\Theta\, x$.*

- *Write the Lagrangian $L(x; \dot{x}, \dot{y}, \dot{\phi})$, and express the sliding constraint as a linear relation among the velocities.*

- *For quasicoordinates* $x^1 = x$, $x^2 = y$, *and* $x^3 = \phi$ *evaluate all coefficients* $c^{\kappa}{}_{\rho\sigma}$.
- *Write the Poincaré equation for* $\phi(t)$. *Solve it for the initial conditions* $\phi(0) = 0$, $\dot{\phi}(0) = \omega_0$,
- *Write the Poincaré equations for* x *and* y. *Solve them assuming the skate is at rest at the origin at* $t = 0$.
- *As time is allowed to increase without limit, give the maximum displacements down the hill and horizontally along the hill.*

Problem 5.3.9. *A spherical marble of unit mass rolls without sliding on the inside of a circular cylinder whose axis is perfectly vertical. If the marble is released from rest it will obviously roll straight down with ever-increasing speed. Assuming it is released with finite initial transverse speed, solve for its subsequent motion. Be sure to allow for the possibility of initial angular velocity about an axis through the point of contact and normal to the surface. You should obtain the (surprising) result that the ball does not "fall out the bottom."*

Problem 5.3.10. *Hospital beds and some carts roll on wheels attached by casters that swivel at one end and are fixed at the other as shown in Fig. 5.8. To control the position* (x, y) *and angle* θ *of the cart forces* $\mathbf{F}_f$ *or* $\mathbf{F}_s$ *are applied at the midpoints between the wheels.*

- *Write the equations of motion and constraint equations governing the motion.*
- *Discuss the relative efficacy of pushing the cart from the fixed and swivel ends and explain the way you expect the solutions of the equations of motion to analytically confirm this behavior.*
- *Complete the solution discussed in the previous part.*

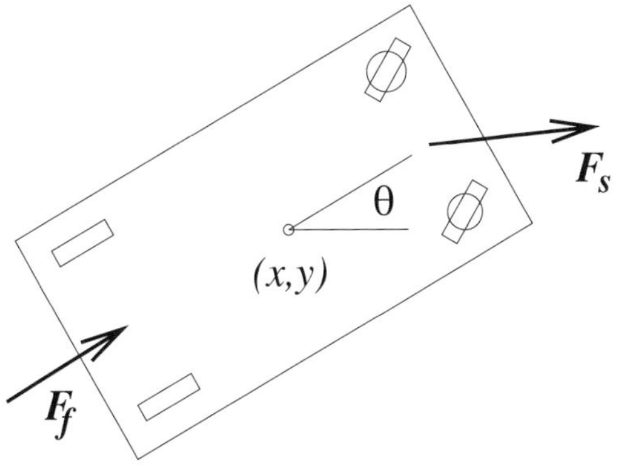

Fig. 5.8 The wheels at one end of a rolling cart or hospital bed are "fixed" while those at the other end are free to swivel. The cart can be propelled by forces $\mathbf{F}_f$ at the fixed end or $\mathbf{F}_s$ at the swivel end.

Bibliography

General References

1 L.D. Landau and E.M. Lifshitz, *Classical Mechanics*, Pergamon, Oxford, 1976, p. 124.

2 E.T. Whittaker, *A Treatise on the Analytical Dynamics of Particles and Rigid Bodies*, Cambridge University Press, Cambridge, UK, 1989.

References for Further Study

Section 5.1

3 N.G. Chetaev, *Theoretical Mechanics*, Springer, Berlin, 1989.

4 H. Poincaré, *C. R. Hebd. Séances Acad. Sci.* **132**, 369(1901).

Section 5.2

5 M. Born and E. Wolf, *Principles of Optics*, 4th ed., Pergamon, Oxford, 1970.

6 R. Courant and D. Hilbert, *Methods of Mathematical Physics*, Vol. I., Interscience, New York, 1953.

7 L.D. Landau and E.M. Lifshitz, *Mechanics*, Pergamon, Oxford, 1976

8 H. Rund, *The Hamilton–Jacobi Theory in the Calculus of Variations*, Van Nostrand, London, 1966.

9 B. F. Schutz, *Geometrical Methods of Mathematical Physics*, Cambridge University Press, Cambridge, UK, 1995.

Section 5.3.7.4

10 V.I. Arnold, V.V. Kozlov, and A.I. Neishtadt, *Mathematical Aspects of Classical and Celestial Mechanics*, Springer, Berlin, 1997,p.13.

11 D.Hopkins and J.Patterson, *Bowling frames: Paths of a bowling ball*, in *The Physics of Sport*, American Institute of Physics, New York, 1992.

Section 5.3.1

12 M. Hamermesh, *Group Theory and its Application to Physical Problems*, Addison-Wesley, Reading, MA, 1962.

Section 5.3.7.2

13 N.G. Chetaev, *Theoretical Mechanics*, Springer, Berlin, 1989.

Section 5.3.7.4

14 V.I. Arnold, V.V. Kozlov, and A.I. Meishtadt, *Dynamical Systems III*, Springer, Berlin, 1990.

6 Newtonian/Gauge Invariant Mechanics

Geometry as the basis of mechanics is a theme of this textbook. Though it may not have been recognized at the time, the importance of geometry was already made clear in freshman mechanics by the central role played by vectors. The purpose of this chapter is to develop a similar, but more powerful, algebraic/geometric basis for mechanics. However, unlike the chapter just completed, the approach will be Newtonian, with no artificial Lagrangian or Hamiltonian-like functions being introduced and no variational principles. The description of motion in noninertial frames of reference will be of central importance. Though this approach is very old, it continues to influence modern thinking, especially through the topic of "gauge invariance," which restricts theories to those that naturally support freedom in coordinate choices. This has been significant both for general relativity and string theory (and even more so in quantum field theory). The issue of "geometric phases" enters Newtonian mechanics similarly. To indicate the intended style we begin by reviewing vector mechanics.

6.1 Vector Mechanics

6.1.1 Vector Description in Curvilinear Coordinates

In its simplest form Newton's law for the motion of a point particle with mass m (an inertial quantity) subject to a force $\mathbf{F}$ (a dynamical quantity) yields the acceleration (a kinematical quantity);

$$\mathbf{a} = \frac{\mathbf{F}}{m}. \tag{6.1}$$

This is hypothesized to be valid only in an inertial frame of reference. In such a frame the acceleration vector is given by

$$\mathbf{a} = \frac{d^2\mathbf{r}}{dt^2} \equiv \ddot{\mathbf{r}}, \tag{6.2}$$

Geometric Mechanics: Toward a Unification of Classical Physics. 2nd Edition. Richard Talman

ISBN: 978-3-527-40683-8

where $\mathbf{r}(t)$ is the radius vector from the origin. The traditional notation has been used of replacing d/dt, the "total derivative" taken along the actual particle trajectory, by an overhead dot.

For actual computation it is often appropriate to introduce unit vectors such as $(\widehat{\mathbf{x}}, \widehat{\mathbf{y}}, \widehat{\mathbf{z}})$ or $(\widehat{\mathbf{r}}, \widehat{\boldsymbol{\theta}}, \widehat{\boldsymbol{\phi}})$ with the choice depending, for example, on the symmetry of the problem. With Euclidean geometry being assumed implicitly, these are "unit vector"[1] triads, mutually orthogonal and each having unit length. The "components" of $\mathbf{r}$ are then given (in rectangular and spherical coordinates) as the coefficients in

$$\mathbf{r} = x\,\widehat{\mathbf{x}} + y\,\widehat{\mathbf{y}} + z\,\widehat{\mathbf{z}} = r\,\widehat{\mathbf{r}}. \tag{6.3}$$

The component-wise differentiation of this vector is simple in the rectangular form, because the unit vectors are constant, but it is more complicated for other coordinate systems. In the case of spherical coordinates (see Fig. 6.2), as the particle moves, a local unit vector, $\widehat{\mathbf{r}}$ for example, varies. As a result, the velocity $\mathbf{v} = d\mathbf{r}/dt$ is given by

$$\mathbf{v} \equiv v^r\widehat{\mathbf{r}} + v^\theta\widehat{\boldsymbol{\theta}} + v^\phi\widehat{\boldsymbol{\phi}} = \dot{r}\,\widehat{\mathbf{r}} + r\,\dot{\widehat{\mathbf{r}}}. \tag{6.4}$$

Already at this stage there are minor complications. One is notational: in this text, when $(r, \theta, \phi) \equiv (q^1, q^2, q^2)$ are taken as "generalized coordinates," we refer to $(\dot{q}^1, \dot{q}^2, \dot{q}^2)$ as their "generalized velocities" and these are not the same as v^r, v^θ, and v^ϕ. Furthermore, symbolizing the time derivative of $\mathbf{r}$ by $\mathbf{v}$, we have to accept the fact that the components of the time derivative are not equal to the time derivatives of the components (except in rectangular components.) Finally, formula (6.3), which is intended to give the velocity components, still depends on the rate of change of a basis vector.

In general a vector can vary both in magnitude and direction, but a unit vector can vary only in direction – with its tail fixed, the most that can be happening to it is that it is rotating about some axis $\widehat{\boldsymbol{\omega}}$ with angular speed ω; together $\boldsymbol{\omega} = \omega\widehat{\boldsymbol{\omega}}$. Consider the radial unit vector $\widehat{\mathbf{r}}$ illustrated in Fig. 6.1. Since its change in time Δt is given (in the limit) by $\boldsymbol{\omega}\Delta t \times \widehat{\mathbf{r}}$ we have

$$\frac{d\widehat{\mathbf{r}}}{dt} = \boldsymbol{\omega} \times \widehat{\mathbf{r}}. \tag{6.5}$$

1) In this chapter, and only in this chapter, "unit vectors" are defined to have unit length. In other chapters a unit vector (or, preferably, a basis vector) is usually a vector pointing along the curve on which its corresponding coordinate varies, while the other coordinates are held fixed and "unit" implies that (in a linearized sense) the coordinate increases by one unit along the curve. To reduce the likelihood of confusion, the overhead "hat" symbol will be used only for vectors having unit length.

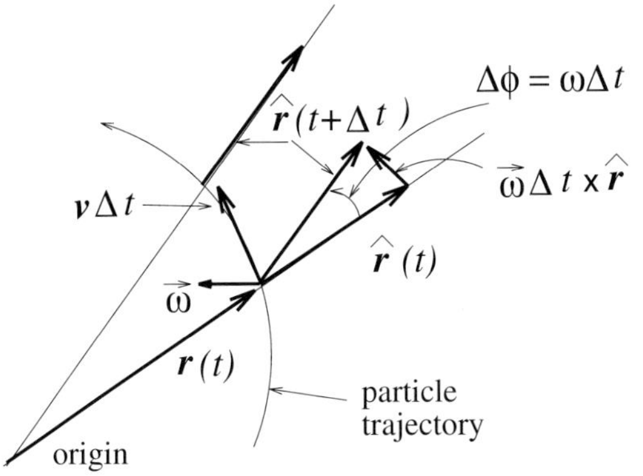

Fig. 6.1 For a particle moving instantaneously in the plane of the paper, the direction of the radial unit vector $\widehat{\mathbf{r}}(t)$ at its instantaneous location varies with time but its length remains constant. Instantaneously, $\widehat{\mathbf{r}}$ is rotating with angular velocity ω about the axis $\boldsymbol{\omega}$ normal to the paper.

This same formula holds with $\widehat{\mathbf{r}}$ replacing $\widehat{\boldsymbol{\theta}}$ or $\widehat{\boldsymbol{\phi}}$, and hence for any vector $\mathbf{u}$ or unit vector $\widehat{\mathbf{u}}$ fixed relative to the coordinate triad;

$$\frac{d\widehat{\mathbf{u}}}{dt} = \boldsymbol{\omega} \times \widehat{\mathbf{u}}. \tag{6.6}$$

The change in orientation of the unit triad is due to the motion of the particle; from the geometry of Fig. 6.1 one infers

$$\boldsymbol{\omega} = \frac{\widehat{\mathbf{r}} \times \mathbf{v}}{r}. \tag{6.7}$$

Combining this with Eq. (6.6) yields

$$\frac{d\widehat{\mathbf{u}}}{dt} = \frac{1}{r}\left(\widehat{\mathbf{r}} \times \mathbf{v}\right) \times \widehat{\mathbf{u}} = \left(\widehat{\mathbf{r}} \cdot \widehat{\mathbf{u}}\right) \frac{\mathbf{v}}{r} - \left(\mathbf{v} \cdot \widehat{\mathbf{u}}\right) \frac{\widehat{\mathbf{r}}}{r}. \tag{6.8}$$

When this formula is applied to each of the three spherical coordinate unit vectors, with coordinates defined as in Fig. 6.2, using $\mathbf{v} \cdot \widehat{\boldsymbol{\theta}} = r\dot{\theta}$ and $\mathbf{v} \cdot \widehat{\boldsymbol{\phi}} = r\sin\theta\dot{\phi}$, the results are

$$\dot{\widehat{\mathbf{r}}} = \dot{\theta}\widehat{\boldsymbol{\theta}} + \sin\theta\,\dot{\phi}\,\widehat{\boldsymbol{\phi}}, \quad \dot{\widehat{\boldsymbol{\theta}}} = -\dot{\theta}\,\widehat{\mathbf{r}}, \quad \dot{\widehat{\boldsymbol{\phi}}} = -\sin\theta\,\dot{\phi}\,\widehat{\mathbf{r}}. \tag{6.9}$$

Substituting the first of these into Eq. (6.3) yields

$$\mathbf{v} = \dot{r}\,\widehat{\mathbf{r}} + r\,\dot{\theta}\,\widehat{\boldsymbol{\theta}} + r\,\sin\theta\,\dot{\phi}\,\widehat{\boldsymbol{\phi}}. \tag{6.10}$$

This has been a circuitous route to obtain a result that seems self-evident (see Fig. 6.2) but, if one insists on starting by differentiating Eq. (6.3), it is hard to see how to derive the result more directly. The reason Eq. (6.10) seems

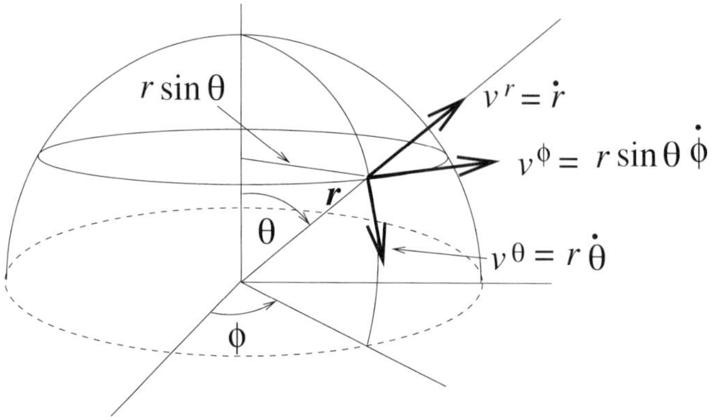

Fig. 6.2 Components of the velocity vector in a spherical coordinate system.

self-evident is that it is taken for granted that velocity is a *true* vector whose spherical and Cartesian components are related as if they belonged to a displacement vector.

Problem 6.1.1. *The acceleration can be calculated similarly, starting by differentiating Eq. (6.10). In this way confirm the calculations of Section 3.2.1.*

We have seen then that calculating kinematic quantities in curvilinear coordinates using vector analysis and the properties of vectors is straightforward though somewhat awkward.

6.1.2 The Frenet–Serret Formulas

Describing the evolution of a triad of unit basis vectors that are naturally related to a curve in space is one of the classic problems of the subject of *differential geometry*. It is done compactly using the formulas of Frenet and Serret. If the curve in question represents the trajectory of a particle these formulas describe only variation in space and contain nothing concerning the time rate of progress along the curve. Also, the case of free motion (in a straight line) is degenerate and needs to be treated specially. For these reasons (and a more important reason to be mentioned later) traditional treatments of mechanics usually re-derive the essential content of these elegant formulas explicitly rather than using them as a starting point.

A vector $\mathbf{x}(t)$ pointing from some origin to a point on its trajectory locates a particle's position P at time t. But because time t is to be suppressed from this treatment we take arc length s along the curve as an independent variable and represent the curve as $\mathbf{x}(s)$. To represent differentiation with respect to s a prime (as in $\mathbf{x}'$) will be used in the way that a dot is used to represent differentiation with respect to t.

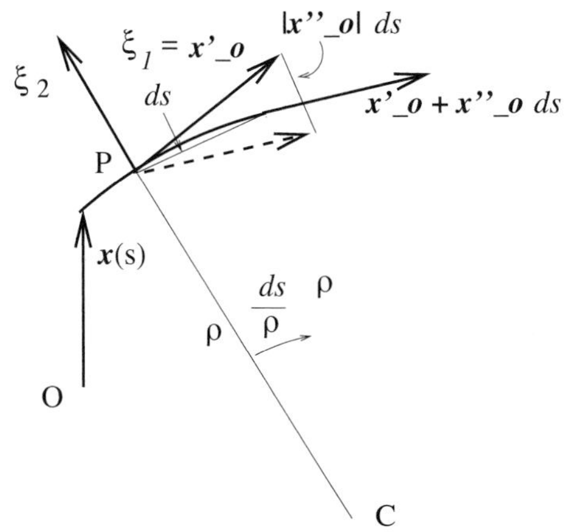

Fig. 6.3 Vector construction illustrating the derivation of the Frenet–Serret formulas. $\boldsymbol{\xi}_1$ is the unit tangent vector; $\boldsymbol{\xi}_2$ is the unit principal normal vector.

Any three disjoint points on a smoothly curving space curve define a plane and the limiting plane when these points approach each other is known as the "osculating plane." (It is because this plane is not unique for a straight line that free motion has to be regarded as special.) Clearly the velocity vector $\mathbf{v}$ defined in the previous section lies in this plane. But, depending as it does on speed v, it is not a unit vector, so it is replaced by the parallel vector

$$\boldsymbol{\xi}_1 = \frac{\mathbf{v}}{v} = \frac{d\mathbf{x}}{ds} \equiv \mathbf{x}', \tag{6.11}$$

where $\boldsymbol{\xi}_1$ is known as the "unit tangent vector." The unique vector $\boldsymbol{\xi}_2$ that also lies in the osculating plane but is perpendicular to $\boldsymbol{\xi}_1$ and points "outward" is known as the "principal normal" to the curve. From the study of circular motion in elementary mechanics one knows that the trajectory is instantaneously circular, with the center C of the circle being "inward" and lying in the osculating plane as well. Letting ρ stand for the radius of curvature of this circle, we know that the acceleration vector is $-(v^2/\rho)\boldsymbol{\xi}_2$, but we must again eliminate references to time (see Fig. 6.3).

If $\mathbf{x}_0'$ is the tangent vector at the point P in question then the tangent vector at a distance ds further along the curve is given by Taylor expansion to be $\mathbf{x}_0' + \mathbf{x}_0'' ds + \cdots$. Denoting the angle between these tangents by $\theta(s)$ the radius of curvature is defined by

$$\frac{1}{\rho} = \lim^{ds\to 0} \frac{d\theta}{ds} = \sqrt{\boldsymbol{\xi}_1' \cdot \boldsymbol{\xi}_1'}. \tag{6.12}$$

From the figure it can be seen that $\boldsymbol{\xi}_2$ is parallel to $\mathbf{x}_0''$ and that $|\mathbf{x}_0''| = 1/\rho$.

Since $\boldsymbol{\xi}_2$ is to be a unit vector it follows that

$$\boldsymbol{\xi}_2 = \rho \mathbf{x}''. \tag{6.13}$$

To make a complete orthonormal triad of basis vectors at the point C, we also define the "unit binormal" $\boldsymbol{\xi}_3$ by

$$\boldsymbol{\xi}_3 = \boldsymbol{\xi}_1 \times \boldsymbol{\xi}_2. \tag{6.14}$$

Proceeding by analogy with the introduction of the radius of curvature, the angle between $\boldsymbol{\xi}_3|_P$ and $\boldsymbol{\xi}_3(s+ds)$ is denoted by $\phi(s)$ and a new quantity, the "torsion" $1/\tau$ is defined by

$$\frac{1}{\tau} = \overset{ds\to 0}{\lim} \frac{d\phi}{ds} = \sqrt{\boldsymbol{\xi}_3' \cdot \boldsymbol{\xi}_3'}. \tag{6.15}$$

This relation does not fix the sign of τ. It will be fixed below. The Frenet–Serret formulas are first-order (in s) differential equations governing the evolution of the orthonormal triad $(\boldsymbol{\xi}_1, \boldsymbol{\xi}_2, \boldsymbol{\xi}_3)$ (to be called "Frenet vectors") as the point P moves along the curve. The first of these equations, obtained from Eqs. (6.11) and (6.13), is

$$\boldsymbol{\xi}_1' = \frac{\boldsymbol{\xi}_2}{\rho}. \tag{6.16}$$

Because the vector $\boldsymbol{\xi}_3$ is a unit vector, its derivative $\boldsymbol{\xi}_3'$ is normal to $\boldsymbol{\xi}_3$ and hence expandable in $\boldsymbol{\xi}_1$ and $\boldsymbol{\xi}_2$. But $\boldsymbol{\xi}_3'$ is in fact also orthogonal to $\boldsymbol{\xi}_1$. To see this differentiate the equation that expresses the orthogonality of $\boldsymbol{\xi}_1$ and $\boldsymbol{\xi}_3$;

$$0 = \frac{d}{ds}(\boldsymbol{\xi}_1 \cdot \boldsymbol{\xi}_3) = \boldsymbol{\xi}_1' \cdot \boldsymbol{\xi}_3 + \boldsymbol{\xi}_1 \cdot \boldsymbol{\xi}_3' \tag{6.17}$$

where, using Eq. (6.16) and the orthogonality of $\boldsymbol{\xi}_2$ and $\boldsymbol{\xi}_3$, the first term must vanish. We have therefore that $\boldsymbol{\xi}_3'$ is parallel to $\boldsymbol{\xi}_2$ and the constant of proportionality is obtained from Eq. (6.15)

$$\boldsymbol{\xi}_3' = -\frac{\boldsymbol{\xi}_2}{\tau}; \tag{6.18}$$

the sign of τ has been chosen to yield the sign shown. From Eq. (6.14) we obtain $\boldsymbol{\xi}_2 = \boldsymbol{\xi}_3 \times \boldsymbol{\xi}_1$ which can be differentiated to obtain $\boldsymbol{\xi}_2'$. Collecting formulas, we have obtained the Frenet–Serret formulas

$$\boldsymbol{\xi}_1' = \frac{\boldsymbol{\xi}_2}{\rho}, \quad \boldsymbol{\xi}_2' = -\frac{\boldsymbol{\xi}_1}{\rho} + \frac{\boldsymbol{\xi}_3}{\tau}, \quad \boldsymbol{\xi}_3' = -\frac{\boldsymbol{\xi}_2}{\tau}. \tag{6.19}$$

Problem 6.1.2. *Show that*

$$\frac{1}{\tau} = \frac{\mathbf{x}' \cdot (\mathbf{x}'' \times \mathbf{x}''')}{\mathbf{x}'' \cdot \mathbf{x}''}. \tag{6.20}$$

Problem 6.1.3. *If the progress of a particle along its trajectory is parameterized by time t, show that the curvature ρ and torsion $1/\tau$ are given by*

$$
\begin{aligned}
\frac{1}{\rho^2} &= \frac{(\dot{\mathbf{x}} \times \ddot{\mathbf{x}}) \cdot (\dot{\mathbf{x}} \times \ddot{\mathbf{x}})}{(\dot{\mathbf{x}} \cdot \dot{\mathbf{x}})^3}, \\
\frac{1}{\tau} &= \frac{\dot{\mathbf{x}} \cdot (\ddot{\mathbf{x}} \times \dddot{\mathbf{x}})}{(\dot{\mathbf{x}} \times \ddot{\mathbf{x}}) \cdot (\dot{\mathbf{x}} \times \ddot{\mathbf{x}})}.
\end{aligned}
\tag{6.21}
$$

As they have been defined, both ρ and τ are *inverse* in the sense that the trajectory becomes more nearly straight as they become large. For this reason their inverses $1/\rho$, known as "curvature" and $1/\tau$, known as "torsion" are more physically appropriate parameters for the trajectory. Loosely speaking, curvature is proportional to $\ddot{\mathbf{x}}$ and torsion is proportional to $\dddot{\mathbf{x}}$. It might seem to be almost accurate to say that in mechanics the curvature is more important than the torsion "by definition." This is because, the curvature being proportional to the transverse component of the applied force, the instantaneously felt force has no component along the binormal direction. This is also why the leading contribution to the torsion is proportional to $\dddot{\mathbf{x}}$. The only circumstance in which the torsion can be appreciable is when the instantaneous force is small but strongly dependent on position. Unfortunately, in this case, the direction of the principal normal can change rapidly in this case even when the force is weak. If the motion is essentially free except for a weak transverse force, the principal normal tracks the force even if the force is arbitrarily small, no matter how its direction is varying. In this case the Frenet frame is simply inappropriate for describing the motion as its orientation is erratically related to the essential features of the trajectory. Furthermore the torsion is also, in a sense, redundant, because the specification of instantaneous position and velocity at any moment, along with a force law giving the acceleration, completely specifies the entire subsequent motion of a particle (including the instantaneous torsion.) Perhaps these considerations account for the previously mentioned lack of emphasis on the Frenet–Serret formulas in most accounts of mechanics?

Since the triad $(\boldsymbol{\xi}_1, \boldsymbol{\xi}_2, \boldsymbol{\xi}_3)$ remains orthonormal it is related to the triad of inertial frame basis vectors $(\hat{\mathbf{x}}, \hat{\mathbf{y}}, \hat{\mathbf{z}})$ by a pure rotation and, instantaneously, by a pure angular velocity vector $\boldsymbol{\omega}$ such as was introduced just before Eq. (6.5). This being the case, the Frenet vectors should satisfy Eq. (6.6). Combined with the Frenet equations this yields

$$
\boldsymbol{\omega} \times \boldsymbol{\xi}_1 = \frac{v\boldsymbol{\xi}_2}{\rho}, \quad \boldsymbol{\omega} \times \boldsymbol{\xi}_2 = -\frac{v\boldsymbol{\xi}_1}{\rho} + \frac{v\boldsymbol{\xi}_3}{\tau}, \quad \boldsymbol{\omega} \times \boldsymbol{\xi}_3 = -\frac{v\boldsymbol{\xi}_2}{\tau}. \tag{6.22}
$$

Furthermore, $\boldsymbol{\omega}$ should itself be expandable in terms of the Frenet vectors, and this expansion must be

$$
\boldsymbol{\omega} = \frac{v}{\tau}\boldsymbol{\xi}_1 + \frac{v}{\rho}\boldsymbol{\xi}_3, \tag{6.23}
$$

as can be quickly checked. Normalized curvature v/ρ measures the rate of rotation of the Frenet frame about the principle normal and normalized torsion v/τ measures its rate of rotation around the tangent vector. Previously in this text, relations specifying the relative orientations of coordinate frames at different positions have been known as "connections" so curvature and torsion can be said to parameterize the connection between the fixed and moving frames of reference.

6.1.3
Vector Description in an Accelerating Coordinate Frame

Another important problem in Newtonian dynamics is that of describing motion using coordinates that are measured in a noninertial frame. Two important applications of this are the description of trajectories using coordinates fixed relative to the (rotating) earth and description of the angular motion of a rigid body about its centroid. These examples are emphasized in this and the following sections. Though frames in linear acceleration relative to each other are also important, the concepts in that case are fairly straightforward, so we will concentrate on the acceleration of rotation. The treatment in this section is not appreciably different and probably not clearer than the excellent and clear corresponding treatment in Symon's book.

Though we will not describe rigid body motion at this time, we borrow terminology appropriate to that subject, namely *space frame* K and *body frame* $\overline{K}$. Also it will seem natural in some contexts to refer to frame $\overline{K}$ as "the laboratory frame" to suggest that the observer is at rest in this frame. The frame K, which will also be known as "the inertial frame," has coordinates $\mathbf{r} = (r^1, r^2, r^3)$ which are related to $\overline{K}$ coordinates $\bar{\mathbf{r}} = (\bar{r}^1, \bar{r}^2, \bar{r}^3)$ by rotation matrix $\mathbf{O}(t)$;

$$\mathbf{r} = \mathbf{O}(t)\,\bar{\mathbf{r}}, \quad \text{or} \quad r^j = O^j_{\ k}(t)\,\bar{r}^k. \tag{6.24}$$

The "inertial" designation has been interjected at this point in preparation for writing Newton's equations in an inertial frame. Much more will be said about the matrix $\mathbf{O}(t)$ but for now we only note that it *connects* two different frames of reference. Unfortunately there is nothing in its notation that specifies what frames $\mathbf{O}(t)$ connects and it is even ambiguous whether or not it deserves to have an overhead bar.[2] It would be possible to devise a notation codifying this information but our policy is to leave the symbol $\mathbf{O}$ unembellished, planning to explain it in words as the need arises.

2) It has been mentioned before, and it will again become clear in this chapter, that when one attempts to maintain a parallelism between "intrinsic appearing" formulas like the first of Eq. (6.24) and coordinate formulas like the second, there is an inevitable notational ambiguity that can only be removed by accompanying verbal description.

In this section the point of view of an observer fixed in the $\overline{K}$ frame will be emphasized (though all vector diagrams to be exhibited will be plotted in the inertial system unless otherwise indicated.) A $\overline{K}$-frame observer locates a particular particle by a vector $\overline{\mathbf{r}}_P$, where the overhead bar connotes that its elements $\overline{x}^i_P$ refer to frame $\overline{K}$. If this frame is accelerating or rotating, the motion will be described by Newton's law expressed in terms of $\overline{x}^i_P$ and the effects of frame rotation are to be accounted for by including fictitious forces, to be called "centrifugal force" and "Coriolis force." The absolute coordinates of P in inertial frame K are then given by the second of Eqs. (6.24).[3]

Equation (6.24) could have been interpreted *actively*, with $\overline{\mathbf{r}}$ being, for example, the initial position of a particular point mass and $\mathbf{r}$ its position at time t. This would be a convenient interpretation for describing, in a single frame of reference, motion starting at $\overline{\mathbf{r}}$ and evolving to $\mathbf{r}$. We *do not* allow this interpretation of Eq. (6.24) however, at least for now. Our policy is explained in the following digression.

When one vector, $\overline{\mathbf{V}}$, has an overhead bar and the other, $\mathbf{V}$, does not, the equation $\mathbf{V} = \mathbf{O}\overline{\mathbf{V}}$ will always be regarded *passively*. That is, the symbols $\mathbf{V}$ and $\overline{\mathbf{V}}$ stand for the same arrow and the equation is an abbreviation for the equation $V^j = O^j_{\ k}\overline{V}^k$ that relates the components of the arrow in two different coordinate frames.[4] It is unattractive to have an *intrinsic*, bold-face, symbol modified to key it to a particular frame, but it is a price that has to be paid to maintain an abbreviated matrix-like notation. It is important to remember this feature of the notation. So $\overline{\mathbf{V}}$ is that true vector (or arrow) whose $\overline{K}$-frame components are $\overline{V}^i$ and $\mathbf{V}$ is the *same* arrow, with K-frame components x^i.[5]

Note that it is only the equation relating apparently intrinsic (because they are in bold face type) quantities for which the notation has to be strained in this way – the relation among components $V^j = O^j_{\ k}\overline{V}^k$ is unambiguous. Whenever the "matrix" $\mathbf{O}$ appears in a formula like $\mathbf{V} = \mathbf{O}\overline{\mathbf{V}}$ that links a barred and an unbarred quantity, the quantity $\overline{\mathbf{O}}$ will not be used because it would become ambiguous later on. Also, with $\boldsymbol{\omega}$ being the angular velocity of frame $\overline{K}$ relative to frame K, we must resist the temptation to use $\overline{\boldsymbol{\omega}}$ to signify the

3) It might seem artificial to describe motion from the point of view of a rotating frame were it not for the fact that, living on a rotating earth, we do it all the time.

4) One must fear ambiguity whenever a frame-specific notation, such as an overhead bar, is attached to a (bold-face) vector symbol. This ambiguity is *intrinsic* to the *intrinsic* nature of a *true* vector, since such a vector has an existence that transcends any particular frame of reference. There is no such ambiguity when a notation such as an overhead bar is attached to the *components* of a vector.

5) We have to accept the unsettling feature of this notation that, though we say $\overline{\mathbf{V}}$ and $\mathbf{V}$ are the same arrow, it would not be good form to say $\overline{\mathbf{V}} = \mathbf{V}$ since that would make the equation $\mathbf{V} = \mathbf{O}\overline{\mathbf{V}}$ seem silly. If we slip into writing such an equation it should be written as $\overline{\mathbf{V}} \stackrel{q}{=} \mathbf{V}$, or as an intermediate step in an algebraic simplification where the situation is to be repaired in a subsequent step.

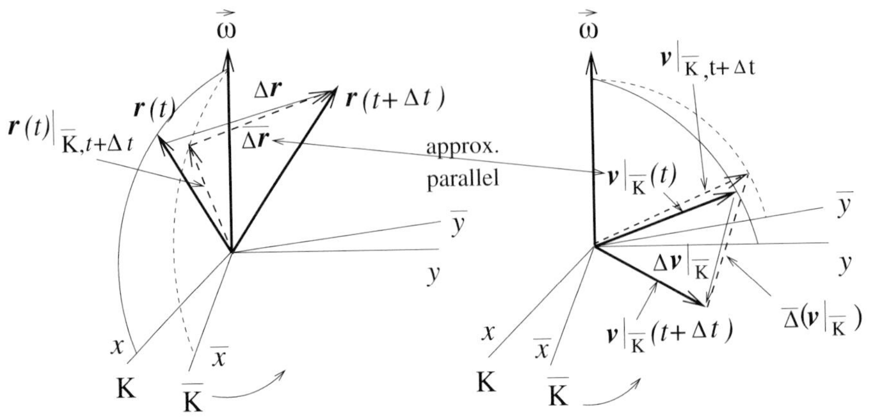

(a) $\overline{v}$ determination, K's plot (b) $\overline{a}$ determination, K's plot

Fig. 6.4 Body frame $\overline{K}$ rotates with angular velocity $\boldsymbol{\omega}$ about its common origin with inertial frame K. All arrows shown are plotted in frame K. (a) Two points $\mathbf{r}(t)$ and $\mathbf{r}(t+\Delta t)$ on the trajectory of a moving particle are shown. For an observer in frame $\overline{K}$ at time $t+\Delta t$ the radius vector $\mathbf{r}(t+\Delta t)$ is the same arrow as the K-frame arrow at that time; but $\mathbf{r}(t)|_{\overline{K},t+\Delta t}$, the $\overline{K}$-frame observer's recollection at $t+\Delta t$ of where the particle was at time t, is different from its actual location at time t, which was $\mathbf{r}(t)$. (b) A similar construction permits determination of acceleration $\mathbf{a}|_{\overline{K}}$.

angular velocity of frame K relative to frame $\overline{K}$ as that would clash with our later definition of $\overline{\boldsymbol{\omega}}$.

For the time being (until Section 6.2) this discussion will have been academic since equations of the form $\mathbf{V} = \mathbf{O}\overline{\mathbf{V}}$ will not appear and, for that matter, neither will vectors symbolized as $\overline{\mathbf{V}}$.

The description of the motion of a single particle by observers in frames K and $\overline{K}$ is illustrated in Fig. 6.4. (One can apologize for the complexity of this figure without knowing how to make it simpler.) Heavy lines in this figure are the images of arrows in a double exposure snapshot (at t and $t+\Delta t$) taken in the inertial frame. Like all arrows, the arrows in this figure illustrate *intrinsic* vectors. Body frame $\overline{K}$ rotates with angular velocity $\boldsymbol{\omega}$ about the common origin. Like $\mathbf{O}$, $\boldsymbol{\omega}$ is called a *connecting* quantity since it connects two different frames. Mainly for convenience in drawing the figure, (x,y) and $(\overline{x},\overline{y})$ axes are taken orthogonal to $\boldsymbol{\omega}$ which is therefore a common z- and $\overline{z}$-axis, and all axes coincide at $t=0$. (The axes will not actually be used in the following discussion.) At any time t the position of the moving particle is represented by an arrow $\mathbf{r}(t)$, which is necessarily the same arrow whether viewed from K or $\overline{K}$. But at time $t+\Delta t$ an observer in frame $\overline{K}$ "remembers" the position of the particle at time t as having been at a point other than where it actually was – in the figure this is indicated by the dashed arrow labeled $\mathbf{r}(t)|_{\overline{K},t+\Delta t}$. As a result, the actual displacement $\Delta\mathbf{r}$ occurring during time interval Δt and the apparent-to-$\overline{K}$ displacement (shown dashed) $\overline{\Delta\mathbf{r}}$ are different.

Our immediate task is to relate the velocities observed in the two frames. Since the vectors $\Delta\mathbf{r}$ and $\overline{\Delta\mathbf{r}}$ stand for unambiguous arrows, plotted in the same frame K, it is meaningful to add or subtract them. From the figure, in the limit of small Δt,

$$\mathbf{r}(t) = \mathbf{r}(t)|_{\overline{K},t+\Delta t} - \boldsymbol{\omega}\Delta t \times \mathbf{r}(t), \quad \text{or} \quad \Delta\mathbf{r} = \overline{\Delta\mathbf{r}} + \boldsymbol{\omega}\Delta t \times \mathbf{r}(t). \tag{6.25}$$

From this we obtain

$$\frac{d\mathbf{r}}{dt} = \overline{\frac{d}{dt}}\mathbf{r} + \boldsymbol{\omega} \times \mathbf{r}, \quad \text{or} \quad \mathbf{v} = \mathbf{v}|_{\overline{K}} + \boldsymbol{\omega} \times \mathbf{r}, \tag{6.26}$$

where, transcribing the geometric quantities into algebraic quantities, we have defined

$$\mathbf{v}|_{\overline{K}} \equiv \overline{\frac{d}{dt}}\mathbf{r} \equiv \lim_{\Delta t \to 0} \frac{\overline{\Delta\mathbf{r}}}{\Delta t}, \tag{6.27}$$

and thereby assigned meaning to the operator $\overline{d/dt}$. The $\overline{K}$ components of $\overline{d/dt}\,\mathbf{r}$ are $(d\overline{x}^1/dt, d\overline{x}^2/dt, \dots)$.

Since Eq. (6.26) is a vector equation, it is valid in any coordinate frame, but if it is to be expressed in components, it is essential that components on the two sides be taken in the same frame. Since we are trying to describe motion from the point of view of a $\overline{K}$ observer, we will eventually use components in that frame. First though, to apply Newton's law, we must calculate acceleration.

Though the derivation so far was based on the displacement vector $\mathbf{r}$, any other *true vector* $\mathbf{V}$, being an equivalent geometric object, must satisfy an equivalent relation, namely

$$\frac{d\mathbf{V}}{dt} = \overline{\frac{d}{dt}}\mathbf{V} + \boldsymbol{\omega} \times \mathbf{V}. \tag{6.28}$$

In particular this can be applied to velocity $\mathbf{v}$, with the result

$$\frac{d\mathbf{v}}{dt} = \overline{\frac{d}{dt}}\mathbf{v} + \boldsymbol{\omega} \times \mathbf{v} = \overline{\frac{d^2}{dt^2}}\mathbf{r} + \overline{\frac{d}{dt}}(\boldsymbol{\omega} \times \mathbf{r}) + \boldsymbol{\omega} \times \mathbf{v}|_{\overline{K}} + \boldsymbol{\omega} \times (\boldsymbol{\omega} \times \mathbf{r}), \tag{6.29}$$

where the extra step of using Eq. (6.26) to replace $\mathbf{v}$ has been taken.

Though the formal manipulations have been simple we must be sure of the meaning of every term in Eq. (6.29). The term on the left is the well-known inertial frame acceleration; for it we will use the traditional notation $\mathbf{a} \equiv d\mathbf{v}/dt$. In the terms $\boldsymbol{\omega} \times \mathbf{v}|_{\overline{K}}$ and $\boldsymbol{\omega} \times (\boldsymbol{\omega} \times \mathbf{r})$ only standard vector multiplication operations are performed on arrows illustrated in Fig. 6.4. All except $\mathbf{v}|_{\overline{K}}$ are shown in the (a) part of the figure and $\mathbf{v}|_{\overline{K}}$ is shown in the (b) part. $\mathbf{v}|_{\overline{K}}$ is the apparent velocity where "apparent" means "from the point of view of an observer stationary in the $\overline{K}$ frame who is (or pretends to be) ignorant of being in a noninertial frame." The $\overline{K}$ frame components of $\mathbf{v}|_{\overline{K}}$ are $(d\overline{x}^1/dt, d\overline{x}^2/dt, \dots)$. It is shown as "approximately parallel" to $\overline{\Delta \mathbf{r}}$ because average and instantaneous velocities over short intervals are approximately parallel – in the limit of small Δt this becomes exact. For simplicity let us assume that $\boldsymbol{\omega}$ is time independent, (this restriction will be removed later,) in which case $\frac{\overline{d}}{dt}(\boldsymbol{\omega} \times \mathbf{r}) = \boldsymbol{\omega} \times \mathbf{v}|_{\overline{K}}$. The only remaining term in Eq. (6.29) deserves closer scrutiny. Defining

$$\mathbf{a}|_{\overline{K}} \equiv \frac{\overline{d}}{dt}(\mathbf{v}|_{\overline{K}}) \equiv \frac{\overline{d^2}}{dt^2}\,\mathbf{r}, \tag{6.30}$$

it can be said to be the *apparent* acceleration from the point of view of $\overline{K}$. Its $\overline{K}$ components are $(d^2\overline{x}^1/dt^2, d^2\overline{x}^2/dt^2, \dots)$. The (b) part of Fig. 6.4 continues the construction from the (a) part to determine **a**.

Combining these results we obtain

$$\mathbf{a} = \mathbf{a}|_{\overline{K}} + 2\boldsymbol{\omega} \times \mathbf{v}|_{\overline{K}} + \boldsymbol{\omega} \times (\boldsymbol{\omega} \times \mathbf{r}). \tag{6.31}$$

At the risk of becoming repetitious let it again be stressed that, even though all terms on the right-hand side of this equation are expressed in terms of quantities that will be evaluated in the $\overline{K}$ frame, the arrows they stand for are all plotted in the K frame in Fig. 6.4 and are hence commensurable with **a** – otherwise Eq. (6.31) could not make sense. On the other hand, since Eq. (6.31) is a *vector* equation, it can be expressed in component form in any frame – for example in the $\overline{K}$ frame. The resulting $\overline{K}$-components are related to the K-components in the well-known way vectors transform, namely the component form of Eq. (6.24).[6] The point of introducing fictitious forces has been to validate an analysis that describes kinematics purely in terms of the vectors shown as heavy arrows in Fig. 6.5.

Since the second and third terms of Eq. (6.31), though artifacts of the description, appear to augment (negatively) the inertial acceleration, they are known as "fictitious" accelerations. With the inertial acceleration related to the "true force" $\mathbf{F}(\text{true})$ by Eq. (6.1), the "fictitious" forces are

$$\begin{aligned} \mathbf{F}(\text{centrifugal}) &= -\,m\boldsymbol{\omega} \times (\boldsymbol{\omega} \times \mathbf{r}) = m\omega^2\,\mathbf{r} - m(\boldsymbol{\omega}\cdot\mathbf{r})\boldsymbol{\omega}, \\ \mathbf{F}(\text{Coriolis}) &= -\,2m\boldsymbol{\omega} \times \mathbf{v}|_{\overline{K}}, \end{aligned} \tag{6.32}$$

and the equation of motion becomes

$$\mathbf{a}|_{\overline{K}} = \frac{1}{m}\big(\mathbf{F}(\text{true}) + \mathbf{F}(\text{centrifugal}) + \mathbf{F}(\text{Coriolis})\big). \tag{6.33}$$

For practical calculations in component form each of the terms is decomposed into $\overline{K}$-frame components. In the case of $\mathbf{F}(\text{true})$ this exploits the fact that $\mathbf{F}(\text{true})$ is in fact *a true vector*.

6) A vector construction analogous to that of Fig. 6.4 can be performed in the $\overline{K}$ frame, as shown by dashed vectors in Fig. 6.5. These vectors are only shown to make this point though; the noninertial frame description describes the motion using only the heavy arrows.

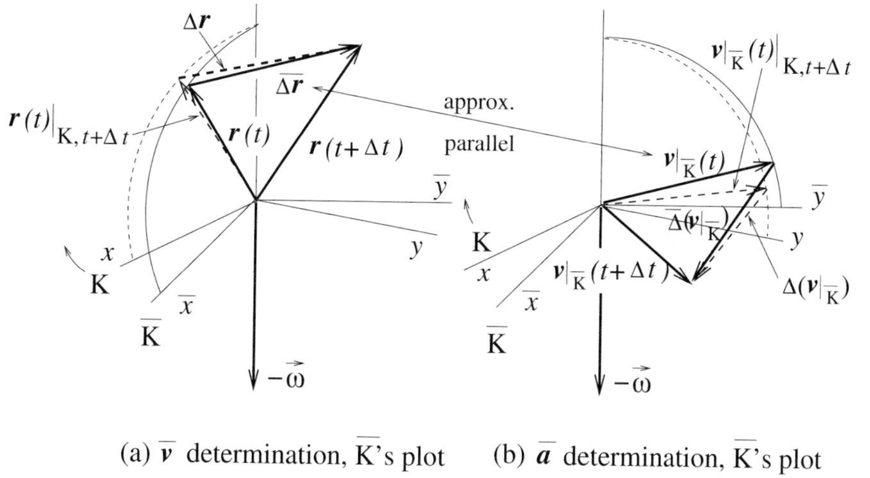

Fig. 6.5 In this figure vectors entering into the determination of $\mathbf{v}|_{\overline{K}}$ and $\mathbf{a}|_{\overline{K}}$ are plotted in a plot that is stationary in the $\overline{K}$ frame. $\mathbf{a}|_{\overline{K}}$ is given by $\overline{\Delta}\mathbf{v}|_{\overline{K}}/\Delta t$ in the limit of small Δt. This figure violates our convention that *all* figures be drawn in an inertial frame.

After some examples illustrating the use of these formulas, this analysis will be re-expressed in different terms, not because anything is wrong with the derivation just completed, but in preparation for proceeding to more complicated situations.

Problem 6.1.4. *Express in your own words the meaning of the symbols* $\Delta\mathbf{r}$ *and* $\Delta(\mathbf{v}|_{\overline{K}})$ *in Fig. 6.5. If that figure seems obscure to you, feel free to redraw it in a way to make it seem clearer.*

Problem 6.1.5. *The radial force on a mass* m*, at radius* r *relative to the center of the earth (mass* M_E*) is* $\mathbf{F} = -mM_E G\hat{\mathbf{r}}/r^2$*. Ignoring the motion of the earth about the sun, but not the rotation of the earth, the motion of a satellite of the earth can be described in inertial coordinates with the earth at the origin or in terms of* (r, θ, ϕ)*, which are the traditional radial distance, co-latitude, and longitude that are used for specifying geographical objects on earth.*

(a) It is possible for the satellite to be in a "geosynchronous" orbit such that all of its coordinates (r, θ, ϕ) *are independent of time. Give the conditions determining this orbit and find its radius* r_S *and latitude* θ_S*.*

(b) Consider a satellite in an orbit just like that of part (a) except that it passes over the North and South poles instead of staying over the equator. Give (time dependent) expressions for the coordinates (r, θ, ϕ)*, as well as for the Coriolis and centrifugal forces, and show that Newton's law is satisfied by the motion.*

6.1.4 Exploiting the Fictitious Force Description

The mental exertion of the previous section is only justified if it simplifies some physical calculation. The reader has probably encountered discussions of the influence of the Coriolis force on weather systems in the earth's atmosphere. (e.g., Kleppner and Kolenkow, p. 364.) Here we will only give enough examples to make clear the practicalities of using Eq. (6.33). Working Problem 6.1.5 goes a long way in this direction. Finally the historically most significant example, the Foucault pendulum, is analyzed.

Though Eq. (6.33) was derived by working entirely with vectors drawn in the K frame, since it is a vector equation, it can be used working entirely with vector calculations in the $\overline{K}$ frame. In this frame the Coriolis and centrifugal forces are every bit as effective in producing acceleration as is the force which to this point has been called "true."

For terrestrial effects the angular velocity is

$$\omega_E = \frac{2\pi}{24 \times 3600} = 0.727 \times 10^{-4}\,\mathrm{s}^{-1}. \tag{6.34}$$

On the earth's equator, since the centrifugal force points radially outward parallel to the equatorial plane, the acceleration it produces can be compared directly to the "acceleration of gravity" $g = 9.8\,\mathrm{m/s}$. The relative magnitude is

$$\frac{{\omega_E}^2 R_E}{g} = 3.44 \times 10^{-3}. \tag{6.35}$$

Though appreciable, this is comparable with the fractional variation of g over the earth's surface due to local elevation. Furthermore, this effect can be included by "renormalizing" the force of gravity slightly in magnitude and direction. This has no appreciable meteorological consequence. The relative magnitude of the Coriolis and centrifugal forces tends to be dominated by the extra factor of ω in F (centrifugal) relative to F (Coriolis). The centrifugal force would be expected to make itself most effectively visible through the force difference occurring over an altitude difference comparable with the height of the earth's atmosphere; let a typical value be $\Delta r = 10\,\mathrm{km}$. The Coriolis force can be estimated as being due to the velocity of a particle having "fallen" through such a change of altitude. For a particle accelerating through distance Δr under the influence of the earth's gravity the velocity v is $\sqrt{g\,\Delta r}$ and a typical value for the ratio $v/\Delta r$ is $\sqrt{g/\Delta r} \approx \sqrt{9.8/10^4} = 0.03\,\mathrm{s}^{-1}$. This is for a "large" fall; for a smaller fall the ratio would be greater. Since this $v/\Delta r$ is already much greater than ω_E, the Coriolis force tends to be more significant than the centrifugal force in influencing terrestrial phenomena.

The Coriolis force has the property of depending on the velocity of the moving particle. Some precedents for velocity dependence in elementary mechanics are friction and viscous drag. These forces are dissipative however, while the Coriolis force clearly is not, since it resulted from a change of reference. With regard to depending on velocity, but being lossless, the Coriolis force resembles the force on a moving charged particle in a magnetic field, with the magnetic field and $\boldsymbol{\omega}$ playing roughly analogous roles. The characteristic qualitative feature of motion of a charged particle in a magnetic field is that it tends to move in a roughly circular helix wrapping around the magnetic field lines. This suggests, at least in some ranges of the parameters, that the Coriolis force will cause qualitatively similar motion around $\boldsymbol{\omega}$. Of course the presence of the earth's surface acting as a boundary that is not normal to $\boldsymbol{\omega}$ tends to invalidate this analogy, but it should not be surprising that the Coriolis force can lead to atmospheric motion in "vortices."

Example 6.1.1. A particle falling freely close to the earth's surface. *Newton's law for free fall with the earth's curvature neglected is* $\dot{\mathbf{v}} = -g\hat{\mathbf{z}}$, *where* g *is the "acceleration of gravity" and* $\hat{\mathbf{z}}$ *points in the local "vertical" direction. Starting with velocity* $\mathbf{v}_0$, *after time* t *the particles velocity is* $\mathbf{v}_0 - gt\,\hat{\mathbf{z}}$. *Including the Coriolis force the equation of "free fall" becomes*

$$\dot{\mathbf{v}} + g\hat{\mathbf{z}} = 2\omega_E \mathbf{v} \times \hat{\boldsymbol{\omega}}, \tag{6.36}$$

where $\hat{\boldsymbol{\omega}}$ *is directed along the earth's axis of rotation. As a matter of convention, to be followed frequently in this text and elsewhere, the terms describing an idealized, solvable, system have been written on the left-hand side of this equation and the "perturbing" force that makes the system deviate from the ideal system has been written on the right-hand side. If the perturbing term is "small" then it can be estimated by approximating the factor* $\mathbf{v}$ *by its "unperturbed" value which is obtained by solving the equation with right-hand side neglected. This procedure can be iterated to yield high accuracy if desired. In the present case, the equation in first iteration is*

$$\dot{\mathbf{v}} + g\hat{\mathbf{z}} = 2\omega_E(\mathbf{v}_0 - gt\hat{\mathbf{z}}) \times \hat{\boldsymbol{\omega}} = 2\omega_E \mathbf{v}_0 \times \hat{\boldsymbol{\omega}} - 2g\omega_E \sin\theta\,\hat{\phi}\,t, \tag{6.37}$$

where θ *is the "co-latitude" angle (away from North) and* $\hat{\phi}$ *is a unit vector pointing from east to west along a line of latitude. Since all the force terms are now functions only of* t, *(6.37) can be integrated easily. Starting from rest, the falling object veers eastward because of the Coriolis force.*

Problem 6.1.6. *Integrate Eq. (6.37) (twice) to find* $\mathbf{r}(t)$ *for a freely falling mass subject to gravity and the Coriolis force. For the case* $\mathbf{v}_0 = 0$, *find the approximate spatial trajectory by using the relation between time and altitude appropriate for unperturbed motion.*

Problem 6.1.7. *Using the velocity obtained from Eq. (6.37), perform a second iteration and from it write an equation of motion more accurate than Eq. (6.37).*

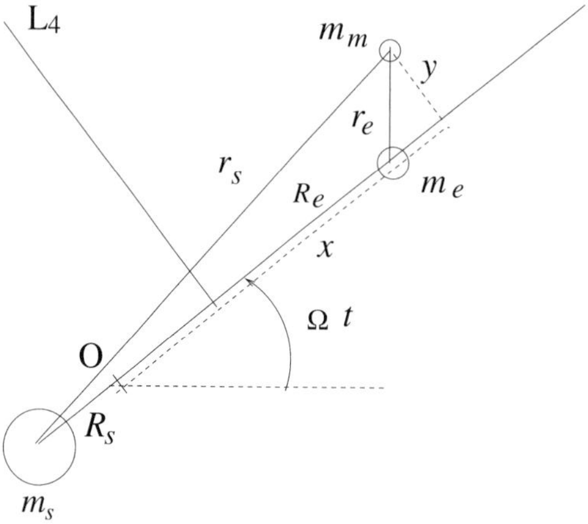

Fig. 6.6 The sun, the earth, the moon system, with all orbits assumed to lie in the same plane. The bisector of the sun–earth line is shown and the point L_4 makes an equilateral triangle with the sun and the moon.

6.1.4.1 **The Reduced Three-Body Problem**

Though the problem of three bodies subject to each other's gravitational attraction is notoriously nonintegrable in general, there are simplifying assumptions that can be made which simplify the problem while still leaving it applicable to realistic celestial systems. In the so-called "reduced three-body" problem the three masses are taken to be $1-\mu$, μ, and 0 (where mass being zero should better be stated as "mass is negligible" and implies that the position of the third mass has no effect on the motions of the first two.) In this case the motion of the first two is integrable and they move *inexorably*, independent of the third. This inexorable motion causes the gravitational potential sensed by the third particle to be time varying. Since the problem is still complicated one also assumes that all three orbits lie in the same plane; call it the (x, y) plane. For further simplification one also assumes that the orbits of the first two masses around each other are circular.

All the approximations mentioned so far are applicable to the system consisting of the sun, the earth, and the moon, so let us say this is the system we are studying. Symbols defining the geometry are shown in Fig. 6.6. Formulated in this way, the problem still has interest apart from its mundane, everyday, observability. One can, for example, inquire as to what stable orbits the earth's moon might have had, or what are the possible orbits of satellites around other binary systems. For that reason, though it would be valid to assume $m_m \ll m_e \ll m_s$, we will only assume $m_m \ll m_e$ and $m_m \ll m_s$. Also we will not assume $r_e \ll r_s$ even though it is true for the earth's moon. As mentioned above, the gravitational potential at the moon depends on time.

But if the system is viewed from a rotating coordinate system this feature can be removed. Defining the "reduced mass" m of the sun–earth system by $m \equiv m_s m_e/(m_s + m_e)$, R as their separation distance, M as their angular momentum, one knows that this system is rotating about their centroid with *constant* angular velocity Ω given by

$$\Omega = \frac{M}{mR^2}. \tag{6.38}$$

It was the requirement that Ω and R be constant that made it appropriate to require the sun and the earth orbits to be circular. The other constant distances satisfy $R_s = Rm/m_s$, $R_e = Rm/m_e$, and $R_s + R_e = R$. Viewed from a system rotating with angular velocity Ω about the centroid, both the sun and the earth appear to be at rest so the gravitational potential has been rendered time independent. The potentials due to the sun and the earth are

$$V_s = -\frac{m_s G}{\sqrt{(x + R_s)^2 + y^2}}, \quad \text{and} \quad V_e = -\frac{m_e G}{\sqrt{(x - R_e)^2 + y^2}}. \tag{6.39}$$

The centrifugal force can be included by including a contribution to the potential energy given by

$$V_{\text{cent}} = -\frac{1}{2}\Omega^2(x^2 + y^2). \tag{6.40}$$

Combining all potentials we define

$$V_{\text{eff}} = V_s + V_e + V_{\text{cent}}. \tag{6.41}$$

Including the Coriolis force, the equations of motion are

$$\begin{aligned} \ddot{x} &= 2\Omega\dot{y} - \frac{\partial V_{\text{eff}}}{\partial x}, \\ \ddot{y} &= -2\Omega\dot{x} - \frac{\partial V_{\text{eff}}}{\partial y}. \end{aligned} \tag{6.42}$$

Problem 6.1.8. *The quantity $h = V_{eff} + v^2/2$ where $v^2 = \dot{x}^2 + \dot{y}^2$ would be the total energy of the moon, which would be conserved (because the total energy of the sun–earth system is constant) except (possibly) for the effect of being in a rotating coordinate system. Show, by manipulating Eqs. (6.42) to eliminate the Coriolis terms, that h is, in fact, a constant of the motion. It is known as the "Jacobi integral."*

Problem 6.1.9. *Find a Lagrangian for which Eqs. (6.42) are the Lagrange equations. (It is not necessary for a Lagrangian to have the form $L = T - V$, and if it is written in this form it is legitimate for $V(\mathbf{r}, \dot{\mathbf{r}})$ to depend on both velocities and positions.)*

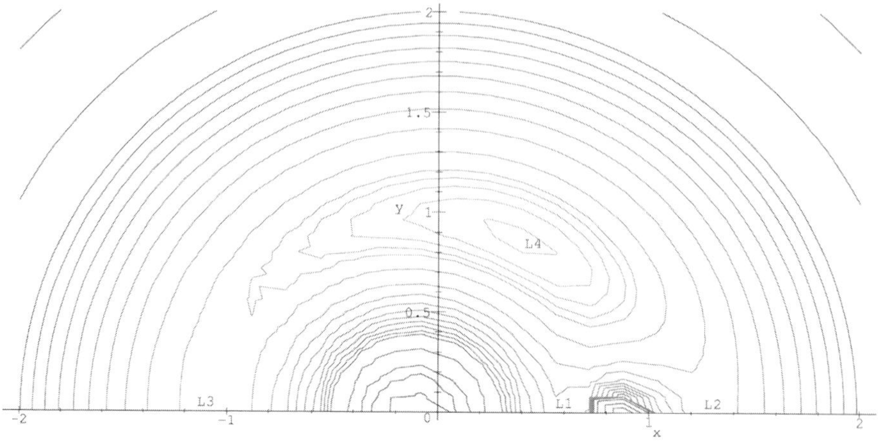

Fig. 6.7 Contour plot of V_{eff} for $R = 1$, $G = 1, \Omega = 1, \mu = 0.1, m_s = 1 - \mu, m_e = \mu$. The contours shown are for constant values of V_{eff} given by $-10, -5, -4, -3, -2.5, -2.4, -2.3, -2.2, -2.1, -2, -1.9, -1.8, -1.7, -1.6, -1.54, -1.52, -1.5, -1.48, -1.46, -1.44, -1.42, -1.4, -1.38, -1.36, -1.3, -1.2, -1.1, -1, -.5, -.2, -.1$. The order of these contours can be inferred by spot calculation and the observation that there is a maximum running roughly along a circle of radius 1. Only the positive y region is shown; V_{eff} is an even function of y.

On the basis of the constancy of the Jacobi integral h, some things can be inferred about possible motions of the moon from a contour plot of V_{eff}. For a particular choice of the parameters, such a contour plot is shown in Fig. 6.7. Some of these contours are approximate trajectories, in particular the "circles" close to and centered on the sun, but otherwise the relations between these contours and valid orbits is less clear. For each of these contours, if it were a valid trajectory, since both h and V_{eff} are constant, so also would be v. For an orbit that is temporarily tangent to one of these contours the tangential components of both the Coriolis force and the force due to V_{eff} vanish so v is temporarily stationary. Presumably the "generic" situation is for v to be either a maximum or a minimum as the orbit osculates the contour. For orbits that are approximate elliptical Kepler orbits around the sun, these two cases correspond approximately to the maximum and minimum values of v as the moon (in this case it would be more appropriate to say "other planet") moves more or less periodically between a smallest value (along a semi-minor axis) and a largest value (along a semi-major axis). In this case then, the orbit stays in a band between a lowest and a highest contour, presumably following a rosetta-shaped orbit that, though resembling a Kepler ellipse does not quite close. If the moon's velocity matches the speed v required by the osculating contour then this band is slender.[7] In greater generality, at any point in the space, by

7) There are remarkable "ergodic theorems" (due originally to Poincaré) that permit heuristically plausible statements such as these to be turned into rigorous results.

judicious choice of initial conditions it should similarly be possible to launch a satellite with the correct speed and direction so it will follow the particular contour passing through the launch point for an appreciable interval. It will not stay on the contour indefinitely though since the transverse acceleration deviates eventually from that required to remain on the contour.

The points labeled L1, L2, and L3, are known as "Lagrange unstable fixed points" and L4, with its symmetric partner L5, are known as "Lagrange stable fixed points." These are the points for which $\partial V_{\text{eff}}/\partial x = \partial V_{\text{eff}}/\partial y = 0$. If a "moon" is placed at rest at one of these points, since the Coriolis force terms and the V_{eff} force terms vanish, the moon would remain at rest.

The most interesting points are L4 and L5. Since there are closed curves surrounding these points there appears to be the possibility of satellite orbits "centered" there. In modern jargon one would say that Lagrange "predicted" the presence of satellites there. Some 100 years later it was discovered that the asteroid Achilles resides near L4 in the sun, Jupiter system, and numerous other asteroids have been discovered subsequently near L4 and L5.

Problem 6.1.10. *On a photocopy of Fig. 6.7 sketch those contours passing through the Lagrange fixed points, completing the lower half of the figure by symmetry and taking account of the following considerations. At "generic" points (x, y) the directional derivative of a function $V(x, y)$ vanishes in one direction (along a contour) but not in directions transverse to this direction. (On the side of a hill there is only one "horizontal" direction.) In this case adjacent contours are more or less parallel and hence cannot cross each other. At particular points though (fixed points) both derivatives vanish and contours can cross (saddle points) or not (true maxima or minima). It is easy to see that L1 is a saddle point and from the figure it appears that L4 and L5 are either maxima or minima. For the parameter values given, test which is the case and see if this agrees with Eq. (6.45) below. For L2 and L3, determine if they are stable or unstable, and, if the latter whether they are saddle points or maxima. Sketching contours that either cross or not, as the case may be, at these points. It should be possible to follow each such contour back to its starting point, wherever that is. Also, in general, one would not expect two fixed points to lie on the same contour.*

The linearized equations of motion, valid near one of the fixed points, say L4, are

$$\begin{aligned} \ddot{x} &= 2\Omega\dot{y} - V_{xx}x - V_{xy}y, \\ \ddot{y} &= -2\Omega\dot{x} - V_{yx}x - V_{yy}y, \end{aligned} \tag{6.43}$$

where partial derivatives are indicated by subscripts and the origin has been placed at the fixed point. Conjecturing a solution of the form $x = Ae^{\lambda t}$, $y = Be^{\lambda t}$, these equations become

$$\begin{pmatrix} \lambda^2 + V_{xx} & -2\Omega\lambda + V_{xy} \\ 2\Omega\lambda + V_{xy} & \lambda^2 + V_{yy} \end{pmatrix} \begin{pmatrix} A \\ B \end{pmatrix} = 0. \tag{6.44}$$

The condition for such linear homogeneous equations to have nontrivial solutions is that the determinant of coefficients vanishes;

$$\lambda^4 + (4\Omega^2 + V_{xx} + V_{yy})\lambda^2 + (V_{xx}V_{yy} - V_{xy}^2) = 0. \qquad (6.45)$$

This is a quadratic equation in λ^2. The condition for stable motion is that both possible values of λ be pure imaginary. This requires λ^2 to be real and negative.

Problem 6.1.11. *Evaluate the terms of Eq. (6.45) for the Lagrange fixed point L4, and show that the condition for stable motion in the vicinity of L4 is*

$$27\mu(1-\mu) < 1. \qquad (6.46)$$

where mu $= m_e/m_s$. *For the sun–Jupiter system* $\mu \approx 10^{-3}$ *which satisfies the condition for stability, consistent with the previously mentioned stable asteroids near L4 and L5.*

Problem 6.1.12. Larmor's Theorem

(a) *The force* $\mathbf{F}_m$ *on a particle with charge* q *and velocity* $\mathbf{v}$ *in a constant and uniform magnetic field* $\mathbf{B}$ *is given by* $\mathbf{F}_m = q\mathbf{v} \times \mathbf{B}$. *Write the equation of motion of the particle in a frame of reference that is rotating with angular velocity* $\boldsymbol{\omega}$ *relative to an inertial frame. Assume that* $\boldsymbol{\omega}$ *is parallel to* $\mathbf{B}$. *Show, if the magnetic field is sufficiently weak, that the magnetic and fictitious forces can be made to cancel by selecting the magnitude of the angular velocity. Give a formula expressing the "weakness" condition that must be satisfied for this procedure to provide a good approximation.*

(b) *Consider a classical mechanics model of an overall neutral atom consisting of light, negatively charged, electrons circulating around a massive, point nucleus. Known as the "Zeeman effect," placing an atom in a magnetic field* $\mathbf{B}$ *shifts the energy levels of the electrons. In the classical model each electron is then subject to electric forces from the nucleus and from each of the other electrons as well as the magnetic force. Assuming the weakness condition derived above is satisfied, show that the electron orbits could be predicted from calculations in a field free rotating frame of reference.*

6.2 Single Particle Equations in Gauge Invariant Form

The term "gauge invariant," probably familiar from electromagnetic theory, has recently acquired greater currency in other fields of theoretical physics. In colloquial English a "gauge" is a device for measuring a physical quantity – a thermometer is a temperature gauge, a ruler is a length gauge. In electromagnetic theory "gauge invariant" describes a kind of freedom of choice of scalar or vector potentials, but it is hard to see why the word "gauge" is thought to call such freedom to mind in that case. In the context of geometric mechanics, the term "gauge invariant" more nearly approximates its colloquial meaning. When one describes a physical configuration by coordinates that refer to inertial, fixed, orthonormal, Euclidean axes, one is committed to choosing a single measuring stick, or gauge, and locating every particle by laying off distances along the axes using the same stick. A theory of the evolution of such a system expressed in these coordinates would not be *manifestly* gauge invariant, because it is explicitly expressed in terms of a particular gauge. But this does not imply that the same theory *cannot* be expressed in gauge invariant form. An example of this sort of mathematical possibility (Gauss's theorem) was discussed in Section 4.3.5. Though Gauss's theorem is commonly expressed in Euclidean coordinates, it is expressed in coordinate-independent form in that section. In this chapter the term "gauge-invariant" will have the similar meaning of "coordinate-frame invariance." The gauge could, in principle, depend on position, but since that will not be the case here, we have to deal with only a much simplified form of gauge-invariance.

Much of the analysis to follow can be described operationally as the effort to derive equations in which all quantities have overhead bars (or all do not.) Such an equation will then be said to be *form invariant*. If the frame in which the equation was derived is itself general then the equation will have the powerful attribute of being applicable in any coordinated system having the corresponding degree of generality. An example of equations having this property are Maxwell's equations; they have the same form in all frames traveling at constant speed relative to a base frame.

6.2.1 Newton's Force Equation in Gauge Invariant Form

A particle of mass m[8], situated at a point P with coordinates x^i, is subject to Newton's equation,

$$m\frac{d^2x^i}{dt^2} = f^i(\mathbf{r}, \dot{\mathbf{r}}, t), \tag{6.47}$$

where f^i is the force[9] (possibly dependent on $\mathbf{r}$, $\dot{\mathbf{r}}$, and t). We are interested in descriptions of motions in two relatively rotating frames. Since the f^i are

8) We talk of a point mass m even though it will often be the mass dm contained in an infinitesimal volume dV, perhaps fixed in a rigid body, that is being discussed.

9) Since we will use only Euclidean coordinates for now, it is unnecessary to distinguish between covariant and contravariant components of the force.

components of a vector they are subject to the same transformation (6.24) as **r**;

$$\mathbf{f} = \mathbf{O}(t)\,\overline{\mathbf{f}}, \quad f^j = O^j{}_k(t)\,\overline{f}^k. \tag{6.48}$$

Recall that, by our conventions, this is to be regarded as a *passive* transformation, relating the components of **f** and $\overline{\mathbf{f}}$, which stand for the same arrow. Our mission is to write Newton's equation in "gauge invariant" form where, in this case, choice of "gauge" means choice of a coordinate system, with rotating coordinate systems being allowed. (The ordinary Newton equations are gauge invariant in this sense when the choice is restricted to inertial frames; this is known as "Galilean invariance.") Formally, the task is to re-express Newton's equation entirely in terms of quantities having overhead bars.

The introduction of Coriolis and centrifugal forces has gone a long way toward realizing our goal – they are fictitious forces which, when added to true forces, make it legitimate to preserve the fiction that a rotating frame is inertial. This "fictitious force" formulation has amounted to evaluating inertial frame quantities entirely in terms of moving frame quantities. Once the gauge invariant formulation has been established there will be no further need for more than one frame and it will be unnecessary to distinguish, say, between $\overline{d/dt}$ and d/dt. For want of better terminology, we will use the terms "fictitious force formalism" and "gauge invariant formalism" to distinguish between these two styles of description even though the terminology is a bit misleading. It is misleading because, not only are the formulations equivalent in content, they are similarly motivated.

The present treatment strives to express Newton's equations in such a way that they have the same form in any reference frame. It is somewhat more general than the simple introduction of fictitious centrifugal and Coriolis forces because the axis and velocity of rotation of the rotating frame will now not necessarily be constant. At some point a sense of *déja vu* may develop, as the present discussion is very similar to that contained in Section 3.2 which dealt with the application of the absolute differential in mechanics, though the present situation is somewhat more general because time-dependent frames are now to be allowed. In that earlier section an operator D was introduced with the property that position **r** and its time derivative were related by $\dot{\mathbf{r}} = D\mathbf{r}$ but the curvilinear effect was shown to cause the absolute acceleration vector $\mathbf{a} = D^2\mathbf{r}$ to differ from $\ddot{\mathbf{r}}$. In the present case, an analogous differential operator D_t will be defined; in terms of it the absolute velocity is $D_t\mathbf{r}$. As in the curvilinear chapter it will be true that $\ddot{\mathbf{r}} \neq D_t^2\mathbf{r}$ but now the time-dependent relation between frames will cause $\dot{\mathbf{r}}$ to differ from $D_t\mathbf{r}$ as well.

Differentiating Eq. (6.24), the inertial-frame velocity $\mathbf{v} \equiv \dot{\mathbf{r}}$ is given by

$$\mathbf{v} = \mathbf{O}\dot{\overline{\mathbf{r}}} + \dot{\mathbf{O}}\overline{\mathbf{r}} = \mathbf{O}(\dot{\overline{\mathbf{r}}} + (\mathbf{O}^T\dot{\mathbf{O}})\overline{\mathbf{r}}) = \mathbf{O}(\dot{\overline{\mathbf{r}}} + \boldsymbol{\Omega}\overline{\mathbf{r}}), \tag{6.49}$$

where[10]

$$\overline{\boldsymbol{\Omega}} \equiv \mathbf{O}^T \dot{\mathbf{O}}, \tag{6.50}$$

and the *orthogonality* of $\mathbf{O}$ has been used; $\mathbf{O}^{-1} = \mathbf{O}^T$. It is easy to be too glib in manipulations such as those just performed in Eq. (6.49). Having said that $\mathbf{r}$ and $\bar{\mathbf{r}}$ are in some sense the same but d/dt and $\overline{d/dt}$ are different, what is the meaning of $\dot{\bar{\mathbf{r}}}$? We mean it as $\dot{\bar{\mathbf{r}}} \equiv \overline{d/dt}\,\bar{\mathbf{r}}$, a quantity with components $(d\bar{x}^1/dt, d\bar{x}^2/dt, \dots)$ and we mean $\mathbf{O}\dot{\bar{\mathbf{r}}}$ as an abbreviated notation for the array of elements $O^j{}_k\, d\bar{x}^k/dt$.

We introduce a vector $\bar{\mathbf{v}}$, related to $\mathbf{v}$ in the same way $\bar{\mathbf{r}}$ is related to $\mathbf{r}$;

$$\mathbf{v} = \mathbf{O}(t)\,\bar{\mathbf{v}}. \tag{6.51}$$

From Eq. (6.49), $\bar{\mathbf{v}}$ is therefore given by

$$\bar{\mathbf{v}} = \dot{\bar{\mathbf{r}}} + \overline{\boldsymbol{\Omega}}\,\bar{\mathbf{r}}. \tag{6.52}$$

(This shows that $\bar{\mathbf{v}}$ *is not equal* to the quantity $\dot{\bar{\mathbf{r}}}$ which might have seemed to deserve being called the $\bar{K}$-frame velocity of the moving point, and reminds us that transformation between frames can be tricky.) We introduce a "time derivative" operator

$$\overline{D_t} = \overline{\frac{d}{dt}} + \overline{\boldsymbol{\Omega}}, \tag{6.53}$$

dependent on "gauge" $\overline{\boldsymbol{\Omega}}$, that relates $\bar{\mathbf{r}}$ and $\bar{\mathbf{v}}$ as in (6.52);

$$\bar{\mathbf{v}} = \overline{D_t}\,\bar{\mathbf{r}}. \tag{6.54}$$

This equation has the desirable feature that all quantities have overhead bars and hence is "gauge invariant."[11] There was no meaningful way to introduce initially a transformation $\overline{\mathbf{O}}$ as distinct from $\mathbf{O}$ but, by being the product of a "forward" and a "backward" transformation, it is natural to associate $\overline{\boldsymbol{\Omega}}$ with the barred frame.

A way of calculating $\bar{\mathbf{v}}$ equivalent to Eq. (6.54) is to first calculate the space-frame displacement $\mathbf{O}\bar{\mathbf{r}}$, find its time derivative $(d/dt)\mathbf{O}\bar{\mathbf{r}}$, and then transform back;

$$\bar{\mathbf{v}} = \mathbf{O}^T \frac{d}{dt}\,\mathbf{O}\bar{\mathbf{r}} = \mathbf{O}^T\mathbf{O}\,\frac{d\bar{\mathbf{r}}}{dt} + \mathbf{O}^T\dot{\mathbf{O}}\bar{\mathbf{r}} = \left(\overline{\frac{d}{dt}} + \overline{\boldsymbol{\Omega}}\right)\bar{\mathbf{r}} = \overline{D_t}\,\bar{\mathbf{r}}. \tag{6.55}$$

10) The quantity $\mathbf{O}^T\dot{\mathbf{O}}$ is known in differential geometry as "the Cartan matrix." It was introduced by him in his "méthode du repère mobile" or "moving frame method."

11) When a tensor equation expresses a relationship between components in the same frame of reference using only invariant operations such as contraction on indices it is said to be *manifestly covariant*. The concept of gauge invariance currently under discussion is therefore a similar concept.

This shows that $\overline{D_t}$ can also be written as

$$\overline{D_t} = \mathbf{O}^T \frac{d}{dt} \mathbf{O}. \tag{6.56}$$

Remembering that $\mathbf{O}^T = \mathbf{O}^{-1}$, this formula shows that $\overline{D_t}$ and d/dt, though they operate in different spaces, can be regarded as being related by a "similarity transformation." In this case the two spaces are related by $\mathbf{O}$ and the $\overline{K}$-frame evolution operator $\overline{D_t}$ is "similar" to the K-frame evolution operator d/dt. This sequence of operations, depending as it does on absolute time, or Galilean relativity, would not be valid in relativistic mechanics.

We now write Newton's equation using the operator $\overline{D_t}$;

$$m\overline{D_t}^2 \overline{\mathbf{r}} = \overline{\mathbf{f}}(\overline{\mathbf{r}}, \overline{\mathbf{v}}). \tag{6.57}$$

This has the sought-for property of being expressed entirely in terms of quantities with overhead bars. As in Eq. (6.48), the vectorial property of force $\mathbf{f}$, has been assumed. (The force can also depend on time, but that dependence is not shown since it does not affect the present discussion.) To check that this equation is correct we need to see that it agrees with Eq. (6.47), which is Newton's equation in an inertial frame;

$$\begin{aligned} m\overline{D_t}^2 \overline{\mathbf{r}} &= m\left(\mathbf{O}^T \frac{d}{dt}\mathbf{O}\right)\left(\mathbf{O}^T \frac{d}{dt}\mathbf{O}\right)\overline{\mathbf{r}} = m\mathbf{O}^T \ddot{\mathbf{r}} = \mathbf{O}^T \mathbf{f}(\mathbf{r}, \mathbf{v}) = \mathbf{O}^T \mathbf{f}(\mathbf{O}\overline{\mathbf{r}}, \mathbf{O}\overline{\mathbf{v}}) \\ &= \overline{\mathbf{f}}(\overline{\mathbf{r}}, \overline{\mathbf{v}}); \end{aligned} \tag{6.58}$$

These manipulations have been formally motivated by the goal of eliminating quantities without overhead bars. The only remaining evidence of the fact that the body frame is rotating is that the operations depend on the gauge $\overline{\boldsymbol{\Omega}}$. When expanded more fully, the "acceleration" term of Newton's equation becomes

$$\overline{D_t}^2 \overline{\mathbf{r}} = \overline{D_t}(\dot{\overline{\mathbf{r}}} + \overline{\boldsymbol{\Omega}}\overline{\mathbf{r}}) = \ddot{\overline{\mathbf{r}}} + \dot{\overline{\boldsymbol{\Omega}}}\,\overline{\mathbf{r}} + 2\overline{\boldsymbol{\Omega}}\,\dot{\overline{\mathbf{r}}} + \overline{\boldsymbol{\Omega}}^2\,\overline{\mathbf{r}}. \tag{6.59}$$

The term $-2m\overline{\boldsymbol{\Omega}}\,\dot{\overline{\mathbf{r}}}$ is the Coriolis force, $-m\overline{\boldsymbol{\Omega}}^2\,\overline{\mathbf{r}}$ is the centrifugal force, and $-m\dot{\overline{\boldsymbol{\Omega}}}\,\overline{\mathbf{r}}$ accounts for nonconstancy of the relative angular velocities of the frames.[12]

6.2.2
Active Interpretation of the Transformations

Essentially the same transformations that have just been discussed can also be given an *active* interpretation. This is important in further understanding

12) It has been explained repeatedly why one need not be concerned by the absence of an overhead bar on $\dot{\overline{\boldsymbol{\Omega}}}$ as it appears in Eq. (6.59).

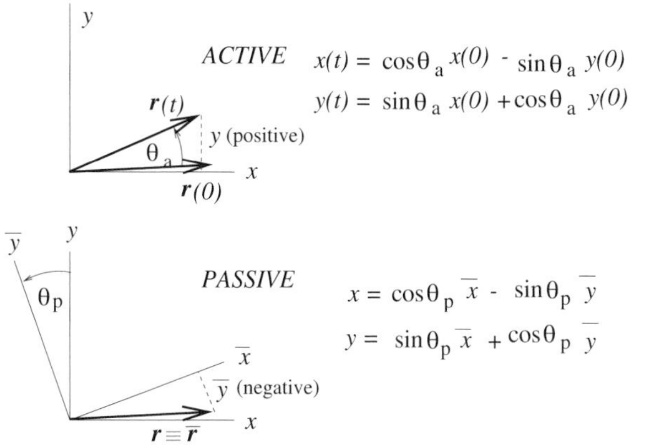

Fig. 6.8 Pictorial representations of Eqs. (6.60) and (6.61) exhibiting the active and passive effects of the same matrix. The figures have been arranged so that the coefficients of the inset equations are element-by-element equal (for the case $\theta_a = \theta_p$).

the Cartan matrix $\boldsymbol{\Omega}$ that figured prominently in the previous section. The Cartan matrix was also introduced in Section 4.2.7 where it was used to relate bivectors and infinitesimal rotations. As bivectors and infinitesimal rotations are discussed there, a fixed vector is related to a moving vector. Instead of Eq. (6.24), we therefore start with an orthogonal matrix $\mathbf{O}'(t)$ and transformation

$$\mathbf{r}(t) = \mathbf{O}'(t)\,\mathbf{r}(0). \tag{6.60}$$

Now the natural interpretation is *active* with $\mathbf{O}'(t)$ acting on constant vector $\mathbf{r}(0)$ to yield rotating vector $\mathbf{r}(t)$. This is illustrated in the upper part of Fig. 6.8. This interpretation can be contrasted with the *passive* interpretation we insisted on for Eq. (6.24),

$$\mathbf{r} = \mathbf{O}(t)\,\bar{\mathbf{r}}, \tag{6.61}$$

where $\mathbf{r}$ and $\bar{\mathbf{r}}$ stand for the same arrow.

These vectors are illustrated in the lower part of Fig. 6.8. Any relation between $\mathbf{O}'(t)$ and $\mathbf{O}(t)$ would depend on the relation between *active* angle θ_a and *passive* angle θ_p. In general there is no such relation since the relative orientation of frames of reference and the time evolution of systems being described are unrelated. Commonly though, the passive view is adopted in order to "freeze" motion that is rotating in the active view. To achieve this, after setting $\mathbf{r}(0) = \bar{\mathbf{r}}(0)$, we combine Eqs. (6.60) and (6.61)

$$\bar{\mathbf{r}}(t) = \left(\mathbf{O}^{-1}(t)\,\mathbf{O}'(t)\right)\bar{\mathbf{r}}(0), \tag{6.62}$$

In order for $\bar{\mathbf{r}}(t)$ to be constant we must have $\mathbf{O}'(t) = \mathbf{O}(t)$.

Though it is logically almost equivalent, essentially the same argument applied to linear motion seems much easier to comprehend – to freeze the active motion $x = a + vt$, we define $x = \overline{x} + b + vt$ so that $\overline{x} = a + vt - b - vt = a - b$.

We now continue with the Cartan analysis. Differentiating Eq. (6.60), the velocity of a moving point P is

$$\mathbf{v}(t) = \dot{\mathbf{O}}(t)\,\mathbf{r}(0). \tag{6.63}$$

Using the inverse of Eq. (6.60), $\mathbf{v}(t)$ can therefore be expressed in terms of $\mathbf{r}(t)$;

$$\mathbf{v}(t) = \dot{\mathbf{O}}\mathbf{O}^T\mathbf{r}(t), \tag{6.64}$$

where $\mathbf{O}^T = \mathbf{O}^{-1}$ because $\mathbf{O}$ is an orthogonal matrix. It was shown in Section 4.2.7 (from the requirement $\mathbf{r}\cdot\mathbf{v} = 0$) that the quantity $\dot{\mathbf{O}}\mathbf{O}^T$ is antisymmetric. It then follows from $d/dt(\mathbf{O}\mathbf{O}^T) = 0$ that $\mathbf{O}^T\dot{\mathbf{O}}$, for which we introduce the symbol $\overline{\boldsymbol{\Omega}}$, is also antisymmetric;[13]

$$\mathbf{O}^T\dot{\mathbf{O}} \equiv \overline{\boldsymbol{\Omega}} = \begin{pmatrix} 0 & -\varrho^3 & \varrho^2 \\ \varrho^3 & 0 & -\varrho^1 \\ -\varrho^2 & \varrho^1 & 0 \end{pmatrix}. \tag{6.65}$$

This meets the requirement of antisymmetry but for the time being the quantities ϱ^1, ϱ^2, and ϱ^3 are simply undetermined parameters; the signs have been chosen for later convenience. We next introduce a quantity $\overline{\mathbf{v}}$ that is related to $\mathbf{v}$ by

$$\mathbf{v} = \mathbf{O}\,\overline{\mathbf{v}}, \tag{6.66}$$

which is to say, the same passive way vectors introduced previously have been related in the two frames of reference; that is, $\mathbf{v}$ and $\overline{\mathbf{v}}$ stand for *the same arrow* but with components to be taken in different frames. Combining these formulas;

$$\overline{\mathbf{v}} = \mathbf{O}^T\mathbf{v} = \mathbf{O}^T\dot{\mathbf{O}}\mathbf{O}^T\,\mathbf{r}(t) = \overline{\boldsymbol{\Omega}}(t)\,\overline{\mathbf{r}}. \tag{6.67}$$

The essential feature of $\overline{\boldsymbol{\Omega}}$ is that it relates the instantaneous position vector and the instantaneous velocity vector "as arrows in the same frame." This is the basis for the phrase "Cartan's moving frame." If we now allow the point P to move with velocity $\dot{\overline{\mathbf{r}}}$ in the moving frame this becomes

$$\overline{\mathbf{v}} = \dot{\overline{\mathbf{r}}} + \overline{\boldsymbol{\Omega}}(t)\,\overline{\mathbf{r}}. \tag{6.68}$$

We have rederived Eq. (6.52) though the components of $\overline{\boldsymbol{\Omega}}$ are as yet undetermined.

13) The calculation $\dot{\mathbf{O}}\mathbf{O}^T$ has generated an element of the *Lie algebra* of antisymmetric matrices from the *Lie group* of orthogonal matrices. This generalizes to arbitrary continuous symmetries.

Clearly the "fictitious-force" and the "gauge-invariant" descriptions contain the same physics, but their conceptual bases are somewhat different and their interrelationships are subtle. Equation (6.59) is equivalent to Eq. (6.31) (or rather it generalizes that equation by allowing nonuniform rotation) but minor manipulation is required to demonstrate the fact, especially because Eq. (6.59) is expressed by matrix multiplication and Eq. (6.31) is expressed by vector cross products. The needed formula was derived in Eq. (4.37), but to avoid the need for correlating symbols we re-derive it now.

To do this, and to motivate manipulations to be performed shortly in analyzing rigid body motion, the two representations of the same physics can now be juxtaposed, starting with velocities from Eqs. (6.26) and (6.52);

$$\mathbf{v} = \mathbf{v}|_{\overline{K}} + \boldsymbol{\omega} \times \mathbf{r}, \quad \text{fictitious force description,} \tag{6.69}$$

$$\overline{\mathbf{v}} = \dot{\overline{\mathbf{r}}} + \overline{\boldsymbol{\Omega}}\,\overline{\mathbf{r}}, \quad \text{gauge-invariant description.} \tag{6.70}$$

The latter equation, in component form, reads

$$\begin{pmatrix} \overline{v}_x \\ \overline{v}_y \\ \overline{v}_z \end{pmatrix} = \left(\frac{d}{dt} + \begin{pmatrix} 0 & -\overline{\omega}^3 & \overline{\omega}^2 \\ \overline{\omega}^3 & 0 & -\overline{\omega}^1 \\ -\overline{\omega}^2 & \overline{\omega}^1 & 0 \end{pmatrix} \right) \begin{pmatrix} \overline{x} \\ \overline{y} \\ \overline{z} \end{pmatrix}. \tag{6.71}$$

where the components of $\overline{\boldsymbol{\Omega}}$ have been chosen to make the two equations match term-by-term. The mental pictures behind Eqs. (6.69) and (6.70) are different. For the former equation one has *two* coordinate frames explicitly in mind, and the equation yields the inertial-frame quantity $\mathbf{v}$ from quantities evaluated in a moving frame. In the gauge-invariant description one "knows" only one frame, the frame one inhabits, and that frame has a gauge $\overline{\boldsymbol{\Omega}}$ externally imposed upon it. In Eq. (6.70), $\overline{\boldsymbol{\Omega}}$ acts as, and is indistinguishable from, an externally imposed field. That the quantity $\overline{\mathbf{v}}$ is more deserving than, say $\dot{\overline{\mathbf{r}}}$, of having its own symbol is just part of the formalism. (There is a similar occurrence in the Hamiltonian description of a particle in an electromagnetic field; in that case the mechanical momentum is augmented by a term proportional to the vector potential. Furthermore, recalling Problem 6.1.12, one knows that $\overline{\boldsymbol{\Omega}}$ has somewhat the character of a magnetic field.)

In Eq. (6.70), or more explicitly in Eq. (6.71), all coordinates have bars on them in the only frame that is in use. Except that it would introduce confusion while comparing the two views, we could simply remove all the bars in Eqs. (6.70) and (6.71).

For these two views to correspond to the same physics, there must be an intimate connection between the quantities $\boldsymbol{\omega}$ and $\overline{\boldsymbol{\Omega}}$. Identifying $\mathbf{v}|_{\overline{K}} \stackrel{q}{=} (\frac{d}{dt}\overline{x}, \frac{d}{dt}\overline{y}, \frac{d}{dt}\overline{z})^T$ (This is necessarily a "qualified" equality, since the quantity on one side is intrinsic and on the other side it is in component form.) and equating corresponding coefficients, it is clear that the quantities ω^1, ω^2, and ω^3

entering the definition of $\overline{\boldsymbol{\Omega}}$ in Eq. (6.65) are in fact the components of $\boldsymbol{\omega}$. The two formalisms can then be related by replacing vector cross product multiplication by $\boldsymbol{\omega}$ with matrix multiplication by $\overline{\boldsymbol{\Omega}}$;

$$\boldsymbol{\omega}\times \;\to \overline{\boldsymbol{\Omega}}\,\cdot \tag{6.72}$$

Spelled out in component form this is the well-known cross-product expansion of ordinary vector analysis;

$$\begin{pmatrix} 0 & -\omega^3 & \omega^2 \\ \omega^3 & 0 & -\omega^1 \\ -\omega^2 & \omega^1 & 0 \end{pmatrix}\begin{pmatrix} x^1 \\ x^2 \\ x^3 \end{pmatrix} \stackrel{\text{q}}{=} \boldsymbol{\omega}\times\mathbf{r}. \tag{6.73}$$

Like numerous previous formulas, this has to be regarded as a "qualified" equality since it equates an intrinsic and a nonintrinsic quantity. It is valid in any frame as long as the appropriate components are used in each frame.

Accelerations, as given in the two approaches by Eqs. (6.31) and (6.59) can also be juxtaposed;

$$\mathbf{a} = \mathbf{a}|_{\overline{K}} + 2\boldsymbol{\omega}\times\mathbf{v}|_{\overline{K}} + \boldsymbol{\omega}\times(\boldsymbol{\omega}\times\mathbf{r}), \quad \text{fictitious force description,} \tag{6.74}$$

$$\begin{pmatrix} \overline{a}_x \\ \overline{a}_y \\ \overline{a}_z \end{pmatrix} = \left(\overline{\frac{d}{dt}}^{2} + 2\begin{pmatrix} 0 & -\overline{\omega}^3 & \overline{\omega}^2 \\ \overline{\omega}^3 & 0 & -\overline{\omega}^1 \\ -\overline{\omega}^2 & \overline{\omega}^1 & 0 \end{pmatrix}\overline{\frac{d}{dt}} + \begin{pmatrix} 0 & -\overline{\omega}^3 & \overline{\omega}^2 \\ \overline{\omega}^3 & 0 & -\overline{\omega}^1 \\ -\overline{\omega}^2 & \overline{\omega}^1 & 0 \end{pmatrix}^2\right)\begin{pmatrix} \overline{x} \\ \overline{y} \\ \overline{z} \end{pmatrix},$$

$$\text{gauge-invariant description.} \tag{6.75}$$

One identifies acceleration components according to $\mathbf{a}|_{\overline{K}} \stackrel{\text{q}}{=} (\overline{\frac{d}{dt}}^2\overline{x}, \overline{\frac{d}{dt}}^2\overline{y}, \overline{\frac{d}{dt}}^2\overline{z})^T$. The quantities on the left side of Eq. (6.75) are "dynamical" in the sense that they can be inferred from the applied force and Newton's law; the quantities on the right side can be said to be "kinematical" as they are to be inferred from the particle's evolving position. Clearly the matrix and vector equations are equivalent representations of the same physics.

6.2.3
Newton's Torque Equation

For analyzing rotational motion of one or more particles it is useful to introduce "torques" and to write Newton's equation for a particle, with radius vector $\mathbf{r}$ (relative to O) and velocity $\mathbf{v}$, in terms of the angular momentum $\mathbf{L}$ (relative to O), which is defined by

$$\mathbf{L} \equiv \mathbf{r}\times m\mathbf{v}. \tag{6.76}$$

By the rules of vector analysis this is a true vector (actually pseudo-vector) since both $\mathbf{r}$ and $\mathbf{v}$ are true vectors. The torque about O due to force $\mathbf{F}$ acting at position $\mathbf{r}$ is defined by

$$\boldsymbol{\tau} \equiv \mathbf{r} \times \mathbf{F}. \tag{6.77}$$

As it applies to $\mathbf{L}$, Newton's "torque law," valid in inertial frames, is

$$d\frac{\mathbf{L}}{dt} = \frac{d\mathbf{r}}{dt} \times m\mathbf{v} + \mathbf{r} \times m\frac{d\mathbf{v}}{dt} = \mathbf{r} \times \mathbf{F} = \boldsymbol{\tau}. \tag{6.78}$$

Consider next a point B, also at rest in the inertial frame, with radius vector $\mathbf{r_B}$ relative to O, and let $\mathbf{r} - \mathbf{r}_B = \mathbf{x}$ such that $\mathbf{x}$ is the displacement of mass m relative to point B. The angular momentum $\mathbf{L}_B$ and torque $\boldsymbol{\tau}_B$, both relative to B, are defined by

$$\mathbf{L}_B \equiv \mathbf{x} \times m\mathbf{v}, \quad \boldsymbol{\tau}_B \equiv \mathbf{x} \times \mathbf{F}. \tag{6.79}$$

We have therefore

$$\frac{d\mathbf{L}_B}{dt} = \left(\frac{d}{dt}(\mathbf{r} - \mathbf{r}_B)\right) \times m\mathbf{v} + \mathbf{x} \times m\frac{d\mathbf{v}}{dt} = \boldsymbol{\tau}_B - \mathbf{v}_B \times m\mathbf{v}, \tag{6.80}$$

where the final term vanishes because point B has (so far) been taken to be at rest. We have not dropped this correction term explicitly to cover the possibility that point B is in fact moving.

This formula is especially useful when the system is constrained by forces incapable of applying torque or when two or more particles are rigidly connected so the torques due to their internal forces cancel out in pairs. Then, if there are external torques but no net external force, the point of application of the forces does not matter. But, for the time being, all forces are to be applied directly to a single particle.

Next let us consider similar quantities reckoned in the rotating frame $\overline{K}$. We revert to the fictional force formalism, in which all vector equations equate vectors in an inertial system. From the rules of vector analysis and from our algebraic conventions, the following relations have to be true:

$$\overline{\mathbf{L}}_B = \overline{\mathbf{x}} \times m\overline{\mathbf{v}}, \quad \overline{\boldsymbol{\tau}}_B = \overline{\mathbf{x}} \times \overline{\mathbf{F}}, \quad \frac{d}{dt}\overline{\mathbf{L}}_B = \overline{\frac{d}{dt}}\,\overline{\mathbf{L}}_B + \overline{\boldsymbol{\omega}} \times \overline{\mathbf{L}}_B. \tag{6.81}$$

What is not yet clear is the relation between $\overline{d/dt}\,\overline{\mathbf{L}}_B$ and $\overline{\boldsymbol{\tau}}_B$. One complication is that, even if the point B is at rest in the K frame, it will be moving in the $\overline{K}$ frame and vice versa. We need to evaluate

$$\begin{aligned}
\overline{\frac{d}{dt}}\,\overline{\mathbf{L}}_B &= \overline{\frac{d}{dt}}\left((\overline{\mathbf{r}} - \overline{\mathbf{r}}_B) \times m\overline{\mathbf{v}}\right) \\
&= -\overline{\mathbf{v}}_B \times \overline{\mathbf{v}} + \overline{\mathbf{x}} \times \overline{\frac{d}{dt}}\, m\overline{\mathbf{v}} \\
&= -\overline{\mathbf{v}}_B \times \overline{\mathbf{v}} + \overline{\mathbf{x}} \times (\overline{\mathbf{F}} + \overline{\boldsymbol{\omega}} \times m\overline{\mathbf{v}}) \\
&= -\overline{\mathbf{v}}_B \times \overline{\mathbf{v}} + \overline{\boldsymbol{\tau}}_B - \overline{\boldsymbol{\omega}} \times \overline{\mathbf{L}}_B.
\end{aligned} \tag{6.82}$$

This can be rearranged as

$$\left(\overline{\frac{d}{dt}} + \overline{\boldsymbol{\omega}} \times\right) \overline{\mathbf{L}}_B = \overline{\boldsymbol{\tau}}_B - \overline{\mathbf{v}}_B \times m\overline{\mathbf{v}}. \qquad (6.83)$$

6.2.4 The Plumb Bob

What could be simpler than a plumb bob, a point mass hanging at the end of a light string or rigid rod, and not even swinging? It hangs straight down "in the laboratory." But what is "down"? The fact that the earth rotates makes this system not quite so simple. But its apparent simplicity makes it a good system to exercise the methods under discussion. It will be approached in several ways. Because the bob appears to be at rest there is no Coriolis force, even when working in the laboratory system. The earth plumb-bob system is illustrated in Fig. 6.9.

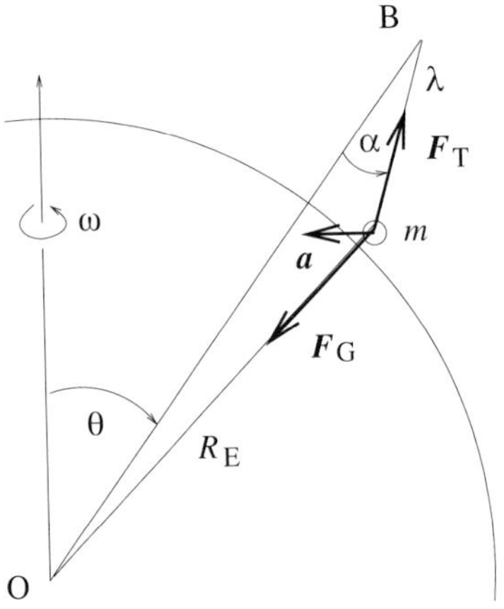

Fig. 6.9 Mass m, hanging at rest at the end of a light string of length λ constitutes a plumb bob. Its length is much exaggerated relative to the earth's radius R_E.

Example 6.2.1. Inertial frame force method. *The mass m hanging, apparently at rest, at the end of a light string of length* λ *is subject to a gravitational force* $\mathbf{F}_E$ *directed toward the center of the earth and tension force* $\mathbf{F}_T$ *along the string. The resultant of these forces causes the mass to accelerate toward the axis of rotation of the earth. Of course, this is just the acceleration needed to stay on a circular path of radius* $R_E \sin\theta$ *and keep up with the earth's rotation at angular velocity* ω. *Its radial acceleration is* $a = R_E \sin\theta\, \omega^2$ *where* θ *is the co-latitude of the bearing point B on the earth from which the bob is suspended. The angle by which the plumb bob deviates* *from the line to the earth's center is α. Since the mass is at rest axially, we have*

$$F_T \cos(\theta - \alpha) = F_G \cos\left(\theta + \alpha\,\frac{\lambda}{R_E}\right). \tag{6.84}$$

An extremely conservative approximation has already been made in the second argument, and the term $\alpha\lambda/R_E$, of order λ/R_E relative to α, will immediately be dropped in any case. Taking $F_G = mg$ we have therefore

$$F_T = mg\,\frac{\cos\theta}{\cos(\theta - \alpha)}. \tag{6.85}$$

Equating the radial components yields

$$\cos\theta\,\tan(\theta - \alpha) - \sin\theta = -\frac{R_E}{g}\sin\theta\,\omega^2, \tag{6.86}$$

which simplifies to

$$\left(1 - \frac{R_E\omega^2}{g}\right)\tan\theta = \tan(\theta - \alpha) \approx \tan\theta - \frac{\alpha}{\cos^2\theta}. \tag{6.87}$$

Finally, this reduces to

$$\alpha \approx \frac{R_E\omega^2}{2g}\sin 2\theta \overset{\text{typ.}}{=} 1.6 \times 10^{-3}\ \text{radians}, \tag{6.88}$$

where the Greenwich co-latitude of $\theta = 38.5^\circ$ has been used.

It is customary to incorporate this tiny angle by redefining what constitutes "down" so that the plumb bob points "down" and not toward the earth's center. (Actually, the bob would not point toward the center of the earth in any case since the earth is not a perfect sphere. Its major deviation from being spherical is itself due to the centrifugal force which, acting on the somewhat fluid earth, has caused it to acquire an ellipsoidal shape.) Once this has been done the centrifugal force can (to good approximation) be ignored completely. It is customary, therefore, to define an "effective" gravitational acceleration vector $\mathbf{g}(\theta)$

$$\mathbf{g}(\theta) = -g\hat{\mathbf{r}} - \boldsymbol{\omega} \times (\boldsymbol{\omega} \times R_E\mathbf{r}), \tag{6.89}$$

which will permit the centrifugal force to be otherwise neglected. This constitutes an approximation, but since the deviation is so small it is a good one. If the earth were spinning sufficiently fast the plumb bob would end up pointing sideways, and treating the centrifugal force would be more difficult, though not as difficult as our other problems.

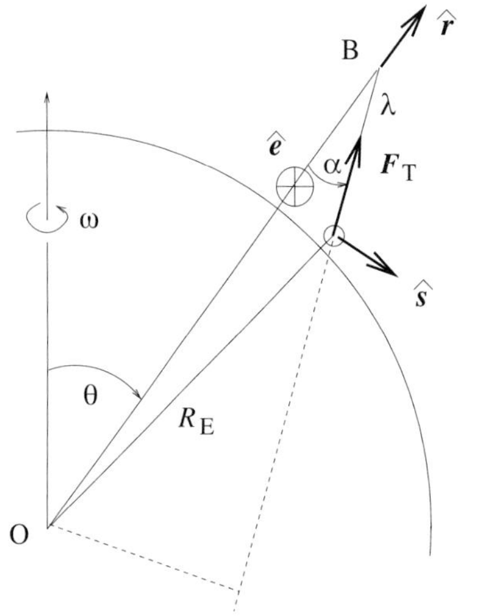

Fig. 6.10 The same plumb bob as in the previous figure is shown along with local basis vectors $\hat{\mathbf{r}}$ radial, $\hat{\mathbf{e}}$ eastward and $\hat{\mathbf{s}}$ southward.

Example 6.2.2. Inertial frame torque/angular-momentum method. *To evaluate the torque acting on m about point O we define local unit vectors* $\hat{\mathbf{r}}$ *radial,* $\hat{\mathbf{e}}$ *eastward and* $\hat{\mathbf{s}}$ *southward, as shown in Fig. 6.10. Still working in the inertial frame, the angular momentum of the mass m relative to point O is*

$$\mathbf{L} = R_E\hat{\mathbf{r}} \times (mR_E \sin\theta\,\omega\,\hat{\mathbf{e}}) = -mR_E^2\omega\sin\theta\,\hat{\mathbf{s}}, \tag{6.90}$$

and its time rate of change is given by

$$\frac{d\mathbf{L}}{dt} = -mR_E^2\omega^2\sin\theta\,\hat{\mathbf{e}}\cos\theta. \tag{6.91}$$

As shown in the figure, the only force applying torque about O is the string tension, which is approximately given by $F_T = mg$. *Its torque is*

$$\boldsymbol{\tau} = R_E\hat{\mathbf{r}} \times \mathbf{F}_T = -R_E mg\sin\alpha\,\hat{\mathbf{e}}. \tag{6.92}$$

Equating the last two expressions we obtain

$$\sin\alpha = \frac{R_E\omega^2}{2g}\sin 2\theta, \tag{6.93}$$

in approximate agreement with Eq. (6.88) since α *is an exceedingly small angle.*

Example 6.2.3. Fictitious force method. *If we act as if the earth is at rest and continue to use the center of the earth as origin, the situation is as illustrated in Fig. 6.11. There is an outward directed fictitious force* $\mathbf{F}_{cent}$ *with magnitude* $mR_E\omega^2\sin\theta$ *which*

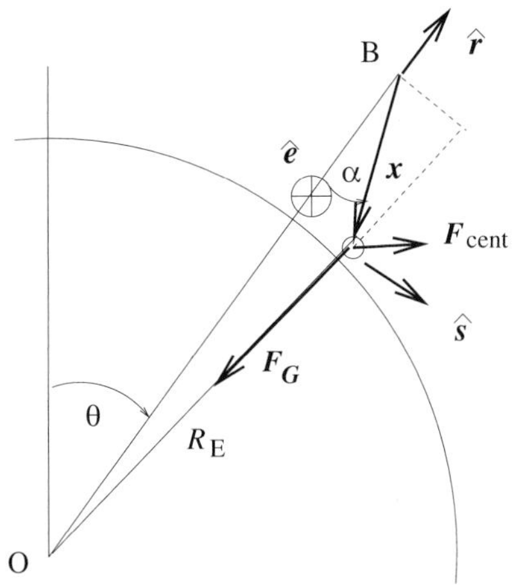

Fig. 6.11 When viewed "in the laboratory," the earth, the bearing point B and the mass m all appear to be at rest, but there is a fictitious centrifugal force F_{cent}.

has to balance the gravitational force F_G in order for the bob to remain at rest. Since both of these forces are applied directly to m we can equate their components normal to the bob, which amounts also to equating their torques about point B. The condition for this balance is

$$mg\alpha \approx mR_E\omega^2 \sin\theta \cos\theta, \tag{6.94}$$

which agrees with the previous calculations. In this case it has been valid to work with torques about B, but this would be risky in general because B is not fixed in an inertial frame.

Example 6.2.4. Transformation of angular momentum; origin at O. *Alternatively we can use transformation formulas to infer inertial frame quantities. Since m is at rest relative to B we can use $R_E\hat{\mathbf{r}}$ both as its (approximate) position vector and for calculating its velocity using Eq. (6.26). The inertial frame angular momentum of m about O is given by*

$$\mathbf{L} = mR_E\hat{\mathbf{r}} \times \left(\overline{\frac{d}{dt}} R_E\hat{\mathbf{r}} + \boldsymbol{\omega} \times R_E\hat{\mathbf{r}} \right) = -mR_E^2\omega \sin\theta\, \hat{\mathbf{s}}. \tag{6.95}$$

Note that the $\overline{d/dt}$ term has vanished because the mass appears *to be at rest. The time rate of change of angular momentum is given by*

$$\frac{d\mathbf{L}}{dt} = \overline{\frac{d}{dt}}\mathbf{L} + \boldsymbol{\omega} \times \mathbf{L} = -mR_E^2(\hat{\mathbf{r}}\cdot\boldsymbol{\omega})(\boldsymbol{\omega}\times\hat{\mathbf{r}}) = -mR_E^2\omega^2 \cos\theta\, \sin\theta\, \hat{\mathbf{e}}, \tag{6.96}$$

where the $\overline{d/dt}$ term has again vanished because the angular momentum appears *to be constant. The torque is again given by Eq. (6.93) and the result (6.88) is again obtained.*

Example 6.2.5. Gauge invariant method. *Referring to Fig. 6.11 and substituting into Eq. (6.83), we have*

$$\left(\overline{\frac{d}{dt}} + \overline{\boldsymbol{\omega}}\times\right)(\overline{\mathbf{x}}\times\overline{\mathbf{v}}) = \boldsymbol{\omega}\times(\mathbf{x}\times\mathbf{v}). = \mathbf{x}\times\mathbf{F}_G. \tag{6.97}$$

(In evaluating expressions such as this a mistake that is hard to avoid making is to set $\overline{\mathbf{v}}$ to zero, because the bob is not moving – it is $\overline{d/dt}\overline{\mathbf{x}}$ that vanishes, not $\overline{\mathbf{v}}$.) The term $\mathbf{v}_B\times\mathbf{v}$ in Eq. (6.83) vanishes because, though the support point B is moving in the inertial system, the velocities of the bob and the support point are parallel. The needed vectors are

$$\mathbf{x}\approx-\lambda\hat{\mathbf{r}},\quad \boldsymbol{\omega}=\omega(\cos\theta\hat{\mathbf{r}}-\sin\theta\hat{\mathbf{s}}),\quad \overline{\boldsymbol{\tau}}_B=\overline{\mathbf{x}}\times\overline{\mathbf{F}}_G\approx\lambda mg\alpha\hat{\mathbf{e}}, \tag{6.98}$$

and Eq. (6.83) becomes

$$R_E\omega^2\lambda\sin\theta\cos\theta\hat{\mathbf{e}} = \lambda mg\alpha\hat{\mathbf{e}}, \tag{6.99}$$

which agrees with the previous determinations.

No great significance should be placed on which of the above plumb bob equations have been indicated as equalities and which as approximations. Most of the equations are approximate in one way or another. The first, inertial frame, method, though the most elementary, has the disadvantage compared to all the other methods, because of the need to equate components parallel to the plumb line, that more careful approximation is required.

6.3
Gauge Invariant Description of Rigid Body Motion

Much of the material in this section is repetitive of material in earlier chapters in which the Lagrange–Poincaré approach is taken. This is partly to permit this chapter to be accessible whether or not that material has been mastered and partly to compare and contrast the two approaches.

The main example will be the description of angular motion of a rigid body by the "Euler equations" describing rigid body motion. These are equations governing the time evolution of the body's angular velocity components, as reckoned in the body frame, $(\overline{\omega}^1, \overline{\omega}^2, \overline{\omega}^3)$. The equations will be derived using Lie algebraic methods. One purpose of this discussion is to practice with the commutator manipulations that are basic to the application of Lie algebraic methods. The method to be developed is applicable to any system with symmetry describable as invariance under a Lie group of transformations.

Every physicist has intuitively assimilated Newton's equations as giving a valid description of mechanics (at least in the nonrelativistic, nonquantal domain.) Every physicist who has advanced beyond freshman level mechanics has acquired a similar (perhaps less deeply held) confidence in the Lagrange equations as containing the same physics and leading economically to correct equations of motion. And many physicists have followed a derivation of the Euler equations and their application to describe rigid body motion. Probably far fewer physicists can answer the question "Why is it that the Euler equations can be derived from Newton's law, but not from the Lagrange equations?" For our theory to be regarded as satisfactorily powerful it should be possible to derive the Euler equations by straightforward manipulations. The Poincaré equation derived in Section 5.1 provided exactly that capability, making it possible to derive the Euler equations just by "turning the crank."

The situation was still somewhat unsatisfactory, in the way that Lagrangian mechanics often seems, because the formulas provide little visualizable content. This may make it hard to make sensible approximations or to see how symmetries or simple features of the physical system can be exploited to simplify the equations, or how the method can be applied to other problems. This justifies working on the same problem with different methods. Here this will mean algebraic representations of geometry. Commutation relations again play an important role, but now the noncommuting elements will be 2×2 or 3×3 matrices, rather than the vector fields prominent in Lagrange–Poincaré mechanics.

As well as the material just covered concerning rotating reference systems and Coriolis and centrifugal forces, the description of rotational motion of rigid bodies and the mathematics of infinitesimal rotations deserve review at this point. It is entirely intentional that Hamilton's equations and "canonical methods" have not been, and will not be used in this discussion. Hamiltonian formulation is of no particular value for clarifying the issues under discussion, though of course the present discussion will have to be reviewed in that context later on.

6.3.1
Space and Body Frames of Reference

We wish to describe rigid body motion in much the way single particle motion was described in the preceding sections. At this point familiarity with the inertia tensor and its use in expressing the kinetic energy of a rotating body is assumed. Position and orientation of a rigid body can be specified in an inertial "space frame" K, or a "body frame" $\overline{K}$ whose origin is fixed at the centroid of the body, and whose axes are fixed in the body. Another inertial frame might be introduced with origin at the centroid and axes aligned with those of K, but for simplicity from now on, we ignore centroid motion and assume the centroid remains at rest. The inertial-frame rotational, kinetic energy T_{rot}

can be written in terms of K-space variables as

$$T_{\text{rot}} = \frac{1}{2}\boldsymbol{\omega}^T(t) \cdot \mathbf{I}(t) \cdot \boldsymbol{\omega}(t), \tag{6.100}$$

where the angular velocity vector $\boldsymbol{\omega}(t)$ is *time dependent* (if the body is tumbling) and the inertia tensor $\mathbf{I}(t)$ is time dependent because the mass distribution varies relative to an inertial frame (if the body is tumbling). In Eq. (6.100) the matrix multiplications are indicated by the $\cdot$ symbol. This is purely artificial since the spelled-out form

$$T_{\text{rot}} = \frac{1}{2}\sum_{i,j} \omega^i(t)\, I_{ij}(t)\, \omega^j(t), \tag{6.101}$$

is the same as is implied by ordinary matrix multiplication. The dot or "dyadic" notation is intended to discourage the interpretation of I_{ij} as having any geometric significance whatsoever – it is preferably regarded as the array of coefficients of a quadratic form since there are quite enough transformation matrices without introducing another one.[14]

The kinetic energy T_{rot} can be written alternatively in terms of body variables but $\bar{\mathbf{I}}$ is *time independent* (because, the body being rigid, its particles are not moving relative to one another in the $\overline{K}$ frame),[15]

$$T_{\text{rot}} = \frac{1}{2}\overline{\boldsymbol{\omega}}^T(t) \cdot \bar{\mathbf{I}} \cdot \overline{\boldsymbol{\omega}}(t). \tag{6.102}$$

The inertia tensor also relates $\boldsymbol{\omega}$ to angular momentum $\mathbf{l}$

$$\mathbf{l} = \mathbf{I} \cdot \boldsymbol{\omega}. \tag{6.103}$$

Here, as in Eqs. (6.100) and (6.102), we use dyadic notation in which expressions like these treat $\mathbf{I}$ as a matrix that multiplies a vector by the normal rules of matrix multiplication. Equation (6.103) is simplest in the body frame where $\bar{\mathbf{I}}$ is time independent. Since $\bar{\mathbf{I}}$ is diagonal it can be diagonalized by an appropriate choice of axes, in which case

$$\bar{\mathbf{I}} = \begin{pmatrix} \bar{I}_1 & 0 & 0 \\ 0 & \bar{I}_2 & 0 \\ 0 & 0 & \bar{I}_3 \end{pmatrix} \quad \text{and} \quad \bar{\mathbf{l}}_i = \bar{I}_i \omega^i. \tag{6.104}$$

Like other vectors, the angular momentum components in different coordinate frames are related by a rotation matrix $\mathbf{O}$, as in Eq. (6.24). The matrix $\mathbf{O}$,

14) Of course there is a useful "moment of inertia ellipsoid" which is a kind of geometric significance that I_{ij} has, but this has nothing to do with the geometry of transformation between coordinate frames.

15) It is still true, that vectors $\boldsymbol{\omega}$ and $\overline{\boldsymbol{\omega}}$ signify the same arrow, and are best regarded as simple algebraic abbreviations for the arrays of elements $\omega^1, \omega^2, \ldots$ and $\overline{\omega}^1, \overline{\omega}^2, \ldots$.

being antisymmetric, can itself be expanded in terms of basis matrices defined in Eq. (4.38);

$$J_1 = \begin{pmatrix} 0 & 0 & 0 \\ 0 & 0 & -1 \\ 0 & 1 & 0 \end{pmatrix}, \quad J_2 = \begin{pmatrix} 0 & 0 & 1 \\ 0 & 0 & 0 \\ -1 & 0 & 0 \end{pmatrix}, \quad J_3 = \begin{pmatrix} 0 & -1 & 0 \\ 1 & 0 & 0 \\ 0 & 0 & 0 \end{pmatrix}. \tag{6.105}$$

They satisfy commutation relations

$$[J_i, J_j] = \epsilon_{ijk} J_k. \tag{6.106}$$

For rotation through angle ϕ about axis $\widehat{\boldsymbol{\phi}}$, defining $\boldsymbol{\phi} = \phi\widehat{\boldsymbol{\phi}}$, the formula for $\mathbf{O}$ is[16]

$$\mathbf{O} = e^{\boldsymbol{\phi}\cdot\mathbf{J}}. \tag{6.107}$$

Example 6.3.1. *Let us check this for J_3 (which satisfies the equation $J_3^2 = -\mathbf{1}$ after the third row and the third column have been suppressed.)*

$$e^{\phi J_3} = 1 + \frac{\phi J_3}{1!} + \frac{(\phi J_3)^2}{2!} + \frac{(\phi J_3)^3}{3!} + \cdots = \begin{pmatrix} \cos\phi & -\sin\phi \\ \sin\phi & \cos\phi \end{pmatrix}. \tag{6.108}$$

After restoring the suppressed rows and columns, when acting on a radius vector, this clearly produces rotation by angle ϕ about the z-axis.

Problem 6.3.1. *Derive Eq. (6.107) by "exponentiating" Eq. (4.40).*

As explained previously, since the angular velocity is a *true* vector (actually pseudovector), the same rotation matrix relates angular velocity vectors $\boldsymbol{\omega}$ and $\overline{\boldsymbol{\omega}}$,

$$\boldsymbol{\omega} = \mathbf{O}(t)\,\overline{\boldsymbol{\omega}}, \quad \omega^j = O^j_{\ k}(t)\overline{\omega}^k; \tag{6.109}$$

similarly the angular momentum vectors $\mathbf{l}$ and $\overline{\mathbf{l}}$ are related by

$$\mathbf{l} = \mathbf{O}(t)\,\overline{\mathbf{l}}, \quad l^j = O^j_{\ k}(t)\overline{l}^k. \tag{6.110}$$

In terms of $\mathbf{O}$, known to be an orthogonal matrix, the inertia tensors of Eqs. (6.100) and (6.102) are related by

$$\overline{\mathbf{I}} = \mathbf{O}^T \cdot \mathbf{I} \cdot \mathbf{O}. \tag{6.111}$$

To confirm this, substitute it into Eq. (6.102) and use Eq. (6.109)

$$T_{\text{rot}} = \frac{1}{2}\overline{\boldsymbol{\omega}}^T\mathbf{O}^T \cdot \mathbf{I} \cdot \mathbf{O}\overline{\boldsymbol{\omega}} = \frac{1}{2}\left(\mathbf{O}\overline{\boldsymbol{\omega}}\right)^T \cdot \mathbf{I} \cdot \left(\mathbf{O}\overline{\boldsymbol{\omega}}\right) = \frac{1}{2}\boldsymbol{\omega}^T \cdot \mathbf{I} \cdot \boldsymbol{\omega}, \tag{6.112}$$

16) Because rotations do not commute, it is not legitimate to factorize this as a product of three exponentials, $e^{-\phi_1 J_1}e^{-\phi_2 J_2}e^{-\phi_3 J_3}$ though, of course, angles can be found to make such a factorization correct.

which agrees with Eq. (6.100). This manipulation has used the fact that the dot operation is performed by standard matrix multiplication. From here on the dot will be suppressed.

6.3.2 Review of the Association of 2×2 Matrices to Vectors

This section consists mainly of a series of problems that review material developed in Chapter 4. But they are supposed to be intelligible even without having studied that material. Furthermore, it is only a digression since the association between vectors and 2×2 matrices that is derived here will not actually be used for the analysis of rigid body motion. The purpose is to refresh the essential ideas. In the subsequent section following this one a corresponding association to 3×3 matrices will be developed and *that* will be the basis for analyzing rigid body motion.

A concept to be used again is that of "similarity transformation." Consider two arrows $\mathbf{a}$ and $\mathbf{b}$ and suppose that the pure rotation of $\mathbf{a}$ into $\mathbf{b}$ is symbolized by $\mathbf{b} = \mathbf{T}\mathbf{a}$. Imagine further an azimuthal rotation by some angle (such as one radian) around $\mathbf{a}$; let it be symbolized by $\mathbf{\Phi}_a$. The result of this rotation about $\mathbf{a}$ of a vector $\mathbf{x}$ is a vector $\mathbf{x}' = \mathbf{\Phi}_a\,\mathbf{x}$. Are we in a position to derive the operator $\mathbf{\Phi}_b$ that rotates $\mathbf{x}$ by the same angle azimuthally around the vector $\mathbf{b}$? The answer is yes, because we can first rotate $\mathbf{b}$ into $\mathbf{a}$ using $\mathbf{T}^{-1}$ then rotate around $\mathbf{a}$ using $\mathbf{\Phi}_a$ and then rotate back using $\mathbf{T}$. The result is

$$\mathbf{\Phi}_b = \mathbf{T}\mathbf{\Phi}_a\mathbf{T}^{-1}. \tag{6.113}$$

This is known as a similarity transformation. The rationale for the terminology is that transformations $\mathbf{\Phi}_a$ and $\mathbf{\Phi}_b$ are "similar" transformations around different axes – the word "similar" is used here as it is in the "high school" or "synthetic" Euclidean geometry of rulers and compasses. The same argument would be valid if $\mathbf{\Phi}_a$ and $\mathbf{\Phi}_b$ designated reflections in planes orthogonal to $\mathbf{a}$ and $\mathbf{b}$ respectively.

Consider the following associations (introduced in Section 4.4.3, where upper case letters stand for matrices and lower case letters stand for vectors:

$$\mathbf{X} = \begin{pmatrix} x^3 & x^1 - ix^2 \\ x^1 + ix^2 & -x^3 \end{pmatrix} = x^1\sigma_1 + x^2\sigma_2 + x^3\sigma_3 \equiv \mathbf{x}\cdot\boldsymbol{\sigma}, \tag{6.114}$$

$$\sigma_1 = \begin{pmatrix} 0 & 1 \\ 1 & 0 \end{pmatrix}, \quad \sigma_2 = \begin{pmatrix} 0 & -i \\ i & 0 \end{pmatrix}, \quad \sigma_3 = \begin{pmatrix} 1 & 0 \\ 0 & -1 \end{pmatrix}. \tag{6.115}$$

$$\mathbf{x} \to \mathbf{X}, \quad \mathbf{y} \to \mathbf{Y}, \quad 2i(\mathbf{x}\times\mathbf{y}) \to [\mathbf{X},\mathbf{Y}] \equiv \mathbf{XY} - \mathbf{YX}. \tag{6.116}$$

Though these associations were derived previously, they can be re-derived by solving the following series of problems, thereby obviating the need to review that material. Another relation that was derived earlier, $\mathbf{x}\cdot\mathbf{y} \to (1/2)(\mathbf{XY}+\mathbf{YX})$, will not be needed.

In transforming between frames, as for example from inertial frame to body frame, vectors transform by matrix multiplication as in Eqs. (6.24), (6.48), (6.109), and (6.110). Because the matrices associated with these vectors themselves represent transformations, the transformation of these matrices between frames are similarity transformations. This will now be spelled out explicitly.

The matrices $\sigma_1, \sigma_2, \sigma_3$ are the *Pauli spin matrices*; they satisfy the algebraic relations

$$\begin{aligned}\sigma_j\sigma_k &= \delta_{jk}\mathbf{1} + i\epsilon_{jkl}\sigma_l,\\ \left[\sigma_j, \sigma_k\right] &= \sigma_j\sigma_k - \sigma_k\sigma_j = 2i\epsilon_{jkl}\sigma_l.\end{aligned} \tag{6.117}$$

Problem 6.3.2. *Show that*

$$(\mathbf{a}\cdot\boldsymbol{\sigma})\,(\mathbf{b}\cdot\boldsymbol{\sigma}) = \mathbf{a}\cdot\mathbf{b}\,\mathbf{1} + i(\mathbf{a}\times\mathbf{b})\cdot\boldsymbol{\sigma}. \tag{6.118}$$

Problem 6.3.3. *Show that*

$$e^{i\frac{\theta}{2}\,\widehat{\mathbf{n}}\cdot\boldsymbol{\sigma}} = \cos\frac{\theta}{2}\,\mathbf{1} + i\sin\frac{\theta}{2}\,\widehat{\mathbf{n}}\cdot\boldsymbol{\sigma}. \tag{6.119}$$

Shortly, this matrix will be symbolized by S^{-1}; it appeared previously in Eq. (4.117).

Problem 6.3.4. *According to Eq. (6.114), the matrices σ_1, σ_2, and σ_3 are "associated with" the unit vectors $\widehat{\mathbf{e}}_1$, $\widehat{\mathbf{e}}_2$, and $\widehat{\mathbf{e}}_3$, respectively. Derive the following similarity transformations:*

$$\begin{aligned}e^{-i\frac{\gamma}{2}\sigma_3}\sigma_1 e^{i\frac{\gamma}{2}\sigma_3} &= \cos\gamma\;\sigma_1 + \sin\gamma\;\sigma_2\\ e^{-i\frac{\gamma}{2}\sigma_3}\sigma_2 e^{i\frac{\gamma}{2}\sigma_3} &= -\sin\gamma\;\sigma_1 + \cos\gamma\;\sigma_2\\ e^{-i\frac{\gamma}{2}\sigma_3}\sigma_3 e^{i\frac{\gamma}{2}\sigma_3} &= \sigma_3.\end{aligned} \tag{6.120}$$

A coordinate frame related to the original frame by a rotation of angle γ around the x^3-axis has unit vectors given by $\cos\gamma\,\widehat{\mathbf{e}}_1 + \sin\gamma\,\widehat{\mathbf{e}}_2$, $-\sin\gamma\,\widehat{\mathbf{e}}_1 + \cos\gamma\,\widehat{\mathbf{e}}_2$, and $\widehat{\mathbf{e}}_3$. The right-hand sides of Eq. (6.120) are the matrices "associated" with these unit vectors. This demonstrates, in a special case, that when vectors transform by an ordinary rotation, their associated matrices transform by a similarity transformation based on the corresponding matrix.

Problem 6.3.5. *Making the association $\mathbf{x}\to\mathbf{X}\equiv\mathbf{x}\cdot\boldsymbol{\sigma}$ show that*

$$\det|X| = -\mathbf{x}\cdot\mathbf{x}. \tag{6.121}$$

Problem 6.3.6. *Show that the inverse association $\mathbf{X}\to\mathbf{x}$ can be written in component form as*

$$x^i = \frac{1}{2}\,\mathrm{tr}(\mathbf{X}\sigma_i). \tag{6.122}$$

Problem 6.3.7. *Compute* $\mathbf{X}' = e^{-i(\theta/2)\hat{\mathbf{n}}\cdot\boldsymbol{\sigma}}\mathbf{X}e^{i(\theta/2)\hat{\mathbf{n}}\cdot\boldsymbol{\sigma}}$ *and show that*

$$\mathbf{x}' = (\hat{\mathbf{n}} \cdot \mathbf{x})\,\hat{\mathbf{n}} + \cos\theta\,\big((\hat{\mathbf{n}} \times \mathbf{x}) \times \hat{\mathbf{n}}\big) + \sin\theta\,\big(\hat{\mathbf{n}} \times \mathbf{x}\big). \tag{6.123}$$

Note that this is the same as Eq. (4.120).

Problem 6.3.8. *Show that*

$$\mathbf{x} \cdot \mathbf{y} \to \frac{\mathbf{XY} + \mathbf{YX}}{2} \quad \text{and} \quad \mathbf{x} \times \mathbf{y} \to -\frac{i}{2}\,\big[\mathbf{X}, \mathbf{Y}\big].$$

6.3.3 "Association" of 3×3 Matrices to Vectors

We now set up a similar association between vectors $\mathbf{x}$ and 3×3 matrices $\mathbf{X}$. The use of the same upper case symbol for both 2×2 and 3×3 matrices should not be too confusing since only 3×3 matrices will occur for the remainder of this chapter. Using the triplet of matrices $\mathbf{J}$ defined in Eq. (6.105), the association is

$$\mathbf{x} \to \mathbf{X} \equiv \mathbf{x} \cdot \mathbf{J}. \tag{6.124}$$

Observe that the matrix infinitesimal rotation operator $\boldsymbol{\Omega}$ and the angular velocity vector $\boldsymbol{\omega}$ introduced previously are *associated* in this sense, and their symbols were chosen appropriately to indicate the same.

Problem 6.3.9. *Show, with this association, that*

$$\mathbf{x} \times \mathbf{y} \to \big[\mathbf{X}, \mathbf{Y}\big]; \tag{6.125}$$

i.e., vector cross products map to matrix commutators.

Problem 6.3.10. *By analogy with Eq. (6.120), one anticipates the following equations:*

$$\begin{aligned} e^{-\phi J_3} J_1 e^{\phi J_3} &= \cos\phi\ J_1 + \sin\phi\ J_2 \\ e^{-\phi J_3} J_2 e^{\phi J_3} &= -\sin\phi\ J_1 + \cos\phi\ J_2 \\ e^{-\phi J_3} J_3 e^{\phi J_3} &= J_3. \end{aligned} \tag{6.126}$$

Prove this result.

Problem 6.3.11. *Compute* $\mathbf{X}' = e^{-\boldsymbol{\phi}\cdot\mathbf{J}}\mathbf{X}e^{\boldsymbol{\phi}\cdot\mathbf{J}}$ *and show that*

$$\mathbf{x}' = \mathbf{O}\,\mathbf{x}, \tag{6.127}$$

where $\mathbf{O}$ *is given by Eq. (6.107).*

A coordinate frame related to the original frame by a rotation of angle ϕ around the x^3-axis has unit vectors given by $\cos\phi\,\widehat{\mathbf{e}}_1 + \sin\phi\,\widehat{\mathbf{e}}_2$, $-\sin\phi\,\widehat{\mathbf{e}}_1 + \cos\phi\,\widehat{\mathbf{e}}_2$, and $\widehat{\mathbf{e}}_3$. The right-hand sides of Eq. (6.126) are the matrices "associated" with these unit vectors. This demonstrates, in a special case, that when vectors transform by an ordinary rotation, their associated matrices transform by a similarity transformation based on the corresponding matrix.

6.3.4
Derivation of the Rigid Body Equations

We now apply the associations defined in the previous section to the generalized Newton's equation derived before that. For reasons that should become clear gradually, rather than studying the evolution of displacement vector $\mathbf{x}$, we study the evolution of its associated "displacement matrix" $\mathbf{X}$.

The fixed-frame and rotating-frame "displacement matrices" are $\mathbf{X} \equiv \mathbf{x}\cdot\mathbf{J}$ and $\overline{\mathbf{X}} \equiv \overline{\mathbf{x}}\cdot\mathbf{J}$, respectively; they are related by

$$\mathbf{X} = \mathbf{O}\overline{\mathbf{X}}\mathbf{O}^T. \tag{6.128}$$

As we have seen, $\mathbf{X}$ and $\overline{\mathbf{X}}$ have geometric interpretations as transformation matrices for infinitesimal rotation around the vectors $\mathbf{x}$ and $\overline{\mathbf{x}}$. This conforms with the remark made previously that the operators $\mathbf{X}$ and $\overline{\mathbf{X}}$ are related by similarity transformation. By analogy with our earlier treatment the "time derivative operator" $\overline{D_t}$ should be defined so that "velocity" matrices are related by

$$\mathbf{V} = \mathbf{O}\overline{\mathbf{V}}\mathbf{O}^T = \mathbf{O}(\overline{D_t}\,\overline{\mathbf{X}})\mathbf{O}^T, \tag{6.129}$$

where the parentheses indicate that $\overline{D_t}$ does not operate on the final factor $\mathbf{O}^T$. Differentiating Eq. (6.124) with respect to t and using $(d/dt)(\mathbf{O}^T\mathbf{O}) = 0$ yields

$$\begin{aligned}\mathbf{V} = \dot{\mathbf{X}} &= \frac{d}{dt}(\mathbf{O}\overline{\mathbf{X}}\mathbf{O}^T) = \dot{\mathbf{O}}\overline{\mathbf{X}}\mathbf{O}^T + \mathbf{O}\dot{\overline{\mathbf{X}}}\mathbf{O}^T + \mathbf{O}\overline{\mathbf{X}}\dot{\mathbf{O}}^T \\ &= \mathbf{O}\left(\dot{\overline{\mathbf{X}}} + \left[\overline{\mathbf{\Omega}}, \overline{\mathbf{X}}\right]\right)\mathbf{O}^T.\end{aligned} \tag{6.130}$$

This conforms with Eq. (6.129) if

$$\overline{\mathbf{V}} = \overline{D_t}\,\overline{\mathbf{X}} = \dot{\overline{\mathbf{X}}} + \left[\overline{\mathbf{\Omega}}, \overline{\mathbf{X}}\right], \quad \text{or} \quad \overline{D_t} = \overline{\frac{d}{dt}} + \left[\overline{\mathbf{\Omega}}, \cdot\,\right]. \tag{6.131}$$

Here the same qualifications mentioned previously have to be made concerning the meaning of derivatives with respect to time. Also the $\cdot$ in $\left[\overline{\mathbf{\Omega}}, \cdot\,\right]$ is to be replaced by the quantity $\overline{\mathbf{X}}$ being operated upon.[17] Yet another dispensation is required in that the symbol $\overline{D_t}$ has acquired a new meaning that can be inferred only from the context.[18]

We can also define a matrix $\mathbf{L}^{(i)}$ associated with angular momentum $\mathbf{l}^{(i)} = \mathbf{x}^{(i)} \times m^{(i)}\mathbf{v}^{(i)}$ defined in Eq. (6.103). Here, in anticipation of analyzing multiparticle systems, the notation has been generalized by introducing the superscript (i) which is a particle index. The space and moving frame angular momenta "matrices" for the ith particle are given by

$$\mathbf{L}^{(i)} = m^{(i)}\left[\mathbf{X}^{(i)}, \mathbf{V}^{(i)}\right], \quad \overline{\mathbf{L}}^{(i)} = \mathbf{O}^T\mathbf{L}^{(i)}\mathbf{O} = m^{(i)}\left[\overline{\mathbf{X}}^{(i)}, \overline{\mathbf{V}}^{(i)}\right]. \tag{6.132}$$

Newton's torque equation (6.80), expressed in an arbitrary frame as in Eq. (6.83), relative to a point on the axis of rotation, becomes

$$\overline{D_t\mathbf{L}}^{(i)} = \dot{\overline{\mathbf{L}}}^{(i)} + \left[\overline{\boldsymbol{\Omega}}, \overline{\mathbf{L}}^{(i)}\right] = \overline{\boldsymbol{\mathcal{T}}}^{(i)}. \tag{6.133}$$

Expressed here in "associated" matrices, this is the gauge invariant equation of rotation of a rigid body that consists of a single point mass $m^{(i)}$ subject to applied torque. Since the centroid of a one particle system is coincident with the mass itself, this equation so far gives a useful and complete description only for a spherical pendulum, with the mass attached to the origin by a light rod (or for an unconstrained cheerleader's baton which amounts to the same thing.)

One wishes to employ Eq. (6.133) to obtain $\dot{\overline{\mathbf{L}}}^{(i)}$ and eventually the evolution of the angular momentum. This is simplest to do in the body frame, where $\dot{\overline{\mathbf{X}}}^{(i)}$ vanishes. Working in that frame, where it is also true that $\overline{\mathbf{V}}^{(i)} = \left[\overline{\boldsymbol{\Omega}}, \overline{\mathbf{X}}^{(i)}\right]$, using Eq. (6.132) one obtains a formula for the "angular momentum" in terms of the "position" (and mass) of the particle;

$$\overline{\mathbf{L}}^{(i)} = -m^{(i)}\left[\overline{\mathbf{X}}^{(i)}\left[\overline{\mathbf{X}}^{(i)}, \overline{\boldsymbol{\Omega}}\right]\right]. \tag{6.134}$$

As required it has dimensions $[ML^2/T]$. Substituting this into Eq. (6.133), Newton's torque equation becomes

$$\dot{\overline{\mathbf{L}}}^{(i)} = \overline{\boldsymbol{\mathcal{T}}}^{(i)} + m^{(i)}\left[\overline{\boldsymbol{\Omega}}, \left[\overline{\mathbf{X}}^{(i)}, \left[\overline{\mathbf{X}}^{(i)}, \overline{\boldsymbol{\Omega}}\right]\right]\right]. \tag{6.135}$$

17) With $\overline{\boldsymbol{\Omega}}$ regarded to be a member of a Lie algebra, the operator $\left[\overline{\boldsymbol{\Omega}}, \cdot\,\right]$ is known as its *adjoint operator* and $\overline{\boldsymbol{\Omega}} = \left[\overline{\boldsymbol{\Omega}}, \cdot\,\right]$. We refrain from using this notation since formal results from the theory of Lie algebra will not be used.

18) A computer scientist would say $\overline{D_t}$ is "overloaded;" a mathematician might call it "abuse of notation" (but use it nevertheless).

This equation will be exploited in the next section. There has been a large investment in establishing formalism. Equation (6.135) represents the first return on this "overhead." With a remarkable thrice-nested matrix commutation multiplication the torque is augmented by a fictitious torque that accounts for the rotation of the moving frame relative to the inertial frame.

6.3.5
The Euler Equations for a Rigid Body

Consider a rigid body made up of masses $m^{(i)}$. Any one of these masses $m^{(i)}$ can be considered as contributing $\mathbf{L}^{(i)}$ (given by Eq. (6.134)) to the total angular momentum of the body, and hence an amount $\mathbf{I}^{(i)}$ to the moment of inertia tensor;

$$\overline{\mathbf{L}}^{(i)} = \overline{\mathbf{I}}^{(i)}(\overline{\boldsymbol{\Omega}}), \quad \text{where} \quad \overline{\mathbf{I}}^{(i)}(\cdot\,) = -m^{(i)}\Big[\overline{\mathbf{X}}^{(i)}, \big[\overline{\mathbf{X}}^{(i)}, \cdot\,\big]\Big]. \tag{6.136}$$

Here the (per particle) "moment of inertia" tensor has been generalized to be a function that generates the "angular momentum" $\mathbf{l}$ linearly from the "angular velocity" $\overline{\boldsymbol{\Omega}}$. Then the total moment of inertia tensors $\overline{\mathbf{I}}$ and the total angular momentum $\overline{\mathbf{L}}$ are

$$\overline{\mathbf{I}}(\cdot\,) = -\sum_i m^{(i)}\Big[\overline{\mathbf{X}}^{(i)}, \big[\overline{\mathbf{X}}^{(i)}, \cdot\,\big]\Big], \quad \text{and} \quad \overline{\mathbf{L}} = \overline{\mathbf{I}}(\overline{\boldsymbol{\Omega}}). \tag{6.137}$$

When the moment of inertia is a symmetric tensor, as in standard theory, we know that it can be diagonalized with orthogonal transformations, yielding three orthogonal principal axes and corresponding principal moments of inertia, call them $\overline{I}_i$. In the body frame these are independent of time. The same algebra assures the existence of principal-axes determined by our new, generalized, moment of inertia tensor. (See Problem 1.6.2.) The argument of $\overline{\mathbf{I}}$, namely $\overline{\boldsymbol{\Omega}}$, can itself be expanded in terms of the "basis" matrices J_i defined in Eq. (6.105) and each of these, being in turn associated with an angular rotation vector aligned with its respective axis, is the transformation matrix for an infinitesimal rotation around that axis. Superposition is applicable because $\overline{\mathbf{I}}$ is a linear operator. Supposing that these axes were judiciously chosen to start with to be these principal axes we must have

$$\overline{\mathbf{I}}(J_1) = \overline{I}_1 J_1, \quad \overline{\mathbf{I}}(J_2) = \overline{I}_2 J_2, \quad \overline{\mathbf{I}}(J_3) = \overline{I}_3 J_3. \tag{6.138}$$

Clearly it is advantageous to express $\overline{\boldsymbol{\Omega}}$ in terms of components along these axes.

$$\overline{\boldsymbol{\Omega}} = \sum_{i=1}^{3} \overline{\omega}^i J_i. \tag{6.139}$$

Of course the coefficients are the body frame, principal-axis components of the instantaneous angular velocity of the rigid body. The total angular momentum is then given by

$$\overline{\mathbf{L}} = \overline{\mathbf{I}}\Big(\sum_{j=1}^{3} \overline{\omega}^j J_j\Big) = \sum_{j=1}^{3} \overline{\omega}^j \overline{\mathbf{I}}(J_j) = \sum_{j=1}^{3} \overline{I}_j \overline{\omega}^j J_j. \tag{6.140}$$

Substitution into Newton's equation (6.133) (with vanishing torque for simplicity) yields

$$\sum_{j=1}^{3} \overline{I}_j \dot{\overline{\omega}}_j J_j = -\Big[\sum_{i=1}^{3} \overline{\omega}^i J_i, \sum_{j=1}^{3} \overline{I}_j \overline{\omega}^j J_j\Big] = -\sum_{i,j} \overline{\omega}^i \overline{\omega}^j I_j \epsilon_{ijk} J_k, \tag{6.141}$$

since the J_j satisfy the commutation relations of Eq. (6.106). The equations of motion become

$$\overline{I}_1 \dot{\overline{\omega}}_1 = \sum_{j=1}^{3} \overline{I}_j \overline{\omega}^i \overline{\omega}^j \epsilon_{ij1} = (\overline{I}_2 - \overline{I}_3)\overline{\omega}^2 \overline{\omega}^3, \tag{6.142}$$

and cyclic permutations. Once again, these are the Euler equations. Once the machinery was in place their derivation has been remarkably brief.

6.4 The Foucault Pendulum

Seen by every one who visits a science museum, the Foucault pendulum is one of the best experiments of all time. It rarely gets the credit it deserves. It is cheap to construct (though it behaves improperly if it is implemented too flimsily) and requires no more sophisticated data acquisition apparatus, even for quantitatively accurate measurements, than a patient "nurse" willing to look at it every few hours for a few days. If the base has a readable scale, permitting one to note the pendulum's advance over a couple of hours, one can check the advance of the set-up to perhaps ten percent accuracy. If one is prepared to spend all day at the science museum one can do better yet.[19] Yet these observations have profound implications. For example, the experiment can be regarded as experimental confirmation of the parallel transport formalism of Levi-Civita, which is fundamental to Einstein's theory of general relativity.

If one starts the pendulum swinging, say at noon, say parallel to a nearby wall, then leaves and comes back a few hours later, the pendulum is no longer swinging parallel to the wall. "It is the earth's rotation" you say, "check it

19) The most professional set-up I have seen is at the Science Museum, London, England, where the co-latitude is 38.5° and plane of oscillation rotates 11.8° /h.

when the earth has made a complete revolution" so everything is back where it started. Coming back at noon the next day everything is back except the pendulum. The wall is presumably back (in just one day the earth's orbit around the sun has introduced only a one degree reorientation of the earth) but the pendulum is not.[20]

Since the Foucault pendulum at rest is nothing other than the plumb bob analyzed earlier, we know that it hangs down along the direction of *effective* gravity and not precisely aimed toward the earth's center. As suggested there, we will take the rest orientation of the pendulum as defining the effective direction of the acceleration of gravity $\hat{\mathbf{g}}_{\rm eff}$ and continue to use the traditional symbol g for its magnitude. Once this is done we can from then on, to a good approximation, ignore the centrifugal force. Furthermore, we will use $\hat{\mathbf{g}}_{\rm eff} = -\hat{\mathbf{r}}$ even though that neglects the small angular deviation relative to the earth's center.

Using the Foucault pendulum, we can illustrate the discussion of evolving orientation and at the same time introduce the curious concept of *holonomy*, or more interestingly, *anholonomy*. You are instructed to perform the *gedanken experiment* illustrated in Fig. 6.12. Supporting a simple bob-pendulum of mass m, length λ, by an ideal bearing (swivel bearing, not compass bearing) you are to walk west to east with angular velocity ω_E, once around a nonrotating earth, radius R, along the θ line of latitude – say the one passing through New York City. Here "ideal bearing" means that the bearing cannot apply any torque component parallel to the support wire. (In practice, if the support wire is sufficiently long and slender, and the bob sufficiently massive, this condition can be adequately met even without the support point being an actual rotating bearing, but this analysis will not be described here.) The practical realization of this experiment with a very long, very heavy bobbed pendulum is known as the Foucault experiment.

6.4.1
Fictitious Force Solution

At this point we "solve" the Foucault pendulum problem using "fictitious force" arguments. Though extremely efficient this solution does not explain how the motion is consistent with the conservation of angular momentum. In the following section the motion will be studied in greater detail which will also serve to illustrate "gauge invariant" reasoning.[21]

20) Even such an excellent text as Kleppner and Kolenkov seems to get this wrong when they say "The plane of motion tends to stay fixed in inertial space while the earth rotates beneath it." And the literature describing the London Science Museum apparatus mentioned above is similar. (Not to gloat, though; the updated edition of this text has corrected errors as egregious as this.)

21) We continue the somewhat artificial distinction between the "fictitious force" and "gauge invariant" formulations.

With centrifugal force accounted for by the redefinition of the gravitational "up" direction, the forces acting on the pendulum bob are gravity $-mg\hat{\bar{r}}$ and the Coriolis force $-2m\omega_E(\cos\theta\,\hat{\bar{\mathbf{r}}} - \sin\theta\,\hat{\bar{\mathbf{s}}}) \times (\dot{\bar{s}}\,\hat{\bar{\mathbf{s}}} + \dot{\bar{e}}\,\hat{\bar{\mathbf{e}}})$, where "south" has been assigned coordinate $\bar{s}$ and "east" $\bar{e}$. In the usual approximation of small pendulum oscillations[22] the equations of motion are

$$\ddot{\bar{s}} - 2\omega_E \cos\theta\,\dot{\bar{e}} + \frac{g}{\lambda}\bar{s} = 0,$$
$$\ddot{\bar{e}} + 2\omega_E \cos\theta\,\dot{\bar{s}} + \frac{g}{\lambda}\bar{e} = 0. \tag{6.143}$$

The earth's rotation frequency ω_E was given numerically by Eq. (6.34) to be $0.727 \times 10^{-4}\,\mathrm{s}^{-1}$. Knowing the pendulum swinging frequency to be $\omega_0 = \sqrt{g/l}$, it is meaningful to compare the magnitudes of three frequencies in the problem. Recalling several seconds as being a typical period of a swinging Foucault pendulum, and something less than one revolution of secular advance to its plane of oscillation, it is clear that

$$\omega_F \leq \omega_E \ll \sqrt{\frac{g}{\lambda}}, \tag{6.144}$$

where $\omega_F = \cos\theta\,\omega_E$ is the "Foucault frequency."

As a general rule the presence of velocity terms in the second-order linear equations such as this reflects the presence of damping or antidamping (very weak according to the numerical estimate just given) and in that case a solution by Laplace transform would be appropriate. But the fact that the coefficients in the two equations are equal and opposite (and our expectation that the solution should exhibit no damping when none has been included in the model) makes a special method of solution appropriate.[23] Introducing the "complex displacement"

$$\bar{\xi} = \bar{s} + i\,\bar{e} \tag{6.145}$$

the equations of motion become

$$\ddot{\bar{\xi}} + 2i\,\omega_F\,\dot{\bar{\xi}} + \omega_0^2\bar{\xi} = 0. \tag{6.146}$$

Assuming the pendulum is started from rest with a southerly amplitude a, $(\xi(t=0) = a)$, the solution of this equation is

$$\bar{\xi} = \frac{a}{2\omega}\Big((\omega + \omega_F)e^{i(\omega - \omega_F)}(\omega - \omega_F)e^{-i(\omega + \omega_F)}\Big), \tag{6.147}$$

22) Approximate solution of large amplitude pendulum motion is treated in Chapter 16.

23) Conditions under which velocity-dependent forces do not destroy the Hamiltonian nature of mechanical systems are treated in Chapter 16.

where

$$\omega^2 = \omega_0^2 + \omega_F^2. \tag{6.148}$$

Substituting

$$\overline{\xi} = e^{-i\omega_F t}\overline{\zeta} \tag{6.149}$$

in Eq. (6.147), we get

$$\overline{\zeta} = a\cos\omega t + i\frac{\omega_F}{\omega}\,a\sin\omega t \approx a\cos\omega t. \tag{6.150}$$

With the "complex coordinate" defined as in Eq. (6.145), with south defined by the real axis and east defined by the imaginary axis, the transformation (6.149) amounts to viewing the pendulum from a frame of reference rotating with an angular velocity ω_F. Be sure to notice though, from the first inequality in (6.144), that the rotation period is not the same as the earth's rotation period (except at the equator.) When the pendulum is viewed for only a few periods, the motion as given by Eq. (6.150) is just what one expects for a pendulum, oscillating with frequency $\sqrt{g/l}$, because the time dependent factor in transformation (6.149) is varying so slowly. Coming back some hours later and viewing a few periods of oscillation one sees the same thing, but now the plane of oscillation is altered, according to (6.149). All the observations have been accounted for.

6.4.2
Gauge Invariant Solution

We now consider the same system in somewhat greater detail, using the gauge invariant formulas. Assuming the trip starts in the x, z plane, the "trip equation" of the support point is $\phi = \omega_E t$ and the location of the point of support $\big(x(t), y(t), z(t)\big)$ can be related to its initial position by

$$\begin{pmatrix} x(t) \\ y(t) \\ z(t) \end{pmatrix} = \begin{pmatrix} \cos\omega_E t & -\sin\omega_E t & 0 \\ \sin\omega_E t & \cos\omega_E t & 0 \\ 0 & 0 & 1 \end{pmatrix} \begin{pmatrix} R\sin\theta \\ 0 \\ R\cos\theta \end{pmatrix}, \quad \text{or} \quad \mathbf{r}(t) = \mathbf{O}(t)\,\mathbf{r}(0). \tag{6.151}$$

Because of attraction by the earth's mass, which is treated as if it were concentrated at the origin (on the earth's axis, but slightly south of its center to account for centrifugal force if we are in the Northern hemisphere), the pendulum bob always points more or less toward the origin. It is the orientation of its instantaneous swing plane that will be of primary interest. The gravitational force between the earth and the bob applies torque to the pendulum

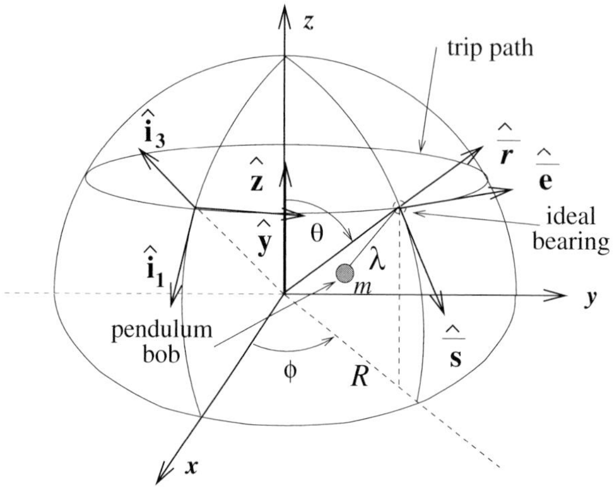

Fig. 6.12 Illustration of parallel transport of an ideally mounted pendulum around a line of latitude. Note that the fixed frame is specific to the particular latitude along which the pendulum is carried.

– that is what makes the pendulum oscillate – but the torque about the bearing point due to gravitational attraction to the earth has no component parallel to $\hat{\mathbf{r}}$. We expect the radial component of angular momentum therefore to be either exactly or approximately conserved, where the "approximate" reservation has to be included until we are sure that the Coriolis force is correctly incorporated.

The motion of the pendulum could be analyzed in the inertial space frame K, with unit vectors being $(\hat{\imath}_1, \hat{\mathbf{y}}, \hat{\imath}_3)$ as shown in the figure. (Note that $\hat{\imath}_1$ would be "south" and $\hat{\mathbf{y}}$ would be "east," but, not being attached to the earth, they are fixed in an inertial frame.) We will instead analyze the pendulum in a moving frame $\overline{K}$, $(\hat{\bar{\mathbf{s}}}, \hat{\bar{\mathbf{e}}}, \hat{\bar{\mathbf{r}}})$, with the origin at the support point. The moving-frame axes satisfy $\hat{\bar{\mathbf{e}}} = \widehat{\mathbf{z} \times \bar{\mathbf{r}}}$, and $\hat{\bar{\mathbf{s}}} = \hat{\bar{\mathbf{e}}} \times \hat{\bar{\mathbf{r}}}$. The angular velocity of frame $\overline{K}$ is

$$\boldsymbol{\omega}_E = \omega_E \hat{\mathbf{z}} = \overline{\boldsymbol{\omega}}_E = \omega_E \big(\cos\theta \, \hat{\bar{\mathbf{r}}} - \sin\theta \, \hat{\bar{\mathbf{s}}} \big). \tag{6.152}$$

(Since this vector is constant, its K-frame components are constant, and its $\overline{K}$-frame components are constant also, even though the body axes are varying. Also we know by now that we can get away with not distinguishing between $\overline{\boldsymbol{\omega}}_E$ and $\boldsymbol{\omega}_E$, as the identification of the same intrinsic arrow, but use the overhead symbol to indicate which frame is intended for their components to be calculated.) Notice that the moving frame in this case is *not* the body frame of the pendulum (unless the pendulum happens to be in its neutral position) so the use of body-fixed axes is contraindicated since the gravitational force applies a torque that complicates the motion in that frame. The bob location relative to the support position is

$$\overline{\mathbf{x}} = -\lambda\hat{\overline{\mathbf{r}}} + \overline{\mathbf{x}}_\perp. \tag{6.153}$$

Here a small term, quadratic in the pendulum swing angle, has been neglected in the first term and a "transverse" $\overline{K}$ frame displacement vector $\overline{\mathbf{x}}_\perp$ has been introduced: it satisfies $\overline{\mathbf{x}}_\perp \cdot \hat{\overline{\mathbf{r}}} = 0$. The angular momentum of the bob about the point of support is

$$\overline{\mathbf{L}} = \overline{\mathbf{x}} \times m\overline{\mathbf{v}} = m\overline{\mathbf{x}} \times \left(\overline{\frac{d}{dt}}\overline{\mathbf{x}} + \overline{\boldsymbol{\omega}}_E \times \overline{\mathbf{x}}\right), \tag{6.154}$$

and its radial component is

$$\overline{\mathbf{L}}_{\overline{r}} = m\hat{\overline{\mathbf{r}}} \cdot \left(\overline{\mathbf{x}}_\perp \times \overline{\frac{d}{dt}}\overline{\mathbf{x}} + \overline{x}_\perp^2\, \overline{\boldsymbol{\omega}}_E - (\overline{\mathbf{x}}_\perp \cdot \overline{\boldsymbol{\omega}}_E)\,\overline{\mathbf{x}}\right). \tag{6.155}$$

This equation contains the essence of the calculation. It follows from Eq. (6.81), which derives the $\overline{K}$-frame angular momentum from the absolute velocity $\overline{\mathbf{v}}$. $\overline{\frac{d}{dt}}\overline{\mathbf{x}}$, is an "apparent velocity" obtained by differentiating the moving frame coordinates with respect to time. We know this quantity is not itself a vector, but that $\overline{\frac{d}{dt}}\overline{\mathbf{x}} + \overline{\boldsymbol{\omega}}_E \times \overline{\mathbf{x}}$, the true velocity vector expressed in terms of moving frame variables, *is* a true vector. Also, since the calculation is being performed in the $\overline{K}$ frame, the angular velocity vector has been written as $\overline{\boldsymbol{\omega}}_E$.

In the $\overline{K}$ frame, the equation of motion is given by Eq. (6.83), simplified by the fact that the support point is at rest.

$$\overline{\frac{d}{dt}}\overline{\mathbf{L}} = \overline{\boldsymbol{\tau}}. \tag{6.156}$$

The force acting on mass m is $-mg(\hat{\overline{\mathbf{r}}} + \overline{\mathbf{x}}/R)$, directed along the line pointing from the gravitational center toward m

$$\overline{\boldsymbol{\tau}} = -mg\overline{\mathbf{x}} \times \left(\hat{\overline{\mathbf{r}}} + \frac{\overline{\mathbf{x}}}{R}\right), \tag{6.157}$$

which has no component parallel to $\hat{\overline{\mathbf{r}}}$. As a result the angular momentum component along $\hat{\overline{\mathbf{r}}}$ is conserved. Taking $\overline{L}_r(0)$ as its initial value and substituting from Eqs. (6.152) and (6.153) into Eq. (6.156),

$$\begin{aligned}\overline{L}_r(0) = \overline{\mathbf{L}} \cdot \hat{\overline{r}} &= m\left(\overline{\mathbf{x}}_\perp \times \overline{\frac{d}{dt}}\overline{\mathbf{x}}_\perp\right) \cdot \hat{\overline{\mathbf{r}}} + m\overline{x}_\perp^2\, \omega_E \cos\theta + m\lambda(\overline{\mathbf{x}} \cdot \overline{\boldsymbol{\omega}}_E) \\ &= m\left(\overline{s}\,\dot{\overline{e}} - \overline{e}\,\dot{\overline{s}} + (\overline{e}^2 + \overline{s}^2)\omega_E \cos\theta - \lambda\omega_E \sin\theta\, \overline{s}\right),\end{aligned} \tag{6.158}$$

where we have neglected the radial velocity and have used $\overline{\mathbf{x}}_\perp \cdot \hat{\overline{\mathbf{s}}} = \overline{s}$ and $\overline{\mathbf{x}}_\perp \cdot \hat{\overline{\mathbf{e}}} = \overline{e}$.

The pendulum could be set swinging initially in an elliptical orbit, but for simplicity we suppose that it is released from rest from an initial displacement a along the $\bar{s}$-axis. As a result $\overline{L}_r(0) = 0$. Casually viewed, the pendulum will continue to oscillate in this plane, but we will allow for the possibility that the plane changes gradually (because that is what is observed.) Let us then conjecture a solution such that the components of $\overline{\mathbf{x}}_\perp$ are given by

$$\begin{aligned} \bar{s} &= a \sin \omega_0 t \cos \psi(t) = \frac{a}{2}\Big(\sin\left(\omega_0 t + \psi(t)\right) + \sin\left(\omega_0 t - \psi(t)\right)\Big), \\ \bar{e} &= a \sin \omega_0 t \sin \psi(t) = \frac{a}{2}\Big(-\cos\left(\omega_0 t + \psi(t)\right) + \cos\left(\omega_0 t - \psi(t)\right)\Big). \end{aligned} \tag{6.159}$$

Here a is an amplitude factor, $\psi(t)$ defines the axis of the elliptical orbit, and $\omega_0 = \sqrt{g/\lambda}$ is the pendulum frequency. The small, quadratic-in-amplitude vertical displacement is neglected. This form of solution is likely to be valid only if the angle $\psi(t)$ is slowly varying compared to $\omega_0 t$. We are also implicitly assuming that amplitude a remains constant; this relies on the fact that it is an *adiabatic invariant*, a fact that will be explained in Chapter 14. Substitution from Eq. (6.159) into Eq. (6.158) yields

$$\begin{aligned} &\frac{1}{2} m a^2 (1 - \cos(2\omega_0 t))\dot{\psi} + m a^2 \omega_E \cos\theta \sin^2(\omega_0 t) \\ &\qquad = m\omega_E \lambda a \sin \omega_0 t \cos \psi(t) \sin\theta. \end{aligned} \tag{6.160}$$

Note that the *ansatz* (6.159) has assigned zero initial radial angular momentum to the pendulum. In detail the motion implied by Eq. (6.160) is complicated but, since we are assuming $|\dot{\psi}| \ll \omega_0$, it is possible to distinguish between *rapidly varying* terms like $\sin \omega_0 t$ and *slowly varying* terms like $\dot{\psi}$ in Eq. (6.160). This permits the equation to be averaged over the rapid variation while treating the slow variation as constant. Recalling that $\langle \sin^2 \omega_0 t \rangle = 1/2$ this yields

$$\langle \dot{\psi} \rangle = -\omega_E \cos\theta. \tag{6.161}$$

(The sort of averaging by which Eq. (6.160) has been derived will be considered further and put on a firmer foundation in Chapter 16.)

This shows that carrying the support point along a line of latitude with angular velocity ω_E causes the plane of oscillation of the pendulum to rotate relative to axes $\hat{\bar{\mathbf{s}}}, \hat{\bar{\mathbf{e}}}$, with rotation rate $-\omega_E \cos\theta$. The presence of the $\cos\theta$ factor gives a nontrivial dependence on latitude; at the end of one earth-rotation period $T = 2\pi/\omega_E$, the support point has returned to its starting point, but the plane of oscillation has deviated by angle $-\omega_E \cos\theta 2\pi/\omega_E = -2\pi\cos\theta$.

If one performs this experiment at the North pole,[24] the support point never moves and it is clear that the plane of oscillation remains fixed in space. This

24) The distinction between magnetic axis and rotation axis is being ignored.

agrees with Eq. (6.160) which, because $\cos\theta = 1$, predicts a $-\omega_E$ rotation rate about the earth's axis. This just compensates the $+\omega_E$ rotation rate of the earth.

If you were performing this experiment at the North pole it would not be necessary to rely on the earth's rotation to perform the experiment. Rather, holding the pendulum at arm's length, you could move the support point around a tiny circle (radius equal to the length of your arm) centered over the North pole. In this case ω_E could be chosen at will, say fast compared to the earth's rotation rate, but still slow compared to ω_0. Since the plane of oscillation of the pendulum would be perceived to be invariant, the plane of oscillation would return to its original orientation after your hand had returned to its starting point. But this special case is misleading since it suggests that the pendulum necessarily recovers its initial spatial orientation after one complete rotation. In fact, the pendulum orientation suffers secular change, as observation of an actual Foucault pendulum after 24 h confirm experimentally.

For latitudes other than at the poles it is more complicated and the only experimentally easy value for ω_E is $(2\pi/24)\,\mathrm{h}^{-1}$. At the equator there is no apparent rotation of the plane of oscillation – formula (6.160) gives that result and it is just as well, since symmetry requires it, especially in the case that the pendulum plane is parallel to the equatorial plane.

Since the Foucault precession rate is proportional to the earth's rotation rate, the angle of precession after one revolution is independent of the earth rotation rate. Furthermore, the precession is independent of gravitational constant g – the same experiment on the moon would yield the same precession angle (after one moon rotation period.) These features show that the effect is geometric rather than dynamic. Two masses joined by an ideal spring and supported on a frictionless horizontal table and oscillating losslessly along the line joining them would exhibit the same result – the line joining them would precess. A geometric analysis of the phenomenon will be pursued in the next section.

Problem 6.4.1. *Analyze the average motion of a pendulum that is performing circular motion about a "vertical" axis as the support point moves along a line of latitude θ due to the earth's rotation at frequency ω_E.*

Problem 6.4.2. *The circularly swinging pendulum of the previous problem can be used as a clock, with one "tick" occurring each time the bob completes a full circle. Suppose there are two such clocks, initially synchronized, and at the same place. Suppose further that one of the clocks remains fixed in space (or rather it stays on the earth's orbit about the sun) while the other, fixed on earth, comes side-by-side with the other clock only once per day. Which clock "runs slow" and by how much? This is a kind of "twin paradox" for Foucault clocks.*

Problem 6.4.3. *Using steel as the material for both the suspension wire and the bob of the Foucault pendulum, after choosing plausible values for wire length and radius, calculate the torsional frequency of the pendulum in its straight up configuration. It is important for this frequency to be large compared to ω_F. It is also important for lossiness of this oscillation to be small enough to cause only an acceptably small reduction of quality factor Q of the dominant pendulum oscillation. Since this damping (as well as air resistance) is usually not negligible, practical set-ups usually use active control mechanisms that maintain the main oscillation without applying twist.*

6.4.3
"Parallel" Translation of Coordinate Axes

The fact that, after one circumnavigation of the earth, the Foucault pendulum plane does not return to its original orientation is an example of *anholonomy*. Though one tried one's best to avoid "twisting it," by supporting the pendulum with an ideal bearing, its plane is found to be "twisted" when the pendulum returns to its home position. Since all practical implementations of the Foucault pendulum rely on the slenderness of the suspension wire, rather than the presence of an ideal bearing at the top, a complete analysis also requires the analysis of the torsional pendulum formed from the wire and the ball; see Problem 6.4.3. Because no external torque is applied to the pendulum, its radial angular momentum is conserved, but this does not prevent rotational displacement around the radial axis from accumulating.[25] Recall the two triads $(\hat{\imath}_1, \hat{\mathbf{y}}, \hat{\imath}_3)$ and $(\hat{\hat{\mathbf{s}}}, \hat{\hat{\mathbf{e}}}, \hat{\hat{\mathbf{r}}})$ used to analyze the Foucault pendulum. (In this section we will refer to the latter triad as $(\hat{\mathbf{s}}, \hat{\mathbf{e}}, \hat{\mathbf{r}})$ and restrict the description to rest frame quantities.) The following comments and questions arise:

- In 3D, (x, y, z) space the two triads are manifestly *not* parallel, except initially.
- Can meaning be assigned to "parallel translation" of such a triad in the 2D surface of the sphere?
- If so, are $(\hat{\imath}_1, \hat{\mathbf{y}}, \hat{\imath}_3)$ and $(\hat{\mathbf{s}}, \hat{\mathbf{e}}, \hat{\mathbf{r}})$ parallel in this sense?

These questions relate to the concept of parallel displacement of a vector in differential geometry. This concept was introduced by the Italian mathematician Levi-Civita toward the end of the nineteenth century. The importance of the concept in physics is discussed, for example, by M. V. Berry, *The Quantum Phase, Five Years After.* He describes requirements to be satisfied for the "parallel transport" of an initially orthonormal triad of unit vectors $(\hat{\mathbf{e}}_1, \hat{\mathbf{e}}_2, \hat{\mathbf{r}})$ that is

25) A relevant observation (I am unaware of its actually having been reported) based on viewing a mark on the equator of the bob, would be to confirm that, at the end of the day, the ball stays aligned with the swing plane.

attached to the tip of a radius vector $\mathbf{r}(t)$ pointing from the center of a sphere of unit radius to a point moving on the surface of a sphere:

- $\mathbf{r}(t)$ is to remain a unit vector (so the origin of the triad stays on the surface of the sphere.)
- $\hat{\mathbf{r}}(t)$ is to remain parallel to $\mathbf{r}(t)$.
- $\hat{\mathbf{e}}_1 \cdot \hat{\mathbf{r}}$ is to remain zero. That is, $\mathbf{e}_1$ remains tangent to the sphere. With $\hat{\mathbf{r}}$ normal to the surface, and $\mathbf{e}_2$ normal to $\hat{\mathbf{r}}$, $\mathbf{e}_2$ also remains tangential.
- The triad is not to "twist" about $\hat{\mathbf{r}}$, i.e., $\boldsymbol{\omega} \cdot \hat{\mathbf{r}} = 0$, where $\boldsymbol{\omega}$ is the instantaneous angular velocity of the triad.

To visualize the meaning of the final requirement imagine a single-gimbel-mounted globe with bearings at the North and the South pole. Such a mount allows only pure rotations with $\boldsymbol{\omega}$ parallel to the North–South-axis and an arbitrary point A can be rotated into an arbitrary point B if and only if they are on the same latitude. The path followed by A is a circle with center on the earth's axis but not, in general, coincident with the earth's center and the condition $\boldsymbol{\omega} \cdot \hat{\mathbf{r}} = 0$ is not met unless both A and B to lie on the equator; only in that case would the motion of the triad be said to be twist-free. Next suppose the globe has a double-gimbel mounting. Then any point A can be rotated into any point B by a twist-free rotation – to obtain a pure rotation about a single axis, one has to seize the globe symmetrically with both hands and twist them in synchronism about the desired axis. Point A is then said to be taking the "great circle" route to B. The center of such a circle necessarily coincides with the earth's center. Since the path taken by the Foucault pendulum is not a great circle path, the triads $(\hat{\imath}_1, \hat{\mathbf{y}}, \hat{\imath}_3)$ and $(\hat{\mathbf{s}}, \hat{\mathbf{e}}, \hat{\mathbf{r}})$ used to analyze that system are *not* parallel in the sense being discussed.

To meet the first requirement of parallel transport, the evolution of $\mathbf{r}(t)$ must be describable by an orthogonal matrix $\mathbf{O}(t)$ as in Eq. (6.151),

$$\mathbf{r}(t) = \mathbf{O}(t)\,\mathbf{r}(0). \tag{6.162}$$

and our task then, for arbitrary evolution $\mathbf{r}(t)$, is to find out how $\mathbf{O}(t)$ evolves with t. In the differential evolution occurring during time dt the simplest rotation carrying $\mathbf{r}$ to $\mathbf{r} + \dot{\mathbf{r}}dt$ is around the axis $\mathbf{r} \times \dot{\mathbf{r}}$ – the motion remains in a plane through the center of the sphere. The angular speed being $\dot{\mathbf{r}}$ and the sphere having unit radius, the angular velocity vector of this rotation is $\boldsymbol{\omega} = \mathbf{r} \times \dot{\mathbf{r}}$ which implies

$$\dot{\mathbf{r}} = \boldsymbol{\omega} \times \mathbf{r}. \tag{6.163}$$

This rotation does not determine $\mathbf{O}(t)$ uniquely however, since there remains the possibility of further (pre or post) rotation of the globe around $\hat{\mathbf{r}}$. Still, this *is* the twist-free motion being sought since it satisfies

$$\boldsymbol{\omega} \cdot \mathbf{r} = 0, \tag{6.164}$$

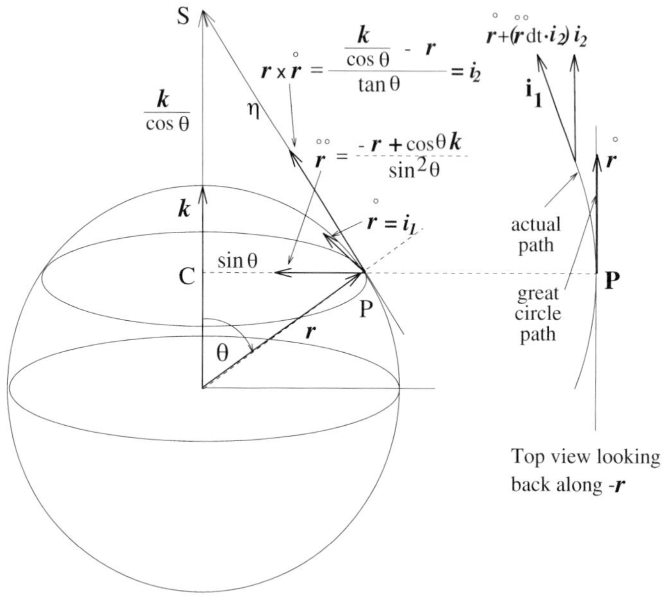

Fig. 6.13 Vector geometry illustrating the rate of accumulation of angular deviation ϕ_T between a twist-free frame and a frame with one axis constrained to line up with $\dot{\mathbf{r}}$. To reduce clutter it is assumed that $v = 1$. At any instant the figure can be drawn to look like the Foucault trajectory along a circular arc as shown here, but in general the point P moves along any smooth closed path on the surface of the sphere.

which is the no-twist condition listed above. As in Eq. (6.163) the time rates of change of unit vectors $\hat{\mathbf{e}}_1$ and $\hat{\mathbf{e}}_1$ due to angular rotation velocity $\boldsymbol{\omega}$ are

$$\dot{\hat{\mathbf{e}}}_1 = \boldsymbol{\omega} \times \hat{\mathbf{e}}_1 \quad \text{and} \quad \dot{\hat{\mathbf{e}}}_2 = \boldsymbol{\omega} \times \hat{\mathbf{e}}_2. \tag{6.165}$$

The moving origin is constrained to stay on the sphere, but it can otherwise be specified arbitrarily by specifying $\mathbf{r}(t)$ and then $\boldsymbol{\omega}$ follows from Eq. (6.163). An example of a "trip plan" for $\mathbf{r}(t)$ is that taken by the Foucault pendulum in Fig. 6.12, but notice that $\boldsymbol{\omega}$ as given by Eq. (6.163) is not parallel to the North–South-axis and hence differs from the angular velocity vector of the Foucault experiment.

From Eq. (6.50) we know the antisymmetric matrix "associated" with $\boldsymbol{\omega}$ is $\mathbf{J} \cdot \boldsymbol{\omega} = \mathbf{O}^T\dot{\mathbf{O}}$ and hence that

$$\dot{\mathbf{O}} = \mathbf{O}\mathbf{J} \cdot \boldsymbol{\omega} = \mathbf{O}\mathbf{J} \cdot (\mathbf{r} \times \dot{\mathbf{r}}). \tag{6.166}$$

This differential equation is to be integrated to obtain the twist-free rotation matrix $\mathbf{O}(t)$. Not surprisingly, the solution turns out to depend on the path taken by $\mathbf{r}(t)$. The geometry used to investigate this is indicated in Fig. 6.13.

As P, the point at the tip $\mathbf{r}$ moves on the unit sphere, its velocity $\dot{\mathbf{r}}$ lies in the tangent plane to the sphere. Requiring the speed v of the point's motion to be constant, the vector $\mathbf{i}_1 = \dot{\mathbf{r}}/v$ is a unit vector parallel to the motion. (This vector was referred to as the unit tangent vector $\mathbf{t}$ in the Frenet–Serret description of a curve in space.) It can be taken as the first axis of a local coordinate system. It is specialized to the particular motion being studied and is not intended to be useful for any other purpose, but it *does* have the property, *by definition*, assuming the point P moves on a smooth, kink-free closed path, of *returning to its starting value* after one complete circuit along a closed path. The other two local orthonormal axes can be taken to be $\mathbf{i}_3 = \hat{\mathbf{r}}$ and $\mathbf{i}_2 = \mathbf{r} \times \dot{\mathbf{r}}/v$. The latter is given by

$$\mathbf{i}_2 = \mathbf{r} \times \frac{\dot{\mathbf{r}}}{v} = \frac{\frac{\mathbf{k}}{\cos\theta} - \mathbf{r}}{\tan\theta}, \tag{6.167}$$

where θ and $\mathbf{k}$ are to be defined next. There is a best fit (osculating) circle with center at point C, lying in the local orbit plane, and having radius of curvature $\rho \equiv \sin\theta$ equal to the local curvature. The unit vector $\mathbf{k}$ is directed from the origin toward point C. From the elementary physics of circular motion one knows that the acceleration vector has magnitude v^2/ρ and points toward C. Explicitly, it is given by

$$\ddot{\mathbf{r}} = \frac{v^2}{\sin\theta}\frac{-\mathbf{r} + \cos\theta\mathbf{k}}{\sin\theta}. \tag{6.168}$$

The component of $\ddot{\mathbf{r}}$ lying in the tangential plane is

$$-v\frac{d\phi_T}{dt} = \ddot{\mathbf{r}} \cdot \left(\mathbf{r} \times \frac{\dot{\mathbf{r}}}{v}\right) = \frac{v^2}{\sin\theta}\frac{-\mathbf{r} + \cos\theta\mathbf{k}}{\sin\theta} \cdot \frac{\frac{\mathbf{k}}{\cos\theta} - \hat{\mathbf{r}}}{\tan\theta}. \tag{6.169}$$

Here the result has been used that in circular motion with speed v on a circle of radius η, the rate of angular advance $d\phi_T/dt$ satisfies $a = -v^2/\eta = -v\,d\phi_T/dt$. From the top view in Fig. 6.13 looking back along $-\mathbf{r}$, it can be seen that the axis $\mathbf{i}_1$ twists relative to an axis pointing along the tangential great circle through P, and that $d\phi_T/dt$ measures the time rate of twist. The radius η in this case is the distance from P to S, but this distance does not appear explicitly in the formulas. The accumulated twist in making a complete circuit is

$$\phi_T = -\oint \frac{\ddot{\mathbf{r}}}{v^2} \cdot \left(\mathbf{r} \times \frac{\dot{\mathbf{r}}}{v}\right) vdt = \oint (\mathbf{r} \times \mathbf{r}'') \cdot d\mathbf{r}, \tag{6.170}$$

where $ds = vdt$ and primes indicate differentiation with respect to s. Let us apply this formula to the trip taken by the Foucault pendulum in Fig. 6.12. Using Eq. (6.169),

$$\phi_T = -\oint \frac{1}{\sin\theta}\frac{-\mathbf{r} + \cos\theta\mathbf{k}}{\sin\theta} \cdot \frac{\frac{\mathbf{k}}{\cos\theta} - \mathbf{r}}{\tan\theta} 2\pi\sin\theta = 2\pi\cos\theta. \tag{6.171}$$

We have calculated the twist of $\mathbf{i}_1$ relative to the no-twist frame. But since we know that $\mathbf{i}_1$ returns to its starting value, it follows that the no-twist frame returns rotated by $-\phi_T = -2\pi\cos\theta$. This is the same twist we calculated (and observed) for the Foucault pendulum. This implies that the pendulum frame and the no-twist frame are the same thing. As far as I can see there is no *a priori* guaranteed equivalence of no-twist of geometry and no-twist caused by the Foucault support bearing, but the observation supports this equivalence experimentally.

No-twist displacement of the axes is also known as "parallel displacement" of the axes. It is clear that the orientation of the oscillation of a Foucault pendulum would satisfy Eq. (6.170) for a trip plan more complicated than simply along a line of latitude.

Problem 6.4.4. *Making any assumptions you wish concerning the orientations of path direction, twist direction, and solid angle orientation, show that the accumulated twist accompanying an arbitrary smooth closed path on the surface of a sphere can be expressed as* 2π *minus the solid angle enclosed by the path.*

6.5
Tumblers and Divers

An intriguing question has to do with falling cats. Everyone "knows" that a cat released from an upside down position from a height of less than a meter still manages to land on its feet. Everyone also knows that angular momentum is conserved. Many people believe these two statements are contradictory. One "explanation" is that the cat "pushes off" giving itself some initial angular momentum. Anyone who, one hopes in youth, has investigated this issue experimentally is certain to doubt this explanation. In any case this explanation requires the cat to be very good at mechanics to know how hard to push, and prescient to know its initial height.

The stunts performed by divers and gymnasts are as amazing as that by falling cats. Again the maneuvers appear to violate the laws of nature. Human inability to register exactly what is happening makes one doubt one's eyes. Does the trampoline artist push off with a twist? Otherwise how can she or he be facing one way on take off, and the other way on landing? Is the diver erect or bent when taking off from the diving board? And so on. These ambiguities introduce enough confusion to prevent resolution of observations that appear to violate the laws of nature.

Once one has unraveled one of these "paradoxes" one is less troubled by all the rest (and can perhaps advance to harder problems like "does the curve ball really curve?") The moves of divers and gymnasts are more controlled, less ambiguous, and more subject to experimentation, than are the gyrations

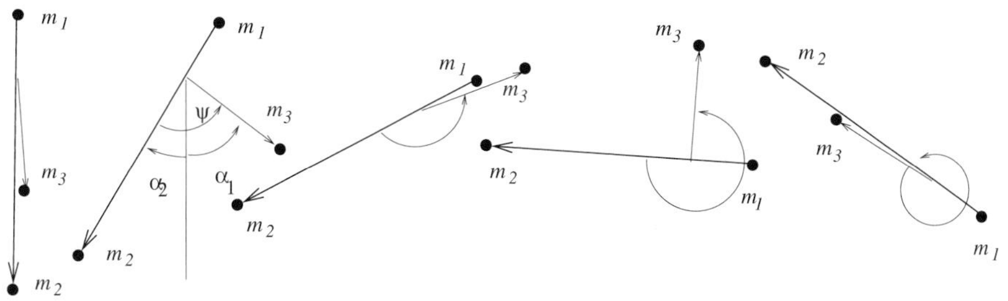

Fig. 6.14 Successive shapes and orientations of an astronaut performing the exercise of completing a full rotation of both arms.

of falling cats. They are therefore better subject to disciplined analysis using mechanics. The article by Frohlich, listed at the end of the chapter, describes these things clearly and there is no reason to repeat his explanations in any detail.

Only a simple example will be analyzed here. Imagine an astronaut (in gravity-free space) performs the exercise indicated (highly schematically) in Fig. 6.14. Fully erect initially, with arms straight down ($\alpha = \alpha_1 = \alpha_2 = 0$) the astronaut slowly (or quickly for that matter) raises her arms toward the front and on through one full revolution, ending therefore with the initial shape. The question is "what is her final orientation?"

To simplify discussion let us simplify the model by lumping head and shoulder into one mass, m_1, pelvis and and legs into m_2, and arms into m_3. This is not quite right for an actual human being but it should be "good enough for government work." It would be a lucky coincidence if the centroid of the astronaut coincided with the rotation axis of her arms. Nevertheless, in the spirit of the discussion so far.

Proceeding in steps, let us suppose the astronaut pauses at $\alpha = \pi/2$ – what is her configuration? pauses at $\alpha = \pi$ – what is her configuration? pauses at $\alpha = 3\pi/2$ – what is her configuration? ends at $\alpha = \pi/2$ – what is her orientation? These configurations are illustrated in Fig. 6.14. In the first step the astronaut has to apply the torque needed to rotate her arms forward and the torque of reaction pushes her torso back. Once her arms are straight forward her orientation is therefore something like that shown in the second figure. Much the same action occurs in the next step and is illustrated in the third figure. The shoulders of most men would make the next step difficult, but this is irrelevant because the astronaut is a woman with exceptionally supple shoulders. The torque she applies to keep her arms going in the same direction has to be accompanied by a pull on the rest of her body and this causes her torso to continue rotating in the same direction as in the first two steps. The final step leaves her as shown in the final figure, with orientation very different from her original orientation.

Since the astronaut is in free space, her angular momentum is presumably preserved at all times in the exercise just described, but this fact has not so far been used explicitly in predicting the motion. Let the moments of inertia (about the shoulder) of the arms and the assumed-rigid, rest of the body be I_1 and I_2, respectively. Let the angles of arms and torso in inertial space be α_1 and α_2 as shown, and let $\psi = \alpha_1 + \alpha_2$ be the angle between arms and torso. One notes, in passing, a possibly unexpected feature – the angle ψ does not in fact advance through 2π in the exercise. But let us work with angular momenta. The angular momentum of the arms is $I\dot{\alpha}_1$ and of the torso $-I_2\dot{\alpha}_2$. The total angular momentum is zero initially and presumably stays that way throughout. By conservation of angular momentum we conclude that

$$I_1\dot{\alpha}_1 = I_2\dot{\alpha}_2, \quad \text{or} \quad I_2\dot{\alpha}_2 = I_1\big(\dot{\psi}(t) - \dot{\alpha}_2\big). \tag{6.172}$$

Solving this differential equation with appropriate initial condition produces

$$\alpha_2 = \frac{I_1}{I_1 + I_2}\,\psi(t). \tag{6.173}$$

This corroborates the result obtained intuitively while describing the figure. For any reasonable estimate of the ratio I_1/I_2 for an actual person the figure clearly exaggerates the degree of reorientation occurring after one cycle. But there is no doubt that finite reorientation is consistent with conservation of angular momentum. This is another example of anholonomy.

Bibliography

General References

1 D. Kleppner and R.J. Kolenkow, *An Introduction to Mechanics,* McGraw-Hill, New York, 1973, p. 355.

2 K.R. Symon, *Mechanics*, 3rd ed., Addison-Wesley, Reading, MA, 1971, p. 271.

References for Further Study

Section 6.1.4

3 V.I. Arnold, V. Kozlov, and A. Neishtadt, *Mathematical Aspects of Classical and Celestial Mechanics,* 2nd ed., Springer, Berlin, 1997, p. 69.

4 G. Pascoli, *Elements de Mécanique Céleste,* Masson, Paris, 1997, p. 150.

Section 6.3

5 D.H. Sattinger and O.L. Weaver, *Lie Groups and Algebras with Applications to Physics, Geometry, and Mechanics,* Springer, New York, 1993.

Section 6.4.3

6 M.V. Berry, in S. Shapere and F. Wilczek, eds., *Geometric Phases in Physics,* pp. 7–28, World Scientific, Singapore, 1989.

Section 6.5

7 C. Frohlich, *Am. J. Phys.,* **47**, 583 (1979); **54**, 590 (1986).

8 C. Frohlich, *Sci. Am.,* **242**, 154 (1980).

7
Hamiltonian Treatment of Geometric Optics

It is not uncommon for Hamiltonian mechanics to be first encountered in the waning days of a course emphasizing Lagrangian mechanics. This has made it natural for the next course to start with and emphasize Hamiltonian mechanics. Here, though standard Hamiltonian arguments enter, they do so initially via a geometric, Hamilton–Jacobi route. Later, in Chapter 17, the full geometric artillery developed earlier in the text is rolled out under the name of symplectic mechanics, which is just another name for Hamiltonian mechanics with its geometric structure emphasized. Especially important is Liouville's theorem and its generalizations. Because of their fundamental importance, for example in accelerator physics, and because of their connection with quantum mechanics, adiabatic invariants and Poisson brackets, central to both traditional "canonical" formalism and the more geometric treatment are also stressed.

In his formulation of classical mechanics, Hamilton was motivated by geometrical optics. For that reason we briefly review this subject. But only those aspects that can be reinterpreted as results or methods of classical mechanics will be discussed, though not in detail. It would be somewhat pointless to formulate mechanics in terms of a partial differential equation (which is what the Hamilton–Jacobi equation is) without reviewing a context in which that mathematics is familiar, namely physical optics. In this chapter traditional vector analysis – gradients, divergences, curls – will be used.

Particle trajectories are the analogs of the *rays* of geometric optics. In the process we will also find the analog of *wavefronts*. The equation analogous to the "eikonal," or wavefront, equation will be the Hamilton–Jacobi equation. The *action S* is the analog of the *optical path length* and the principle of least action, also known as Hamilton's principle, is the analog of Fermat's principle of least time. Not attempting to justify it *a priori*, we will take Hamilton's principle as a postulate leading to equations whose correctness is to be confirmed later.

Methods developed in this chapter will also be needed, in Chapter 13, to analyze the "gravitational lensing" predicted by general relativity.

Geometric Mechanics: Toward a Unification of Classical Physics. 2nd Edition. Richard Talman

ISBN: 978-3-527-40683-8

7.1 Analogy Between Mechanics and Geometric Optics

Our initial purpose is to recast mechanics to more nearly resemble optics. To start one is to visualize a congruence of space filling and nonintersecting valid trajectories, with time t parameterizing each curve (Fig. 7.1). This very picture already represents a deviation from Newtonian mechanics toward a description like the beam and wave-oriented discussion of physical optics. This formulation emphasizes the importance of *boundary conditions* satisfied by initial and final configuration space coordinates, and contrasts with Newtonian mechanics, which concerns itself more naturally with matching *initial conditions* for both configuration and velocity space coordinates. Also, while Newtonian mechanics concentrates its attention on the particular trajectory of a solitary system under study, it is more natural in optics to consider whole "beams" of rays, and the corresponding fields.

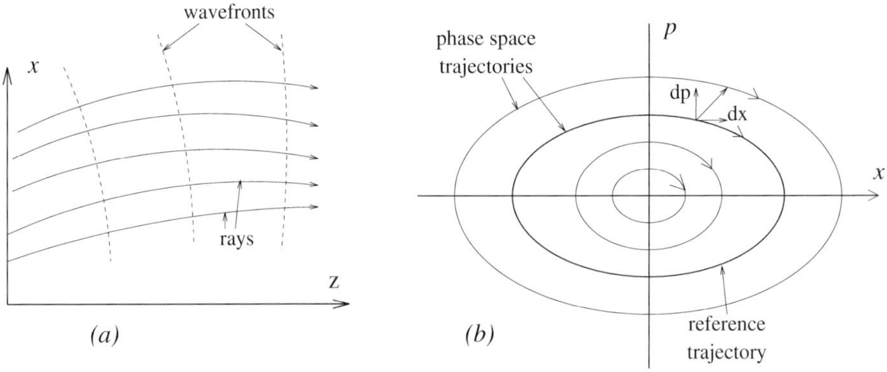

Fig. 7.1 (a) Configuration space curves, transverse coordinate x versus longitudinal coordinate z, natural for describing optical rays. They can usefully be parameterized by arc length s. (b) Phase space trajectories, p versus x. They cannot cross. Modulo an arbitrary additive constant, they can best be regarded as parameterized by time t. It is sometimes useful to refer orbits to a single reference orbit as shown.

One way in which the analogy between mechanics and optics is imperfect comes from the fact that in variational integrals like Eq. (1.21) the curves are parameterized by independent variable t whereas in geometric optics, it is the path taken by a ray rather than the rate of progress along it that is of interest. This can perhaps be understood by the historical fact that the existence of photons was not even contemplated when geometric optics was developed, the principle of least time notwithstanding. This made it natural to parameterize rays with arc length s or, in the paraxial case, with coordinate z along some straight axis. See Eq. (1.28). In mechanics we parameterize trajectories by time t and by treating velocities, or perhaps momenta, on the same footing as displacements, keep track of progress along trajectories. This makes it natural to visualize trajectories in "phase space," such as those in Fig. 7.1(b). An invaluable property of phase space is that trajectories cannot cross – this follows because the instantaneous values of positions and velocities uniquely specify the subsequent evolution of the system.

Another way the analogy of mechanics with optics is defective is that, in systems describable by generalized coordinates, the concept of orthogonality does not exist in general. While rays are perpendicular to wavefronts in optics, the absence of metric – distances and angles – requires the relation between trajectories and "surfaces of constant phase" to be specified differently in mechanics. This leads eventually to the so-called "symplectic geometry." To a physicist who is unwilling to distinguish between "geometry" and "high school geometry" this might better be called "symplectic nongeometry" since the hardest step toward understanding it may be jettisoning of much of the geometric intuition acquired in high school. Stated differently, it may not seem particularly natural to a physicist to impose a geometric interpretation on Lagrangians and Hamiltonians that have previously been thought to play only formal roles as artificial functions whose only purposes were to be formally differentiated.

7.1.1 Scalar Wave Equation

To study geometric optics in media with spatially varying index of refraction $n{=}n(\mathbf{r})$ one should work with electric and magnetic fields, but to reduce complication (without compromising the issues to be analyzed) we will work with *scalar* waves. The simplest example is a plane wave in a medium with constant index of refraction n,

$$\Psi(\mathbf{r},t) = a\,\Re\, e^{i(\mathbf{k}\cdot\mathbf{r}-\omega t)} = a\,\Re\, e^{ik_0(n\hat{\mathbf{k}}\cdot\mathbf{r}-ct)}. \tag{7.1}$$

Here $\Re$ stands for real part, a is a constant amplitude, c is the speed of light in vacuum, ω is the angular frequency, and $\mathbf{k}$, the "wave vector," satisfies $\mathbf{k} = k_0 n\hat{\mathbf{k}}$, where $\hat{\mathbf{k}}$ is a unit vector pointing in the wave direction and k_0 is the "vacuum wave number." (That is, $k_0 \equiv 2\pi/\lambda_0 = \omega/c$, where λ_0 is the vacuum wavelength for the given frequency ω; linearity implies that all time variation has the same frequency everywhere.)

The index of refraction n is a dimensionless number, typically in the range from 1 to 2 for the optics of visual light. Because n is the wavelength in free space divided by the local wavelength, the product $n\,dr$, or distance "weighted" by n, where dr is a path increment along $\hat{\mathbf{k}}$, is said to be the "optical path length." The "phase velocity" is given by

$$v = \frac{\omega}{|\mathbf{k}|} = \frac{c}{n}, \tag{7.2}$$

and the "group velocity" will not be needed. A result to be used immediately, that is valid for constant n, is

$$\nabla(n\,\hat{\mathbf{k}}\cdot\mathbf{r}) = n\,\nabla(\hat{\mathbf{k}}\cdot\mathbf{r}) = n\,\hat{\mathbf{k}}. \tag{7.3}$$

A wave somewhat more general than is given by Eq. (7.1) but which has the same frequency is required if $n(\mathbf{r})$ depends on position;

$$\Psi(\mathbf{r},t) = a(\mathbf{r})\,\Re e^{ik_0\left(\phi(\mathbf{r})-ct\right)} \equiv a(\mathbf{r})\,\Re e^{i\psi}. \tag{7.4}$$

(In four dimensional analysis (i.e., time included) in later chapters, $\psi(t,\mathbf{r})$ is known as the "eikonal.") The importance of ψ in wave theory comes from this definition. The wavelength and wave direction (or rather, equivalently, the local wave vector $\mathbf{k}$) and the wave frequency ω of a sinusoidally varying wave are given by

$$\mathbf{k} = \nabla\psi, \quad \text{and} \quad \omega = -\frac{\partial\psi}{\partial t}. \tag{7.5}$$

In three-dimensional analysis, applicable for the rest of this chapter, $\phi(\mathbf{r})$ is also called the "eikonal." Since this wave function must satisfy the wave equation the (weak) spatial variation of the amplitude $a(\mathbf{r})$ is necessarily position-dependent, so that Ψ can satisfy

$$\nabla^2\Psi \equiv \nabla\cdot\nabla\Psi = \frac{n^2(\mathbf{r})}{c^2}\frac{\partial^2\Psi}{\partial t^2}, \tag{7.6}$$

which is the wave equation for a wave of velocity n/c. Rays and wavefronts for n constant are shown in Fig. 7.2(a) and for n variable in Fig. 7.2(b).

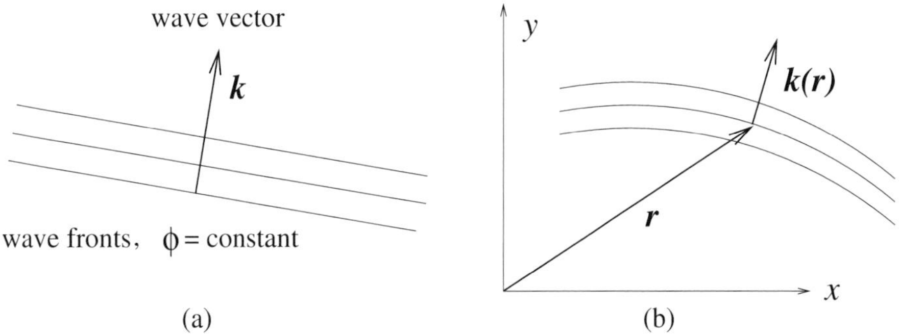

Fig. 7.2 (a) The wave vector $\mathbf{k}$ is normal to the wavefronts of a plane wave. (b) Wavefronts of light wave in a medium with nonconstant index of refraction.

7.1.2
The Eikonal Equation

Since the function $\phi(\mathbf{r})$ in Eq. (7.4) takes the place of $n\hat{\mathbf{k}} \cdot \mathbf{r}$ in Eq. (7.1), it generalizes the previously mentioned optical path length; ϕ is known as the "eikonal," a name with no mnemonic virtue whatsoever to recommend it. One can think of ϕ as a "wave phase" advancing by 2π and beyond as one moves a distance equal to one wavelength and beyond along a ray.

The condition characterizing "geometric optics" is for wavelength λ to be short compared to distances x over which $n(\mathbf{r})$ varies appreciably in a fractional sense. More explicitly this is $\frac{dn/n}{dx/(\lambda/2\pi)} \ll 1$, or

$$\frac{1}{n}\frac{dn}{dx} \ll k. \tag{7.7}$$

This is known as an "adiabatic" condition. (This condition is violated at boundaries, for example at the surfaces of lenses, but this can be accommodated by matching boundary conditions.) This approximation will permit dropping terms proportional to $|dn/dx|$. By matching exponents of Eqs. (7.4) and (7.1) locally, one can define a local wave vector $\hat{\mathbf{k}}$ such that

$$\phi(\mathbf{r}) = n(\mathbf{r})\,\hat{\mathbf{k}}(\mathbf{r}) \cdot \mathbf{r}. \tag{7.8}$$

This amounts to best-approximating the wave function locally by the plane wave solution of Eq. (7.1). Because n and $\hat{\mathbf{k}}$ are no longer constant, Eq. (7.3) becomes

$$\boldsymbol{\nabla}\boldsymbol{\phi} = (\boldsymbol{\nabla} n(\mathbf{r}))\hat{\mathbf{k}}(\mathbf{r}) \cdot \mathbf{r} + n(\mathbf{r})\boldsymbol{\nabla}\left(\hat{\mathbf{k}}(\mathbf{r}) \cdot \mathbf{r}\right) \approx n(\mathbf{r})\hat{\mathbf{k}}(\mathbf{r}), \tag{7.9}$$

where inequality (7.7) has been used to show that the first term is small compared to the second. Also spatial derivatives of $\hat{\mathbf{k}}(\mathbf{r})$ have been dropped because deviation of the local plane wave solution from the actual wave are necessarily proportional to $|dn/dx|$. (A simple rule of thumb expressing the approximation is that all terms that are zero in the constant-n limit can be dropped. Equation (7.9) shows that $\boldsymbol{\nabla}\boldsymbol{\phi}$ varies slowly, even though ϕ varies greatly (i.e., of order 2π) on the scale of one wavelength.

One must assure, with ϕ given by Eq. (7.8), that Ψ, as given by Eq. (7.4), satisfies the wave equation (7.6). Differentiating Eq. (7.4) twice, the approximation can be made of neglecting the spatial variation of $\mathbf{r}$-dependent factors, $n(\mathbf{r})$ and $a(\mathbf{r})$. relative to that of eikonal $\phi(\mathbf{r})$.

$$\boldsymbol{\nabla}\Psi \approx ik_0\boldsymbol{\nabla}\boldsymbol{\phi}(\mathbf{r})\,\Psi, \quad \text{and} \quad \nabla^2\Psi = \boldsymbol{\nabla}\cdot\boldsymbol{\nabla}\Psi \approx -k_0^2|\boldsymbol{\nabla}\boldsymbol{\phi}|^2\Psi. \tag{7.10}$$

With this approximation, substituting Eq. (7.6) becomes

$$|\boldsymbol{\nabla}\boldsymbol{\phi}(\mathbf{r})|^2 = n^2(\mathbf{r}), \quad \text{or} \quad |\boldsymbol{\nabla}\boldsymbol{\phi}(\mathbf{r})| = n(\mathbf{r}), \tag{7.11}$$

which is known as "the eikonal equation." It can be seen to be equivalent to Eq. (7.9), provided $\phi(\mathbf{r})$ and $\hat{\mathbf{k}}(\mathbf{r})$ are related by Eq. (7.8) and then the eikonal equation can be written as a vector equation that fixes the direction as well as the magnitude of $\nabla\boldsymbol{\phi}$,

$$\nabla\boldsymbol{\phi} = n\hat{\mathbf{k}}. \tag{7.12}$$

The real content of this equation is twofold: it relates $\nabla\boldsymbol{\phi}$, in magnitude, to the local index of refraction and, in direction, to the ray direction. Since this equation might have been considered obvious, and written down without apology at the start of this section, the discussion to this point can be regarded as a review of the wave equation and wave theory in the short wavelength limit.

7.1.3 Determination of Rays from Wavefronts

Any displacement $\mathbf{dr}_{(f)}$ lying in a surface, $\phi(\mathbf{r}) = \text{constant}$, satisfies

$$0 = \frac{\partial\phi}{\partial x^i} dx^i_{(f)} = \nabla\boldsymbol{\phi}\cdot\mathbf{dr}_{(f)}. \tag{7.13}$$

This shows that the vector $\nabla\boldsymbol{\phi}$ is orthogonal to the surface of constant ϕ. (Which is why it is called the "gradient" – $\phi(\mathbf{r})$ varies most rapidly in that direction.) From Eq. (7.12) we then obtain the result that $\hat{\mathbf{k}}(\mathbf{r})$ is locally orthogonal to a surface of constant $\phi(\mathbf{r})$. "Wavefronts" are, by definition, surfaces of constant $\phi(\mathbf{r})$, and rays are directed locally along $\hat{\mathbf{k}}(\mathbf{r})$.[1] It has been shown then that "rays" are curves that are normal everywhere to wave fronts. If the displacement $\mathbf{dr}$ lies along the ray and ds is its length then $\mathbf{dr}/ds$ is a unit vector and hence

$$\hat{\mathbf{k}} = \frac{\mathbf{dr}}{ds}. \tag{7.14}$$

Combining Eqs. (7.12) and (7.14), we obtain a *differential equation for the ray*,

$$\frac{\mathbf{dr}}{ds} = \frac{1}{n}\nabla\boldsymbol{\phi}. \tag{7.15}$$

1) The unit vectors $\hat{\mathbf{k}}(\mathbf{r})$ defined throughout some region, and having the property that smooth rays are to be drawn everywhere tangent to them, is a good picture to keep in mind when contemplating the "vector fields" of geometric mechanics. In the jargon of "dynamical systems" the entire pattern of rays is known as a "flow." Another mathematical expression for them is "a congruence" of rays. Though perfectly natural in optics, such a congruence may seem artificial in mechanics, but it may be the single most important concept differentiating between the dynamical systems approach and Newtonian mechanics.

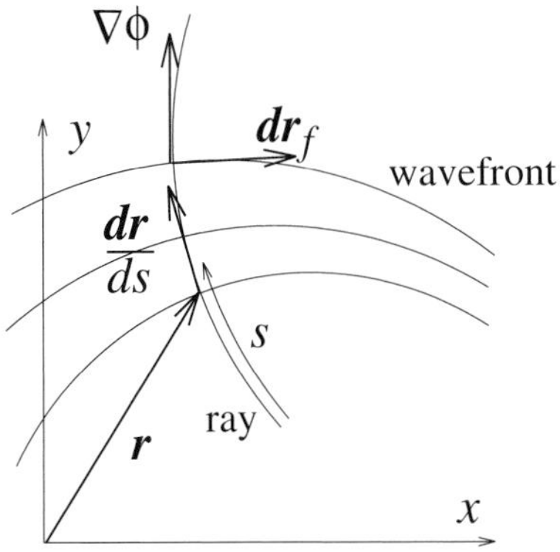

Fig. 7.3 Geometry relating a ray to the wavefronts it crosses.

7.1.4
The Ray Equation in Geometric Optics

Equation (7.15) is a *hybrid* equation containing two unknown functions, $\mathbf{r}(s)$ and $\phi(\mathbf{r})$, and as such is only useful if the wave-front function $\phi(\mathbf{r})$ is already known. But we can convert this equation into a differential equation for $\mathbf{r}(s)$ alone. Expressing Eq. (7.15) in component form, differentiating it, and then re-substituting from it, yields

$$\frac{d}{ds}\left(n\frac{dx^i}{ds}\right) = \frac{d}{ds}\frac{\partial\phi}{\partial x^i} = \frac{\partial^2\phi}{\partial x^j \partial x^i}\frac{dx^j}{ds} = \frac{\partial^2\phi}{\partial x^j \partial x^i}\frac{1}{n}\frac{\partial\phi}{\partial x^j}. \tag{7.16}$$

The final expression can be re-expressed using Eq. (7.11);

$$\frac{\partial^2\phi}{\partial x^j \partial x^i}\frac{1}{n}\frac{\partial\phi}{\partial x^j} = \frac{1}{2n}\frac{\partial}{\partial x^i}\sum_j\left(\frac{\partial\phi}{\partial x^j}\right)^2 = \frac{1}{2n}\frac{\partial}{\partial x^i}|\boldsymbol{\nabla\phi}|^2 = \frac{1}{2n}\frac{\partial n^2}{\partial x^i} = \frac{\partial n}{\partial x^i}. \tag{7.17}$$

Combining results yields the vector equation,

$$\frac{d}{ds}\left(n\frac{\mathbf{dr}}{ds}\right) = \boldsymbol{\nabla}\mathbf{n}(\mathbf{r}). \tag{7.18}$$

This is "the ray equation." A second order, ordinary differential equation, it is the analog for "light trajectories" of the Newton equation for a point particle. In this analogy arc length s plays the role of time and the index of refraction $n(\mathbf{r})$ is somewhat analogous to the potential energy function $U(\mathbf{r})$. The analogy will be made more precise shortly.

All of the geometric optics of refraction of light in the presence of position-dependent optical media, such as lenses, can be based on the ray equation.

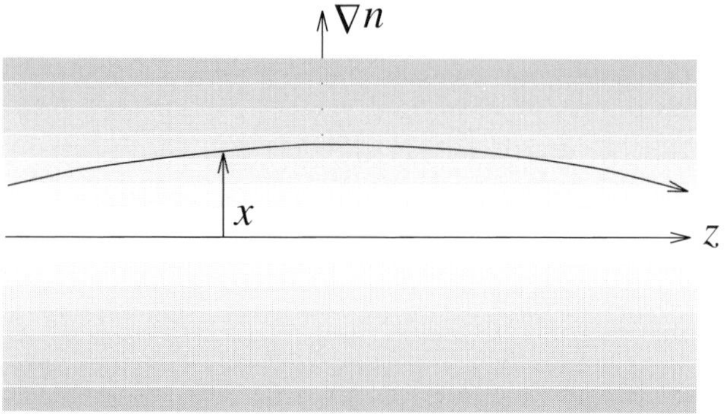

Fig. 7.4 A ray of light being guided by a graded fiber having an axially symmetric index of refraction n. Darker shading corresponds to lower index of refraction.

Problem 7.1.1.

(a) **Light rays in a lens-like medium.** *Consider paraxial, that is to say, almost parallel to the z-axis, rays in a medium for which the index of refraction is a quadratic function of the transverse distance from the axis;*

$$n(x,y) = n_0(1 + B\, r^2), \tag{7.19}$$

where $r^2 = x^2 + y^2$ and B is a constant. Given initial values (x_0, y_0) and initial slopes (x'_0, y'_0) at the plane $z = 0$, using Eq. (7.18) find the space curve followed by the light ray. See Yariv, Quantum Electronics, *or any book about fibre optics for discussion of applications of such media.*

(b) *Next suppose the coefficient B in part (i) depends on z (arbitrarily though consistent with short wavelength approximation (7.7)) but x and y can be approximated for small r as in part (i). In that case the "linearized" ray equation becomes*

$$\frac{d}{dz}\left(n(z)\,\frac{dx}{dz}\right) + k_B(z)x = 0, \quad or \quad p' + k_B x = 0, \tag{7.20}$$

where $p(z) \equiv n(z)(dx/dz)$, prime stands for d/dz, and there is a similar equation for y. Consider any two (independent) solutions $x_1(z)$ and $x_2(z)$ of this equation. For example, $x_1(z)$ can be the "cosine-like" solution with $C(0) = 1, C'(0) = 0$ and $x_2(z) \equiv S(z)$, the "sine-like" solution with $S(0) = 0$, $n(0)S'(0) = 1$. Show that the propagation of any solution from $z = z_0$ to $z = z_1$ can be described by a matrix equation

$$\begin{pmatrix} x(z_1) \\ p(z_1) \end{pmatrix} = M \begin{pmatrix} x(z_0) \\ p(z_0) \end{pmatrix}, \tag{7.21}$$

where M is a 2×2 matrix called the "transfer matrix." Identify the matrix elements of M with the cosine-like and sine-like solutions. Show also, for sufficiently small values of r, that the expression obtained from two separate rays,

$$x_1(z)p_2(z) - x_2(z)p_1(z), \tag{7.22}$$

is conserved as z varies. Finally, use this result to show that $\det|M| = 1$. *The analog of this result in mechanics is Liouville's theorem. In the context of optics it would not be difficult to make a more general proof by removing assumptions made in introducing this problem.*

Problem 7.1.2. *Consider an optical medium with spherical symmetry (e.g., the earth's atmosphere), such that the index of refraction $n(r)$ is a function only of distance r from the center. Let d be the perpendicular distance from the center of the sphere to any tangent to the ray. Show that the product nd is conserved along the ray. This is an analog of the conservation of angular momentum.*

7.2 Variational Principles

7.2.1 The Lagrange Integral Invariant and Snell's Law

In this section we will work with a "congruence of curves," which is the name given to families of curves, like the rays just encountered, that accompany a definite, single-valued wavefront function ϕ; it is a family of nonintersecting smooth curves that "fill" the region of space under study, with one and only one curve passing through each point. Returning to Eq. (7.15), it can be seen that the quantity $n\hat{\mathbf{k}}$ is the gradient of a scalar function, namely ϕ. This causes the field $n\hat{\mathbf{k}}$ to be related to ϕ in the same way that an electric field is related to an electric potential, and similar "conservation" properties follow. We therefore define an integral,[2]

$$\text{L.I.I.}(C) = \int_{P_1}^{P_2} n\,\hat{\mathbf{k}} \cdot \mathbf{dr}, \tag{7.23}$$

known as the "Lagrange invariant integral," whose invariance is equivalent to the existence of a single-valued function ϕ. The integral has the same value for any path C joining any two points P_1 and P_2, whether or not they are not on the same ray. One such path is shown in Fig. 7.6. As the integration path crosses any particular ray, such as at point P in the figure, the unit vector $\hat{\mathbf{k}}$

2) In the context of mechanics an analogous integral is called the Poincaré–Cartan integral invariant.

corresponding to that ray is used in the integrand. Hence, though the integral is independent of path it depends on the particular steady wave field under discussion. The same result can be expressed as the vanishing of an integral around any closed path,

$$\oint n\hat{\mathbf{k}} \cdot \mathbf{dr} = 0, \tag{7.24}$$

or as the vector equation[3]

$$\nabla \times (n\hat{\mathbf{k}}) = 0. \tag{7.25}$$

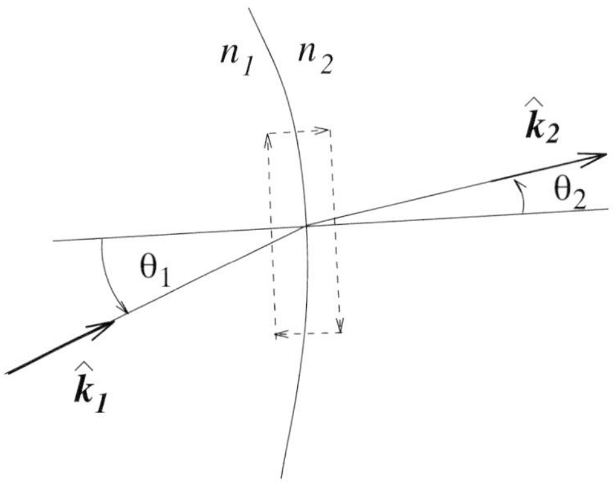

Fig. 7.5 Refraction of a ray as it passes from a medium with index of refraction n_1 to a medium with index of refraction n_2. The Lagrange invariant integral is evaluated over the closed, broken line.

Example 7.2.1. Snell's Law. *Consider the situation illustrated in Fig. 7.5, with light leaving a medium with index n_1 and entering a medium with index n_2. It is assumed that the transition is abrupt compared to any microscopic curvature of the interface but gradual compared to the wavelength of the light. (The latter assumption should not really be necessary but it avoids distraction.) An actual ray is illustrated in the figure and the Lagrange invariant can be evaluated along the broken line shown. The result, obtained by neglecting the short end segments as the integration path shrinks onto the interface (as one has probably done before in electricity and magnetism), is*

$$n_1 \sin\theta_1 = n_2 \sin\theta_2, \tag{7.26}$$

which is, of course, Snell's law.

3) A possibility that is easily overlooked and is not being addressed satisfactorily here, but which will actually become important later on, is the case where $n\hat{\mathbf{k}}$ is the gradient of a scalar function everywhere except at a single point at which it diverges. In that case the integral for curves enclosing the divergent point, though somewhat independent of path, may or may not vanish, depending on whether the path encloses the singular point. This problem will be discussed in a later chapter.

7.2.2
The Principle of Least Time

The optical path length of an interval $\mathbf{dr}$ has previously been defined to be $n|\mathbf{dr}| \equiv nds$. The optical path length O.P.L.($\mathcal{C}$) of the same curve $\mathcal{C}$ illustrated in Fig. 7.6 and used for the Lagrange invariant integral is

$$\text{O.P.L.}(\mathcal{C}) = \int_{P_1}^{P_2} n|\mathbf{dr}|; \tag{7.27}$$

this need not be a path light actually follows, but if a photon did travel that path the time taken would be O.P.L.($\mathcal{C}$)/c because

$$\frac{n\,ds}{c} = \frac{n\,v}{c}\,dt = dt. \tag{7.28}$$

If the path taken *is* an actual ray $\mathcal{R}$, which is only possible if P_1 and P_2 lie on the same ray $\mathcal{R}$, as in Fig. 7.6, then the optical path length is

$$\text{O.P.L.}(\mathcal{R}) = \int_{\mathcal{R}} n\hat{\mathbf{k}}\cdot\mathbf{dr} = \int_{\mathcal{R}} \boldsymbol{\nabla}\boldsymbol{\phi}\cdot\mathbf{dr} = \phi(P_2) - \phi(P_1). \tag{7.29}$$

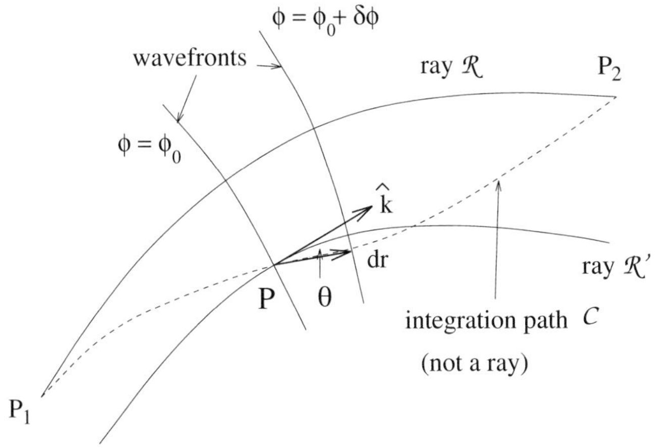

Fig. 7.6 Ray and nonray curves used in the derivation of the principle of least time. Note that $\hat{\mathbf{k}}\cdot\widehat{\mathbf{dr}} = \cos\theta < 1$.

We can calculate both the L.I.I. and the O.P.L. for both the ray joining P_1 and P_2 and for the nonray $\mathcal{C}$. The following series of inferences are each simple to derive:

$$\text{O.P.L.}(\mathcal{R}) = \text{L.I.I.}(\mathcal{R}), \tag{7.30}$$

$$\text{L.I.I.}(\mathcal{R}) = \text{L.I.I.}(\mathcal{C}), \tag{7.31}$$

$$\text{L.I.I.}(\mathcal{C}) < \text{O.P.L.}(\mathcal{C}), \tag{7.32}$$

and hence

$$\text{O.P.L.}(\mathcal{R}) < \text{O.P.L.}(\mathcal{C}). \tag{7.33}$$

Part (7.30) is the same as Eq. (7.29). Part (7.31) follows because L.I.I. is in fact invariant. Part (7.32) follows because the integrands of L.I.I. and O.P.L. differ only by a factor $\cos\theta$, where θ is the angle between curve $\mathcal{C}$ and the particular (different) ray $\mathcal{R}'$ passing through the point P. Since $\cos\theta < 1$ the inequality follows. The conclusion is known as *Fermat's principle.* Spelled out more explicitly for ray $\mathcal{R}$ and nonray $\mathcal{C}$ joining the same points, the principle is

$$\int_{\mathcal{R}} n\,ds < \int_{\mathcal{C}} n\,ds. \tag{7.34}$$

Except for a factor c the optical path length is the same as the time taken by a "photon" would take traveling along the ray, and for that reason the condition is also known as the *Principle of Least Time* – light gets from point P_1 to point P_2 by that path which takes the least time. (Under some conditions this can become an *extremum* condition, and not necessarily a *minimum.*)

7.3 Paraxial Optics, Gaussian Optics, Matrix Optics

In this section we will consider a light beam traveling almost parallel to and close to an axis – call it the z-axis. These conditions constitute the *paraxial* conditions. This set-up lends itself to linearization and discussion using matrices, and is also known as *Gaussian optics.* For full generality the refractive index $n(x, y, z)$ depends on the *transverse* coordinates x and y as well as on the longitudinal coordinate z. But since the purpose of this section is only to be an example illustrative of Hamilton's original line of thought we will consider 1D optics only and assume that $n = n(x, z)$ is independent of y.

Consider Fig. 7.7 in which a ray $\mathcal{R}$ propagates from input plane z_1 in a region with index of refraction n_1 to output plane z_2 in a region with index of refraction n_2. In the paraxial approximation the intervals such as from z to z' are neglected. The two outgoing rays are shown separate to show the sort of effect that is to be neglected. Also points such as z, z', and z'' are to be treated as essentially the same point. The fact that this is manifestly *not* the case reflects the fact that, to support this discussion, the figure illustrates a case somewhat beyond where the approximation would normally be judged appropriate.

As the ray $\mathcal{R}$ passes z_1 its displacement is x_1 and its angle is θ_1 but, for reasons to be explained later, the quantity $p_1 = n_1\theta_1$ will be used as the coor-

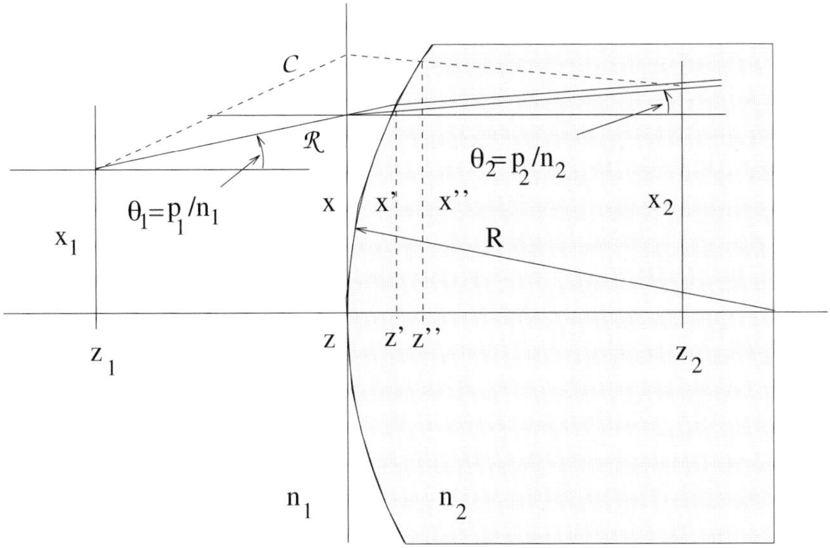

Fig. 7.7 Paraxial ray $\mathcal{R}$ propagates from z_1 in a region with index of refraction n_1 to z_2 in a region with index of refraction n_2. A spherical surface of radius R separates the two regions. Curve $\mathcal{C}$ is tentatively, but in fact not, a ray.

dinate fixing the slope of the straight line.[4] Sometimes p may even be called a "momentum" in anticipation of later results. Curve $\mathcal{C}$ is a tentative trajectory, close to $\mathcal{R}$ but not, in fact, a physically possible ray. The quantity to be emphasized is $\phi(z_1, z_2)$ the optical path length from the input plane at z_1 to the output plane at z_2.

All important points will be made in the following series of problems, adapted from V. Guillemin, and S. Sternberg, *Symplectic Techniques in Physics.*

Problem 7.3.1. *Referring to Fig. 7.7, the ray $\mathcal{R}$ leads from the input plane at z_1 to the output plane at z_2. With coordinate $p \equiv n\theta$ defined as above:*

(a) Show that propagation from z_1 to z is described by

$$\begin{pmatrix} x \\ p \end{pmatrix} = \begin{pmatrix} 1 & \frac{z-z_1}{n_1} \\ 0 & 1 \end{pmatrix} \begin{pmatrix} x_1 \\ p_1 \end{pmatrix} + O(x^2). \tag{7.35}$$

(b) Using Snell's law, the approximation $x' \approx x$, and related approximations, show that propagation from z to z' of ray $\mathcal{R}$ is described by

$$\begin{pmatrix} x' \\ p' \end{pmatrix} = \begin{pmatrix} 1 & 0 \\ -\frac{n_2-n_1}{R} & 1 \end{pmatrix} \begin{pmatrix} x \\ p \end{pmatrix}. \tag{7.36}$$

The point z' is to be interpreted as $z'+$, just to the right of the interface.

4) The ray equation, (7.18), suggests that the variable $n \sin\theta$ is preferable to $n\theta$, but the distinction will not be pursued here.

(c) *The 2 × 2 matrices in the previous parts are called "transfer matrices." Find the transfer matrix $M^{(21)}$ for propagation from z_1 to z_2 as defined by*

$$\begin{pmatrix} x_2 \\ p_2 \end{pmatrix} = \begin{pmatrix} A & B \\ C & D \end{pmatrix} \begin{pmatrix} x_1 \\ p_1 \end{pmatrix}. \tag{7.37}$$

Confirm that $\det|M^{(21)}| = AD - BC = 1$.[5] *Suppose, instead of regarding this as an initial value problem for which the output variables are to be predicted, one wishes to formulate it as a boundary value problem in which the input and output displacements are known but the input and output slopes are to be determined. Show that* $p_1 = (x_2 - Ax_1)/B$ *and* $p_2 = (-x_1 + Dx_2)/B$ *can be used to obtain* p_1 *and* p_2 *from* x_1 *and* x_2. *Under what circumstance will this fail, and what is the optical interpretation of this failure?*

Problem 7.3.2. *You are to calculate the optical path lengths for the three path segments of tentative ray* $\mathcal{C}$.

(a) *Show that the optical path length of the segment of* $\mathcal{C}$ *from* z_1 *to* z *is given approximately by*

$$\phi(z_1, z) = n_1(z - z_1) + n_1 \frac{1}{2} \frac{(x'' - x_1)^2}{z - z_1}. \tag{7.38}$$

You are to treat z, z', *and* z'' *as being essentially the same point. As drawn in the figure, this optical path length is somewhat longer than the corresponding segment of the true ray* $\mathcal{R}$. *Defining* $p_1'' = n_1(x'' - x_1)/(z - z_1)$ *as above, show that* $\phi(z_1, z)$ *also can be written as*

$$\phi(z_1, z) = n_1(z - z_1) + \frac{1}{2}(x'' p_1'' - x_1 p_1). \tag{7.39}$$

(b) *There is one phase advance for which the separation of* z, z', *and* z'' *cannot be neglected. The tiny segment of path from the transverse plane at* z *to the spherical surface occurs in a region with index* n_1 *while, on-axis, this region has index* n_2. *Show that this discrepancy can be accounted for in the optical path length of* $\mathcal{C}$ *by*

$$\phi(z, z'') = -\frac{1}{2} \frac{n_2 - n_1}{R} x''^2. \tag{7.40}$$

Defining $p_2'' = n_2(x_2 - x'')/(z_2 - z)$, *show that* $\phi(z, z'')$ *can be written as*

$$\phi(z, z'') = \frac{1}{2}(x'' p_2'' - x'' p_1''). \tag{7.41}$$

5) For the simple 1D case under study the condition $\det|M| = 1$ is necessary and sufficient for M to be a physically realizable transfer matrix. For higher dimensionality this condition is necessary but not sufficient.

(c) *Having accounted for the gap region, the rest of the optical path length can be calculated as in part (a) and the complete optical path length of C is*

$$\phi(z_1, z_2) = n_1(z - z_1) + n_2(z_2 - z) + \frac{1}{2}\left(n_1 \frac{(x'' - x_1)^2}{z - z_1} - \frac{n_2 - n_1}{R} x''^2 + n_2 \frac{(x_2 - x'')^2}{z_2 - z}\right). \quad (7.42)$$

(d) *By differentiating Eq. (7.42) with respect to x'', regarded as its only variable, express the condition that the optical path length of C be an extremum. Show that this condition, reexpressed in terms of p_1'' and p_2'' is*

$$p_2'' = p_1'' - \frac{n_2 - n_1}{R} x'', \quad (7.43)$$

which is the same condition as was obtained using Snell's law in part (b) of the previous problem.

Problem 7.3.3. *Equations (7.39) and (7.41), have been written to facilitate* concatenation, *which is to say successively accumulating the effects of consecutive segments. A similar expression can be written for the third segment.*

(a) *Using this remark and reverting to the case of true ray $\mathcal{R}$, show that $\phi(z_1, z_2)$ also can be written as*

$$\phi(z_1, z_2) = \text{constant} + \frac{1}{2}(x_2 p_2 - x_1 p_1). \quad (7.44)$$

where "constant" means independent of x and p. The motivation behind this form is that it depends only on displacements *at the end points and not on the "momenta."*

(b) *When expressed entirely in terms of x_1 and x_2, $\phi(z_1, z_2)$ is known as "Hamilton's point characteristic," $W(x_1, x_2)$. Substituting from part (c) of the first problem of this series, show that*

$$W(x_1, x_2) = \text{constant} + \frac{A x_1^2 + D x_2^2 - 2 x_1 x_2}{2B}. \quad (7.45)$$

Repeating it for emphasis, $W(x_1, x_2)$ is the optical path length expressed in terms of x_1 and x_2. As such it has an intuitively simple qualitative content, whether or not it is analytically calculable. Show finally that

$$p_1 = -\frac{\partial W}{\partial x_1}, \quad \text{and} \quad p_2 = \frac{\partial W}{\partial x_2}. \quad (7.46)$$

These are "Hamilton's equations" in embryonic form. The reason Hamilton considered $W(x_1, x_2)$ valuable was that it could be obtained from $\phi(z_1, z_2)$ which was obtainable from the simple geometric constructions of the previous two problems.

7.4
Huygens' Principle

Because rays of light respect the principle of least time there is a construction due to Huygens' principle by which their trajectories can be plotted. (This construction is also much used in heuristic analyses of diffraction patterns, for example due to multiple thin slits.) This principle is illustrated in Fig. 7.8. In this case it is assumed that the medium in which the light is traveling is anisotropic in such a way that light travels twice as fast in the y direction as in the x direction. (None of the formulas appearing earlier in this chapter apply to this case since the most general possibility to this point has been the variability with position of the index of refraction.) As well as making it easier to draw a figure illustrating Huygens' principle, this anisotropic possibility more nearly represents the degree of complexity encountered when one proceeds from optics to mechanics in the next chapter.

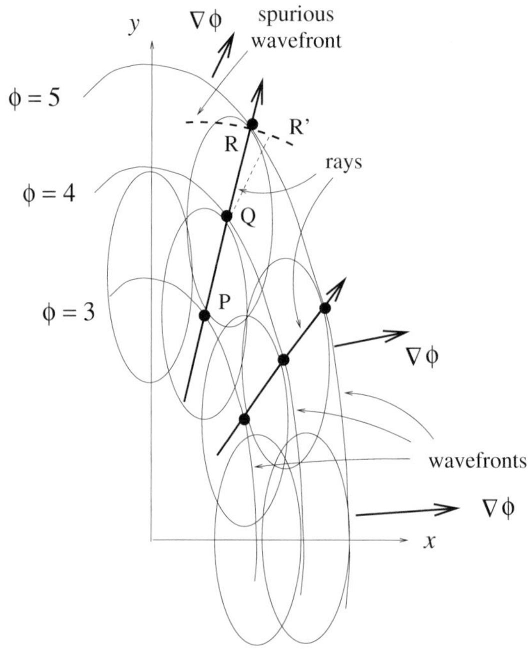

Fig. 7.8 Rays and wavefronts in a medium in which the velocity of propagation of light traveling in the y direction is twice as great as for light traveling in the x direction. Three "snapshots" of a wavefront taken at equal time intervals are shown.

It is not really obvious that a medium such as this is possible or if it *is*, that the principle of least time is actually valid but we accept both to be true. The little ellipses in the figure indicate the progress rays starting from their centers would make in one unit of time. Recall that, by definition, a "wavefront" is a surface of constant ϕ where ϕ is the "optical path length" which is the flight time to that point along the trajectory taken by the light. The Huygens construction is to center numerous little ellipses on one wavefront and call the envelope they define the "next" wavefront. A few of the centers are marked by heavy dots, two on each of three wavefronts, or rather on the same wavefront photographed at equal time intervals.

The emphasized points such as P, Q, and R are chosen so that the point on the next curve is at the point of tangency of the little ellipse and the next envelope. If light actually travels along this path it gets to P at time 3, to Q at time 4, and to R at time 5, where "time" is the value of ϕ. Huygens' principle declares that the light actually *does* travel along the curve PQR constructed as has been described. (In this case the rays are *not* normal to the wavefronts.)

To make this persuasive, a *spurious* wavefront supposedly appropriate for rays emerging from the point Q is drawn with a heavy broken line. Being a supposed "wavefront" this has the property that light leaving point Q gets to it in one unit of time. We consider the path PQ as the *correct* path taken by a ray under consideration. If the wavefront labeled "spurious" were correct the ray would proceed from Q to point R' along the faint broken line. This means the light will have followed a path PQR' with a kink at point Q. But in that case light could have got to R' even more quickly by a path not including Q. Since this contradicts our hypothesis the ray must actually proceed to point R. In other words any tentative "next" wavefront is spurious unless it is tangential to the envelope of little ellipses centered on the earlier wavefront.

As already noted, because of the anisotropy of the medium, the light rays are not normal to the wavefronts of constant ϕ. On the other hand, the vector

$$\mathbf{p} = \nabla\boldsymbol{\phi} \tag{7.47}$$

is normal to these wavefronts. This quantity has been called $\mathbf{p}$ in anticipation of a similar formula that will be encountered in mechanics; the momentum $\mathbf{p}$ will satisfy an analogous equation.

Letting $\dot{\mathbf{x}}$ be the velocity vector of a "photon" following a ray, we have seen graphically that $\dot{\mathbf{x}}$ and $\mathbf{p}$ are not generally parallel. More striking yet, as Fig. 7.8 shows, their magnitudes are, roughly speaking, inverse. This is shown by the lengths of the vectors labeled $\nabla\boldsymbol{\phi}$; they are large when directed along the "slow" axis. The gradient is greatest in the direction in which the ray speed is least. For this reason Hamilton himself called $\mathbf{p}$ *the vector of normal slowness to the front*.

Quite apart from the fact that this phrase grates on the ear, the picture it conjures up may be misleading because of its close identification of ϕ with time-of-flight. Whether or not one is willing to visualize a ray as the trajectory followed by a "material" photon, one must not think of a wavefront as derived from this picture. In preparation for going on to mechanics it is better to to suppress the interpretation of ϕ as flight time and to concentrate on the definition of the surface ϕ = constant as a surface of constant phase of some wave. Anyone who has understood wave propagation in an electromagnetic waveguide will have little trouble grasping this point. In a waveguide the "rays" (traveling at the speed of light) reflect back and forth at an angle off the sides of the guide, but surfaces of constant phase are square with the waveguide. Furthermore, the speed with which these fronts advance (the phase velocity) differs from the speed of light – in this case it is greater. (Another velocity that enters in this context is the *group velocity* but what has been discussed in this section bears at most weakly on that topic.) What has been intended is to make the point that $\mathbf{p}$ as defined in Eq. (7.47), need not be proportional, either in magnitude or direction, to the velocity of a "particle" along its trajectory.

Bibliography

General References

1 M. Born and E. Wolf, *Principles of Optics*, 4th ed., Pergamon, Oxford, 1970.

2 L.D. Landau and E.M. Lifshitz, *Classical Theory of Fields*, Pergamon, Oxford, 1971.

References for Further Study

Section 7.3

3 V. Guilleman and S. Sternberg, *Symplectic Techniques in Physics*, Cambridge University Press, Cambridge, UK, 1984.

8
Hamilton–Jacobi Theory

8.1
Hamilton–Jacobi Theory Derived from Hamilton's Principle

To develop mechanics based on its analogy with optics we work initially in $\mathbf{q}$-only configuration space rather than $(\mathbf{q}, \mathbf{p})$ phase space. Because they are both integrals on which variational principles are based, it is natural to regard the *action*

$$S = \int_{P_1}^{P} L\,dt \tag{8.1}$$

as the analog of the *eikonal* ϕ. The Lagrange/Poincaré equations were derived in Section 5.2 by applying Hamilton's principle to S. The present discussion will deviate primarily by replacing the upper limit by a point P that is variable in the vicinity of P_2. For fixed lower limit P_1, and any choice of upper limit $P = P_2 + \delta P$ in Eq. (8.1), after the extremal path has been found by solving the Lagrange equations, the action $S(P_1, P) \equiv S(\mathbf{q}, t)$ is a well-defined function of $(\mathbf{q}, t) = (\mathbf{q}_2, t_2) + (\boldsymbol{\delta}\mathbf{q}, \delta t)$, the coordinates of P. Three particular variations are illustrated in Fig. 8.1.

The variation δS accompanying change δx with $\delta t = 0$ as illustrated in Fig. 8.1(a) can be obtained directly from Eq. (1.24) for which an upper boundary contribution was calculated but then set to zero at that time. The result is $\delta S = (\partial L/\partial \dot{x})\delta x \equiv p_x \delta x$. For multiple variables the result is

$$\delta S_{(a)} = p_i \delta q^i, \quad \text{or} \quad \left.\frac{\partial S}{\partial q^i}\right|_{t_2} = p_i(t_2). \tag{8.2}$$

With the upper position held fixed, $\boldsymbol{\delta}\mathbf{q} = 0$, but with its time varied by δt as indicated in Fig. 8.1(b), the change of action is

$$\delta S_{(b)} = \frac{\partial S}{\partial t}\,\delta t. \tag{8.3}$$

The case in which the motion is identical to the reference motion over the original time interval, but is followed for extra time is illustrated in Fig. 8.1(c). In this case the path of integration is unchanged and the dependence of S comes

Geometric Mechanics: Toward a Unification of Classical Physics. 2nd Edition. Richard Talman

ISBN: 978-3-527-40683-8

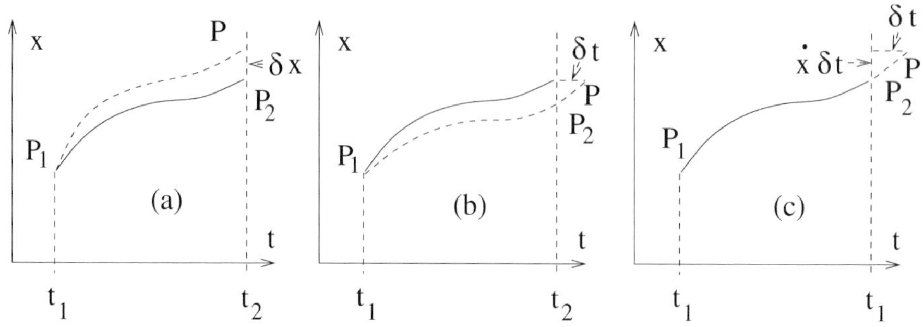

Fig. 8.1 Possible variations of the location upper end point P for extremal paths from P_0 to points P close to P_2. The "reference trajectory" is the solid curve.

entirely from the path's upper end extension. Differentiating (1.24) with respect to the upper limit yields

$$\delta S_{(c)} = \frac{dS}{dt}\delta t = L\delta t, \quad \text{or} \quad \frac{dS}{dt} = L(\mathbf{q}, t). \tag{8.4}$$

Using Fig. 8.1(c) and combining results,

$$-\frac{\partial S}{\partial t} = p_i \dot{q}^i - L. \tag{8.5}$$

The final expression $p_i\dot{q}^i - L$ is of course equal to the Hamiltonian H, but one must be careful to specify the arguments unambiguously in order for the final result to be usable.[1] Once this has been done, a preliminary form of the Hamilton–Jacobi equation is the result;

$$-\frac{\partial S}{\partial t}(\mathbf{q}, t) = H(\mathbf{q}, \mathbf{p}, t). \tag{8.6}$$

"Solving" this H–J equation is not immediately practical since $H(\mathbf{q}, \mathbf{p}, t)$ depends on $\mathbf{p}$ which itself depends on the motion. This dependency can be eliminated using Eq. (8.2) to give

$$-\frac{\partial S}{\partial t}(\mathbf{q}, t) = H\left(\mathbf{q}, \frac{\partial S}{\partial \mathbf{q}}, t\right). \tag{8.7}$$

This is the Hamilton–Jacobi Equation. It is a first-order (only the first derivatives occur) partial differential equation for the *action* function $S(\mathbf{q}, t)$. It is the analog of the *eikonal* equation. Momentum variables do not appear explicitly but, for given $S(\mathbf{q}, t)$, the momentum can be inferred immediately from

1) The reader is assumed to be familiar with the Hamiltonian function but, logically, it is being introduced here for the first time.

Eq. (8.2);

$$\mathbf{p} = \frac{\partial S}{\partial \mathbf{q}}. \tag{8.8}$$

The coordinates $\mathbf{q}$ used so far are any valid Lagrangian generalized coordinates. In the common circumstance that they are Euclidean displacements this equation becomes

$$\mathbf{p} = \boldsymbol{\nabla} S. \tag{8.9}$$

This result resembles Eq. (7.12), which relates rays to wavefronts in optics.

8.1.1 The Geometric Picture

We have not yet stated what it means to "solve" the Hamilton–Jacobi equation, nor even what good it would do us to have solved it. The latter is easier. A function $S(\mathbf{q}, t)$ satisfying the Hamilton–Jacobi equation over all configuration space includes descriptions of the evolution of a family of systems from consistent (but not all possible) initial conditions. Actual system trajectories are "transverse"[2] to "wavefronts" of constant S in the following sense. From Eqs. (8.2) and (8.6) the dependencies of S near point P are given by

$$dS = p_i dq^i - H dt. \tag{8.10}$$

If we treat $\mathbf{q}$ and t together as Cartesian coordinates, then a pair of dynamical variables $(\mathbf{A}, B)$ can be said to be "transverse" to $(\mathbf{dq}, dt)$ if

$$A_i dq^i + B dt = 0. \tag{8.11}$$

Suppose the displacement (dq^i, dt) lies in a surface of constant S, i.e., $dS = 0$. From Eq. (8.10) one can then say that the "vector" $(p_i, -H)$ is transverse to the surface of constant S. More simply one can say that $(p_i, -H)$ is a "generalized gradient" of S

$$p_i = \frac{\partial S}{\partial q^i}, \quad -H = \frac{\partial S}{\partial t}. \tag{8.12}$$

This can be regarded as the analog of the hybrid ray equation $n(\mathbf{dr}/ds) = \boldsymbol{\nabla}\phi$.

2) The geometric discussion in this section is not "intrinsic." This means the picture depends on the coordinates being used. An intrinsic description will be given in chapter 17. It is the notion that a vector is "transverse" that makes the present discussion *nonintrinsic*. Especially for constrained systems, inferences drawn from this discussion may therefore be suspect.

Since there is now an extra independent variable, t, the geometry is more complicated than it is in optics. If the system configuration is specified by a point in configuration space and if, as usual, many noninteracting identical systems are all indicated by points on the same figure, then the hyperplane of constant t yields a "snapshot" of all systems at that time. (At each point in this space the further specification of momentum $\mathbf{p}$ uniquely specifies the subsequent system evolution – another snapshot taken infinitesimally later, at time $t + dt$, would capture this configuration and determine the subsequent evolution.) In the time t snapshot, consider a curve of constant S. By Eq. (8.8), momentum $\mathbf{p}$ is transverse to that surface. This resembles the relation between rays and wavefronts in optics. But, since system velocity and system momentum are not necessarily proportional, the system velocity $\dot{\mathbf{q}}$ is not necessarily transverse to surfaces of constant S, (though it often will be.) The important case of a charged particle in an electromagnetic field is an example for which $\mathbf{p}$ and $\dot{\mathbf{q}}$ are not parallel; this is discussed in Section 1.2.1.

8.1.2 Constant S Wavefronts

We now develop a qualitative picture of the connection between "wavefronts" and surfaces of constant S shown in Fig. 8.2. for a system with just two coordinates (x, y) plus time t. At an initial time $t = t_0$ suppose the function $S = S(x, y, t_0)$ is specified over the entire $t = t_0$ plane. One can attempt to solve an *initial value* problem with these initial values. Loosely speaking, ac-

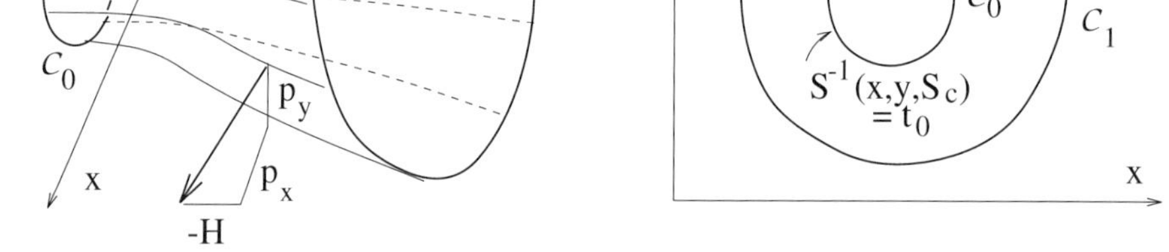

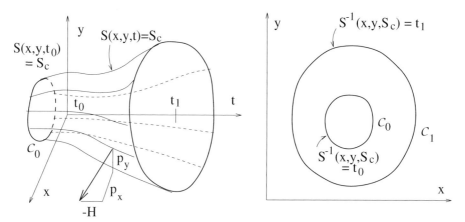

Fig. 8.2 Wavefront-like curves in the right figure are intersections with planes of constant t of the surface of constant S shown in the left figure. Trajectories are not in general orthogonal to curves of constant S in the right figure, but $\mathbf{p}, -H$ *is* orthogonal to surfaces of constant S on the left.

cording to the theorem of Cauchy and Kowaleski, discussed for example in Courant and Hilbert, Vol. II, the Hamilton–Jacobi equation (8.7) uniquely determines the time evolution of $S(x,y,t)$ as t varies away from $t = t_0$. On the $t = t_0$ plane the out-of-plane partial derivative $\partial S/\partial t$ needed to propagate S away from $t = t_0$ is given by Eq. (8.12). In principle then, $S(x,y,t)$ can be regarded as known throughout the (x,y,t) space when it is known at an initial time and satisfies the H–J equation.

One can consider a contour of constant S in the $t = t_0$ plane such as

$$\mathcal{C}_0 : S(x,y,t_0) = S_c = \text{ constant.} \tag{8.13}$$

As shown in Fig. 8.2, there is a surface on which $S = S_c$ and its intersection with the plane $t = t_0$ is $\mathcal{C}_0$. Intersections of the constant S surface with planes of other constant t determine other "wavefronts." The equation $S(x,y;t) = S_c$ can be inverted, $t = S^{-1}(x,y;S_c)$, in order to label curves with corresponding values of t.

A solution $S(x,y;t;\overline{S}(x,y))$ of the Hamilton–Jacobi equation like this, that is able to match arbitrary initial condition $S(x,y,t_0) = \overline{S}(x,y)$, is known as a *general integral*. It is satisfying to visualize solving the Hamilton–Jacobi equation as an initial value problem in this way, and using the solution to define wavefronts and hence trajectories. But that is not the way the Hamilton–Jacobi equation has been applied to solve practical problems. Rather there is a formal, operational, procedure that makes no use whatsoever of the geometric picture just described. This will be developed next.

8.2 Trajectory Determination Using the Hamilton–Jacobi Equation

8.2.1 Complete Integral

For solving practical problems in mechanics, it is not necessary to find the *general integral* discussed in the previous section. Rather, one starts by finding a so-called "complete integral" of the Hamilton–Jacobi equation. This is a solution containing as many free parameters as there are generalized coordinates of the system. Though there are ways of using such a complete integral to match initial conditions, that is not the profitable approach. Rather, there is an operational way of solving the mechanics problem of interest without ever completing the solution of the Hamilton–Jacobi initial value problem.

Though this method is completely general, we will continue to work with just x, y, and t. We seek a solution for x, y, p_x, and p_y as functions of t, satisfying given initial conditions. If we have a solution with four "constants of integration," we can presumably find values for them such that initial values x_0, y_0, p_{x0}, and p_{y0} are matched.[3]

8.2.2
Finding a Complete Integral by Separation of Variables

The Hamilton–Jacobi equation is

$$\frac{\partial S}{\partial t} + H\left(x, y, \frac{\partial S}{\partial x}, \frac{\partial S}{\partial y}, t\right) = 0. \tag{8.14}$$

Recall that its domain is (x, y, t)-space, with momentum nowhere in evidence. Assume a variable-separated solution of the form[4]

$$S(x, y, t) = S^{(x)}(x) + S^{(y)}(y) + S^{(t)}(t). \tag{8.15}$$

For this to be effective, substitution into Eq. (8.14) should cause the Hamilton–Jacobi equation to take the following form:

$$f^{(x)}\left(x, \frac{dS^{(x)}}{dx}\right) + f^{(y)}\left(y, \frac{dS^{(y)}}{dy}\right) + f^{(t)}\left(t, \frac{dS^{(t)}}{dt}\right) = 0. \tag{8.16}$$

By straightforward argument this form assures the validity of introducing two arbitrary constants, α_1 and α_2 such that

$$f^{(x)}\left(x, \frac{dS^{(x)}}{dx}\right) = \alpha_1, \quad f^{(y)}\left(y, \frac{dS^{(y)}}{dy}\right) = \alpha_2, \quad f^{(t)}\left(t, \frac{dS^{(t)}}{dt}\right) = -\alpha_1 - \alpha_2. \tag{8.17}$$

Being first-order ordinary differential equations, each of these supplies an additional constant of integration, but only one of these, call it α, is independent since $S^{(x)}$, $S^{(y)}(y)$, and $S^{(t)}$ are simply added in Eq. (8.15).

Whether found this way by separation of variables or any other way, suppose then that one has a *complete integral* of the form

$$S = S(x, y, t, \alpha_1, \alpha_2, t) + \alpha. \tag{8.18}$$

That there is an additive constant α is obvious since the H–J equation depends only on the first derivatives of S. It is required that α_1 and α_2 be *independent*.

3) Without saying it every time, we assume that mathematical pathologies do not occur. In this case we assume that four equations in four unknowns have a unique solution.

4) Since S is a "phase-like" quantity, its simple behavior is additive, in contrast to a quantity like $\phi = \exp i\psi$, whose corresponding behavior is multiplicative. This accounts for the surprising appearance of an additive form rather than the multiplicative form that appears when "separation of variables" is applied to the Schrödinger equation in quantum mechanics.

For example it would not be acceptable for α_2 to be a definite function of α_1, even with position or time dependent coefficients. For reasons to be become clear shortly, the constants $\alpha_1, \alpha_1, \ldots$ appearing in such a complete integral are sometimes called "new momenta." For the time being the symbol α (and later P), with indices as necessary, will be reserved for these new momenta. Before showing how (8.18) can be exploited in general, we consider an example.

8.2.3 Hamilton–Jacobi Analysis of Projectile Motion

Consider the motion of a projectile of mass m in a uniform gravitational field for which

$$V = mgy, \quad L = \frac{m}{2}(\dot{x}^2 + \dot{y}^2) - mgy, \quad H = \frac{1}{2m}(p_x^2 + p_y^2) + mgy, \tag{8.19}$$

and the H–J equation is

$$\frac{\partial S}{\partial t} + \frac{1}{2m}\left(\frac{\partial S}{\partial x}\right)^2 + \frac{1}{2m}\left(\frac{\partial S}{\partial y}\right)^2 + mgy = 0. \tag{8.20}$$

Neither x nor t appears explicitly in this equation. The reason for this is that there is no explicit dependence of the Lagrangian on x or t. This equation is therefore simple enough that the variable separation approach of the previous section can be performed mentally to yield

$$S = \alpha_1 t + \alpha_2 x + S^{(y)}(y) + \alpha. \tag{8.21}$$

Substitution back into Eq. (8.20) yields

$$\begin{aligned}
&\alpha_1 + \frac{\alpha_2^2}{2m} + \frac{1}{2m}\left(\frac{dS^{(y)}}{dy}\right)^2 + mgy = 0;\\
&\frac{dS^{(y)}}{dy} = \pm\sqrt{2m}\sqrt{-\alpha_1 - \frac{\alpha_2^2}{2m} - mgy};\\
&S^{(y)} = \pm\sqrt{2m}\int_{y_0}^{y} dy'\sqrt{-\alpha_1 - \frac{\alpha_2^2}{2m} - mgy'}\\
&\quad\;\; = \mp\sqrt{\frac{2}{m}\frac{2}{3g}}\left(-\alpha_1 - \frac{\alpha_2^2}{2m} - mgy\right)^{3/2}\Bigg|_{y_0}^{y}.
\end{aligned} \tag{8.22}$$

Arbitrarily picking $y_0 = 0$ and merging additive constants yields

$$S = \alpha_1 t + \alpha_2 x \mp \sqrt{\frac{2}{m}\frac{2}{3g}}\left(-\alpha_1 - \frac{\alpha_2^2}{2m} - mgy\right)^{3/2} + \alpha. \tag{8.23}$$

One can check Eq. (8.12);

$$p_x = \frac{\partial S}{\partial x} = \alpha_2, \quad p_y = \frac{\partial S}{\partial y} = \pm\sqrt{2m}\sqrt{-\alpha_1 - \frac{\alpha_2^2}{2m} - mgy}. \tag{8.24}$$

From the first of these it is clear that $\alpha_2 = p_{x0} = m\dot{x}_0$, the initial horizontal momentum, and that it is a conserved quantity. Rearranging the second equation and substituting $p_y = m\dot{y}$ yields

$$\frac{1}{2}m\dot{y}^2 + \frac{1}{2}m\dot{x}_0^2 + mgy = -\alpha_1, \tag{8.25}$$

a result clearly interpretable as energy conservation, with $-\alpha_1$ being the (conserved) total energy E_0. Based only on theory presented so far the interpretation is only that the function of variables y and $\dot{y}$ appearing on the left side of Eq. (8.25) is conserved.

8.2.4 The Jacobi Method for Exploiting a Complete Integral

By "solving" a mechanics problem at one hand, one usually means finding explicit expressions for $x(t)$ and $y(t)$. If velocities or momenta are also required they can then be found by straightforward differentiation. It was just seen by example, that a complete integral $S(x, y, \alpha_1, \alpha_2, t)$ yields immediately

$$p_x = \frac{\partial S}{\partial x}(x, y, \alpha_1, \alpha_2, t), \quad p_y = \frac{\partial S}{\partial y}(x, y, \alpha_1, \alpha_2, t). \tag{8.26}$$

(In the previous section's example they were Eqs. (8.24).) This can perhaps be regarded as having completed one level of integration since it yields p_x and p_y as functions of x, y and t. There are only two free parameters, α_1 and α_2 but that is all that is needed for matching initial momenta. This leaves another level of integration to be completed to obtain $x(t)$ and $y(t)$.

Where are we to get two more relations giving $x(t)$ and $y(t)$? One relation available for use is

$$H(x, y, p_x, p_y, t) = -\frac{\partial S}{\partial t}(x, y, t), \tag{8.27}$$

but this is just another complicated implicit relation among the variables. The remarkable discovery of Jacobi was, starting from a complete integral as in Eq. (8.18), that expressions for $x(t)$ and $y(t)$ can be written down mechanically, uncluttered with dependency upon p_x and p_y. His procedure was to regard equations obtained by inverting Eqs. (8.26) as transformation equations defining new dynamical variables α_1 and α_2 as functions of x,y,p_x,p_y and t. Then Jacobi defined two further dynamical variables β_1 and β_2 by

$$\beta_1 = \frac{\partial S}{\partial \alpha_1}(x, y, \alpha_1, \alpha_2, t), \quad \text{and} \quad \beta_2 = \frac{\partial S}{\partial \alpha_2}(x, y, \alpha_1, \alpha_2, t). \tag{8.28}$$

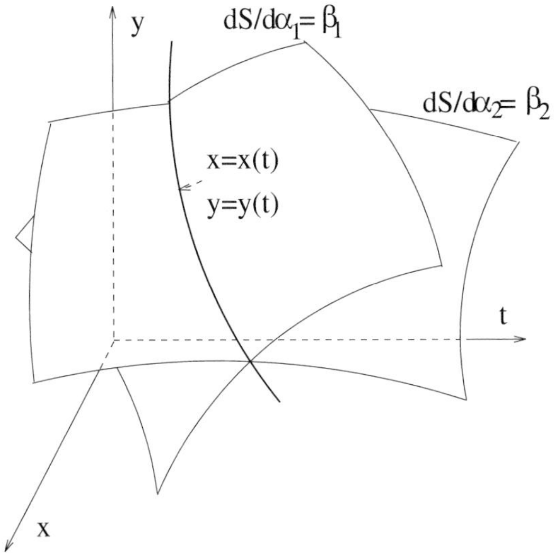

Fig. 8.3 A particle trajectory satisfying Hamilton's equations is found as the intersection of two surfaces derived from a complete integral of the Hamilton–Jacobi equation.

These are to be regarded as a coordinate transformation $(x, y) \rightarrow (\beta_1, \beta_2)$. The amazing result then follows that all four of the dynamic variables α_1, α_2, β_1, and β_2, defined by $S(x, y, \alpha_1, \alpha_2, t)$ and Eq. (8.28), are constants of the motion. This can be visualized geometrically as in Fig. 8.3. The newly introduced dynamical variables $\beta_1, \beta_2, \ldots$ will be known as "new coordinates" and the symbol β (and later Q) with appropriate indices, will be reserved for them.

To demonstrate Jacobi's result, first differentiate Eqs. (8.28) with respect to t;

$$\frac{\partial^2 S}{\partial \alpha_1 \partial x}\dot{x} + \frac{\partial^2 S}{\partial \alpha_1 \partial y}\dot{y} = -\frac{\partial^2 S}{\partial \alpha_1 \partial t}, \quad \text{and} \quad \frac{\partial^2 S}{\partial \alpha_2 \partial x}\dot{x} + \frac{\partial^2 S}{\partial \alpha_2 \partial y}\dot{y} = -\frac{\partial^2 S}{\partial \alpha_2 \partial t}. \tag{8.29}$$

Terms proportional to $\dot{\alpha}_1$, $\dot{\alpha}_2$, $\dot{\beta}_1$, and $\dot{\beta}_2$ have vanished by hypothesis. The order of differentiations in these equations has been reversed for convenience in the next step. Next, partially differentiate the H–J equation itself (8.27) with respect to α_1 (respectively α_2) holding x, y, and t fixed, after substituting for p_x and p_y from Eq. (8.26);

$$\begin{aligned} \frac{\partial H}{\partial p_x}\frac{\partial^2 S}{\partial \alpha_1 \partial x} + \frac{\partial H}{\partial p_y}\frac{\partial^2 S}{\partial \alpha_1 \partial y} &= -\frac{\partial^2 S}{\partial \alpha_1 \partial t}, \quad \text{and} \\ \frac{\partial H}{\partial p_x}\frac{\partial^2 S}{\partial \alpha_2 \partial x} + \frac{\partial H}{\partial p_y}\frac{\partial^2 S}{\partial \alpha_2 \partial y} &= -\frac{\partial^2 S}{\partial \alpha_2 \partial t}. \end{aligned} \tag{8.30}$$

Subtracting Eq. (8.30) from (8.29) yields

$$\begin{pmatrix} \frac{\partial^2 S}{\partial\alpha_1\partial x} & \frac{\partial^2 S}{\partial\alpha_1\partial y} \\ \frac{\partial^2 S}{\partial\alpha_2\partial x} & \frac{\partial^2 S}{\partial\alpha_2\partial y} \end{pmatrix} \begin{pmatrix} \dot{x} - \frac{\partial H}{\partial p_x} \\ \dot{y} - \frac{\partial H}{\partial p_y} \end{pmatrix} = 0. \tag{8.31}$$

Unless the determinant formed from the coefficients vanishes this equation implies

$$\dot{x} = \frac{\partial H}{\partial p_x}, \quad \text{and} \quad \dot{y} = \frac{\partial H}{\partial p_y}, \tag{8.32}$$

But the vanishing of the determinant would imply that α_1 and α_2 were functionally dependent, contrary to hypotheses. It has been shown therefore that half of Hamilton's equations are satisfied.

Similar manipulations show that the remaining Hamilton equations are satisfied. Differentiate Eqs. (8.26) with respect to t, again under the hypothesis that α_1 and α_2 are constant,

$$\dot{p}_x = \frac{\partial^2 S}{\partial x^2}\dot{x} + \frac{\partial^2 S}{\partial x \partial y}\dot{y} + \frac{\partial^2 S}{\partial x \partial t}. \tag{8.33}$$

Also partially differentiate the H–J equation with respect to x to obtain

$$\frac{\partial^2 S}{\partial x \partial t} = -\frac{\partial H}{\partial x} - \frac{\partial H}{\partial p_x}\frac{\partial^2 S}{\partial x^2} - \frac{\partial H}{\partial p_y}\frac{\partial^2 S}{\partial y \partial x}. \tag{8.34}$$

Using Eq. (8.32) and subtracting these equations

$$\dot{p}_x = -\frac{\partial H}{\partial x}, \tag{8.35}$$

and $\dot{p}_y = -\partial H/\partial y$ follows similarly. Hence all of Hamilton's equations are satisfied by Jacobi's hypothesized solution.

8.2.5
Completion of Projectile Example

To continue the Jacobi prescription for the projectile example of Section 8.2.3, define β_1 by substituting Eq. (8.23) into the first of Eqs. (8.28) to obtain

$$\beta_1 = t \pm \frac{1}{g}\sqrt{\frac{2}{m}}\sqrt{E_0 - \frac{p_{x0}^2}{2m} - mgy}. \tag{8.36}$$

It was noted earlier that $-\alpha_1$ and α_2 could be identified with the total energy E_0 and p_{x0} and those replacements have been made. The expression inside the final square root is the vertical contribution to the kinetic energy. It vanishes as the projectile passes through "zenith," the highest point on its trajectory. This makes it clear that β_1 can be interpreted as the time of passage through that point – clearly a constant of the motion. It is necessary to take the positive sign for $t < \beta_1$ and the negative sign otherwise. Equation (8.36) can be inverted to give $y(t)$ directly, since x has dropped out. Superficially β_1 appears to increase linearly with t but this time dependence is precisely cancelled by the variation of the spatial quantities in the second term, in this case y. The remaining equation is the other of Eqs. (8.28);

$$\beta_2 = x \pm \frac{p_{x0}}{g}\sqrt{\frac{2}{m^3}}\sqrt{E_0 - \frac{p_{x0}^2}{2m} - mgy}. \tag{8.37}$$

Eliminating the square root expression using Eq. (8.36) yields

$$\beta_2 = x + \frac{p_{x0}}{m}(\beta_1 - t). \tag{8.38}$$

Since β_1 is the time of passage through zenith the second term vanishes at that instant and β_2 is the x-coordinate of that point – obviously also a constant of the motion. Again the superficial dependence (on x) in the first term is cancelled by the second term. This example has shown (and it is the same in general) that the constants of motion $\beta_1, \beta_2, \ldots$ in the Jacobi procedure have a kind of "who's buried in Grant's tomb?" character. In this case "How do the coordinates of the highest point on the trajectory, that will be (or was) reached, vary as the projectile moves along its trajectory?" When $\beta_1, \beta_2, \ldots$ are expressed in terms of the evolving coordinates they may not *look* constant, but they *are* nevertheless.

8.2.6
The Time-Independent Hamilton–Jacobi Equation

Though this problem has been very simple and special, the features just mentioned are common to all the cases where H is independent of t. The H–J equation can then be seen to be at least partially separable, with the only time dependence of S being an additive term of the form $S^{(t)} = \alpha t$. Furthermore, since $H = E =$ constant when the Hamiltonian is independent of time, it can be seen that $-\alpha$ can be identified as the energy E so

$$S(x, y; t) = -Et + S_0(x, y; E). \tag{8.39}$$

$S_0(x, y; E)$ contains the spatial variation of S with E as a parameter but is independent of t. It satisfies the "time-independent" Hamilton–Jacobi equation;

$$H\left(x, y, \frac{\partial S_0}{\partial x}, \frac{\partial S_0}{\partial y}\right) = E. \tag{8.40}$$

The Jacobi coordinate definition yields the "constant" β_1 according to

$$\beta_1 = \frac{\partial S}{\partial(-E)} = t - \frac{\partial S_0}{\partial E}. \tag{8.41}$$

Re-ordering this equation, it becomes

$$\frac{\partial S_0}{\partial E} = t - \beta_1 \equiv t - t_0. \tag{8.42}$$

Since β_1 subtracts from t, the symbol β_1 is commonly replaced by t_0, which is commonly then called the "initial" time. Translating the origin of time gives a corresponding shift in β_1. (With S_0 expressed in terms of x and y it is again not obvious that it can also be expressed as a linear function of t in this way though.)

We will develop shortly a close analogy between the Hamilton–Jacobi equation and the Schrödinger equation of quantum mechanics. Equation (8.40) will then be the analog of the time-independent Schrödinger equation.

8.2.7
Hamilton–Jacobi Treatment of 1D Simple Harmonic Motion

Though it is nearly the most elementary conceivable system, the one-dimensional simple harmonic oscillator is basic to most oscillations and provides a simple illustration of the Jacobi procedure. This formalism may initially seem a bit "heavy" for such a simple problem, but the entire theory of adiabatic invariance follows directly from it and nonlinear oscillations cannot be satisfactorily analyzed without this approach. The Hamiltonian is

$$H(q,p) = \frac{p^2}{2m} + \frac{1}{2} m\omega_0^2 q^2. \tag{8.43}$$

This yields as the (time-independent) H–J equation

$$\frac{1}{2m}\left(\frac{dS_0}{dq}\right)^2 + \frac{1}{2} m\omega_0^2 q^2 = E, \tag{8.44}$$

which can be solved to give

$$S_0(q,E) = m\omega_0 \int_0^q \sqrt{\frac{2E}{m\omega_0^2} - q'^2}\, dq'. \tag{8.45}$$

(The lower limit has been picked arbitrarily.) It will be necessary to handle the ± 1 ambiguity coming from the square root on an *ad hoc* basis; here the positive sign has been chosen. This is a *complete integral* in that it depends on E, which we now take as the first (and only) "Jacobi momentum" that would previously have been denoted by α_1 (or $-\alpha_1$). Following the Jacobi procedure we next find β_1, but which we will now call Q, or Q_E since it is to be the "new coordinate" corresponding to E. (If we were to insist on conventional terminology we would also introduce a "new momentum" $P \equiv E$.) That is, we are performing a transformation of phase space variables $(x,p) \rightarrow (Q,P)$. Since the main purpose of $S_0(q,E)$ is to be differentiated, explicit evaluation of the integral in Eq. (8.45) may not be necessary, but for definiteness, the result is

$$S_0(q,E) = \frac{m\omega_0}{2} q \sqrt{\frac{2E}{m\omega_0^2} - q^2} + \frac{E}{\omega_0} \sin^{-1}\left(\sqrt{\frac{m\omega_0^2}{2E}}\, q\right). \tag{8.46}$$

By Jacobi's defining equation for Q_E we have

$$Q_E = \frac{\partial S_0}{\partial E} = \frac{1}{\omega_0} \int_0^q \frac{1}{\sqrt{\frac{2E}{m\omega_0^2} - q'^2}}\, dq' = \frac{1}{\omega_0} \sin^{-1}\left(\sqrt{\frac{m\omega_0^2}{2E}}\, q\right). \tag{8.47}$$

As previously warned, it is not obvious that Q_E is a linear function of t but from the general theory of the previous section, in particular Eq. (8.42), we know this to be the case;

$$Q_E = t - t_0. \tag{8.48}$$

Combining Eqs. (8.47) and (8.48) yields

$$q = \sqrt{\frac{2E}{m\omega_0^2}} \sin \omega_0(t - t_0), \tag{8.49}$$

which begins to look familiar. The corresponding variation of p is given by

$$p = \frac{\partial S_0}{\partial q} = \pm m\omega_0 \sqrt{\frac{2E}{m\omega_0^2} - q^2} = \sqrt{2mE} \cos \omega_0(t - t_0). \tag{8.50}$$

Phase space plots of the motion are shown in Fig. 8.4. From considerations of continuity in this figure it has been necessary to restore the $\pm$ options for the square root that entered in the first place. The trajectory equation is

$$E = \frac{p^2}{2m} + \frac{1}{2} m\omega_0^2 q^2. \tag{8.51}$$

8.3 The Kepler Problem

We now take up the Kepler problem from the point it was left in Problem 1.5.1. It is important both for celestial mechanics and as a classical precursor to the theory of the atom. The latter topic is introduced nicely by Ter Haar.

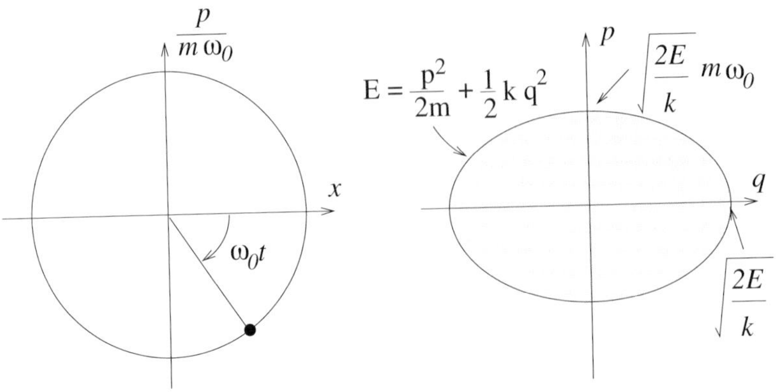

Fig. 8.4 The phase space trajectory of simple harmonic motion is a circle traversed at constant angular velocity ω_0 if the axes are q and $p/(m\omega_0)$. The shaded area enclosed within the trajectory for one cycle of the motion in q and p *phase space* is $2\pi I$ where I is the "action."

8.3.1 Coordinate Frames

Since one is dealing with formulas derived by astronomers one may as well use their frames of reference, referring to Fig. 8.5. For studying earth satellites it is natural to use the *equatorial plane* as the x, y plane. For studying solar planetary orbits, such as that of the earth, it is natural to use the *ecliptic plane* which is the plane of the earth's orbit around the sun. (Recall that the equatorial plane is inclined by about 23° relative to the plane of the earth's orbit.) In either case it remains to fix the orientation of the x-axes, and if com-

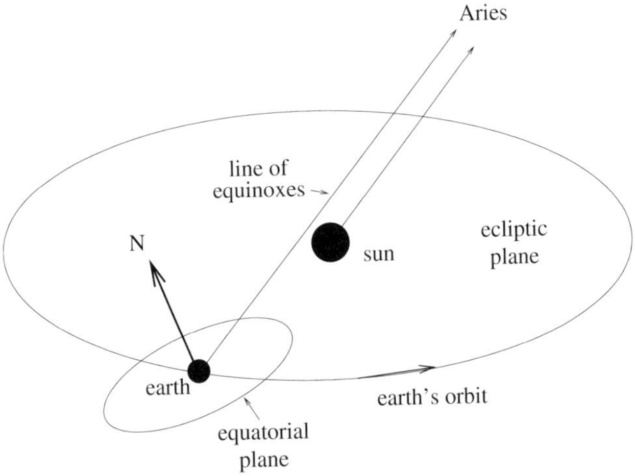

Fig. 8.5 Astronomical reference frames.

parisons are to be made between the two frames, to fix also the rule relating these choices. By convention the x-axis in both frames is chosen to be *the line of equinoxes*, which is to say the line joining the earth to the sun on the day of the year the sun is directly over the equator so night and day are equal in duration everywhere on earth. This line necessarily lies in both the equatorial and ecliptic planes and is therefore their line of intersection. It happens that a "distant" star called Aries lies approximately on this line, and it can be used to "remember" this direction at other times in the year.

8.3.2 Orbit Elements

Having chosen one or the other of these frames, specification of the three-dimensional motion of a satellite can be discussed using Fig. 8.6 in which the instantaneous satellite position is projected onto a sphere of constant radius. The trace of the orbit has the polar coordinates θ,ϕ of the true orbit. The satellite lies instantaneously along ray C and the orbit plane is defined by this line and the line OA of the "ascending node."

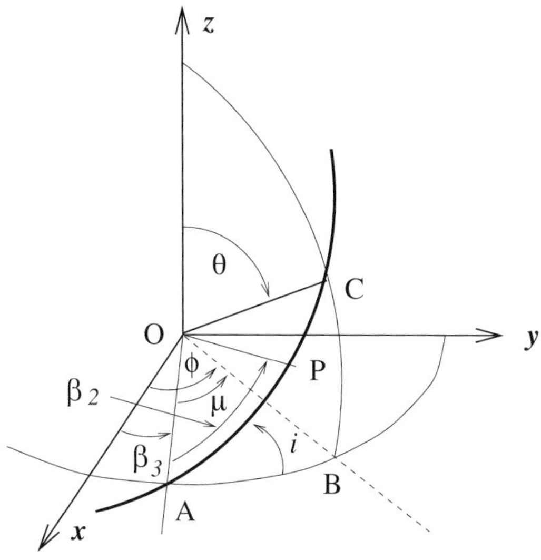

Fig. 8.6 A Kepler orbit is projected onto a sphere centered on the center of gravity of the binary system. The true orbit emerges from the x, y plane, passes through perigee, and is instantaneously situated along the lines OA, OP, and OC, respectively.

Fixing three initial positions and three initial velocities fixes the subsequent three-dimensional trajectory of the satellite. But there are other way of specifying the orbit. The orbit plane can be specified by the "azimuth β_3 of the line OA, and the "inclination" i which is the polar angle between the normal to the orbit plane and the z-axis. Two coordinates locate the point of nearest approach ("perigee") along the line OP in the orbit plane plus the particle speed as it passes perigee (necessarily at right angles to OP). But it is more conventional to choose parameters that characterize the geometric shape and orientation of the orbit, which is known to be to be elliptical as in Fig. 1.3. The semimajor axis is a and the eccentricity is ϵ. The angle β_2 between OA and OP specifies the orientation of the ellipse. Finally the location C can be located relative to P by specifying the "time of passage through perigee" β_1. The parameters introduced in this way are known as "orbit elements."

Other parameters are sometimes introduced for convenience. The parameters β_1, β_2, and β_3 (which depend on choice of coordinates and initial time) have already been named in anticipation of the way they will appear in the H–J theory, but it remains to introduce parameters α_1, α_2, and α_3, as functions of a, ϵ and i. But there are always six independent parameters in all.

8.3.3 Hamilton–Jacobi Formulation.

Using polar coordinates, the Lagrangian for a particle of mass m moving in three dimensions in an inverse square law potential is

$$L = \frac{1}{2}m(\dot{r}^2 + r^2\dot{\theta}^2 + r^2 \sin^2\theta\, \dot{\phi}^2) + \frac{K}{r}. \tag{8.52}$$

The canonical momenta are

$$p_r = m\dot{r}, \quad p_\theta = mr^2\dot{\theta}, \quad p_\phi = mr^2 \sin^2\theta\, \dot{\phi}, \tag{8.53}$$

and the Hamiltonian is

$$H = \frac{p_r^2}{2m} + \frac{p_\theta^2}{2mr^2} + \frac{p_\phi^2}{2mr^2 \sin^2\theta} - \frac{K}{r}. \tag{8.54}$$

Preparing to look for a solution by separation of the variables, the time-independent H–J equation is

$$E = \frac{1}{2m}\left(\left(\frac{dS^{(r)}}{dr}\right)^2 + \frac{1}{r^2}\left(\frac{dS^{(\theta)}}{d\theta}\right)^2 + \frac{1}{r^2 \sin^2\theta}\left(\frac{dS^{(\phi)}}{d\phi}\right)^2\right) - \frac{K}{r}. \tag{8.55}$$

Since ϕ does not appear explicitly we can separate it immediately in the same way t has already been separated;

$$S = -Et + \alpha_3\phi + S^{(\theta)}(\theta) + S^{(r)}(r). \tag{8.56}$$

Here α_3 is the second "new momentum" of Jacobi. (E is the first.) It is interpretable as the value of a conserved angular momentum around the z-axis because

$$p_\phi = \frac{\partial S}{\partial \phi} = \alpha_3. \tag{8.57}$$

Substituting this into Eq. (8.55) and multiplying by $2mr^2$ yields

$$2mEr^2 + 2mKr - r^2\left(\frac{dS^{(r)}}{dr}\right)^2 = \left(\frac{dS^{(\theta)}}{d\theta}\right)^2 + \frac{1}{\sin^2\theta}\left(\frac{dS^{(\phi)}}{d\phi}\right)^2 = \alpha_2^2, \tag{8.58}$$

where the equality of a pure function of r to a pure function of θ implies that both are constant; this has permitted a third Jacobi parameter α_2 to be introduced. The physical meaning of α_2 can be inferred by expanding M^2, the square of the total angular momentum;

$$M = \sqrt{(mr^2\dot{\theta})^2 + (mr^2\sin\theta\,\dot{\phi})^2} = \sqrt{p_\theta^2 + \frac{p_\phi^2}{\sin^2\theta}} = \alpha_2. \tag{8.59}$$

From the interpretation α_3 as the z component of α_2 it follows that

$$\alpha_3 = \alpha_2 \cos i. \tag{8.60}$$

Determination of the other terms in S has been "reduced to quadratures" since Eqs. (8.58), gives expressions for $dS^{(\theta)}/d\theta$ and $dS^{(r)}/dr$ that can be re-arranged to yield $S^{(\theta)}(\theta)$ and $S^{(r)}(r)$ as indefinite integrals;

$$\begin{aligned} S_2 &= -\int^{\theta}\sqrt{\alpha_2^2 - \frac{\alpha_3^2}{\sin^2\theta'}}\,d\theta', \\ S_3 &= \int^{r}\sqrt{2mE + \frac{2mK}{r'} - \frac{\alpha_2^2}{r'^2}}\,dr'. \end{aligned} \tag{8.61}$$

Instead of using E as the first Jacobi "new momentum" it is conventional to use a function of E, namely

$$\alpha_1 = \sqrt{\frac{-K^2m}{2E}}, \quad E = \frac{-K^2m}{2\alpha_1^2}. \tag{8.62}$$

Like α_2 and α_3, α_1 has dimensions of angular momentum. Referring to Fig. 1.3, the semimajor axis a and the orbit eccentricity ϵ are given by

$$a = \frac{\alpha_1^2}{Km}, \quad 1 - \epsilon^2 = \left(\frac{\alpha_2}{\alpha_1}\right)^2, \tag{8.63}$$

with inverse relations

$$\alpha_1^2 = Kma, \quad \alpha_2^2 = (1-\epsilon^2)Kma. \tag{8.64}$$

Combining results, the complete integral of the H–J equation is

$$S = \frac{mK^2}{2\alpha_1^2} t + \alpha_3 \phi - \int_{\pi/2}^{\theta} \sqrt{\alpha_2^2 - \frac{\alpha_3^2}{\sin^2\theta'}}\, d\theta' + \int_{a(1-\epsilon^2)}^{r} \sqrt{-\frac{m^2K^2}{\alpha_1^2} + \frac{2mK}{r'} - \frac{\alpha_2^2}{r'^2}}\, dr'. \quad (8.65)$$

The lower limits and some signs have been chosen arbitrarily so that the Jacobi "new momenta" β_1, β_2, and β_3 will have conventional meanings. To define them requires the following *tour de force* of manipulations from spherical trigonometry. They are the work of centuries of astronomers.

Starting with β_3 and using Eq. (8.60) we obtain

$$\beta_3 = \phi + \int_{\pi/2}^{\theta} \frac{d\theta'}{\sin\theta' \sqrt{\frac{\sin^2\theta'}{\cos^2 i} - 1}} = \phi - \tan^{-1} \frac{\cos\theta \cos i}{\sqrt{\sin^2\theta - \cos^2 i}}$$
$$= \phi - \sin^{-1}(\cot\theta \cot i) = \phi - \mu. \quad (8.66)$$

The second to last step is justified by the trigonometry of Fig. 8.7(a) and the last step by the spherical trigonometry of Fig. 8.7(b) which can be used to show that $\cot\theta \cot i = \sin\mu$. Referring back to Fig. 8.6 one sees that β_3 is indeed the nodal angle and, as such, is a constant of the motion.

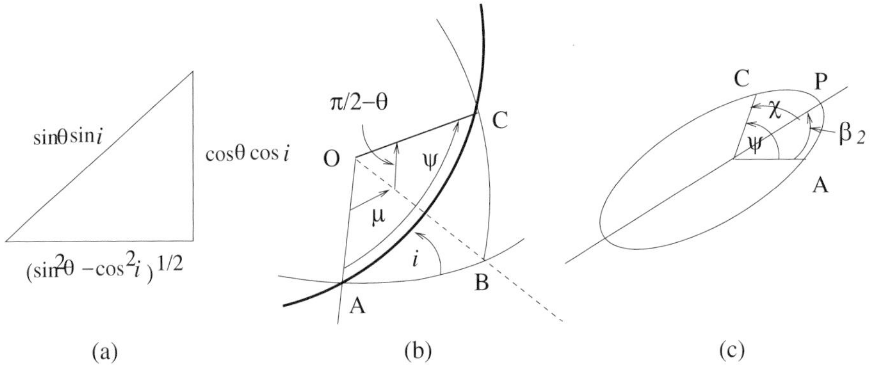

Fig. 8.7 Figures illustrating the trigonometry and spherical trigonometry used in assigning meaning to the Jacobi parameters β_1, β_2, and β_3. Angle χ in part (c) is the same angle as in Fig. 1.3.

We next consider $\beta_2 = \partial S/\partial \alpha_2$.

$$\beta_2 = -\int_{\pi/2}^{\theta} \frac{\alpha_2 \sin\theta' \, d\theta'}{\sqrt{\alpha_2^2 \sin^2\theta' - \alpha_3^2}} - \alpha_2 \left(\frac{a}{Km}\right)^2 \int_{a(1-\epsilon)}^{r} \frac{-\alpha_2 dr'/r'^2}{\sqrt{\alpha_1^2 - \frac{\alpha_2^2}{r'^2} - \left(\frac{mK}{\alpha_1} - \frac{\alpha_1}{r'}\right)^2}}$$

$$= -\int_{\pi/2}^{\theta} \frac{\alpha_2 \sin\theta' \, d\theta'}{\sqrt{\alpha_2^2 \sin^2\theta' - \alpha_3^2}} - \alpha_2 \sqrt{\frac{a}{Km}} \int_{a(1-\epsilon)}^{r} \frac{dr'/r'}{\sqrt{\epsilon^2 a^2 - (r'-a)^2}} \qquad (8.67)$$

$$= \sin^{-1}\left(\frac{\cos\theta}{\sin i}\right) - \chi = \psi - \chi.$$

The second integral was evaluated using Eq. (1.70). The first integral was performed using Eq. (8.60) and changing the variable of integration according to

$$\cos\theta = \sin i \, \sin\psi, \quad \text{or} \quad \psi = \sin^{-1}\left(\frac{\cos\theta}{\sin i}\right). \qquad (8.68)$$

Using spherical trigonometry again on Fig. 8.7(b), the variable ψ can be seen to be the angle shown both there and in Fig. 8.7(c). That is, ψ is the angle in the plane of the orbit from ascending node A to the instantaneous particle position. It follows that β_2 is the difference between two fixed points, A and P, and is hence a constant of the motion (as expected.)

Since α_1 is a function only of E we expect its conjugate variable β_1 to be a linear function of t. It is given by

$$\beta_1 = \frac{\partial S}{\partial \alpha_1} = -\frac{mK^2}{\alpha_1^3} t + \frac{m^2K^2}{\alpha_1^3} \int_{a(1-\epsilon)}^{r} \frac{dr'}{\sqrt{-\frac{m^2K^2}{\alpha_1^2} + \frac{2mK}{r'} - \frac{\alpha_2^2}{r'^2}}}$$

$$= -\frac{mK^2}{\alpha_1^3} t + \frac{1}{a} \int_{a(1-\epsilon)}^{r} \frac{dr'}{\sqrt{a^2\epsilon^2 - (r'-a)^2}} \qquad (8.69)$$

$$= -\sqrt{\frac{K}{m\,a^3}}\, t + u - \epsilon \sin u.$$

The integral has been performed by making the substitution $r = a(1 - \epsilon \cos\mu)$ and the result of Eq. (1.78) has been replicated, with β_1 being proportional to the time since passage through perigee.

We will return to this topic again in Section 14.6.3 as a (degenerate) example of conditionally periodic motion and as example of the use of action-angle variables.

Problem 8.3.1. *In each of the following problems a Lagrangian function* $L(\mathbf{q}, \dot{\mathbf{q}}, t)$ *is given. In every case the Lagrangian is appropriate for some practical physical system, but that is irrelevant to doing the problem. You are to write the Lagrange equations and, after defining momenta* $\mathbf{p}$*, give the Hamiltonian* $H(\mathbf{q}, \mathbf{p}, t)$ *and write the*

Hamilton–Jacobi equation. In each case find a "complete integral"; (in all cases except one this can be accomplished by separation of variables.) Finally use the complete integrals to solve for the motion, given initial conditions. Leave your answers as definite integrals, in some cases quite ugly. Confirm that there are enough free parameters to match arbitrary initial conditions, but do not attempt to do it explicitly. Figure out from the context which symbols are intended to be constants and which variables.

(a) $L = \frac{1}{2}m\dot{x}^2 - \frac{1}{2}kx^2$.

(b) $L = \frac{1}{2}\dot{x}^2 + Atx$. *(Something to try here for solving the H–J equation is to make a change of dependent variable $S \to S'(S, x, t)$ such that the equation for S' is separable. An alternative thing to try is to "cheat" by solving the Lagrange equation and working from that solution. This latter approach is more instructive, but also trickier to do correctly.)*

(c) $L = \frac{1}{2}m(R^2\dot{\theta}^2 + R^2\sin^2\theta\,\omega^2 - mgR(1-\cos\theta)$. *$\theta$ is the only variable,*

(d) $L = \frac{1}{2}m(\dot{r}^2 + r^2\dot{\theta}^2) - V(r)$. *($r$ and θ are cylindrical coordinates.)*

(e) $L = A(\dot{\theta}^2 + \sin^2\theta\,\dot{\phi}^2) + C(\dot{\psi} + \dot{\phi}\cos\theta)^2 - Mg\ell\cos\theta$. *(Euler angles.)*

(f) $L = m_0c^2(1 - \sqrt{1 - |\dot{\mathbf{r}}|^2/c^2}) + e\mathbf{A}(\mathbf{r})\cdot\dot{\mathbf{r}} - V(r)$, *where $\mathbf{A}(\mathbf{r})$ is a vector function of position $\mathbf{r}$ and $V(r)$ is a scalar function of radial coordinate r. In this case write the H–J equation only for the special case $\mathbf{A} = 0$ and assume the motion is confined to the $z = 0$ plane, with r and θ being cylindrical coordinates. This problem contains much of the content of relativistic mechanics.*

8.4
Analogies Between Optics and Quantum Mechanics

8.4.1
Classical Limit of the Schrödinger Equation

In the Hamilton–Jacobi formulation of mechanics one proceeds by solving a partial differential equation for the "wavefront" quantity S. This makes the mathematics of mechanics closely analogous to the mathematics of waves, as shown in Fig. 8.8, and therefore also analogous to wave (i.e. quantum) mechanics.

It seemed natural to Schrödinger to pursue the possibility that the Hamilton–Jacobi equation was itself the short wave approximation to a more general wave equation. We know that, for the case of a single particle of mass m in a potential $V(\mathbf{r})$, Schrödinger was led to the equation

$$i\hbar\frac{\partial\psi}{\partial t} = -\frac{\hbar^2}{2m}\nabla^2\psi + V(\mathbf{r})\psi. \tag{8.70}$$

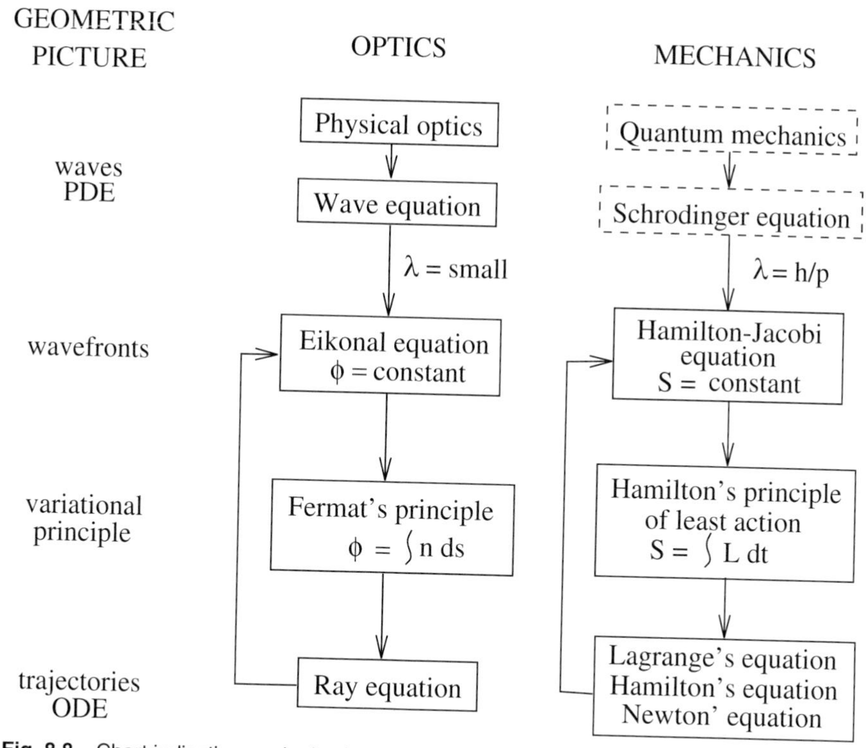

Fig. 8.8 Chart indicating analogies between optics and mechanics. Topics not discussed are in broken line boxes and derivation paths discussed are indicated by arrows.

As in Section 7.1.1 we can seek a solution to this equation that approximates a plane wave locally;

$$\psi(\mathbf{r}) = Ae^{iS(\mathbf{r},t)/\hbar}. \tag{8.71}$$

The constant h, as introduced here, has the same units as S but remains to be determined. This establishes h to have units of *action* which partially accounts for its having the name "quantum of action." Substitution into Eq. (8.70) yields

$$\frac{\partial S}{\partial t} + \frac{1}{2m}|\nabla \mathbf{S}|^2 + V(\mathbf{r}) - \frac{i\hbar}{2m}\nabla^2 S = 0. \tag{8.72}$$

The final term vanishes in the limit $\hbar \to 0$, and in that limit the Schrödinger equation becomes

$$\frac{\partial S}{\partial t} + \frac{1}{2m}|\nabla \mathbf{S}|^2 + V(\mathbf{r}) = 0. \tag{8.73}$$

This is precisely the Hamilton–Jacobi equation, since the Hamiltonian for this system is

$$H = \frac{\mathbf{p}^2}{2m} + V(\mathbf{r}). \tag{8.74}$$

In Eq. (8.71), to make the local wavelength of wave function ψ be λ, the spatial dependence at fixed time of $S(\mathbf{r}, t)$ should be

$$S(\mathbf{r}) = \frac{2\pi\hbar\hat{\mathbf{k}} \cdot \mathbf{r}}{\lambda}, \quad \text{or} \quad \nabla \mathbf{S} = \frac{h}{\lambda}\hat{\mathbf{k}}. \tag{8.75}$$

But, in the H–J formalism, the momentum is given by $\mathbf{p} = \nabla \mathbf{S}$, and hence self consistency requires

$$\mathbf{p} = \frac{h}{\lambda}\hat{\mathbf{k}}, \tag{8.76}$$

which is the de Broglie relation between momentum and wavelength.

Since the momentum can be inferred from mechanics and the wavelength λ can be measured – for example by electron diffraction from a crystal of known lattice spacing – the numerical value of h can be determined. This value can be compared with the value appearing in $E = h\nu$, as measured, for example, using photoelectric measurements. This provides a significant, almost definitive, test of the validity of quantum mechanics.

Bibliography

General References

1 R. Courant and D. Hilbert, *Methods of Mathematical Physics*, Vol. 2, Interscience, New York, 1962.

2 D. Ter Haar, *Elements of Hamiltonian Mechanics*, 2nd ed., Pergamon, Oxford, 1971.

References for Further Study

Section 8.2

3 L.A. Pars, *Analytical Dynamics*, Ox Bow Press, Woodbridge, CT, 1979.

Section 8.3

4 F.T. Geyling and H.R. Westerman, *Introduction to Orbital Mechanics*, Addison-Wesley, Reading, MA, 1971.

9
Relativistic Mechanics

The treatment of relativity here follows Landau and Lifshitz, *Classical Theory of Fields*, and is largely equivalent to the treatment in Jackson's *Classical Electrodynamics*. There are notational problems in comparing with books on general relativity or string theory, such as Schutz, *A First Course in General Relativity*, or Zwiebach, *A First Course in String Theory*. These authors distinguish between a gravitation-free, special relativistic metric $\eta_{\mu\nu}$, and a metric $g_{\mu\nu}$ that is similar, except for including gravitational effects;

$$g_{\mu\nu} = \pm\eta_{\mu\nu} + \text{ gravitational effect.} \tag{9.1}$$

Further possible confusion results when the sign in this equation is taken to be negative or when all the signs in $g_{\mu\nu}$ are reversed from what is conventional in electromagnetic theory. In this text, at least until later chapters, only the symbol $g_{\mu\nu}$ will be used.

Another notational convention concerns the alphabet used for tensor indices. We use Latin letters $i, j, k, \dots$ to span the four coordinates of Minkowski space and Greek letters $\alpha, \beta, \gamma, \dots$ to span the three coordinates of Euclidean space. Other authors reverse these choices.

9.1
Relativistic Kinematics

9.1.1
Form Invariance

Form invariance is the main idea of relativity: *All in-vacuum equations should have the same* form *in all coordinate frames. If a scalar quantity such as* $c = 3 \times 10^{10}$ *cm/s occurs in the equations in one frame of reference, then that same quantity c, with the same value, must appear in the equations in any other coordinate system.* If it should turn out, as it does, that c is the speed of propagation of waves (e.g., light) predicted by the equations, then the speed of light must be the same in all frames. (In systems of units such as SI units, c does not appear explicitly in Maxwell's equations, but it is derivable from other constants and the same

Geometric Mechanics: Toward a Unification of Classical Physics. 2nd Edition. Richard Talman

ISBN: 978-3-527-40683-8

conclusion follows.) It is found that there are wave solutions to Maxwell's equations and that their speed is c. Maxwell evaluated c from electrical and magnetic measurements and, finding the value close to the speed of light, conjectured that light is an electromagnetic phenomenon.

Putting these things together, we can say that light travels with the same speed in all inertial frames. This conclusion was corroborated by the Michelson–Morley experiment.

Of course, the numerical values of physical quantities such as position, velocity, electric field, and so on, can have different values in different frames. In Galilean relativity, velocities measured by two relatively moving observers are necessarily different. Hence the constancy of the speed of light is not consistent with Galilean relativity.

Einstein also introduced the concept that *time* need not be the same in different frames. Treating time and space similarly, he stressed the importance of *world points* in a four-dimensional plot with almost symmetric treatment of time as one coordinate and space as the other three.

9.1.2
World Points and Intervals

A *world event* is some occurrence, at position $\mathbf{x}$, and at time t, such as a ball dropping in Times Square, precisely at midnight, New Year's eve. Such a world event is labeled by its time and space coordinates $(t, \mathbf{x})$. To describe the trajectory of the ball requires a *world line* (which may be curved) that describes where the ball is, $\mathbf{x}(t)$, at time t.

If this world line describes a light pulse sent from world point $(t_1, \mathbf{x}_1)$ and received at world point $(t_2, \mathbf{x}_2)$, then, because it *is* light,

$$|\mathbf{x}_2 - \mathbf{x}_1| = c(t_2 - t_1). \tag{9.2}$$

When described in a different coordinate frame, designated by primes, the same two events are related by

$$|\mathbf{x}'_2 - \mathbf{x}'_1| = c(t'_2 - t'_1), \tag{9.3}$$

with c having the *same* value. The interval between *any* two such world events (not necessarily lying on the world line of a light pulse) is defined to be

$$s_{12} = \sqrt{c^2(t_2 - t_1)^2 - (\mathbf{x}_2 - \mathbf{x}_1)^2}. \tag{9.4}$$

Since $c^2(t_2 - t_1)^2 - (\mathbf{x}_2 - \mathbf{x}_1)^2$ can have either sign, s_{12} can be real or purely imaginary. If real, the interval is said to be *time like*,[1] if imaginary it is *space like*. From Eq. (9.2) it can be seen that the interval vanishes if the two ends are on the world line of a light pulse. Either point is then said to be "on the light cone" of the other. This condition is frame independent since, from Eq. (9.3), the same interval reckoned in any other frame also vanishes.

A differential interval ds is defined by

$$ds^2 = c^2dt^2 - dx^2 - dy^2 - dz^2 = c^2dt^2 - |\mathbf{dx}|^2. \tag{9.5}$$

It has been seen that the vanishing of the value ds in one frame implies the vanishing of the value ds' in any other. This implies that ds^2 and ds'^2 (both being homogeneous (quadratic) differentials) are proportional; say $ds = Ads'$, whether or not their end points lie on a single light pulse world line. With space assumed to be homogeneous and isotropic, and time homogeneous, the proportionality factor A can at most depend on the absolute value of the relative velocity of the two frames. *A priori* the factor A could depend on the relative speed of the frames. But, by considering three noncollinear frames, each with the same speed relative to the other two, the same nonunity factor would apply to each pair, which would lead to a contradiction. Hence

$$ds = ds', \tag{9.6}$$

From the equality of differential intervals it follows that finite intervals are also invariant;

$$s_{12} = s'_{12}. \tag{9.7}$$

9.1.3
Proper Time

In special relativity the word "proper" has a very definite technical meaning that is fairly close to its conventional colloquial meaning – as in "If you do not do it properly it may not come out right." Hence the "proper" way to time a mile run is to start a stopwatch at the start of the race and and stop it when the leader crosses the finish line. The stopwatch is chosen for accuracy, but that is not the point being made here – the point is that the watch is present at both world events, start and finish. (This assumes the race is four laps of a 1/4-mile track.) In this sense the traditional method of timing a 100 yard

1) It is *not* mnemonically helpful that s, which seems like an appropriate symbol for a *distance*, is actually better thought of as a *time*, or rather as c times a time. For this reason s will usually appear in the form $\tau = s/c$. In some contexts τ is referred to as "proper time." The fact that modern workers in general relativity favor a metric with sign opposite to that used here may be because they consider it "natural" for $\sqrt{ds^2}$ to be real for space-like separations.

dash is not "proper," because the same watch is not present at start and finish; (unless the winner of the race is carrying it.) In practice, if the timing is not done "properly," it may still be possible to compensate the measurement so as to get the right answer. In timing the 100 yard dash, allowance can be made for the time it takes for sound to get from the starting gun to the finish line.

Hence, a "proper time" in relativity is the time between two world events occurring at the same place. A "proper distance" is the distance between two world events occurring at the same time; this requires the use of a meter stick that is at rest in a frame in which the two events are simultaneous.

The world line of a particle moving with speed v is described, in a fixed, unprimed frame, by coordinates $(t, \mathbf{x}(t))$, where $v = |\mathbf{dx}|/dt$. Consider differential motion along this world line. The *proper* time advances not by dt, which is the time interval measured in the unprimed frame, but by dt', the time interval measured by a clock carried along with the particle. The same interval can be worked out in the fixed frame, yielding ds, and in the frame of the particle, yielding ds':

$$ds^2 = c^2 dt^2 - v^2 dt^2, \quad ds'^2 = c^2 dt'^2. \tag{9.8}$$

Since these quantities are known to be equal we obtain, for the proper time,

$$dt' = d(s/c) = dt\sqrt{1 - \frac{v^2}{c^2}}. \tag{9.9}$$

These equations include the result that, except for factor c, proper time and invariant interval are the same thing in the case of differential intervals between points on the world line of a moving particle. For a finite path from a point P_1 to P_2, with $\mathbf{v}$ not necessarily constant, the proper time is obtained by integration;

$$t'_2 - t'_1 = \int_{t_1}^{t_2} dt\sqrt{1 - \frac{v^2}{c^2}} = \int_{P_1}^{P_2} \frac{ds}{c}. \tag{9.10}$$

It will turn out that the use of proper time, rather than t, is the appropriate independent variable for describing the dynamics of the motion of a particle; i.e., the relativistic generalization of Newton's law.

Recall the twin paradox, according to which a moving clock, carried away from and then returned to a stationary clock, gains less time than does the stationary clock. This leads to a seemingly paradoxical "principle of greatest time" according to which, of the motions from the initial to the final world-point, free motion takes the *greatest* proper time. Superficially, it seems difficult to reconcile this with the principle that a straight line is the shortest distance between two points, but that's relativity!

9.1.4
The Lorentz Transformation

In the so-called Galilean relativity, time is universal, the same in all coordinate frames. In Einstein relativity, space and time coordinates transform jointly. From an unprimed frame K, at rest, to a primed, identically oriented, frame K', moving with uniform speed V along the x-axis, the coordinates transform according to the Lorentz transformation, derived as follows.

Except for signs, the *metric* given in Eq. (9.5) is the same as the Pythagorean formula in Euclidean geometry. In Euclidean geometry the transformations that preserve distances are rotations or translations or combinations of the two. Here we are concerned with the relation between the coordinates in frames K and K'. By insisting that the origins coincide initially we exclude translations and are left with "rotations" of the form

$$\begin{aligned} x &= x' \cosh \psi_V + ct' \sinh \psi_V, \\ ct &= x' \sinh \psi_V + ct' \cosh \psi_V. \end{aligned} \tag{9.11}$$

Substituting into Eq. (9.5) verifies that indeed $ds = ds'$. The occurrence of hyperbolic functions instead of trigonometric functions is due to the negative sign in Eq. (9.5).

Consider the origin of the K' system, $x' = 0$. From Eq. (9.11) the motion of this point in frame K is described by

$$x = ct' \sinh \psi_V, \quad ct = ct' \cosh \psi_V, \tag{9.12}$$

and hence

$$\frac{x}{ct} = \tanh \psi_V, \quad \text{or} \quad \tanh \psi_V = \frac{V}{c}. \tag{9.13}$$

Using properties of hyperbolic functions, this yields

$$\sinh \psi_V = \gamma_V \frac{V}{c}, \quad \cosh \psi_V = \gamma_V, \quad \text{where} \quad \gamma_V = \frac{1}{\sqrt{1-(V/c)^2}}. \tag{9.14}$$

Substituting back into Eq. (9.11), we obtain the Lorentz transformation equations,

$$ct = \gamma_V(ct' + \beta_V x'), \quad x = \gamma_V(\beta_V ct' + x'), \quad y = y', \quad z = z', \tag{9.15}$$

where

$$\beta_V = \frac{V}{c}, \quad \gamma_V = \frac{1}{\sqrt{1-\beta_V^2}}. \tag{9.16}$$

The inverse Lorentz transformation can be worked out algebraically; by symmetry the result must be equivalent to switching primed and unprimed variables and replacing β_V by $-\beta_V$,

$$ct' = \gamma_V(ct - \beta_V x), \quad x' = \gamma_V(-\beta_V ct + x), \quad y' = y, \quad z' = z. \tag{9.17}$$

These transformation equations can be expressed as a matrix multiplication,

$$x'^i = \Lambda^i{}_j x^j, \quad \text{or} \quad \underline{x}' = \Lambda\,\underline{x}, \tag{9.18}$$

where $\underline{x} = (ct, x, y, z)^T$ and

$$\Lambda = \begin{pmatrix} \gamma & -\gamma\beta & 0 & 0 \\ -\gamma\beta & \gamma & 0 & 0 \\ 0 & 0 & 1 & 0 \\ 0 & 0 & 0 & 1 \end{pmatrix}. \tag{9.19}$$

9.1.5 Transformation of Velocities

The incremental primed coordinates of a particle moving with velocity $\mathbf{v}'$ in the moving frame are given by $dx' = v'_x dt'$, $dy' = v'_y dt'$, $dz' = v'_z dt'$. Using Eq. (9.15), the unprimed coordinates are

$$\begin{aligned} cdt &= \gamma_V(cdt' + \beta_V v'_x dt'), \\ dx &= \gamma_V(\beta_V cdt' + v'_x dt'), \quad dy = v'_y dt', \quad dz = v'_z dt'. \end{aligned} \tag{9.20}$$

The fixed frame, unprimed, velocity components are obtained by dividing the last three of these equations by the first;

$$v_x = \frac{v'_x + V}{1 + v'_x V/c^2}, \quad v_y = \frac{v'_y/\gamma_V}{1 + v'_x V/c^2}, \quad v_z = \frac{v'_z/\gamma_V}{1 + v'_x V/c^2}. \tag{9.21}$$

In the special case that the particle is moving along the x'-axis with speed v' this becomes

$$v = \frac{v' + V}{1 + v'V/c^2}. \tag{9.22}$$

Of all the formulas of relativity, this is, to me, the most counterintuitive, since the truth of the same formula, but without the denominator, seems so "obvious." It is easy to see from these formulas that the particle velocity cannot exceed c in any frame.

9.1.6 4-Vectors and Tensors

The formulas of relativity are made much more compact by using the four-component tensor notation introduced by Einstein. The basic particle coordinate 4-vector is given by $x^i, i = 0, 1, 2, 3$, where

$$x^0 = ct; \quad x^1 = x; \quad x^2 = y; \quad x^3 = z. \tag{9.23}$$

These will be referred to as the world-coordinates of the particle. An abstract world point can be referred to by its coordinates x^i, and its spatial part as $\mathbf{x}$, so $x^i = (ct, \mathbf{x})^T$ (where the $=$ sign is being used very loosely). When used to indicate the functional dependence on position x^i of some function $f(x^i) = f(ct, \mathbf{x})$ the dangling superscript i can be confusing, especially in formulas with other tensor indices, because it is not meaningful to replace i by a particular integer. To avoid this problem, the symbol $\underline{x}$ will sometimes be used to stand for $(ct, \mathbf{x})^T$. The symbol $\underline{x}$ can be considered as representing an intrinsic geometric quantity whose coordinates are $(ct, \mathbf{x})$ in the same way that, in vector analysis, the vector $\mathbf{x}$ is a geometric object whose coordinates are $(x, y, z)^T$.

Any other four-component object whose components in different frames are related by Eqs. (9.15) is also called a 4-vector. Hence the 4-vector components A^i (lower case Roman letters are always assumed to range over 0, 1, 2, and 3) and A'^i are related by

$$A^0 = \gamma_V(A'^0 + \beta_V A'^1), \; A^1 = \gamma_V(\beta_V A'^0 + A'^1), \; A^2 = A'^2, \; A^3 = A'^3. \tag{9.24}$$

These components A^i are called *contravariant*. Also introduced are *covariant* components given by

$$A_0 = A^0, \; A_1 = -A^1, \; A_2 = -A^2, \; A_3 = -A^3. \tag{9.25}$$

This is referred to as "lowering the index." The same algebra that assured the invariance of s_{12}, (see Eq. (9.7)) assures the invariance of the combination

$$(A^0)^2 - (A^1)^2 - (A^2)^2 - (A^3)^2 = \sum_{i=0}^{3} A^i A_i \equiv A^i A_i. \tag{9.26}$$

Because of its invariance, one calls $A^i A_i$ a 4-scalar. A more general scalar, called the scalar product, can be formed from two 4-vectors A^i and B^i;

$$A^i B_i \equiv A_i B^i = A^0 B_0 + A^1 B_1 + A^2 B_2 + A^3 B_3. \tag{9.27}$$

Its invariance is assured by the same algebra.

A 16-component object, called a 4-tensor, can be formed from all the products of the components of two 4-vectors

$$T^{ij} = A^i B^j \quad i, j = 0, 1, 2, 3. \tag{9.28}$$

Not all tensors are "factorizable" into two 4-vectors like this but a two index tensor can always be written as a sum of products of 4-vectors. Any 16-component object transforming by the same formulas as $A^i B^j$ is also called a

4-tensor. If $T^{ij} = T^{ji}$, as would be true of A^iB^j if A^i and B^i happened to be equal, then the tensor is said to be "symmetric." If $T^{ij} = -T^{ji}$ it is "antisymmetric."

The Lorentz transformation of a two index tensor can be expressed compactly in terms of the matrix Λ introduced in Eq. (9.19);

$$A'^{ij} = \Lambda^i_{\ k}\, A^{kl}\, \Lambda^j_{\ l}, \quad \text{or} \quad A' = \Lambda\, A\, \Lambda^T. \tag{9.29}$$

The particular order of the factors in the first step has been chosen for convenience in the next step and the matrix transposition in the second step is required because the indices of the factor $\Lambda^j_{\ l}$ are in the "wrong" order for matrix multiplication. (The index notation for tensors has the *formal* advantage that the order of the factors is irrelevant, but the *calculational* disadvantage is that the more compact matrix-like representation of the factors becomes progressively more complicated for tensors of higher order.)

The operation of lowering or raising indices can be accomplished by tensor multiplication by the so-called *metric tensor*,

$$g^{ij} = g_{ij} = \begin{pmatrix} 1 & 0 & 0 & 0 \\ 0 & -1 & 0 & 0 \\ 0 & 0 & -1 & 0 \\ 0 & 0 & 0 & -1 \end{pmatrix}. \tag{9.30}$$

Thus,

$$A_i = g_{ij} A^j. \tag{9.31}$$

The indices of tensors of any order can be raised and lowered the same way. Note that g^{ij} and g_{ij} themselves are consistent with this. Also the *mixed* (one index lowered) tensor $g^i_{\ j}$ is equal to the "Kronecker delta,"

$$g^i_{\ j} = \delta^i_j = \begin{cases} 0, & i \neq j, \\ 1, & i = j. \end{cases} \tag{9.32}$$

The terminology "metric" for g^{ij} is justified by the fact that Eq. (9.5) can be written as

$$ds^2 = g_{ij} dx^i dx^j. \tag{9.33}$$

The tensor g^{ij} has the same components in all coordinate frames (having constant relative velocities.) You should confirm that this is consistent with g^{ij} transforming as a tensor by working Problem 9.3.6.

9.1.7
Three-Index Antisymmetric Tensor

We now return temporarily to Cartesian, three-component, spatial, geometry, and define a 27-component, antisymmetric, three-index tensor, $\varepsilon_{\alpha\beta\gamma}$, which vanishes if any two indices are the same, $\varepsilon_{123} = 1$, and cases for which all indices are different are equal to ± 1, depending on whether the permutation of indices from 123 is even or odd. $e_{\alpha\beta\gamma}$ is known as the Levi-Civita antisymmetric symbol. It is assumed that all readers have encountered this symbol while studying 3×3 determinants. The antisymmetric symbol can be used to represent the *cross product* $\mathbf{C} = \mathbf{A} \times \mathbf{B}$ of two polar vectors $\mathbf{A}$ and $\mathbf{B}$ in component form;

$$C_\alpha = \varepsilon_{\alpha\beta\gamma} A^\beta B^\gamma. \tag{9.34}$$

When the orientations of all coordinate axes are reversed, the signs of all the components of both $\mathbf{A}$ and $\mathbf{B}$ switch. But, from Eq. (9.34), this transformation leaves the signs of components of $\mathbf{C}$ unchanged. $\mathbf{C}$ is therefore said to be an *axial* vector, (or pseudovector). (Note the convention being employed that lower case Greek letters have the range 1, 2, 3, while Roman letters have the range 0, 1, 2, 3.) C_α can be related to an antisymmetric tensor $C^{\beta\gamma} = A^\beta B^\gamma - A^\gamma B^\beta$ according to

$$C_\alpha = \frac{1}{2} e_{\alpha\beta\gamma} C^{\beta\gamma}. \tag{9.35}$$

9.1.8
Antisymmetric 4-Tensors

It is confirmed in Problem 9.3.1 that, from the components p_x, p_y, and p_z of a polar 3-vector, and the components a_x, a_y, and a_z of an axial 3-vector, it is possible to construct an antisymmetric 4-tensor according to

$$(A^{ij}) = \begin{pmatrix} 0 & p_x & p_y & p_z \\ -p_x & 0 & -a_z & a_y \\ -p_y & a_z & 0 & -a_x \\ -p_z & -a_y & a_x & 0 \end{pmatrix}. \tag{9.36}$$

This is a very important form in electrodynamics.

A useful step in analyzing the transformation of a tensor like this is to partition the matrices in Eq. (9.29) appropriately for a transformation that mixes the x^1, x^2 and x^3 coordinates, but leaves the x^0 coordinate unchanged;

$$\begin{pmatrix} 0 & P'^T \\ -P' & A' \end{pmatrix} = \begin{pmatrix} 1 & 0 \\ 0 & R \end{pmatrix} \begin{pmatrix} 0 & P^T \\ -P & A \end{pmatrix} \begin{pmatrix} 1 & 0 \\ 0 & R^T \end{pmatrix}, \tag{9.37}$$

where R is a 3D spatial rotation matrix. Completing the multiplications, the results are

$$P' = -RP, \quad \text{and} \quad A' = RAR^T, \tag{9.38}$$

for the Lorentz transformation of an antisymmetric, two-index tensor under spatial rotation. The transformation of a general, not necessarily antisymmetric, tensor, can be performed similarly. In analyzing transformations between frames it is always advantageous to label the axes and to choose the partitioning to take advantage of any special feature of the configuration.

9.1.9
The 4-Gradient, 4-Velocity, and 4-Acceleration

One can form a four-component object called the four-gradient by differentiating a four-scalar function $\phi(ct, x, y, z)$ with respect to its four arguments. These derivatives appear naturally in the expression for the differential $d\phi$:

$$d\phi = \frac{\partial \phi}{\partial x^i}\, dx^i. \tag{9.39}$$

From Eq. (9.27) it can be seen that for $d\phi$ to be a scalar quantity, as it is by definition, $\partial\phi/\partial x^i$ must be a covariant tensor. A compact notation is

$$\phi_{,i} \equiv \frac{\partial \phi}{\partial x^i} = \left(\frac{\partial \phi}{\partial (ct)}, \nabla\phi\right). \tag{9.40}$$

(The final notation is ambiguous, and should be used carefully or avoided. The choice of sign of the spatial part depends on whether the listed entries are to be interpreted as co- or as contravariant. Often when the components of a vector are listed within parentheses they are assumed to be contravariant. But here they are covariant. In this text, contravariant listings like this are expressed as $x^i = \begin{pmatrix} ct \\ \mathbf{x} \end{pmatrix}$.)

The 4-velocity u_i is defined by

$$u^i = \frac{dx^i}{ds/c}, \tag{9.41}$$

where the factor c has been included to give u_i the dimensions of velocity. Comparing with Eq. (9.9) it can be seen that

$$u^i = \begin{pmatrix} \dfrac{c\,dt}{dt\sqrt{1 - v^2/c^2}} \\ \dfrac{\mathbf{dx}}{dt\sqrt{1 - v^2/c^2}} \end{pmatrix} = \begin{pmatrix} \gamma_v c \\ \gamma_v \mathbf{v} \end{pmatrix}, \tag{9.42}$$

where $\mathbf{v}$ is the ordinary particle velocity. Because $ds^2 = dx^i dx_i$, the 4-scalar formed from u^i is constant,

$$u^i u_i = c^2. \tag{9.43}$$

The fact that $u^i u_i$ is independent of the particle's three velocity makes it inappropriate to interpret $u^i u_i$ as any useful function of the particle's speed. The 4-acceleration w^i is defined similarly;

$$w^i = \frac{d^2 x^i}{ds^2/c^2} = \frac{du^i}{ds/c}. \tag{9.44}$$

Differentiating Eq. (9.43), the 4-velocity and 4-acceleration are seen to be mutually "orthogonal";

$$u^i w_i = 0. \tag{9.45}$$

9.2 Relativistic Mechanics

9.2.1 The Relativistic Principle of Least Action

It is straightforward to generalize the principle of least action in such a way as to satisfy the requirements of relativity while at the same time leaving nonrelativistic relationships (i.e., Newton's law) valid when speeds are small compared to c. Owing to the homogeneity of both space and time, the relativistically generalized action S cannot depend on the particle's coordinate 4-vector x^i. Furthermore, it must be a relativistic scalar since otherwise it would have directional properties, forbidden by the isotropy of space.

Though a Lagrangian depends on both position and velocity, owing to Eq. (9.43), it is impossible to form a scalar other than a constant using the 4-vector u^i. The only possibility, therefore, for the action of a free particle (i.e., one subject to no force) is

$$S = (-mc) \int_{t_0}^{t} ds = (-mc^2) \int_{t_0}^{t} \sqrt{1 - \frac{v^2}{c^2}}\, dt, \tag{9.46}$$

where the invariant interval ds is the proper time multiplied by c defined in Eq. (9.10). As always, the dimensions of S are momentum$\times$distance or, equivalently, as energy$\times$time. Though the first expression for S is manifestly invariant, the second depends on values of $v \equiv |\dot{\mathbf{x}}|$ and t in the particular frame of reference in which Hamilton's principle is to be applied. *A priori* the multiplicative factor could be any constant, but it will be seen shortly why the factor has to be $(-mc^2)$. The negative sign is significant. It corresponds to the seemingly paradoxical result mentioned above that the free particle path from position P_1 to position P_2 *maximizes* the proper time taken. Comparing with the standard definition of the action in terms of the Lagrangian, it can be seen that the free particle Lagrangian is

$$L(\mathbf{x}, \dot{\mathbf{x}}) = -mc^2\sqrt{1 - \frac{|\dot{\mathbf{x}}|^2}{c^2}}. \tag{9.47}$$

As always, the Lagrangian has the dimensions of an energy.

9.2.2
Energy and Momentum

In Lagrangian mechanics, once the Lagrangian is specified, the equations of motion follow just by "turning the crank." Slavishly following the Lagrangian prescriptions, the momentum $\mathbf{p}$ is *defined* by

$$\mathbf{p} = \frac{\partial L}{\partial \dot{\mathbf{x}}} = \frac{m\mathbf{v}}{\sqrt{1 - v^2/c^2}}. \tag{9.48}$$

For v small compared to c, this gives the nonrelativistic result $\mathbf{p} \simeq m\mathbf{v}$. This is the relation that fixed the constant factor in the initial definition of the Lagrangian. Using Eqs. (9.47) and (9.48), one obtains the Hamiltonian H and hence the energy $\mathcal{E}$ by

$$H = \mathbf{p} \cdot \mathbf{v} - L = \frac{mc^2}{\sqrt{1 - v^2/c^2}}. \tag{9.49}$$

For v small compared to c, and the numerical value of H symbolized by $\mathcal{E}$, this gives

$$\mathcal{E} \simeq \mathcal{E}_0 + \frac{1}{2}mv^2, \tag{9.50}$$

which is the classical result for the kinetic energy, except for the additive constant $\mathcal{E}_0 = mc^2$, known as the rest energy. An additive constant like this has no effect in the Lagrangian description. From Eqs. (9.48) and (9.49) come the important identities

$$\mathcal{E}^2 = \mathbf{p}^2c^2 + m^2c^4, \quad \mathbf{p} = \frac{\mathcal{E}\mathbf{v}}{c^2}. \tag{9.51}$$

For massless particles like photons these reduce to $v = c$ and

$$p = \frac{\mathcal{E}}{c}. \tag{9.52}$$

This formula also becomes progressively more valid for a massive particle as its total energy becomes progressively large compared to its rest energy. As stated previously, m is the "rest mass," a constant quantity, and there is no question of "mass increasing with velocity" as occurs in some descriptions of relativity, such as the famous "$\mathcal{E} = mc^2$," which is incorrect in modern formulations.

Remembering to express it in terms of $\mathbf{p}$, the relativistic Hamiltonian is given by

$$H(\mathbf{p}) = \sqrt{p^2c^2 + m^2c^4}. \tag{9.53}$$

9.2.3
4-Vector Notation

Referring back to Eq. (9.42), it can be seen that $\mathbf{p}$, as given by Eq. (9.48), and $\mathcal{E}$, as given by Eq. (9.49), are closely related to the 4-velocity u^i. We define a momentum 4-vector p^i by

$$p^i = mu^i = \frac{m}{\sqrt{1 - v^2/c^2}} \begin{pmatrix} c \\ \mathbf{v} \end{pmatrix} = \begin{pmatrix} \mathcal{E}/c \\ \mathbf{p} \end{pmatrix}. \tag{9.54}$$

We expect that $p^i p_i$, the scalar product of p^i with itself should, like all scalar products, be invariant. The first of Eqs. (9.51) shows this to be true;

$$p^i p_i = \mathcal{E}^2/c^2 - p^2 = m^2c^2. \tag{9.55}$$

Belonging to the same 4-vector, the components of $\mathbf{p}$ and $\mathcal{E}/c$ in different coordinate frames are related according to the Lorentz transformation, Eq. (9.15).

9.2.4
Forced Motion

If the 4-velocity is to change, it has to be because force is applied to the particle. It is natural to define the 4-force G^i by the relation

$$G^i = \frac{dp^i}{ds/c} = \gamma \left(\frac{d\mathcal{E}/c}{dt}, \frac{d\mathbf{p}}{dt} \right)^T = \left(\frac{\mathbf{F}\cdot\mathbf{v}/c}{\sqrt{1 - v^2/c^2}}, \frac{\mathbf{F}}{\sqrt{1 - v^2/c^2}} \right)^T \tag{9.56}$$

where

$$\mathbf{F} = \frac{d\mathbf{p}}{dt} \tag{9.57}$$

is the classically defined force. Since this formula is valid both relativistically and nonrelativistically it is least error-prone 3D form of Newton's law. The energy/time component G^0 is related to the rate of work done on the particle by the external force. See Problem 9.3.9. Note that this component vanishes in the case that $\mathbf{F} \cdot \mathbf{v} = 0$, as is true, for example, for a charged particle in a purely magnetic field.

9.2.5
Hamilton–Jacobi Formulation

All this is quite elementary, and it is all that one really needs to remember in order to proceed with relativistic dynamics. The following more formal development will be needed shortly. When the minimized function $S(x_0, t_0, x, t)$ is expressed as a function of x and t at the upper spatial end point, holding the lower end point fixed, S satisfies the so-called "Hamilton Jacobi equation." This equation is rarely used for single particle dynamics but the following definitions (associated with the equation) are important for subsequent theory:

$$\mathbf{p} = \frac{\partial S}{\partial \mathbf{x}}, \quad H = -\frac{\partial S}{\partial t}. \tag{9.58}$$

Corresponding to the Hamiltonian of Eq. (9.53), the Hamilton–Jacobi equation is

$$\left(\frac{\partial S}{\partial t}\right)^2 = c^2\left(\frac{\partial S}{\partial x}\right)^2 + c^2\left(\frac{\partial S}{\partial y}\right)^2 + c^2\left(\frac{\partial S}{\partial z}\right)^2 + m^2c^4. \tag{9.59}$$

Since the relations (9.58) can be derived for arbitrary S purely from the calculus of variations, without reference to the physical interpretation of the quantities, they must remain valid in relativistic mechanics. Nevertheless we will re-derive these relations for practice in using abbreviated manipulations for the calculus of variations. The action integral (9.46), expressed in terms of $ds = \sqrt{dx_i dx^i}$, is

$$S = -mc \int_{P_0}^{P} \sqrt{dx_i dx^i}. \tag{9.60}$$

The variation δS in the action that accompanies a variation $\delta x^i(t)$ away from the true world trajectory, is what establish the equations of motion. Here, $\delta x^i(t)$ is an arbitrary function. Variation of the integrand yields

$$\begin{aligned}
\delta\sqrt{dx_i dx^i} &= \frac{(\delta dx_i)dx^i + dx_i(\delta dx^i)}{2\sqrt{dx_i dx^i}} \\
&= \frac{dx_i}{ds} d\delta x^i = \frac{u_i}{c}\, d\delta x^i \\
&= d\left(\frac{u_i}{c}\,\delta x^i\right) - \frac{du_i/c}{ds}\delta x^i\, ds.
\end{aligned} \tag{9.61}$$

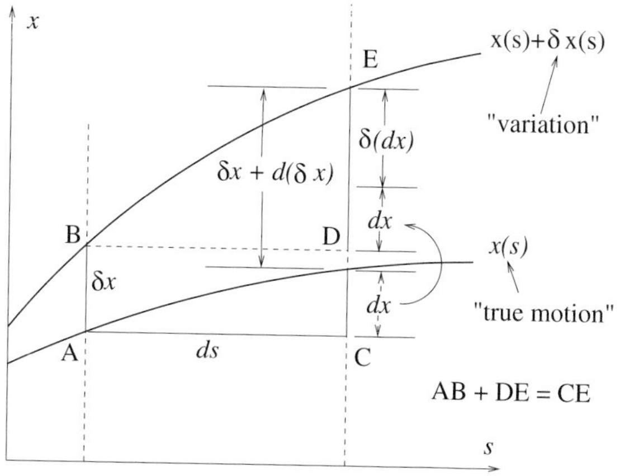

Fig. 9.1 The equation shown, AB+DE=CE, when expressed in terms of dx, δx, $d\delta x$, and δdx, shows that $d\delta x = \delta dx$.

The relation $\delta dx = d\delta x$, whose validity is exhibited in Fig. 9.1, has been used. The last line is preparatory to integration by parts. The action variation is

$$\delta S = -mc \int_{P_0}^{P} \left(d\left(\frac{u_i}{c}\delta x^i\right) - \frac{du_i/c}{ds}(\delta x^i)ds \right). \tag{9.62}$$

In this form the upper integration limit can be held fixed or varied as we wish. If both end points are held fixed, the first term in the integral vanishes, in which case, since the principle of least action requires the vanishing of δS, and since δx^i is arbitrary, the vanishing of the 4-acceleration $w_i = du^i/(ds/c)$ follows. This is appropriate for force-free motion.

When the upper end point of the integral in Eq. (9.62) is varied, but with the requirement that the trajectory be a true one, then the second term in the integral vanishes, leaving

$$\delta S = -mu_i \delta x^i. \tag{9.63}$$

Substituting into Eq. (9.58) yields

$$-p_i = \frac{\partial S}{\partial x^i} = -mu_i = \left(-\frac{\mathcal{E}}{c}, \mathbf{p} \right), \tag{9.64}$$

which confirms the validity of those equations. Remembering that the spatial covariant and contravariant 4-vector components have opposite signs, notice that the contravariant and covariant indices magically take care of the signs. Also the result is consistent with Eq. (9.40); taking the 4-gradient of a scalar yields a *covariant* 4-vector.

9.3 Introduction of Electromagnetic Forces into Relativistic Mechanics

9.3.1 Generalization of the Action

A relativistically satisfactory generalization of the free space Lagrangian is now to be hypothesized. It will be found to describe the force on a charged particle just like the forces that we ascribe to electric and magnetic forces. The action for a free particle was adopted because, except for an arbitrary multiplicative factor, ds was the only first-order-differential, origin-independent, 4-scalar that could be constructed.

We now generalize this by introducing an initially arbitrary 4-vector function of position, with *contravariant*, position-dependent components $A^i(\underline{x}) \equiv (\phi, \mathbf{A})$, and take for the action

$$S = \int_{t_0}^{t} \left(-mc\, ds - \frac{e}{c} A_i(\underline{x}) dx^i \right). \tag{9.65}$$

The integrand certainly satisfies the requirement of being a relativistic invariant. Like the factor $-mc^2$, that was chosen to make free motion come out right, the factor $-e/c$ is chosen to make this action principle lead to the forces of electromagnetism, with e being the charge on the particle (in Gaussian units). Also, anticipating the correlation with electromagnetic theory, the factors ϕ and $\mathbf{A}$ are called scalar and vector potentials, respectively. Spelling out the integrand more explicitly, and making the differential be dt, to enable extraction of the Lagrangian, the action is

$$S = \int_{t_0}^{t} \left(-mc^2 \sqrt{1 - \frac{v^2}{c^2}} + \frac{e}{c} \mathbf{A} \cdot \mathbf{v} - e\phi \right) dt. \tag{9.66}$$

This shows that the Lagrangian is

$$L = -mc^2 \sqrt{1 - \frac{v^2}{c^2}} + \frac{e}{c} \mathbf{A}(\underline{x}) \cdot \mathbf{v} - e\phi(\underline{x}). \tag{9.67}$$

(Another candidate for the action that would be consistent with relativistic invariance is $\int A(\underline{x})\, ds$ where $A(\underline{x})$ is a scalar function, but that would *not* lead to electromagnetism.)

Once the Lagrangian has been selected one must mechanically follow the prescriptions of Lagrangian mechanics in order to introduce a 3-vector $\mathbf{P}$, which is the "momentum" conjugate to $\mathbf{x}$, and then to obtain the equations of motion. This newly introduced (uppercase) momentum will be called the generalized momentum, to distinguish it from the previously introduced "ordinary momentum" or "mechanical momentum" $\mathbf{p}$. You should continue to think of the (lower case) quantity $\mathbf{p}$ as the generalization of the familiar mass times velocity of elementary mechanics. The generalized momentum $\mathbf{P}$ has a more formal significance connected with the Lagrange equations. It is given by

$$\mathbf{P} = \frac{\partial L}{\partial \mathbf{v}} = \frac{m\mathbf{v}}{\sqrt{1 - v^2/c^2}} + \frac{e}{c}\mathbf{A} = \mathbf{p} + \frac{e}{c}\mathbf{A}(\underline{x}). \tag{9.68}$$

Notice in particular that, unlike $\mathbf{p}$, the generalized momentum $\mathbf{P}$ depends explicitly on position (and time) $\underline{x}$.

We need only follow the rules to define the Hamiltonian by

$$H = \mathbf{v} \cdot \mathbf{P} - L = \frac{mc^2}{\sqrt{1 - v^2/c^2}} + e\phi, \tag{9.69}$$

which must still, however, be expressed in terms of $\mathbf{P}$ rather than $\mathbf{v}$. The first term energy can be referred to as "kinetic" the second, "potential." According to Eq. (9.55), the rest mass m, the mechanical momentum $\mathbf{p}$, and the ordinary or mechanical energy $\mathcal{E}_{\text{kin}} = mc^2/\sqrt{1 - v^2/c^2}$ are related by

$$\mathcal{E}_{\text{kin}}^2 = p^2c^2 + m^2c^4. \tag{9.70}$$

Here, we have used the symbol $\mathcal{E}_{\text{kin}}$ which, since it includes the rest energy, differs by that much from being a generalization of the "kinetic energy" of Newtonian mechanics. Nevertheless it is convenient to have a symbol for the energy of a particle that accompanies its very existence and includes its energy of motion but does not include any "potential energy" due to its position in a field of force. Using Eqs. (9.68) and (9.69) this same relation can be expressed in terms of $\mathbf{P}$ and H,

$$(H - e\phi)^2 = \left(\mathbf{P} - \frac{e}{c}\mathbf{A}\right)^2 c^2 + m^2c^4. \tag{9.71}$$

Solving for H yields

$$H(\mathbf{x}, \mathbf{P}) = \sqrt{m^2c^4 + \left(\mathbf{P} - \frac{e}{c}\mathbf{A}(\underline{x})\right)^2 c^2} + e\phi(\underline{x}). \tag{9.72}$$

Remember that the Hamiltonian is important in two ways. One is formal; differentiating it appropriately leads to Hamilton's equations. The other deals with its numerical value, which is called the energy, at least in those cases where it is conserved.

Equation (9.72) should seem entirely natural; the square root term gives the mechanical energy (remember that the second term under the square root is just c^2 times the ordinary momentum) and the other term gives the energy that a particle has by virtue of its having charge e and location at position $\underline{x}$

where the potential function is $\phi(\underline{x})$. Corresponding to this Hamiltonian, the H–J equation is

$$\left(\frac{\partial S}{\partial t} + e\phi\right)^2 = \left(\nabla S - \frac{e}{c}\mathbf{A}\right)^2 c^2 + m^2c^4. \tag{9.73}$$

9.3.2 Derivation of the Lorentz Force Law

To obtain the equations of motion for our charged particle we write the Lagrange equations with L given by Eq. (9.67). One term is

$$\nabla L = \frac{e}{c}\nabla\left(\mathbf{A}(\underline{x})\cdot\mathbf{v}\right) - e\nabla\phi(\underline{x}). \tag{9.74}$$

Remembering that the very meaning of the partial derivative symbol in the Lagrange equation is that $\mathbf{v}$ is to be held constant, the first term factor becomes

$$\nabla(\mathbf{A}\cdot\mathbf{v}) = (\mathbf{v}\cdot\nabla)\mathbf{A} + \mathbf{v}\times(\nabla\times\mathbf{A}), \tag{9.75}$$

where a well-known vector identity has been used. The meaning of the expression $(\mathbf{v}\cdot\nabla)\mathbf{A}$ is certainly unambiguous in Cartesian coordinates. Its meaning may be ambiguous in curvilinear coordinates, but we assume Cartesian coordinates without loss of generality, since this term will be eliminated shortly.

With Eq. (9.75), and using Eq. (9.68), the Lagrange equation becomes

$$\frac{d}{dt}\mathbf{p} + \frac{e}{c}\frac{d}{dt}\mathbf{A} = \frac{e}{c}(\mathbf{v}\cdot\nabla)\mathbf{A} + \frac{e}{c}\mathbf{v}\times(\nabla\times\mathbf{A}) - e\nabla\phi. \tag{9.76}$$

At this point a great bargain appears. For any function $F(\mathbf{x}, t)$, the total derivative and its partial derivative are related by

$$\frac{d}{dt}F = \frac{\partial}{\partial t}F + (\mathbf{v}\cdot\nabla)F. \tag{9.77}$$

The first term gives the change of F at a fixed point in space, and the second term gives the change due to the particle's motion. This permits a hard-to-evaluate term on the left-hand side, $d\mathbf{A}/dt$, and a hard-to-evaluate term on the right-hand side, $(\mathbf{v}\cdot\nabla)\mathbf{A}$, to be combined to make an easy-to-evaluate term, yielding

$$\frac{d\mathbf{p}}{dt} = -\frac{e}{c}\frac{\partial\mathbf{A}}{\partial t} - e\nabla\phi + \frac{e}{c}\mathbf{v}\times(\nabla\times\mathbf{A}). \tag{9.78}$$

At this point, we introduce the *electric field vector* $\mathbf{E}$ and the *magnetic field vector* $\mathbf{B}$ *defined* by

$$\begin{aligned}\mathbf{E} &= -\frac{1}{c}\frac{\partial\mathbf{A}}{\partial t} - \nabla\phi,\\ \mathbf{B} &= \nabla\times\mathbf{A}.\end{aligned} \tag{9.79}$$

Finally, we obtain the so-called *Lorentz force law*;

$$\frac{d\mathbf{p}}{dt} = e\mathbf{E} + e\frac{\mathbf{v}}{c} \times \mathbf{B}. \tag{9.80}$$

Since A^i was arbitrary, the electric and magnetic fields are completely general, consistent with Eq. (9.79).

From its derivation, the Lorentz force law, though not *manifestly* covariant, has unquestionable relativistic frame invariance. It describes the evolution of the spatial components. One can look for the corresponding time-component evolution equation. It is

$$\frac{d\mathcal{E}_{\text{kin}}}{dt} = \frac{d}{dt}\left(\frac{mc^2}{\sqrt{1 - v^2/c^2}}\right) = \mathbf{v} \cdot \frac{d\mathbf{p}}{dt}. \tag{9.81}$$

Recognizing $d\mathbf{p}/dt$ as the applied force, this shows the rate of change in mechanical energy to be the applied power.

This is entirely consistent with the Newtonian result that the rate of change of energy is the rate at which the external force (given by $d\mathbf{p}/dt$) does work. Under the Lorentz force law, since the magnetic force is normal to $\mathbf{v}$, it follows that a magnetic field can never change the particle energy. Rather the rate of change of energy is given, as expected, by

$$\frac{d\mathcal{E}_{\text{kin}}}{dt} = e\mathbf{E} \cdot \mathbf{v}. \tag{9.82}$$

9.3.3 Gauge Invariance

Though the 4-potential $A^i \equiv (\phi, \mathbf{A})^T$ was introduced first, it is the electric and magnetic fields $\mathbf{E}$ and $\mathbf{B}$ that manifest themselves physically through the forces acting on charged particles. They must be determinable uniquely from the physical conditions. But because $\mathbf{E}$ and $\mathbf{B}$ are obtained from A^i by differentiation, there is a lack of uniqueness in A^i, much like the "constant of integration" in an indefinite integral. In electrostatics this indeterminacy has already been encountered; adding a constant to the electric potential has no observable effect. With the 4-potential the lack of determinacy can be more complicated, because a change in ϕ can be compensated by a change in $\mathbf{A}$. For mathematical methods that are based on the potentials, this can have considerable impact on the analysis, though not on the (correctly) calculated $\mathbf{E}$ and $\mathbf{B}$ fields. The invariance of the answers to transformations of the potentials is called "gauge invariance."

The gauge invariance of the present theory follows immediately from the action principle of Eq.(9.65). Suppose the (covariant) components in that equation are altered according to

$$A_i(\underline{x}) \rightarrow A_i(\underline{x}) + \frac{\partial f(\underline{x})}{\partial x^i}, \tag{9.83}$$

where $f(\underline{x})$ is an arbitrary function of position. Expressed in components, the changes are

$$\phi \rightarrow \phi + \frac{\partial f}{\partial (ct)}, \qquad -\mathbf{A} \rightarrow -\mathbf{A} + \nabla f. \tag{9.84}$$

As a result of these changes, the action integral acquires an extra sum of terms

$$-\frac{e}{c}\int_{t_0}^{t} \frac{\partial f(\underline{x})}{\partial x^i}\, dx^i. \tag{9.85}$$

Each of these terms, and hence their sum, is expressible in terms of values of f on the boundaries. They are unaffected by variation of the particle trajectory and they therefore do not affect the extremal determination or the equation of motion. As a result the physics is unaffected by this "change of gauge." It is also instructive to confirm that this change in A^i has no effect on **E** and **B** evaluated by Eq. (9.79). (Problem 9.3.10.)

This is one of many instances in which an alteration of the Lagrangian is permissable without altering the extremal determination. In practice, this makes it legitimate to place another, presumably simplifying, condition for the Lagrangian to satisfy. In electromagnetic theory, gauge conditions such as the Lorentz gauge, the Coulomb gauge, or the radiation gauge, can be imposed that simplify those equations that are most important for some particular phenomenon under study.

In developing a Lagrangian theory of (nonquantum-mechanical) relativistic strings a very similar gauge invariance will provide great simplification. Historically, as quantum field theory developed, and an appropriately generalized gauge invariance principle developed, far more fundamental consequences ensued, beginning with the so-called Yang–Mills gauge field theory. From that theory evolved what is now considered to be the "standard model" of fundamental particles.

Problem 9.3.1. *The four-dimensional tensor transformation formalism has to incorporate purely three-dimensional transformations. For example, consider a rotation through angle θ in the x, y plane. A pure spatial vector having components $0, p_x, p_y, 0$ in one frame has components $0, p'_x, p'_y, 0$ in the rotated frame. Confirm the statement associated with Eq. (9.36) according to which certain elements of a two-index antisymmetric 4-tensor transform like the components of a three-dimensional spatial vector. That is, show that the rotation-in-a-plane transformation is correctly subsumed into the four-dimensional formalism for appropriately placed elements of an antisymmetric, two-index tensor. Also check pseudovector $0, a_x, a_y, 0$.*

Problem 9.3.2. *From two 4-vectors,* $x^i_{(1)} = (ct_{(1)}, x_{(1)}, y_{(1)}, z_{(1)})^T$ *and* $x^i_{(2)} = (ct_{(2)}, x_{(2)}, y_{(2)}, z_{(2)})^T$ *one can form a symmetric two-index tensor whose elements in the primed frame are*

$$S'^{ij} = x'^i_{(1)} x'^j_{(2)} + x'^i_{(2)} x'^j_{(1)}.$$

You are to find Lorentz transformation formulas analogous to Eqs. (9.17), but for two-index tensors instead of 4-vectors, as follows. For each particular element, such as S'^{00}, *after performing the Lorentz substitution (9.17) for the* $x'^i_{(1)}$ *and* $x'^j_{(2)}$ *components, eliminate the* $x^i_{(1)}$ *and* $x^j_{(2)}$ *components in favor of the* S^{ij} *components. Only diagonal and above-diagonal formulas are needed and only diagonal and above-diagonal components should appear in the formulas.*

Problem 9.3.3. *Use Eq. (9.29) to obtain the same results as in the previous problem – the transformation equations for the elements of a symmetric tensor. Before performing the matrix multiplications it is appropriate to partition the matrices appropriately for a transformation that mixes* x^0 *and* x^1 *but leaves* x^2 *and* x^3 *unchanged.*

Problem 9.3.4. *From the same two 4-vectors as in Problem 9.3.2 one can form a two-index antisymmetric tensor*

$$A^{ij} = x^i_{(1)} x^j_{(2)} - x^i_{(2)} x^j_{(1)}.$$

Repeat the steps taken in that problem to find the Lorentz transformation formulas for an antisymmetric two-index tensor. Only above-diagonal formulas are needed and only above-diagonal components should appear in the formulas.

Problem 9.3.5. *Use Eq. (9.29) to obtain the same results as in the previous problem – the transformation equations for the elements of an antisymmetric tensor.*

Problem 9.3.6. *The quantities* g^{ij} *introduced in Eq. (9.30) are* defined *to have the same values in all coordinate frames. Confirm that, in spite of their having the same values in all frames, the* g^{ij} *"transform like a two-index tensor."*

Problem 9.3.7. *Use the fully antisymmetric "Levi-Cevita" symbol* $\varepsilon_{\alpha\beta\gamma}$ *defined in Section 9.1.7 to prove the relations*

(a) $(\mathbf{A} \times \mathbf{B})_\alpha = \varepsilon_{\alpha\beta\gamma} A_\beta B_\gamma$

(b) $\mathbf{A} \cdot (\mathbf{B} \times \mathbf{C}) = \varepsilon_{\alpha\beta\gamma} A_\alpha B_\beta C_\gamma$

(c) $\varepsilon_{\alpha\beta\gamma}\varepsilon_{\delta\epsilon\gamma} = \delta_{\alpha\delta}\delta_{\beta\epsilon} - \delta_{\alpha\epsilon}\delta_{\beta\delta}$

(d) $\mathbf{A} \times (\mathbf{B} \times \mathbf{C}) = (\mathbf{A} \cdot \mathbf{C})\mathbf{B} - (\mathbf{A} \cdot \mathbf{B})\mathbf{C}$

where **A**, **B**, *and* **C** *are ordinary 3D space vectors and summation convention is used. (Remember that all indices can be written as subscripts in Euclidean geometry – there is no need to distinguish between contravariant and covariant components.)*

Problem 9.3.8. *It is valid and useful to regard* $\varepsilon_{\alpha\beta\gamma}$ *as a three-index tensor in ordinary 3D space. Interpret the product* $\mathbf{A} \cdot (\mathbf{B} \times \mathbf{C})$ *appearing in Problem 9.3.7(b) as a* geometric property *of the structure defined by vectors* **A**, **B**, *and* **C** *that is invariant to spatial rotation of the frame of reference. What is the geometric property in question? Use this invariance to confirm the legitimacy of treating* $\varepsilon_{\alpha\beta\gamma}$ *as a three-index tensor in spite of the fact that its elements are unchanged by rotation transformations. (There is a mathematical theorem to the effect that a quantity that yields an invariant upon contraction with a general tensor must itself be a tensor.) You could also confirm this by explicitly performing the transformation, as in Problem 9.3.6.*

Problem 9.3.9. *The formula for the energy component of the relativistic force was stated without proof in Eq. (9.56). Supply the proof. Also check Eq. (9.81).*

Problem 9.3.10. *Directly from their definitions in Eqs. (9.79), show that the electric and magnetic fields* **E** *and* **B** *are, in fact, unchanged, when* A^i *is subjected to gauge transformation (9.83).*

Bibliography

General References

1 L.D. Landau and E.M. Lifshitz, *The Classical Theory of Fields*, Pergamon, Oxford, 1971.

2 J.D. Jackson, *Classical Electrodynamics*, Wiley, New York, 1999.

10
Conservation Laws and Symmetry

Conservation laws can be derived using any of the formalisms of mechanics. The treatment here emphasizes the Poincaré approach. Arguments applicable to Lagrangian mechanics also apply automatically to Poincaré mechanics. One topic that is strikingly simplified by the Poincaré approach is that of "integrability," which is briefly discussed. Discussion of Noether's theorem, which provides the most fundamental description of conservation laws, is also discussed briefly here, even though this discussion might more appropriately be delayed until after the development of symplectic mechanics in Chapter 17. Since symmetry groups provided the original motivation for the Poincaré equation and, since the mathematical description of symmetry is based on groups, it is natural to use the Poincaré equation while investigating the effect of symmetry on mechanical systems.

The latter half of the chapter is devoted to conservation laws in classical field theory. Since field theory has only been introduced in a preliminary way in Chapter 1, and will only be applied seriously in later chapters, it may be appropriate to skim this material for now, intending to return to it more seriously while reading later chapters on Electromagnetic Theory, String Theory, and General Relativity. The treatment of fields in this chapter is intended to emphasize the great generality of total field energy and momentum and their derivation from the energy–momentum tensor T^{ij}. For readers primarily interested in field theory it is the first half of the chapter that can be safely skipped.

10.1
Conservation of Linear Momentum

The kinetic energy of a system of N particles is

$$T = \frac{1}{2}\sum_{i=1}^{N} m_{(i)}(\dot{x}_{(i)}^2 + \dot{y}_{(i)}^2 + \dot{z}_{(i)}^2) \equiv \frac{1}{2}\sum m(\dot{x}^2 + \dot{y}^2 + \dot{z}^2). \tag{10.1}$$

Here, and in future, replacements like $\sum_{i=1}^{N} x_{(i)} \to \sum x$ will be made to reduce clutter. The presence of the summation sign is the only reminder that there

Geometric Mechanics: Toward a Unification of Classical Physics. 2nd Edition. Richard Talman

ISBN: 978-3-527-40683-8

is one term for each of the N particles. An essential simplifying feature of these rectangular coordinates, but which is not valid in general, is that the coefficients of the quadratic velocity terms are independent of position.

The particle-(i)-specific infinitesimal displacement generators are $\boldsymbol{\partial}/\boldsymbol{\partial} x_{(i)}$, $\boldsymbol{\partial}/\boldsymbol{\partial} y_{(i)}$, and $\boldsymbol{\partial}/\boldsymbol{\partial} z_{(i)}$. The Poincaré equation for $v^x_{(i)} = \dot{x}_{(i)}$ is

$$\frac{d}{dt}\frac{\partial T}{\partial v^x_{(i)}} = m_{(i)}\dot{v}^x_{(i)} = -\frac{\partial}{\partial x_{(i)}} U(\mathbf{x}_{(1)}, \mathbf{x}_{(2)}, \dots, \mathbf{x}_{(N)}), \tag{10.2}$$

not different from the Lagrange equation, or for that matter from Newton's equation. This has used the fact that $\boldsymbol{\partial} T_{(i)}/\boldsymbol{\partial}\mathbf{x}_{(i)} = 0$, which is a manifestation of the invariance of kinetic energy under pure translation. Defining total mass, centroid displacement and velocity by

$$M = \sum m, \quad M\mathbf{X} = \sum m\mathbf{x}, \quad M\mathbf{V} = \sum m\mathbf{v}, \tag{10.3}$$

and summing Eqs. (10.2) yields

$$M\dot{\mathbf{V}} = -\frac{\boldsymbol{\partial}}{\boldsymbol{\partial}\mathbf{x}} U = \mathbf{F}_{\text{tot}}, \tag{10.4}$$

where the operator

$$\sum_{i=1}^{N} \frac{\boldsymbol{\partial}}{\boldsymbol{\partial}\mathbf{x}_{(i)}} \equiv \frac{\vec{\boldsymbol{\partial}}}{\boldsymbol{\partial}\mathbf{x}} \tag{10.5}$$

is the infinitesimal (vector) generator that translates all particles equally. The somewhat *ad hoc* inclusion of an overhead arrow is to indicate that there is one operator for each component. Operating on $-U$, the vector of operators $\vec{\boldsymbol{\partial}}/\boldsymbol{\partial}\mathbf{x}$ yields the components of the total force $\mathbf{F}_{\text{tot}}$.

Suppose translation of the mechanical system parallel to the (x, y) plane generates a "congruent" system. This would be true for a displacement parallel to the earth's surface (when treating the earth as flat) with vertical gravitational acceleration g. In this case

$$\frac{\boldsymbol{\partial}}{\boldsymbol{\partial} x} U = \frac{\boldsymbol{\partial}}{\boldsymbol{\partial} y} U = 0, \tag{10.6}$$

with the result that MV^x and MV^y are "constants of the motion." But MV^z is not constant in this case.

The case of linear momentum conservation just treated has been anomalously simple because of the simple relation between velocity and momentum. A more general formulation would have been to define the linear momentum vector $\mathbf{P}$ by

$$\mathbf{P}_{(i)} = \frac{\partial T}{\partial \dot{\mathbf{x}}_{(i)}}, \quad \text{and} \quad \mathbf{P} = \sum \mathbf{P}_{(i)}, \tag{10.7}$$

which would have led to equivalent results.

10.2
Rate of Change of Angular Momentum: Poincaré Approach

For a rigid body rotating with one point fixed, the kinetic energy summation of Eq. (10.1) can be re-expressed in terms of particle-specific angular velocities about the fixed point, $(\omega^x_{(i)}, \omega^y_{(i)}, \omega^z_{(i)})$,

$$T_{\text{rot}} = \frac{1}{2}\sum m\big((y^2+z^2)\omega^{x2} + (x^2+z^2)\omega^{y2} + (x^2+y^2)\omega^{z2} - 2yz\omega^y\omega^z - 2xz\omega^x\omega^z - 2xy\omega^x\omega^y\big). \tag{10.8}$$

This formula is valid even though quasivelocities $(\omega^x_{(i)}, \omega^y_{(i)}, \omega^z_{(i)})$ are not valid Lagrangian velocities. Define particle-specific operators

$$\mathcal{R}_{(i)x} = y_{(i)}\frac{\partial}{\partial z_{(i)}} - z_{(i)}\frac{\partial}{\partial y_{(i)}},$$
$$\mathcal{R}_{(i)y} = z_{(i)}\frac{\partial}{\partial x_{(i)}} - x_{(i)}\frac{\partial}{\partial z_{(i)}}, \qquad \mathcal{R}_{(i)z} = x_{(i)}\frac{\partial}{\partial y_{(i)}} - y_{(i)}\frac{\partial}{\partial x_{(i)}}. \tag{10.9}$$

We have seen previously that these are equivalent to

$$\mathcal{R}_{(i)x} = \frac{\partial}{\partial \phi^x_{(i)}}, \qquad \mathcal{R}_{(i)y} = \frac{\partial}{\partial \phi^y_{(i)}}, \qquad \mathcal{R}_{(i)z} = \frac{\partial}{\partial \phi^z_{(i)}}, \tag{10.10}$$

where $\boldsymbol{\phi} = (\phi^x, \phi^y, \phi^z)$ is a "vector" of quasiangles. Since appearance of these angles would not be valid in Lagrangian mechanics, the following development would not be valid there (though the same results can be obtained from Newton's equations.) It does represent a considerable emancipation to be able to work, guilt free, with angular velocity components. We also define

$$\mathcal{R}_x = \frac{\partial}{\partial \phi^x} = \sum \frac{\partial}{\partial \phi^x_{(i)}},$$
$$\mathcal{R}_y = \frac{\partial}{\partial \phi^y} = \sum \frac{\partial}{\partial \phi^y_{(i)}}, \qquad \mathcal{R}_z = \frac{\partial}{\partial \phi^z} = \sum \frac{\partial}{\partial \phi^z_{(i)}}. \tag{10.11}$$

The Poincaré equations using these variables are like Eqs. (10.2) but with linear velocities replaced by ω-velocities and Cartesian derivative operators replaced by $\mathcal{R}$ operators. The first term needed for substitution in the equation for $\omega^x_{(i)}$ is

$$\frac{1}{m}\mathcal{R}_x T_{\text{rot}} = yz\omega^{y2} - yz\omega^{z2} + (-y^2+z^2)\omega^y\omega^z - xy\omega^x\omega^z + xz\omega^x\omega^y. \tag{10.12}$$

Also, using structure constants from Eq. (5.154),

$$\frac{1}{m}(-\omega^z)\frac{\partial T_{\text{rot}}}{\partial \omega^y} = -(x^2+z^2)\omega^y\omega^z + yz\omega^{z2} + xy\omega^x\omega^z,$$
$$\frac{1}{m}(\omega^y)\frac{\partial T_{\text{rot}}}{\partial \omega^z} = (x^2+y^2)\omega^y\omega^z - yz\omega^{y2} - xz\omega^x\omega^y. \tag{10.13}$$

On substituting into the Poincaré equation, all these terms cancel. Defining

$$L_x = \sum_{i=1}^{N} \frac{\partial T_{\text{rot}}}{\partial \omega_{(i)}^x}, \quad L_y = \sum_{i=1}^{N} \frac{\partial T_{\text{rot}}}{\partial \omega_{(i)}^y}, \quad L_z = \sum_{i=1}^{N} \frac{\partial T_{\text{rot}}}{\partial \omega_{(i)}^z}, \tag{10.14}$$

and realizing that the vanishing of $\mathcal{R}_x T_{\text{rot}}$ follows from the invariance of T_{rot} under rotation, the Poincaré equation for L_x is

$$\frac{d}{dt} L_x = -\mathcal{R}_x U = -\frac{\partial U}{\partial \phi^x} \tag{10.15}$$

with similar equations for L_y and L_z. These are the three components of angular momentum vector $\mathbf{L}$ and

$$\dot{\mathbf{L}} = -\frac{\vec{\partial}}{\partial \phi} U \equiv \mathbf{K}, \tag{10.16}$$

where $\mathbf{K}$ is the applied torque.

If the potential U is "azimuthally" symmetric around some axis, say the z-axis, then $\partial U / \partial \phi_z = 0$ and it follows that the component of $\mathbf{L}$ along that axis is conserved. If U is independent of direction ("isotropic") then all components of the angular momentum vector $\mathbf{L}$ are conserved.

Problem 10.2.1. *In the derivation of angular momentum conservation just completed no account was taken of "internal" forces of one mass within the rotating object acting on another. Show that including such forces does not alter the result.*

10.3 Conservation of Angular Momentum: Lagrangian Approach

Proof of the conservation of angular momentum in the absence of external forces is also easy using ordinary Lagrangian methods. Under an infinitesimal rotation $\boldsymbol{\Delta\phi}$ each particle radius vector is shifted $\mathbf{r} \to \mathbf{r} + \boldsymbol{\Delta\phi} \times \mathbf{r}$ and its velocity is shifted similarly $\mathbf{v} \to \mathbf{r} + \boldsymbol{\Delta\phi} \times \mathbf{v}$. With no external forces the Lagrangian (equal to the kinetic energy) is unchanged by the rotation;

$$0 = \sum \Big(\frac{\partial T}{\partial \mathbf{r}} \cdot (\boldsymbol{\Delta\phi} \times \mathbf{r}) + \frac{\partial T}{\partial \mathbf{v}} \cdot (\boldsymbol{\Delta\phi} \times \mathbf{v}) \Big) = \boldsymbol{\Delta\phi} \cdot \sum (\mathbf{r} \times \dot{\mathbf{p}} + \mathbf{v} \times \mathbf{p}). \tag{10.17}$$

In the last step both the defining relation for $\mathbf{p}$ and the Lagrange equation for $\dot{\mathbf{p}}$ have been used. Defining angular momentum

$$\mathbf{L} = \sum \mathbf{r} \times \mathbf{p}, \tag{10.18}$$

it follows from the vanishing of (10.17) that

$$\frac{d\mathbf{L}}{dt} = 0. \tag{10.19}$$

Problem 10.3.1. *The particles making up an electro-mechanical system under study have arbitrary masses and charges. This system is subjected to electric or magnetic fields caused by the various electrical configurations listed below. In each case, indicate what total linear and angular momentum components of* **P** *and* **L** *of the system under study are conserved. Forces due to the fields generated by the charges of the system under study can be ignored (because they are internal). In each case indicate which choice of axes best takes advantage of the symmetry and assume that choice has been made.*

(a) A plane is uniformly charged.
(b) An infinite circular cylinder is uniformly charged.
(c) The surface of a noncircular infinite cylinder is uniformly charged.
(d) Two parallel infinite lines have equal, uniform, charge densities.
(e) Two equal point charges.
(f) A uniformly charged infinite cone.
(g) A uniformly charged torus, circular in both cross sections.
(h) An infinite, uniformly charged, uniform pitch, solenoid. The pitch is such that the distance along its axis at which the helix has made one revolution is Δ. In this case, the only conserved momentum is a combination of two of the elementary momenta.

10.4 Conservation of Energy

Suppose that both T and U have no explicit time dependence. Multiplying Poincaré equation (5.147) by ω^ρ, summing over ρ, and utilizing the antisymmetry of $c^\lambda_{\ \mu\rho}$ to justify setting $\omega^\rho c^\lambda_{\ \mu\rho}\omega^\mu = 0$, yields

$$\omega^\rho \frac{d}{dt}\frac{\partial T}{\partial \omega^\rho} = \frac{d}{dt}\left(\omega^\rho \frac{\partial T}{\partial \omega^\rho}\right) - \dot{\omega}^\rho \frac{\partial T}{\partial \omega^\rho} = \omega^\rho X_\rho (T - U). \tag{10.20}$$

The last two terms can be merged using

$$\frac{d}{dt}(T - U) = \frac{\partial T}{\partial \omega^\rho}\dot{\omega}^\rho + \omega^\rho X_\rho (T - U). \tag{10.21}$$

(This equation would acquire an extra term $(\partial/\partial t)(T - U)$ if we were not assuming this quantity vanishes. This extra term would otherwise have to be included in the following equations.) We therefore obtain

$$\frac{d}{dt}(T - U) = \frac{d}{dt}\left(\omega^\rho \frac{\partial T}{\partial \omega^\rho}\right). \tag{10.22}$$

Defining a new function

$$h(\boldsymbol{\omega}, \mathbf{q}) = \omega^\rho \frac{\partial T}{\partial \omega^\rho} - T + U, \tag{10.23}$$

and integrating Eq. (10.22) yields

$$h(\boldsymbol{\omega}, \mathbf{q}) = \text{constant} = E_0. \tag{10.24}$$

This formula says that the function h remains equal to its initial value E_0. In our present approach the function h, obviously related to the "Hamiltonian," makes its appearance here for the first time in the Poincaré approach. It is a dynamical variable whose numerical value remains constant, equal to initial energy E_0. Technically the function h, though equal in value to the energy, cannot legitimately be called the Hamiltonian however, since its functional dependency is incorrect.

Problem 10.4.1. *Starting from the Euler equations of a freely moving rigid body with one point fixed,*

$$\begin{aligned} &I_1\dot{\omega}^1 = (I_2 - I_3)\omega^2\omega^3, \\ &I_2\dot{\omega}^2 = (I_3 - I_1)\omega^3\omega^1, \qquad I_3\dot{\omega}^3 = (I_1 - I_2)\omega^1\omega^2, \end{aligned} \tag{10.25}$$

exhibit explicitly the constancy of both energy h, and the total angular momentum squared $L^2 = L^{x2} + L^{y2} + L^{z2}$.

10.5
Cyclic Coordinates and Routhian Reduction

One of the important problems in "Dynamical System Theory" is that of "reduction," which means exploiting some symmetry of the system to reduce the dimensionality of the problem. Normally the term reduction also includes the requirement that the equations retain the same form, be it Lagrangian, Poincaré, or Hamiltonian, in the reduced number of unknowns. Even when a constant of the motion is known it is not necessarily easy to express the problem explicitly in terms of a reduced set of variables. This section considers the simplest example of reduction. This is not essentially different from a procedure due to Routh for reducing the Lagrange equations to take advantage of an *ignorable* coordinate. Within the Poincaré formalism the procedure is quite analogous. The procedure can also be regarded as a generalization of the procedure for deriving Hamilton's equations.

Suppose that one coordinate, say q^1, along with its matching velocity v^1, have the property that $(d/dt)(\partial T/\partial v^1)$ is the only nonvanishing term in their Poincaré equation. This equation can be integrated immediately;

$$\frac{\partial T}{\partial v^1} = \beta_1, \tag{10.26}$$

where β_1 is an integration constant. Two sufficient conditions for this simplification to occur are that all relevant commutators vanish, $[\mathcal{X}_1, \mathcal{X}_\mu] = 0$, and that $\mathcal{X}_1(T - V) = 0$. In this case the variable q^1 is said to be "cyclic."

Elimination of q^1 and v^1 begins by solving Eq. (10.26) for v^1;

$$v^1 = v^1(q^2, \ldots, q^n; v^2, \ldots, v^n). \tag{10.27}$$

For the method to work this has to yield an explicit formula for v^1. The "Routhian" is then defined by

$$R(q^2, \ldots, q^n; \beta_1, v^2, \ldots, v^n) = T - V - v^1 \frac{\partial T}{\partial v^1}. \tag{10.28}$$

The absence of q^1 is due to the fact that q^1 is cyclic, and the absence of v^1 is due to the fact it will have been replaced using Eq. (10.27). The Poincaré-based Routhian reduction continues by writing the Poincaré equations for the remaining variables;

$$\frac{d}{dt}\frac{\partial R}{\partial v^\rho} - \sum_{\mu=2}^{n}\sum_{\lambda=2}^{n} c^\lambda{}_{\mu\rho} v^\mu \frac{\partial R}{\partial v^\lambda} - \sum_{\mu=2}^{n} c^1{}_{\mu\rho} v^\mu \beta_1 - X_\rho R = 0, \quad \rho = 2, \ldots, n. \tag{10.29}$$

The quantity β^1 can be treated as a constant parameter as these equations are being solved. After they *have* been solved, v^1 can be found by substituting into Eq. (10.28) and then differentiating with respect to β_1;

$$v^1 = -\frac{\partial R}{\partial \beta^1}. \tag{10.30}$$

This follows from substituting Eq. (10.26) into Eq. (10.28).

Example 10.5.1. Symmetric top. *Consider the axially symmetric top shown in Fig. 10.1, rotating with its tip fixed at the origin. Let its equal moments of inertia be I_1 and its moment of inertia about the symmetry axis be I_3. Its body axis angular velocities $s^{x'}, s^{y'}, s^{z'}$ are related to the Euler angular velocities $\dot\phi$, $\dot\theta$, and $\dot\psi$ by the relations*

$$\begin{pmatrix} s^{x'} \\ s^{y'} \\ s^{z'} \end{pmatrix} = \begin{pmatrix} \sin\theta\,\sin\psi & \cos\psi & 0 \\ \sin\theta\,\cos\psi & -\sin\psi & 0 \\ \cos\theta & 0 & 1 \end{pmatrix} \begin{pmatrix} \dot\phi \\ \dot\theta \\ \dot\psi \end{pmatrix}. \tag{10.31}$$

The kinetic energy is given by

$$\begin{aligned} T &= \frac{1}{2} I_1 (s^{x'2} + s^{y'2}) + \frac{1}{2} I_3 s^{z'2} \\ &= \frac{1}{2} I_1 (\sin^2\theta\,\dot\phi^2 + \dot\theta^2) + \frac{1}{2} I_3 (\cos^2\theta\,\dot\phi^2 + 2\cos\theta\,\dot\phi\dot\psi + \dot\psi^2), \end{aligned} \tag{10.32}$$

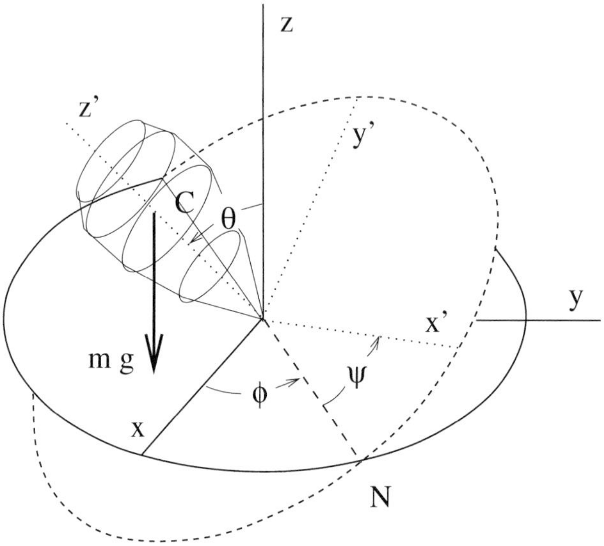

Fig. 10.1 An axial symmetric top rotating with its tip fixed. Its orientation is determined by Euler angles ϕ, θ, and ψ. The vertical force of gravity mg acts at its centroid C, at distance ℓ from the tip. "Space axes" are x, y, z. "Body axes" are x', y', z'.

and the potential energy by

$$V = mg\ell\cos\theta. \tag{10.33}$$

Neither of these depend on ψ*, which is therefore chosen as the variable* q^1*, and* $\dot\psi = v^1$ *in the Routhian reduction. (This step was made possible by a cancellation depending on the equality of* I_1 *and* I_2 *and by the absence of "commutator terms" in the Poincaré equation which, in this case is simply the Lagrange equation because legitimate generalized coordinates are being used.) The relations corresponding to Eqs. (10.25) and (10.27) are*

$$\frac{\partial T}{\partial \dot\psi} = I_3(\cos\theta\dot\phi + \dot\psi) = \beta_1, \quad or \quad \dot\psi = \frac{\beta_1}{I_3} - \cos\theta\dot\phi. \tag{10.34}$$

The Routhian reduces to

$$\begin{aligned} R &= T - V - \dot\psi\frac{\partial T}{\partial\dot\psi} \\ &= \frac{1}{2}I_1(\sin^2\theta\,\dot\phi^2 + \dot\theta^2) + \beta_1\cos\theta\dot\phi - mg\ell\cos\theta - \frac{\beta^{1\,2}}{2I_3}. \end{aligned} \tag{10.35}$$

The final term being constant, it drops out of the "Lagrange" equations obtained treating R as a Lagrangian.

Problem 10.5.1. *After the Routhian reduction just performed for the symmetric top, the Routhian R is independent of the coordinate ϕ. Perform another step of Routhian reduction to obtain a one-dimensional equation in Lagrangian form. Any such equation can be "reduced to quadratures."*

Problem 10.5.2. *For coordinates other than rectangular the kinetic energy acquires a general form; $T = \frac{1}{2} A_{rs}(\mathbf{q})\,\dot{q}^r \dot{q}^s$, and the potential energy is $V = V(\mathbf{q})$. Defining matrix $B = A^{-1}$, find the Hamiltonian, write Hamilton's equations and show that the (conserved) value of H is the total energy $E = T + V$. Only if matrix A_{rs} is diagonal are the momentum components proportional to the velocity components.*

Problem 10.5.3. *Recall Problem 7.1.1 which described rays approximately parallel to the z-axis, with the index of refraction given by $n = n_0(1 + B(x^2 + y^2))$. Generalizing this a bit, allow the index to have the form $n(\rho)$, where $\rho = \sqrt{x^2 + y^2}$. Using (ρ, ϕ) coordinates, where ϕ is an azimuthal angle around the z-axis, write the Lagrangian $L(\rho, \rho', \phi, \phi', z)$ appropriate for use in Eq. (1.28). (As in that equation primes stand for d/dz.) Find momenta $p_\rho = \partial L/\partial \rho'$ and $p_\phi = \partial L/\partial \phi'$, and find the functions f^i defined in Eq. (1.12). Find an ignorable coordinate and give the corresponding conserved momentum. Write the Hamiltonian H according to Eq. (1.13). Why is H conserved? Take $H = E$. Solve this for $\dot{\rho}$ and eliminate $\dot{\phi}$ using the conserved momentum found earlier. In this way the problem has been "reduced to quadratures." Write the integral that this implies.*

10.5.1
Integrability; Generalization of Cyclic Variables

The Routhian reduction just studied was made possible partially by the absence of "commutator terms" in the Poincaré equation. But reduction may be possible even if some of these terms are nonvanishing. It is possible for all terms in the Poincaré equation except $(d/dt)(\partial T/\partial v^1)$ to cancel even when some of the $c^\lambda_{\ \mu 1}$ coefficients are nonvanishing. With T given by

$$T = \frac{1}{2} I_{ij} v^i v^j, \tag{10.36}$$

and assuming $\mathcal{X}_1(T - V) = 0$, the unwanted terms are

$$c^\lambda_{\ \mu 1} v^\mu \frac{\partial T}{\partial v^\lambda} = (c^\lambda_{\ \mu 1} I_{i\lambda})(v^i v^\mu), \tag{10.37}$$

which vanishes if

$$c^\lambda_{\ \mu 1} I_{i\lambda} = 0. \tag{10.38}$$

Example 10.5.2. *Consider the same axially symmetric top, subject to gravity, spinning with its lower point fixed. With body axes, $T = (1/2)(I_1 \omega^{1^2} + I_2 \omega^{2^2} + I_3 \omega^{3^2})$.*

If $I_1 = I_2$ the Poincaré equation for $\phi^3 \equiv \psi$, which is a quasiangle of rotation around the instantaneous axis of the top, is

$$I_3 \dot{\omega}^3 = -\mathcal{X}_3 U = 0. \tag{10.39}$$

Since rotation around the top axis does not change the potential energy, $\mathcal{X}_3 U = 0$, and as a result ω^3 is a constant of motion.

Note that this cancellation is not "generic" in that $I_1 = I_2$ cannot be *exactly* true in practice. On the other hand, in the absence of this cancellation the equation of motion is nonlinear, which almost always leads to bizarre behavior at large amplitudes. Hence, one can say that bizarre behavior is a *generic* property of the top.

There are a small number of other known choices of the parameters in rigid body motion with one point fixed that lead to completely integrable systems. See Arnold *et al.*, *Mathematical Aspects of Classical and Celestial Mechanics*, p. 120.

10.6
Noether's Theorem

In this section (and only this section) we use notation $\partial/\partial t$ instead of d/dt for the "total time derivative." The reason for this is that a new subsidiary variable s will be introduced and the main arguments have to do with functions $f(t,s)$ and derivatives holding either t or s constant.

For much of Lagrangian mechanics the Lagrangian can be regarded as a purely formal construct whose only role is to be differentiated to yield the Lagrange equations. For this purpose it is not necessary to have more than an operational understanding of the meaning of the dependence of the Lagrangian on $\dot{\mathbf{q}}$. But here we insist on treating $L(\mathbf{q}, \dot{\mathbf{q}})$ as a regular scalar function on the configuration space M of coordinates $\mathbf{q}$ augmented by the tangent spaces of velocities. We must define carefully the meaning of the dependence on $\dot{\mathbf{q}}$. Based on curves $\mathbf{q}(t)$ in configuration space, parameterized by time t, one constructs the so-called tangent spaces $\mathrm{TM}_{\mathbf{q}}$ at every point $\mathbf{q}$ in the configuration space. $\mathrm{TM}_{\mathbf{q}}$ is the space of possible instantaneous system velocities at $\mathbf{q}$. This space has the same dimensionality n as does the configuration space M itself. The union of these spaces at every point is known as the "tangent bundle" TM. It has dimensionality $2n$ with a possible choice of coordinates being $q^1, q^2, \dots, q^n$ and the remaining n coordinates being the corresponding "natural" velocity components introduced in Eq. (3.9). The Lagrangian $L(\mathbf{q}, \dot{\mathbf{q}})$ is a scalar function on TM.

We have seen many examples in which symmetries of the Lagrangian are reflected in conserved dynamical variables. The simplest of these involve "ignorable coordinate." For example, the absence of z in the Lagrangian for particles in Euclidean space subject to potential V, depending on the x_i and y_i components but not on the z_i components,

$$L = \frac{1}{2} \sum_i m_i (\dot{x}_i^2 + \dot{y}_i^2 + \dot{z}_i^2) - V, \tag{10.40}$$

implies the conservation of total momentum p_z; this follows trivially within the customary Lagrangian formalism. The *ad hoc* nature of such conclusions should seem mildly troubling and one wishes to be able to have a more general theoretical construct from which all such conserved quantities, or at least broad classes of them, can be derived. Such a construct would exhibit simple invariance properties (symmetries) of the Lagrangian and associate a definite conserved dynamical variable with each such symmetry. This is what Noether's theorem accomplishes.

Since the Lagrangian is a scalar function, constancy under particular transformation of its arguments is the only sort of symmetry to which it can be subject. For example, the Lagrangian of Eq. (10.40) is invariant under the transformation

$$x \to x, \quad y \to y, \quad z \to z + s, \tag{10.41}$$

where s is an arbitrary parameter, provided this configuration space transformation is accompanied by the following tangent space transformation:

$$\dot{x} \to \dot{x}, \quad \dot{y} \to \dot{y}, \quad \dot{z} \to \dot{z}. \tag{10.42}$$

From a physical point of view the fact that these configuration space and tangent space transformations have to go together in this way is an obvious requirement on descriptions of the same physics in two frames that differ only by constant displacement s along the z-axis.

From a mathematical point of view, a smooth transformation $\mathbf{f} : \mathrm{M} \to \mathrm{M}$ mapping $\mathbf{q}$ to $\mathbf{f}(\mathbf{q})$ implies a transformation $\mathbf{f}_{*\mathbf{q}} : \mathrm{TM}_{\mathbf{q}} \to \mathrm{TM}_{\mathbf{f}(\mathbf{q})}$ from the tangent space at $\mathbf{q}$ to the tangent space at $\mathbf{f}(\mathbf{q})$. For any particular curve through $\mathbf{q}$ this maps its instantaneous velocity $\dot{\mathbf{q}}$ inferred at $\mathbf{q}$ into $\mathbf{f}_{*\mathbf{q}}(\dot{\mathbf{q}})$ which is equal to the instantaneous velocity inferred at $\mathbf{f}(\mathbf{q})$. "Infer" here means "go through the usual limiting procedure by which instantaneous velocity is extracted from a system trajectory." $\mathbf{f}_{*\mathbf{q}}$ is an n-dimensional transformation. Geometrically the output of this transformation is a vector which is the reason a bold face symbol has been used for $\mathbf{f}$. Combining the transformations $\mathbf{f}_{*\mathbf{q}}$ at all points $\mathbf{q}$ we obtain an n-dimensional transformation $\mathbf{f}_*$ at every point in an n-dimensional domain; it maps all tangent spaces.

The coordinate map can depend on a parameter s and therefore be symbolized by $\mathbf{f}^s$ and the corresponding map of velocities is symbolized by $\mathbf{f}^s_*$. Consider a valid system trajectory $\mathbf{q}(t)$. At every time t the point $\mathbf{q}(t)$ can be

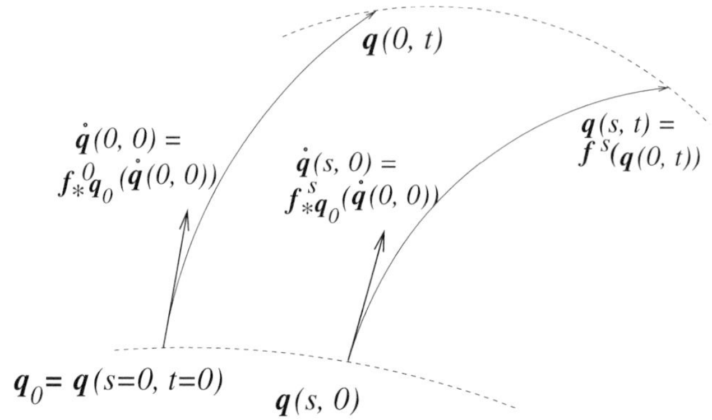

Fig. 10.2 A family of curves, each parameterized by time t, with the different curves differentiated by s. The curve with $s = 0$ is a valid system trajectory, but the curve with $s \neq 0$ need not be.

mapped by $\mathbf{f}^s$ to yield a point $\mathbf{q}(s,t)$. This yields a family of curves that are individually parameterized by t with different curves distinguished by parameter s (as illustrated in Fig. 10.2). By definition the $s = 0$ curve satisfies the equations of motion but the curves for other values of s will not be valid system trajectories except under conditions to be considered next.

The Lagrangian system is said to be invariant under a mapping $\mathbf{f}$ if

$$L(\mathbf{f}(\mathbf{q}), \mathbf{f}_{*\mathbf{q}}(\dot{\mathbf{q}})) = L(\mathbf{q}, \dot{\mathbf{q}}). \tag{10.43}$$

Theorem 10.6.1. Noether's theorem. *If a Lagrangian system is invariant for the transformations* $\mathbf{f}^s$ *along with the induced transformations* $\mathbf{f}^s_*$ *for all values of the parameter* s *then the quantity*

$$I(\mathbf{q}, \dot{\mathbf{q}}) = \left\langle \frac{\partial L}{\partial \dot{\mathbf{q}}}, \frac{\partial \mathbf{f}^s(\mathbf{q})}{\partial s}\bigg|_{s=0} \right\rangle = \sum_{i=1}^{n} \frac{\partial L}{\partial \dot{q}^i} \frac{\partial \mathbf{f}^{s\,i}(\mathbf{q})}{\partial s}\bigg|_{s=0} \tag{10.44}$$

is a constant of the motion.

The middle term in Eq. (10.44) uses the notation first introduced in Eq. (2.3). Since $\partial L/\partial \dot{\mathbf{q}}$ is covariant and $\partial \mathbf{f}^s(\mathbf{q})/\partial s|_{s=0}$ is contravariant the quantity $I(\mathbf{q}, \dot{\mathbf{q}})$ is invariantly defined, independent of the choice of coordinates. But its practical evaluation requires the use of coordinates as spelled out in the rightmost term of Eq. (10.44). The former notation will be used in the proof that follows.

Proof. The analytic statement of the invariance of L under $\mathbf{f}^s$ is

$$\frac{\partial L(\mathbf{q}, \dot{\mathbf{q}})}{\partial s} = \left\langle \frac{\partial L}{\partial \mathbf{q}}, \frac{\partial \mathbf{q}}{\partial s}\bigg|_{s=0} \right\rangle + \left\langle \frac{\partial L}{\partial \dot{\mathbf{q}}}, \frac{\partial \dot{\mathbf{q}}}{\partial s}\bigg|_{s=0} \right\rangle = 0. \tag{10.45}$$

When this equation is expressed in terms of the function $\mathbf{f}^s$ it becomes

$$\left\langle \frac{\partial L}{\partial \mathbf{q}}, \frac{\partial \mathbf{f}^s(\mathbf{q})}{\partial s}\bigg|_{s=0} \right\rangle + \left\langle \frac{\partial L}{\partial \dot{\mathbf{q}}}, \frac{\partial(\partial \mathbf{f}^s(\mathbf{q})/\partial t)}{\partial s}\bigg|_{s=0} \right\rangle = 0. \tag{10.46}$$

With the assumed invariance the same variational calculation showing that $\mathbf{q}(t)$ satisfies the equations of motion shows that $\mathbf{q}(t,s)$ satisfies the equations of motion also, for all s. As a result, the Lagrange equations are satisfied as identities in s;

$$\frac{\partial}{\partial t}\frac{\partial L}{\partial \dot{q}^i}(\mathbf{q}(s,t), \dot{\mathbf{q}}(s,t)) = \frac{\partial L}{\partial q^i}(\mathbf{q}(s,t), \dot{\mathbf{q}}(s,t)). \tag{10.47}$$

Proceeding to calculate $\partial I/\partial t$ (while recalling that the notation $\partial/\partial t$ is a total time derivative with s held constant) and using Eq. (10.47) we obtain

$$\begin{aligned}\frac{\partial I}{\partial t} &= \frac{\partial}{\partial t}\left\langle \frac{\partial L}{\partial \dot{\mathbf{q}}}, \frac{\partial \mathbf{f}^s(\mathbf{q})}{\partial s}\bigg|_{s=0} \right\rangle \\ &= \left\langle \frac{\partial L}{\partial \mathbf{q}}, \frac{\partial \mathbf{f}^s(\mathbf{q})}{\partial s}\bigg|_{s=0} \right\rangle + \left\langle \frac{\partial L}{\partial \dot{\mathbf{q}}}, \frac{\partial}{\partial t}\frac{\partial \mathbf{f}^s(\mathbf{q})}{\partial s}\bigg|_{s=0} \right\rangle = 0.\end{aligned} \tag{10.48}$$

In the final step the order of s and t derivatives has been reversed and Eq. (10.46) used. This completes the proof. □

Example 10.6.1. *For the transformation given by Eqs. (10.41), and (10.42) we have*

$$\frac{\partial \mathbf{f}^s}{\partial s}\bigg|_{s=0} = \mathbf{e}_z, \tag{10.49}$$

where $\mathbf{e}_z$ is a unit vector along the z-axis. With the Lagrangian of (10.40) we obtain

$$I = \sum_i m_i(\dot{x}_i\mathbf{e}_x + \dot{y}_i\mathbf{e}_y + \dot{z}_i\mathbf{e}_z)\cdot \mathbf{e}_z = \sum_i m_i\dot{z}_i = p_z. \tag{10.50}$$

The assumed Euclidean geometry and the orthonormal property of the coordinates have been used here (for the first time) in evaluating the invariant product as an ordinary dot product of two vectors.

Example 10.6.2. *Consider next rotation about a particular axis, say the $\hat{\boldsymbol{\omega}}$-axis, through angle s. According to Eq. (4.36), keeping only the term linear in s, we have*

$$\mathbf{x}_i \rightarrow \mathbf{x}_i + s\,\hat{\boldsymbol{\omega}} \times \mathbf{x}_i + \cdots, \tag{10.51}$$

and therefore that

$$\frac{\partial \mathbf{f}^s}{\partial s}\bigg|_{s=0} = \hat{\boldsymbol{\omega}} \times \mathbf{x}_i. \tag{10.52}$$

If the Lagrangian is invariant under transformation (10.51) (as (10.40) would be if V is invariant to rotation around $\hat{\boldsymbol{\omega}}$*) then the Noether invariant is given by*

$$I = \left\langle \frac{\partial L}{\partial \dot{\mathbf{q}}}, \frac{\partial \mathbf{f}^s(\mathbf{q})}{\partial s}\bigg|_{s=0} \right\rangle = \sum_i m_i \dot{\mathbf{x}} \cdot (\hat{\boldsymbol{\omega}} \times \mathbf{x}_i) = \hat{\boldsymbol{\omega}} \cdot \sum_i \mathbf{x}_i \times (m_i \dot{\mathbf{x}}). \tag{10.53}$$

This shows that the component of angular momentum about the $\hat{\boldsymbol{\omega}}$*-axis is conserved.*

Problem 10.6.1. *Formulate each of the invariance examples of Problem 10.3.1 as an example of Noether's theorem and express the implied conserved quantities in the form of Eq. (10.44).*

10.7 Conservation Laws in Field Theory

Formulas in the remainder of this chapter are intended to serve two purposes. *First* they are intended to describe conservation laws satisfied by general field theories. But, as these laws are spelled out in examples, they are intended to, *second*, be directly applicable to electromagnetic theory. Results that seem obscure or ill-motivated may become clearer while applying these results to electromagnetic theory in Chapter 11. A simpler interpretation of the formulas is obtained by applying them to the field theory of a nonrelativistic stretched string as discussed in Section 1.8. The treatment here is explicitly relativistic, however. It is later adapted to string theory and to general relativity.

10.7.1 Ignorable Coordinates and the Energy Momentum Tensor

A general field theory will have one or more "potential functions," such as $A^{(\alpha)}(\underline{x})$ which depend on position and time. For example, there might be four, $A^{(0)}$, $A^{(1)}$, $A^{(2)}$, $A^{(3)}$, which may or may not be the components of a 4-vector $A^i(\underline{x})$. The purpose for the parentheses on the upper indices is to "protect" them from the summation convention. In other words, summation over the index α can only be expressed using the explicit summation symbol $\sum_\alpha$. These functions $A^{(\alpha)}(\underline{x})$ will serve as the analogs of generalized coordinates (conventionally symbolized by $q_{(0)}, q_{(1)}, \dots$, in regular mechanics) as the Lagrangian formulation is being extended to field theory.

The system under study is assumed to satisfy Hamilton's principle, with the Lagrangian density being given by

$$\mathcal{L}(A^{(\alpha)}, A^{(\alpha)}_{,0}, A^{(\alpha)}_{,1}, A^{(\alpha)}_{,2}, A^{(\alpha)}_{,3}), \qquad \alpha = 1, 2, 3, 4. \tag{10.54}$$

Here the "comma, index" notation is being used to specify partial derivatives;

$$A^{(\alpha)}_{,i} \equiv \frac{\partial A^{(\alpha)}}{\partial x^i}. \tag{10.55}$$

As listed in Eq. (10.54), $\mathcal{L}$ has four coordinate arguments, analogs of q, and 16 "first derivative" arguments, analogs of $\dot{q}$. What is more important, in the present context, is that $\mathcal{L}$, *does not* depend on the world coordinates $x^i = (ct, x, y, z)$. In mechanics such coordinates are said to be "ignorable." Such independence of origin is applicable when the system is invariant to translations in space or time. Most commonly this invariance is due to a system being free in space, not subject to external forces.

Reviewing point particle mechanics, the *absence* of dependence on a coordinate is always accompanied by the *presence* of a conserved quantity. The most fundamental quantity of this sort is the total energy E of the system; conservation of E corresponds to the time t being ignorable. Formally, this invariance is guaranteed by the *absence* of t from the argument list of $L = L(q, \dot{q})$. The procedure for determining a conserved quantity in that case started by defining the Hamiltonian $H(q, p)$. Then, as shown in Chapter 1, starting with the fact that H has no explicit dependence on t when L does not, and applying the Lagrange equations, it follows that H is conserved; its value is defined to be the system energy E.

Here we wish to derive the quantities whose conservation is guaranteed by the *absence* of ct, x, y, and z from the argument list of $\mathcal{L}$. Defining 4-volume element $d\Omega = c\,dt\,dV$, and using Hamilton's principle, the evolution equations for the system are derivable from the principle of least action;

$$\delta S = \delta \int \mathcal{L}\left(A^{(0)}, A^{(0)}_{,0}, A^{(0)}_{,1}, \ldots, A^{(1)}, A^{(1)}_{,0}, \ldots, A^{(3)}_{,3}\right) \frac{d\Omega}{c} = 0, \tag{10.56}$$

with the integral running over all of space-time. The argument list consists of all the potentials and all their first derivatives, and nothing else. Performing the variation:

$$\sum_\alpha \int \left(\frac{\partial \mathcal{L}}{\partial A^{(\alpha)}} \delta A^{(\alpha)} + \frac{\partial \mathcal{L}}{\partial A^{(\alpha)}_{,i}} \delta A^{(\alpha)}_{,i} \right) \frac{d\Omega}{c} = 0. \tag{10.57}$$

As usual, we prepare for integration by parts by writing

$$\frac{\partial}{\partial x^i}\left(\frac{\partial \mathcal{L}}{\partial A^{(\alpha)}_{,i}} \delta A^{(\alpha)} \right) = \frac{\partial}{\partial x^i}\left(\frac{\partial \mathcal{L}}{\partial A^{(\alpha)}_{,i}} \right) \delta A^{(\alpha)} + \frac{\partial \mathcal{L}}{\partial A^{(\alpha)}_{,i}} \delta A^{(\alpha)}_{,i}. \tag{10.58}$$

In this equation there is summation over the i index, but *not* over the α index. Substitution from Eq. (10.58) into Eq. (10.57) yields

$$\sum_\alpha \int \left(\frac{\partial \mathcal{L}}{\partial A^{(\alpha)}} \delta A^{(\alpha)} + \frac{\partial}{\partial x^i}\left(\frac{\partial \mathcal{L}}{\partial A^{(\alpha)}_{,i}} \delta A^{(\alpha)} \right) - \frac{\partial}{\partial x^i}\left(\frac{\partial \mathcal{L}}{\partial A^{(\alpha)}_{,i}} \right) \delta A^{(\alpha)} \right) \frac{d\Omega}{c} = 0. \tag{10.59}$$

The central term here is a divergence which, using Gauss's theorem, can be expressed as a "surface"-integral at infinity. Under suitable assumptions about the behavior at infinity this term can be dropped, yielding

$$\sum_{\alpha} \int \delta A^{(\alpha)} \left(\frac{\partial \mathcal{L}}{\partial A^{(\alpha)}} - \frac{\partial}{\partial x^i} \left(\frac{\partial \mathcal{L}}{\partial A^{(\alpha)}_{,i}} \right) \right) \frac{d\Omega}{c} = 0. \tag{10.60}$$

As usual, since the factor $\delta A^{(\alpha)}$ is arbitrary, one concludes that

$$\frac{\partial}{\partial x^i} \left(\frac{\partial \mathcal{L}}{\partial A^{(\alpha)}_{,i}} \right) = \frac{\partial \mathcal{L}}{\partial A^{(\alpha)}}, \qquad \alpha = 0, 1, 2, 3. \tag{10.61}$$

These are the Lagrange equations for the system. As noted previously these are partial differential equations and $\mathcal{L}$ is a density with dimensions *energy* per *spatial volume*. Note also that there is no longer a summation so, as in ordinary mechanics, there is one equation for each generalized coordinate. It was only as an example that α ranges over four values. There can be any number of fields. But, when applied to electromagnetic theory, the $A^{(\alpha)}$ will actually be the four components of the vector potential.

There may seem to be only an imperfect analogy of this Lagrangian *field* formulation with the ordinary Lagrange *particle* formulation: the partial derivative $\partial/\partial x^i$, on the left-hand side of Eq. (10.61) in the field formulation, has to play the role of the "total" derivative d/dt (which applies to actual solution trajectories) in the particle Lagrange equations. This blemish is partially removed by noting that a partial derivative $\partial \mathcal{L}/\partial A^{(\alpha)}$ has a very different character from a partial derivative such as $\partial/\partial x^i$. In the former it is the remaining arguments of the Lagrangian that are being held constant. In the latter it is the other three of the coordinates ct, x, y, and z that are being held constant. These latter derivatives, as they appear in Eq. (10.61), are *also* "total" derivatives, which are to be evaluated along solution trajectories. But, with the solution space being multidimensional, there is no single solution trajectory. For want of a clearer notation, the partial derivative symbol is used to distinguish rates of change in the various directions in space, of the functions solving the variational problem. Some writers prefer to use d/dx^i for these derivatives.

Returning to the task at hand, we wish to establish conservation laws corresponding to the ignorable coordinates. Though the Lagrange density $\mathcal{L}$ plays primarily a formal role, when its arguments are replaced by validly evolving system coordinates, the evolution of $\mathcal{L}$ is also determined, and can be used to define a Hamiltonian-like function or functions that can potentially be conserved. Derivatives of $\mathcal{L}$ are given by

$$\begin{aligned}\mathcal{L}_{,i} &= \sum_\alpha \left(\frac{\partial \mathcal{L}}{\partial A^{(\alpha)}} A^{(\alpha)}_{,i} + \frac{\partial \mathcal{L}}{\partial A^{(\alpha)}_{,j}} A^{(\alpha)}_{,i,j} \right) \\ &= \sum_\alpha \left(\frac{\partial}{\partial x^j} \left(\frac{\partial \mathcal{L}}{\partial A^{(\alpha)}_{,j}} \right) A^{(\alpha)}_{,i} + \frac{\partial \mathcal{L}}{\partial A^{(\alpha)}_{,j}} A^{(\alpha)}_{,i,j} \right) \\ &= \sum_\alpha \frac{\partial}{\partial x^j} \left(\frac{\partial \mathcal{L}}{\partial A^{(\alpha)}_{,j}} A^{(\alpha)}_{,i} \right),\end{aligned} \tag{10.62}$$

where the substitution in the second step used Lagrange equations (10.61).

The first factor within parenthesis in the final form of Eq. (10.62), namely $\partial\mathcal{L}/\partial A^{(\alpha)}_{,j}$, can be defined to be a "generalized momentum," conjugate to "generalized coordinate" $A^{(\alpha)}_{,j}$. The structure of each term inside the same parentheses is then much like a term $p\dot{q}$ which is subtracted from the Lagrangian in forming the Hamiltonian in particle mechanics – in that case one sums over all such $p_i\dot{q}^i$ terms. For an analogous field definition one must integrate $\sum_\alpha (\partial\mathcal{L}/\partial A^{(\alpha)}_{,j})\, A^{(\alpha)}_{,i}$ over all space. Furthermore, the structure of Eq. (10.62), with each term being expressed as a 4-divergence, suggests the eventual usefulness of converting the volume integral into a surface integral.

Continuing to develop a Hamiltonian-like object in the field formalism, we need to "subtract off" the Lagrangian density from the quantity envisaged in the previous paragraph. But that is impossible, since one cannot subtract a scalar from a two-index tensor. One can, however, introduce an appropriate tensor, closely related to $\mathcal{L}$. Define a two-index tensor by

$$T^j_{\ i} = \sum_\alpha \frac{\partial \mathcal{L}}{\partial A^{(\alpha)}_{,j}} A^{(\alpha)}_{,i} - \delta^j_{\ i}\, \mathcal{L}, \tag{10.63}$$

and replace the left-hand side of Eq. (10.62) using the trivially valid relation

$$\frac{\partial \mathcal{L}}{\partial x^i} = \delta^j_{\ i} \frac{\partial \mathcal{L}}{\partial x^j}. \tag{10.64}$$

By this construction we have produced a tensor density T^{ij} satisfying

$$\frac{\partial T^{ij}}{\partial x^i} = 0, \quad j = 0, 1, 2, 3. \tag{10.65}$$

Here the order of i and j indices does not matter and it is justified to treat them both as contravariant since the vanishing of all contravariant components follows from the vanishing of all covariant components.

This construction has not determined T^{ij} uniquely. T^{ij} can be augmented by a term $\partial\psi^{ijl}/\partial x^l$, where ψ^{ijl} is a function that is arbitrary except for being antisymmetric in the j,l indices. This freedom is pursued below and in a problem.

10.8 Transition From Discrete to Continuous Representation

Before using T^{ij} to develop conservation laws it is appropriate to consider the simpler, and one hopes familiar, case of charge conservation. The discussion of field theory up to this point has intentionally been sufficiently abstract to encompass diverse areas of physics. At this point, in the interest of concreteness, the treatment will focus more narrowly on the description of charged particle motions under circumstances in which it is sensible to extend the continuum treatment, previously restricted to force fields, also to particle distributions. The description has to respect relativity.

10.8.1 The 4-Current Density and Charge Conservation

We need to express the locations and motions of discrete charged particles by the continuum charge distributions and current densities they are equivalent to, assuming that the charges on individual particles (typically electrons) are so small, and the charges so numerous, that it is appropriate to represent them by continua.

The charge e of an isolated particle is defined to be a relativistic scalar that has the same value in every coordinate frame. This is the simplest expression of the assumed *conservation of charge* principle. Taking a charge to be a point, the corresponding charge density of a particle located at $\mathbf{r}_0$ is $\rho(\mathbf{r}) = e\delta(\mathbf{r} - \mathbf{r}_0)$. Notice that $\rho(\mathbf{r})$ is *not* a relativistic scalar. On the other hand, when integrating over a charge distribution, the differential contribution ρdV, which is the quantity of charge within a defined volume element, has to be scalar to satisfy the charge conservation principle just annunciated.

Often the flow of charge consists of a charge density ρ moving with a velocity which is the spatial part of dx^i/dt, for example as illustrated in Fig. 10.3. One defines "4-current density" J^i by

$$J^i = \rho\frac{dx^i}{dt}. \tag{10.66}$$

It is necessary to confirm that J^i is, in fact, a 4-vector. Multiplying the right-hand side of this equation by the factor $dVdt$, which is known (see Fig. 10.3) to be a 4-scalar, produces the quantity $(\rho dV)\,dx^i$. This quantity is a 4-vector

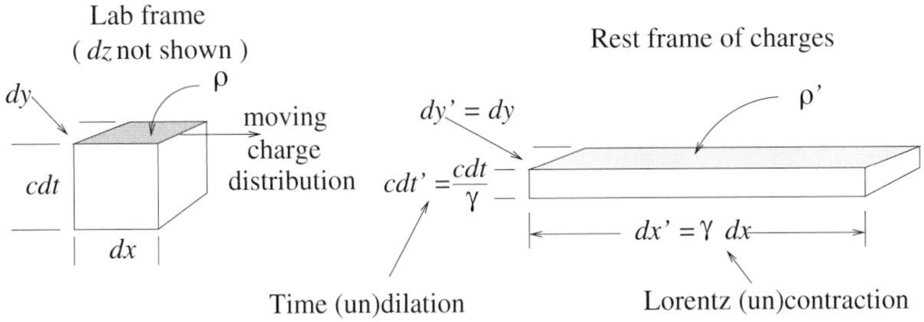

Fig. 10.3 A bunch of charge, long in its rest frame, is observed foreshortened in the laboratory. Since total charge is invariant, the spatial charge density is correspondingly altered; $\rho' = \rho/\gamma$. So charge density transforms like the time component of a 4-vector. The figure also shows that $d\Omega = cdt\,dx\,dy\,dz$ is invariant.

like dx^i because, by charge conservation, ρdV is a scalar. It follows that J^i is a 4-vector. The spatial part of J^i coincides with the conventional definition of the current density $\mathbf{J} = \rho\mathbf{v}$, and we have

$$J^i = \begin{pmatrix} c\rho \\ \mathbf{J} \end{pmatrix}. \tag{10.67}$$

One quickly verifies that the well-known *continuity equation*

$$\nabla \cdot \mathbf{J} + \frac{\partial \rho}{\partial t} = 0, \tag{10.68}$$

which expresses the conservation of charge in conventional electromagnetic theory, can be written in 4-notation using the summation convention, as

$$\frac{\partial J^i}{\partial x^i} = 0. \tag{10.69}$$

Problem 10.8.1. *Use the Lorentz transformation formulas to confirm the statement, made in connection with Eq. (10.66), that the quantity $dVdt$ transforms as a 4-scalar.*

The total system charge Q can be obtained by integrating ρ over a volume V containing all the charge; $Q = \int_V \rho dV$. In four dimensions the spatial volume V can be interpreted as a hyperplane $S^{(3)}$ at fixed time t_0 with the "surface area" element (which is actually a volume) being dV. Expressed as a covariant vector, the element orthogonal to this hyperplane has components $dS_i^{(3)} = (dV, 0, 0, 0)$. This permits Q to be expressed as

$$Q = \int_V \rho dV = \frac{1}{c}\int_{t=t_0} J^0\,dV = \frac{1}{c}\int_{S^{(3)}} J^i dS_i^{(3)}. \tag{10.70}$$

For the special $t = t_0$ surface $S^{(3)}$ just defined, Q can be interpreted as being a count of all charges whose world lines pass through $S^{(3)}$. For this surface, the final form of Eq. (10.70) is trivially valid.

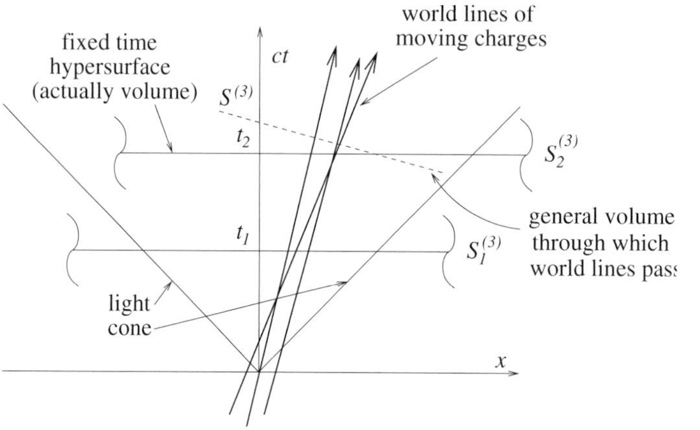

Fig. 10.4 A current consisting of three charged particles (possibly with charge extended in the y and z directions) traveling more or less parallel to the x-axis but projected onto the (ct, x) plane. The integral over $S_1^{(3)}$ or $S_2^{(3)}$ (fixed time "surfaces") $(1/c)\int_S J^i dS_i$ can be interpreted as the charge-weighted count of the charges passing through the "volumes" they represent. The same count can be performed by integrating over a surface $S^{(3)}$ that is not at fixed time.

An inference from this result can be drawn from the world diagram shown in Fig. 10.4, which shows the world lines of some charged particles traveling more or less parallel to the x-axis. The total system charge at fixed time is obtained by summing their charges. In the figure this amounts to forming the charge-weighted count of the particles intersecting surface (actually volume) $\mathcal{S}_1^{(3)}$ or $\mathcal{S}_2^{(3)}$. This integral can be generalized by permitting the three-dimensional surface $S^{(3)}$ to be finite, and not necessarily at fixed time. Still the integral can be interpreted as a count of the charges whose world lines pass through $S^{(3)}$.

This result can be obtained more formally as shown in Fig. 10.5 in which a closed 4-volume Ω is formed from two of the hypersurfaces in Fig. 10.4 enclosed by "cylindrical" sides well outside the locations of any charges. Calling the enclosing surface Σ, and applying the four-dimensional version of Gauss's law and Eq. (10.69), one obtains

$$\int_S J^i dS_i - \int_{S_1^{(3)}} J^i (dS_1)_i = \int_\Sigma J^i dS_i = \int \frac{\partial J^i}{\partial x^i}\, d\Omega = 0. \tag{10.71}$$

This confirms the conservation of Q.

It is often the case that the charge distribution consists of a localized bunch of charges, for example in an accelerator, all traveling in more or less the same direction. For some purposes it is useful to treat the collection as a single particle like bunch. The gross bunch motion can be described by an average

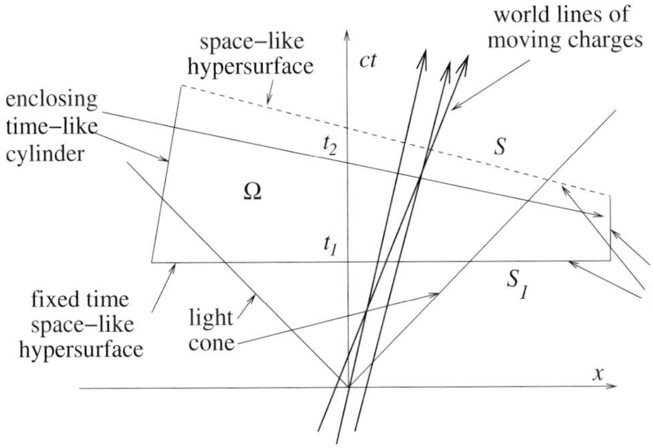

Fig. 10.5 A closed 4-volume Ω is formed by closing "sides" which are a constant-time hypersurface and a space-like hypersurface from Fig. 10.4 and by an enclosing hypercylinder formed from space-like curves. Call the enclosing surface Σ.

velocity $\mathbf{v}$ and an average 4-velocity $u^i = \langle \gamma dx^i/dt \rangle$. Multiplying Eq. (10.66) by dV and integrating over space, the time component integral is the same as in Eq. (10.70). The other integrals can be approximated by moving the 4-velocity outside the integral and replacing it by its average;

$$\int_V J^i dV \approx \begin{pmatrix} c \\ \mathbf{v} \end{pmatrix} \int_V \rho\, dV = \frac{1}{\gamma} Q u^i. \tag{10.72}$$

This shows that, for an arbitrary charge distribution in free space, and fully enclosed in volume V, the integrals $\gamma \int_V J^i dV$ need the factor γ to be the components of a 4-vector.

To further interpret the spatial components of J^i we drop back temporarily into three spatial dimensions. When expressed using vector analysis, to find the charge per second flowing through a (two-dimensional) element of area dS, one defines a vector $\mathbf{dS} = dS\,\hat{\mathbf{n}}$, where $\hat{\mathbf{n}}$ is a unit vector normal to the surface in question. Then the charge per second (i.e., current) I is given by $\mathbf{J} \cdot \mathbf{dS}$.

To stay more in the spirit of this text, it is more appropriate to express this relation using tensor analysis, in which the current density is a contravariant vector J^α. Then, to produce a scalar quantity such as charge per second, the surface element has to be represented by a covariant vector dS_α. In Cartesian all this is just pedantic, since the components are the same whether expressed as tensor or as vector components. But, to make it coordinate independent, we express the total current I through surface $S^{(2)}$ by

$$I = \int_{S^{(2)}} J^\alpha\, dS_\alpha. \tag{10.73}$$

10.8.2
Energy and Momentum Densities

Except for having a free index j, Eq. (10.65) for T^{ij} is just like Eq. (10.69) for J^i. One can therefore use it to define 16 densities much like the current densities just discussed. Consider one of the fixed time hyperplanes (actually volumes) $S_1^{(3)}$ introduced there, and illustrated in Fig. 10.4, or a general hyperplane such as in Fig. 10.5, and define

$$P^i = \frac{1}{c}\int T^{ij}\, dS_j^{(3)}. \tag{10.74}$$

As in the discussion of charge conservation, if $dS^{(3)}$ is at fixed time, we can represent the volume differential as $dS_j^{(3)} = (dV, 0, 0, 0)$, the components of a 4-vector orthogonal to $S^{(3)}$. We then obtain

$$\begin{pmatrix} E/c \\ \mathbf{P} \end{pmatrix} = P^i = \int \frac{T^{i0}}{c}\, dV. \tag{10.75}$$

Being subject to conservation laws like those provided by current density, the quantity T^{i0}/c is defined to be the 4-momentum density for the field. In the 0,0 element of T^{ij} it is the Lagrangian density itself that is being subtracted in definition (10.63). Clearly then it is the T^{00} that generalizes the Hamiltonian and can be expected to be the energy density. Then the total system energy is given by integrating T^{00} over space.

It remains to be confirmed that the conserved quantities E and $\mathbf{P}$ given by Eq. (10.75) agree with our existing concepts of energy and momentum. The mathematical description of mass distribution μ is like that of charge distribution ρ. Just as point charge e corresponds to $\rho = e\,\delta(\mathbf{r} - \mathbf{r}_0)$, for a point mass m, at position $\mathbf{r}_0$, the mass density is $\mu = m\,\delta(\mathbf{r} - \mathbf{r}_0)$. Recall also, from Eq. (10.66), the definition of the current density 4-vector as $J^i = \rho dx^i/dt = (c\rho, \mathbf{J})^T$, which is the analog of T^{i0}/c. Making these analogies, and recognizing that the relativistic energy acquires an extra factor $\gamma = (1 - v^2/c^2)^{-1/2}$ corresponding to particle velocity v, Eq. (10.72) can be recast as

$$\int_V \frac{T^{i0}}{c}\, dV \approx \gamma \begin{pmatrix} c \\ \mathbf{v} \end{pmatrix} \int_V \mu\, dV = m\, u^i. \tag{10.76}$$

From this we have $T^{i0} = c\mu u^i$ (where u^i is the proper velocity) which can be expanded to a full two-index tensor

$$T^{ij} = c\mu u^i \frac{u^j}{\gamma c} = \frac{\mu}{\gamma} u^i u^j. \tag{10.77}$$

For $j = 0$ the final factor in the middle expression is 1, which gives the correct value of T^{i0}. As given, T^{ij} is manifestly a two-index tensor and is therefore correct also for $j = 1, 2$, and 3. The mass density μ can be either a continuous distribution function or a sum of delta functions describing point particles. Clearly T^{ij} is symmetric.

Since this derivation has neglected any energy of random motion it is only valid in reference frames in which the systematic particle motion is large compared to the random motion. This is more or less equivalent to assuming the systematic velocity is large compared to typical velocities in what Schutz refers to as the MCRF (momentarily comoving reference frame). Incidentally Schutz also *defines* the mass density μ to be applicable in the MCRF. With this definition $T^{ij} = \mu\, u^i u^j$.

Identification of the conserved quantities E and $\mathbf{P}$ will be done for electrodynamics in Chapter 11 , with the fields A^i being the components of the electromagnetic vector potential.

A similar interpretation is to be applied to T^{ij} for other possible choices of the fields $A^{(0)}, A^{(1)}, \ldots$. For example, the same formulas will be applied to relativistic strings in Chapter 12.

A common modern terminology, based on its analogy with J^i in Eq. (10.69), is to refer to T^{ij} as a *conserved current*, even though *conserved current density* might seem to be more appropriate. Also, as Zwiebach points out, it is not really the T^{ij} elements that are conserved. Rather, as with Q in Eq. (10.70), it is the "charges" obtained by volume integrations at fixed time that are conserved. Zwiebach, p. 148, uses the symbol $j^{\alpha}{}_{\beta}$ for our $T^{\alpha}{}_{\beta}$. The interpretation of T^{ij} within string theory is pursued in Chapter 12.

The tensor T^{ij} is also known as the energy–momentum–stress tensor. In the presence of material media it is possible for transverse stresses to be applied from one section of medium to the adjacent section. This is especially true in elasticity theory. It is even true in Maxwell theory, once dielectric and magnetic media, with their $\mathbf{D}$ and $\mathbf{H}$ fields, have been allowed into the theory. Any useful theory of media or of cosmology must, minimally, include, either theoretically or phenomenologically, a realistic version of T^{ij}.

This text, restricted as it is to classical physics, makes no pretense of understanding or describing such things. From our pristine point of view such extensions of the theory are on the other side of the interface between classical and quantum physics. In this view T^{ij} does not deserve to have "stress" in its name. The only material version of T^{ij} is given by Eq. (10.77) and the only electromagnetic version by Eq. (11.43). In general relativity these two contributions to T^{ij} will be summed.

In fact there are other conservation laws involving the $j \neq 0$ elements of T^{ij}. Of these we will consider only the conservation of angular momentum, since the others would carry us too deeply into the areas of physics just dismissed.

10.9
Angular Momentum of a System of Particles

Consider a collection of particles, having positions $x^i_{(\alpha)}$ and momenta $p^i_{(\alpha)}$. The particles may interact with each other, but they are otherwise in free space. In 3D notation the angular momentum of this system is given by the "moment of momentum,"

$$\mathbf{M} = \sum_\alpha \mathbf{x}_{(\alpha)} \times \mathbf{p}_{(\alpha)} = \sum_\alpha \begin{pmatrix} \hat{\mathbf{x}} & \hat{\mathbf{y}} & \hat{\mathbf{z}} \\ x & y & z \\ p_x & p_y & p_z \end{pmatrix}. \tag{10.78}$$

Using 4D notation, a spatial rotation of the entire system, around some axis, at fixed time, can be described by shifting the particle positions according to

$$x'^i = x^i + \delta x^i = x^i + \delta\Omega^{ij}\, x_j. \tag{10.79}$$

For example, for infinitesimal rotation around one axis by angle $\delta\theta$, the relevant elements of $\delta\Omega^{ij}$ are arrayed as $\begin{pmatrix} 0 & \delta\theta \\ -\delta\theta & 0 \end{pmatrix}$. Being a purely spatial rotation, this transformation preserves the time component of x^i and the particle invariant measures; $x'_i x'^i = x_i x^i$. Dropping higher differentials, and reversing up–down locations of contracted variables, this imposes the requirement

$$x^i x^j \delta\Omega_{ij} = 0. \tag{10.80}$$

Since the first factor is unchanged by interchanging i and j for arbitrary x^i, the factor $\delta\Omega_{ij}$ has to be an antisymmetric 4-tensor. The rotation angle corresponding to $\delta\Omega_{ij}$ can be taken to be a generalized coordinate and, because the system is free, this coordinate is *ignorable.*

In Eq. (9.63) the variation of action δS for evolution along valid system trajectories is given as

$$\delta S = -\sum_\alpha p_{(\alpha)i} \delta x^i_{(\alpha)}. \tag{10.81}$$

This formula, which is one way of writing the Hamilton–Jacobi equation, gives the particle momenta to be $p_{(\alpha)i} = -\partial S/\partial(\delta x^i_{(\alpha)})$. In the special case of displacement along a single axis, $\delta x^i_{(\alpha)} = \delta x$, the factor multiplying δx in this formula can be recognized to be the total momentum along that axis; $P_i = \sum_\alpha p_{(\alpha)i}$. For the displacements $\delta x^i_{(\alpha)}$ given by Eq. (10.79), the variation is

$$\delta S = -\delta\Omega_{ij} \sum_\alpha p^i_{(\alpha)} x^j_{(\alpha)} = -\delta\Omega_{ij} \frac{1}{2} \sum_\alpha (p^i_{(\alpha)} x^j_{(\alpha)} - p^j_{(\alpha)} x^i_{(\alpha)}). \tag{10.82}$$

This relation can be interpreted in terms of angular momentum much the way Eq. (10.81) is interpreted in terms of ordinary momentum. The factor multiplying $\delta\Omega^{ij}$ is proportional to the angular momentum. Because the rotation angles are ignorable, the derivatives

$$\frac{\partial S}{\partial \Omega_{ij}} = -\frac{1}{2}\sum_{\alpha}(p^i_{(\alpha)}x^j_{(\alpha)} - p^j_{(\alpha)}x^i_{(\alpha)}). \tag{10.83}$$

are constants of the motion. To match the spatial components defined in Eq. (10.78), define the relativistic, angular momentum 4-tensor by

$$M^{ij} = \sum_{\alpha}(x^i_{(\alpha)}p^j_{(\alpha)} - x^j_{(\alpha)}p^i_{(\alpha)}). \tag{10.84}$$

All components of M^{ij} are conserved. Spelling out all the components, suppressing subscript α, and matching element symbols with Eq. (10.78),

$$M^{ij} = \sum_{\alpha}\begin{pmatrix} 0 & ctp_x - xE/c & ctp_y - yE/c & ctp_z - zE/c \\ -ctp_x + xE/c & 0 & M_z & -M_y \\ -ctp_y + yE/c & -M_z & 0 & M_x \\ -ctp_z + zE/c & M_y & -M_x & 0 \end{pmatrix}. \tag{10.85}$$

With M^{ij} having 16 elements it may appear superficially that there are that many conservation laws. But, being antisymmetric, there are far fewer. In fact, like Eq. (10.78), there are only three independent components making up the 3×3 spatial partition. Furthermore, the constancy of the elements in the upper row and left column follows from the conservation of momentum and energy. In particular, the combination $\sum_\alpha(\mathbf{p}c\,t - \mathbf{x}E/c)$ is conserved. Using constancy of the total energy $\sum E$, this can be expressed as

$$\frac{\sum \mathbf{p}c^2}{\sum E}t = \frac{\sum \mathbf{x}E}{\sum E}. \tag{10.86}$$

In this equation the right-hand side can be used to define the position vector $\mathbf{X}$ of the system centroid and the ratio $\sum \mathbf{p}c^2/\sum E$ is the system velocity $\mathbf{V}$. Equation (10.86) then reduces to $\mathbf{X} = \mathbf{V}t$, which describes the uniform motion of the centroid.

10.10 Angular Momentum of a Field

The same angular variables $\delta\Omega^{ij}$ are ignorable for free space fields. With $T^{ij}dS^{(3)}_j/c$ being the momentum of fields in the volume $dS^{(3)}$ and replacing

$p^i_{(\alpha)}$ in Eq. (10.84), the total angular momentum tensor of the field is

$$M^{ij} = \frac{1}{c}\int (x^i T^{jl} - x^j T^{il})\, dS_l^{(3)}. \tag{10.87}$$

The constancy of these elements can be discussed in the same way that the constancy of charge Q was discussed in Eq. (10.71) and Fig. 10.5. The differences between their values on two hypersurfaces can be expressed as integrals over the 4-volume Ω enclosed by the two hypersurfaces and an enclosing "cylinder" (on which the integrand vanishes appropriately).

$$\delta M^{ij} = \frac{1}{c^2}\int_\Omega \frac{\partial(x^i T^{jl} - x^j T^{il})}{\partial x^l}\, d\Omega. \tag{10.88}$$

Using Eq. (10.65), one obtains

$$\frac{\partial(x^i T^{jl} - x^j T^{il})}{\partial x^l} = \delta^i{}_l T^{jl} + x^i \frac{\partial T^{jl}}{\partial x^l} - \delta^j{}_l T^{il} - x^j \frac{\partial T^{il}}{\partial x^l} = T^{ji} - T^{ij}, \tag{10.89}$$

which vanishes if and only if T^{ij} is symmetric. Using the freedom mentioned below Eq. (10.65), it is possible to adjust T^{ij} to make it symmetric. This is what is needed to assure the conservation of the angular momentum components. In the case of electromagnetic theory this alteration is pursued in problems at the end of Chapter 11. With the fields belonging to the electromagnetic vector potential A^i, after symmetrization of T^{ij}, this theory produces standard definitions of the field energy, momentum, and angular momentum.

Bibliography

General References

1 L.D. Landau and E.M. Lifshitz, *The Classical Theory of Fields*, Pergamon, Oxford, 1971.

References for Further Study

Section 10.5.1

2 N.G. Chetaev, *Theoretical Mechanics*, Springer, Berlin, 1989.

Section 10.6

3 V.I. Arnold, *Mathematical Methods of Classical Mechanics*, Springer, Berlin, 1978, p. 88.

4 V.I. Arnold, V.V. Kovlov, and A.I. Neishtadt, *Mathematical Aspects of Classical and Celestial Mechanics*, Springer, Berlin, 1997, p. 120.

Sections 10.7,

5 L.D. Landau and E.M. Lifshitz, *The Classical Theory of Fields*, Pergamon, Oxford, 1971.

11
Electromagnetic Theory

Chapter 9 has shown how charged particles respond to externally established force fields. In the simplest case these forces have been interpreted as being electric and magnetic. It remains to be seen to what extent these fields are interrelated in the ways embodied in Maxwell's equations.

Historically, Maxwell invented his equations before special relativity was discovered. As a result his equations, though consistent with special relativity, are not "manifestly invariant" (which means that their consistency with Einstein is not obvious.) It is only when the dynamics of material particles in electromagnetic fields is addressed that the *manifestly* invariant treatment becomes essential. This chapter develops this theory.

This task will be approached by generalizing the principle of least action and the Lagrangian formalism. So far, starting from the description of one-dimensional motion, which is described by a single Lagrange equation for dependent variable $x(t)$, generalization has already proceeded to three equations for $x(t)$, $y(t)$, $z(t)$, for motion of a single particle in space, and to $\mathbf{x}_i(t)$, the positions of a countable number of particles. In Chapter 1 the theory was extended to the one-dimensional continuum needed to describe waves on a string. This was the first extension to *field theory*.

To describe electromagnetism the Lagrangian formalism has to be further generalized to describe the time evolution of fields $\mathbf{E}(\mathbf{x}, t)$ and $\mathbf{B}(\mathbf{x}, t)$. These are the *dependent variables* of the theory. The dependence on position $\mathbf{x}$ is to be treated as the generalization from a countable index, such as the integer i indexing particles, to a continuous "indexing" by the position $\mathbf{x}$ at which the fields are to be determined.

Two of the Maxwell equations govern the interrelations of electric and magnetic fields in charge-free (and therefore current-free) space. These will be referred to as the "homogeneous" Maxwell equations since they are linear in the electric and magnetic field components. The two remaining Maxwell equations describe the way that charges and currents act as *sources* of electric and magnetic fields. The presence of *source terms* cause the equations to be "inhomogeneous."

Geometric Mechanics: Toward a Unification of Classical Physics. 2nd Edition. Richard Talman

ISBN: 978-3-527-40683-8

The generalization from countable index i to continuous index $\mathbf{x}$ requires the generalization from a countable number of ordinary differential Lagrange equations for particle positions indexed by i, to a finite number (more or less equal to the number of independent electric and magnetic field components) of partial differential Lagrange equations. These Maxwell equations have to be derived from an assumed Lagrangian. To the extent that trusted principles lead to a unique Lagrangian, this can be interpreted as a "derivation" of the Maxwell equations.

Electromagnetism is not the only form of classical physics that can be subsumed into mechanics in this way. Fluid mechanics, elasticity, and magnetohydrodynamics are other examples. These important areas of physics require little more fundamental input than will be encountered in this chapter, but the resulting Lagrange equations are too complicated for treatment at the level of this text. Paradoxically, other, more fundamental extensions, such as general relativity and string theory, are more consistent with the superficial treatment attempted here. These topics will be pursued in later chapters.

As stated in the Preface, the *theme* of this text is to exhibit the *unification* of classical physics within Lagrangian mechanics. Of course any actual *grand unification of fundamental physics* will require quantum field theory that goes far beyond anything considered in this text. But, as far as I know, any such grand unification currently conceived of will be consistent with the classical unification described here.

Even with this qualification, the treatment in this chapter is biased toward being elementary rather than rigorous. The purpose is to present the sort of general principles/arguments that have been important in developing fundamental physics. The quotation marks around "derivation" when discussing the incorporation of electromagnetic theory into classical mechanics were intended to convey some qualifications. For one thing the arguments will "developed conjectures" rather than mathematical theorems. Most of the arguments in this chapter have been extracted from Landau and Lifshitz's book, *The Classical Theory of Fields*, and are more fully explained there. Even that source is not as authoritative as the original sources.

It should become obvious to the reader that, in deriving Maxwell's equation, what is actually being derived is the functional form of an *action* quantity that leads to the known Maxwell equations. The fact that this turns out to be possible is what justifies the term *unification*. Repeating the unification principle for emphasis: "All classical physics is expressible in the form of Hamilton's principle of least action." Part of the attractiveness of this unification is the relative simplicity of the action, which distills an entire field of physics, in this case electromagnetism, into a single formula.

To prepare for the extra complexity accompanying the transition to partial differential equations it is appropriate to reformulate the discrete analysis in a more abbreviated and, regrettably therefore, in a more abstract form.

11.1
The Electromagnetic Field Tensor

11.1.1
The Lorentz Force Equation in Tensor Notation

Previously, in Eq. (9.80), the Lorentz law, the relativistic generalization of Newton's law for a particle in an electromagnetic field, was derived by substituting the Lagrangian into the Lagrange equations. Here, exploiting the power of the tensor formulation, we skip the extraction of an explicit Lagrangian and go directly from the principle of least action to the equation of motion.

Revising Eq. (9.65) slightly, the action takes the form

$$S = -\int_{P_0}^{P} \left(mc\sqrt{dx_i dx^i} + \frac{e}{c} A_i dx^i \right). \tag{11.1}$$

A possible small deviation from the true orbit is symbolized by $\delta x(s)$. Except for being infinitesimal and vanishing at the end points, $\delta x(s)$ is an arbitrary function, and the symbol $\delta\dot{x}(s)$ is subject to the same interpretation as has been given earlier. When the x^i are varied to $x^i + \delta x^i$, the differential ds changes to $\sqrt{(dx_i + d\delta x_i)(dx^i + d\delta x^i)}$. We continue to use s rather than t as independent variable, so the fundamental total derivative describing evolution along the paths will be d/ds. A partial derivative continues to imply formal differentiation with respect to one of the formally listed arguments of a function. As before $d\delta x^i$ and δdx^i mean the same thing.

The variation in S that accompanies the replacement $x^i \Rightarrow x^i(s) + \delta x^i$ and $u_i \Rightarrow u_i(s) + \delta u_i$ is,

$$\delta S = -\int_{P_0}^{P} \left(mu_j d\delta x^j + \frac{e}{c} A_j \delta dx^j + \frac{e}{c} \Big(\sum_j \frac{\partial A_i}{\partial x^j} \delta x^j \Big) dx^i . \right) \tag{11.2}$$

The first term was obtained previously in Eq. (9.61) and, for the moment, the summation sign has been explicitly included for emphasis; in this case it comes from the rules of calculus rather than from a scalar product. Using the summation convention, the summation will be left implicit from here on. Notice that the differentiation $\partial/\partial x^j$ generates the needed covariant (lower) index for this summation.

As usual we integrate by parts to pull out a common factor δx^j from all terms in the integrand; for this we use

$$\begin{aligned} d(u_j \delta x^j) &= du_j \delta x^j + u_j d\delta x^j, \quad \text{and} \\ d(A_j \delta x^j) &= A_j d\delta x^j + dA_j \delta x^j \\ &= A_j \delta dx^j + \left(\frac{\partial A_j}{\partial x^i} dx^i \right) \delta x^j, \end{aligned} \tag{11.3}$$

to obtain

$$0 = \int_{P_0}^{P} \delta x^j \left(m\, du_j + \frac{e}{c}\frac{\partial A_j}{\partial x^i} dx^i - \frac{e}{c}\frac{\partial A_i}{\partial x^j} dx^i \right). \tag{11.4}$$

The terms coming from the total differentials on the left-hand sides of Eqs. (11.3) vanished because the end points are being held fixed. This integral can now be written in more conventional form as an integral over s;

$$0 = \int_{s_0}^{s} \delta x^j \left(m\frac{du_j}{ds} + \frac{e}{c}\frac{\partial A_j}{\partial x^i}\frac{u^i}{c} - \frac{e}{c}\frac{\partial A_i}{\partial x^j}\frac{u^i}{c} \right) ds. \tag{11.5}$$

The vanishing of this integral for arbitrary differentials δx^j implies the vanishing of the integrand factor, and we obtain

$$m\frac{du_j}{d(s/c)} = \frac{e}{c}\left(\frac{\partial A_i}{\partial x^j} - \frac{\partial A_j}{\partial x^i} \right) u^i. \tag{11.6}$$

This gives the force in terms of the 4-vector potential A_i and the 4-velocity u^i. Correlating with what we already know about electromagnetism, it is clear from Eq. (11.6) that the electric and magnetic fields are tied up in the terms having derivatives of A^i. We introduce the notation

$$F_{ij} = -F_{ji} = \frac{\partial A_j}{\partial x^i} - \frac{\partial A_i}{\partial x^j}, \tag{11.7}$$

which is an antisymmetric 4-tensor with both indices being covariant, and call F_{ij} the *electromagnetic field tensor*. The contravariant versions of Eq. (11.7) are

$$F^{ij} = -F^{ji} = \frac{\partial A^j}{\partial x_i} - \frac{\partial A^i}{\partial x_j}. \tag{11.8}$$

One way of remembering the relation between both-indices-up and both-indices-down values of such a tensor is to realize that the raising of the second index corresponds to multiplying the $j = 1,2,3$ columns by -1, while raising the first index corresponds to multiplying the $i = 1,2,3$ rows by -1. This introduces a pattern of sign changes,

$$\begin{pmatrix} + & - & - & - \\ - & + & + & + \\ - & + & + & + \\ - & + & + & + \end{pmatrix}. \tag{11.9}$$

The equation of motion becomes

$$m\frac{du_j}{d(s/c)} = eF_{ji}\,\frac{u^i}{c}, \quad \text{or} \quad m\frac{du^j}{d(s/c)} = eF^{ji}\,\frac{u_i}{c}. \tag{11.10}$$

Interpreting the components of mu^i as $(\mathcal{E}/c, \mathbf{p}) = (\gamma mc, \gamma m\mathbf{v})$ as in Eq. (9.54), and comparing with Eq. (9.80), the tensor F_{ij} can be interpreted as describing electric and magnetic fields by assigning the matrix elements to be

$$F_{ij} = \begin{pmatrix} 0 & E_x & E_y & E_z \\ -E_x & 0 & -B_z & B_y \\ -E_y & B_z & 0 & -B_x \\ -E_z & -B_y & B_x & 0 \end{pmatrix}, \quad F^{ij} = \begin{pmatrix} 0 & -E_x & -E_y & -E_z \\ E_x & 0 & -B_z & B_y \\ E_y & B_z & 0 & -B_x \\ E_z & -B_y & B_x & 0 \end{pmatrix}. \tag{11.11}$$

As an example of this identification, consider the motion of a particle of mass m traveling with speed v_x along the x-axis in electric field $E_y\hat{\mathbf{y}}$ and magnetic field $B_y\hat{\mathbf{y}}$. Using $d(s/c) = dt/\gamma$, as shown by Eq. (9.9), the second of Eqs. (11.10) yields

$$\gamma \begin{pmatrix} d(\mathcal{E}/c)/dt \\ dp_x/dt \\ dp_y/dt \\ dp_z/dt \end{pmatrix} = e \begin{pmatrix} 0 & 0 & -E_y & 0 \\ 0 & 0 & 0 & B_y \\ E_y & 0 & 0 & 0 \\ 0 & -B_y & 0 & 0 \end{pmatrix} \begin{pmatrix} \gamma \\ -\gamma v_x/c \\ 0 \\ 0 \end{pmatrix}. \tag{11.12}$$

These examples agree with Eq. (9.80).

11.1.2
Lorentz Transformation and Invariants of the Fields

Because the newly introduced quantity F_{ij} is a tensor, its components in different coordinate systems are related to each other by very definite transformations. For two frames at rest with respect to each other the transformations are spatial rotations and the relationships are just those appropriate for **E** and **B** being 3-vectors. Between relatively moving frames (for simplicity let us say that the primed frame has the same orientation as an unprimed frame and moves with velocity V along the positive x-axis) the relation is a Lorentz transformation. Using the fully antisymmetric, 4-index tensor ε^{ijkl} one can construct invariants $F_{ij}F^{ij}$ and $\varepsilon^{ijkl}F_{ij}F_{kl}$ from the tensor F^{ij}. Assuming that $\varepsilon^{0123} = 1$, problems below show that

$$F_{ij}F^{ij} = 2(B^2 - E^2) \quad \text{and} \quad \varepsilon^{ijkl}F_{ij}F_{kl} = -4\mathbf{E}\cdot\mathbf{B}. \tag{11.13}$$

The invariant nature of these equations can be used to provide simple proofs of two general transformation constraints. For example, the equality of the magnitudes of E and B in one frame implying their equality in all other frames. Also the orthogonality of **E** and **B** in one frame implies their orthogonality in all other frames. Both of these are well-known properties of electromagnetic waves in free space.

11.2 The Electromagnetic Field Equations

11.2.1 The Homogeneous Pair of Maxwell Equations

The field tensor F_{ij} is related to the 4-vector potential A_i by Eq. (11.7). It is easy to see from this definition that F_{ij} satisfies the equation

$$\frac{\partial F_{ij}}{\partial x^k} + \frac{\partial F_{jk}}{\partial x^i} + \frac{\partial F_{ki}}{\partial x^j} = 0. \tag{11.14}$$

(Separating into a case where two indices are the same and another case where no two indices are the same simplifies the task.) Actually there are many equations, but the only nontrivial ones have all of i, j, and k different. For example, consider a pure electrostatic field for which

$$F_{ij} = \begin{pmatrix} 0 & E_x & E_y & E_z \\ -E_x & 0 & 0 & 0 \\ -E_y & 0 & 0 & 0 \\ -E_z & 0 & 0 & 0 \end{pmatrix}, \tag{11.15}$$

and consider the $(i = 0, j = 1, k = 2)$ case;

$$\frac{\partial F_{01}}{\partial x^2} + \frac{\partial F_{12}}{\partial x^0} + \frac{\partial F_{20}}{\partial x^1} = \frac{\partial E_x}{\partial y} - \frac{\partial E_y}{\partial x} = -[\nabla \times \mathbf{E}]_z = 0. \tag{11.16}$$

showing that the z-component of the curl vanishes. Working out all cases one finds that Eq. (11.14) is equivalent to the homogeneous Maxwell equations

$$\nabla \times \mathbf{E} + \frac{1}{c}\frac{\partial \mathbf{B}}{\partial t} = 0, \quad \nabla \cdot \mathbf{B} = 0. \tag{11.17}$$

11.2.2 The Action for the Field, Particle System

The inhomogeneous pair of Maxwell equations can also be derived from a principle of least action. As usual in the application of Maxwell's equations we assume the system of charges and fields is self-consistent, with the motions of all the charges responding to all the fields produced.

To this point the action has been a sum of one-dimensional time integrals, one for each coordinate of each particle being described. The *dependent* variables have been the spatial coordinates of these particles and the problem has been to express these dependent variables as functions of the *independent* variable t, which ranges over a single infinity of values. This formalism must be generalized if it is to express the dependent variables **E** and **B** as functions of the quadruple infinity of values that define all positions and times.

In going from one to many particles one simply summed over the actions of all of the particles. We do the same now except, since we have a continuum of different points in space, we define the action as the time integral of a Lagrangian that is itself defined as a volume integral over all space. The action is therefore expressible as an integral over a four-dimensional region. Spatially this region is all of space, and the temporal bounds are defined by "hyperplanes" of initial and final times. The differential for this integration is $d\Omega = dVd(ct)$. Though this integration sounds formidable it really is not, since it is only formal. As usual in the calculus of variations, it is the vanishing of the integrand that yields the equations of motion. All that will be done here is to show that these equations are the inhomogeneous Maxwell equations.

First, though, one must conjecture a form for the Lagrangian (actually one must start with a Lagrangian "density" since it needs to be integrated over volume to have dimensions proper for a Lagrangian.) There is surprisingly little freedom in choosing this density. The new arguments of the Lagrangian density are to be the components of **E** and **B** which are the analogs of generalized coordinates in the purely mechanical Lagrangian. They are the dependent variables whose evolutions are to be described by the Lagrange equations. The density can also depend on time derivatives $\dot{\mathbf{E}}$ and $\dot{\mathbf{B}}$, which are the analogs of velocities in the purely mechanical formulation. But these quantities now have to be expressed as $\partial\mathbf{E}/\partial t$ and $\partial\mathbf{B}/\partial t$, because the derivatives are to be taken at fixed spatial locations.

The Lagrangian cannot depend on spatial derivatives of the components of **E** and **B**.[1] [2] Furthermore, the Lagrangian density cannot depend on the potentials ϕ and **A** (except through their previously introduced interaction-with-charge terms) because, being unobservable, they cannot appear in the electromagnetic field equations. For the field equations to be linear in **E** and **B** the density has to depend quadratically on them. Finally, the density has to be a relativistic invariant.

The only quantity meeting all these requirements is $F_{ij}F^{ij}$. (Another candidate might seem to be the quantity $\epsilon^{ijkl}F_{ij}F_{kl}$ worked out in Eq. (11.13). But this quantity is shown in Problem 11.2.9 to be a divergence which would not influence the determination of extremal trajectories. As it happens this quantity, because it is a pseudovector, might also be rejectable on the grounds that it would lead to parity nonconservation that seems not to be observed in nature.) Accepting this, and combining all terms we now know must occur in

1) In an elastic medium forces on material in one region due to its neighboring region would be represented by spatial derivatives in the action, but no such forces are present in vacuum electromagnetism.

2) Even though spatial derivatives of **E** and **B** do not occur in the Lagrangian density, such terms can appear in the derived Lagrange equation, just as they do in purely mechanical Lagrange equations.

the Lagrangian, we define the action to be

$$S = -\sum \int mcds - \sum \int \frac{e}{c} A_k dx^k - \frac{1}{16\pi c} \int F_{ij} F^{ij} d\Omega, \tag{11.18}$$

where the incremental 4-volume element is $d\Omega = c\,dt\,dx\,dy\,dz$. No explicit indexing of the possibly multiple particles is shown. Such indexing is intended to be implied by the summation signs shown. As before, the precise constant factor $-(16\pi c)^{-1}$ has been preselected to yield a choice of units that will later be shown to be conventional. Using Eq. (11.13), the new part of the Lagrangian, ascribable purely to the electromagnetic field, is

$$L_{\text{em}} = \frac{1}{8\pi} \int (E^2 - B^2)\, dx\, dy\, dz. \tag{11.19}$$

As has been emphasized repeatedly, it is necessary to specify explicitly what the arguments of the Lagrangian, or in this case the Lagrangian density, are to be. We will regard the A^i functions as the dependent variables for which equations of motion are sought, and only later express the equations in terms of **E** and **B**, which are expressible as derivatives of the A^i.

11.2.3
The Electromagnetic Wave Equation

Problem 11.2.1. *The last term in Lagrangian density (11.18) is not very different from the Lagrangian density of a stretched string given in Eq. (1.117). It depends only on the first-order derivatives of the dependent variables, and it is a quadratic function of those terms, which are $\partial A^\alpha / \partial x_\beta \equiv A^a{}_{,b}$. The Lagrange equations are therefore "wave equations" that are not very much more complicated than Eq. (1.80). To derive them start by showing that*

$$F_{ij} F^{ij} = (g_{jk} A^k{}_{,i} - g_{ik} A^k{}_{,j})(g^{il} A^j{}_{,l} - g^{jm} A^i{}_{,m}). \tag{11.20}$$

Then show that

$$\frac{\partial (F_{ij} F^{ij})}{\partial A^a{}_{,b}} = 4(g_{ja} g^{bl} A^j{}_{,l} - A^b{}_{,a}). \tag{11.21}$$

Then show that the Lagrange equations are

$$g_{ja} g^{bl} \frac{\partial^2 A^j}{\partial x^c \partial x^l} - \frac{\partial^2 A^b}{\partial x^c \partial x^a} = 0. \tag{11.22}$$

By manipulating the indices this can be written in various forms. For example, show that

$$\frac{\partial^2 A^c}{\partial x^b \partial x_b} - \frac{\partial}{\partial x_c} \left(\frac{\partial A^b}{\partial x^b} \right) = 0. \tag{11.23}$$

It would clearly be advantageous to suppress the second term, because it causes the equations to be coupled. *As it happens, by appropriate choice of* gauge, *the last term* can *be suppressed. Then, compared to the equation of waves on a string, this set has four times as many equations, because there are four components of A^b, and each equation has four terms instead of two, because there are three spatial coordinates (x, y, z) instead of just x.*

11.2.4
The Inhomogeneous Pair of Maxwell Equations

In the definition of the action in Eq. (11.18), if individual particles are to be represented by charge distributions rather than as point particles, then the summation over particles in the second term has to be replaced by a volume integral. That is, the factor $\sum e$ is replaced by $\int \rho dV$. Then, using Eq. (10.66), the factor ρdx^k can be replaced by $J^k dt$. The volume element dV, taken with the element cdt, form the same element $d\Omega$, used in the part of the action ascribable to just the fields. Finally, we obtain

$$S = -\sum \int mcds - \frac{1}{c^2} \int A_k J^k d\Omega - \frac{1}{16\pi c} \int F_{ij} F^{ij} d\Omega. \tag{11.24}$$

The principle of least action is to be applied to S, defined this way, to find the equations of motion of the dependent variables, which are the components of the vector potential, whose first derivatives produce $\mathbf{E}(\mathbf{x})$ and $\mathbf{B}(\mathbf{x})$. Using Eq. (11.11), these field variables will be obtained from F^{ij}, which can temporarily be regarded as satisfactory Lagrange function (first derivative) arguments in terms of which the Lagrange equations can be expressed. These functions are given arbitrary "variations" δF^{ij}, but 4-vector J^i, describing the "sources," which are charges and currents, is assumed to be given, at least instantaneously, and is not allowed to vary. Performing the variation yields

$$\delta S = -\int \left(\frac{1}{c^2} (\delta A_i)\, J^i + \frac{1}{8\pi c}\, F^{ij} (\delta F_{ij}) \right) d\Omega = 0. \tag{11.25}$$

The usual strategy in applying the calculus of variation is to express all terms in the integrand as multiples of the (arbitrary) factor δA_i. In this case, because $F_{ij} = \partial A_j / \partial x^i - \partial A_i / \partial x^j$, a first step in this direction converts the integrand to

$$-\frac{1}{c^2} J^i \delta A_i + \frac{1}{4\pi c} F^{ij} \frac{\partial}{\partial x^j} (\delta A_i). \tag{11.26}$$

The validity of commuting the δ and $\partial / \partial x^j$ operations has been justified before, and the equality of the two terms coming from δF_{ij} follows from the antisymmetry of F^{ij}. In further preparation for integration by parts, the integrand can be written as

$$-\left(\frac{1}{c^2}J^i + \frac{1}{4\pi c}\frac{\partial F^{ij}}{\partial x^j}\right)\delta A_i + \frac{1}{4\pi c}\delta(F^{ij}A_i). \tag{11.27}$$

The integral in Eq. (11.25) is four-dimensional, but the integration region can be broken into long parallelepipeds at fixed x, y, and z, parallel to the t-axis. Being a total differential, the last term in Eq. (11.27) can be dropped as usual, since its contribution to each of these parallelepipeds vanishes. Then, since δA_i is arbitrary, we obtain

$$\frac{\partial F^{ij}}{\partial x^j} = -\frac{4\pi}{c}J^i. \tag{11.28}$$

When the individual components of F^{ij} are identified with the components of $\mathbf{E}(\mathbf{x})$ and $\mathbf{B}(\mathbf{x})$ and are substituted into this equation, the results are the inhomogeneous Maxwell equations

$$\nabla \times \mathbf{B} - \frac{1}{c}\frac{\partial \mathbf{E}}{\partial t} = \frac{4\pi}{c}\mathbf{J}, \quad \nabla \cdot \mathbf{E} = 4\pi\rho. \tag{11.29}$$

11.2.5
Energy Density, Energy Flux, and the Maxwell Stress Energy Tensor

It would be possible (but would go deeply into electromagnetic theory) to continue the development by interpreting all 16 elements of the energy–momentum density of electromagnetism at this time. The spatial part of T^{ij} is the Maxwell stress tensor.

The most elementary description of stress in an electric or magnetic field (due to Faraday) is that the field lines are under tension. The magnitude of the tension can be inferred, for example, by calculating the pressure exerted by the field lines as they terminate at a surface, for example separating a conductor from free space.

We will consider only the T^{00} element and only in the purely 3D Maxwellian formulation rather than in the four-dimensional, manifestly relativistic, context. This is intended to provide a minimal, though satisfactory, bridge from familiar mechanics to familiar electromagnetic theory – the Poynting vector in particular. Starting from the two Maxwell equations

$$\begin{aligned}\nabla \times \mathbf{B} &= \frac{1}{c}\frac{\partial \mathbf{E}}{\partial t} + \frac{4\pi}{c}\mathbf{J},\\ \nabla \times \mathbf{E} &= -\frac{1}{c}\frac{\partial \mathbf{B}}{\partial t},\end{aligned} \tag{11.30}$$

taking a dot product of the second with $\mathbf{B}$, the first with $\mathbf{E}$, and subtracting, one obtains

$$\frac{1}{c}\mathbf{E}\cdot\frac{\partial \mathbf{E}}{\partial t} + \frac{1}{c}\mathbf{B}\cdot\frac{\partial \mathbf{B}}{\partial t} = -\frac{4\pi}{c}\mathbf{J}\cdot\mathbf{E} - \Big(\mathbf{B}\cdot(\nabla\times\mathbf{E}) - \mathbf{E}\cdot(\nabla\times\mathbf{B})\Big). \tag{11.31}$$

The left-hand side can be recognized as the time derivative of $(E^2 + B^2)/(2c)$. Using a vector identity,

$$\nabla \cdot (\mathbf{E} \times \mathbf{B}) = \mathbf{B} \cdot (\nabla \times \mathbf{E}) - \mathbf{E} \cdot (\nabla \times \mathbf{B}), \tag{11.32}$$

the quantity in square brackets can be expressed in terms of $\mathbf{E} \times \mathbf{B}$; this vector is of sufficient importance to deserve its own symbol

$$\mathbf{S} = \frac{c}{4\pi} \mathbf{E} \times \mathbf{B}, \tag{11.33}$$

which is called the *Poynting vector*, named after the physicist who first introduced it. Combining formulas, we have

$$\frac{\partial}{\partial t} \frac{E^2 + B^2}{8\pi} = -\mathbf{J} \cdot \mathbf{E} - \nabla \cdot \mathbf{S}. \tag{11.34}$$

If this equation is integrated over all space, using Gauss's theorem to evaluate the second term on the right, and assuming that the fields vanish at infinity, the result is

$$\frac{\partial}{\partial t} \int \frac{E^2 + B^2}{8\pi} dV = - \int \mathbf{J} \cdot \mathbf{E} \, dV. \tag{11.35}$$

The right-hand side can be related to the energy of the particles in the system using formula (9.82) which reads

$$\frac{d\mathcal{E}_{\text{kin}}}{dt} = e\mathbf{E} \cdot \mathbf{v}. \tag{11.36}$$

Converting from the point particle description to the charge density description using Eq. (10.66) and taking $\sum \mathcal{E}_{\text{kin}}$ to be the mechanical energy of all particles in the system, Eq. (11.35) becomes

$$\frac{\partial}{\partial t} \left(\int \frac{E^2 + B^2}{8\pi} dV + \sum \mathcal{E}_{\text{kin}} \right) = 0. \tag{11.37}$$

This formula makes it natural to interpret the quantity

$$W = \frac{E^2 + B^2}{8\pi} \tag{11.38}$$

as the *energy density* of the electromagnetic field, so that Eq. (11.37) can be interpreted as the *conservation of energy*. In a problem it is shown that this energy density W is equal to the T^{00} element of the energy–momentum tensor introduced in the previous section. As the particles move, their increase in mechanical energy cancels the decrease in field energy.

Instead of integrating Eq. (11.34) over all space, it can be integrated only over a finite volume V bounded by a closed surface $\mathcal{S}$. In that case, the term involving the Poynting vector **S** no longer vanishes. The volume integral it contributes can however, using Stokes's theorem, be converted to a surface integral over S. This yields

$$\frac{\partial}{\partial t}\left(\int_V \frac{E^2+B^2}{8\pi} dV + \sum \mathcal{E}_{\text{kin}}\right) = -\int_{\mathcal{S}} \mathbf{S}\cdot d\mathbf{A}, \tag{11.39}$$

where $d\mathbf{A}$ is an outward-directed area element of surface $\mathcal{S}$. This formula now permits us to interpret S as the energy flux density – the amount of electromagnetic field energy passing unit area per unit time. In Eq. (11.39) it accounts for the energy leaving V.

Problem 11.2.2. *Consider the fully antisymmetric 4-index symbol ε^{iklm} introduced in Eq. (11.13).*

(a) Assuming it is legitimate (as it is) to treat ε^{ijkl} as a 4-index tensor with $\varepsilon^{0123} = 1$, evaluate ε_{0123} and the quadruple sum $\varepsilon^{ijkl}\varepsilon_{ijkl}$.

(b) (Summing over both k and l) show that $\varepsilon^{ijkl}\varepsilon_{mnkl} = -2(\delta^i_m\delta^j_n - \delta^i_n\delta^j_m)$.

Problem 11.2.3. *Prove Eqs. (11.13). That is*

$$(a)\quad F_{ij}F^{ij} = 2(B^2 - E^2)$$

$$(b)\quad \varepsilon^{ijlm}F_{ij}F_{lm} = -8\mathbf{E}\cdot\mathbf{B}.$$

One can evaluate $F_{ij}F^{jk}$ by matrix multiplication and then complete part (i) by taking the trace of the result. For tensor products expressed purely with indices the order of the factor does not matter.

Problem 11.2.4. *Fill in the steps leading from Eq. (11.28) to Eq. (11.29).*

Problem 11.2.5. *The Lagrangian density corresponding to electromagnetic fields (not including charges and currents) was found, in Eq. (11.18) to be*

$$\mathcal{L} = -\frac{1}{16\pi} F_{ij}F^{ij}. \tag{11.40}$$

Show that the energy–momentum tensor (Eq. (10.63)) is given by

$$T^{ij} = -\frac{1}{4\pi}\frac{\partial A^l}{\partial x^i} F^j{}_l + \frac{1}{16\pi} g^{ij} F_{kl} F^{kl}. \tag{11.41}$$

These relations, along with an entire development of their implications for electromagnetic theory, are derived in section 33 of Landau and Lifshitz, The Classical Theory of Fields. *That is the best place for further study of the material in this chapter.*

Problem 11.2.6. *The tensor T^{ij} just derived is not symmetric. Show that addition of the term*

$$\frac{1}{4\pi}\frac{\partial A^i}{\partial x^l} F^j{}_l, \tag{11.42}$$

preserves the validity of Eq. (10.65) while rendering T^{ij} symmetric. The result is

$$T^{ij} = \frac{1}{4\pi}\left(- F^{il}F^{j}_{\;l} + \frac{1}{4} g^{ij} F_{lm}F^{lm}\right). \tag{11.43}$$

With T^{ij} defined in this way, show that

$$T^{00} = \frac{E^2 + B^2}{8\pi}. \tag{11.44}$$

Problem 11.2.7. *Confirm the identifications of elements of F_{ij} in Eq. (11.11) with elements of* **E** *and* **B**. *This amounts to reconciling them with Eqs. (9.79) and (11.7).*

Problem 11.2.8. *Show that Eq. (11.14), in the nondegenerate case for which i, j, and k are all different, can also be expressed as*

$$\epsilon^{ijkl}\frac{\partial}{\partial x^j} F_{kl} = 0. \tag{11.45}$$

Problem 11.2.9. *Use the antisymmetry of ϵ^{ijkl} to show that*

$$\epsilon^{ijkl}F_{ij}F_{kl} = 4\frac{\partial}{\partial x^i}\left(\epsilon^{ijkl}A_j\frac{\partial A_l}{\partial x^k}\right). \tag{11.46}$$

Being a divergence, this quantity can be added to the Lagrangian density without influencing extremal trajectories.

Bibliography

General References

1 L.D. Landau and E.M. Lifshitz, *The Classical Theory of Fields*, Pergamon, Oxford, 1971.

2 J.D. Jackson, *Classical Electrodynamics*, Wiley, New York, 1999.

12
Relativistic Strings

The purpose of this chapter is to show how a mechanical system of very modern interest, the relativistic string, can fit into the framework of classical mechanics. In preparation for this subject, nonrelativistic string were analyzed using (classical) Lagrangian field theory in Chapter 1. In particular, waves on the string were analyzed. That material should be reviewed before starting this chapter. Our mission in this chapter is to obtain a similar description of waves on a relativistic string. One curious feature to be understood is the suppression of longitudinal waves (which is needed because the string model cannot support them.) Another important formula to be derived is an expression for string's rest mass-squared m^2 as a quadratic function of its mode amplitudes. This is an appropriate final classical result since it is needed as the starting point for a quantum mechanical treatment.

12.1
Introduction

Most of the material in this chapter has been learned, and in some cases virtually copied, from the text *A First Course in String Theory* by Barton Zwiebach – mainly chapters 6 through 9. But the material presented here merely scratches the surface of the material in Zwiebach's book, and the thrust is entirely different. The intention here is to give yet another example illustrating how classical physics is subsumed into classical mechanics. So what is a concluding example here is a starting point for Zwiebach, who goes on to discuss the quantum field theory of relativistic strings – "String Theory" in modern parlance.

12.1.1
Is String Theory Appropriate?

As most physicists know, in recent years there has been a major push by theoretical physicists toward a "grand unification" of fundamental physics. Usually it is the four forces, electromagnetism, the strong and weak nuclear forces, and gravity that are considered to be "fundamental." The incorporation of

Geometric Mechanics: Toward a Unification of Classical Physics. 2nd Edition. Richard Talman

ISBN: 978-3-527-40683-8

gravity has apparently been the most difficult part of this (as yet incomplete) unification program. Much of this program has centered on the development of the quantum field theory of relativistic strings. Part of the motivation for this concentration is that the gravitational field has been successfully quantized using string theory.

In spite of the concentrated efforts of large numbers of theoreticians, there seems to have been relatively little advance in the field of string theory in the last few years and certainly there is no experimental confirmation of the theory. Early optimism christened string theory "The Theory of Everything" but modern critics have been known to refer to it as "The Theory of Nothing." Apparently there is an embarrassment of riches in string theory, and ways have not yet been found to distill the range of theories down to one that applies uniquely to the actual world we live in.

The present text provides an introduction to this field, but it makes no pretense of contributing to these controversial issues. The presence of string theory here is justified by the fact that string theory is so *geometric*. This makes the subject singularly appropriate as an example for inclusion in a text on geometric classical mechanics.

This text discusses only classical theory with no quantum mechanics. Even with this limitation, relativistic effects make the treatment of strings far different than is the treatment of nonrelativistic strings. Nevertheless the analysis can be patterned to a large extent after the treatment in Chapter 1. It must be acknowledged though, that there are *absolutely* no realizable physical systems describable as relativistic free strings. If relativistic strings have any role in real life it is only after their quantization. In spite of these reservations, we persevere in the development of string theory because it so clearly exemplifies the geometric approach to classical mechanics. One important property of Lagrangian mechanics – parameterization invariance – is especially well demonstrated by classical string theory.

Over the century since the invention of special relativity there have been any number of unexpected behaviors predicted by relativity (such as the pole vaulter with pole entirely inside a building shorter than the pole) or weird properties (such as being able to see the back side of an object). Classical relativistic string theory brings in curiosities like these, but in a more serious context than suffices for these frivolous examples.

So the answer to the question in the title to this section is "String theory is appropriate for illustrating the geometric character of classical mechanics whether or not the quantum field theory of strings eventually proves its value in explaining nature."

12.1.2
Parameterization Invariance

A significant dividend accompanying the principle of least action in Lagrangian mechanics is the coordinate independence of the Lagrange equations. The system trajectory minimizing the action integral is obviously independent of the particular coordinates used in the trajectory's description. It follows that the Lagrange equations should be (and are) formally independent of the particular generalized coordinates used to describe system configurations. This property is implicitly exploited every time generalized coordinates are cleverly chosen to simplify the description of a mechanical system. For example, coordinates that are constants of the motion are greatly favored. As already mentioned, parameterization-invariance plays an especially important role in relativistic string theory. It can even be regarded as a form of gauge invariance, since the advance of one coordinate along a coordinate curve of another can be obtained by laying down measuring sticks along the curve. A choice of parameters is frequently, therefore, referred to as a *choice of gauge*.

12.1.3
Postulating a String Lagrangian

As always, in preparation for applying Hamilton's principle, the first thing to do is to find the appropriate Lagrangian, or rather, because a string is continuous, the Lagrangian density. The Lagrangian for a nonrelativistic beaded string was easily obtained in Section 1.7, and the Lagrangian density was obtained by proceeding to the continuum limit in Section 1.8. But we have seen in Chapter 9 that the principle of relativity imposes strict conditions on the Lagrangian, even for a system as simple as a point particle.

In Section 9.2.1 the free point particle action was found to be

$$S = -mc \int_{t_0}^{t} ds = (-mc^2) \int_{t_0}^{t} \sqrt{1 - \frac{v^2}{c^2}}\, dt, \tag{12.1}$$

where ds is the proper length interval along the world line of the particle. Just as ds is an (invariant) geometric property of the particle's world trajectory, for the string Lagrangian density one is motivated to seek an invariant property derived from the world trajectories of all the points making up the string, perhaps starting with a beaded string of relativistic particles. This would be hopelessly complicated. The only well-understood relativistic force is electromagnetic and even the motion of just two particles subject to electromagnetic interaction is quite difficult. In any case it would not be persuasive to model the tension force in a string as being electromagnetic.

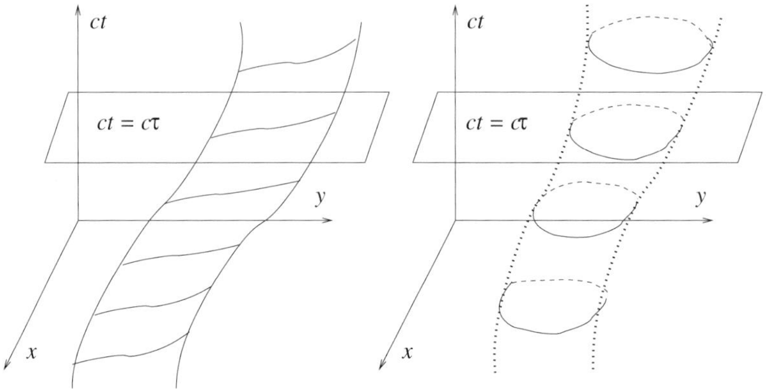

Fig. 12.1 World sheets of an *open* string on the left and a *closed* string on the right. Transverse curves are intersections of the world sheet with planes of constant time. To help in intuiting figures like these it may be helpful to pretend that the string always lies instantaneously in a plane perpendicular to the z-axis; this makes it unimportant that z is not exhibited.

To get a manageable theory it is necessary to postulate the existence of a mechanical system that is describable by a far simpler Lagrangian. For starters we give up any pretense of keeping track of the positions of identifiable points interior to the string, which is something that is possible with an ordinary nonrelativistic string.[1] We also intend to avail ourselves of a theoretical physicist's luxury of being guided by simplicity in making choices (but *we* will not accept the responsibility of deriving *all* the implications flowing from these choices.)

Clearly the choice is to be based on the geometry of the string's motion. It is appropriate to distinguish between *closed* rubber-band-like strings, and *open* strings, whose ends may be free or may be attached to something else. In either case, the relativistic motion has to be represented by world sheets, such as those shown in Fig. 12.1. Because the dimensionality of the string is one higher than that of a point particle, it is natural to suppose that the dimensionality of contributions to the Lagrangian should also be one higher – hence incremental areas on the world sheet rather than incremental arc lengths along the world trajectory.

1) The inability to keep track of interior points of the string is related to parameterization invariance. A point on the string identified by a longitudinal coordinate σ cannot be regarded as being the same material point at a later time just because σ is the same. In other words the string cannot be regarded as having any permanent substructure.

12.2 Area Representation in Terms of the Metric

The parameterization of a 2D surface in 3D Euclidean, (x^0, x^1, x^2) space, by parameters ξ^1 and ξ^2 is illustrated in Fig. 12.2. A general point on the surface

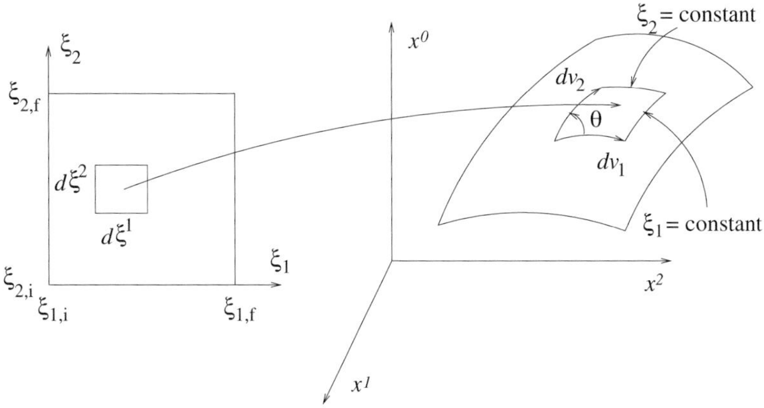

Fig. 12.2 Parameterization of a 2D surface in Euclidean space by ξ^1 and ξ^2.

can be specified by a vector $\mathbf{x}(\xi^1, \xi^2)$. Proceeding as in Eq. (2.102), the metric induced on the surface can be expressed as

$$ds^2 = g_{ij} d\xi^i d\xi^j, \quad \text{where} \quad g_{ij} = \frac{\partial \mathbf{x}}{\partial \xi^i} \cdot \frac{\partial \mathbf{x}}{\partial \xi^j}. \tag{12.2}$$

Let vector $\mathbf{dv}_1$, tangent to the surface, corresponds to change $d\xi^1$, with ξ^2 held constant, and $\mathbf{dv}_2$ corresponds in the same way to $d\xi^2$. They are given by

$$\mathbf{dv}_1 = \frac{\partial \mathbf{x}}{\partial \xi^1} d\xi^1, \quad \text{and} \quad \mathbf{dv}_2 = \frac{\partial \mathbf{x}}{\partial \xi^2} d\xi^2. \tag{12.3}$$

The area they delineate is given by

$$\begin{aligned} dA &= |\mathbf{dv}_1||\mathbf{dv}_2| \sin\theta = |\mathbf{dv}_1||\mathbf{dv}_2| \sqrt{1-\cos^2\theta} \\ &= \sqrt{|\mathbf{dv}_1|^2 |\mathbf{dv}_2|^2 - (\mathbf{dv}_1 \cdot \mathbf{dv}_2)^2}, \end{aligned} \tag{12.4}$$

and the total surface area is given by

$$A = \iint d\xi^1 d\xi^2 \sqrt{\left(\frac{\partial \mathbf{x}}{\partial \xi^1} \cdot \frac{\partial \mathbf{x}}{\partial \xi^1}\right), \left(\frac{\partial \mathbf{x}}{\partial \xi^2} \cdot \frac{\partial \mathbf{x}}{\partial \xi^2}\right) - \left(\frac{\partial \mathbf{x}}{\partial \xi^1} \cdot \frac{\partial \mathbf{x}}{\partial \xi^2}\right)^2} \tag{12.5}$$

$$= \iint d\xi^1 d\xi^2 \sqrt{g_{11} g_{22} - g_{12}^2}$$

$$= \iint d\xi^1 d\xi^2 \sqrt{g}, \quad \text{where} \quad g = \begin{vmatrix} g_{11} & g_{12} \\ g_{21} & g_{22} \end{vmatrix}. \tag{12.6}$$

Problem 12.2.1. *The area calculation just completed could have been performed using other parameters,* $\overline{\xi}_1$ *and* ξ_2*, by introducing metric tensor* $\overline{g}_{ij}$ *and retracing the same steps. This would be known as "reparameterization" and Eq. (12.6) shows the area to be manifestly invariant to this change. Alternatively, the integral could have been performed by using the rules of integral calculus and introducing the Jacobean matrix* $J = |\mathbf{J}((\xi^1,\xi^2)/(\overline{\xi}^1,\overline{\xi}^2))|$ *to change the integration variables. Show that* $\overline{g} = J^2 g$ *and thereby reconfirm the parameterization invariance.*

12.3 The Lagrangian Density and Action for Strings

12.3.1 A Revised Metric

Up to this point we have always used the same $1,-1,-1,-1$ relativistic metric, $ds^2 = c^2dt^2 - |\mathbf{dx}|^2$. For *time-like* displacements $ds^2 > 0$. This continues to be, by and large, the metric used in electromagnetic theory and in accelerator and elementary particle physics. But modern workers in cosmology and string theory tend to use a different metric and a more compact notation. This alteration may be partly due to the complication, introduced by general relativity, of the metric being affected by nearby masses. As well as adapting to this complication by introducing a "free space metric,"

$$\eta_{\mu\nu} = \begin{pmatrix} -1 & 0 & 0 & 0 \\ 0 & 1 & 0 & 0 \\ 0 & 0 & 1 & 0 \\ 0 & 0 & 0 & 1 \end{pmatrix}, \tag{12.7}$$

the new notation also introduces a "dot product" notation for the scalar product;

$$a \cdot b = a^\mu b_\mu = \eta_{\mu\nu} a^\mu b^\nu = -a^0 b^0 + \mathbf{a} \cdot \mathbf{b}, \tag{12.8}$$

where $a = (a^0, \mathbf{a})$ and $b = (b^0, \mathbf{b})$ are two 4-vectors. This notation, mimicking the dot product notation of ordinary 3D geometry, tends to rein in the "débauche d'indices" deplored by Cartan. The overall change of signs to $-1,1,1,1$ in the metric has the effect of matching the sign of the 4D dot product to the sign of its spatial 3D part. By defining the invariant displacement along the world line of a particle by

$$ds^2 = -dx \cdot dx, \tag{12.9}$$

one preserves the rule that $ds^2 > 0$ for *time-like* displacements.

12.3.2
Parameterization of String World Surface by σ and τ

We wish to parameterize the world surfaces of strings such as those illustrated in Fig. 12.1 using a space-like parameter σ and a time-like parameter τ. Curves lying in planes of constant time are exhibited in that figure. A possible parameterization is to use the time t corresponding to each of these planes as the τ parameter. Being at fixed time, this is known as a "static gauge." The σ parameter can be chosen to be any monotonically advancing parameterization of the curve in each fixed time hyperplane.

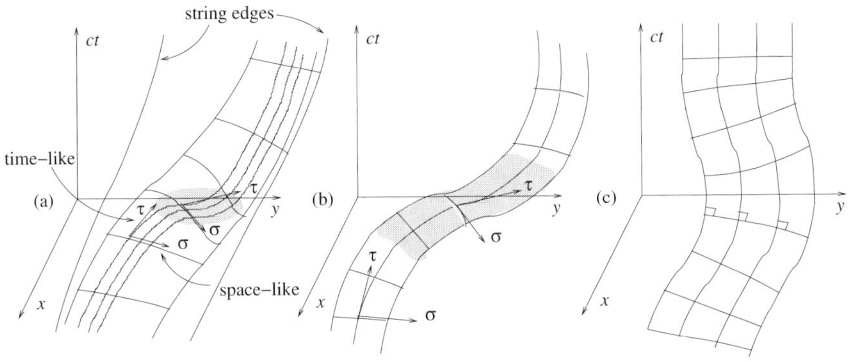

Fig. 12.3 Candidate parameterizations of the world sheets of open strings. Case (a) is *ill-advised* in that tangents to coordinate curves switch character from time-like to space-like on entry to or exit from the shaded region. Case (b) is *unphysical* because the sheet edges are space-like in the shaded region which violates relativity. The string in case (c) is *physical* and the coordinate curves are mutually *orthogonal*.

Other candidate parameterizations are illustrated in Fig. 12.3, but only case (c) is sensible. The end point of a string, like any point particle cannot travel faster than the speed of light. The angle between the ct-axis and a tangent to the end point world line can therefore not exceed $45\,^\circ$. On these grounds case (b) is therefore unphysical. The string illustrated in case (a) is physically valid, but the parameterization shown is ill-advised because the character of individual coordinate curves switch between time-like and space-like in the shaded region.

12.3.3
The Nambu–Goto Action

With the metric re-defined as in Eq. (12.9), the dot product re-defined as in Eq. (12.8), and the parameters ξ^1 and ξ^2 replaced by σ and τ, formula (12.5) becomes

$$A = \iint d\sigma d\tau \sqrt{\left(\frac{\partial X}{\partial \tau} \cdot \frac{\partial X}{\partial \sigma}\right)^2 - \left(\frac{\partial X}{\partial \tau} \cdot \frac{\partial X}{\partial \tau}\right)\left(\frac{\partial X}{\partial \sigma} \cdot \frac{\partial X}{\partial \sigma}\right)}. \tag{12.10}$$

In this formula X is the 4-position of an arbitrary point on the surface, and the sign under the square root has been reversed for a reason to be justified in Problem 12.3.1.

To abbreviate further let us introduce the notations

$$\dot{X}^\mu \equiv \frac{\partial X^\mu}{\partial \tau}, \quad \text{and} \quad X^{\mu\prime} \equiv \frac{\partial X^\mu}{\partial \sigma}. \tag{12.11}$$

The elements $\gamma_{\alpha\beta}$ of the metric induced on the world sheet can then be expressed as 4-dot-products:

$$\gamma_{\alpha\beta} = \begin{pmatrix} \dot{X} \cdot \dot{X} & \dot{X} \cdot X' \\ \dot{X} \cdot X' & X' \cdot X' \end{pmatrix}, \quad \text{and} \quad \gamma = \det(\gamma_{\alpha\beta}). \tag{12.12}$$

Combining all these formulas, and postulating the world-surface area to be the Lagrangian density, we obtain the so-called "Nambu–Goto Action:"[2]

$$\begin{aligned} S &= -\frac{T_0}{c} \int_{\tau_i}^{\tau_f} d\tau \int_0^{\sigma_1} d\sigma \sqrt{(\dot{X} \cdot X')^2 - (\dot{X} \cdot \dot{X})(X' \cdot X')} \\ &= -\frac{T_0}{c} \iint d\tau d\sigma \sqrt{-\gamma}. \end{aligned} \tag{12.13}$$

In this form the action of a point particle (as given, for example, in the first form of Eq. (12.1)) has been generalized to apply to an extended object, namely a string. The σ parameterization has been specialized so that the string ends are at 0 and σ_1. Also, as always, there is an undetermined overall, dimensional, multiplicative factor, whose value $-T_0/c$, remains to be fixed.

In the static gauge (defined earlier) τ is chosen to be the time t on a fixed-time hyperplane (which, in this case, is the volume of all ordinary 3D space). In this gauge, the coordinates of a general point on the surface are

$$X = (c\tau, \mathbf{X}(\tau, \sigma)), \tag{12.14}$$

with partial derivatives given by

$$X' = (0, \mathbf{X}'), \quad \dot{X} = (c, \dot{\mathbf{X}}), \tag{12.15}$$

and 4-dot products given by

$$X' \cdot X' = \mathbf{X}' \cdot \mathbf{X}', \quad \dot{X} \cdot \dot{X} = -c^2 + \dot{\mathbf{X}} \cdot \dot{\mathbf{X}}, \quad X' \cdot \dot{X} = \mathbf{X}' \cdot \dot{\mathbf{X}}. \tag{12.16}$$

2) The Nambu–Goto action is named after the two physicists who introduced it.

Problem 12.3.1. *Working in the static gauge, and using formulas just derived, justify the choice of sign of the square root argument in Eq. (12.10) by showing that it is nonnegative everywhere on a physically valid string world sheet. Exclude unphysical world surfaces such as the one shown in Fig. 12.3(b). For confirmation and, likely a more rigorous proof, see Zwiebach, p. 99.*

12.3.4 String Tension and Mass Density

One is free to assign dimensions arbitrarily to the parameters in the string action. But the only mnemonically helpful choice is to treat σ as a length and τ as a time. With these choices $\sqrt{-\gamma}/c$ is dimensionless. The dimension of action S is *always* energy times time. It follows from Eq. (12.13) that T_0 times a length is an energy. Hence T_0 is a force or a tension, which accounts for its symbol – it seems natural to associate a tension with a string.

The strings under discussion are "ideal" (and unlike any spring encountered in the laboratory) in that their tension is independent of their length.[3] This being the case, the work done on a string as it is being stretched is proportional to the change in the string's length. If the entire string rest energy is ascribed to this work, then the string of zero length has zero rest energy and the energy of a string of length l is equal to $T_0 l$. Ascribing this energy to the rest mass of string of mass density μ_0, we have

$$T_0 = \mu_0 c^2. \tag{12.17}$$

In other words, T_0 and μ_0 are essentially equivalent parameters.

Problem 12.3.2. *When expressed in terms of the new metric, the action for a free point particle (Eq. (12.1)) is given by*

$$S = -mc \int_{t_i}^{t_f} d\tau \sqrt{-\eta_{\mu\nu} \frac{dx^\mu}{d\tau} \frac{dx^\mu}{d\tau}}. \tag{12.18}$$

3) A spring whose tension is independent of its length is unthinkable classically – it extends Hooke's law unconscionably. But elementary particle physicists are not at all troubled by such behavior in the context of quarks. Though the evidence for the existence of quarks is overwhelming, the fact that no free quark has ever been detected is troubling. The explanation is that the field lines of the (strong nuclear force) field acting on an individual quark are confined to a tube such that the force on one quark is independent of its distance from the quark at the other end of the tube. Neither quark can ever escape this attractive force to become free and individually detectable. This is known as "quark confinement." The closest classical analog is the *retractable* leash attaching a dog to its owner. The force exerted by the leash on the dog is independent of the length of the leash; the dog can roam, but never escape.

For this action the Lagrange equation reduces to

$$\frac{du^\nu}{d\tau} = 0, \quad \textit{where} \quad \dot{u}^\nu = \frac{x^\nu}{\sqrt{-\eta_{\mu\nu}\dot{x}^\mu \dot{x}^\nu}}. \tag{12.19}$$

The square root in Eq. (12.18) makes the Lagrangian unwieldy. A trick for getting rid of it is to permit the action (call it S') to depend on an artificial function $\zeta(\tau)$ whose form, like those of the other variables, is to be determined by causing S' to be extreme. The modified action is

$$S' = \frac{c}{2} \int \left(\frac{\eta_{\mu\nu}\dot{x}^\mu \dot{x}^\nu}{\zeta(\tau)} - \zeta(\tau)\, m^2 \right) d\tau. \tag{12.20}$$

Requiring $\delta S' = 0$, for a variation in which ζ is varied to $\zeta + \delta\zeta$, find an expression for $\zeta(\tau)$. Substitute this expression back into S' and show that the integrand reduces to the integrand of S. As with a Lagrange multiplier, the Lagrangian has been made manageable at the cost of introducing another function to be determined. One can consider $\zeta(\tau)$ to be a (monotonic) function of time defining an arbitrary reparameterization of the particle's world curve.

Problem 12.3.3. *A procedure like that in the previous problem can be used to get rid of the square root in the Nambu–Goto action for a relativistic string, which is Eq. (12.13). For the 2D metric on the world surface of the string, instead of accepting the metric induced from the standard 4D metric tensor $\eta_{\mu\nu}$ of the embedding space, one can introduce a 2D metric tensor $\zeta_{\alpha\beta}(\tau, \sigma)$. Its determinant is $\zeta = \det(\zeta_{\alpha\beta})$ and its covariant and contravariant components are related by the usual index raising and lowering operations giving, for example,*

$$\zeta_{\alpha\beta}\zeta^{\beta\gamma} = \delta^\gamma_\alpha, \tag{12.21}$$

where δ^γ_α is the Kronecker delta. When the elements of $\zeta_{\alpha\beta}$ are subjected to variations $\delta\zeta_{\alpha\beta}$, show that

$$\delta\zeta = \zeta\, \zeta^{\alpha\beta}\delta\zeta_{\alpha\beta} = -\zeta\, \zeta_{\alpha\beta}\delta\zeta^{\alpha\beta}. \tag{12.22}$$

An alternate action, call it S', can be written as

$$S' = -\frac{T_0}{c} \int\!\!\int d\tau\, d\sigma\, (-\zeta)^{1/2} \zeta^{\alpha\beta}\, h_{\alpha\beta}, \tag{12.23}$$

where

$$h_{\alpha\beta} = \eta_{\mu\nu}\frac{\partial X^\mu}{\partial x^\alpha}\frac{\partial X^\nu}{\partial x^\beta}, \quad \textit{and} \quad h = \det(h_{\alpha\beta}). \tag{12.24}$$

As was done with ζ in the previous problem, the elements of $\zeta_{\alpha\beta}$, in an initial step, are to be varied, and hence determined. Show that, in this step, the extremal condition $\delta S' = 0$ leads to the relation

$$\zeta^{\alpha\beta} h_{\alpha\beta} = 2(-h)^{1/2}(-\zeta)^{-1/2}. \tag{12.25}$$

Finally show that substituting this result into S' recovers the Nambu–Goto action. The action S' is known as the "Polyakov Action." It is apparently more suitable than the Nambu–Goto action for some purposes in quantum field theory.

12.4
Equations of Motion, Boundary Conditions, and Unexcited Strings

Part of the lore of Lagrangian mechanics is that once a Lagrangian has been found the problem is essentially solved. After defining momentum variables, one writes the Lagrange equations and then "turns the crank." Since we have found the Lagrangian we have, in that sense, finished the first part of our program. We will find though that relativity continues to impose complications.

We start by mimicking the steps taken in analyzing nonrelativistic strings in Chapter 1. Momentum densities on a nonrelativistic string were derived in Eqs. (1.122) and the wave equation was expressed in terms of those densities in Eq. (1.123). From Eq. (12.13) our relativistic Lagrangian density is

$$\mathcal{L} = -\frac{T_0}{c}\sqrt{(\dot{X}\cdot X')^2 - (\dot{X}\cdot\dot{X})(X'\cdot X')}. \tag{12.26}$$

This Lagrangian density shares some simplifying features with the nonrelativistic string Lagrangian of Eq. (1.117). Being a function only of first derivatives of X, $\mathcal{L}$ is independent of X itself, and of the coordinates τ and σ. (For the time being we do not investigate the conservation laws these imply.)

Following standard Lagrangian practice, canonical momentum densities are defined by

$$\begin{aligned}
\mathcal{P}^{(\tau)}_{\mu} &= \frac{\partial\mathcal{L}}{\partial\dot{X}^{\mu}} = -\frac{T_0}{c}\frac{(\dot{X}\cdot X')X'_{\mu} - (X'\cdot X')\dot{X}_{\mu}}{\sqrt{(\dot{X}\cdot X')^2 - (\dot{X}\cdot\dot{X})(X'\cdot X')}}\\
\mathcal{P}^{(\sigma)}_{\mu} &= \frac{\partial\mathcal{L}}{\partial X^{\mu\prime}} = -\frac{T_0}{c}\frac{(\dot{X}\cdot X')\dot{X}_{\mu} - (\dot{X}\cdot\dot{X})X_{\mu}{}'}{\sqrt{(\dot{X}\cdot X')^2 - (\dot{X}\cdot\dot{X})(X'\cdot X')}}.
\end{aligned} \tag{12.27}$$

Problem 12.4.1. *Formulas (12.27) for the canonical momentum densities appear somewhat less formidable when written in the static gauge. Show, for example, that the (σ) component is*

$$\begin{pmatrix}\mathcal{P}^{(\sigma)}_0\\ \vec{\mathcal{P}}^{(\sigma)}\end{pmatrix} = -\frac{T_0}{c\sqrt{}}\left(\dot{\mathbf{X}}\cdot\mathbf{X}'\begin{pmatrix}-c\\ \dot{\mathbf{X}}\end{pmatrix} - (-c^2 + \dot{\mathbf{X}}\cdot\dot{\mathbf{X}})\begin{pmatrix}0\\ \mathbf{X}'\end{pmatrix}\right); \tag{12.28}$$

here the blank square root symbol stands for the denominators of the preceding formulas.

Applying variations δX^μ to S, the variation $\delta\mathcal{L}$ can be expressed in terms of $\mathcal{P}_\mu^{(\tau)}$ and $\mathcal{P}_\mu^{(\sigma)}$. Permuting ∂ and δ as required and, as always, preparing for integration by parts,

$$\begin{aligned}\delta\mathcal{L} &= \mathcal{P}_\mu^{(\tau)}\delta\dot{X}^\mu + \mathcal{P}_\mu^{(\sigma)}\delta X^{\mu\prime} = \mathcal{P}_\mu^{(\tau)}\overbrace{\delta X^\mu}^{\cdot} + \mathcal{P}_\mu^{(\sigma)}\overbrace{\delta X^\mu}^{\prime} \\ &= \overbrace{\mathcal{P}_\mu^{(\tau)}\delta X^\mu}^{\cdot} + \overbrace{\mathcal{P}_\mu^{(\sigma)}\delta X^\mu}^{\prime} - (\overbrace{\mathcal{P}_\mu^{(\tau)}}^{\cdot} + \overbrace{\mathcal{P}_\mu^{(\sigma)}}^{\prime})\,\delta X^\mu.\end{aligned} \tag{12.29}$$

Being total derivatives, the first two terms, after integration, are expressible in terms of functions evaluated at the limits of integration. One way or another they will be made to vanish. We have control of the arbitrary factor δX_μ and can require it to vanish at initial and final times. This permits us to drop the first term immediately.

Treatment of the second term in the final form of Eq. (12.29) depends on boundary conditions at the ends of the string. This term will be made to vanish shortly. We are then left with only the final term which, as usual, yields the Lagrange equation which, in this case, is the wave equation;

$$\frac{\partial\mathcal{P}_\mu^{(\tau)}}{\partial\tau} + \frac{\partial\mathcal{P}_\mu^{(\sigma)}}{\partial\sigma} = 0. \tag{12.30}$$

Except for indices, this is the same as Eq. (1.125). A few comments concerning these indices should suggest some of the issues still to be faced.

The easiest comment concerns the $\partial/\partial\tau$ derivative in Eq. (12.30). At least in the static gauge, the parameter τ has been identified with the time t, which brings this term closer to conformity with Eq. (1.125). When it becomes necessary to introduce a different gauge this comment can be revisited.

The next easiest comment concerns the $\partial/\partial\sigma$ derivative in Eq. (12.30). Evidently this is analogous to $\partial/\partial x$ in Eq. (1.125) and measures rate of change along the string. But now, since the σ parameterization has been left open so far, this derivative is not yet uniquely defined. In any case $\mathcal{P}_\mu^{(\sigma)}$ describes the distribution along the string of a μ-component of momentum, whose total for the string will eventually be obtained by integrating over σ. A corresponding comment regarding dependence on τ has to be deferred to Section 12.9. The most serious relativistic complication is that, μ, being a tensor index, now takes on four values rather than the three values it can have in Eq. (1.125). This means that there are (superficially) four possible components of "string displacement."

String with fixed ends and at rest: The simplest possible configuration for a string is to be at rest and aligned with, say, the x-axis, with one end fixed at $x = 0$ and the other at $x = a$. From Eq. (12.14), in the static gauge, the string configuration is given by

$$X = (ct, f(\sigma), 0, 0), \quad \dot{X} = (c, 0, 0, 0), \quad X' = (0, f', 0, 0), \tag{12.31}$$

where $f(\sigma)$ is a monotonic, but otherwise arbitrary, parameterization function running from $f(0) = 0$ to $f(\sigma_1) = a$. The terms appearing in action S are

$$\dot{X} \cdot \dot{X} = -c^2, \quad X' \cdot X' = f'^2, \quad \dot{X} \cdot X' = 0, \quad \surd = \sqrt{0 - (-c^2)f'^2} = cf'. \tag{12.32}$$

Substitution into Eq. (12.13) produces

$$S = \int_{t_i}^{t_f} dt(-T_0 a). \tag{12.33}$$

The integrand can be interpreted as the Lagrangian in a system with no kinetic energy but potential energy V equal to the work done by tension T_0 in stretching the string from length zero to length a. This reconciles the formal equations with the same result obtained more heuristically above.

Problem 12.4.2. *Continuing with this simplest possible configuration, evaluate $\mathcal{P}_\mu^{(\tau)}$ and $\mathcal{P}_\mu^{(\sigma)}$ and show that wave equation (12.30) is satisfied for each value of μ.*

Fixed ends, general excitation: The conditions $\partial \mathbf{X}/\partial\tau = 0$ at both $\sigma = 0$ and $\sigma = \sigma_1$ are enough to kill the $\mu = 1,2,3$ boundary terms in Eq. (12.29) as required. But the δX^0 variation must be explicitly set to zero at the ends since (as in Eq. (12.31)), the X^0 component increases inexorably with t. In quantum field theory the fixed end possibility is not without problems as it begs the question "fixed to what." The answer to this question involves esoteric aspects of string theory such as *branes*, which we have no hope of understanding. We will either ignore the question or work with strings having free ends.

String with free ends: From Eq. (12.29) the conditions on free ends causing the boundary terms to vanish are

$$\mathcal{P}_\mu^{(\sigma)}(\tau, 0) = \mathcal{P}_\mu^{(\sigma)}(\tau, \sigma_1) = 0, \quad \mu = 0, 1, 2, 3. \tag{12.34}$$

These conditions will shortly be seen to have surprising consequences. But, before that, it is useful to transform the Lagrangian into a more convenient form.

12.5 The Action in Terms of Transverse Velocity

The action, as written so far, is too complicated to be used easily. Referring again to a point particle Lagrangian, instead of an implicit form, such as in the second form of Eq. (12.1), we need explicit velocity dependence, like that in the third form.

The local string velocity is $\dot{\mathbf{X}}(\tau, \sigma)$. But, as mentioned in an earlier footnote, because of the arbitrary nature of σ, this velocity is not ascribable to any identifiable particle. In particular, except for the end points, which are identifiable points, no meaningful interpretation whatsoever can be given to the velocity component parallel to $\mathbf{X}'$, which is parallel to the string. This component can be adjusted arbitrarily by changing the sigma parameterization. The velocity $\dot{\mathbf{X}}$, though not necessarily orthogonal to the string as in Fig. 12.3(c), necessarily has an orthogonal component, call it $\mathbf{v}_{\perp}$. Two snapshots of the string at times t and $t + dt$ show the string at two closely spaced positions. A plane drawn perpendicular to the first curve at position σ intersects the second curve at some position σ' (arbitrarily different from σ). $\mathbf{v}_{\perp}$ is the velocity inferred by pretending these points belong to an identifiable particle.

According to Frenet formula (6.11), a unit tangent vector to the string at fixed time is given by

$$\hat{\mathbf{t}} = \frac{d\mathbf{X}}{d\ell} = \mathbf{X}' \frac{d\sigma}{d\ell}, \tag{12.35}$$

where ℓ is arc length along the curve. Similarly, the vector $\dot{\mathbf{X}} \equiv \partial \mathbf{X}/\partial t$ is directed along a curve of constant σ. As already mentioned, this curve is not necessarily orthogonal to the string. But $\mathbf{v}_{\perp}$ can be obtained from $\dot{\mathbf{X}}$ by subtracting the component of $\dot{\mathbf{X}}$ parallel to the string. That is

$$\mathbf{v}_{\perp} = \dot{\mathbf{X}} - (\dot{\mathbf{X}} \cdot \hat{\mathbf{t}})\, \hat{\mathbf{t}} = \dot{\mathbf{X}} - \left(\dot{\mathbf{X}} \cdot \frac{d\mathbf{X}}{d\ell} \right) \frac{d\mathbf{X}}{d\ell}. \tag{12.36}$$

The dot product of this equation with itself gives

$$v_{\perp}^2 = \dot{\mathbf{X}} \cdot \dot{\mathbf{X}} - \left(\dot{\mathbf{X}} \cdot \frac{d\mathbf{X}}{d\ell} \right)^2. \tag{12.37}$$

Along with Eqs. (12.16), this permits the square root in the expression for S to be written in terms of $v_{\perp}$;

$$\begin{aligned}(\dot{X} \cdot X')^2 - (\dot{X} \cdot \dot{X})(X' \cdot X') &= (\dot{\mathbf{X}} \cdot \mathbf{X}')^2 - (-c^2 + \dot{\mathbf{X}} \cdot \dot{\mathbf{X}})|\hat{\mathbf{t}}|^2 \left(\frac{d\ell}{d\sigma} \right)^2 \\ &= \left(\frac{d\ell}{d\sigma} \right)^2 \left(\left(\dot{\mathbf{X}} \cdot \frac{d\mathbf{X}}{d\ell} \right)^2 + c^2 - \dot{\mathbf{X}} \cdot \dot{\mathbf{X}} \right) \\ &= c^2 \left(\frac{d\ell}{d\sigma} \right)^2 \left(1 - \frac{v_{\perp}^2}{c^2} \right).\end{aligned} \tag{12.38}$$

Substituting this into Eq. (12.26), and continuing to identify τ with time t, the Lagrangian density is

$$\mathcal{L} = -T_0 \frac{d\ell}{d\sigma} \sqrt{1 - \frac{v_{\perp}^2}{c^2}}. \tag{12.39}$$

Then the Lagrangian expressed as an integral over ℓ is,

$$L = -\int_0^l (T_0 d\ell)\sqrt{1 - \frac{v_\perp^2}{c^2}}. \tag{12.40}$$

Though σ runs over a fixed range, from 0 to σ_1, the string length l is, in general, time dependent. Recalling Eq. (12.17), the parenthesized factor can be expressed as $T_0 d\ell = \mu_0 c^2 d\ell$. Comparing again with the Lagrangian for a point particle (see Eq. (12.1)) one sees that the string Lagrangian can be interpreted as the sum of the Lagrangians for the intervals making up the string, treated as points. In this interpretation, the rest energy of a length $d\ell$ is $\mu_0 c^2 d\ell$ (which is consistent with the previous interpretation of the parameter μ_0) and the appropriate speed is $v_\perp$.

String With Free Ends (continued): With the denominator factor simplified by Eq. (12.38), the momentum densities in Eq. (12.28) simplify further. By Eq. (12.32) they must vanish at both end points of a free string;

$$0 = \begin{pmatrix} \mathcal{P}_0^{(\sigma)} \\ \vec{\mathcal{P}}^{(\sigma)} \end{pmatrix} = -\frac{T_0/c^2}{\sqrt{1 - v_\perp^2/c^2}} \frac{d\sigma}{d\ell} \left(\dot{\mathbf{X}} \cdot \mathbf{X}' \begin{pmatrix} -c \\ \dot{\mathbf{X}} \end{pmatrix} + (c^2 - \dot{\mathbf{X}} \cdot \dot{\mathbf{X}}) \begin{pmatrix} 0 \\ \mathbf{X}' \end{pmatrix} \right) \tag{12.41}$$

Though the denominator can approach zero it cannot become large and the $d\sigma/d\ell$ factor cannot vanish. It follows, therefore, that the parenthesized expression must vanish at free string ends.

Vanishing of the upper component requires $\dot{\mathbf{X}} \cdot \mathbf{X}' = 0$. This implies, since $\mathbf{X}'$ is directed along the string, that the velocity of the string end is purely orthogonal to the string. Essentially the same condition would apply at the end of a free nonrelativistic string; (for example think of the skater at the end of a crack-the-whip formation.)

The other implication of Eq. (12.41) is more troubling. Since $\mathbf{X}'$ cannot vanish, it follows that $|\dot{\mathbf{X}}| = c$. This means that the free string end has to move with the speed of light. As a consequence the present theory *cannot* be continued smoothly to the theory of a nonrelativistic string, at least for strings with free ends. This consideration alone would force a classical physicist to reject the theory, but it has not prevented quantum physicists from continuing.

Problem 12.5.1. *For a string with* fixed *ends there can be a smooth transition from nonrelativistic to relativistic motion. Consider a string stretched between $x = 0$ and $x = a$ as in Eq. (12.31), but distorted into the (x, y)-plane. For nonrelativistic $v_\perp \ll c$ and small amplitude $|dy/dx| \ll 1$ motion:*

(a) Determine $\mathbf{v}_\perp$ and string length interval $d\ell$ in terms of x, y and t.

(b) *Derive the nonrelativistic limit of the static gauge action (12.40) and obtain the nonrelativistic tension and mass density in terms of the parameters of the relativistic theory.*

(c) *Using the parameterization $x = f(\sigma) = a\sigma/\sigma_1$, repeat the previous two parts starting from the first of Eqs. (12.13).*

Problem 12.5.2. *It is rarely considered appropriate to study the motion of a nonrelativistic string in which the* paraxial *approximation, $|dy/dx| \ll 1$, is inapplicable. However, the relativistic string Lagrangian density should be applicable even for amplitudes large enough to violate this condition. Show that, in the nonrelativistic limit, the Lagrangian density valid for large amplitudes is*

$$\mathcal{L} = -T_0 \int_0^a dx \sqrt{1 + \left(\frac{\partial y}{\partial x}\right)^2 - \left(\frac{\partial y}{\partial ct}\right)^2}. \tag{12.42}$$

Problem 12.5.3. *An ordinary (certainly nonrelativistic) rubber band is stretched and at rest in the form of a circle of radius R_0 where it has mass density $\mu_0 = m/(2\pi R_0)$. After release of the band, the shape remains circular with radius R. The mechanical system can therefore be described by a single generalized coordinate $R(t)$. Now make the assumption (unrealistic for actual rubber bands) that the tension is independent of R. Write the Lagrangian for this system, find the Lagrange equation and solve it to find the band's time evolution after its release at $t = 0$. Don't be distressed if the string has to pass through a seemingly impossible configuration. [This problem could just as well have been in Chapter 1 since you are intended to use only nonrelativistic mechanics.]*

12.6 Orthogonal Parameterization by Energy Content

The strategy has been to delay specification of the string world surface's parameterization. So far, only the τ parameterization has been fixed by working in the static gauge. (Later even this choice will be revisited.) But, to make detailed calculations, it is necessary to have fully specified parameterizations.

One natural choice is to require the curves of constant σ to be *orthogonal* to curves of constant τ. This condition, illustrated in Fig. 12.3(c), can be expressed by

$$\mathbf{X}' \cdot \dot{\mathbf{X}} = 0, \quad \text{or} \quad \dot{\mathbf{X}} = \mathbf{v}_\perp. \tag{12.43}$$

Note that these are relations between 3-vectors, not among the 4-vectors they belong to. The latter constraint, already known from Eq. (12.41) to be valid at string ends, now applies at every point on the string. Comparing with

Eq. (12.41), it can be seen that these relations provide important simplification.

$$\begin{pmatrix} \mathcal{P}_0^{(\sigma)} \\ \vec{\mathcal{P}}^{(\sigma)} \end{pmatrix} = -T_0 \frac{\sqrt{1 - v_\perp^2/c^2}}{d\ell/d\sigma} \begin{pmatrix} 0 \\ \mathbf{X}' \end{pmatrix}, \tag{12.44}$$

$$\begin{pmatrix} \mathcal{P}_0^{(\tau)} \\ \vec{\mathcal{P}}^{(\tau)} \end{pmatrix} = \frac{T_0}{c^2} \frac{d\ell/d\sigma}{\sqrt{1 - v_\perp^2/c^2}} \begin{pmatrix} -c \\ \dot{\mathbf{X}} \end{pmatrix}. \tag{12.45}$$

The second result has come from making the same substitutions in the first of Eqs. (12.27). Substitution of the time-like components into the wave equation (12.30) yields

$$\frac{\partial(\mathcal{P}^{(\tau)0} d\sigma)}{\partial \tau} = \frac{\partial}{\partial \tau}\left(\frac{T_0 d\ell}{\sqrt{1 - v_\perp^2/c^2}} \right) = 0, \tag{12.46}$$

where cross-multiplying the $d\sigma$ factor is valid because the evolution has previously been arranged to be parallel to lines of constant σ. This equation makes it valid to identify $dE = \mathcal{P}^{(\tau)0} d\sigma$ as a conserved quantity and natural to call dE the total energy of that segment of string. With rest energy $c^2 dm$ as previously defined, we have

$$\frac{dE}{c^2 dm} = \frac{T_0 d\ell / \sqrt{1 - v_\perp^2/c^2}}{T_0 d\ell} = \frac{1}{\sqrt{1 - v_\perp^2/c^2}}. \tag{12.47}$$

This ratio is the usual "relativistic gamma factor," as it applies to point masses, though here calculated from only the transverse component of string velocity.

At this point, comparing with Section 1.9, one might be tempted to claim the behavior of the relativistic string to be simpler than that of a nonrelativistic string. For the classical string, though the total energy is conserved, the local energy density is not. For the relativistic string even the local energy density (as defined here) is constant. This is something of a mathematical artifact however, since the relativistic density is reckoned per unit of σ, while the nonrelativistic density is reckoned per unit of x. So the present formulas approach the nonrelativistic limit smoothly.

The σ parameterization has not yet been established. Equations (12.44) and (12.45) beg to have this parameterization fixed by the relation

$$d\sigma = \frac{d\ell}{\sqrt{1 - v_\perp^2/c^2}} = \frac{dE}{T_0}, \quad \text{and hence} \quad \sigma_1 = \frac{E}{T_0}. \tag{12.48}$$

Based on the previous discussion this can be referred to as "parameterization by energy content" since, except for a constant factor, $d\sigma$ is equal to the energy content of the segment. This consideration has fixed the total energy E in terms of σ_1 and T_0.

At this point we make a brief digression. This parameterization by energy content is not manifestly invariant. For present purposes this will not matter, but it will become important in later sections. Instead of concentrating on total string energy E one can introduce both p, as the total string 4-momentum, and a 4-vector $n_\mu = (1,0,0,0)$, which is directed along the fixed σ, changing τ, coordinate curve. With $\mathcal{P}^{(\tau)}$ being the energy/momentum density, in terms of these quantities the parameterization by energy content can be expressed as the requirement that $n \cdot \mathcal{P}^{(\tau)}$ be constant. Since the first component of p is E/c, this parameterization is the same (except for a possible numerical factor depending on the choice of units and the range of σ) to that of Eq. (12.48). Spelled out more explicitly, the parameterization could be expressed as

$$\sigma = \pi \frac{n \cdot \int_0^\sigma \mathcal{P}^{(\tau)}(\tau, \sigma')\, d\sigma'}{n \cdot p}. \tag{12.49}$$

The integral is calculated along a curve of constant τ. Except that the range of σ here has been arranged to be $0 \leq \sigma \leq \pi$, this parameterization is equivalent to Eq. (12.48). This formula obviously establishes σ as increasing monotonically from 0 at one end to π at the other end of the string. For the time being this rearrangement of formulas is purely pedantic. In a later section, to simplify the formulas, a different choice will be made for the vector n. But Eq. (12.49) will still be valid.

After this digression, and continuing from Eqs. (12.48), the momentum densities are given by

$$\begin{pmatrix} \mathcal{P}_0^{(\sigma)} \\ \vec{\mathcal{P}}^{(\sigma)} \end{pmatrix} = -T_0 \begin{pmatrix} 0 \\ \partial \mathbf{X}/\partial \sigma \end{pmatrix}, \quad \begin{pmatrix} \mathcal{P}_0^{(\tau)} \\ \vec{\mathcal{P}}^{(\tau)} \end{pmatrix} = \frac{T_0}{c} \begin{pmatrix} -c \\ \partial \mathbf{X}/\partial t \end{pmatrix}. \tag{12.50}$$

Substituting these expressions into Eq. (12.30), the partial differential equation governing excitations of the string is

$$\frac{1}{c^2} \frac{\partial^2 \mathbf{X}}{\partial t^2} = \frac{\partial^2 \mathbf{X}}{\partial \sigma^2}. \tag{12.51}$$

This obviously reduces to the wave equation for small oscillations, but, for large amplitudes, the complexity has been masked by the σ-parameterization introduced in Eq. (12.48).

12.7 General Motion of a Free Open String

Equation (12.51) can be solved by mimicking the steps taken in Section 1.10. Here these steps will be expressed in the form of a series of problems, interrupted only to consider the complications brought on by the parameterization conditions. Let us assume the string is open, with free ends and the longitudinal parameter σ is defined to run from 0 to σ_1, so the boundary conditions are

$$\mathbf{X}'(t,0) = \mathbf{X}'(t,\sigma_1) = 0. \tag{12.52}$$

Problem 12.7.1. *Show that the general solution of Eq. (12.51) can be written as*

$$\mathbf{X}(t,\sigma) = \frac{1}{2}\left(\mathbf{F}(ct+\sigma) + \mathbf{F}(ct-\sigma)\right), \tag{12.53}$$

where the function $\mathbf{F}(u)$*, like the function illustrated in Fig. 1.8, satisfies*

$$\frac{d\mathbf{F}}{du}(u+2\sigma_1) = \frac{d\mathbf{F}}{du}(u), \tag{12.54}$$

and which therefore satisfies

$$\mathbf{F}(u+2\sigma_1) = \mathbf{F}(u) + 2\sigma_1\frac{\mathbf{v}_0}{c}. \tag{12.55}$$

The constant of integration $\mathbf{v}_0$ is determined by initial conditions; its accompanying factor has been chosen for dimensional convenience. Unlike the solution in Section 1.10, because the present string is free, this term cannot be dropped. This solution is *too general* however, as no account has yet been taken of the orthogonal and energy content σ-parameterization constraints.

Problem 12.7.2. *Show that the orthogonal constraint (Eq. (12.43) second version) along with the energy content constraint (Eq. (12.48) expressed in terms of* $\mathbf{X}'$ *and* $\dot{\mathbf{X}}$*) can be combined into the constraint equations*

$$(\mathbf{X}' \pm \dot{\mathbf{X}})^2 = 1. \tag{12.56}$$

Problem 12.7.3. *Substituting Eq. (12.53) into constraint (12.56), show that the constraint is met by requiring*

$$\left|\frac{dF(u)}{du}\right| = 1. \tag{12.57}$$

With this requirement included, Eq. (12.53) provides the general solution. The motion of the σ=0 end of the string is especially simple – by Eq. (12.53),

$$\mathbf{X}(t,0) = \mathbf{F}(ct). \tag{12.58}$$

At time $t + 2\sigma_1/c$, because of the periodicity condition, the displacement $\mathbf{X}(t + 2\sigma_1/c, 0)$ will have shifted by the amount $2\sigma_1\mathbf{v}_0/c$. The average velocity of this point (and every other point on the string) is therefore $\mathbf{v}_0$.

12.8
A Rotating Straight String

A uniformly rotating string of length l, centered on the origin, is shown in Fig. 12.4. Since there is no net string motion $\mathbf{v}_0 = 0$. The motion of the $\sigma = 0$ end, and from it, $\mathbf{F}$, are given by

$$\mathbf{X}(t,0) = \frac{l}{2}\,(\cos\omega t, \sin\omega t) = \mathbf{F}(ct). \tag{12.59}$$

Then, by Eq. (12.58),

$$\mathbf{F}(u) = \frac{l}{2}\left(\cos\frac{\omega u}{c},\ \sin\frac{\omega u}{c}\right). \tag{12.60}$$

Periodicity condition (12.55) requires

$$\frac{\omega}{c} = m\,\frac{\pi}{\sigma_1}, \tag{12.61}$$

where m is an integer. This condition synchronizes the externally visible angular frequency ω with the frequency needed to satisfy the wave equation and boundary conditions. At fixed time, again using Eq. (12.53), the string shape is given by

$$\mathbf{X}(0,\sigma) = \frac{1}{2}\,(\mathbf{F}(\sigma) + \mathbf{F}(-\sigma)) = \frac{l}{2}\left(\cos m\pi\frac{\sigma}{\sigma_1},\ 0\right), \quad \text{where} \quad m = 1. \tag{12.62}$$

Here the requirement that parameter σ and arc length along string be monotonically related has selected the value $m = 1$.

So far the description is consistent with the string motion being nonrelativistic, with freedom to choose its physical observables, but there are still more conditions to be met. Combining Eqs.(12.48) and (12.61) yield

$$\frac{\omega}{c} = \frac{\pi T_0}{E}, \tag{12.63}$$

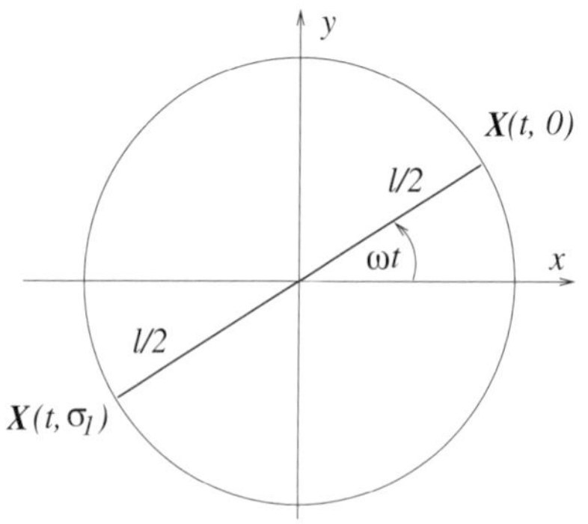

Fig. 12.4 A uniformly rotating relativistic string.

showing that ω varies inversely with E. Finally, it is necessary to satisfy Eq. (12.57), which requires

$$l = \frac{2c}{\omega}, \quad \text{or equivalently,} \quad E = \frac{\pi}{2} T_0 l, \tag{12.64}$$

thereby specifying both l and E. This also clearly reproduces the result that both ends of the string travel at speed c.

Problem 12.8.1. *Show that the fully parameterized expression for the string's evolution is*

$$\mathbf{X}(t,\sigma) = \frac{\sigma_1}{\pi} \cos\frac{\pi\sigma}{\sigma_1}\left(\cos\frac{\pi c t}{\sigma_1}, \sin\frac{\pi c t}{\sigma_1}\right). \tag{12.65}$$

Problem 12.8.2. *For the uniformly rotating string just analyzed, let ℓ be a coordinate measuring (signed) distance from the origin, with $\ell = -l/2$ corresponding to $\sigma = 0$. The energy density at position ℓ on a uniformly rotating string is defined by $\mathcal{E} = dE/d\ell$.*

(a) Find $v_\perp$ as a function of ℓ.

(b) With T_0 given, find and then plot the function $\mathcal{E}(\ell)$.

(c) Find the total string energy E (confirming that the singularities at the end points are integrable) and from it find the average energy density $\langle\mathcal{E}\rangle = E/l$.

(d) Find the central fraction of the total string length that contains all but one percent of the total string energy.

Problem 12.8.3.

(a) Repeat problem 12.5.3 for a relativistic string having fixed tension T_0, collapsing from rest at radius $R(0) = R_0$ and remaining circular with radius $R(t)$. It is useful to find $v_\perp$ as a function of dR/dt, and from that to find the total energy E (which is conserved).

(b) Find the function $\mathbf{F}(u)$ which, substituted into Eq. (12.53), describes the motion of part (a).

Problem 12.8.4. *One end of a relativistic "jumping rope" is attached at the origin and the other at z=a. The rope lies in a plane that coincides with the (x,z)-plane at $t = 0$ but which rotates with angular velocity ω around the z-axis. The vector function*

$$\mathbf{F}'(u) = \sin\gamma\cos\omega t\hat{\mathbf{x}} + \sin\gamma\sin\omega t\hat{\mathbf{y}} + \cos\gamma\hat{\mathbf{z}}, \tag{12.66}$$

meets the requirement of Eq. (12.57). Find the angle the rope makes with the z-axis at the ends of the rope. Also find the total energy of the string and find how it is distributed along the string.

12.9 Conserved Momenta of a String

So far one has concentrated mainly on deformations of a relativistic string. These deformations are necessarily restricted to the world sheet of the string. If a string of finite length is to eventually serve as a model for an elementary particle, it is important to infer the dynamical properties manifested by the string as a whole. These are the same properties whose constancy reflects the symmetries of the string Lagrangian – the linear and angular momenta. Since spin angular momentum is an inherent property of an elementary particle, if the eventual goal is the modeling of elementary particles by strings, then the angular momentum of the string is of special interest,

As defined in Eq. (12.27), the densities $\mathcal{P}_\mu^{(\tau)}$ and $\mathcal{P}_\mu^{(\sigma)}$ were introduced following the standard recipe of Lagrangian mechanics. But we know, from the translational invariance of the Lagrangian, that these quantities are conserved or, at least that they are central to the identification of conserved quantities. (One reason they may not themselves be conserved is that they involve derivatives with respect to the somewhat arbitrary parameters σ and τ.)

In the static gauge, with τ identified with time t, and $\mathcal{P}_\mu^{(\tau)}$ being a derivative of the Lagrangian with respect to a velocity, it is natural to define

$$p_\mu(\tau) = \int_0^{\sigma_1} \mathcal{P}_\mu^{(\tau)}(\tau,\sigma)\, d\sigma, \tag{12.67}$$

as a candidate for total string momentum. Being the integral over σ of a density with respect to σ, the quantity thus defined has the possibility of being independent of the σ parameterization. To test whether $p_\mu(\tau)$ is conserved, which is to say, independent of τ, we differentiate with respect to τ;

$$\frac{dp_\mu}{d\tau} = \int_0^{\sigma_1} \frac{\partial \mathcal{P}_\mu^{(\tau)}}{\partial \tau}\, d\sigma = -\int_0^{\sigma_1} \frac{\partial \mathcal{P}_\mu^{(\sigma)}}{\partial \sigma}\, d\sigma = -\mathcal{P}_\mu^{(\sigma)}\Big|_0^{\sigma_1}. \tag{12.68}$$

Equation (12.30), which is the wave equation for waves on the string, was used in the intermediate step. The inferences to be drawn from this equation depend on the string configuration:

(a) Closed string: since $\sigma = 0$ and $\sigma = \sigma_1$ are the same point, $p_\mu|_\tau$ is conserved.

(b) Open string, free ends: By Eq. (12.32), $\mathcal{P}_\mu^{(\sigma)}$ vanishes at both ends, so $p_\mu|_\tau$ is conserved.

(c) Open string, attached end: $p_\mu|_\tau$ is not, in general, conserved.

With discussion being restricted to cases (a) and (b), p_μ will be a conserved. In fact, from now on, we will limit the discussion case (b), free open strings.

When based on the static gauge, as it has been, it has been natural to integrate the momentum density at fixed time and interpret the result as momentum at that time. With a more general gauge, an integration over σ at fixed τ brings in contributions at different values of the time t. Nevertheless, using conserved charge considerations like those in Section 10.8.1 of the present text, Zwiebach shows how the string momentum can be obtained for arbitrary string parameterization.

With formulas from Section 10.9, the angular momentum of a string can be obtained as an integral of "moments of momentum" over the string. Copying from Eq. (10.84), the tensor of angular momentum density is given by $\mathcal{M}^{(\tau)}_{\mu\nu}(\sigma,\tau) = X_\mu \mathcal{P}^{(\tau)}_\nu - X_\nu \mathcal{P}^{(\tau)}_\mu$ and the total angular momentum is

$$M_{\mu\nu} = \int_0^{\sigma_1} \left(X_\mu \mathcal{P}^{(\tau)}_\nu - X_\nu \mathcal{P}^{(\tau)}_\mu \right) d\sigma. \tag{12.69}$$

Though the right-hand side of this equation depends formally on τ, conservation of $M_{\mu\nu}$ follows as in Section 10.9.

As discussed so far, the momentum densities $\mathcal{P}^{(\tau)}_\mu$ and $\mathcal{P}^{(\sigma)}_\mu$ are necessarily tangent to the 2D world surface of the string. The "indices" (τ) and (σ) define two directions in this surface. For these momentum densities to be geometric quantities in four dimensions they would need to have four indices. A possible parameterization of the 4D space by coordinates $(\xi^0, \xi^1, \xi^2, \xi^3)$ *could* be chosen such that $\xi_0 \equiv \tau$ and $\xi_1 \equiv \sigma$. Then, in all formulas appearing so far, the superscripts (τ) and (σ) would be 0 and 1, respectively. The momentum densities would then be symbolized $\mathcal{P}^\nu_\mu$, $\nu = 0, 1, 2, 3$, with the upper indices no longer in parentheses. We will not make this change of notation however.

12.9.1 Angular Momentum of Uniformly Rotating Straight String

We can apply the formula just derived to find the angular momentum of the uniformly rotating string described in Section 12.8. The fully parameterized string evolution was given in Eq. (12.65), and the momentum density therefore by

$$\mathcal{P}^{(\tau)} = \frac{T_0}{c^2}\dot{\mathbf{X}} = \frac{T_0}{c}\cos\frac{\pi\sigma}{\sigma_1}\left(-\sin\frac{\pi c t}{\sigma_1}, \cos\frac{\pi c t}{\sigma_1} \right). \tag{12.70}$$

Since the string is rotating in the x, y-plane, the only nonvanishing spatial components of M are $M_{12} = -M_{21}$, with the angular momentum $J = |M_{12}|$. Substituting into Eq. (12.69) and performing the integration yields

$$J = \int_0^{\sigma_1} \left(X_1 \mathcal{P}^{(\tau)}_2 - X_2 \mathcal{P}^{(\tau)}_1 \right) d\sigma = \frac{\sigma_1}{\pi}\frac{T_0}{c}\frac{\sigma_1}{2} = \frac{E^2}{2\pi T_0 c}. \tag{12.71}$$

The final simplification has used the result $\sigma_1 = E/T_0$, where E is the total string energy.

As expected, the angular momentum J is a constant of the motion. Reflecting on this formula, the dependence of J on E is curious. For a string of fixed length J would be proportional to ω which would be proportional to $\sqrt{E}$. The far stronger dependence on E shown by Eq. (12.71) reflects the increase of string length with increasing E.

From the point of view of classical physics we are now finished with this system. But, from the point of view of elementary particle physics, a few more points can be made. For one thing our string now has spin angular momentum; this meets one requirement for modeling a particle with spin. But it is curious (unless compensated by the T_0 dependence) for the spin to be proportional to the particle mass-squared (since E is to be interpreted as mc^2). On the contrary it was actually this dependence that first suggested the importance of analyzing strings. (If I am not mistaken) it was a dynamical particle model in which a series of elementary particles had angular momentum proportional to mass-squared that caused a physicist (Susskind, I believe) to first introduce strings as models of elementary particles.

Accepting as "physics" the proportionality $J \sim E^2$, with α' (a symbol adopted for historical reasons) being the constant of proportionality, we have

$$J = \alpha' E^2, \quad \text{where} \quad \alpha' = \frac{1}{2\pi T_0 c}. \tag{12.72}$$

As "physics" this constitutes a major advance, because the entirely artificial parameter T_0 has been replaced by a quantity that is, in principle, measurable, or at least can be estimated within elementary particle physics. To see the orders of magnitude involved one should refer to Zwiebach's book. Of course, this includes the introduction of Planck's constant $\hbar$; this being an angular momentum, it is natural to express J in units of $\hbar$.

If you ask me, this motivation for continuing the development seems pretty lame but, so far, no one has asked me. In any case, if the string we have analyzed so far were a violin string, it would not yet make a noise. In other words, internal deformations of the string have not yet been described. More explicitly the modes of oscillation have to be determined, and the corresponding system energy found. These are the last remaining tasks to be completed within classical physics, as preparation for launching into quantum theory.

12.10 Light Cone Coordinates

When expressed in so-called "natural" units, with $\hbar = c = 1$, with T_0 eliminated in favor of α', and with τ and σ now taken to be dimensionless, the

Nambu–Goto action of Eq. (12.13) is

$$S = -\frac{1}{2\pi\alpha'} \int_{\tau_i}^{\tau_f} d\tau \int_0^{\sigma_1} d\sigma \sqrt{(\dot{X} \cdot X')^2 - (\dot{X} \cdot \dot{X})(X' \cdot X')}. \tag{12.73}$$

There is a quite close correspondence between the description of waves on a relativistic string in Section 12.7 and waves on an ordinary sting in Section 1.10. In mimicking the subsequent steps from Section 1.10, there are two formulas that are going to give trouble, namely Eqs. (12.43) and (12.56) which read

$$\mathbf{X}' \cdot \dot{\mathbf{X}} = 0, \quad \text{and} \quad (\mathbf{X}' \pm \dot{\mathbf{X}})^2 = 1. \tag{12.74}$$

The second equation can be simplified a bit by substituting from the first, but we are still left with a kind of Pythagorean constraint between the slope and the velocity components of a wave on a string. From a numerical standpoint such a constraint can be routinely handled, for example by solving the quadratic equation satisfied by, say, X'^0, to fix X'^0 as a function of the remaining components of $\mathbf{X}'$ and all the components of $\dot{\mathbf{X}}$. But we are talking here about the components of 4-vectors, for which only linear relationships can be invariantly described. So such a quadratic relationship cannot be tolerated. Clearly, for a manifestly invariant theory, relations (12.74) have to be expressed as 4-scalar relationships. The trick for doing this is to work in the "light-cone gauge." This is the subject to be addressed next.

Up to this point a distinction has been made between the parameter τ and the time t (which is the same as X^0/c).[4] In reality, the distinction between t and τ has been purely academic up to this point since all results were obtained in the static gauge. But with τ being taken proportional to a new coordinate X^+, as we now intend, we must, definitively, leave the static gauge behind. Fortunately, it is not too difficult to update the formulas accordingly.

The static gauge choice, $X^0(\tau, \sigma) = c\tau$, could have been expressed as $n \cdot X = c\tau$ where $n = (1, 0, 0, 0)$ is the same constant 4-vector that was introduced in the introductory comments preceding Eq. (12.49). That equation anticipated the possibility of a different choice for n, and that choice is now to be considered. Let us take the τ-parameterization as

$$n \cdot X = 2\alpha' \, (n \cdot p) \, \tau, \tag{12.75}$$

where p is the string total 4-momentum vector, and the choice of n is left open for the moment. This choice amounts to requiring τ to be constant along the

4) Zwiebach also makes a valid distinction between (lower case) coordinate 4-vector x^μ and (upper case) string displacement X^μ, but we are suppressing this distinction.

curve formed by the intersection of the string's world surface with a hyperplane $n \cdot X = 2\alpha' (n \cdot p)\, \tau$. Pictorially this parameterization amounts to tipping the coordinate hyperplanes in Fig. 12.1. The constant of proportionality $2\alpha' (n \cdot p)$ has been chosen for later convenience.

For the same choice of n, the σ-parameterization will now be taken to be that given in Eq. (12.49). With that parameterization $n \cdot \mathcal{P}^{(\tau)}(\tau, \sigma)$ is constant as a function of σ. Furthermore, $n \cdot \mathcal{P}^{(\tau)}$ is, in fact, constant on the entire string world sheet, since the range of σ is independent of τ. Then Eq. (12.49) reduces to

$$n \cdot p = \pi\, n \cdot \mathcal{P}^{(\tau)}, \tag{12.76}$$

which relates the string momentum to its momentum density. Through the wave equation (12.30), this constancy also has an important implication for $n \cdot \mathcal{P}^{(\sigma)}$. Operating with $n\cdot$ on that equation gives $\partial (n \cdot \mathcal{P}^{(\sigma)})/\partial\sigma = 0$. For open strings with free ends, which is the only case we are considering, according to Eq. (12.68), $\mathcal{P}^{(\sigma)}$ vanishes at the string ends. We therefore have

$$n \cdot \mathcal{P}^{(\sigma)} = 0, \tag{12.77}$$

everywhere on the string's world sheet. When this formula is expressed in terms of the string deformation using Eq. (12.27) the result is

$$0 = n \cdot \mathcal{P}^{(\sigma)} = -\frac{1}{2\pi\alpha'} \frac{(\dot{X} \cdot X')\, \partial (n \cdot X)/\partial\tau}{\sqrt{(\dot{X} \cdot X')^2 - (\dot{X} \cdot \dot{X})(X' \cdot X')}}; \tag{12.78}$$

it was Eq.(12.75), according to which $n \cdot X$ is independent of σ, that caused the $\partial/\partial\sigma$ term in the numerator to vanish. Finally, this gives the manifestly invariant generalization of Eq. (12.74) that we have been looking for, namely,

$$\dot{X} \cdot X' = 0. \tag{12.79}$$

Working in a different "gauge" means using a particular set of basis vectors other than the standard ones. In the "light-cone gauge" in terms of the original (X^0, X^1, X^2, X^3) the new coordinates (X^+, X^-, X^2, X^3) for a 4-vector X are given by

$$X^+ = \frac{1}{\sqrt{2}}(X^0 + X^1), \quad X^- = \frac{1}{\sqrt{2}}(X^0 - X^1), \tag{12.80}$$

with X^2 and X^3, the so-called "transverse" coordinates, are common to both sets. By choosing $n_\mu = (1/\sqrt{2})(1, 1, 0, 0)$ the time-like component of vectors like X and p are given by

$$X^+ = n \cdot X \quad \text{and} \quad p^+ = n \cdot p. \tag{12.81}$$

Substituting these values into Eq. (12.75) yields

$$X^+(\tau,\sigma) = 2\alpha'\, p^+\, \tau. \tag{12.82}$$

This is a rather satisfying result in that it gives the coordinate that we wish to treat as the "time" coordinate as being proportional to τ, independent of σ.

Even though there are only two transverse components, it is useful to introduce an indexing convention that allows summations over them to be expressed by the summation convention – upper case Roman letters $I, J, \ldots$, will be used for this purpose. In these new, so-called "light cone coordinates," X^+ is to be treated as the "time-coordinate" and the metric is

$$ds^2 = 2\,dX^+dX^- - (dX^I)^2, \tag{12.83}$$

where $(dX^I)^2 \equiv dX^2dX^2 + dX^3dX^3$. You can easily check this by substituting into Eq. (12.9).

These light-cone coordinates amount to choosing space-time axes that are orthogonal to each other and slanted at 45° relative to the usual axes. It is not difficult to correlate the evolution of particle world lines when expressed in these coordinates with their description in standard coordinates, but the details will not be discussed here. When expressed with X^+ as independent "time-like" coordinate the world trajectories may seem bizarre, but, for particles within the light cone (β<1) the transformation $(X^0, X^1) \rightarrow (X^+, X^-)$ is single valued and well behaved. You should refer to Zwiebach to become familiar with particle relationships when they are expressed using these new coordinates.

After making all these choices, the string equations for open strings have become:

(a) Wave equation: $\ddot{X}^\mu - X^{\mu''} = 0$.

(b) Orthonormal constraint: $\left(\dot{X} \pm X'\right)^2 = 0$.

(c) Boundary conditions: $X'(\tau, 0) = X'(\tau, \pi) = 0$.

(d) Momentum densities: $\mathcal{P}^{(\tau)\mu} = \frac{1}{2\pi\alpha'}\,\dot{X}^\mu, \quad \mathcal{P}^{(\sigma)\mu} = -\frac{1}{2\pi\alpha'}\,X^{\mu\prime}$.

In all the cases the functions depend on σ and τ.

After all this weakly motivated build-up we are finally in a position to see what has been the benefit of working in the light-cone gauge. Using the metric as given by Eq. (12.83) to expand the self dot products of the expressions in constraint (b), the results are

$$2(\dot{X}^+ \pm X'^+)(\dot{X}^- \pm X'^-) = \left(\dot{X}^I \pm X'^I\right)^2. \tag{12.84}$$

Both terms in the first factor can be worked out using Eq. (12.82). Dividing through by this result yields

$$\dot{X}^- \pm X'^- = \frac{1}{4\alpha' p^+}\left(\dot{X}^I \pm X'^I\right)^2. \tag{12.85}$$

We are now in a position to answer a question raised earlier. With a wave represented by a four component object, what do the four amplitudes represent? Since the first component is essentially an independent variable, like a time, it doesn't count as an amplitude. If the strings were nonrelativistic, one would say that, of the three remaining amplitudes, two are transverse, and one is longitudinal. But, for relativistic strings we have insisted that no longitudinal disturbance is supportable, because points internal to the string have no distinguishable identity. The constraints just obtained in Eq. (12.85) are consistent with this picture. They can be interpreted as showing that the X^- waves are "phony," in that they are derivable from the X^2 and X^3 waves. They have no detectable physical significance. This suppression of one mode should be reminiscent of electromagnetic waves. In that case the amplitudes are three of the four components of the vector potential. As now, in that case, the longitudinal wave is suppressed, leaving only two (transverse) amplitudes.

Problem 12.10.1. *Combine formulas from the previous section to obtain the two orthonormal constraint formulas listed as part (b).*

12.11
Oscillation Modes of a Relativistic String

Finding the normal modes of a classical stretched string is very easy; it was done, for example, in Chapter 1. Essentially, the same calculation is to be repeated now for a relativistic string, with σ standing in for longitudinal position along the string and τ standing in for time. But there are still relativistic complications that have to be faced.

One is familiar with the Fourier expansion of waves on a string, for example from Section 1.10. The range of σ, from 0 to π, has been chosen to simplify this expansion. Here, we simply write down the solution of (a) as a Fourier series, though using complex exponentials this time instead of sines and cosines, intending to justify the various coefficients by *ex post facto* comments;

$$X^\mu(\tau,\sigma) = x_0^\mu + 2\alpha' p^\mu \tau - i\sqrt{2\alpha'}\sum_{n=1}^{\infty}\left(a_n^{\mu *}e^{in\tau} - a_n^\mu e^{-in\tau}\right)\frac{\cos n\pi}{\sqrt{n}}. \qquad (12.86)$$

The leading term obviously represents the initial position of the string; it can be set to zero with impunity. The second term represents systematic uniform translation of the string as a whole. The coefficient is proportional to the string 4-momentum p. This can be confirmed, along with its numerical factor, by substituting Eq. (12.86) into the first of the (d) equations above, and integrating over σ. The remaining terms in the expansion all vanish in this integration. However, the value of p itself depends on the state of string excitation. Setting this term to zero would amount to working in the rest system of the string.

As regards the Fourier expansion, the fact that the n and $-n$ coefficients are complex conjugates is required for the result to be real, and the factors $-i\sqrt{2\alpha'}$ and $1/\sqrt{n}$ have been introduced for later convenience.

As it stands, Eq. (12.86) satisfies wave equation (a), and it also satisfies boundary conditions (c). But it does not, as yet, satisfy orthonormalization condition (b). To reduce the Fourier series to the sum of one (instead of two) terms, while allowing n to range over all nonzero integers, Zwiebach introduces equivalent coefficients

$$\alpha_0^\mu = \sqrt{2\alpha'}\, p^\mu, \quad \text{and} \quad \alpha_n^\mu = a_n^\mu \sqrt{n}, \quad \alpha_{-n}^\mu = a_n^{\mu *} \sqrt{n}, \quad n > 0. \tag{12.87}$$

In terms of these coefficients the expansion becomes

$$X^\mu(\tau,\sigma) = x_0^\mu + \sqrt{2\alpha'}\,\alpha_0^\mu\,\tau + i\sqrt{2\alpha'} \sum_{n\neq 0}^{\infty} \frac{1}{n}\,\alpha_n^\mu e^{-in\tau} \cos n\pi. \tag{12.88}$$

Notice that the time dependence factor has been arranged to be $e^{-in\tau}$, with a negative sign, which is the usual physicist's convention. To be consistent with Eq. (12.82), for the X^+ amplitude, only the second term can survive; all the other coefficients vanish. As shown in Eq. (12.85) and the next problem, all the X^- coefficients are derivable from the X^2 and X^3 coefficients.

Problem 12.11.1. *Using the result*

$$\dot{X}^\mu \pm X'^\mu = \sqrt{2\alpha'} \sum_n \alpha_n^\mu\, e^{-in(\tau\pm\sigma)}, \tag{12.89}$$

(obtained by substituting from Eq. (12.88) into constraint (b)) with the sum being over all integers, but only transverse amplitudes, show that

$$\sqrt{2\alpha'}\alpha_n^- = \frac{1}{2p^+} \sum_p \left(\alpha_{n-p}^2\,\alpha_p^2 + \alpha_{n-p}^3\,\alpha_p^3\right), \tag{12.90}$$

or look it up in Zwiebach. For what it is worth, this is known as a Virasoro expansion – named after a pure mathematician working in a field utterly removed from string theory. Regrettably, since the light-cone coordinates are not *related to laboratory coordinates by a Lorentz transformation, the transverse coordinates are* not *the same as laboratory frame transverse coordinates.*

There is only one more thing to be done. It is to calculate the mass m (which, with $c = 1$, is the same as rest energy) of a string in an arbitrarily excited state. From the light-cone metric (12.83), m is given by

$$m^2 = 2p^+p^- - p^2p^2 - p^3p^3. \tag{12.91}$$

By Eq. (12.87), $p^- = \alpha_0^-/\sqrt{2\alpha'}$, and, by Eq. (12.90), $\alpha_0^- = \frac{1}{2p^+\sqrt{2\alpha'}} \sum_p \alpha_{-p}^I\,\alpha_p^I$, so

$$2p^+p^- = \frac{1}{2\alpha'} \sum_p \alpha_{-p}^I\,\alpha_p^I. \tag{12.92}$$

Substituting into Eq. (12.91) yields

$$m^2 = \frac{1}{\alpha'} \sum_{n=1}^{\infty} n \left(\alpha_n^{2*} \alpha_n^2 + \alpha_n^{3*} \alpha_n^3\right). \tag{12.93}$$

The complex conjugate replacements have come from the reality requirement mentioned earlier. This is the closest we will come to the quantum mechanical treatment. In his Chapter 12, Zwiebach generalizes the α coefficients into operators that raise or lower harmonic oscillator states.

At the level of classical physics this model of an elementary particle as a string has many of the properties that one would have to insist on. The particle has mass, momentum, energy and angular momentum, and all the laws of special relativity are respected. As far as I know, no other classical model can make the same claims. Whether or not the quantized string model corresponds to real world particles remains to be seen.

Bibliography

General References

1 B. Zwiebach, *A First Course in String Theory*, Cambridge University Press, Cambridge, UK, 2004.

13
General Relativity

13.1
Introduction

It has to be admitted, up front, that the present chapter just scratches the surface of the subject of general relativity. Furthermore, the presentation will not quite follow the pattern, established in previous chapters, in which "all" of classical physics was said to be unified into mechanics. That pattern consists of using general principles to restrict additions to the Lagrangian, and to make such additions as sparingly as possible. Everything else then follows from Hamilton's principle. Here, though we follow this pattern, to reduce technically difficulties, we justify only partially the zeroing in on a uniquely satisfactory Lagrangian density. Furthermore, because gravity alters the underlying geometry, a theory containing gravity as well as other physics, such as electromagnetic theory, cannot, in principle, be obtained by the simple addition of the separate actions. Here we consider mainly weak gravity, in which this sort of superposition *is* valid. The chapter therefore provides only the briefest of introductions to the concepts of general relativity. The real bread and butter of the subject, namely solving the Einstein equations in the presence of strong gravitational fields and studying its implications for cosmology, are not described.

"General Relativity" is curiously misnamed. It is really two subjects in one. One of these subjects extends special relativity to relate observations in relatively accelerating reference frames. The name "general relativity" is appropriate for this aspect of the theory. But most people regard general relativity as being primarily a theory of gravity. For this aspect, "Theory of Gravity" would be a more appropriate name.

General relativity includes a great deal of physics even in the absence of masses. To make this point, generalized transformations, which extend relativity beyond Lorentz transformations to more general transformations, are studied first. The representation of gravity by geometry is what is to be understood next, but this has to be preceded by a substantial amount of geometry. Some of this reprises subjects discussed earlier in the test but some, especially

Geometric Mechanics: Toward a Unification of Classical Physics. 2nd Edition. Richard Talman

ISBN: 978-3-527-40683-8

concerning the curvature tensor, is new. Next the so-called Einstein equation, which amounts to being Newton's law in the presence of masses, are derived. Finally, some experimental tests of the theory in the presence of weak gravity are described.

Curiously, the mass parameter m of a particle plays two roles in Newtonian mechanics. It plays an *inertial* role, which causes the acceleration of a particle subject to a given force to depend inversely on m. It also plays a *gravitational role*, causing the force on a particle in a gravitational force to be proportional to m. In this latter role m acts like a gravitational "charge," much like electron charge e. It was this dual nature of mass, more than anything else, that drove Einstein to invent general relativity.

As well as its mass parameter, a particle also has kinetic energy $\mathcal{E}$ and momentum p that influence its motion. From special relativity we know that $\mathcal{E}$ and momentum p depend on the observing frame of reference, but that the relation $\mathcal{E}^2 = p^2c^2 + m^2c^4$ is always true. Continuing to reflect on the curious nature of mass m, its direct gravitational effect and its inverse inertial effect cancel in situations where only gravitational forces are present. In this circumstance, instead of having a dual role, the mass can be said to have *no* role. That being the case, and *other things being equal*, the trajectories of all particles should be identical. Galileo's observations already confirmed this experimentally for all particles he had access to. But, for consistency, the equivalence should include even massless particles such as photons. Einstein also insisted on meeting this requirement. (Because of energy dependence, it is hard, experimentally, to meet the "other things being equal" condition accompanying this principle. The challenge is to reconcile the parabolic orbit of a massive particle in the earth's gravitational field with the apparently straight line trajectory of a photon.)

Even before beginning a quantitative analysis one can note some qualitative ways in which light has been observed experimentally to be influenced by gravity. The first such observation, in 1919, concerned the apparent shift in the position of stars observed close to the edge of the sun during a total eclipse. In recent years this phenomenon has been abundantly reconfirmed in the so-called "gravitational lensing" observed in astronomy. A quite different confirmation that light "accelerates" when subjected to gravity was the Pound–Rebka experiment in 1960. In that experiment a difference was observed between the interactions of x-rays (of extremely well-defined wavelength at the top of a tower) with crystalline atoms, depending on the vertical location of the crystal in the tower. A detailed formal theory is required before either of these phenomena can be understood quantitatively.

The "equivalence principle" of general relativity includes the considerations mentioned so far plus a far more significant physical principle. The principle is that, no matter what the source of gravitational field, there is some (presumably accelerating) frame of reference in which, at least locally, a particle moves as if it is free. In some simple situations, such as the trajectory of a particle in the earth's gravitational field, as viewed from a freely falling elevator, the equivalence principle is obviously valid. But the equivalence is now postulated to be *always* valid. The central role played by transformations between different frames is how "relativity" gets into the title of the subject.

There is one respect in which the equivalence principle is closer to mathematical artifact than to physical principle. It is never intended for the *entire* distribution of masses to be representable by identifying a single, appropriately accelerating frame in which there are no masses. For masses distributed over a finite region of space the gravitational force must vanish at infinity, while no uniformly accelerating frame (especially one that, because it is rotating, gives centrifugal forces that increase with radius) could have that property. On the other hand, a force that appears to be gravitational may, in fact, be only due to the choice of coordinates, in which case it would vanish after transformation to an inertial frame.

In the context of this text there is an equally important step to be taken. So far we have discussed metric geometry in the context only of using curvilinear coordinates in Euclidean geometry. This includes, especially, the metric tensor g_{ij}, but also the affine-connecting Christoffel symbols, $\Gamma^i{}_{jk}$. Without changing any of the equations developed so far, we have to change their physical interpretation. The simplest example of this reinterpretation concerns the motion of a "free" particle, where "free" means "subject to no forces other than gravity." We know that the particle trajectory is a geodesic, for which, by Eq. (3.51), the equation of motion is

$$\frac{d^2x^i}{ds^2} = -\Gamma^i{}_{jk}\frac{dx^j}{ds}\frac{dx^k}{ds}. \tag{13.1}$$

As complicated as it is, up to this point, since the solution of this equation has been known to be a straight line, the complexity of this equation has been due to only an ill-advised choice of curvilinear coordinate system. Now we are to interpret the terms on the right-hand side of the equation as representing the effect of gravity in "our" reference system. In this frame the solution of Eq. (13.1) has ceased to be a straight line. By hypothesis there *is* a "freely falling" coordinate system in which the solution is a straight line, but in our frame the motion is the geodesic solving Eq. (13.1) with appropriate initial conditions.

With transformations more general than Lorentz transformations newly allowed, and with photons assumed to follow geodesics, it is important to reconfirm the assumption, from special relativity, that light always travels with speed c. This also makes it necessary to revisit issues such as the synchronization of clocks.

With these reinterpretations we are finished, except for finding the ${\Gamma^i}_{jk}$ coefficients and solving the equations. Unfortunately these two tasks have occupied armies of theoretical physicists for close to a century. Einstein himself produced the (presumably correct) formulas giving ${\Gamma^i}_{jk}$ in terms of the distribution of masses. But the formulas are formidably complicated and Einstein completed the calculation for only a few simple configurations, where the gravitational effects are weak. These examples were, however, sufficiently realistic to predict measurable effects that could be used to test and corroborate the theory. What made these examples calculable is that the gravitational effects could be treated as perturbations of conventional Newtonian theory.

Much of the complexity of the full theory is due to the fact that the equations are nonlinear (which makes superposition invalid) and need to be solved self-consistently. Even after the ${\Gamma^i}_{jk}$ coefficients are known, solving for system evolution is still difficult, especially if analytic solutions are sought.

13.2 Transformation to Locally Inertial Coordinates

The mathematical expression of the equivalence principle is that there is a "inertial" system of coordinates ξ^i, in terms of which Eq. (13.1) reduces to

$$\frac{d^2\xi^i}{ds^2} = 0, \tag{13.2}$$

which is consistent with free particles moving in straight lines. This equation now applies to 4D spacetime, with s/c being proper time in a particle's motion.

The fact that the ${\Gamma^i}_{jk}$ coefficients vanish locally *does not* imply that they vanish elsewhere. Their derivatives with respect to the ξ^i coordinates do not, in general, vanish even locally. Newtonian mechanics makes a similar assumption when it assumes that the force on a particle can depend on its instantaneous velocity, but not on its instantaneous acceleration. Lagrangian mechanics also makes an equivalent assumption: the Lagrangian can depend on first, but not second, time derivatives. Though the assumption in general relativity is similar in principle, its implications are far harder to apply because of the far greater complexity of the ${\Gamma^i}_{jk}$ coefficients.

There have to be functions $\xi^i(x^j)$ giving the inertial coordinates in terms of our x^i coordinates. These, and the inverse functions, satisfy

$$\frac{\partial x^l}{\partial \xi^i}\frac{\partial \xi^i}{\partial x^j} = \delta^l_j. \tag{13.3}$$

Starting to reexpress Eq. (13.2) in "our" coordinates,

$$0 = \frac{d}{ds}\left(\frac{\partial \xi^i}{\partial x^j}\frac{dx^j}{ds}\right) = \frac{\partial \xi^i}{\partial x^j}\frac{d^2x^j}{ds^2} + \frac{\partial^2 \xi^i}{\partial x^k \partial x^j}\frac{dx^k}{ds}\frac{dx^j}{ds}. \tag{13.4}$$

Multiplying this equation by $\partial x^l / \partial \xi^i$ and using Eq. (13.3) yields

$$0 = \frac{d^2x^l}{ds^2} + \left(\frac{\partial x^l}{\partial \xi^i}\frac{\partial^2 \xi^i}{\partial x^k \partial x^j}\right)\frac{dx^k}{ds}\frac{dx^j}{ds}. \tag{13.5}$$

Comparing with Eq. (13.1), and rearranging indices, one obtains

$$\Gamma^i{}_{jk} = \frac{\partial x^i}{\partial \xi^m}\frac{\partial^2 \xi^m}{\partial x^j \partial x^k}. \tag{13.6}$$

This expresses the Christoffel coefficients in terms of the transformation functions between our coordinates and inertial coordinates.

If all the components of a *true* tensor vanish in one frame of reference then the components of the tensor have to vanish in all frames of reference. The fact that the $\Gamma^i{}_{jk}$ components vanish in the inertial frame, but not in our frame, shows that $\Gamma^i{}_{jk}$ is *not* a tensor.

In purely mathematical terms, the nonvanishing of the $\Gamma^i{}_{jk}$ coefficients is ascribed to "curvature" of the underlying space. This account for the statement that, according to Einstein, space is *curved*. (This usage of the term "curved" is, of course, not to be confused with a particle trajectory being "curved," which is the normal state of affairs in Newtonian mechanics). Before continuing with incorporating gravity, it is appropriate to study the geometry of curved spaces.

13.3 Parallel Transport on a Surface

Consider a circular cylinder of radius ρ_0 with axis coinciding with the z-axis. The metric in cylindrical coordinates is $ds^2 = d\rho^2 + \rho^2 d\phi^2 + dz^2$ which, when restricted to the surface of the cylinder becomes $ds^2 = d(\rho_0\phi)^2 + dz^2$. Introducing a new coordinate $\sigma = \rho_0\phi$, the metric for geometry on the surface of the cylinder is

$$ds^2 = d\sigma^2 + dz^2. \tag{13.7}$$

Since this metric is the same as the Euclidean metric in a plane, the local geometric properties on the cylinder and on the plane are the same. If a geometric figure, such as a parallelogram is inked on the surface of a plane, and the cylinder is rolled over the plane with the ink still wet, the pattern of the parallelogram will be transferred onto the surface of the cylinder. In this transfer

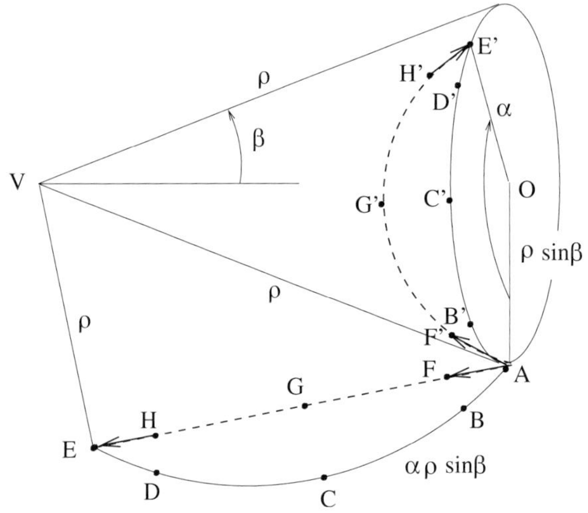

Fig. 13.1 A cone with vertex fixed at V rolls onto the plane AEV. Circular rim AB′C′D′E′ rolls onto the circle ABCDE lying in the plane.

both the lengths of the sides and the angles at the corners will be preserved. This is known as "isometry." The "parallelogram" sides on the cylinder will be curved, not straight, however, unless they happen to be parallel to the z-axis.

A similar, but slightly more general transfer is shown in Fig. 13.1. In this case a cone with vertex fixed at V rolls onto the plane AEV. Circular rim AB′C′D′E′e rolls onto the circle ABCDE lying in the plane. The length of arc AB′C′D′E′ is $\alpha\,\rho \sin\beta$ which is therefore also the length of arc ABCDE.

One knows that, on the plane, the shortest distance from point A to point E follows the straight line AFGHE. From the figure it can be seen that this line transfers to the curve AF′G′H′E on the cone. This curve must therefore be the shortest route (i.e., the geodesic) leading from A to E on the cone. (If there were a shorter route it would transfer to some curve on the plane shorter than AFGHE, which would be a contradiction.) Under parallel translation in the plane, if an arrow initially at AF slides along the straight line, always parallel to the line, ending up at HE, it will certainly be parallel to AF. Similarly, after sliding the arrow AF′ always parallel to the geodesic AF′G′H′E, it ends up at H′E. Contrary to ones possible expectation of AB′ and D′E′ being parallel, one finds instead that AF′ and H′E′ are parallel.

Next consider Fig. 13.2 (a) in which the same cone as in the previous figure is placed, like a dunce's cap, on the top of the earth with its vertex V on the earth's axis, such that the cone is tangent to the earth along the entire line of colatitude β. This time an arrow originally at point A, and pointing toward the North pole, is parallel translated along the line of latitude, back to its original location at A'. Figure 13.2 (b) shows this translation being performed in the

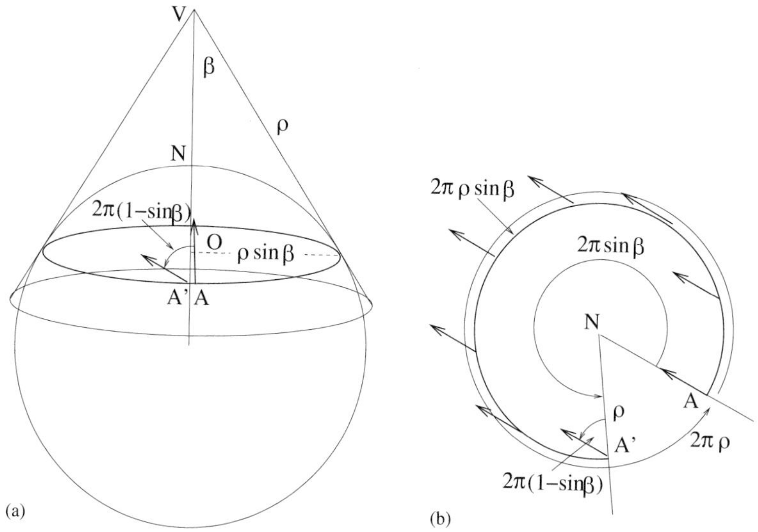

Fig. 13.2 (a) A cone with vertex V on the axis through the North pole is tangent to a line of colatitude β on the earth's surface. (b) An arrow at A, pointing toward the North pole, is parallel translated around its line of latitude back to its original position at A$'$, as shown in both figures.

surface of the cone after the cone has been unrolled (or "developed") into a plane. In this plane parallel translation has its ordinary Euclidean significance. But the full arc of the unrolled cone has a length of only $2\pi\rho\sin\beta$, which is less than the full circle circumference of $2\pi\rho$. The translated arrow, still lying in the plane, fails to point toward point N by an angle equal to $2\pi(1-\sin\beta)$.

Problem 13.3.1. *Compare the parallel displacement just analyzed and the "no-twist" displacement analyzed in Section 6.4.3.*

Problem 13.3.2. *A surface of revolution is to be formed, by rotating around the z-axis a curve lying in the x, z plane and defined parametrically, with parameter u, by* $(x,y,z) = \big(g(u),0,h(u)\big)$, *where* $g(u)$ *and* $h(u)$ *are arbitrary functions. Points at which this curve crosses the z-axis will be "singular" points of the surface of revolution. After rotating the plane containing the curve around the z-axis by an angle* $0 \le v < 2\pi$, *the moved curve is defined parametrically by* $(x,y,z) = \big(g(u)\cos v, g(u)\sin v, h(u)\big)$. *The parameters u and v can serve as coordinates on the surface of revolution swept out by the curve for all values of v. A coordinate curve on which* $u = k =$ *constant, while v varies is* $\big(g(k)\cos v, g(k)\sin v, h(k)\big)$. *Such a curve is known as a "parallel" or as a "line of latitude." A coordinate curve on which* $v = c =$ *constant, while u varies is* $\big(g(u)\,k, g(u)\sqrt{1-k^2}, h(u)\big)$. *Such a curve is* *known as a "meridian." The following metrics can be used to confirm the result of the previous problem.*

(a) *Determine functions $g(u)$ and $h(u)$ that describe the cone shown in Fig. 13.2 (a), and describe the resulting coordinate curves. For this surface, write the metric ds^2 induced from the 3D Euclidean space, in terms of u and v.*

(b) *Determine functions $g(u)$ and $h(u)$ that describe the sphere shown in Fig. 13.2 (a), and describe the coordinate curves. Write the induced metric ds^2 in terms of u and v.*

13.3.1 Geodesic Curves

The invariant differential was defined in terms of Christoffel symbols in Eq. (3.29), which is repeated here, though in terms of covariant components;

$$DA_{\;i} = dA_i - \Gamma^j_{\;ik} A_j \, dx^k. \tag{13.8}$$

Comparing components at two nearby locations, $A_i(x^k)$ and $A_i(x^k + dx^k)$, this formula gives the covariant components $DA_{\;i}$ of the arrow obtained by subtracting the Euclidean-parallel-translated original arrow from the arrow at the displaced location. Expressing the differential as an absolute derivative,

$$A_{i;k} = \frac{\partial A_i}{\partial x^k} - \Gamma^j_{\;ik} A_j. \tag{13.9}$$

The Christoffel symbols were determined as functions of the metric coefficients in Eq. (3.21). After adjusting the indices and multiplying that equation by g^{im} and summing on m, the result is

$$\Gamma^i_{\;jk} = \frac{1}{2} g^{im} \left(\frac{\partial g_{mj}}{\partial x^k} + \frac{\partial g_{mk}}{\partial x^j} - \frac{\partial g_{jk}}{\partial x^m} \right). \tag{13.10}$$

In Euclidean geometry all the Christoffel coefficients clearly vanish.

If the arrow at the displaced location is, in fact, parallel translated from the original, then $A_{i;k} = 0$. Therefore, the condition for parallel translation is

$$\frac{\partial A_i}{\partial x^k} = \Gamma^j_{\;ik} A_j. \tag{13.11}$$

A curve in space has the form $x^i = x^i(s)$, where s is, for example, the arc length along the curve. The vector $u^i = dx^i/ds$ is tangent to the curve. A curve having the property that tangent vector u^i is parallel translated along the curve is known as a "geodesic." The condition for this to be true is $Du^i = 0$. As in Eq. (3.51), the differential equation satisfied by such a curve is

$$\frac{d^2x^i}{ds^2} + \Gamma^i_{\;jk} \frac{dx^j}{ds} \frac{dx^k}{ds} = 0. \tag{13.12}$$

Problem 13.3.3. *Returning to the terminology of Section 13.2, as well as "our," unprimed, coordinate frame and the inertial frame, suppose there is another, primed, frame, in which the Christoffel coefficients are* ${\Gamma'}^{i}{}_{jk}$. *Show that the transformation formulas of Christoffel coefficients from primed frame to unprimed frame are*

$$\Gamma^{i}{}_{jk} = {\Gamma'}^{l}{}_{mn} \frac{\partial x^{i}}{\partial x'^{l}} \frac{\partial x'^{m}}{\partial x^{j}} \frac{\partial x'^{n}}{\partial x^{k}} + \frac{\partial x^{i}}{\partial x'^{m}} \frac{\partial^{2} x'^{m}}{\partial x^{j} \partial x^{k}}. \tag{13.13}$$

Do this by exploiting, in Eq.(13.8), the fact that the left-hand side is known to be a true *tensor, and transforming the right-hand side as in Section 13.2.*

The first term on the right-hand side of Eq. (13.13), derived in the previous problem, gives the transformation law for $\Gamma^{i}{}_{jk}$ if it were a *true* tensor. The second term shows that the nonvanishing of Christoffel coefficients can be ascribed to nonlinearity (i.e., nonvanishing second derivatives) in the transformation formulas from "flat" space, where the Christoffel coefficients vanish, to "curved" space, where they do not. For the same reasons, with the primed frame taken to be the inertial frame, Eq. (13.13) is consistent with Eq. (13.6).

Problem 13.3.4. *Suppose that, by gravitational measurements made near the origin in our unprimed frame, the Christoffel coefficients have been determined to be* $(\Gamma^{i}{}_{jk})_0$. *Show that* ξ^i*-coordinates defined by transformation relations*

$$\xi^{i} = x^{i} + \frac{1}{2} \left(\Gamma^{i}{}_{jk}\right)_0 x^{j} x^{k}, \tag{13.14}$$

are, in fact, inertial frame coordinates.

13.4 The Twin Paradox in General Relativity

The goal of this section is to use general relativity to analyze the so-called twin paradox. That is, to calculate the extent to which an itinerant twin has aged less upon his return from a circuitous trip than his twin, who has remained stationary in an inertial frame.

The equations of general relativity apply even in cases where there are no masses present (as we are now assuming) but where the particular coordinate system being used causes effects similar to those due to gravity. Two points that are relatively fixed in one frame of reference may, for example due to the use of a rotating frame of reference, be moving relative to each other in another frame. This causes even their separation to be uncertain because of ambiguity in the time at which the separation is to be determined. The timing ambiguity will be resolved by using the fact that the speed of light is constant and the separation of the particles will be taken to be the speed of light multiplied by half the time it takes light to go from one particle to the other and back.

In general relativity a "constant" metric has the property that the metric coefficients are independent of x^0;

$$ds^2 = g_{00}(\mathbf{x})\,dx^{0^2} + 2g_{0\alpha}(\mathbf{x})\,dx^0\,dx^\alpha + g_{\alpha\beta}(\mathbf{x})\,dx^\alpha dx^\beta, \tag{13.15}$$

where Greek indices $\alpha, \beta, \ldots$ range over 1, 2, 3 and the metric coefficients are arbitrary functions of $\mathbf{x} = (x^1, x^2, x^3)$, but are independent of x^0. One important assumption, carried over from special relativity, is that a clock at fixed spatial position $\mathbf{x}$ measures *proper time* $d\tau = ds/c$ at that location. This implies

$$dx^0 = \frac{c\,d\tau}{\sqrt{g_{00}(\mathbf{x})}}. \tag{13.16}$$

Because of the $\mathbf{x}$-dependence of $g_{00}(\mathbf{x})$, the scaling between x^0 and τ depends on position $\mathbf{x}$. This makes it inappropriate to regard x^0 as the time measured by a clock at fixed spatial location $\mathbf{x}$, as one does in special relativity.

Two things (at least) prevent Eq. (13.15) from being treated simply as a metric from special relativity with curvilinear spatial coordinates being used. The "crossed" term, $dx^0\,dx^\alpha$, because it depends on $dx^0\,dx^\alpha$, is not present in special relativity. Also, even if it were, because clocks at different positions keep time differently, for world events occurring at different positions, the meaning of dx^0 is, *a priori*, unclear.

By analogy with the definition of proper time in Eq. (13.16), one would like to define the proper distance between two stationary spatial points, A and B, as their spatial separation at fixed time. But this definition requires the clocks at A and B to be synchronized. Let us assume initially that the separation of these two points is infinitesimal, with dx^α being the spatial displacement of point A relative to point B. How can A and B clocks be synchronized? There will be a synchronization correction dx^0 that is proportional to dx^β so, after correction, both of the final two terms in Eq. (13.15) will be proportional to $dx^\alpha dx^\beta$.

Suppose that particle B emits a light pulse at world event (a) which gets to particle A at world event (b) and is sent back immediately, arriving at particle B at world event (c). World lines of the particles and of the light pulses are shown in Fig. 13.3. Another important feature carrying over from special relativity is that $ds = 0$ along the world line of a pulse of light. Therefore, as measured with the clock of particle A, the timings are

$$\begin{aligned} x_A^0(a) &= x_A^0(b) + dx_A^{0(1)}, \\ x_A^0(c) &= x_A^0(b) + dx_A^{0(2)}, \end{aligned} \tag{13.17}$$

where $dx_A^{0(1)}$ is the more negative root of the quadratic equation in dx_A, obtained from Eq. (13.15) with $ds = 0$,

$$0 = g_{00} dx_A^{0\,2} + 2g_{0\alpha} dx_A^0\, dx^\alpha + g_{\alpha\beta} dx^\alpha dx^\beta, \tag{13.18}$$

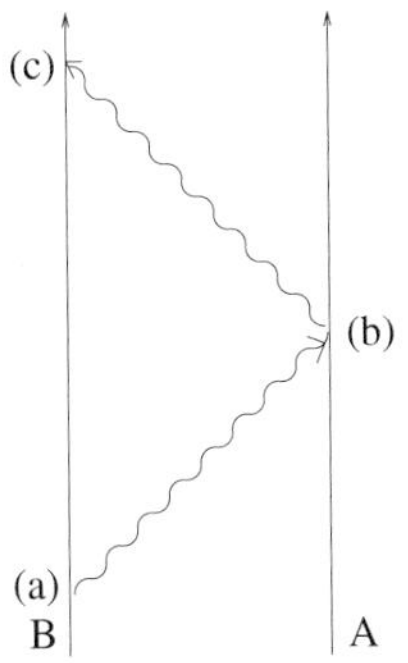

Fig. 13.3 World lines of particles A and B are both purely vertical because both particles are at rest. A light signal is transmitted from B at (a) and, immediately on its arrival (b), is retransmitted, arriving at A at (c).

and $dx_A^{0(2)}$ is the other root. When the times on the left-hand side of Eq. (13.17) are expressed in terms of times measured on the B clock, the results are

$$\begin{aligned} x_B^0(a) + dx_{AB}^0 &= x_A^0(b) + dx_A^{0(1)}, \\ x_B^0(c) + dx_{AB}^0 &= x_A^0(b) + dx_A^{0(2)}, \end{aligned} \tag{13.19}$$

where dx_{AB}^0 is a timing offset to be determined next. The condition for A and B clocks to be synchronized is

$$x_A^0(b) = \frac{x_B^0(a) + x_B^0(c)}{2}. \tag{13.20}$$

In other words, this condition assures that the turn around time, which is directly available from the A clock, is the average of the start and end times made available by the B clock. Any error caused by this assumption would be quadratic in the differentials and hence can be neglected in the limit.

Solving these equations, the results are

$$\begin{aligned} dx_A^{0(2)} - dx_A^{0(1)} &= \frac{2}{g_{00}} \sqrt{(g_{0\alpha}g_{0\beta} - g_{\alpha\beta}g_{00})dx^\alpha dx^\beta}, \\ dx_{AB}^0 &= -\frac{g_{0\alpha}dx^\alpha}{g_{00}}. \end{aligned} \tag{13.21}$$

The first of these equations gives double the time light takes in going from B to A (at speed c), so the spatial separation dl of these two points satisfies

$$dl^2 = \left(-g_{\alpha\beta} + \frac{g_{0\alpha}g_{0\beta}}{g_{00}} \right) dx^\alpha dx^\beta. \tag{13.22}$$

This formula can therefore be regarded as being the spatial metric tensor applicable to the region near A. Combining this with Eq. (13.16), the invariant measure of world intervals near A can be expressed as

$$ds^2 = c^2 d\tau_A^2 - dl^2. \tag{13.23}$$

Allowing particle A to move, the procedure just described permits the sequential synchronization of all the clocks along a world curve followed by A, relative to a clock at A's starting location. The second of Eqs. (13.21) gives the discrepancy from synchronism corresponding to the interval dx^α, so the accumulated discrepancy (assuming the path returns to its starting position) is

$$\Delta x^0 = -\oint \frac{g_{0\alpha} dx^\alpha}{g_{00}}. \tag{13.24}$$

If the area enclosed by the curve is finite, a clock carried around the closed curve will not necessarily agree with a clock that remained stationary at A. This situation should be reminiscent of the "twin paradox" already understood in special relativity.

Consider ordinary inertial space with no masses (or anything else) present, with metric $ds^2 = c^2 dt^2 - x^2 - y^2 - z^2$. World points can be located by cylindrical coordinates (t, r, ϕ, z) and the metric is

$$ds^2 = c^2 dt^2 - dr^2 - r^2 d\phi^2 - dz^2. \tag{13.25}$$

We assume that synchronized clocks are distributed everywhere in this inertial frame, which makes t available to all observers, independent of their velocities. Also consider cylindrical coordinates (t, r', ϕ', z') in a different frame of reference that is rotating about the z-axis with constant angular velocity Ω, so that $r = r'$, $z = z'$, and $\phi = \phi' + \Omega t$. Note, however, that the time coordinate continues to be t. Substituting into Eq. (13.25), one obtains the so-called "Langevin metric,"

$$ds^2 = (1 - \Omega^2 r'^2 / c^2)\, c^2 dt^2 - 2\Omega r'^2 d\phi' dt - d\sigma'^2, \tag{13.26}$$

where $d\sigma'^2 = dr'^2 + r'^2 d\phi'^2 + dz'^2$ is the apparent spatial metric in the rotating frame. (These coordinates are only applicable for values of r small enough so that the first coefficient remains positive. This is clearly associated with the prohibition of velocities exceeding the speed of light.) Setting $z' = 0$ for convenience, and matching the coefficients in this expression to the symbols g'_{ij} in Eq. (13.15), now applied in the new frame, the nonvanishing metric coefficients are

$$g'_{00} = 1 - \Omega^2 r^2 / c^2, \quad g'_{0\phi} = g'_{\phi 0} = -\Omega r^2 / c, \quad g'_{\phi\phi} = -r^2, \quad g'_{rr} = -1. \tag{13.27}$$

Then, from Eq. (13.22),

$$d\ell'^2 = dr'^2 + \frac{r'^2 d\phi'^2}{1 - \Omega^2 r'^2/c^2}. \tag{13.28}$$

In the geometry with lengths given by the metric dl'^2 just obtained, ignoring time, the circumference C of a circle of radius $r = R$, centered on the origin is given by

$$C = \int_0^{2\pi} \frac{R|d\phi'|}{\sqrt{1 - \Omega^2 R^2/c^2}} = \frac{2\pi R}{\sqrt{1 - \Omega^2 R^2/c^2}}. \tag{13.29}$$

The integral in Eq. (13.24) evaluates to

$$\Delta x^0 = \frac{\Omega R^2}{c} \oint \frac{d\phi'}{1 - \Omega^2 R^2/c^2} = \pm \frac{2\pi R^2}{c} \frac{\Omega}{1 - \Omega^2 R^2/c^2}, \tag{13.30}$$

where the sign depends on the direction of integration.

Problem 13.4.1. *A clockwise-traveling and a counter-clockwise traveling signal are emitted simultaneously from a source moving on the circle just discussed. The* differ-ence *of the two values of $\Delta x^0 = c\Delta t$ just calculated can be interpreted to be the difference in arrival times (times c) of these two signals back at the source. The counter-traveling signal will arrive first. Working in the inertial frame of reference, work out this time difference and show that your result agrees with Eq. (13.30). This time difference, known as the Sagnac effect, was observed, as a shift of interferometer fringes, by Sagnac in 1914.*

Consider an observer P at rest in the inertial frame at position $(r, \phi) = (R, 0)$ and another observer P′ that is initially coincident with P, but stationary in the rotating frame. Clocks of both observers are initially set to zero. In the rotating frame the point P will be moving at constant speed along the same circle of radius R centered on the origin that P′ is on so, at later times, the points will coincide periodically.

The parameter x'^0 is being interpreted as ct which is (c times) the time observer P′ reads off the particular P-frame synchronized clock that happens to be present at his immediate location. In particular, when P and P′ again coincide, t will have advanced by $2\pi/\Omega$, since that is the revolution period according to inertial frame observer P. But the clock carried by P′ measures proper time τ. According to Eq. (13.16)

$$ct = \frac{c\tau}{\sqrt{1 - \Omega^2 R^2/c^2}} = \frac{c\tau}{\sqrt{1 - V^2/c^2}}, \tag{13.31}$$

where $V = \Omega R$ is the speed of P′ as observed in the inertial system. "Twin" P′ will therefore have aged less than twin P by the amount given, which agrees with special relativity.

As observed in the P′ frame, the distance P travels between encounters with P′ is greater than $2\pi R$, and the time between encounters is less than $2\pi/\Omega$. So the proper speed of P observed in the P′ reference frame is greater than the proper speed of P′ observed in the P reference frame.

Suppose light is emitted tangentially to the circle of radius R and is channeled along the circle by repeated reflections in a gigantic cylindrical mirror of diameter $2R$. If the reflections are sufficiently glancing the light can be treated as if it is traveling on a circle of radius R. The proper velocity of the light will, as always, be c. But, applying Eq. (13.30), to adjust from proper time to time t, the speed, reckoned as distance traveled divided by t, will depend on whether the light is traveling with, or against, the rotation direction.

13.5
The Curvature Tensor

After a sequence of parallel translations, according to Eq. (13.8), the changes ΔA_k in the components of a vector satisfy

$$0 = \Delta A_j - \int \Gamma^i{}_{ja} A_i \, dx^a. \tag{13.32}$$

If the sequence returns to the starting point then

$$\Delta A_j = \oint \Gamma^i{}_{ja} A_i \, dx^a. \tag{13.33}$$

Since they are the components of the difference of two vectors at the same point in space, these ΔA_j values are the covariant components of a true vector. (Or, if one prefers, they are the coefficients of a 1-form.) For an infinitesimal loop the vector "almost" returns to its starting orientation, and the differential ΔA_j is of the same order of smallness as the area of the loop traversed. With a limiting process, one can therefore use ΔA_j to define, at every point in the space, a "curl-like" entity, called the "curvature tensor." Like the curl, the curvature tensor has multiple components. Along with a surface-defining two-tensor the curvature operation acts on one vector to define another. Unlike the curl, which is specific to Euclidean geometry. the curvature operation depends on the geometry of the space, which is represented by the metric tensor.) A differentially small, possible integration path is shown in Fig. 13.4. It is based on two differential displacements $\mathbf{\Delta x}_{(1)}$ and $\mathbf{\Delta x}_{(2)}$. A differential area tensor Δf^{ab} can be formed from these vectors

$$\Delta f^{ab} = \Delta x^a_{(1)} \Delta x^b_{(2)} - \Delta x^b_{(1)} \Delta x^a_{(2)}. \tag{13.34}$$

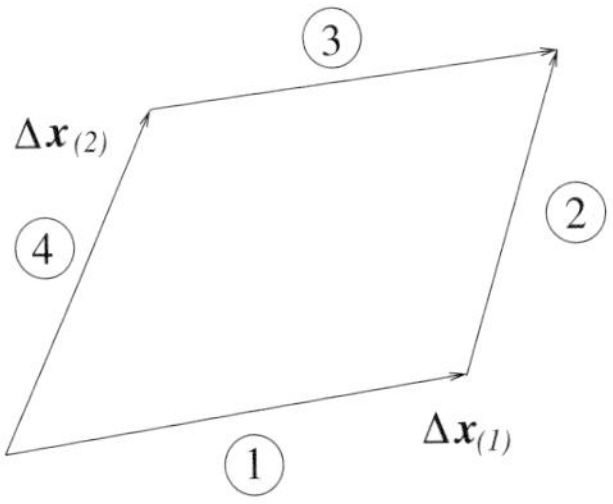

Fig. 13.4 Differential integration path used to define the curvature tensor. The enclosed differential area can be represented by the 2-index tensor Δf^{ab} defined in Eq. (13.34).

Evaluation of the integral around this loop is performed just as in the definition of the curl in vector analysis, following the pattern of Section 4.3.4;

$$\Delta A_j = \oint \Gamma^i{}_{ja} A_i dx^a = \int_1 - \int_3 + \int_2 - \int_4$$
$$\approx \frac{1}{2}\left(\frac{\partial(\Gamma^i{}_{jb}A_i)}{\partial x^a} - \frac{\partial(\Gamma^i{}_{ja}A_i)}{\partial x^b}\right)\Delta f^{ab}$$
$$\equiv \frac{1}{2} R^i{}_{jab}\,\Delta f^{ab}\,A_i, \tag{13.35}$$

where the limits of the integrals are labeled as in the figure. The path is traversed in the order, 1, 2, 3, 4, and the signs of the 3 and 4 legs are reversed because they are traversed in the negative direction. Only leading terms have been retained in the Taylor expansion of the integrand. Also neglected during paths after the first, is the effect of parallel translation on the earlier paths. All these effects become negligible in the limit of small displacements. In the last step of Eq. (13.35) it has been recognized that, where derivatives $\partial A_i/\partial x^j$ appear they can be replaced using Eq. (13.11). That is what has permitted pulling out the A_i factor. The remaining coefficient has *defined* the "curvature tensor" or "Riemann tensor." Manipulating indices, $R^k{}_{jab}$ is given by

$$R^k{}_{jab} = \frac{\partial \Gamma^k{}_{jb}}{\partial x^a} - \frac{\partial \Gamma^k{}_{ja}}{\partial x^b} + \Gamma^k{}_{na}\Gamma^n{}_{jb} - \Gamma^k{}_{nb}\Gamma^n{}_{ja}. \tag{13.36}$$

13.5.1 Properties of Curvature Tensor, Ricci Tensor, and Scalar Curvature

By construction $R^k{}_{jab}$ is a *true* tensor. Note that the four indices are best regarded as forming two pairs. The (unusual) choice of a and b as indices has anticipated their special role, which is to identify displacements defining an integration path. It may be conceptually useful to visualize the a and b indices having been suppressed by defining a 2-form which is "waiting for" arguments $\mathbf{\Delta x}_{(1)}$ and $\mathbf{\Delta x}_{(2)}$, to find the coefficients in the expansion of ΔA_j.

The curvature tensor is formidably complicated. It is all but necessary to attempt to simplify the formula before attempting to evaluate $R^i{}_{jab}$, for example as in the following problems.

Problem 13.5.1. *Following the pattern of Section 4.3.4 to approximate ΔA_j, as given by Eq. (13.33), over the integration path shown in Fig. 13.4, derive Eqs. (13.35) and (13.36).*

Problem 13.5.2. *By lowering the first index in Eq. (13.36) and using relations (13.10), show that*

$$R_{ijab} = \frac{1}{2}\left(\frac{\partial^2 g_{ib}}{\partial x^a \partial u^j} - \frac{\partial^2 g_{jb}}{\partial x^a \partial u^i} - \frac{\partial^2 g_{ia}}{\partial x^b \partial u^j} + \frac{\partial^2 g_{ja}}{\partial x^b \partial u^i}\right) + g_{kl}\left(\Gamma^k{}_{ja}\Gamma^l{}_{ib} - \Gamma^l{}_{jb}\Gamma^k{}_{ia}\right). \tag{13.37}$$

In inertial coordinates, for which all Christoffel symbols vanish, the second term vanishes.

Problem 13.5.3. *Prove the following algebraic relationships satisfied by the fully covariant curvature tensor elements R_{ijab} derived in the previous problem:*

(a) R_{ijab} is antisymmetric under the interchange of i and j. This indicates a kind of reciprocity in which the ith component of the change in the jth component is the negative of the jth component of the change in the ith component.

(b) R_{ijab} is antisymmetric under the interchange of a and b. From Eq. (13.34), reversing a and b is equivalent to reversing the direction of traversal of the integration path.

(c) R_{ijab} is symmetric under the interchange of the pair (i, j) with the pair (a, b) (with their orders separately preserved).

(d) The sum of the terms formed by cyclic permutation of any three indices vanishes. For example,

$$R_{ijab} + R_{iabj} + R_{ibja} = 0. \tag{13.38}$$

Since all other such triple sums can be obtained from this example using results (a), (b), and (c), this adds only one further algebraic constraint on the coefficients.

Problem 13.5.4. *Covariant derivatives of curvature tensor elements are most easily evaluated using Eq. (13.36), while working with inertial coordinates for which all $\Gamma^1{}_{jk}$*

elements vanish. Since the last term is quadratic in the Γ's, its first derivatives all vanish. Show therefore that

$$R^k{}_{jab;m} = \frac{\partial^2 \Gamma^k{}_{jb}}{\partial x^m \partial x^a} - \frac{\partial^2 \Gamma^k{}_{ja}}{\partial x^m \partial x^b}. \tag{13.39}$$

Problem 13.5.5. *Use the result of the previous problem to derive the "Bianchi identity," which gives a sum of covariant derivatives of the curvature tensor with cyclically permuted indices;*

$$R^n_{ija;b} + R^n_{iab;j} + R^n_{ibj;a} = 0. \tag{13.40}$$

By contracting on a pair of indices of R_{ijab} one can form a 2-index tensor. But contraction on the (i,j) pair or the (a,b) pair gives zero since R_{ijab} is anti-symmetric in those pairs. Hence, we can contract only on (i,a), (i,b), (j,a), or (j,b). We *define*

$$R_{jb} \equiv \sum_{m=a} g^{mi} R_{ijab} = \sum_{m=a} R^m{}_{jab} = R^m{}_{jmb}. \tag{13.41}$$

From Eq. (13.36) this tensor is given by

$$R_{jb} = \frac{\partial \Gamma^k{}_{jb}}{\partial x^k} - \frac{\partial \Gamma^k{}_{jk}}{\partial x^b} + \Gamma^k{}_{nk}\Gamma^n{}_{jb} - \Gamma^k{}_{nb}\Gamma^n{}_{jk}. \tag{13.42}$$

Except for overall sign, the other three nonvanishing contractions yield the same tensor. R_{jb} is known as the "Ricci tensor." It is clearly symmetric in its two indices.

The only other possible invariant quantity that can be defined is obtained by contracting the Ricci tensor to produce a scalar invariant $\mathcal{R}$ called the "scalar curvature";

$$\mathcal{R} = g^{bj} R_{jb} = g^{bj} g^{mi} R_{ijmb}. \tag{13.43}$$

Problem 13.5.6. *In Problem 13.3.2 the metrics for two surfaces of revolution were determined. Using these metrics, show:*

(a) On the conical surface the scalar curvature is $\mathcal{R} = 0$.

(b) On the surface of a sphere of radius ρ the scalar curvature is $\mathcal{R} = -2/\rho^2$.

13.6 The Lagrangian of General Relativity and the Energy–Momentum Tensor

In the previous section it was shown that there is only one scalar that is invariantly definable in metric geometry, namely the scalar curvature $\mathcal{R}$.[1] Clearly $\mathcal{R}$ has to be selected as the Lagrangian density.

1) Note the unfortunate choice of this symbol $\mathcal{R}$, which suggests "radius" when, in fact, the scalar curvature is proportional to inverse radius squared.

But, in metric geometry, even the definition of the action has to be generalized somewhat. The formula for the action S needs to be modified from that in Eq. (10.56) to

$$S = \frac{1}{c} \int \mathcal{R}(x^i) \sqrt{-g}\, d\Omega. \tag{13.44}$$

For special relativity the factor $\sqrt{-g}$ is equal to 1, so this is also the value $\sqrt{-g}$ must approach in the limit of weak gravity. In Eq. (13.44), as always, the action is defined only up to a multiplicative factor whose choice depends on the units of the physical quantities entering the theory. For now this factor has been set to 1.

For general classical field theory in the context of special relativity, the energy–momentum tensor T_i^j was defined, in Eq. (10.63), as a function of the Lagrangian density. Unfortunately, this definition also has to be updated to be consistent with the metric geometry. Recalling that the conservation laws involving T_i^j were obtained from the invariance following from the existence of ignorable coordinates in the action, it is this derivation that needs to be generalized, and which is, therefore, the immediate task.

Consider a transformation like that described in Problem 3.1.8, in which the world coordinates suffer an infinitesimal displacement, $x^i \rightarrow x^i + \xi^i \equiv x'^i$. When this displacement is applied to the integrand of Eq. (13.44), because there are no explicit dependences on x^i, the variation can be applied as if it is the metric coefficients themselves that are being varied; $g_{ij} \rightarrow g_{ij} + \delta g_{ij}$. As a result,

$$\begin{aligned}
\delta S &= \frac{1}{c} \int \left(\frac{\partial(\mathcal{R}\sqrt{-g})}{\partial g^{ij}} \delta g^{ij} + \frac{\partial(\mathcal{R}\sqrt{-g})}{\partial(\partial g^{ij}/\partial x^l)}\, \delta \frac{\partial g^{ij}}{\partial x^l} \right) d\Omega \\
&= \frac{1}{c} \int \left(\frac{\partial(\mathcal{R}\sqrt{-g})}{\partial g^{ij}} - \frac{\partial}{\partial x^l} \frac{\partial(\mathcal{R}\sqrt{-g})}{\partial(\partial g^{ij}/\partial x^l)} \right) \delta g^{ij}\, d\Omega
\end{aligned} \tag{13.45}$$

where the usual integration by parts step has been taken. We now abbreviate the integrand by introducing T_{ij} defined by

$$\frac{1}{2}\sqrt{-g}\, T_{ij} = \frac{\partial(\mathcal{R}\sqrt{-g})}{\partial g^{ij}} - \frac{\partial}{\partial x^l} \frac{\partial(\mathcal{R}\sqrt{-g})}{\partial(\partial g^{ij}/\partial x^l)}. \tag{13.46}$$

As a result,

$$\delta S = \frac{1}{2c} \int \sqrt{-g}\, T_{ij}\, \delta g^{ij}\, d\Omega \tag{13.47}$$

With the integrand being proportional to the discretionary factor δg^{ij} this formula superficially suggests that T_{ij} must vanish. But that inference is incorrect because it is only the four ξ^i factors that are arbitrary rather than the 16 δg^{ij} factors.

To make progress we therefore have to eliminate the δg^{ij} factors in favor of ξ^i factors. The first step in accomplishing this was taken in Problem 3.1.8, the result of which was Eq. (3.39), $\delta g^{ij} = \xi^{i;j} + \xi^{j;i}$. Substituting this into Eq. (13.47), and exploiting the fact that T_{ij} is symmetric, we get

$$\delta S = \frac{1}{c}\int \sqrt{-g}\, T_{ij}\, \xi^{i;j}\, d\Omega = \frac{1}{c}\int \sqrt{-g}\, T_i^j\, \xi^i_{;j} d\Omega. \tag{13.48}$$

With ξ^i now explicit, this is closer to what is needed to infer properties of the other factor in the integrand. But it is only derivatives of ξ^i present at this point. As a next step we prepare for another integration by parts;

$$\delta S = \frac{1}{c}\int \sqrt{-g}\, \frac{D(T_i^j \xi^i)}{dx^j}\, d\Omega - \frac{1}{c}\int \xi^i\, \frac{DT_i^j}{dx^j} d\Omega. \tag{13.49}$$

We are now in a position to use Eq. (4.83) which defined the divergence in invariant fashion. As in Eq. (4.84), the first integral can be converted to an integral over an enclosing surface, of $T_i^j \xi^i$. This integral can be assumed to vanish, even for ξ^i constant, assuming the fields fall off sufficiently rapidly at infinity.

Hamilton's principle will, in the next section, be applied to the action S that is being worked on. The variation δS has to vanish for arbitrary infinitesimal displacements from the true motion. In particular it has to vanish for the displacements $x^i \to x^i + \xi^i$ assumed in this section. The only term remaining in δS, which is the second term of Eq. (13.49), must therefore vanish. Exploiting the arbitrariness of ξ^i, one therefore has

$$T^j_{i;j} = 0. \tag{13.50}$$

This equation should be compared with Eq. (10.65). Applied within Euclidean geometry the equations are identical. The quantity T_i^j defined in Eq. (10.63) was interpreted to be the energy–momentum tensor and its properties were investigated in Chapter 10. The quantity T_i^j defined in Eq. (13.46) in this section will therefore share the properties described there, and will therefore be called the energy–momentum tensor. The arbitrary multiplicative factor in Eq. (13.46) was chosen so that the two definitions of T_i^j coincide, for example for electromagnetic theory, in the limit of flat space.

Problem 13.6.1. *The Lagrangian density for electromagnetic theory defined to be $\mathcal{L} = -F_{ij}F^{ij}/(16\pi)$ in Eq. (11.40), and the corresponding T^{ij} (symmetric in its indices) was worked out in Eq. (11.43). Setting $\sqrt{-g} = 1$ in Eq. (13.46), show that the same energy–momentum tensor T^{ij} is obtained.*

13.7
"Derivation" of the Einstein Equation

The foundation has been laid for the derivation of gravitational equations with sources. As always, the equations are to be derived using the principle of least action. Previous to general relativity, the vanishing of δS, where S is the total system action, has depended on the cancellation of a field action and a material action. In functional terms, for single particle motion, the cancellation occurs between a potential energy component and the kinetic energy component of the Lagrangian. In the electromagnetic case the separate terms of S, including also an interaction energy, were spelled out in Eq. (11.18). Maxwell's equations with sources were derived by applying Hamilton's principle to this action. In functional terms in that case, the two functions involved in the cancellation are the vector potential and the mass/charge distribution function.

Because masses both cause, and are affected by, gravitation, the action cannot be separated in the same way as in electromagnetic theory. In the present theory, the Lagrangian density depends implicitly (via the energy–momentum tensor) on the distribution of masses. So what is the other function whose variation can lead to the cancellation of δS? Based on the geometric emphasis so far, it is clear that this other function is the metric tensor, $g_{ij}(x^n)$. This is new to general relativity; in special relativity, because the metric coefficients were constant, they could contribute no such variability to the theory.

Though there is a single action function S, we can define two action variations δS_m, corresponding to variation of material distributions, and δS_g, corresponding to variation of the metric tensor. Working out δS_m has essentially already been done in the previous section, Eq. (13.47);

$$\delta S_m = \frac{1}{2c}\int \sqrt{-g}\, T_{ij}\, \delta g^{ij}\, d\Omega \tag{13.51}$$

Working out δS_g will appear superficially similar to the determination of δS_m in that δg^{ij} will again appear but, this time, the coefficient variations δg^{ij} will be independent, rather than dependent, variables. Making the replacement $\mathcal{R} = g^{ij}R_{ij}$ in the integrand of Eq. (13.44) and forming the variation of the integrand, three terms result, of which the third is shown to vanish in Problem 13.7.2. This leaves

$$\begin{aligned}\delta S_g &= \frac{1}{c}\int \delta(\sqrt{-g}\,\mathcal{R})\, d\Omega \\ &= \frac{1}{c}\int \Big((\delta\sqrt{-g})\, g^{ij}\, R_{ij} + \sqrt{-g}(\delta g^{ij})\, R_{ij}\Big)\, d\Omega.\end{aligned} \tag{13.52}$$

The factor $\delta\sqrt{-g} = -\delta g/(2\sqrt{-g})$ can be worked out using Eq. (12.22), according to which

$$\delta g = -g\, g_{ij}\delta g^{ij}. \tag{13.53}$$

We therefore obtain

$$\delta S_g = \frac{1}{c} \int \left(-\frac{1}{2} g_{ij}\, \mathcal{R} + R_{ij} \right) \sqrt{-g}\, \delta g^{ij}\, d\Omega. \tag{13.54}$$

Having succeeded in expressing the integrands of both δS_m and δS_g proportional to the arbitrary factor δg^{ij}, we are finally in a position to apply Hamilton's principle by cancelling their contributions. There is, however, as always, an undetermined numerical factor. Since the curvature of the space is *caused* by mass, one expects this constant, after multiplication by other physical constants entering the theory, to give the Newton gravitational constant G. Continuing to follow Landau and Lifshitz, a dimensional constant k is introduced by expressing Hamilton's principle as

$$\delta S_g = -\frac{c^3}{16\pi k} \int \left(R_{ij} - \frac{1}{2} g_{ij}\, \mathcal{R} - \frac{8\pi k}{c^4} T_{ij} \right) \sqrt{-g}\, \delta g^{ij}\, d\Omega = 0. \tag{13.55}$$

The constant k has been introduced in two places so as to cancel and preserve the definition of T_{ij} as it was introduced in Eq. (13.46). With the δg^{ij} factor being arbitrary, we conclude that

$$R_{ij} - \frac{1}{2} g_{ij}\, \mathcal{R} = \frac{8\pi k}{c^4} T_{ij}. \tag{13.56}$$

This is known as Einstein's equation and the tensor on the left-hand side is known as the Einstein tensor G_{ij};

$$G_{ij} = R_{ij} - \frac{1}{2} g_{ij}\, \mathcal{R}. \tag{13.57}$$

In free space, where $T_{ij} = 0$, one must therefore have $G_{ij} = 0$.

To determine k in terms of G (which is measurable in the Cavendish experiment) we need to solve the Einstein equation under physical circumstances applicable to the Cavendish experiment. This requires, first of all, an interpretation of what it means to "solve" the Einstein equation. Knowing that treating space as flat seems to be an excellent approximation in ordinary life, one expects only tiny deviations to be caused by the presence of scalar curvature $\mathcal{R}$. As a result one expects the factor k to be "small," which will permit the third term of Eq. (13.56) to be treated perturbatively. This condition can be referred to as "weak gravity." Ignoring gravity altogether amounts to setting $k = 0$, in which case the Minkowski metric, which is independent of position and time, causes Eq. (13.56) to be trivially satisfied.

Problem 13.7.1. *Consider Eq. (13.8) as it stands, and also as it would be interpreted with* $\Gamma^i{}_{jk}$ *replaced by* $\Gamma^i{}_{jk} + \delta\Gamma^i{}_{jk}$, *where the* $\delta\Gamma^i{}_{jk}$ *are the infinitesimal variations that appear in the term that was dropped in going from the first to the second line* *of Eq. (13.52). Explain why the factor $\delta\Gamma^{i}{}_{ja}A_{i}dx^{a}$ entering that comparison can be interpreted as the difference, at a single point, of two* true *tensors. Show, therefore, even though the $\Gamma^{i}{}_{jk}$ values are not the components of a tensor, that the $\delta\Gamma^{i}{}_{jk}$* are *the components of a true tensor.*

Problem 13.7.2. *Show that the term dropped in going from the first to the second line of Eq. (13.52) can be expressed as*

$$\int \sqrt{-g}\, g^{ij}\, \delta R_{ij}\, d\Omega = \int \frac{\partial\left(\sqrt{-g}\,(g^{ij}\delta\Gamma^{l}{}_{ij} - g^{il}\delta\Gamma^{j}{}_{ij})\right)}{\partial x^{l}}\, d\Omega. \tag{13.58}$$

Using Eq. (4.83) (with g replaced by $-g$ and X^i replaced by a quantity known from the previous problem to be a true vector), to represent the integrand as the divergence of a vector. Then, using Gauss's law, the integral can be transformed into an integral over an enclosing "surface." Constraining the variations to vanish on this surface, the integral vanishes. This validates having dropped the term in Eq. (13.52).

Much of the complexity of general relativity is due to the complicated formulas giving R_{ij} in terms of the $\Gamma^{i}{}_{jk}$ components, which are themselves complicated sums of derivatives of the metric coefficients. It is sensible, therefore, to simplify the equations to the extent possible. Since the energy–momentum tensor is typically simpler than the curvature tensor, one simplification is to rearrange the Einstein equation, first contracting it on its indices to produce

$$\mathcal{R} = -\frac{8\pi k}{c^4}\mathcal{T}, \tag{13.59}$$

where $\mathcal{T} = T^{i}{}_{i}$. Substituting this back into Eq. (13.56) and rearranging terms produces

$$R_{ij} = \frac{8\pi k}{c^4}\left(T_{ij} - \frac{1}{2}g_{ij}\mathcal{T}\right). \tag{13.60}$$

For weak gravity, which is the only case we will consider, this equation has source terms on the right-hand side of the equation and terms describing the geometric influence of the sources on the left-hand side of the equation. When described in terms of mixed indices this equation becomes

$$R_{i}^{j} = \frac{8\pi k}{c^4}\left(T_{i}^{j} - \frac{1}{2}\delta_{i}^{j}\mathcal{T}\right). \tag{13.61}$$

13.8
Weak, Nonrelativistic Gravity

Essentially all of one's experience with gravity, both in the laboratory and in (early) astronomy, involves only nonrelativistic mechanics. In solving the

Einstein equation we will therefore assume $v \ll c$, where v is the velocity of any particle of any nonzero mass that enters. The components of the 4-velocity of such a particle are $u^i = \gamma(c, \mathbf{v})^T$ where $\gamma = 1$. Wherever u^i appears we can, therefore, set the u^α, $\alpha = 1,2,3$ components to zero, and $u^0 = c$. Here we are following the convention that Greek indices run only over the spatial, $1,2,3$ values.

Another important simplification is that $\partial/\partial x^0$ derivatives, because of the c factor in the denominator, can be neglected compared to $\partial/\partial x^\alpha$ derivatives.

With $\gamma = 1$, as is appropriate nonrelativistically, the momentum-energy tensor T^{ij} needed for the Einstein equation was shown, in Eq.(10.77), to be $T^{ij} = \mu\, u^i u^j$. With u^i approximated as in the previous paragraph this simplifies greatly – the only nonnegligible component is T^{00}, and its value is $T^{00} = \mu c^2$, where $\mu(\mathbf{r})$ is the mass density. We also have $\mathcal{T} = \mu c^2$. The source term of the Einstein equation with mixed components (i.e. the right-hand side of Eq. (13.61)) is therefore given by

$$\frac{8\pi k}{c^4}\left(\mu\, u_i u^j - \frac{1}{2}\,\delta_i^j\, \mu c^2\right). \tag{13.62}$$

Working out the components of R_{ij}, starting with R_{j0}, we have, from Eq. (13.42),

$$R_{j0} \approx \frac{\partial \Gamma^k{}_{j0}}{\partial x^k} - \frac{\partial \Gamma^k{}_{jk}}{\partial x^0} \approx \frac{\partial \Gamma^\alpha{}_{j0}}{\partial x^\alpha}. \tag{13.63}$$

Some of the terms from Eq. (13.42) have been dropped because they are quadratic in the Γ's and hence quadratic in the (small) gravitational constant. Also the second term in the middle form of Eq. (13.63) is negligible because of its $\partial/\partial x^0$ derivative. For the same reason it is only the spatial derivative terms that survive in the final expression.

With all spatial components of u^i being neglected, the geodesic equation (13.12) simplifies greatly, in that only the $\Gamma^i{}_{00}$ terms will survive. Tentatively we will assume we need to keep only this Christoffel coefficient while evaluating the curvature tensor, planning to confirm or improve the validity of this assumption later. We therefore need to work out only the j=0 case of Eq. (13.63). For this we need Eq. (13.10);

$$\Gamma^\alpha{}_{00} = \frac{1}{2} g^{\alpha\beta}\left(\frac{\partial g_{\beta 0}}{\partial x^0} + \frac{\partial g_{\beta 0}}{\partial x^0} - \frac{\partial g_{00}}{\partial x^\beta}\right) \approx -\frac{1}{2} g^{\alpha\beta}\frac{\partial g_{00}}{\partial x^\beta} \approx \frac{1}{2}\,(\nabla \Delta g_{00})^\alpha. \tag{13.64}$$

In the final step $g^{\alpha\beta}$ has been approximated, to relevant accuracy, by its Minkowski limit. Also, defining $g_{00} = 1 + \Delta g_{00}$, the entire gravitational effect has been reduced to $\Delta g_{00}(\mathbf{x})$, which is the local deviation from flat geometry of the time-like component of the metric. Regarded as ordinary 3D vectors, with components related as in Eq. (13.64), $\Gamma_{\ 00}$ is equal to $\nabla \Delta g_{00}/2$.

Before continuing the evaluation of terms in the Einstein equation, let us exploit our just-determined Christoffel coefficients by recalling geodesic equation (13.12), which simplifies now to

$$\frac{d^2x^\alpha}{dt^2} = -\left(\nabla\left(\frac{c^2}{2}\Delta g_{00}\right)\right)^\alpha \equiv -(\nabla\phi_g)^\alpha. \tag{13.65}$$

This is the differential equation satisfied by free particle motion in the curved geometry described by the metric tensor. In Newtonian mechanics this equation would be interpreted as being Newton's law, with the left-hand side being acceleration and the right-hand side being force/mass. Therefore, modulo a constant of proportionality, Δg_{00} can be interpreted as the gravitational potential.

Returning to the Einstein equation by substituting from Eq. (13.64) into Eq. (13.63),

$$R^0_0 = R_{00} = \frac{1}{2}\frac{\partial^2 \Delta g_{00}}{\partial x_\alpha \partial x^\alpha} = \frac{1}{2}\nabla^2 \Delta g_{00}. \tag{13.66}$$

Finally, we are able, equating this to (13.62), to express the metric coefficient deviation in terms of the local mass density;

$$\nabla^2 \Delta g_{00} = \frac{8\pi k \mu}{c^2}. \tag{13.67}$$

This equation, known as the Poisson equation, is familiar both from electrostatics, where it is equivalent to Coulomb's law, and from Newtonian gravity, where it is equivalent to Newton's gravitational law. For a point mass M at the origin, the solution of this equation is

$$\Delta g_{00}(r) = -\frac{2kM}{c^2}\frac{1}{r}. \tag{13.68}$$

where r is the distance from the origin.

A gravitational potential function $\phi_g(r)$ can be defined such that the gravitational force F_g on a mass m is given by $F_g = -m\nabla\phi_g$. If F_g is due to a mass M at the origin, Newton's law of gravitation gives

$$-m\nabla\phi_g = -G\frac{mM}{r^2} = -m\nabla\left(-\frac{GM}{r}\right), \quad \text{or} \quad \phi_g = -\frac{GM}{r}. \tag{13.69}$$

Comparing this equation with Eq. (13.68), one sees that $\Delta g_{00}(r)$ and $\phi_g(r)$ are proportional. This squares well with our observation, below Eq. (13.65), that $\Delta g_{00}(r)$ behaves like a potential function. Eliminating $1/r$ from these two equations produces $\phi_g = (G/k)\Delta g_{00}c^2/2$. One sees that ϕ_g, as introduced here, is made identical to ϕ_g, as introduced in Eq. (13.65), by choosing $k = G$.

In other words, the constant k was introduced in Eq. (13.55) along with appropriate factors to make k be Newton's constant G.

13.9
The Schwarzschild Metric

At this point we have recreated Newtonian gravity within our geometric framework. Not only does this fall short of being a triumph, the formulation as given so far will now be seen to be incomplete. With ϕ_g given by Eq. (13.69), and therefore $\Delta g_{00}(r)$ by Eq. (13.68), using spherical coordinates, the metric tensor corresponding to mass M at the origin is given by

$$ds^2 = \left(1 - \frac{r_g}{r}\right) dx^{0 2} - f(r)\, dr^2 - r^2 d\theta^2 - r^2 \sin^2\theta\, d\phi^2, \tag{13.70}$$

where $r_g = 2GM/c^2$. As in metric (13.26), the interval dx^0 here is a "world time interval" measured by previously distributed, and synchronized, clocks stationary in an inertial frame, and *not* as the (proper) time measured by a local clock at rest. Note, for a nonrelativistic circular Keplerian orbit of radius r, that $r_g/r = 2v^2/c^2$. For planets of the sun and satellites of the earth r_g is therefore much smaller than r. In Eq. (13.70) an undetermined function $f(r)$ has been introduced as the g_{rr} metric coefficient, for reasons to be discussed now. With coordinates taken in the order t, r, θ, ϕ, the nonzero Christoffel coefficients for this metric are

$$\begin{aligned}
\Gamma^0{}_{01} &= \frac{r_g}{2r^2(1 - r_g/r)},\\
\Gamma^1{}_{00} &= -\frac{r_g}{2fr^2}, \quad \Gamma^1{}_{11} = \frac{f'}{2f}, \quad \Gamma^1{}_{22} = \frac{r}{f}, \quad \Gamma^1{}_{33} = \frac{r\sin^2\theta}{f},\\
\Gamma^2{}_{12} &= \frac{1}{r}, \quad \Gamma^2{}_{33} = -\sin\theta\,\cos\theta, \quad \Gamma^3{}_{13} = \frac{1}{r}, \quad \Gamma^3{}_{23} = \frac{\cos\theta}{\sin\theta}.
\end{aligned} \tag{13.71}$$

Problem 13.9.1. *Confirm the Christoffel coefficients given in Eq. (13.71). This can be done by hand or by using MAPLE.*

Problem 13.9.2. *For the same metric, work out the nonzero components of the Einstein tensor, G_{ij}, defined in Eq. (13.57). If you are working out the coefficients by hand, first show that*

$$G_{11} = -\frac{r + f(r - r_g)}{(r - r_g)\, r^2}, \tag{13.72}$$

and continue immediately to the next problem.

Problem 13.9.3. *Show that the vanishing of the Einstein tensor, requires*

$$f(r) = -\frac{1}{1 - r_g/r}, \tag{13.73}$$

and show that, for this choice $G_{ij} = 0$ *for all elements. This is very tedious to do by hand, but very easy using MAPLE.*

Accepting the results of the previous problems, for the formulation to be self-consistent, the metric of Eq. (13.70) needs to take the form

$$ds^2 = \left(1 - \frac{r_g}{r}\right) c^2 dt^2 - \frac{dr^2}{1 - r_g/r} - r^2 d\theta^2 - r^2 \sin^2\theta \, d\phi^2. \tag{13.74}$$

This is known as the Schwarzschild metric. It gives the metric outside a spherically symmetric mass distribution.

13.9.1 Orbit of a Particle Subject to the Schwarzschild Metric

The orbit of any free particle is a geodesic in the geometry of the space it inhabits – a space that we now take to be described by the Schwarzschild metric. In this geometry Newton's law, $a = F/m$, has been replaced by geodesic equation (13.1). It is useful to cast these seemingly very different formulations as nearly parallel as possible. This can start with the relation

$$p^j p_j = m^2 c^2, \tag{13.75}$$

which, based in special relativity, relates the quantities $p_i = (E/c, \mathbf{p})$ to the rest mass m. But now the metric is taken to be Schwarzschild rather than Minkowski. Forming the absolute derivative of this equation yields

$$p^k \frac{\partial p_j}{\partial x^k} = \Gamma^l{}_{jk} \, p^k p_l. \tag{13.76}$$

Also from special relativity one has $p^k = mu^k = mdx^k/d\tau$ which, with τ being proper time, relates momentum and (proper) velocity. One then has $u^k \partial/\partial x^k = d/d\tau$ where, as usual in Newtonian mechanics, $d/d\tau$ represents rate of change as measured by a clock attached to the particle whose dynamics is being studied. Equation (13.76) therefore becomes

$$m \frac{dp_j}{d\tau} = \Gamma^l{}_{jk} \, p^k p_l. \tag{13.77}$$

In this form the geodesic equation more nearly resembles Newton's law, and one begins to see how the right-hand side can be viewed as being "force-like."

For the study of Keplerian orbits, and other problems in mechanics, one has found an alternate approach to be effective – namely the Hamilton–Jacobi method. This method is especially appropriate, once solved analytically for idealized forces, to find the effects of small additional forces not present in the idealized model. (Methods like this are discussed at length in Chapter 16.) Here we wish mainly to introduce and apply the Hamiltonian–Jacobi equation as generalized by the requirements of general relativity.

Recall the main ingredients of H-J theory: (a) From special relativity, the free particle Hamiltonian follows from $E^2 - p^2c^2 = m^2c^4$. (b) The H–J equation is obtained by replacing p by $\partial S/\partial x$ in the Hamiltonian. (c) Hope the H-J equation can be solved by separation of variables. If, for reasons of symmetry, all but one of the coordinates are ignorable, separability is more than a hope, it is *guaranteed*. (d) Use the separation constants as dynamical variables – they have the ideal property of *not* varying. (e) Apply the Jacobi prescription (Section 8.2.4) to identify matching constants of the motion.

To start on step (a), Eq. (13.75) can be rewritten as

$$g^{ij}p_ip_j = m^2c^2. \tag{13.78}$$

To generalize this equation to general relativity all that is necessary is to replace the Minkowski metric by the Schwarzschild metric. This completes step (a). Step (b) gives the H-J equation (a partial differential equation for "action" S) to be

$$g^{ij}\frac{\partial S}{\partial x^i}\frac{\partial S}{\partial x^j} = m^2c^2. \tag{13.79}$$

For the Schwarzschild metric this becomes

$$\frac{1}{1-r_g/r}\left(\frac{\partial S}{\partial ct}\right)^2 - \left(1-\frac{r_g}{r}\right)\left(\frac{\partial S}{\partial r}\right)^2 - \left(\frac{1}{r}\frac{\partial S}{\partial \phi}\right)^2 = m^2c^2, \tag{13.80}$$

where S has been assumed to be independent of θ, based on symmetry and the assumption that the motion is initially in the $\theta = \pi/2$-plane.

The historically first application of Eq. (13.80) was to planetary orbits in the solar system, especially that of Mercury – because of its proximity to the sun the factor r_g/r is fractionally more important for Mercury than for other planets. Some sample numerical values, including an artificial satellite, are shown in the following table, with a being the semimajor axis of the planetary orbit.

System	r_g (m)	a (m)	r_g/a
Sun–Mercury	2.95×10^3	0.579×10^{11}	5.1×10^{-8}
Earth–Moon	0.89×10^{-2}	3.84×10^8	2.3×10^{-11}
Earth–art. sat.	0.89×10^{-2}	6.9×10^6	1.29×10^{-9}

From the smallness of the entries in the final column it can be seen, for almost all planetary systems, that the general relativistic correction is negligible for most purposes. With $r_g = 0$, Eq. (13.80) can be compared to Eq. (8.55), which is the H–J equation for Keplerian orbits. The new equation is subject to the same analysis, which will not be repeated in detail, other than to figure out how the potential energy springs out of the general relativistic formulation.

Because t and ϕ are ignorable in Eq. (13.80), the action takes the form

$$S = -\mathcal{E}t + \alpha\phi + S^{(r)}(r), \tag{13.81}$$

where $\mathcal{E}$ is the "energy," which is known to be constant, but whose physical significance needs to be interpreted, and α_3 is the angular momentum. Substituting this into Eq(13.80) and solving for $dS^{(r)}/dr$, gives $dS^{(r)}$ as an indefinite integral:

$$S^{(r)}(r) = \int dr \sqrt{\frac{\mathcal{E}^2}{c^2}\left(1 - \frac{r_g}{r}\right)^{-2} - \left(m^2c^2 + \frac{\alpha^2}{r^2}\right)\left(1 - \frac{r_g}{r}\right)^{-1}}. \tag{13.82}$$

Because of the c^2 factor in the second term, it is not legitimate to set $r_g = 0$, even for $r_g \ll r$. But, introducing the nonrelativistic energy E, the sum of potential energy $V(r)$ and kinetic energy $mv^2/2$, such that $\mathcal{E} = mc^2 + E$, and assuming $v \ll c$ and $r_g \ll r$, one obtains

$$S^{(r)}(r) \approx \int dr \sqrt{2mE - \frac{\alpha^2}{r^2} + \frac{2GMm^2}{r}}. \tag{13.83}$$

This clearly reduces to S_3, as given by Eq. (8.61). In other words, Einstein gravity reproduces Newton gravity at least for nonrelativistic elliptical orbits.

Continuing to the next order of accuracy in r_g, we proceed to step (e) of the H–J prescription. With α being a constant of the motion, so also is $\partial S/\partial \alpha$. Differentiating Eq. (13.81) then produces

$$\phi = -\frac{\partial S^{(r)}}{\partial \alpha} = 2\int_{r_{\text{min.}}}^{r_{max.}} \frac{\alpha\, dr}{r^2\sqrt{\frac{\mathcal{E}^2}{c^2} - \left(m^2c^2 + \frac{\alpha^2}{r^2}\right)\left(1 - \frac{r_g}{r}\right)}}. \tag{13.84}$$

as the orbit equation or rather, with definite limits having been assigned to the integral, as the formula for the phase advance during one complete "revolution." The reason for the quotation marks is that, with the general relativistic correction, the orbit does not, in fact, close. However, in the limit where Eq. (13.83) is valid the orbit is known to close and the value of the expression in Eq. (13.84) is known to be 2π.

Problem 13.9.4. *The integral in Eq. (13.84) can be evaluated two ways; first, in the limit in which it reduces to Eq. (13.83) and the radial minima and maxima are* *calculated accordingly. The orbit is known to close in that case and the result is known to be 2π. Second, the integrand can be approximated to linear order in r_g, and the integral evaluated using the same limits as in the first integral. Performing the appropriate subtraction, show that the precession of the orbit per revolution is given by*

$$\Delta\phi \approx \frac{6\pi G^2 m^2 M^2}{c^2 \alpha^2}. \tag{13.85}$$

An alternate approach to finding the orbit precession, using the Lagrange planetary equations, is given in Section 16.2.

Problem 13.9.5. *Referring, if you wish, to Schutz,* A First Course in General Relativity, *p. 186, with ${\Gamma^l}_{jk}$ evaluated for the Schwarzschild metric, show that Eq. (13.77) reduces to Newton's law for planetary orbits in the limit of small r_g. Your result should include the introduction of "gravitational potential energy" $V(r) = GMm/r$, a relation that is already implicit in Eq. (13.83).*

13.10 Gravitational Lensing and Red Shifts

The initial guiding principle of general relativity was that the equations governing the orbit of a particle (even including massless particles) should be valid for massless particles. With m being explicitly present in Eqs. (13.76) and (13.79) and subsequent equations, our formulation so far appears not to have satisfied this requirement. It is not that these equations become wrong for $m = 0$, it is that they become *indeterminate*. The reason for this is that the proper time τ does not advance along the geodesic of a massless particle. Hence, for example on the left-hand side of Eq. (13.76), both numerator and denominator vanish.

From the parochial, obsessively classical, perspective of this text, one can even object to the concept of a particle having zero mass. One has to explain light, but not necessarily by the introduction of photons. If Einstein had invented general relativity before he had explained photons he would have had to discuss the gravitational bending of electromagnetic waves rather than of zero mass particles. These gratuitous, unhistorical comments are intended to motivate the following approach (also, of course, due to Einstein) toward discussing the trajectories of massless particles.

Because the Hamilton–Jacobi equation is, itself, a wave equation, it is the natural formalism on which to base a description of the gravitational bending of waves. Furthermore one knows, for example from Chapter 7, that one needs to study wave propagation in the short wavelength limit in order to bridge the gap between optics and mechanics. This evolution, from geometric optics, to physical optics, to particle mechanics is charted in Fig. 8.8.

The tangent to any geometric curve is given by $k^i = dx^i/d\lambda$, where λ is a parameter identifying points continuously along the curve. In wave optics the 4-vector k^i (introduced in Section 7.1.1) is referred to as the "wave vector." The relation of k^i (or rather its spatial component) to wavefronts is illustrated in Fig. 7.2.

A hint toward treating wave propagation in general relativity is the observation that Eqs. (13.75) and (13.78) both continue smoothly to the $m = 0$ limit. The former of these, for $m = 0$, and with p^i replaced by k^i as is customary for massless particles, expressed in terms of covariant components is

$$g^{ij}k_ik_j = 0, \tag{13.86}$$

with, for the moment, g^{ij} still in its Minkowski form. Upon differentiation, this produces $dk_i = dk^i = 0$. (Waves in free space don't change direction or wavelength.) In general relativity this generalizes to $Dk^i = 0$, where D is the absolute differential operator defined in Section 3.1.3. In expanded form this becomes

$$\frac{dk^i}{d\lambda} = \Gamma^i{}_{jk}k^jk^k, \tag{13.87}$$

where the metric is now assumed to be that given by general relativity. (Strictly speaking the left-hand side only needs to be proportional to the right-hand side, but λ can evidently be chosen to convert this to an equality.) The equation just written, with its parameter λ, serves as the ray equation for the wave. In the limit of geometric optics (i.e., short wavelengths) the direction of propagation of the wave gets better and better defined as being along the ray. Equation (13.87) resembles Eq. (7.18);

$$\frac{d}{ds}\left(\frac{\mathbf{dr}}{ds}\right) = \frac{1}{n}\boldsymbol{\nabla}\mathbf{n}(\mathbf{r}). \tag{13.88}$$

(Here we have cheated a bit, by moving the factor n from (inside the derivative) on the left-hand side, to the right-hand side. This is likely to be valid in the cases when it is the variation of the index of refraction *transverse* to the ray that is important. This assumption is valid for astronomical lensing observations.) To complete the analogy $\mathbf{dr}/ds$ has to be replaced by $\mathbf{k}$ and the right-hand side of Eq. (13.87) describes a spatially varying or "focusing" medium. When the right-hand side of Eq. (13.87) is evaluated as in Eq. (13.65), an analogy between gravitational potential ϕ_g and index of refraction n becomes apparent.

One of the two initial confirmations of general relativity came from observing shifts of the apparent position of those stars that happened to be in the "limb" of the sun during a total eclipse. In that case the sun is acting more like a prism than as a lens. The optical path length of a ray coming to the eye from a distant star depends on the "impact parameter" of the ray and the sun. Just like a ray passing off-axis through a lens, the effect is a net bend that shifts the apparent position of the point source. In spite of the sun's great mass, the angular shift is so small as to be barely detectable, so the assumptions of weak gravity are applicable.

In modern astrophysics the use of gravitational lensing in experimental cosmology has far surpassed its earlier importance as confirmation of general relativity. It is possible for a single distant source to show up as more than one image, because the prismatic effect has opposite signs on opposite sides of an intervening massive object. (Multiple images of a single distant light as seen in the refraction/reflection of a wine glass is a not dissimilar phenomenon.) As far as I know it is valid to use the same weak gravitational analysis for calculating all such earth-bound lensing observations. The same can obviously *not* be said for the propagation of light in the vicinity of massive objects such as black holes.

Problem 13.10.1. *Approximating the right-hand side of Eq. (13.87) as in Eq. (13.65), complete the analogy with Eq. (13.88) and find the "effective index of refraction" in the vicinity of mass M – call it the sun – taken to be point like. Two initially parallel rays of light pass the sun with impact parameters b and $b + \Delta b$. The difference of their optical path lengths (which can be approximated by integrating along straight lines) causes a relative deflection of the rays, and their point of intersection downstream is calculable by the principle of least time. From this result calculate the apparent shift of a star viewed near the edge of the sun.*

As in the $m \neq 0$ case, the Hamilton–Jacobi equation is obtained by the replacement $k_i \to \partial\psi/\partial x^i$ in Eq. (13.86);[2]

$$g^{ij}\frac{\partial\psi}{\partial x^i}\frac{\partial\psi}{\partial x^j} = 0. \tag{13.89}$$

This is the analog of the eikonal equation (7.11) of physical optics and ψ is referred to as the "eikonal" though, now, ψ includes both space and time sinusoidal variation. As noted previously this is the same as Eq. (13.79) with $m = 0$. For the Schwarzschild metric the eikonal equation is

$$\frac{1}{1-(2GM/c^2)/r}\left(\frac{\partial\psi}{\partial ct}\right)^2 - \left(1-\frac{2GM/c^2}{r}\right)\left(\frac{\partial\psi}{\partial r}\right)^2 - \left(\frac{1}{r}\frac{\partial\psi}{\partial\phi}\right)^2 = 0. \tag{13.90}$$

Consider a distant source, radiating at frequency ω_0 the light described by Eq. (13.90). Here ω_0 is the frequency referred to coordinate time (a.k.a. world time) $t = x^0/c$. Following Eq. (7.5), to obtain ω_0 as $\omega_0 = -\partial\psi/\partial t$, because the coefficients of Eq. (13.90) do not depend on x^0, ω_0 will be constant throughout

2) In Eq. (13.89) ψ is used instead of the action variable S only in order to have a symbol that is specific to propagation of light.

space. This is like ordinary electromagnetic theory which, because it is perfectly linear, the response everywhere in space (not counting Doppler effect) has the same frequency as the frequency of the source.

In general relativity it is necessary to distinguish the just-defined ω_0 from "frequency" ω, which is referenced to proper time τ, measured, say, by a stationary local atomic clock. At a fixed point in the vicinity of a (not excessively massive) point mass M, substituting from Eq. (13.68) into Eq. (13.16), coordinate time t and proper time τ are related by

$$\tau = t\sqrt{1 - \frac{2GM/c^2}{r}} \approx t\left(1 - \frac{GM/c^2}{r}\right). \tag{13.91}$$

The two frequencies are correspondingly related by

$$\omega = \omega_0\left(1 + \frac{GM/c^2}{r}\right). \tag{13.92}$$

Since ω_0 is constant through space, ω is not, and the discrepancy is referred to as the gravitational red shift.

The clocks of the global positioning system (GPS) are so accurate, and so well synchronized, they can be used to check the effects described in this chapter. Because the gravity is so weak the coefficients of the Langevin metric (13.26) describing the earth's rotation, and of the Schwarzschild metric (13.74) can, to adequate accuracy, simply be merged. Many such tests of general relativity are described in the paper by Ashby.

Bibliography

General References

1 L.D. Landau and E.M. Lifshitz, *Classical Theory of Fields*, Pergamon, Oxford, 1976.

References for Further Study

Section 13.2

2 S. Weinberg, *Gravitation and Cosmology*, Wiley, New York, 1972.

Section 13.9.1

3 B.F. Schwarz, *A First Course in General Relativity*, Cambridge University Press, Cambridge, UK, 1985.

Section 13.10

4 N. Ashby, *Relativity in the Global Positioning System*, in Living Reviews in Relativity, Max Planck Institute, Germany, Vol. 6, 2003.

14
Analytic Bases for Approximation

Once equations of motion have been found they can usually be solved by straightforward numerical methods. But numerical results rarely provide much general insight. This makes it productive to develop analytic results to the extent possible. Since it is usually believed that "fundamental physics" is Hamiltonian, considerable effort is justified in advancing the analytic formulation to the extent possible without violating Hamiltonian requirements. One must constantly ask "is it symplectic."

In this chapter the method of canonical transformation will be introduced and then exercised by being applied to nonlinear oscillators. Oscillators of one kind or another are probably the most ubiquitous systems analyzed using classical mechanics. Some, such as relaxation oscillators, are inherently non-sinusoidal, but many exhibit motion that is approximately simple harmonic. Some of the sources of deviation from harmonicity are (usually weak) damping, Hooke's law violating restoring forces, and parametric drive. Hamiltonian methods, and in particular phase space representation, are especially effective at treating these systems, and adiabatic invariance, to be derived shortly, is even more important than energy conservation.

14.1
Canonical Transformations

14.1.1
The Action as a Generator of Canonical Transformations

We have encountered the Jacobi method within the Hamilton–Jacobi theory while developing analogies between optics and mechanics. But it is possible to come upon this procedure more formally while developing the theory of "canonical transformation" (which means transforming the equations in such a way that Hamilton's equations remain valid.) The motivation for restricting the field of acceptable transformations in this way is provided by the large body of certain knowledge one has about Hamiltonian systems, much of it described in previous chapters.

Geometric Mechanics: Toward a Unification of Classical Physics. 2nd Edition. Richard Talman

ISBN: 978-3-527-40683-8

From a Hamiltonian system initially described by "old" coordinates $q^1, q^2, \dots, q^n$ and "old" momenta $p_1, p_2, \dots, p_n$, we seek appropriate transformations

$$(q^1, q^2, \dots, q^n; p_1, p_2, \dots, p_n) \rightarrow (Q^1, Q^2, \dots, Q^n; P_1, P_2, \dots, P_n), \tag{14.1}$$

to "new coordinates" $Q^1, Q^2, \dots, Q^n$ and "new momenta" $P_1, P_2, \dots, P_n$.[1] (Within the Jacobi procedure these would have been known as β-parameters and α-parameters, respectively.)

Within Lagrangean mechanics we have seen the importance of variational principles in establishing the invariance to coordinate transformation of the form of the Lagrange equations. Since we have assigned ourselves essentially the same task in Hamiltonian mechanics it is appropriate to investigate Hamiltonian variational principles. This method will prove to be successful in establishing conditions that must be satisfied by the new $\mathbf{Q}$ and $\mathbf{P}$ variables.

Later, in Eq. (17.99), the Poincaré–Cartan integral invariant I.I. will be defined and will form the basis of symplectic mechanics. For the time being a simpler "Hamiltonian, variational" line integral H.I. is defined by,[2]

$$\text{H.I.} = \int_{\mathcal{P}_1}^{\mathcal{P}_2} \left(p_i dq^i - H(\mathbf{q}, \mathbf{p}, t)\, dt \right); \tag{14.2}$$

Other than starting at $\mathcal{P}_1$ and ending at $\mathcal{P}_2$ (and not being "pathological") the path of integration is arbitrary in the extended phase space q^i, p_i, and t. It is necessary however, for the given Hamiltonian $H(\mathbf{q}, \mathbf{p}, t)$ to be appropriately evaluated at every point along the path of integration. Here we use a new symbol H.I. to indicate that $\mathbf{p}$ and H are not assumed to have been derived from a solution of the H–J equation as they were in Eq. (17.99). In particular, the path of integration is not necessarily a solution path for the system.

H.I. has the dimensions of action and we now subject it to analysis like that used in deriving the Lagrange equations from $\int L\, dt$. In particular we seek the integration path for which H.I. achieves an *extreme* value. In contrast to coordinate/velocity space, where the principle of extreme action was previously analyzed, consider independent smooth phase space variations $(\delta\mathbf{q}, \delta\mathbf{p})$ away from an arbitrary integration path through fixed end points $(\mathcal{P}_1, t_1)$ and $(\mathcal{P}_2, t_2)$. Evaluating the variations of its two terms individually, the condition

1) It would be consistent with the more formally valid mathematical notation introduced previously to use the symbol $\widetilde{\mathbf{p}}_i$ for momentum p_i since the momenta are more properly thought of as *forms*, but this is rarely done.

2) It was noted in earlier chapters that, when specific coordinates are in use, the differential forms $\widetilde{\mathbf{dq}}^i$ are eventually replaced by the old-fashioned differentials dq^i and similarly for the other differential forms appearing in the theory. Since we will not be insisting on *intrinsic* description, we make the replacement from the start.

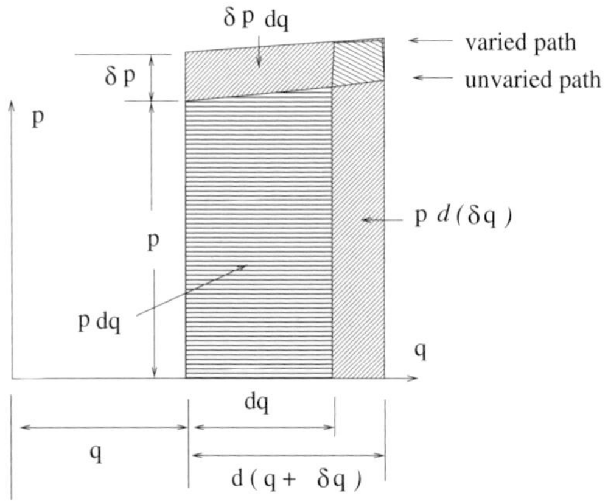

Fig. 14.1 Areas representing terms $\delta p\, dq + pd(\delta q)$ in the Hamiltonian variational integral.

for H.I. to achieve an extreme value is

$$0 = \int_{\mathcal{P}_1,t_1}^{\mathcal{P}_2,t_2} \left(\delta p_i\, dq^i + p_i d(\delta q^i) - \frac{\partial H}{\partial q^i} \delta q^i\, dt - \frac{\partial H}{\partial p_i} \delta p_i\, dt \right). \tag{14.3}$$

The last two terms come from $\int H\, dt$ just the way two terms came from $\int L\, dt$ in the Lagrangian derivation. Where the first two terms come from is illustrated in Fig. 14.1. At each point on the unvaried curve, incremental displacements $\delta q(q)$ and $\delta p(q)$ locate points on the varied curve. Since the end points are fixed the deviation δp vanishes at the ends and $d(\delta q^i)$ must average to zero as well as vanishing at the ends. With a view toward obtaining a common multiplicative factor in the integrand, using the fact that the end points are fixed, the factor $p_i d(\delta q^i)$ can be replaced by $-\delta q^i\, dp_i$ since the difference $d(p_i \delta q^i)$ is a total differential. Then, since the variations δq^i and δp_i are arbitrary, Hamilton's equations follow;

$$\dot{q}^i = \frac{\partial H}{\partial p_i}, \quad \text{and} \quad \dot{p}_i = -\frac{\partial H}{\partial q^i}. \tag{14.4}$$

It has therefore been proved that Hamilton's equations are implied by applying the variational principle to integral H.I. But that has not been our real purpose. Rather, as stated previously, our purpose is to derive canonical transformations. Toward that end we introduce[3] an arbitrary function $G(\mathbf{q}, \mathbf{Q}, t)$ of old coordinates $\mathbf{q}$ and new coordinates $\mathbf{Q}$ and alter H.I. slightly by subtracting

3) Goldstein uses the notation $F_1(\mathbf{q}, \mathbf{Q}, t)$ for our function $G(\mathbf{q}, \mathbf{Q}, t)$.

the total derivative dG from its integrand;

$$\begin{aligned} \text{H.I.}' &= \int_{\mathcal{P}_1}^{\mathcal{P}_2} \Big(p_i dq^i - H dt - dG(\mathbf{q}, \mathbf{Q}, t) \Big) \\ &= \int_{\mathcal{P}_1}^{\mathcal{P}_2} \left(p_i dq^i - H dt - \frac{\partial G}{\partial q^i} dq^i - \frac{\partial G}{\partial Q_i} dQ_i - \frac{\partial G}{\partial t} dt \right). \end{aligned} \tag{14.5}$$

This alteration cannot change the extremal path obtained by applying the same variational principle since the integral over the added term is independent of path. We could subject H.I.′ to a variational calculation like that applied to I but instead we take advantage of the fact that G is arbitrary to simplify the integrand by imposing on it the condition

$$p_i = \frac{\partial G(\mathbf{q}, \mathbf{Q}, t)}{\partial q^i}. \tag{14.6}$$

This simplifies Eq. (14.5) to

$$\text{H.I.}' = \int_{\mathcal{P}_1}^{\mathcal{P}_2} (P_i dQ_i - H' dt), \tag{14.7}$$

where we have introduced the abbreviations

$$P_i = -\frac{\partial G(\mathbf{q}, \mathbf{Q}, t)}{\partial Q^i}, \quad \text{and} \quad H'(\mathbf{Q}, \mathbf{P}, t) = H + \frac{\partial G}{\partial t}. \tag{14.8}$$

The former equation, with Eq. (14.6), defines the coordinate transformation and the latter equation gives the Hamiltonian in the new coordinates. The motivation for this choice of transformation is that Eq. (14.7) has the same form in the new variables that Eq. (14.2) had in the old variables. The equations of motion are therefore

$$\dot{Q}^i = \frac{\partial H'}{\partial P_i}, \quad \text{and} \quad \dot{P}_i = -\frac{\partial H'}{\partial Q^i}. \tag{14.9}$$

Since these are Hamilton's equations in the new variables we have achieved our goal. The function $G(\mathbf{q}, \mathbf{Q}, t)$ is known as the "generating function" of the canonical transformation defined by Eq. (14.6) and the first of Eqs. (14.8). The transformations have a kind of hybrid form (and it is an inelegance inherent to the generating function procedure) with G depending, as it does, on *old* coordinates and *new* momenta. Also there is still "housekeeping" to be done: expressing the new Hamiltonian H' in terms of the new variables. There is no assurance that it will be possible to do this in closed form.

Condition (14.6) that has been imposed on the function G is reminiscent of the formula for $\mathbf{p}$ in the H–J theory, with G taking the place of action function S. Though G could have been any function consistent with Eq. (14.6), if we conjecture that G is a solution of the H–J equation,

$$H + \frac{\partial G}{\partial t} = 0, \tag{14.10}$$

we note from Eq. (14.8) that the new Hamiltonian is given by $H' = 0$. Nothing could be better than a vanishing Hamiltonian since, by Eqs. (14.9), it implies the new coordinates and momenta are constants of the motion. Stated conversely, if we had initially assigned ourselves the task of finding coordinates that were constants of the motion we would have been led to the Hamilton–Jacobi equation as the condition to be applied to generating function G.

The other equation defining the canonical transformation is the first of Eqs. (14.8),

$$P_i = -\frac{\partial G(\mathbf{q}, \mathbf{Q}, t)}{\partial Q^i}. \tag{14.11}$$

Without being quite the same, this relation resembles the Jacobi-prescription formula $\beta = \partial S/\partial \alpha$ for extracting constant of the motion β corresponding to separation constant α in a complete integral of the H–J equation. It is certainly true that if G is a complete integral and the P_i are interpreted as the separation constants in that solution then the quantities defined by Eq. (14.11) *are* constants of the motion. But, relative to the earlier procedure, coordinates and momenta are interchanged. The reason is that the second arguments of G have been taken to be coordinates rather than momenta.

We are therefore motivated to try a different subtraction in the definition of the action. We subtract the total differential of an arbitrary function $dS(\mathbf{q}, \mathbf{P}, t)$[4] (or rather, for reasons that will become clear immediately, the function $d(S - P_i Q^i)$) from the variational integrand;

$$\begin{aligned} \text{H.I.}' &= \int_{\mathcal{P}_1}^{\mathcal{P}_2} \left(p_i dq^i - H dt - \frac{\partial S}{\partial q^i} dq^i - \frac{\partial S}{\partial P_i} dP_i - \frac{\partial S}{\partial t} dt + P_i dQ^i + Q^i dP_i\right) \\ &= \int_{\mathcal{P}_1}^{\mathcal{P}_2} (P_i dQ_i - H' dt), \end{aligned} \tag{14.12}$$

where we have required

$$p_i = \frac{\partial S(\mathbf{q}, \mathbf{P}, t)}{\partial q^i}, \quad Q^i = \frac{\partial S(\mathbf{q}, \mathbf{P}, t)}{\partial P_i}, \quad \text{and} \quad H'(\mathbf{Q}, \mathbf{P}, t) = H + \frac{\partial S}{\partial t}. \tag{14.13}$$

(It was only with the extra subtraction of $d(P_i Q^i)$ that the required final form was obtained.) We have now reconstructed the entire Jacobi prescription. If

4) Goldstein uses the notation $F_2(\mathbf{q}, \mathbf{P}, t)$ for our function $S(\mathbf{q}, \mathbf{P}, t)$. This function is also known as "Hamilton's principal function." Other generating functions, $F_3(\mathbf{p}, \mathbf{Q}, t)$ and $F_4(\mathbf{p}, \mathbf{P}, t)$ in Goldstein's notation , can also be used.

$S(\mathbf{q}, \mathbf{P}, t)$ is a complete integral of the H–J equation, with the P_i defined to be the α_i separation constants, then the $\beta_i \equiv Q^i$ obtained from the second of Eqs. (14.13) are constants of the motion.

To recapitulate, a complete integral of the H–J equation provides a generator for performing a canonical transformation to new variables for which the Hamiltonian has the simplest conceivable form – it vanishes – causing all coordinates and all momenta to be constants of the motion.

14.2
Time-Independent Canonical Transformation

Just as the Hamilton–Jacobi equation is the short-wavelength limit of the Schrödinger equation, the time-independent H–J equation is the same limit of the time-independent Schrödinger equation. As in the quantum case, methods of treating the two cases appear superficially to be rather different even though time independence is just a special case.

When it does not depend explicitly on time, the Hamiltonian is conserved, $H(\mathbf{q}, \mathbf{p}) = E$ and a complete integral of the H–J equation takes the form

$$S(\mathbf{q}, t) = S_0(\mathbf{q}, \mathbf{P}) - E(t - t_0), \tag{14.14}$$

where the independent parameters are listed as $\mathbf{P}$. The term *action*, applied to S up to this point, is also commonly used to refer to S_0.[5] In this case the H–J becomes

$$H\left(\mathbf{q}, \frac{\partial S_0}{\partial \mathbf{q}}\right) = E, \tag{14.15}$$

and a complete integral is defined to be a solution of the form

$$S_0 = S_0(\mathbf{q}, \mathbf{P}) + \text{const.}, \tag{14.16}$$

with as many new parameters P_i as there are coordinates. It is important to recognize though that the energy E can itself be regarded as a Jacobi parameter, in which case the parameter set $\mathbf{P}$ is taken to include E.

In this time-independent case it is customary to use $S_0(\mathbf{q}, \mathbf{P})$ (rather than $S(\mathbf{q}, \mathbf{P}, t)$) as the canonical generating function G. By the general theory, new variables are then related to old by

$$p_i = \frac{\partial S_0}{\partial q^i}, \quad Q^i = \frac{\partial S_0}{\partial P_i}. \tag{14.17}$$

5) Goldstein uses the notation $W(\mathbf{q}, \mathbf{P})$ for our function $S_0(\mathbf{q}, \mathbf{P})$. This function is also known as "Hamilton's characteristic function." The possible basis for this terminology has been discussed earlier in connection with Problem 7.3.3. Landau and Lifshitz call S_0 the "abbreviated action."

In particular, taking E itself as one of the new momentum, its corresponding new coordinate is

$$Q_E = \frac{\partial S_0}{\partial E}, \tag{14.18}$$

which is nonvanishing since the parameter set $\mathbf{P}$ includes E. Defined in this way Q_E is therefore *not* constant. The quantity whose constancy *is* assured by the Jacobi theory is

$$\frac{\partial S}{\partial E} = Q_E - t + t_0 = \text{ constant}. \tag{14.19}$$

This shows that Q_E and time t are essentially equivalent, differing at most by the choice of what constitutes initial time. Equation (14.19) is the basis of the statement that E and t are canonically conjugate variables. Continuing with the canonical transformation, the new Hamiltonian is

$$H'(\mathbf{Q}, \mathbf{P}, t) = H + \frac{\partial S_0}{\partial t} = E. \tag{14.20}$$

We have obtained the superficially curious result that in this simpler, time-independent, case the Hamiltonian is less simple, namely nonvanishing, than in the time-dependent case. This is due to our use of S_0 rather than S as generating function. But H' is constant, which is good enough.[6] We can test one of the Hamilton equations, namely the equation for $\dot{Q}_E$,

$$\dot{Q}_E = \frac{\partial H'}{\partial E} = 1, \tag{14.21}$$

in agreement with Eq. (14.19). For the other momenta, not including E, Hamilton's equations are

$$\dot{P}_i = 0, \quad \text{and} \quad \dot{Q}^i = \frac{\partial E}{\partial P_i} = 0. \tag{14.22}$$

Hence, finding a complete integral of the time-independent H–J equation is tantamount to having solved the problem.

6) When applying the Jacobi prescription in the time-independent case one must be careful not to treat E as functionally dependent on any of the other P_i though.

14.3 Action-Angle Variables

14.3.1 The Action Variable of a Simple Harmonic Oscillator

The Hamiltonian for a simple harmonic oscillator was given in Eq. (8.43);

$$H(q,p) = \frac{p^2}{2m} + \frac{1}{2}m\omega_0^2 q^2. \tag{14.23}$$

Recall that the variation of action S along a true trajectory is given, as in Eq. (8.10), by

$$dS = p_i dq^i - H dt, \quad \text{or} \quad S(P) = \int_{\mathcal{P}_0}^{\mathcal{P}} (p_i dq^i - H dt). \tag{14.24}$$

Applying this formula to the simple harmonic oscillator, since the path of integration is a true particle trajectory, $H = E$, and the second term integrates to $-E(t - t_0)$. Comparing with Eq. (14.13), we obtain, for the abbreviated action,

$$S_0(q) = \int_{q_0}^{q} p(q')dq'. \tag{14.25}$$

The word "action" has already been used to define the basic Lagrangian variational integral and as a name for the function satisfying the H–J equation, but it now now acquires yet another meaning as "$1/2\pi$ times the phase space area enclosed after one cycle." Because this quantity will be used as a dynamic variable it is called the "action variable" I of the oscillator.[7] For simple harmonic motion

$$I = \frac{1}{2\pi}\oint p(q')\,dq' = \frac{1}{2\pi}\iint dp\,dq = \frac{1}{2\pi}\pi\sqrt{2mE}\sqrt{\frac{2E}{m\omega_0^2}} = \frac{E}{\omega_0}. \tag{14.26}$$

The first form of the integral here is a line integral along the phase space trajectory, the second is the area in (q,p) phase space enclosed by that curve. The factor $1/(2\pi)$ entering the conventional definition of I will cause the motion to have the "right" period, namely 2π, when the motion is expressed in terms of "angle variables" (to be introduced shortly).

7) The terminology is certainly strained since I is usually called the "action variable," in spite of the fact that it is constant, but "variable" does not accompany "action" when describing S_0 which actually does vary. Next we will consider a situation in which I might be expected to vary, but will find (to high accuracy) that it does not. Hence, the name "action nonvariable" would be more appropriate. Curiously enough the word "amplitude" in physics suffers from the same ambiguity; in the relation $x = a\cos\omega t$ it is ambiguous whether the "amplitude" is x or a.

14.3.2
Adiabatic Invariance of the Action I

Following Landau and Lifshitz, *Mechanics*, consider a one-dimensional system which is an "oscillator" in the sense that coordinate q returns to its starting point at some time. If the Hamiltonian is time independent, the energy is conserved, and the momentum p returns to its initial value when q does. In this situation, the area within the phase space trajectory is closed and the action variable I just introduced is unambiguously defined.

Suppose however that the Hamiltonian $H(q,p,t)$, and hence the energy $E(t)$ have a weak dependence on time that is indicated by writing

$$E(t) = H(q,p,t) = H(q,p,\lambda(t)). \tag{14.27}$$

The variable $\lambda(t)$ has been introduced artificially to consolidate whatever time dependence exists into a single parameter for purposes of the following discussion. At any time t the energy $E(t)$ is *defined* to have the value it would have if $\lambda(t)$ were held constant at its current instantaneous value. Any non-constancy of $E(t)$ reflects the time dependence of H. The prototypical example of this sort of time dependency is *parametric variation* – for example, the "spring constant" k, a "parameter" in simple harmonic motion, might vary slowly with time, $k = k(t)$. Eventually what constitutes "slow" will be made more precise but, much like short wavelength approximations previously encountered, the fractional change of frequency during one oscillation period is required to be small. Motion with λ fixed/variable will be called "unperturbed/perturbed."

During *perturbed* motion the particle energy,

$$E(t) = H(q,p,\lambda(t)), \tag{14.28}$$

varies, possibly increasing during some parts of the cycle and decreasing during others, and probably accumulating appreciably over many cycles. We are now interested in the systematic or averaged-over-one-cycle variation of quantities like $E(t)$ and $I(t)$. The "time average" $\overline{f(t)}$ of a variable $f(t)$ that describes some property of a periodic oscillating system having period T is defined to be

$$\overline{f(t)} = \frac{1}{T}\int_t^{t+T} f(t')\,dt'. \tag{14.29}$$

From here on we take $t = 0$.

Let us start by estimating the rate of change of E as λ varies. Since $\lambda(t)$ is assumed to vary slowly and monotonically over many cycles, its average rate of change $\overline{d\lambda/dt}$ and its instantaneous rate of change $d\lambda/dt$ differ negligibly, making it unnecessary to distinguish between these two quantities. But the variation of E will tend to be correlated with the instantaneous values of q and p so E can be expected to be above average at some times and below average at others. We seek the time-averaged value $\overline{dE/dt}$. To a lowest approximation we anticipate $\overline{dE/dt} \sim d\lambda/dt$ unless it should happen (which it won't) that dE/dt vanishes to this order of approximation.

Two features that complicate the present calculation are that the perturbed period T is in general different from the unperturbed period and that the phase space orbit is not in general closed. This causes the area enclosed by the orbit to be poorly defined. To overcome this problem the integrals will be recast as integrals over one cycle of coordinate q, since q necessarily returns to its starting value, say $q = 0$. (We assume $\dot{q}(t = 0) \neq 0$.) The action variable

$$I(E, \lambda) = \frac{1}{2\pi} \oint p(q, E, \lambda)\, dq \tag{14.30}$$

is already written in this form. From Eq. (14.28) and energy conservation of the unperturbed motion, the instantaneous rate of change of energy is given by

$$\frac{dE}{dt} = \frac{\partial H}{\partial \lambda} \frac{d\lambda}{dt}, \tag{14.31}$$

and its time average is therefore given by

$$\overline{\frac{dE}{dt}} = \frac{d\lambda}{dt} \frac{1}{T} \int_0^T \frac{\partial H}{\partial \lambda}\, dt. \tag{14.32}$$

Because of the assumed slow, monotonic variation of $\lambda(t)$ it is legitimate to have moved the $d\lambda/dt$ factor outside the integral in this way. To work around the dependence of T on λ we need to recast this expression in terms of phase space line integrals. Using Hamilton's equations, we obtain

$$dq = \left.\frac{\partial H}{\partial p}\right|_{q,\lambda} dt, \quad \text{and hence} \quad T = \oint \frac{1}{\partial H/\partial p|_{q,\lambda}}\, dq. \tag{14.33}$$

Here we must respect the assumed functional form $H(q, p, t)$ and, to emphasize the point, have indicated explicitly what variables are being held constant for the partial differentiation. (To be consistent we should have similarly written $\partial H/\partial \lambda|_{q,p}$ in the integrand of Eq. (14.32).) Making the same substitution (14.33) in the numerator, formula (14.32) can be written as

$$\overline{\frac{dE}{dt}} = \frac{d\lambda}{dt} \oint \frac{\partial H/\partial \lambda|_{q,p}}{\partial H/\partial p|_{q,\lambda}}\, dq \Big/ \oint \frac{1}{\partial H/\partial p|_{q,\lambda}}\, dq. \tag{14.34}$$

Since this expression is already proportional to $d\lambda/dt$ which is the order to which we are working, it is legitimate to evaluate the two integrals using the unperturbed motion. Terms neglected by this procedure are proportional to $d\lambda/dt$ and give only contributions of order $(d\lambda/dt)^2$ to $\overline{dE/dt}$. (This is the sort of maneuver that one always resorts to in perturbation theory.)

The unperturbed motion is characterized by functional relation (14.28) and its "inverse";

$$E = H(q,p,\lambda), \quad \text{and} \quad p = p(q,\lambda,E), \quad \text{or} \quad E = H\big(q, p(q,\lambda,E), \lambda\big). \tag{14.35}$$

From now on, since λ is constant because unperturbed motion is being described, it will be unnecessary to list it among the variables being held fixed during differentiation. Differentiating the third formula with respect to E yields

$$\frac{1}{\partial H/\partial p|_q} = \frac{\partial p}{\partial E}\bigg|_q, \tag{14.36}$$

which provides a more convenient form for one of the factors appearing in the integrands of Eq. (14.34). Differentiating the third of Eqs. (14.35) with respect to λ yields

$$0 = \frac{\partial H}{\partial p}\bigg|_{q,\lambda} \frac{\partial p}{\partial \lambda}\bigg|_{q,E} + \frac{\partial H}{\partial \lambda}\bigg|_{q,p}, \quad \text{or} \quad \frac{\partial H/\partial\lambda|_{q,p}}{\partial H/\partial p|_{q,\lambda}} = -\frac{\partial p}{\partial \lambda}\bigg|_{q,E}. \tag{14.37}$$

Finally, substituting these expressions into Eq. (14.34) yields

$$\overline{\frac{dE}{dt}} = -\frac{d\lambda}{dt}\frac{1}{T}\oint \frac{\partial p}{\partial \lambda}\bigg|_{q,E} dq. \tag{14.38}$$

As stated previously, the integral is to be performed over the presumed-to-be-known unperturbed motion.

We turn next to the similar calculation of $\overline{dI/dt}$. Differentiating Eq. (14.30) with respect to t, using Eq. (14.31) and the first of Eqs. (14.33) yields

$$\begin{aligned} \overline{\frac{dI}{dt}} &= \frac{d\lambda/dt}{2\pi}\oint \left(\frac{\partial p}{\partial E}\bigg|_q \frac{\partial H}{\partial \lambda}\bigg|_{q,p} + \frac{\partial p}{\partial \lambda}\bigg|_{q,E}\right) dq \\ &= \frac{d\lambda/dt}{2\pi}\oint \frac{\partial H/\partial\lambda|_{q,p}}{\partial H/\partial p|_{q,\lambda}}\, dq + \frac{d\lambda/dt}{2\pi}\oint \frac{\partial p}{\partial \lambda}\bigg|_{q,E} dq. \end{aligned} \tag{14.39}$$

From the second of Eqs. (14.37) it can then be seen that

$$\overline{\frac{dI}{dt}} = 0. \tag{14.40}$$

Of course this is only approximate since terms of order $(d\lambda/dt)^2$ have been dropped. Even so this is one of the most important formulas in mechanics. It is usually stated as *the action variable is an adiabatic invariant.* That this is not an *exact* result might be regarded as detracting from its elegance, utility, and importance. In fact the opposite is true since, as we shall see, it is often an *extremely accurate* result, with accuracy in parts per million not uncommon. This would make it perhaps unique in physics – an approximation that is as good as an exact result – except that the same thing can be said for the whole of Newtonian mechanics. It is still possible for I to vary throughout the cycle, as an example in Section 14.3.4 will show, but its average is constant.

There is an important relation between action I and period T (or equivalently frequency $\omega = 2\pi/T$) of an oscillator. Differentiating the defining equation (14.30) for I with respect to E, and using Eqs. (14.36) and (14.33) yields

$$\frac{\partial I}{\partial E} = \frac{1}{2\pi}\oint \frac{\partial p}{\partial E}\bigg|_{q,\lambda} dq = \frac{1}{2\pi}\oint \frac{dq}{\partial H/\partial p}\bigg|_{q,\lambda} = \frac{1}{2\pi}\oint dt = \frac{T}{2\pi} = \frac{1}{\omega}. \tag{14.41}$$

This formula can be checked immediately for simple harmonic motion. In Eq. (14.26) we had $I = E/\omega_0$ and hence

$$\frac{\partial I}{\partial E} = \frac{1}{\omega_0} = \frac{T}{2\pi}. \tag{14.42}$$

Recapitulating, we have considered a system with weakly time-dependent Hamiltonian H, with initial energy E_0 determined by initial conditions. Following the continuing evolution of the motion, the energy, because it is not conserved, may have evolved appreciably to a different value E. Accompanying the same evolution, other quantities such as (*a priori*) action I and oscillation period T also vary. The rates dE/dt, dI/dt, $d\lambda/dt$, etc., are all proportional to $d\lambda/dt$ – doubling $d\lambda/dt$, doubles all rates for small $d\lambda/dt$. Since these rates are all proportional, it should be possible to find some combination that exhibits a first-order cancellation and such a quantity is an "adiabatic invariant" that can be expected to vary only weakly as λ is varied. It has been shown that I itself is this adiabatic invariant.

In thermodynamics one considers "quasistatic" variations in which a system is treated as static even if it is changing slowly and this is what we have been doing here, so "quasistatic" invariant would be slightly more apt than "adiabatic." In thermodynamics "adiabatic" means that the system under discussion is isolated in the sense that heat is neither added nor subtracted from the system. This terminology is not entirely inappropriate since we are considering the effect of purely mechanical external intervention on the system under discussion.

There is an important connection between quantized variables in quantum mechanics and the adiabatic invariants of the corresponding classical system. Suppose a quantum system in a state with given quantum numbers is placed in an environment with varying parameters (such as time varying magnetic field, for example) but that the variation is never quick enough to induce a transition. Let the external parameters vary through a cycle that ends with the same values as they started with. Since the system has never changed state it is important that the physical properties of that state should have returned to their starting values – not just approximately, but exactly. This is what distinguishes an adiabatic invariant. This strongly suggests that *the dynamical variables whose quantum numbers characterize the stationary states of quantum systems have adiabatic invariants as classical analogs*. The Bohr–Sommerfeld atomic theory, that slightly predated the discovery of quantum mechanics, was based on this principle. Though it became immediately obsolete, this theory was not at all *ad hoc* and hence had little in common with what passes for "the Bohr–Sommerfeld model" in modern sophomore physics courses. In short, the fact that the action is an adiabatic invariant makes it no coincidence that Planck's constant is called "the quantum of action."

14.3.3
Action/Angle Conjugate Variables

Because of its adiabatic invariance, the action variable I is an especially appropriate choice as parameter in applying the Jacobi procedure to a system with slowly varying parameters. We continue to focus on oscillating systems. Recalling the discussion of Section 14.2, we introduce the *abbreviated action*

$$S_0(q, I, \lambda) = \int_0^q p(q', I, \lambda)\, dq'. \tag{14.43}$$

Until further notice λ will be taken as constant but it will be carried along explicitly in preparation for allowing it to vary later on. Since λ is constant, both E and I are constant, and either can be taken as the Jacobi "momentum" parameter; previously we have taken E, now we take I, which is why the arguments of S_0 have been given as (q, I, λ). Since holding E fixed and holding I fixed are equivalent,

$$\left.\frac{\partial S_0}{\partial q}\right|_{E,\lambda} = \left.\frac{\partial S_0}{\partial q}\right|_{I,\lambda}. \tag{14.44}$$

Being a function of q through the upper limit of its defining equation, $S_0(q, I, \lambda)$ increases by $2\pi I$ as q completes one cycle of oscillation, since, as in Eq. (14.30),

$$I(E, \lambda) = \frac{1}{2\pi} \oint p(q, E(I), \lambda)\, dq. \tag{14.45}$$

Using $S_0(q, I, \lambda)$, defined by Eq. (14.43), as the generator of a canonical transformation, Eqs. (14.17) become

$$p = \frac{\partial S_0(q, I, \lambda)}{\partial q}, \quad \varphi = \frac{\partial S_0(q, I, \lambda)}{\partial I}. \tag{14.46}$$

where φ, the *new coordinate* conjugate to *new momentum* I, is called an "angle variable." For the procedure presently under discussion to be useful it is necessary for these equations to be reduced to explicit transformation equations $(q, p) \rightarrow (I, \varphi)$, such as Eqs. (14.53) of the next section. By Eq. (14.20) the new Hamiltonian is equal to the energy (expressed as a function of I)

$$H'(I, \varphi, \lambda) = E(I, \lambda), \tag{14.47}$$

and Hamilton's equations are

$$\dot{I} = -\frac{\partial H'}{\partial \varphi} = 0, \quad \text{and} \quad \dot{\varphi} = \frac{\partial E(I, \lambda)}{\partial I} = \omega(I, \lambda), \tag{14.48}$$

where Eq. (14.41) has been used, and the symbol $\omega(I, \lambda)$ has been introduced to stand for the oscillator frequency. Integrating the second equation yields

$$\varphi = \omega(I, \lambda)(t - t_0). \tag{14.49}$$

This is the basis for the name "angle" given to φ. It is an angle that advances through 2π as the oscillator advances through one period.

In these $(q, p) \rightarrow (\varphi, I)$ transformation formulas, λ has appeared simply as a fixed parameter. One way to exploit the concept of adiabatic invariance is now to permit λ to depend on time in a formula such as the second of Eqs. (14.48), $\dot{\varphi} = \omega(I, \lambda(t))$. This formula, giving the angular frequency of the oscillator when λ is constant, will continue to be valid with the value of I remaining constant, even if λ varies arbitrarily, as long as the adiabatic condition is satisfied.

A more robust way of proceeding is to recognize that it is legitimate to continue using Eqs. (14.46) as transformation equations even if λ varies, provided λ is replaced by $\lambda(t)$ *everywhere* it appears. The generating function is then $S_0(q, I, \lambda(t))$ and φ will still be called the "angle variable," conjugate to I. Using Eq. (14.7), and taking account of the fact that the *old* Hamiltonian is now time dependent, the *new* Hamiltonian is

$$H'(\varphi, I, t) = H + \frac{\partial S_0}{\partial t} = E(I, \lambda(t)) + \left.\frac{\partial S_0}{\partial \lambda}\right|_{q,I} \dot{\lambda}. \tag{14.50}$$

The *new* Hamilton equations are

$$\dot{I} = -\frac{\partial}{\partial \varphi}\left(\left.\frac{\partial S_0}{\partial \lambda}\right|_{q,I}\right) \dot{\lambda}, \quad \dot{\varphi} = \frac{\partial E(I, \lambda)}{\partial I} + \frac{\partial}{\partial I}\left(\left.\frac{\partial S_0}{\partial \lambda}\right|_{q,I}\right) \dot{\lambda}, \tag{14.51}$$

Since no approximations have been made these are exact equations of motion provided the function S_0 has been derived without approximation.

14.3.4 Parametrically Driven Simple Harmonic Motion

Generalizing the simple harmonic motion analyzed in Section 8.2.7 by allowing the spring constant $k(t)$ to be time dependent, the Hamiltonian is

$$H(q,p,t) = \frac{p^2}{2m} + \frac{1}{2} m\lambda^2(t)\, q^2. \tag{14.52}$$

Though time dependent, this Hamiltonian represents a *linear* oscillator because the frequency is independent of amplitude. The time-independent transformations corresponding to Eqs. (14.46) can be adapted from Eq. (8.49) by substituting $\omega_0 = \lambda$, $E = I\omega_0 = I\lambda$, and $\omega_0(t - t_0) = \varphi$;

$$\begin{aligned} q(I,\varphi) &= \sqrt{\frac{2E}{m\lambda^2}}\,\sin\varphi = \sqrt{\frac{2I}{m\lambda}}\,\sin\varphi, \\ p(I,\varphi) &= \sqrt{2Im\lambda}\,\cos\varphi. \end{aligned} \tag{14.53}$$

The abbreviated action is given by

$$S_0(q,I,\lambda) = \int^q p' dq' = 2I \int^{\sin^{-1}\left(q\sqrt{\frac{m\lambda}{2I}}\right)} \cos^2\varphi' d\varphi'. \tag{14.54}$$

The dependence on q is through its presence in the upper limit. This dependence can be rearranged as

$$\lambda = \frac{2I}{q^2 m}\,\sin^2\varphi. \tag{14.55}$$

This can be used to calculate the quantity

$$\left.\frac{\partial S_0}{\partial \lambda}\right|_{q,I} = 2I\cos^2\varphi \left.\frac{1}{\partial\lambda/\partial\varphi}\right|_{q,I} = \frac{I}{2\lambda}\sin 2\varphi. \tag{14.56}$$

which can then be substituted into Eqs. (14.51);

$$\begin{aligned} \dot{I} &= -\frac{\partial}{\partial\varphi}\left(\frac{I}{2\lambda}\sin 2\varphi\right)\dot{\lambda} = -I\cos 2\varphi\,\frac{\dot{\lambda}}{\lambda}, \\ \dot{\varphi} &= \omega(I,\lambda) + \frac{\partial}{\partial I}\left(\frac{I}{2\lambda}\sin 2\varphi\right)\dot{\lambda} = \lambda + \sin 2\varphi\,\frac{\dot{\lambda}}{\lambda}. \end{aligned} \tag{14.57}$$

Here the frequency $\omega(I,\lambda)$ has been calculated as if λ were time independent; that is $\omega(I,\lambda) = \lambda$. Since in this case the slowly varying parameter has been chosen as $\lambda = \omega$ one can simply replace λ by ω in Eqs. (14.51), eliminating the artificially introduced λ. The first equation shows that dI/dt is not identically zero, but the fact that $\cos 2\varphi$ averages to zero shows that the equation implies that dI/dt averages to zero to the extent that I *is* constant over one cycle and can therefore be taken outside the averaging. Though this statement may seem a bit circular – if I is constant then I is constant – it shows why I is approximately constant and can be the starting point of an estimate of the accuracy to which this is true. The new Hamiltonian is obtained from Eqs. (14.50) and (14.56),

$$H'(\varphi, I, t) = E(I, \omega(t)) + \left.\frac{\partial S_0}{\partial \omega}\right|_{q,I} \dot{\omega} = I\omega(t) + \frac{I}{2}\sin 2\varphi\,\frac{\dot{\omega}}{\omega}, \tag{14.58}$$

where the time dependence is expressed as the dependence on time (but not amplitude) of the "natural frequency" $\omega(t)$. The *linearity* of the oscillator is here reflected by the fact that H' depends linearly on I. Problems below illustrate how this can be exploited to complete the solution in this circumstance. Equation (14.58) can be used to check Eqs. (14.57) by substituting into Hamilton's equations though that is not different from what has already been done.

The angle φ has appeared in these equations only in the forms $\sin\varphi$, $\cos\varphi$, $\sin 2\varphi$, $\cos 2\varphi$. This is not an accident since, though the abbreviated action is augmented by $2\pi I$ every period, with this subtracted it is necessarily a periodic function of φ. The accumulating part does not contribute to $\partial S_0/\partial\lambda\big|_{q,I}$ because I is being held constant. It follows that H' is a periodic function of φ with period 2π and can therefore be expanded in a Fourier series with period 2π in variable φ. For the particular system under study this Fourier series has a single term, $\sin 2\varphi$.

Problem 14.3.1. *Equation (14.53) gives a transformation $(q, p) \to (I, \varphi)$. Derive the inverse transformation $(I, \varphi) \to (q, p)$. Using a result from Section 17.2.2, show that both of these transformations are symplectic.*

Problem 14.3.2. *Consider a one-dimensional oscillator for which the Hamiltonian expressed in action-angle variables is*

$$H = \omega I + \epsilon I \cos^2\varphi. \tag{14.59}$$

where ω and ϵ are constants (with ϵ not allowed to be arbitrarily large). From Hamilton's equations express the time dependence $\varphi(t)$ as an indefinite integral and perform the integration. Then express $I(t)$ as an indefinite integral.

Problem 14.3.3. *For the system with Hamiltonian given by $H(q, p, t) = p^2/2m + (1/2)m\lambda^2(t)q^2$ as in Eq. (14.52), consider the transformation $(q, p) \to (Q, P)$ given* *by*

$$Q = -\tan^{-1}\left(\frac{r}{q}\left(\frac{rp}{m} - q\dot{r}\right)\right),$$
$$P = \frac{m}{2}\left(\frac{q^2}{r^2} + \left(\frac{rp}{m} - q\dot{r}\right)^2\right), \tag{14.60}$$

where $r(t)$ will be specified more precisely in a later problem. But, for now, $r(t)$ is an arbitrary function of time. Show that this transformation is symplectic.

Problem 14.3.4. *For the same system, in preparation for finding the generating function $G(q, Q, t)$ defined in Eqs. (14.6) and (14.8), rearrange the transformation equations of the previous problem into the form $P = P(q, Q, t)$ and $p = p(q, Q, t)$. Then find $G(q, Q, t)$ such that*

$$p = \frac{\partial G}{\partial q}, \quad P = -\frac{\partial G}{\partial Q}. \tag{14.61}$$

Problem 14.3.5. *In preparation for finding the new Hamiltonian $H'(Q, P, t)$ and expressing it (as is obligatory) explicitly in terms of Q and P, invert the same transformation equations into the form $q = q(Q, P, t)$ and $p = p(Q, P, t)$. Then find $H'(Q, P, t)$ and simplify it by assuming that $r(t)$ satisfies the equation*

$$\ddot{r} + \lambda^2(t)\, r - r^{-3} = 0. \tag{14.62}$$

Then show that Q is ignorable and hence that P is conserved.

Problem 14.3.6. *Assuming that the system studied in the previous series of problems is oscillatory, find its action variable and relate it to the action variable E/ω of simple harmonic motion.*

14.4 Examples of Adiabatic Invariance

14.4.1 Variable Length Pendulum

Consider the variable-length pendulum shown in Fig. 14.2. Tension T holds the string which passes over a frictionless peg (of arbitrarily small diameter), the length of the string below the peg being $l(t)$. Assuming small amplitude motion the "oscillatory energy" of the system E_{osc} is defined so that the potential energy (with pendulum hanging straight down) plus kinetic energy of the system is $-mgl(t) + E_{\text{osc}}$. With fixed l,

$$E_{\text{osc}} = \frac{1}{2}\, mgl\theta_{\text{max}}^2. \tag{14.63}$$

If the pendulum is not swinging, E_{osc} continues to vanish when the length is varied slowly enough that the vertical kinetic energy can be neglected. We assume the length changes slowly enough that $\dot{l}^2$ and $\ddot{l}$ can be neglected throughout. The equation of motion is

$$\ddot{\theta} + \frac{\dot{l}\dot{\theta}}{l} + \frac{g}{l}\sin\theta = 0. \tag{14.64}$$

For "unperturbed" motion the second term is neglected, and the (small amplitude) action is given by

$$I = \sqrt{\frac{l}{g}}\, E_{osc}. \tag{14.65}$$

Change dl in the pendulum length causes change $d\theta_{max}$ in maximum angular amplitude. The only real complication in the problem is that the ratio of these quantities depends on θ. The instantaneous string tension is given by $mg\cos\theta + ml\dot{\theta}^2 - m\ddot{l}$, but we will neglect the last term. The energy change dE_{osc} for length change dl is equal to the work done $-T dl$ by the external agent acting on the system less the change in potential energy;

$$dE_{osc} = -(mg\cos\theta + ml\dot{\theta}^2)\, dl + mg\, dl. \tag{14.66}$$

Continuing to assume small oscillation amplitudes,

$$\frac{dE_{osc}}{dl} = \frac{1}{2} mg\theta^2 - ml\dot{\theta}^2. \tag{14.67}$$

The right-hand side can be estimated by averaging over a complete cycle of the unperturbed motion and for that motion

$$\overline{\theta^2} = \frac{1}{2}\theta_{max}^2 \quad \text{and} \quad \overline{\dot{\theta}^2} = \frac{1}{2}\frac{g}{l}\theta_{max}^2. \tag{14.68}$$

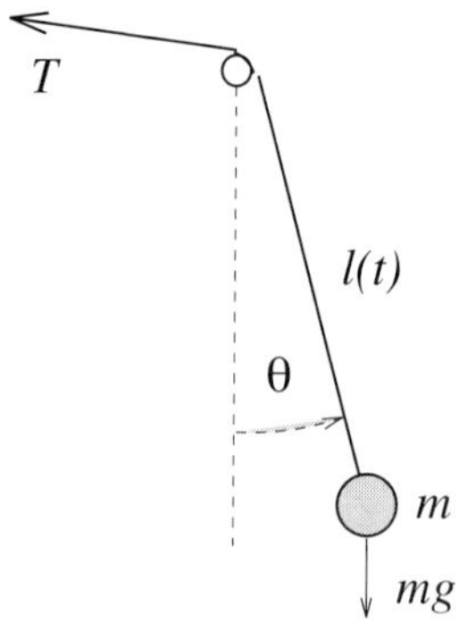

Fig. 14.2 Variable-length pendulum. The fractional change of length during one oscillation period is less than a few percent.

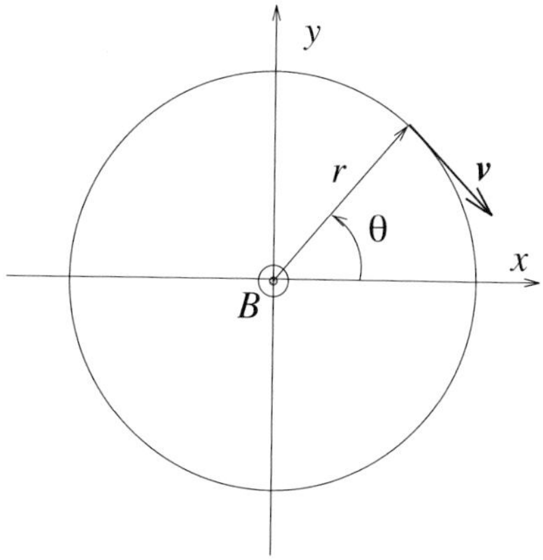

Fig. 14.3 A charged particle moves in a slowly varying, uniform magnetic field.

As a result, using Eq. (14.63), we have

$$\overline{\frac{dE_{\text{osc}}}{dl}} = -\frac{E_{\text{osc}}}{2l}. \tag{14.69}$$

Then from Eq. (14.65)

$$\overline{\frac{dI}{dl}} = \sqrt{\frac{\ell}{g}}\,\overline{\frac{dE_{\text{osc}}}{dl}} + \frac{1}{2}\frac{E_{\text{osc}}}{\sqrt{gl}} = 0. \tag{14.70}$$

Here we have treated both l and E_{osc} as constant and moved them outside the averages. The result is that I is conserved, in agreement with the general theory.

14.4.2 Charged Particle in Magnetic Field

Consider a charged particle moving in a uniform magnetic field $B(t)$ which varies slowly enough that the Faraday's law electric field can be neglected, and also so that the adiabatic condition is satisfied. With coordinate system defined in Fig. 14.3 the vector potential of such a field is

$$A_x = -\frac{1}{2}yB, \quad A_y = \frac{1}{2}xB, \quad A_z = 0, \tag{14.71}$$

since

$$\nabla \times \mathbf{A} = \begin{vmatrix} \hat{\mathbf{x}} & \hat{\mathbf{y}} & \hat{\mathbf{z}} \\ \partial/\partial x & \partial/\partial y & \partial/\partial z \\ A_x & A_y & A_z \end{vmatrix} = B\hat{\mathbf{z}}. \tag{14.72}$$

Introducing cylindrical coordinates, from Eq. (9.67) the (nonrelativistic) Lagrangian is

$$\begin{aligned} L =& \frac{1}{2}\, mv^2 + e\mathbf{A}\cdot\mathbf{v} \\ =& \frac{1}{2}\, mv^2 + \frac{eB}{2}(-y\hat{\mathbf{x}} + x\hat{\mathbf{y}})\cdot\left(\frac{y}{r}\,\hat{\mathbf{x}} - \frac{x}{r}\,\hat{\mathbf{y}}\right)(-r\dot{\theta}) \\ =& \frac{1}{2}\, m(\dot{r}^2 + r^2\dot{\theta}^2 + \dot{z}^2) + \frac{1}{2}\, eB(t)r^2\dot{\theta}. \end{aligned} \tag{14.73}$$

Since this is independent of θ, the conjugate momentum,[8]

$$P_\theta = mr^2\dot{\theta} + \frac{1}{2}\, eB(t)r^2, \tag{14.74}$$

is conserved. With B fixed, and the instantaneous center of rotation chosen as origin, a condition on the unperturbed motion is obtained by equating the centripetal force to the magnetic force;

$$m\dot{\theta} = -eB, \tag{14.75}$$

with the result that

$$P_\theta = \frac{1}{2}\, mr^2\dot{\theta}, \tag{14.76}$$

and the action variable is

$$I_\theta = \frac{1}{2\pi}\oint P_\theta d\theta = P_\theta. \tag{14.77}$$

It is useful to express I_θ in terms of quantities that are independent of the origin using Eq. (14.75),

$$I_\theta = \frac{1}{2}\, m(r\dot{\theta})^2\,\frac{1}{\dot{\theta}} = -\frac{m^2}{2e}\,\frac{v_\perp^2(t)}{B(t)}, \tag{14.78}$$

where $v_\perp$ is the component of particle velocity normal to the magnetic field.

Recapitulating, $v_\perp^2/B$ is an adiabatic invariant. The important result is not that P_θ is conserved when B is constant, which we already knew, but that it is conserved even when B varies (slowly enough) with time. Furthermore, since the change in B is to be evaluated at the particle's nominal position, changes in B can be either due to changes in time of the external sources of B or to spatial variation of B in conjunction with displacement of the moving particle's center of rotation (for example parallel to $\mathbf{B}$). P_θ is one of the important invariants

8) Recall that (upper case) P stands for conjugate momentum which differs from (lower case) p which is the mechanical momentum.

controlling the trapping of charged particles in a magnetic "bottle." This is pursued in the next section.

14.4.3 Charged Particle in a Magnetic Trap

The application *par excellence* of adiabatic invariants describes the trapping of charged particles in an appropriately configured magnetic field. Though the configuration is quite complicated, motion of charges can be understood entirely with adiabatic invariants.

A particle of charge e moves in a time independent, axially symmetric magnetic field $\mathbf{B}(\mathbf{R})$. Symbolizing the component of particle velocity normal to $\mathbf{B}$ by w, the approximate particle motion follows a circle of radius ρ with angular rotation frequency ω_c (known as the "cyclotron frequency"). These quantities are given by

$$\rho = \frac{mw}{eB}, \quad \text{and} \quad \omega_c = 2\pi \frac{w}{2\pi\rho} = \frac{eB}{m}, \tag{14.79}$$

with the latter being independent of the speed of the particle. The field is assumed to be nonuniform but *not too* nonlinear. This is expressed by the condition

$$\rho \frac{|\nabla \mathbf{B}|}{B} \ll 1. \tag{14.80}$$

This condition assures that formulas derived in the previous section are applicable and the particle "gyrates" in an almost circular orbit. The particle retraces pretty much the same trajectory turn after turn. The system is then known as a "magnetic trap." The sort of magnetic field envisaged is illustrated in Fig. 14.4 which also shows typical particle orbits.

In general, the particle also has a component of velocity parallel to $\mathbf{B}$, so the center of the circle (henceforth to be known as the "guiding center") also travels along $\mathbf{B}$. This motion is said to be "longitudinal." There will also be an even slower drift of the guiding center "perpendicular" to $\mathbf{B}$. This is due to the fact that condition (14.80) is not exactly satisfied and the radius of gyration is least in regions where $B = |\mathbf{B}|$ is greatest.

To describe these motions, we introduce the radius vectors shown in Fig. 14.4(a).

$$\mathbf{r} = \mathbf{R} + \boldsymbol{\rho}. \tag{14.81}$$

The three velocities $\mathbf{v} = d\mathbf{r}/dt$, $\mathbf{u} = d\mathbf{R}/dt$, and $\mathbf{w} = d\boldsymbol{\rho}/dt$ satisfy

$$\mathbf{v} = \mathbf{u} + \mathbf{w}. \tag{14.82}$$

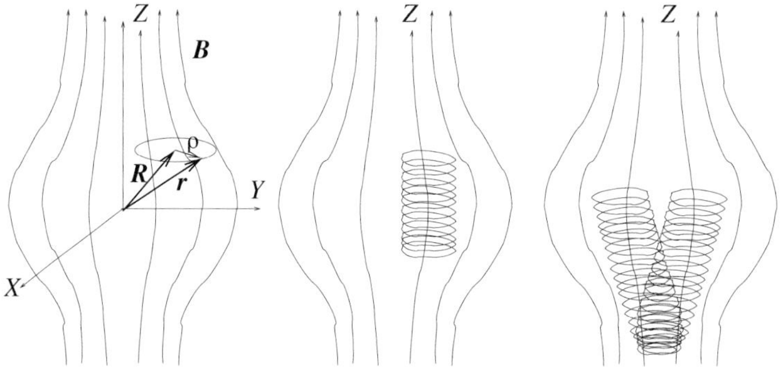

Fig. 14.4 (a) Charged particle gyrating in a nonuniform magnetic field. Its longitudinal and azimuthal motion is exhibited in (b) and (c). The reduction in radius of gyration near the end of the trap is also shown.

Presumably $\mathbf{R}$ and $|\boldsymbol{\rho}|$ are slowly varying compared to $\boldsymbol{\rho}$, which gyrates rapidly. Since particles with large longitudinal velocities can escape out the ends of the bottle (as we shall see) the ones that have not escaped have transverse velocity at least comparable with their longitudinal velocity and it is clear from condition (14.80) that the transverse guiding center drift velocity is small compared to the gyration velocity. These conditions can be expressed as

$$\mathbf{v}_{\parallel} = \mathbf{u}_{\parallel}, \quad \text{and} \quad \mathbf{u}_{\perp} \ll \mathbf{w}, \quad \text{and hence} \quad \mathbf{v}_{\perp} \approx \mathbf{w}. \tag{14.83}$$

General strategy: To start on a detailed description of particle motion in the trap, one can ignore the slow motion of the guiding center in analyzing the gyration. (This part of the problem has already been analyzed in Section 14.4.2, but we will repeat the derivation using the current notation and approximations.) Having once calculated the adiabatic invariant μ for this gyration, it will subsequently be possible to ignore the gyration (or rather to represent it entirely by the value of μ) in following the guiding center. This accomplishes a kind of "averaging over the fast motion." It will then turn out that the motion of the guiding center itself can be similarly treated on two time scales. There is an oscillatory motion of the guiding center parallel to the z-axis in which the azimuthal motion is so slow that it can be ignored. This motion is characterized by adiabatic invariant $I_{\parallel}(\mu)$. As mentioned already, its only dependence on gyration is through μ. Finally, there is a slow azimuthal drift $I_{\perp}(\mu, I_{\parallel})$ that depends on gyration and longitudinal drift only through their adiabatic invariants. In this way, at each stage there is a natural time scale defined by the period of oscillation and this oscillation is described by equations of motion that neglect changes occurring on longer time scales and average over effects that change on shorter time scales.

Gyration: According to Eq. (9.68) the components of the canonical momentum are given by

$$\mathbf{P}_\perp = m\mathbf{w} + e\mathbf{A}_\perp,$$
$$\mathbf{P}_\| = m\mathbf{u}_\| + e\mathbf{A}_\|, \tag{14.84}$$

where approximations (14.83) have been used. The nonrelativistic Hamiltonian is

$$H = \frac{\mathbf{p}^2}{2m} = \frac{(\mathbf{P}_\perp - e\mathbf{A}_\perp^2)}{2m} + \frac{(\mathbf{P}_\| - e\mathbf{A}_\|^2)}{2m}. \tag{14.85}$$

This is the mechanical energy expressed in terms of appropriate variables. There is no contribution from a scalar potential since there is no electric field.

The gyration can be analyzed as the superposition of sinusoidal oscillations in two mutually perpendicular directions in the transverse plane. For adiabatic invariant I_g we can take their average

$$I_g = \frac{1}{4\pi}\oint (P_{\perp x}dx + P_{\perp y}dy) = \frac{1}{4\pi}\oint \mathbf{P}_\perp \cdot \mathbf{dl}_\perp, \tag{14.86}$$

where $\mathbf{dl}_\perp$ is an incremental tangential displacement in the (x, y)-plane, "right handed," with the (x, y, z)-axes being right handed. It is therefore directed opposite to the direction of gyration, as shown in Fig. 14.3, since $\mathbf{B}$ is directed along the (local) positive z-axis. Using Eq. (14.84) we have

$$I_g = \frac{1}{4\pi}\oint m\mathbf{w}\cdot\mathbf{dl}_\perp + \frac{e}{4\pi}\oint \mathbf{A}\cdot\mathbf{dl}_\perp = -\frac{mw\rho}{2} + \frac{e}{4\pi}\int \mathbf{B}\cdot\hat{\mathbf{z}}\,dS, \tag{14.87}$$

where dS is an incremental area in the plane of gyration. The first term is negative because the gyration is directed opposite to $\mathbf{dl}_\perp$. The second term (in particular its positive sign) has been obtained using Stokes's theorem and $\mathbf{B} = \nabla \times \mathbf{A}$. Using Eq. (14.79), we get

$$I_g = -\frac{eB\rho^2}{4}. \tag{14.88}$$

This agrees with Eq. (14.78). I_g can be compared to the "magnetic moment" $\mu = \frac{e^2}{2m}B\rho^2$ of the orbit (which is equal to the average circulating current $e\omega_c/(2\pi)$ multiplied by the orbit area $\pi\rho^2$.) Except for a constant factor, μ and I_g are identical so we can take μ as the adiabatic invariant from here on. If we regard $\boldsymbol{\mu}$ as a vector perpendicular to the plane of gyration, then

$$\boldsymbol{\mu}\cdot\mathbf{B} < 0. \tag{14.89}$$

We also note that the kinetic energy of motion in the perpendicular plane is given by

$$E_\perp = \frac{1}{2} m w^2 = -\boldsymbol{\mu} \cdot \mathbf{B} = \mu B. \tag{14.90}$$

Longitudinal drift of the guiding center: Because of its longitudinal velocity, the particle will drift along the local field line. Since the field is nonuniform this will lead it into a region where B is different. Because the factor $B\rho^2$ remains constant we have $\rho \sim B^{-1/2}$ and (by Eq. (14.79)) $w \sim B^{1/2}$. Superficially this seems contradictory since the speed of a particle cannot change in a pure magnetic field. It *has to be* that energy is transferred to or from motion in the longitudinal direction. We will first analyze the longitudinal motion on the basis of energy conservation and later analyze it in terms of the equations of motion.

The total particle energy is given by

$$E = \mu B(\mathbf{R}) + \frac{1}{2} m u_\parallel^2. \tag{14.91}$$

Since the first term depends only on position $\mathbf{R}$ it can be interpreted as potential energy. It is larger at either end of the trap than in the middle. Since both E and μ are conserved, this equation can be solved for the longitudinal velocity

$$u_\parallel = \pm\sqrt{\frac{2}{m}\left(E - \mu B(\mathbf{R})\right)}. \tag{14.92}$$

In a uniform field $u_\parallel$ would be constant, but in a spatially variable field $u_\parallel$ varies slowly. As the particle drifts toward the end of the trap, the B field becomes stronger and $u_\parallel$ becomes less. At some value Z_{tp} the right-hand side of Eq. (14.92) vanishes. This is therefore a "turning point" of the motion, and the guiding center is turned back to drift toward the center, and then the other end. Perpetual longitudinal oscillation follows. But the motion may be far from simple harmonic, depending as it does on the detailed shape of $\mathbf{B}(\mathbf{R})$ – for example B can be essentially constant over a long central region and then become rapidly larger over a short end region.

In any case an adiabatic invariant $I_\parallel$ for this motion can be calculated (on-axis) by

$$I_\parallel = \frac{1}{2\pi} \oint \mathbf{P}_\parallel \cdot \hat{\mathbf{Z}}\, dZ = \frac{m}{2\pi} \oint u_\parallel dZ, \tag{14.93}$$

where, by symmetry (as in Eq. (14.71)) A_z vanishes on-axis. Then the period of oscillation can be calculated using Eq. (14.41);

$$T_\parallel = 2\pi \frac{\partial I_\parallel}{\partial E_\parallel}. \tag{14.94}$$

Problem 14.4.1. *For the long uniform field magnetic trap with short end regions mentioned in the text, use Eq. (14.94) to calculate the period of longitudinal oscillation $T_{\|}$ and show that the result is the same as one would obtain from elementary kinematic considerations.*

Equation of motion of the guiding center: We still have to study the transverse drift of the guiding center and in the process will corroborate the longitudinal motion inferred purely from energy considerations in the previous paragraph. The equation of motion of the particle is

$$m\,\frac{d(\mathbf{u}+\mathbf{w})}{dt} = e(\mathbf{u}+\mathbf{w}) \times \Big(\mathbf{B} + \boldsymbol{\rho}\cdot\nabla\mathbf{B}\Big)\Big|_0, \tag{14.95}$$

which approximates the magnetic field by its value $\mathbf{B}|_0$ at the guiding center plus the first term in a Taylor expansion evaluated at the same point. We wish to average this equation over one period of the (rapid) gyration which is described relative to local axes by

$$(\rho_x, \rho_y) = w(\cos\theta, \sin\theta), \quad (w_x, w_y) = w(\sin\theta, -\cos\theta). \tag{14.96}$$

When Eq. (14.95) is averaged with all other factors held fixed, the result is

$$m\,\frac{d\mathbf{u}}{dt} = e\mathbf{u}\times\mathbf{B} + e\big\langle(\mathbf{w}\times\boldsymbol{\rho}\cdot\nabla)\mathbf{B}\big\rangle. \tag{14.97}$$

Terms with an odd number of factors of $\boldsymbol{\rho}$ and $\mathbf{w}$ have averaged out to zero. The second term evaluates to

$$\begin{aligned} e\big\langle(\mathbf{w}\times\boldsymbol{\rho}\cdot\nabla)\mathbf{B}\big\rangle &= e\left\langle \det\begin{vmatrix} \hat{\mathbf{x}} & \hat{\mathbf{y}} & \hat{\mathbf{z}} \\ w_x & w_y & 0 \\ \rho_x\frac{\partial B_x}{\partial x} + \rho_y\frac{\partial B_x}{\partial y} & \rho_x\frac{\partial B_y}{\partial x} + \rho_y\frac{\partial B_y}{\partial y} & \rho_x\frac{\partial B_z}{\partial x} + \rho_y\frac{\partial B_z}{\partial y} \end{vmatrix}\right\rangle \\ &= -\frac{ew\rho}{2}\left(\hat{\mathbf{x}}\frac{\partial B_z}{\partial x} + \hat{\mathbf{y}}\frac{\partial B_z}{\partial y}\right) = -\frac{ew\rho}{2}\nabla\mathbf{B_z}, \\ &= -\nabla(\mu\, B), \end{aligned} \tag{14.98}$$

where $\nabla\cdot\mathbf{B} = 0$, and $B_z \approx B$ have been used. The equation of motion is therefore

$$m\,\frac{d\mathbf{u}}{dt} = e\mathbf{u}\times\mathbf{B} - \nabla(\mu\, B). \tag{14.99}$$

When applied to the longitudinal motion of the guiding center, the final term can be seen to be consistent with our interpretation of $\mu\, B$ as a potential energy. Furthermore, the only influence of gyration is through the parameter μ.

The magnitude of the magnetic field presumably falls with increasing R. This causes the gyration to be not quite circular, with its radius increased by

$\Delta\rho$ when the field is reduced by ΔB;

$$\frac{\Delta\rho}{\rho} = -\frac{\Delta B}{B}. \tag{14.100}$$

Along with the cyclotron frequency $\omega_c/(2\pi)$, this can be used to estimate the ratio of the transverse drift velocity $u_\perp$ to w;

$$\frac{u_\perp}{w} \approx \frac{(\omega_c/2\pi)\Delta\rho}{w} \approx \frac{(\omega_c/2\pi)(\partial B/\partial r)\rho^2}{wB} = \frac{1}{2\pi}\frac{\rho}{B}\frac{\partial B}{\partial r} = \frac{1}{2\pi}\frac{\rho}{R_{\text{typ}}}, \tag{14.101}$$

where R_{typ} is a length of the order of the transverse dimensions of the apparatus. Since typical values of the cyclotron radius are much less than this, and since $u_\parallel$ and w have comparable magnitudes, our estimate shows that

$$u_\perp \ll w, \quad \text{and} \quad u_\perp \ll u_\parallel. \tag{14.102}$$

There will nevertheless be a systematic azimuthal motion of the guiding center on a circle of some radius $R_\perp$ centered on the axis. Let the angular frequency of this motion be $\omega_\perp$. We then have

$$\omega_\perp \ll \omega_\parallel \ll \omega_c. \tag{14.103}$$

By a calculation just like that by which I_g was calculated, an adiabatic invariant can also be obtained for this perpendicular drift;

$$I_\perp = -\frac{mu_\perp}{2} + \frac{eBR_\perp^2}{4} = \frac{eR_\perp^2 B}{4}\left(1 - \frac{2\omega_\perp}{\omega_c}\right). \tag{14.104}$$

In practical situations the second term is negligible and we conclude that the third adiabatic invariant $I_\perp$ is proportional to the magnetic flux linked by the guiding center as it makes a complete azimuthal circuit.

14.5 Accuracy of Conservation of Adiabatic Invariants

In order to estimate the accuracy with which the action is invariant as a parameter changes we continue to analyze the oscillator discussed in Section 14.3.4, but with a specific choice of time variation of the natural frequency $\lambda(t)$, namely

$$\lambda(t) = \omega_1\sqrt{\frac{1 + ae^{\alpha t}}{1 + e^{\alpha t}}}. \tag{14.105}$$

As sketched in Fig. 14.5, this function has been carefully tailored to vary smoothly from $\omega_1 = \omega_0 - \Delta\omega$ at $-\infty$ to $\omega_2 = \omega_0 + \Delta\omega$ at ∞, with the main

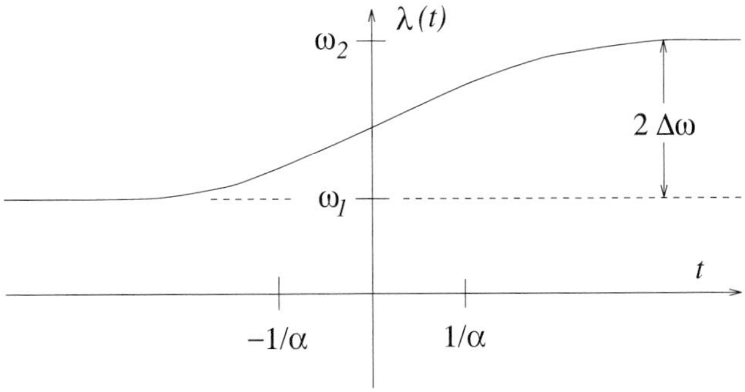

Fig. 14.5 Prototypical adiabatic variation: the natural frequency $\lambda(t)$ of a parametric oscillator varies from ω_1 at $-\infty$ to ω_2 at ∞, with the time range over which the variation occurs being of order $1/\alpha$.

variation occurring over a time interval of order $1/\alpha$, and with $a = (\omega_2/\omega_1)^2$. The adiabatic condition is

$$\frac{1}{\omega_0}\,\dot{\lambda}\,\frac{2\pi}{\omega_0} \approx \frac{1}{\omega_0}\,\alpha\Delta\omega\,\frac{2\pi}{\omega_0} \ll 1. \tag{14.106}$$

With definite parametric variation as given in Eq. (14.105), since the action-angle equations of motion (14.57) are exact, if they can be solved it will supply an estimate of the accuracy with which I is conserved. The second of Eqs. (14.57) yields the deviation of the instantaneous angular frequency from $\lambda(t)$ during one cycle but, averaged over a cycle, this vanishes and the angle variable satisfies

$$\frac{d\varphi}{dt} \approx \lambda(t). \tag{14.107}$$

As shown in Fig. 14.6, the variable φ increases monotonically with t and at only a slowly changing rate. We will change integration variable from t to φ shortly.

In integrating an oscillatory function modulated by a slowly varying function, the frequency of the oscillation is not critical so, for estimation purposes, we accept Eq. (14.107) as an equality. Assuming this variation, one can then obtain $\Delta I = I(+\infty) - I(-\infty)$ by solving the first of Eqs. (14.58). Substituting from the first of Eqs. (14.57) and changing the integration variable from t to φ we obtain

$$\Delta I = \int_{-\infty}^{\infty} \frac{dI}{dt}\,dt = -I\int_{-\infty}^{\infty} \frac{\dot{\lambda}}{\lambda^2}\cos 2\varphi\, d\varphi. \tag{14.108}$$

Here I has been moved outside the integral in anticipation that it will eventually be shown to be essentially constant. With $\lambda(t)$ given by Eq. (14.105),

$$\frac{\dot{\lambda}}{\lambda^2} = \frac{\alpha}{\omega_1}\left(\frac{1}{1+a^{-1}e^{-\alpha t}} - \frac{1}{1+e^{-\alpha t}}\right). \tag{14.109}$$

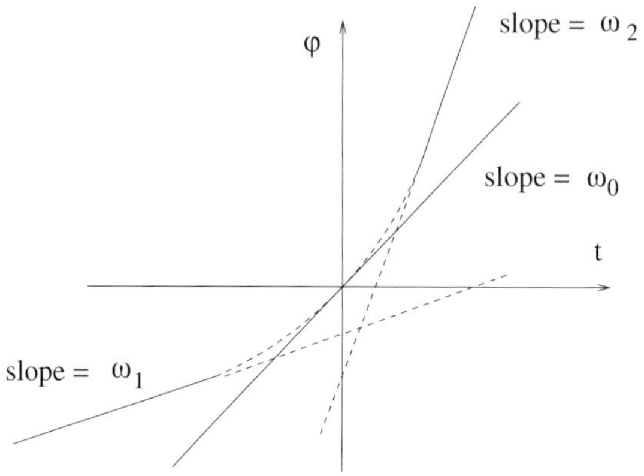

Fig. 14.6 Dependence of angle variable φ on time t as the natural frequency of an oscillator is adiabatically varied from ω_1 to ω_2

When substituting this expression into the integral it is necessary to replace t by $t(\varphi)$. We will approximate this relation by $t = (\varphi - \overline{\varphi})/\overline{\omega}$ where $\overline{\varphi}$ and $\overline{\omega}$ are parameters to be determined by fitting a straight line to the variation shown in Fig. 14.6. The integral becomes

$$\frac{\Delta I}{I} \approx -\frac{1}{2}\frac{\alpha}{\omega_1} Re \int_{-\infty}^{\infty} \left(\frac{1}{1 + a^{-1}e^{-\frac{\alpha(\varphi-\overline{\varphi})}{\overline{\omega}}}} - \frac{1}{1 + e^{-\frac{\alpha(\varphi-\overline{\varphi})}{\overline{\omega}}}} \right) e^{2i\varphi}\, d\varphi. \tag{14.110}$$

The integrand has been made complex to permit its evaluation using contour integration as shown in Fig. 14.7. Because of the $e^{2i\varphi}$ factor and the well-behaved nature of the remaining integrand factor there is no contribution from the arc at infinity. Also the integration path has been deformed to exclude all poles from the interior of the contour.

Our purpose is to show that the integral in Eq. (14.110) is "small." Since accurate evaluation of the contributions of the contour indentations is difficult this demonstration would be difficult if it depended on the cancellation of the two terms, but fortunately the terms are individually small. One can confirm this by looking up the integrals in a table of Fourier transforms, such as F. Oberhettinger, *Tables of Fourier Transforms and Fourier Transforms of Distributions*, Springer, 1990.

Alternatively, continuing to follow Landau and Lifshitz, the integral can be estimated by retaining only the dominant pole. Since the contour is closed on the side of the real axis for which the numerator factor $e^{2i\varphi}$ is a decaying exponential, the integral is dominated by the singularity having the least positive imaginary part; the exponential factor strongly suppresses the relative contribution of the other poles. The first term of Eq. (14.110) has a pole for

$$\varphi = \overline{\varphi} - \frac{\overline{\omega}}{\alpha}(\pm\pi i + \ln a), \quad \text{and hence} \quad \mathrm{Im}\,\varphi_0 = \pi\frac{\overline{\omega}}{\alpha}. \tag{14.111}$$

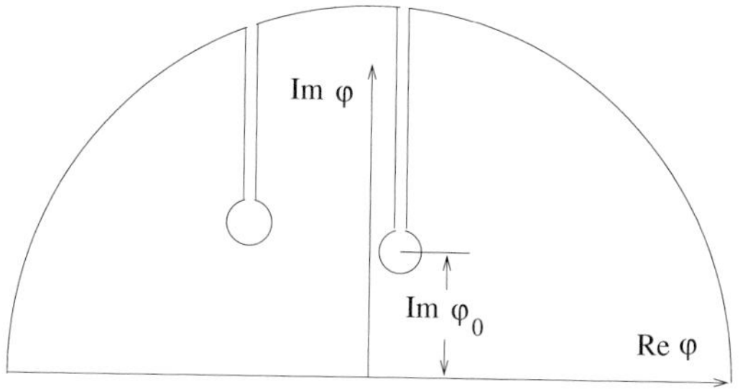

Fig. 14.7 Contour used in the evaluation of integrals in Eq. (14.110). The integrals are dominated by the pole with smallest imaginary part, $\text{Im}\,\varphi_0$.

Since this is just an estimate, the precise value of $\overline{\omega}$ does not matter, but it is approximately the smaller of ω_1 and ω_2. By virtue of the adiabatic condition the ratio in (14.111) is large compared to 1. As a result, apart from the other factors in Eq. (14.110) the deviation ΔI acquires the factor

$$\frac{\Delta I}{I} \sim e^{-2\pi\overline{\omega}/\alpha}. \tag{14.112}$$

This factor is exponentially small and the other term of Eq. (14.110) gives a similar small contribution.

Roughly speaking, if the rate of change of frequency is appreciably less than, say, one cycle out of every 10 cycles, the action I remains essentially constant. For more rapid change it is necessary to calculate more accurately. For slower variation, say ten times slower, the approximation becomes absurdly good.

14.6 Conditionally Periodic Motion

It may be possible to define action-angle variables even in cases where no multiple time scale approximation is applicable. (An example of this is the three-dimensional Kepler satellite problem. It is not entirely typical however, since the orbit is closed so all three independent momentum components vary periodically with the same frequency.) In fact, action-angle variables can be defined for any oscillatory multidimensional system for which the (time independent) H–J equation is separable. The basic theorem on which this approach is based is due to Stäckel. In this section, unless stated otherwise, use of the summation convention will be suspended.

14.6.1 Stäckel's Theorem

Let the system Hamiltonian be

$$H = \frac{1}{2} \sum_{i=1}^{n} c_i(\mathbf{q})\, p_i^2 + V(\mathbf{q}), \tag{14.113}$$

where the c_i are "inverse mass functions" of only the coordinates q^i, so $p_i = \dot{q}^i / c_i$. The time-independent H–J equation, assumed to be separable, is

$$\frac{1}{2} \sum_{i=1}^{n} c_i \left(\frac{\partial S_0}{\partial q^i} \right)^2 + V(\mathbf{q}) = \alpha_1, \tag{14.114}$$

where the first separation constant α_1 has been taken to be the energy E. Let the complete integral (assumed known) be given by

$$S_0 = S^{(1)}(q^1; \boldsymbol{\alpha}) + S^{(2)}(q^2; \boldsymbol{\alpha}) + \cdots + S^{(n)}(q^n; \boldsymbol{\alpha}), \tag{14.115}$$

where $\boldsymbol{\alpha}$ stands for the full set of separation constants $\alpha_1, \alpha_2, \ldots, \alpha_n$, but the individual terms each depend on only one of the q^i. Differentiating Eq. (14.114) partially with respect to each α_j in turn yields

$$\sum_{i=1}^{n} c_i \frac{\partial S_0}{\partial q^i} \frac{\partial^2 S_0}{\partial \alpha_j \partial q^i} \equiv \sum_{i=1}^{n} c_i\, u_{ij}(q^i) = \delta_{1j}. \tag{14.116}$$

Because S_0 has the form (14.115), the function $u_{ij}(q^i)$ introduced as abbreviation for $\partial S_0 / \partial q^i\, \partial^2 S_0 / \partial \alpha_j \partial q^i$ is a functions only of q^i and the same can be said for all j. Rearranging Eq. (14.114) and exploiting the expansion of δ_{1j} given by Eq. (14.116) yields an expansion for the potential energy

$$V = \frac{1}{2} \sum_{i=1}^{n} c_i \left(\alpha_1\, u_{ij}(q^i) - \left(\frac{\partial S_i}{\partial q^i} \right)^2 \right) \equiv \sum_{i=1}^{n} c_i\, w_i(q^i), \tag{14.117}$$

where the newly introduced functions w_i are also functions only of q^i. This has shown that separability of the H–J equation implies this related separability of V – a superposition with the same coefficients c_i as appear in the kinetic energy multiplying functions $w_i(q^i)$. Substituting back into Eq. (14.114) and using Eq. (14.116) again, the H–J equation can therefore be written as

$$\sum_{i=1}^{n} c_i \left(\frac{1}{2} \left(\frac{\partial S_0}{\partial q^i} \right)^2 + w_i(q^i) - \sum_{j=1}^{n} \alpha_j u_{ij}(q^i) \right) = 0. \tag{14.118}$$

Defining $f_i(q^i) = 2\Big(\sum_{j=1}^{n} \alpha_j u_{ij}(q^i) - w_i(q^i) \Big)$, the individual terms in S_0 must satisfy

$$\left(\frac{dS^{(i)}}{dq^i} \right)^2 = f_i(q^i), \quad \text{for} \quad i = 1, \ldots, n. \tag{14.119}$$

Each of these, being a first-order equation, can be reduced to quadratures;

$$S^{(i)}(q^i) = \pm \int^{q^i} \sqrt{f_i(q'^i)}\, dq'^i, \quad \text{for} \quad i = 1, \dots, n. \tag{14.120}$$

Then, according to Hamilton–Jacobi theory, the momentum p_i is given by

$$p_i = \frac{\partial S_0}{\partial q^i} = \frac{dS^{(i)}(q^i;\alpha_i)}{dq^i} = \pm\sqrt{f_i(q^i)}. \tag{14.121}$$

After this equation has been squared to yield $p_i^2 = f_i(q^i)$, it resembles the conservation of energy equation for one-dimensional motion, with $f_i(q^i)$ taking the place of total energy minus potential energy. The function $f_i(q^i)$ can therefore be called "kinetic energy like." The corresponding velocities are given by

$$\dot{q}^i = c_i\, p_i = \pm c_i(\mathbf{q})\, \sqrt{f_i(q^i)}, \tag{14.122}$$

where the second factor depends only on q^i but the first depends on the full set of coordinates $\mathbf{q}$. Values of q^i for which $f_i(q^i) = 0$ have a special qualitative significance as turning points of the motion.

Problem 14.6.1. *With a Hamiltonian known to have the form given by Eq. (14.113) and the potential energy function V therefore necessarily having the form given by Eq. (14.117) write the Lagrange equations for the q^i variables; use the same functions $u_{ij}(q^i)$ and $w_i(q^i)$ as were used in the proof of Stäckel's theorem. Then show that a matrix with elements $v_{ij}(\mathbf{q})$ can be found such that the quantities*

$$\sum_{j=1}^{n} v_{ij}\left(\frac{1}{2}\frac{(\dot{q}^j)^2}{c_j^2} - w_j\right) \quad \textit{for} \quad j = 1, \dots, n, \tag{14.123}$$

are first integrals of the Lagrange equations.

14.6.2 Angle Variables

Equation (14.121) is amazingly simple. Its close similarity to the conservation of energy equation in one dimension implies that the motion in each of the (q^i, p_i) phase space planes is a closed orbit that oscillates repeatedly between the same turning points. This is illustrated in Fig. 14.8. The middle figure shows the function $f_1(q^1)$ for any one of the coordinates (taken to be q^1) and the right figure shows the corresponding q^1, p_1 phase space trajectory. It is easy to overestimate the simplicity of the motion however, for example by incorrectly assuming that the path taken or the time taken in traversing successive phase space orbits will be always the same.

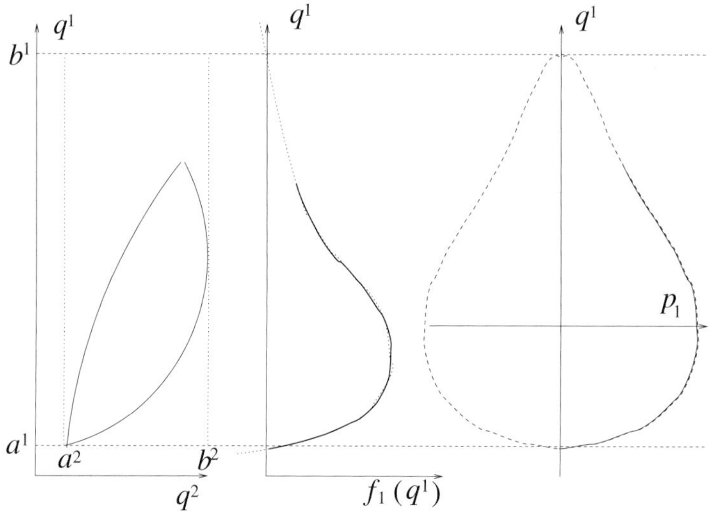

Fig. 14.8 Sample relationships among phase space orbits, regular space orbits, and the "potential energy-like function" in more than one dimension.

In fact, if we use Eq. (14.122) to find the period we obtain

$$t = \int^{q^i} \frac{dq'^i}{c_i(\mathbf{q}')\sqrt{f_i(q'^i)}}. \tag{14.124}$$

Since the integrand depends on all the coordinates, the period in any one q^i, p_i plane depends on the motion in all the others. The sort of motion that is consistent with the motion is shown in the left-most of Fig. 14.8 which shows the motion in the q^1, q^2-plane. With the q^1 motion limited to the range a^1 to b^1 and q^2 limited to a^2 to b^2, the system has to stay inside the rectangle shown. The motion shown is started with both coordinates at one extreme. But, depending on the relative rates of advance of the two coordinates there are an infinity (of which only two are shown) of possible routes the system trajectory can take. The only simple requirement is that the trajectory always "osculates" the enclosing rectangle as it reaches its limiting values.

Each pair (q^1, p_1), $(q^2, p_2), \ldots, (q^n, p_n)$ live a kind of private existence in their own phase space, repeatedly following the same trajectory without reference to time. In this respect the motion resembles the motion of a one-dimensional mechanical system. If the motion is bounded it takes the form of *libration*, as illustrated in Fig. 14.8, and this can be represented as *rotation* as in Problem 1.11.1. Because of the H–J separability this representation is especially powerful.

In Section 14.3.3 the one-dimensional action function S_0 was used as a generator of canonical transformation. We now do the same thing with S_0 as it is given in Eq. (14.115). But first we will replace the Jacobi momenta $\mathbf{a}$ by a set $\mathbf{I} = I_1, I_2, \ldots, I_n$, which are *action variables* defined for each of the phase space pairs (q^i, p_i) that have been under discussion;

$$I_i = \frac{1}{2\pi} \oint p_i(q^i)\, dq^i, \tag{14.125}$$

where the integration is over the corresponding closed phase space orbit. As in the one-dimensional treatment we express the generating function in terms of these action variables;

$$S_0 = \sum_{i=1}^{n} S^{(i)}(q^i; \mathbf{I}). \tag{14.126}$$

The "new momenta" are now to be the $\mathbf{I}$ and the "new coordinates" will be called φ_i. The new Hamiltonian $H = H(\mathbf{I})$ must be independent of the φ_i in order for the first Hamilton equations to yield

$$\dot{I}_i = -\frac{\partial H}{\partial \varphi_i} = 0, \quad i = 1, 2 \ldots, n. \tag{14.127}$$

as must be true since the I_i are constant. The "angle variables" φ_j are defined by

$$\varphi_j = \frac{\partial S_0}{\partial I_j} = \sum_{i=1}^{n} \frac{\partial S^{(i)}(q^i; \mathbf{I})}{\partial I_j}. \tag{14.128}$$

The Hamilton equations they satisfy are

$$\dot{\varphi}_i = \frac{\partial H}{\partial I_i} = \frac{\partial E(\mathbf{I})}{\partial I_i}, \tag{14.129}$$

since $H = E$. These integrate to

$$\varphi_i = \frac{\partial E(\mathbf{I})}{\partial I_i} t + \text{constant}. \tag{14.130}$$

Though new variables $\mathbf{I}, \boldsymbol{\varphi}$ have been introduced, the original variables q^i in terms of which the H–J equation is separable are by no means forgotten. In particular, by their definition in Eq. (14.125), each I_i is tied to a particular q^i and if that variable is allowed to vary through one complete cycle in its q^i, p_i-plane with the other q^j held fixed the corresponding angle change $\Delta\varphi_i$ is given by

$$\Delta\varphi_i = \Delta\left(\frac{\partial S_0}{\partial I_i}\right) = \frac{\partial \Delta S_0}{\partial I_i} = 2\pi. \tag{14.131}$$

Remember though that this variation is quite formal and may not easily relate to a visible periodicity of the entire system.

The transformation relations $\varphi_i = \varphi_i(\mathbf{q}, \mathbf{I})$ therefore have a rather special character. It is not as simple as φ_i depending only on q^i, but should all the variables return to their original values, like the phase of a one-dimensional simple harmonic oscillator, φ_i can only have changed by a multiple of 2π. Stated alternatively, when the angles $\varphi_i(\mathbf{q}, \mathbf{p})$ are expressed as functions of the original variables $\mathbf{q}, \mathbf{p}$ they are not single valued, but they can change only by integral multiples of 2π when the system returns to its original configuration. For this reason, the configuration space is said to be "toroidal" with the toroid dimensionality equal to the number of angle variables and with one circuit around any of the toroids cross sections corresponding to an increment of 2π in the corresponding angle variable.

The greatest power of this development is to generalize to more than one dimension the analysis of not quite time-independent Hamiltonian systems discussed in Section 14.3.3. If this time dependence is described by allowing a previously constant parameter $\lambda(t)$ of the Hamiltonian to be a slowly varying function of time, the earlier analysis generalizes routinely to multiple dimensions. The "variation of constants" equations are

$$\begin{aligned} \dot{I}_i &= -\frac{\partial}{\partial \varphi_i}\left(\left.\frac{\partial S_0}{\partial \lambda}\right|_{\mathbf{q},\mathbf{I}}\right)\dot{\lambda}, \\ \dot{\varphi}_i &= \frac{\partial E(\mathbf{I}, \lambda)}{\partial I_i} + \frac{\partial}{\partial I_i}\left(\left.\frac{\partial S_0}{\partial \lambda}\right|_{\mathbf{q},\mathbf{I}}\right)\dot{\lambda}. \end{aligned} \tag{14.132}$$

The strategy for using these equations perturbatively has been explained earlier.

14.6.3 Action/Angle Coordinates for Keplerian Satellites

All this can be illustrated using the Kepler problem. We now pick up the analysis of Kepler orbits where it was left at the end of Section 8.3, with Jacobi momenta $\alpha_1, \alpha_2, \alpha_3$ and coordinates $\beta_1, \beta_2, \beta_3$ having been introduced and related to coordinates r, θ, ϕ and momenta p_r, p_θ, p_ϕ. Substituting from Eq. (8.61) into Eq. (14.125) we obtain

$$I_\phi = \frac{1}{2\pi}\oint p_\phi \, d\phi = \alpha_3. \tag{14.133}$$

Similarly,

$$I_\theta = \frac{1}{2\pi}\oint p_\theta \, d\theta = \frac{1}{2\pi}\oint \sqrt{\alpha_2^2 - \left(\frac{\alpha_3}{\sin\theta}\right)^2}\, d\theta = \alpha_2 - \alpha_3, \tag{14.134}$$

and

$$I_r = \frac{1}{2\pi}\sqrt{-2mE}\oint\sqrt{(r-r_0)(r_\pi - r)}\,\frac{dr}{r} = \alpha_1 - \alpha_2, \tag{14.135}$$

where the subscripts 0 and π indicate the minimal and maximal radial distances (at the tips of the major axis). They indicate values of u in the formula (1.76), $r = a - a\epsilon\cos u$, giving r in terms of the "eccentric anomaly angle" u, as shown in Fig. 1.3. Transforming the integration variable to u used the relations

$$\sqrt{(r-r_0)(r_\pi - r)} = a\epsilon\sin u, \quad \text{and} \quad dr = a\epsilon\sin u\,du, \tag{14.136}$$

and the integration range is from 0 to 2π. Notice that

$$I_r + I_\theta + I_\phi = \alpha_1 = \sqrt{\frac{-K^2 m}{2E}}, \tag{14.137}$$

which can be reordered as

$$E = -\frac{K^2 m}{2(I_r + I_\theta + I_\phi)^2}. \tag{14.138}$$

Two immediate inferences can be drawn from this form. According to Eq. (14.130) the period of the oscillation of, say, the r variable is

$$T_r = \frac{2\pi}{\partial E/\partial I_r} = 2\pi\sqrt{\frac{ma^3}{K}}. \tag{14.139}$$

The second inference is that T_θ and T_ϕ, calculated the same way, have the same value. The equality of these periods implies that the motion is periodic and *vice versa*.

Bibliography

General References

1 L.D. Landau and E.M. Lifshitz, *Mechanics*, Pergamon, 1976.

References for Further Study

Section 14.5

2 F. Oberhettinger, *Table of Fourier Transforms and Fourier Transforms of Distributions*, Springer, Berlin, 1990.

Section 14.4.3

3 A.J. Lichtenberg, *Phase Space Dynamics of Particles*, Wiley, New York, 1969.

Section 14.6

4 L.A. Pars, *A Treatise on Analytical Dynamics*, Ox Bow Press, Woodbridge, CT, 1979.

15
Linear Hamiltonian Systems

Many systems, though time dependent, are approximately Hamiltonian and approximately linear. Before facing other complications, such as nonlinearity, it is therefore appropriate to find results applicable to systems that are exactly linear, though possibly time dependent. Here we will study some of the more important further properties of such systems. Many more examples and proofs are contained in the first edition of this text. Many more are given in the two books listed at the end of the chapter. The book by Yakubovitch and Starzhinskii is praised faintly by the authors of the other book, Meyer and Hall, as being "well-written but a little wordy" which I think means "mathematically valid but not concise." This makes an ideal combination for a physicist, on account of the fully worked examples.

15.1
Linear Hamiltonian Systems

Under fairly general conditions a multidimensional, linear, Hamiltonian system can be described by a homogeneous matrix equation of the form

$$\left(\mathbf{1}\frac{d^2}{dt^2} + \mathbf{P}(t)\right)\mathbf{e} = 0. \tag{15.1}$$

All time-dependent terms have been lumped into $\mathbf{P}(t)$ and velocity-dependent terms have been transformed away or dropped. This equation can be written in Hamiltonian form as $2n$ equations for the unknowns arrayed as a column vector $\mathbf{z} = (\mathbf{e}, \dot{\mathbf{e}})^T$;

$$\frac{d\mathbf{z}}{dt} = \mathbf{A}(t)\,\mathbf{z}, \tag{15.2}$$

where

$$\mathbf{S} = \begin{pmatrix} \mathbf{0} & -\mathbf{1} \\ \mathbf{1} & \mathbf{0} \end{pmatrix}, \qquad \mathbf{H}(t) = \begin{pmatrix} \mathbf{P}(t) & \mathbf{0} \\ \mathbf{0} & \mathbf{1} \end{pmatrix},$$
$$\text{and} \qquad \mathbf{A}(t) = -\mathbf{S}\mathbf{H}(t) = \begin{pmatrix} \mathbf{0} & \mathbf{1} \\ -\mathbf{P}(t) & \mathbf{0} \end{pmatrix}. \tag{15.3}$$

Geometric Mechanics: Toward a Unification of Classical Physics. 2nd Edition. Richard Talman

ISBN: 978-3-527-40683-8

Problem 15.1.1. *A linear, second-order equation more general than Eq. (15.1) is*

$$\left(\mathbf{1}\frac{d^2}{dt^2} + \mathbf{Q}(t)\,\frac{d}{dt} + \mathbf{P}(t)\right)\mathbf{e} = 0. \tag{15.4}$$

In the simplest, 1D, case, the presence of nonzero coefficient Q is associated with damping, due, for example, to resistance or dissipation, which makes the equation clearly non-Hamiltonian. However, equations such as Eqs. (6.42) have first derivative (sometimes referred to as "gyroscopic") terms, in spite of the fact that they describe a Hamiltonian system.

(a) Defining phase-space coordinates and Hamiltonian by

$$\mathbf{z} = \begin{pmatrix} \mathbf{e} \\ \dot{\mathbf{e}} + \mathbf{Q}\mathbf{e}/2 \end{pmatrix}, \qquad \text{and} \qquad \mathbf{H} = \begin{pmatrix} \mathbf{P} + \mathbf{Q}\mathbf{Q}^T/4 & -\mathbf{Q}^T/2 \\ -\mathbf{Q}/2 & \mathbf{1} \end{pmatrix}, \tag{15.5}$$

derive the (necessarily Hamiltonian) second-order equation satisfied by **e**.

(b) Any matrix **Q** *can be expressed as the sum of a symmetric and an antisymmetric matrix,* $\mathbf{Q} = (\mathbf{Q}+\mathbf{Q}^T)/2 + (\mathbf{Q}-\mathbf{Q}^T)/2$. *Show therefore that Eq. (15.4) can be the equation describing a Hamiltonian system only if* **Q** *is antisymmetric.*

(c) Assuming that **Q** *is antisymmetric, devise a transformation to a new dependent variable which reduces Eq. (15.4) to Eq. (15.1).*

It is possible to group-independent solutions $\mathbf{z}_1(t), \mathbf{z}_2(t), \ldots$ of Eq. (15.2) as the columns of a matrix $\mathbf{Z}(t) = (\mathbf{z}_1(t)\ \mathbf{z}_2(t)\ \cdots)$, which therefore satisfies

$$\frac{d\mathbf{Z}}{dt} = \mathbf{A}(t)\,\mathbf{Z}. \tag{15.6}$$

The matrix **Z** can have as many as $2n$ columns; if it contains $2n$-independent solutions it is known as a "fundamental matrix solution" of Eq. (15.2). The most important matrix of this form is the "transfer matrix" $\mathbf{Z} = \mathbf{M}(t)$ formed from the unique set of solutions for which the initial conditions are given by the identity matrix **1**

$$\mathbf{M}(0) = \mathbf{1}. \tag{15.7}$$

Such transfer matrices were employed in Section 7.1.4. For Hamiltonian systems it will be shown in Chapter 17 that the matrix **M** is symplectic. Some, but not all, of the results in this chapter depend on **M** being symplectic.

If the initial conditions to be imposed on a solution of Eq. (15.1) are arrayed as a column $\mathbf{z}(0)$ of $2n$ values at $t = 0$ then the solution can be written as

$$\mathbf{z}(t) = \mathbf{M}(t)\,\mathbf{z}(0). \tag{15.8}$$

For some purposes it is useful to generalize transfer matrix notation to $\mathbf{M}(t_f, t_i)$, letting $\mathbf{M}$ depend on both an initial time t_i and a final time t_f. Then Eq. (15.8) can be manipulated into the form

$$\mathbf{z}(t) = \mathbf{M}(t,0)\,\mathbf{z}(0) = \mathbf{M}(t,t')\,\mathbf{M}(t',0)\,\mathbf{z}(0), \tag{15.9}$$

where t' is an arbitrary time in the range $0 \le t' \le t$. This again illustrates the accomplishment of "concatenation" of linear transformations by matrix multiplication.

15.1.1
Inhomogeneous Equations

Commonly Eqs. (15.2) are modified by inhomogeneous terms, perhaps due to external forces; these terms can be arrayed as a $2n$-element column matrix and the equations are

$$\frac{d\mathbf{z}}{dt} = \mathbf{A}(t)\,\mathbf{z} + \mathbf{k}(t). \tag{15.10}$$

Such terms destroy the linearity (the constant multiple of a solution is in general not a solution) but the transfer matrix can still be used to obtain a solution satisfying initial conditions $\mathbf{z}(0)$ at $t = 0$. The solution is

$$\mathbf{z}(t) = \mathbf{M}(t)\Big(\mathbf{z}(0) + \int_0^t \mathbf{M}^{-1}(t')\mathbf{k}(t')\,dt'\Big). \tag{15.11}$$

This can be confirmed by direct substitution.

15.1.2
Exponentiation, Diagonalization, and Logarithm Formation of Matrices

Suppose the elements of matrix $\mathbf{A}$ in Eq. (15.2) are constant. The solution with initial values $\mathbf{z}(0)$ can be expressed formally as

$$\mathbf{z} = e^{\mathbf{A}t}\,\mathbf{z}(0). \tag{15.12}$$

The factor $e^{\mathbf{A}t}$ can be regarded as an abbreviation for the power series

$$e^{\mathbf{A}t} = \mathbf{1} + t\mathbf{A} + \frac{t^2}{2}\,\mathbf{A}^2 + \cdots \tag{15.13}$$

in which all terms are well defined. Then, differentiating term-by-term, Eq. (15.2) follows. It is not hard to be persuaded that these manipulations are valid in spite of the fact that $\mathbf{A}t$ is a matrix. If independent solutions of the form (15.12) are grouped as columns of a matrix $\mathbf{Z}$ the result is

$$\mathbf{Z} = e^{\mathbf{A}t}\,\mathbf{Z}(0). \tag{15.14}$$

In particular, for $\mathbf{Z}(0) = \mathbf{1}$ the matrix $\mathbf{Z}$ becomes the transfer matrix $\mathbf{M}$ and Eq. (15.14) becomes

$$\mathbf{M} = e^{\mathbf{A}t}. \tag{15.15}$$

It is similarly possible to define the *logarithm* of a matrix. Recall that the logarithm of complex number $z = re^{i\phi}$ is multiply defined by

$$\ln re^{i\phi} = \ln r + i\phi + 2\pi i\, m, \tag{15.16}$$

where m is any integer. For the logarithm to be an analytic function it is necessary to restrict its domain of definition. Naturally, the same multiple definition plagues the logarithm of a matrix. To keep track of this it is all but necessary to work with diagonalized matrices. This makes it important to understand their eigenvalue structure, especially because the eigenvalues are in general complex. But for problems that are "physical" the elements of $\mathbf{A}$ are real, and this restricts the range of possibilities.

Because the eigenvalues are complex the eigenvectors must be permitted also to have complex elements. There is a way though in which the complete generality that this seems to imply is not needed. It is possible to work only with basis vectors $\mathbf{e}_1, \mathbf{e}_2, \dots,$ that have real components while allowing vectors to have complex expansion coefficients. For example, a complex vector $\mathbf{u}$ may be expressible as $\alpha_1\mathbf{e}_1 + \alpha_2\mathbf{e}_2 + \cdots$ where the coefficients α_i are complex. The complex conjugate of $\mathbf{u}$ is then given by $\mathbf{u}^* = \alpha_1^*\mathbf{e}_1 + \alpha_2^*\mathbf{e}_2 + \cdots$.

It is not necessarily possible to restrict basis elements to be real in this way if vectors are permitted to have arbitrary complex elements – consider for example a two-dimensional space containing both $(1,1)$ and $(i,1)$. But if a vector space is sure to contain $\mathbf{u}^*$ when it contains $\mathbf{u}$, a real basis can be found. All possible arrangements of the eigenvalues of a symplectic matrix are illustrated in Fig. 17.5. Since the eigenvalues are either real or come in complex conjugate pairs, the complex conjugate of an eigenvector is also an eigenvector. It follows that basis vectors can be restricted to be real. (See Meyer and Hall, p. 47, or P. Halmos, *Finite-Dimensional Vector Spaces*, Springer, Berlin, 1987, p. 150, for further explanation.)

Returning to the transfer matrix $\mathbf{M}$, because it is known to be symplectic, according to Eq. (17.49), it satisfies

$$\mathbf{M}^T(t)\,\mathbf{S}\,\mathbf{M}(t) = \mathbf{S}. \tag{15.17}$$

Substituting from Eq. (15.15), differentiating this equation with respect to t, and canceling common factors yields the result

$$\mathbf{A}^T\mathbf{S} = -\mathbf{S}\mathbf{A}. \tag{15.18}$$

A constant matrix $\mathbf{A}$ satisfying this relation is said to be "infinitesimally symplectic" or "Hamiltonian." This equation places strong constraints on the elements of $\mathbf{A}$. (They resemble the relations satisfied by any symplectic matrix according to Eq. (17.49).)

15.1.3
Alternate Coordinate Ordering

The formulas for symplectic matrices take on a different appearance when the coordinates are listed in the order $\mathbf{z} = (q^1, q^2, \ldots, p_1, p_2, \ldots)^T$, With this ordering the matrix $\mathbf{S}$ takes the form

$$\mathbf{S} = \begin{pmatrix} \mathbf{0} & -\mathbf{1} \\ \mathbf{1} & \mathbf{0} \end{pmatrix}. \tag{15.19}$$

Partitioning $\mathbf{M}$ into 2×2 blocks, it and its symplectic conjugate are

$$\mathbf{M} = \begin{pmatrix} \mathbf{A}_a & \mathbf{B}_a \\ \mathbf{C}_a & \mathbf{D}_a \end{pmatrix}, \qquad \overline{\mathbf{M}} = \mathbf{M}^{-1} = \begin{pmatrix} \mathbf{D}_a^T & -\mathbf{B}_a^T \\ \mathbf{B}_a^T & \mathbf{D}_a^T \end{pmatrix}. \tag{15.20}$$

Subscripts a have been added as a reminder of the alternate coordinate ordering. This formula has the attractive property of resembling the formula for the inverse of a 2×2 matrix. With the elements ordered as in Eq. (15.20), condition (15.18) becomes

$$\begin{pmatrix} \mathbf{A}_a^T & \mathbf{C}_a^T \\ \mathbf{B}_a^T & \mathbf{D}_a^T \end{pmatrix} = \mathbf{SAS} = \begin{pmatrix} -\mathbf{D}_a & \mathbf{C}_a \\ \mathbf{B}_a & -\mathbf{A}_a \end{pmatrix}. \tag{15.21}$$

These conditions reduce to the requirements that $\mathbf{B}_a$ and $\mathbf{C}_a$ be symmetric and $\mathbf{A}_a^T = -\mathbf{D}_a$.

15.1.4
Eigensolutions

A standard approach is to seek solutions of Eq. (15.2) in the form

$$\mathbf{z}(t) = e^{\lambda t}\mathbf{a}, \tag{15.22}$$

where λ is a number and $\mathbf{a}$ is a column vector to be obtained. Substitution into Eq. (15.2) yields

$$\mathbf{Aa} = \lambda\mathbf{a}. \tag{15.23}$$

The possible values of λ and the corresponding vectors $\mathbf{a}$ are therefore the eigenvalues and eigenvectors of the matrix $\mathbf{A}$. All eigenvalues and eigenvector elements can be complex. We are, to a large extent, retracing the mathematics of normal mode description. But the present case is not quite identical to that of Problem 1.4.1 since we now have first-order equations and, as a consequence, complex eigenvalues and eigenvectors are more prominent.

For simplicity, we assume the eigenvalues are all distinct, so a set of $2n$ independent solutions is

$$\mathbf{z}_i = e^{\lambda_i t}\mathbf{a}_i. \tag{15.24}$$

Transformation to these eigenvectors as basis vectors proceeds in the well-known way. If this has been done already the matrix $\mathbf{A}$ is diagonal,

$$\mathbf{A} = \text{diag}(\lambda_1, \lambda_2, \ldots, \lambda_{2n}). \tag{15.25}$$

It is sometimes possible to diagonalize (or block diagonalize) a *real* Hamiltonian matrix $\mathbf{A}$ by a similarity transformation

$$\mathbf{A}' = \mathbf{R}^{-1}\mathbf{A}\mathbf{R} \tag{15.26}$$

using a matrix $\mathbf{R}$ that is also *real*, even when the eigenvalues are complex. The general strategy is to simplify the factor $\mathbf{AR}$ in this equation by building $\mathbf{R}$ from column vectors that are eigenvectors of $\mathbf{A}$. One can consider, one-by-one, in the following examples, the various possible eigenvalue arrangements illustrated in Fig. 17.5.

Example 15.1.1. Real, reciprocal eigenvalues. *If $\lambda = e^{\alpha}$ and $1/\lambda = e^{-\alpha}$ are eigenvalues of $e^{\mathbf{A}}$ then α and $-\alpha$ are eigenvalues of $\mathbf{A}$. Taking $\mathbf{A}$ to be a 2×2 matrix, let the eigenvectors be $\mathbf{x}_- = (x_-, p_-)^T$ and $\mathbf{x}_+ = (x_+, p_+)^T$; they satisfy*

$$\mathbf{A}\mathbf{x}_- = -\alpha\mathbf{x}_-, \qquad \textit{and} \qquad \mathbf{A}\mathbf{x}_+ = \alpha\mathbf{x}_+. \tag{15.27}$$

The "symplectic product" of the two eigenvectors is defined by

$$[-,+] = x_- p_+ - p_- x_+, \tag{15.28}$$

and build $\mathbf{R}$ from columns given by $\mathbf{x}_-$ and $\mathbf{x}_+/[-,+]$;

$$\mathbf{R} = \begin{pmatrix} \mathbf{x}_- & \frac{\mathbf{x}_+}{[-,+]} \end{pmatrix}. \tag{15.29}$$

Direct calculation shows that

$$\mathbf{A}' = \begin{pmatrix} -\alpha & 0 \\ 0 & \alpha \end{pmatrix} \tag{15.30}$$

as required.

Example 15.1.2. Pure complex, complex conjugate pairs. *Consider the pair of eigenvalues $\pm i\beta$ with the first eigenvector being $\mathbf{x} = \mathbf{u} + i\mathbf{v}$, where $\mathbf{u}$ and $\mathbf{v}$ are independent and both real. Since we have both*

$$\mathbf{A}\mathbf{x} = i\beta\mathbf{x}, \qquad \textit{and} \qquad \mathbf{A}\mathbf{x}^* = -i\beta\mathbf{x}^*, \tag{15.31}$$

it follows that

$$\mathbf{Au} = -\beta\mathbf{v} \qquad \textit{and} \qquad \mathbf{Av} = \beta\mathbf{u}. \tag{15.32}$$

If the symplectic product $[\mathbf{u},\mathbf{v}]$ *is positive, build* **R** *according to*

$$\mathbf{R} = \begin{pmatrix} \frac{\mathbf{u}}{\sqrt{[\mathbf{u},\mathbf{v}]}} & \frac{\mathbf{v}}{\sqrt{[\mathbf{u},\mathbf{v}]}} \end{pmatrix}. \tag{15.33}$$

Direct calculation shows that

$$\mathbf{A}' = \begin{pmatrix} 0 & -\beta \\ \beta & 0 \end{pmatrix}, \tag{15.34}$$

as required. If necessary to make it positive, change the sign of the symplectic product before taking the square root.

Example 15.1.3. Quartet of two complex conjugate pairs. *Consider a quartet of complex eigenvalues* $\pm\gamma \pm i\delta$. *According to Eq. (15.21), a* 4×4 *Hamiltonian matrix reduced to block-diagonal form must have the following structure:*

$$\mathbf{A}' = \mathbf{R}^{-1}\mathbf{A}\mathbf{R} = \begin{pmatrix} -\mathbf{D}'^{T}_{a} & \mathbf{0} \\ \mathbf{0} & \mathbf{D}'_{a} \end{pmatrix}, \tag{15.35}$$

and the 2×2 *real matrix* $\mathbf{D}'_a$ *must have the form*

$$\begin{pmatrix} \gamma & \delta \\ -\delta & \gamma \end{pmatrix} \tag{15.36}$$

in order for **A** *to have the correct overall set of eigenvalues. Meyer and Hall show that the transformation matrix* **R** *accomplishing this is real, and the manipulations in Section 17.2.3 explicitly performs an equivalent real diagonalization.*

Example 15.1.4. Pure diagonalization. *If one insists on* pure *diagonalization rather than* block *diagonalization, it is necessary for the matrix* **R** *to have complex elements. This is frequently the procedure of choice because it is so much simpler to work with purely diagonal matrices. Let the eigenvalues of* **A** *be* $\pm i\beta$. *Letting the eigenvalue* $\mathbf{x} = \mathbf{u} + i\mathbf{v}$ *with* **u** *and* **v** *both real, Eqs. (15.32) are applicable. The symplectic product of* **x** *and* $\mathbf{x}^*$ *is given by*

$$[\mathbf{x},\mathbf{x}^*] = -2i[\mathbf{u},\mathbf{v}]. \tag{15.37}$$

Build **R** *according to*

$$\mathbf{R} = \frac{1}{|[\mathbf{u},\mathbf{v}]|}\begin{pmatrix} \mathbf{x} & \mathbf{x}^* \end{pmatrix}. \tag{15.38}$$

Direct calculation shows that

$$\mathbf{A}' = \begin{pmatrix} i\beta & 0 \\ 0 & -i\beta \end{pmatrix}. \tag{15.39}$$

The characteristics of the eigenvalues of a Hamiltonian matrix **A** have been discussed in these various examples. In particular, if any one of the eigenvalues is not purely imaginary, then either it or its "mate" yields a factor $e^{\lambda_i t}$ having magnitude greater than 1. By Eq. (15.24) the motion would then diverge at large time. Furthermore, this would be true for any (physically realistic) initial conditions, since the initial motion would contain at least a tiny component of the divergent motion.

After diagonalization in ways such as the examples have illustrated, it is possible to find the logarithm of a matrix by taking the logarithms of the diagonal elements. In most cases the logarithm of a real matrix can be taken to be real and we will assume this to be the case. With **A** in diagonal form as in (15.25) one easily derives the "Liouville formula"

$$\det \left| e^{\mathbf{A}t} \right| = e^{t \operatorname{tr} \mathbf{A}}, \tag{15.40}$$

and this result can be manipulated to derive the same formula whether or not **A** is diagonal.

15.2 Periodic Linear Systems

Suppose the matrix **A** in Hamilton's equation (15.2), though time dependent, is periodic with period T,

$$\mathbf{A}(t+T) = \mathbf{A}(t). \tag{15.41}$$

This condition *does not* imply that the solutions of the equation are periodic, but it *does* greatly restrict the possible variability of the solutions. Condition (15.41) causes the "once-around" or "single-period" transfer matrix[1] $\mathbf{M}_T$, which is the ordinary transfer matrix **M** evaluated at $t = T$,

$$\mathbf{M}_T \equiv \mathbf{M}(T), \tag{15.42}$$

1) There is no consistency concerning the names for matrices **M**, that we call "transfer matrix," and $\mathbf{M}_T$, that we call "single-period transfer matrix." Some of the names used by mathematicians are scarcely suitable for polite company. The term "monodromy matrix" is commonly applied to $\mathbf{M}_T$. Yakubovich and Starzhinskii refer to **M** as the "matrizant" and Meyer and Hall refer to it as the "fundamental matrix solution satisfying $\mathbf{M}(0) = \mathbf{1}$," where any matrix **Z** satisfying Eq. (15.6) is known as a "fundamental matrix solution." This terminology agrees with that used by Pars. But I prefer terminology drawn from electrical engineering and accelerator physics. Since (it seems to me) these fields make more and better use of the formalism it seems their notation should be favored. The term "once-around" comes from circular storage rings that are necessarily periodic in the present sense of the word. However, "single-period transfer matrix" may be more universally acceptable.

to have special properties. A single-period transfer matrix can also be defined for starting times other than $t = 0$ and this is indicated by assigning t as an argument; $\mathbf{M}_T(t)$. Recall that the columns of $\mathbf{M}$ are themselves solutions for a special set of initial conditions, so the columns of $\mathbf{M}_T$ are the same solutions evaluated at $t = T$.

There are two main ways in which equations containing periodically varying parameters arise. One way is that the physical system being described is itself periodic. Examples are crystal lattices, lattices of periodic electrical or mechanical elements, and circular particle accelerator lattices. For particle accelerators it is customary to work with a longitudinal coordinate s rather than time t as independent variable, but the same formulas are applicable. The other way periodic systems commonly arise is while analyzing the effects of perturbations acting on otherwise-closed orbits.

The main theorem's satisfied by solutions of Eq. (15.2) subject to Eq. (15.41) are due to Floquet and Lyapunov. These theorems are essentially equivalent. Lyapunov's contributions were to generalize Floquet's theorem to multiple dimensions and to use it for an especially effective coordinate transformation. For convenience of reference we will muddle the chronology a bit by regarding the multidimensional feature as being included in Floquet's theorem and only the transformation in Lyapunov's. Both of these theorems are valid whether or not the system is Hamiltonian, but the most interesting questions concern the way that Hamiltonian requirements constrain the motion when "conditions," though changing, return periodically to previous values.

15.2.1 Floquet's Theorem

Substituting $t' = T$ in Eq. (15.9), propagation from $t = 0$ to $t + T$ can be described by

$$\mathbf{M}(t+T) = \mathbf{M}(t+T,T)\,\mathbf{M}(T). \tag{15.43}$$

Because of periodicity condition (15.41), propagation from $t = T$ is identical to propagation from $t = 0$, or

$$\mathbf{M}(t+T,T) = \mathbf{M}(t). \tag{15.44}$$

Using definition (15.42), it follows then from Eq. (15.43) that

$$\mathbf{M}(t+T) = \mathbf{M}(t)\,\mathbf{M}_T. \tag{15.45}$$

This is the essential requirement imposed by the periodicity. Since $\mathbf{M}(0) = \mathbf{1}$ this relation is trivially true for $t = 0$ and an equivalent way of understanding it is to recall the definitions of columns of $\mathbf{M}(t)$ as solutions of Eq. (15.2) satisfying special initial conditions. The corresponding columns on both sides of the equation are clearly *the same* solutions.

According to Eq. (15.15), if the coefficient matrix **A** were constant, the single-period matrix $\mathbf{M}_T$ would be related to **A** by $\mathbf{M}_T = e^{T\mathbf{A}}$. Motivated by this equation, and being aware of the considerations of Section 15.1.2, we form a logarithm and call it **K**;

$$\mathbf{K} = \frac{1}{T} \ln \mathbf{M}_T, \qquad \text{which implies} \qquad \mathbf{M}_T = e^{T\mathbf{K}}. \tag{15.46}$$

(Assuming that the elements of **A** are real, the elements of $\mathbf{M}_T$ will also be real, but the matrix **K** may be complex. In any case, because the logarithm is not single valued and because $\mathbf{A}(t)$ is time dependent in general, it is *not* legitimate to identify **K** with **A**. By its definition the matrix **K** is "Hamiltonian" in the sense defined below Eq. (15.18), but it is almost certainly misleading to read this as implying a direct relationship between **K** and the system Hamiltonian (assuming there is a Hamiltonian, that is).

From **K** and transfer matrix $\mathbf{M}(t)$ we form the matrix

$$\mathbf{F}(t) = \mathbf{M}(t)\, e^{-t\mathbf{K}}, \tag{15.47}$$

an equation that we will actually use in the form

$$\mathbf{M}(t) = \mathbf{F}(t)\, e^{t\mathbf{K}}. \tag{15.48}$$

What justifies these manipulations is that $\mathbf{F}(t)$ can be shown to be periodic (obviously with period T). Evaluating (15.47) with $t \rightarrow t + T$ and using condition(15.45) and $e^{-t\mathbf{K}} = \mathbf{M}_T^{-1}$ we obtain

$$\mathbf{F}(t+T) = \mathbf{M}(t)\, \mathbf{M}_T\, e^{-T\mathbf{K}} e^{-t\mathbf{K}} = \mathbf{F}(t). \tag{15.49}$$

We have therefore proved Floquet's theorem which states that transfer matrix $\mathbf{M}(t)$ can be written as the product of a periodic function $\mathbf{F}(t)$ and the "sinusoidal" matrix $e^{t\mathbf{K}}$ as in Eq. (15.48). As stated previously, with the elements of **A** being real, normally the elements of **K** can be constrained to be real, and the elements of $\mathbf{F}(t)$ are therefore also real.

By manipulating Eq. (15.9) and substituting from Eq. (15.48), we obtain a formula for the two argument transfer matrix;

$$\mathbf{M}(t,t') = \mathbf{M}(t)\, \mathbf{M}^{-1}(t') = \mathbf{F}(t)\, e^{(t-t')\mathbf{K}}\, \mathbf{F}^{-1}(t'). \tag{15.50}$$

With the transfer matrix given by Eq. (15.48), solutions take the form

$$\mathbf{z}(t) = \mathbf{F}(t)\, e^{t\mathbf{K}}\, \mathbf{z}(0). \tag{15.51}$$

Such a solution is known as "pseudo-harmonic" because the motion can be regarded as simple-harmonic (that is to say, sinusoidally time-varying) but with "amplitude" being "modulated" by the factor $\mathbf{F}(t)$. By differentiating Eq. (15.47) and rearranging terms, one finds that $\mathbf{F}(t)$ satisfies the equation

$$\dot{\mathbf{F}} = \mathbf{A}\mathbf{F} - \mathbf{F}\,\mathbf{K}. \tag{15.52}$$

Since $\mathbf{F}(t)$ is known to be periodic, it is necessary to select the particular solution of this equation having this property.

15.2.2
Lyapunov's Theorem

We seek the coordinate transformation $\mathbf{z} \to \mathbf{x}$ that best exploits Floquet's theorem to simplify Eq. (15.2). It will now be shown to be

$$\mathbf{z} = \mathbf{F}(t)\,\mathbf{w}. \tag{15.53}$$

There are two ways that $\dot{\mathbf{z}}$ can be worked out in terms of $\mathbf{w}$. On the one hand

$$\dot{\mathbf{z}} = \mathbf{A}\,\mathbf{z} = \mathbf{A}\,\mathbf{F}\,\mathbf{w}. \tag{15.54}$$

On the other hand, differentiating both Eqs. (15.53) and (15.47) and taking advantage of the fact that the transfer matrix satisfies the same equation as $\mathbf{z}$ yields

$$\begin{aligned}\dot{\mathbf{z}} = \dot{\mathbf{F}}\,\mathbf{w} + \mathbf{F}\,\dot{\mathbf{w}} &= (\dot{\mathbf{M}}\,e^{-t\mathbf{K}} - \mathbf{M}\,e^{-t\mathbf{K}}\mathbf{K})\,\mathbf{w} + \mathbf{F}\,\dot{\mathbf{w}} \\ &= (\mathbf{A}\,\mathbf{M}\,e^{-t\mathbf{K}} - \mathbf{M}\,e^{-t\mathbf{K}}\mathbf{K})\,\mathbf{w} + \mathbf{F}\,\dot{\mathbf{w}}.\end{aligned} \tag{15.55}$$

Applying Eq. (15.47) again, the first term of this equation can be seen to be the same as the right-hand side of Eq. (15.54), and the second term can be simplified. Canceling a common factor, it then follows that $\mathbf{w}$ satisfies the equation:

$$\frac{d\mathbf{w}}{dt} = \mathbf{K}\,\mathbf{w}. \tag{15.56}$$

This result is known as *Lyapunov's theorem*. With $\mathbf{K}$ being a constant matrix, this constitutes a major improvement over Eq. (15.2), whose matrix $\mathbf{A}(t)$ depended on time.

15.2.3
Characteristic Multipliers, Characteristic Exponents

An eigenvalue ρ corresponding to eigenvector $\mathbf{a}$ of the single-period transfer matrix $\mathbf{M}_T$ satisfies

$$\mathbf{M}_T\mathbf{a} = \rho\,\mathbf{a}, \qquad \text{and} \qquad \det|\mathbf{M}_T - \rho\,\mathbf{1}| = 0, \tag{15.57}$$

and is known as a "characteristic multiplier" of Eq. (15.2) . If solution $\mathbf{z}(t)$ satisfies the initial condition $\mathbf{z}(0) = \mathbf{a}$, then its value at time t is $\mathbf{M}(t)\mathbf{a}$ and, using Eq. (15.45), its value at $t + T$ is

$$\mathbf{z}(t+T) = \mathbf{M}(t+T)\mathbf{a} = \mathbf{M}(t)\mathbf{M}_T\mathbf{a} = \rho\,\mathbf{M}(t)\mathbf{a} = \rho\,\mathbf{z}(t). \tag{15.58}$$

This relation, true for arbitrary t, is the basis for the name *characteristic multiplier*. It shows that the essential behavior of the solution at large time is controlled by the value of ρ. In particular, the amplitude grows (shrinks) uncontrollably if $|\rho| > 1$ ($|\rho| < 1$). The case of greatest interest from our point of view is therefore $|\rho| = 1$, in which case ρ can be expressed as $\rho = e^{i\alpha\,T}$ (factor T has been included for later convenience) where $\alpha\,T$ is real and can be thought of as an angle in the range from $-\pi$ to π.[2]

When $\mathbf{M}_T$ is expressed in terms of $\mathbf{K}$ as in Eq. (15.46), Eq. (15.57) becomes

$$e^{T\mathbf{K}}\mathbf{a} = \rho\,\mathbf{a}, \quad \text{or} \quad \mathbf{K}\mathbf{a} = \frac{1}{T}\ln\rho\;\mathbf{a}. \tag{15.59}$$

This shows that $\mathbf{K}$ has the same eigenvectors as $\mathbf{M}_T$. Its eigenvalues are $\alpha = \ln\rho/T$. Because of the multiple-valued nature of the logarithm this determines α only *modulo* $2\pi i/T$. The α values are known as "characteristic exponents."[3]

For the case $|\rho| = 1$ that we have declared to be of greatest interest, the α_i can be taken in the range $-\pi < \alpha_i < \pi$. Even limited in this way there is too great a variety of possible arrangements of the α_i to permit a thorough survey in this text of all possibilities. We will assume them to be all unequal. Since, in the cases we will study, they come in equal and opposite pairs $\pm\alpha_i$, we will therefore be excluding even the case $\alpha_i = 0$.

If the coefficient matrix appearing in Eq. (15.2) is a constant matrix $\mathbf{C}$ it can be regarded as periodic for any period T. In this case the single-period transfer matrix is given by

$$\mathbf{M}_T = e^{\mathbf{C}T}. \tag{15.60}$$

Then a characteristic multiplier ρ belonging to $\mathbf{M}_T$ can be associated with an eigenvalue λ of $\mathbf{C}$ according to

$$\rho = e^{\lambda T}. \tag{15.61}$$

If the eigenvalues of $\mathbf{C}$ are expressed as $\lambda_h = \pm i\mu_h$, this equation becomes

$$\rho_h = e^{\pm i\mu_h T}. \tag{15.62}$$

2) One is accustomed to being able to make similar inferences from the eigenvalues of $\mathbf{A}$ in the coefficient-constant case. In this sense at least, $\mathbf{M}_T$ can be thought of as being the constant propagation matrix that "best represents" the time dependence of $\mathbf{A}(t)$.

3) In accelerator physics α (or rather $\alpha/(2\pi)$) is known as the "tune" of its corresponding eigenmotion.

Comparing this with the relation $\alpha = \ln \rho / T$, we see in this case (constant $\mathbf{C}$) that (modulo $2\pi i/T$)

$$\alpha_h = \pm i \mu_h; \tag{15.63}$$

the characteristic exponents are simply the eigenvalues of $\mathbf{C}$.

15.2.4 The Variational Equations

This section may seem initially to be something of a digression from the flow of consideration of periodic systems. What will make it germane is that the significance of 1 as a characteristic multiplier will be illuminated.

The first-order equations of motion for the most general *autonomous* system have the form $\dot{\mathbf{x}} = \mathbf{X}(\mathbf{x})$ or

$$\dot{x}^i = X^i(\mathbf{x}), \tag{15.64}$$

where the $2n$ functions of $X^i(\mathbf{x})$ are arbitrary. Let $\mathbf{x}(t)$ stand for a known actual solution of Eq. (15.64) and let $\mathbf{x}(t) + \delta\mathbf{x}(t)$ be a nearby function that also satisfies (15.64). Sufficiently small values of $\delta\mathbf{x}$ will satisfy an equation obtained by taking the first terms in a Taylor expansion centered on the known solution. This set of equations is

$$\dot{\delta x}^i = \sum_{j=1}^{2n} \left.\frac{\partial X^i}{\partial x_j}\right|_{\mathbf{x}_t} \delta x^j = \sum A_{ij}(t)\delta x^j. \tag{15.65}$$

The matrix of coefficients $\mathbf{A}(t)$ is normally called the Jacobian matrix and these are known as "the variational equations," or sometimes as "the Poincaré variational equations." By construction it is a linear set of equations, but they have become *nonautonomous* since the coefficients depend explicitly of t. If the unperturbed solution is periodic, the coefficients will be periodic functions of t, however. The theory of periodic systems that has been developed can therefore be applied to equations that emerge in this way.

Problem 15.2.1. *For a planar Kepler orbit the Hamiltonian was given in Eq. (8.54) to be*

$$H = \frac{p_r^2}{2m} + \frac{p_\theta^2}{2mr^2} - \frac{K}{r}. \tag{15.66}$$

For an orbit characterized by the (conserved) value of p_θ being α and with the coordinates listed in the order $(r, \theta, p_r, p_\theta)$, show that the matrix of the variational equations is

$$\mathbf{A} = \begin{pmatrix} 0 & 0 & \frac{1}{m} & 0 \\ \frac{-2\alpha}{mr^2} & 0 & 0 & \frac{1}{mr^2} \\ \frac{-3\alpha^2}{mr^4} + \frac{2K}{r^3} & 0 & 0 & \frac{2\alpha}{mr^3} \\ 0 & 0 & 0 & 0 \end{pmatrix}. \tag{15.67}$$

For the special case of the orbit being circular, find the eigenvalues of this matrix.

Problem 15.2.2. *For not quite circular orbits the unperturbed radial motion in the system of the previous problem is given by*

$$r = a(1 - \epsilon \cos \mu), \quad \textit{where} \quad t = \sqrt{\frac{ma^3}{K}}\,(u - \epsilon \sin u), \tag{15.68}$$

where ϵ can be treated as a small parameter. Find the time-dependent variational matrix. Each coefficient should be expressed as a (possibly terminating) Fourier series.

Just by inspection of the Jacobean matrix in Problem 15.2.1 one can see that it has $\lambda = 0$ as an eigenvalue. If the system is Hamiltonian (which it is) then this has to be a double root. This is too bad, since we have said that we would neglect the possibility of double roots. We will, in fact, for want of space and time, not work more with these equations, but we can at least contemplate the source of the vanishing eigenvalues.

If $\mathbf{A}$ has 0 as an eigenvalue, then $\mathbf{M}_T = e^{T\mathbf{A}}$ has 1 as an eigenvalue, and 1 is therefore one of the *multipliers* of the variational equations. If the multiplier is 1 then the corresponding solution is itself periodic, with the same period T as the underlying unperturbed motion. We could have seen *a priori* that such a periodic solution of the variational equations could have been obtained directly from the known unperturbed solution $\mathbf{x}(t)$. Simply differentiating Eqs. (15.64) with respect to t yields

$$\ddot{x}^i = \sum_{j=1}^{2n} \frac{\partial X^i}{\partial x_j}\dot{x}^j = \sum A_{ij}(t)\dot{x}^j. \tag{15.69}$$

This means that $\dot{\mathbf{x}}$ is a solution of the variational equations. But $\dot{\mathbf{x}}$ has the same periodicity as $\mathbf{x}$ and hence has period T. We see therefore that *to have at least one vanishing eigenvalue* is a generic property of the variational equations describing motion perturbed from a periodic orbit. This is something of a nuisance, and we will not pursue it further.

Bibliography

References for Further Study

Section 15.1.2

1 P.R. Halmos, *Finite-Dimensional Vector Spaces*, Springer, New York, 1987, p. 150.

2 K.R. Meyer and R. Hall, *Introduction to Hamiltonian Dynamical Systems and the N-Body Problem*, Springer, New York, 1992.

Section 15.2

3 V.A. Yakubovich and V.M. Starzhinskii, *Linear Differential Equations With Periodic Coefficients*.

16
Perturbation Theory

Nowadays students of physics may be inclined to think of perturbation theory as a branch of quantum mechanics since that is where they have mainly learned about it. For the same reason they may further think that there are precisely two types of perturbation theory – time-independent and time-dependent. It seems to me more appropriate to think of perturbation theory as a branch of applied mathematics with almost as many perturbative methods as there are problems, all similarly motivated, but with details determined by the particular features of the problem. There are methods that arise naturally and repeatedly in classical mechanics and some of them are discussed in this chapter.

One natural category for distinguishing among methods is whether or not they assume the unperturbed motion is Hamiltonian. Since the "purest" mechanical systems are Hamiltonian, we will emphasize methods for which the answer is affirmative.

The next natural categorization is whether the perturbation (a) violates the Hamiltonian requirements or (b) respects them. It is not possible to say which of these possibilities is the more important. (a) The "generic" situation in physics is for perturbations to violate d'Alembert's principle and therefore to be non-Hamiltonian. In fact most systems are treated as Hamiltonian only because non-Hamiltonian terms have been neglected in anticipation of later estimating the small deviations they cause. Lossy or viscous forces such as friction and wind resistance are examples. This case is by far the most important as far as engineering considerations are concerned and the required methods are rather straightforward. (b) The hardest problem, and the one that has tended to be of greatest theoretical interest over the centuries, is the case of perturbations that, though they respect d'Alembert's principle, lead to equations that can only be solved by approximation. It is usually very difficult to insure that an approximation method being employed does not introduce artificial nonsymplectic features into the predicted motion. This difficulty is most pronounced in nearly lossless systems such as high energy particles circulating in the vacuum of particle accelerators or heavenly bodies moving through the sky.

Geometric Mechanics: Toward a Unification of Classical Physics. 2nd Edition. Richard Talman

ISBN: 978-3-527-40683-8

Another categorization is based on whether the Hamiltonian is time-independent or time-dependent. (a) Time-independent systems are said to be "autonomous." They are systems that are so isolated from the rest of the world that there is no possibility of their being influenced by time-dependent external influences. (b) Systems that are not isolated are called "nonautonomous;" in general the external effects influencing them will be time-dependent. Among such systems the time dependence can be periodic or nonperiodic. It might be thought justified to slight the periodic case as being too special, but the opposite is more nearly appropriate. When external conditions return regularly to earlier values any errors that have been made in analyzing the motion are likely to stick out and this imposes serious demands on the methods of approximation. On the other hand, if the external conditions vary in an irregular way, that very irregularity tends to overwhelm any delicate Hamiltonian features of the motion.

Some of the important perturbative themes are as follows: (a) variation of constants, (b) averaging over a cycle of the unperturbed motion, (c) elimination of secular terms, (d) eliminating arbitrariness from the solutions of homogeneous equations, (e) successive approximation ("iterative") methods of approximation, and (f) taking account of the possibility of "resonance." Some of these have already been much discussed and the others will be in this chapter.

16.1 The Lagrange Planetary Equations

16.1.1 Derivation of the Equations

It was Lagrange himself who introduced powerful approximation techniques into celestial mechanics. He developed a procedure for analyzing the effect of perturbing forces on the same Kepler problem that has played such a prominent role in the history of physics as well as in this textbook. Lagrange's method can be characterized as "Newtonian" though Poisson brackets will play a prominent role, but not before the closely related and historically prior "Lagrange brackets" appear.

Copying the Kepler potential energy from Problem 1.5.1 and augmenting it by a perturbing potential energy $mR(\mathbf{r}, t)$ that depends arbitrarily on position and time, but is weak enough to perturb the motion only slightly, we are to find the trajectory of a particle of mass m with potential energy

$$V(\mathbf{r}) = -\frac{K}{r} - mR(\mathbf{r}). \tag{16.1}$$

It is not really essential that the perturbing force be representable by a potential energy function as here, and the force can be time-dependent without seriously complicating the solution, but we simplify the discussion a bit by making these assumptions. Since the motion is assumed to resemble the pure Kepler motion analyzed in earlier chapters it is appropriate to introduce the Jacobi parameters $\alpha_1, \alpha_2, \alpha_3, \beta_1, \beta_2, \beta_3$ of the nearby pure motion. (Recall that these are also known as "orbit elements.") For the time being, since there will be no need to distinguish between the α and the β elements, we will label them as $\alpha_1, \alpha_2, \alpha_3, \alpha_4, \alpha_5, \alpha_6$, and represent them all as $\boldsymbol{\alpha}$. Knowing the initial position and velocity of the mass, one can solve for the orbit elements that would match these conditions if the motion were unperturbed and hence actually *do* match the perturbed motion briefly. Using rectangular coordinates, the equations accomplishing this have the form

$$\begin{aligned} x &= x(t, \boldsymbol{\alpha}), & y &= y(t, \boldsymbol{\alpha}), & z &= z(t, \boldsymbol{\alpha}), \\ \dot{x} &= p_x(t, \boldsymbol{\alpha})/m, & \dot{y} &= p_y(t, \boldsymbol{\alpha})/m, & \dot{z} &= p_z(t, \boldsymbol{\alpha})/m. \end{aligned} \tag{16.2}$$

For later convenience the Cartesian velocity components have been expressed in terms of Cartesian momentum components. Equations (16.2) (actually their inverses) can be employed at any time t to find the instantaneous values of $\boldsymbol{\alpha}$ and $\boldsymbol{\beta}$ that would give the actual instantaneous values of $\mathbf{r}$ and $\dot{\mathbf{r}}$. The (Newtonian) perturbed equations of motion are

$$\ddot{x} + \frac{Kx}{mr^3} = \frac{\partial R}{\partial x}, \qquad \ddot{y} + \frac{Ky}{mr^3} = \frac{\partial R}{\partial y}, \qquad \ddot{z} + \frac{Kz}{mr^3} = \frac{\partial R}{\partial z}. \tag{16.3}$$

A convention that is commonly employed, especially in this text, is to place the terms corresponding to unperturbed motion on the left-hand sides of the equations and to place the perturbing terms on the right-hand sides, as here.

Respecting the functional form of the unknowns as they were introduced in Eqs. (16.2), the unperturbed equations of motion can be written as

$$\left.\frac{\partial^2 x}{\partial t^2}\right|_{\boldsymbol{\alpha}} = -\frac{Kx}{r^3}, \qquad \left.\frac{\partial^2 y}{\partial t^2}\right|_{\boldsymbol{\alpha}} = -\frac{Ky}{r^3}, \qquad \left.\frac{\partial^2 z}{\partial t^2}\right|_{\boldsymbol{\alpha}} = -\frac{Kz}{r^3}. \tag{16.4}$$

In other words these equations state that the functions specified in Eqs. (16.2) satisfy the unperturbed equations if the orbit elements are constants.

The method of "variation of constants" (which is by no means specific to this problem) consists of allowing the "constants" $\boldsymbol{\alpha}$ to vary slowly with time in such a way that the perturbed equations of motion are satisfied, while insisting that the relations (16.2) continue to be satisfied at all times. At any instant the motion will be appropriate to the orbit elements as they are evaluated at that time and they will vary in such a way as to keep this true.

This matching is based on a picture of "osculation" which means that the perturbed and unperturbed orbits not only touch, they "kiss," meaning they

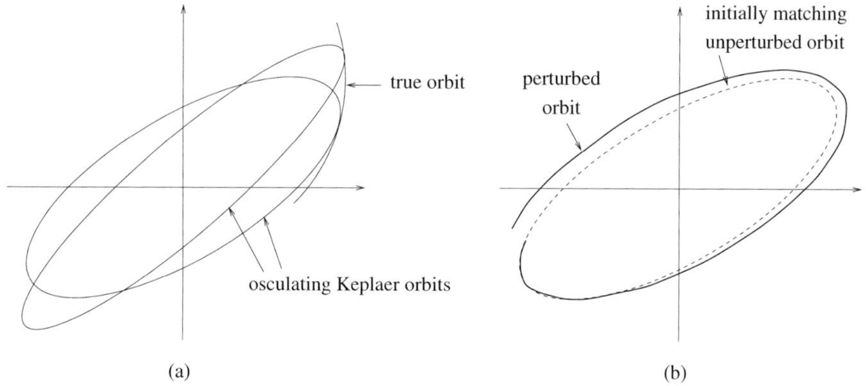

(a) (b)

Fig. 16.1 (a) The true orbit "osculates" the matching unperturbed orbits at successive times. The orbits need not lie in the same plane. (b) The deviation of the true orbit (solid curve) from the unperturbed orbit that matched at the start of the calculation (dashed curve) is based on evaluating all functions on the dashed curve. The perturbed orbit may come out of the plane.

have the same slopes, as in Fig. 16.1. Part (a) of the figure shows that a more or less arbitrary trajectory can be matched by Kepler ellipses, but part (b) more nearly represents the sort of almost-elliptical perturbed orbit we have in mind.

The true instantaneous velocities are obtained (by definition) from

$$\begin{aligned}\frac{dx}{dt} &= \left.\frac{\partial x}{\partial t}\right|_{\boldsymbol{\alpha}} + \sum_s \frac{\partial x}{\partial \alpha_s}\dot{\alpha}_s,\\ \frac{dy}{dt} &= \left.\frac{\partial y}{\partial t}\right|_{\boldsymbol{\alpha}} + \sum_s \frac{\partial y}{\partial \alpha_s}\dot{\alpha}_s,\\ \frac{dz}{dt} &= \left.\frac{\partial z}{\partial t}\right|_{\boldsymbol{\alpha}} + \sum_s \frac{\partial z}{\partial \alpha_s}\dot{\alpha}_s,\end{aligned} \tag{16.5}$$

and the matching unperturbed velocities are given by the first terms on the right-hand sides of these equations. Hence the calculus expression of the osculation condition is

$$\sum_s \frac{\partial x}{\partial \alpha_s}\dot{\alpha}_s = 0, \qquad \sum_s \frac{\partial y}{\partial \alpha_s}\dot{\alpha}_s = 0, \qquad \sum_s \frac{\partial z}{\partial \alpha_s}\dot{\alpha}_s = 0. \tag{16.6}$$

Differentiating the lower of Eqs. (16.2) with respect to t, substituting the result into Eqs. (16.3), and taking advantage of Eqs. (16.4), we obtain

$$\sum_s \frac{\partial p_x}{\partial \alpha_s}\dot{\alpha}_s = m\frac{\partial R}{\partial x}, \qquad \sum_s \frac{\partial p_y}{\partial \alpha_s}\dot{\alpha}_s = m\frac{\partial R}{\partial y}, \qquad \sum_s \frac{\partial p_z}{\partial \alpha_s}\dot{\alpha}_s = m\frac{\partial R}{\partial z}. \tag{16.7}$$

Together, Eqs. (16.6) and (16.7) are six differential equations for the six orbit elements, but they are not yet manageable equations as they depend as well on the Cartesian coordinates and momenta.

This dependency can be removed by the following remarkable manipulations. Multiplying the first of Eqs. (16.7) by $\partial x/\partial\alpha_r$ and subtracting the first of Eqs. (16.6) multiplied by $\partial p_x/\partial\alpha_r$ yields

$$\sum_s X_{rs}\,\dot{\alpha}_s = m\frac{\partial R}{\partial x}\frac{\partial x}{\partial\alpha_r}, \qquad \text{where} \qquad X_{rs} = \frac{\partial x}{\partial\alpha_r}\frac{\partial p_x}{\partial\alpha_s} - \frac{\partial p_x}{\partial\alpha_r}\frac{\partial x}{\partial\alpha_s}. \tag{16.8}$$

Quantities Y_{rs} and Z_{rs} are defined similarly and the same manipulations can be performed on the other equations. We define the "Lagrange bracket" of pairs of orbit elements by

$$L_{rs} \equiv [\alpha_r, \alpha_s] \equiv X_{rs} + Y_{rs} + Z_{rs} \equiv \sum_i^n \left(\frac{\partial q^i}{\partial\alpha_r}\frac{\partial p_i}{\partial\alpha_s} - \frac{\partial p_i}{\partial\alpha_r}\frac{\partial q^i}{\partial\alpha_s}\right), \tag{16.9}$$

where we have introduced $(q^1, q^2, q^3) \equiv (x, y, z)$ and $(p_1, p_2, p_3) \equiv (p_x, p_x, p_z)$ and, though $n = 3$ in this case, similar manipulations would be valid for arbitrary n. The purpose of the duplicate notation for $[\alpha_r, \alpha_s]$ is so that we can regard the Lagrange brackets as the elements of a matrix $\mathbf{L} = (L_{sr})$. Adding the three equations like (16.8) we obtain

$$[\alpha_r, \alpha_s]\,\dot{\alpha}_s = m\left(\frac{\partial R}{\partial x}\frac{\partial x}{\partial\alpha_r} + \frac{\partial R}{\partial y}\frac{\partial y}{\partial\alpha_r} + \frac{\partial R}{\partial z}\frac{\partial z}{\partial\alpha_r}\right). \tag{16.10}$$

After these manipulations the coordinates x, y, and z no longer appear explicitly on the left-hand sides of the equations. This may appear like an altogether artificial improvement since the Lagrange brackets themselves depend implicitly on these quantities. The next stage of the development is to show that there is no such dependence, or rather that the dependence can be neglected in obtaining an approximate solution of the equations. More precisely we will show that $[\alpha_r, \alpha_s]$ is a constant of the unperturbed motion (provided both α_r and α_s are constants of the unperturbed motion which, of course, they are). This is exact for unperturbed orbits but, applying it to the perturbed orbit, we will obtain only an approximate result. We defer this proof and continue to reduce the perturbation equations.

While discussing adiabatic invariants in Chapter 14 we already learned the efficacy of organizing the calculation so that the *deviation* from unperturbed motion can be calculated as an integral over an unperturbed motion. This is illustrated in Fig. 16.1(b). In this case the unperturbed orbit, shown as a dashed curve, is closed while the perturbed orbit may not be, and that is the sort of effect a perturbation is likely to have. But we assume that over a single period the deviation of the perturbed orbit is small on the scale of either orbit. It is implicit in this assumption that the fractional changes in the orbit elements $\boldsymbol{\alpha}$ will also be small over one period. As in the proof of constancy of the action variable in Section 14.3.2, we can approximate the right-hand side of Eq. (16.10) by averaging over one period T;

$$[\alpha_r, \alpha_s]\,\dot{\alpha}_s = \frac{m}{T}\oint\left(\frac{\partial R}{\partial x}\frac{\partial x}{\partial \alpha_r} + \frac{\partial R}{\partial y}\frac{\partial y}{\partial \alpha_r} + \frac{\partial R}{\partial z}\frac{\partial z}{\partial \alpha_r}\right)dt \equiv m\,\overline{\frac{\partial R}{\partial \alpha_r}}. \tag{16.11}$$

These are the "Lagrange planetary equations." Since they can be written in matrix form,

$$\sum_s L_{rs}\dot{\alpha}_s = m\,\overline{\frac{\partial R}{\partial \alpha_r}}, \tag{16.12}$$

they can be solved for the time derivatives;

$$\dot{\alpha}_s = \sum_r P_{sr} m \overline{\frac{\partial R}{\partial \alpha_r}}, \qquad \text{where} \qquad \mathbf{P} = \mathbf{L}^{-1}. \tag{16.13}$$

Since the integrations required to evaluate the averages in (16.11) are taken over the matching unperturbed orbit, the Lagrange brackets are, in principle, known. If they are calculated numerically they are known also in practice and this commonly solves the problem at hand satisfactorily. But we will continue the analytical development and succeed in completing realistic calculations in closed form.

It has already been stated that the coefficients $[\alpha_r, \alpha_s]$ are constants of the motion and, since linear differential equations with constant coefficients are very manageable we can see what a great simplification the Lagrange planetary equations have brought.

One of the nicest applications of the Lagrange planetary equations concerns a calculation of the advance of the perihelion of Mercury predicted by Einstein's general relativity. This calculation is spelled out in some detail in Section 16.2.

16.1.2
Relation Between Lagrange and Poisson Brackets

Because it is needed in Eq. (16.13), we now set about finding $\mathbf{P} = \mathbf{L}^{-1}$, or rather showing that the elements of $\mathbf{P}$ are in fact the Poisson brackets of the orbit elements

$$P_{rs} = \{\alpha_r, \alpha_s\} = \sum_{j=1}^{n}\left(\frac{\partial \alpha_r}{\partial q^j}\frac{\partial \alpha_s}{\partial p_j} - \frac{\partial \alpha_r}{\partial p_j}\frac{\partial \alpha_s}{\partial q^j}\right). \tag{16.14}$$

We are generalizing somewhat by allowing arbitrary generalized coordinates and momenta and Eq. (16.9) has already been generalized in the same way. Recall that there are $2n$ orbit elements α_r since they include both the n Jacobi β_i elements as well as the n Jacobi α_i elements. As a result the matrices under study are $2n \times 2n$; the indices r and s run from 1 to $2n$ while the indices i and j run from 1 to n. Using summation convention and relations like

$$\frac{\partial q^i}{\partial \alpha_r}\frac{\partial \alpha_r}{\partial q^j} = \delta_{ij}, \tag{16.15}$$

we now show that

$$\begin{aligned}
(\mathbf{L}^T\mathbf{P})_{st} &= [\alpha_r, \alpha_s]\{\alpha_r, \alpha_t\} \\
&= \frac{\partial q^i}{\partial \alpha_r}\frac{\partial \alpha_r}{\partial q^j}\frac{\partial p_i}{\partial \alpha_s}\frac{\partial \alpha_t}{\partial p_j} - \frac{\partial q^i}{\partial \alpha_r}\frac{\partial \alpha_r}{\partial p_j}\frac{\partial p_i}{\partial \alpha_s}\frac{\partial \alpha_t}{\partial q^j} - \frac{\partial p_i}{\partial \alpha_r}\frac{\partial \alpha_r}{\partial q^j}\frac{\partial q^i}{\partial \alpha_s}\frac{\partial \alpha_t}{\partial p_j} + \frac{\partial p_i}{\partial \alpha_r}\frac{\partial \alpha_r}{\partial q^j}\frac{\partial q^i}{\partial \alpha_s}\frac{\partial \alpha_t}{\partial q^j} \\
&= \frac{\partial \alpha_t}{\partial p_j}\frac{\partial p_j}{\partial \alpha_s} + \frac{\partial \alpha_t}{\partial q^j}\frac{\partial q^j}{\partial \alpha_s} \qquad (16.16) \\
&= \delta_{st},
\end{aligned}$$

and hence

$$\mathbf{L}^T = \mathbf{P}^{-1}, \tag{16.17}$$

which is the desired result.

It is shown in Chap. 17 that the (Poisson bracket) elements of $\mathbf{P}$ are constants of the unperturbed motion, so we now know that the Lagrange brackets (elements of $\mathbf{P}^{-1}$) are also constants of the motion. This greatly simplifies the Lagrange planetary equations. It also shows, incidentally, that Lagrange himself was aware of the most important properties of these bracket expressions well before Poisson. Lagrange's proof of the invariance of the Lagrange brackets was specific to the Kepler problem and proceeded as in a problem below.

16.2 Advance of Perihelion of Mercury

One of the two experiments that Einstein suggested initially to test general relativity concerned the advance of the perihelion of planet Mercury. In pure Newtonian gravity the orientation of the elliptical orbit is a constant of the motion. But general relativity predicts a small reorientation of the orbit as time advances. See Fig. 16.2. (A "laboratory" demonstration of the advance of an "elliptical" orbit is seen occasionally in shopping malls where one is encouraged to roll coins into a large shallow, curved-wall, funnel. The coin rolls through many revolutions in a rosette-shaped orbit, until it is eventually swallowed by a hole in the center of the funnel.)

From the point of view of studying general relativity *per se*, the following material would more logically be contained in Chapter 13.

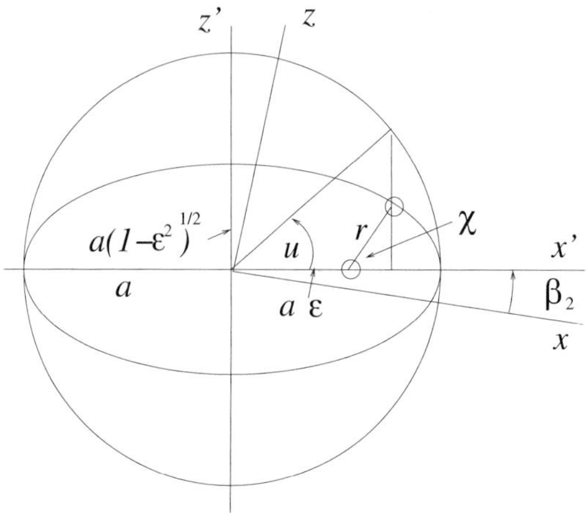

Fig. 16.2 Advance of perihelion is registered as deviation from zero of the coordinate β_2 which is the angle from the major axes at time $t = 0$ to the major axis at time t.

We know from general relativity that the mass of the sun introduces "curvature" into the geometry of the space in its vicinity. This has the effect of modifying the Kepler potential slightly, into the form

$$U(r) = -\frac{K}{r} - \frac{\beta}{r^n}, \qquad \text{or} \qquad mR = \frac{\beta}{r^n}, \tag{16.18}$$

where the parameters β and n are obtained by perturbative expansion of the equations of general relativity. We will calculate the influence of the correction term on the orbits of planets of the sun. The planet for which the effect is most appreciable and the observations least subject to extraneous difficulties is Mercury.[1]

Orbit elements to be used are a, ϵ which are *intrinsic* properties of the unperturbed orbit, along with i and $\beta_1, \beta_2, \beta_3$ which establish its orientation. These quantities were all defined in Section 8.3. We specialize to planar geometry corresponding to the figure by assuming $i = \pi/2$, $\beta_3 = 0$. Generalizing Eqs. (1.77) by introducing β_2 to allow for the possible advance of the peri-

1) In practice the orbit of Mercury *appears* to process at a rate roughly one hundred times greater than the Einstein effect because the coordinate system (described in Section 8.3) is not itself fixed. Furthermore there are perturbations to Mercury's orbit due to other nearby planets and these forces cause precession on the order of ten times greater than the Einstein effect. These precessions are eminently calculable using the Lagrange planetary equations, but we ignore them, or rather treat them as nuisance corrections that need to be made before the result of greatest interest can be extracted.

helion angle β_2, the Cartesian coordinates of the planet are given by

$$x = \cos\beta_2\,(a\cos u - a\epsilon) - \sin\beta_2\, a\sqrt{1-\epsilon^2}\,\sin u$$
$$z = \sin\beta_2\,(a\cos u - a\epsilon) + \cos\beta_2\, a\sqrt{1-\epsilon^2}\,\sin u \tag{16.19}$$

Note that to specify the azimuthal position of the planet we are using the intermediate variable u, known as the "eccentric anomaly," rather than ordinary cylindrical coordinate χ, which is the "true anomaly." A formula relating u to time t is Eq. (1.78);

$$t - \tau = \sqrt{\frac{ma^3}{K}}\,(u - \epsilon\sin u), \tag{16.20}$$

where we have introduced the "time of passage through perigee" τ; it is closely related to β_1 because, according to Eq. (8.69),

$$\tau = -\sqrt{\frac{ma^3}{K}}\,\beta_1. \tag{16.21}$$

For a general perturbing force a replacement like this would be ill-advised since the dependence of the coefficient on one of the other orbit elements, namely a, "mixes" the elements. This would lead at best to complication and at worst to error. But our assumptions have been such that energy E is conserved and the definitions of the orbit elements then imply that a is also conserved for the particular perturbation being analyzed.

Since we need derivatives with respect to t we need the result

$$\frac{du}{dt} = -\frac{du}{d\tau} = \sqrt{\frac{K}{ma^3}}\,\frac{1}{1-\epsilon\cos u}. \tag{16.22}$$

Differentiating Eqs. (16.19) we obtain

$$p_x = (-\cos\beta_2\,\sin u - \sqrt{1-\epsilon^2}\,\sin\beta_2\,\cos u)\,\frac{\sqrt{Km/a}}{1-\epsilon\cos u},$$
$$p_z = (-\sin\beta_2\,\sin u + \sqrt{1-\epsilon^2}\,\cos\beta_2\,\cos u)\,\frac{\sqrt{Km/a}}{1-\epsilon\cos u}. \tag{16.23}$$

Since no further differentiations with respect to t will be required, and since the Lagrange brackets are known to be independent of t, we can set $t = 0$ from here on. Then, by Eq. (16.20), u is a function only of τ. Furthermore, we can set $\beta_2 = 0$ and, after differentiating with respect to τ, we can also set $\tau = 0$ (and hence $u = 0$). These assumptions amount to assuming perigee occurs along the x-axis at $t = 0$, and that the orbit elements are being evaluated at that point. Hence, for example,

$$\left.\frac{\partial x}{\partial \tau}\right|_{\tau=\beta_2=0} = \sin u \left.\frac{\sqrt{K/(ma)}}{1-\epsilon\cos u}\right|_{u=0} = 0,$$
$$\left.\frac{\partial z}{\partial \tau}\right|_{\tau=\beta_2=0} = -\frac{\sqrt{1-\epsilon^2}}{1-\epsilon}\sqrt{\frac{K}{ma}}. \tag{16.24}$$

The vanishing of $\partial x/\partial\tau$ reflects the fact that it is being calculated at perigee. For the Lagrange brackets involving orbit elements other than τ it is legitimate to make the $u = 0$ simplification before evaluating the required partial derivatives:

$$\begin{aligned}
x &= a(1-\epsilon)\cos\beta_2,\\
z &= a(1-\epsilon)\sin\beta_2,\\
p_x &= -\sqrt{\frac{Km}{a}}\sqrt{\frac{1+\epsilon}{1-\epsilon}}\sin\beta_2,\\
p_x &= \sqrt{\frac{Km}{a}}\sqrt{\frac{1+\epsilon}{1-\epsilon}}\cos\beta_2.
\end{aligned} \tag{16.25}$$

Completion of this example is left as an exercise. Einstein's calculation of the constants β and n in Eq. (16.18) leads to the "astronomically *small*" precession rate of 43 s of arc per century. One should be suitably impressed not just by Einstein but also by Newton, who permits such a fantastic "deviation from null" observation, since the unperturbed orbit closes to such high precision.

Problem 16.2.1. *Setting $m = 1$ so $\mathbf{p} = \dot{\mathbf{x}}$, the potential energy that yields a "central force," radial and with magnitude depending only on r is given by $V(r)$. The unperturbed equations of motion in this potential are*

$$\left.\frac{\partial^2 x}{\partial t^2}\right|_{\boldsymbol{\alpha}} = -\frac{\partial V}{\partial x}, \quad \left.\frac{\partial^2 y}{\partial t^2}\right|_{\boldsymbol{\alpha}} = -\frac{\partial V}{\partial y}, \quad \left.\frac{\partial^2 z}{\partial t^2}\right|_{\boldsymbol{\alpha}} = -\frac{\partial V}{\partial z}. \tag{16.26}$$

Let $\boldsymbol{\alpha}$ stand for the orbit elements in this potential. By explicit differentiation show that $d/dt[\alpha_r, \alpha_s] = 0$

Problem 16.2.2. *Check some or all of the following Lagrange brackets for the Kepler problem. They assume as orbit elements a, ϵ, i along with β_2, β_3, all defined in Section 8.3 as well as τ which is closely related to β_1 as in Eq. (16.21).*

$$[a,\epsilon] = 0, \qquad [a,\tau] = \frac{K}{2ma^2}, \qquad [a,\beta_2] = -\frac{1}{2}\sqrt{\frac{(1-\epsilon^2)K}{ma}}, \qquad [a,i] = 0,$$
$$[a,\beta_3] = -\frac{1}{2}\sqrt{\frac{(1-\epsilon^2)K}{ma}}\cos i, \qquad [\epsilon,\tau] = 0, \qquad [\epsilon,\beta_2] = \sqrt{\frac{aK/m}{1-\epsilon^2}}\,\epsilon,$$

$$[\epsilon, i] = 0, \quad [\epsilon, \beta_3] = \sqrt{\frac{aK/m}{1-\epsilon^2}}\,\epsilon\,\cos i, \quad [\tau, \beta_2] = 0, \quad [\tau, i] = 0, \quad [\tau, \beta_3] = 0,$$

$$[i, \beta_3] = \sqrt{(1-\epsilon^2)aK/m}\,\sin i, \qquad [i, \beta_2] = 0, \qquad [\beta_3, \beta_2] = 0. \tag{16.27}$$

Problem 16.2.3. *To obtain orbits that start and remain in the (x, z) plane, assume $i = \pi/2$, $\beta_3 = 0$, $\overline{\partial R/\partial \beta_3} = 0$, and $\overline{\partial R/\partial i} = 0$. Show that the Lagrange planetary equations are*

$$\begin{aligned}
\frac{K}{2ma^2}\,\dot{\tau} - \frac{1}{2}\sqrt{\frac{(1-\epsilon^2)K/m}{a}}\,\dot{\beta}_2 &= \overline{\frac{\partial R}{\partial a}},\\
\sqrt{\frac{aK/m}{1-\epsilon^2}}\,\epsilon\,\dot{\beta}_2 &= \overline{\frac{\partial R}{\partial \epsilon}},\\
-\frac{K}{2ma^2}\,\dot{a} &= \overline{\frac{\partial R}{\partial \tau}},\\
\frac{1}{2}\sqrt{\frac{(1-\epsilon^2)K/m}{a}}\,\dot{a} - \sqrt{\frac{aK/m}{1-\epsilon^2}}\,\epsilon\,\dot{\epsilon} &= \overline{\frac{\partial R}{\partial \beta_2}}.
\end{aligned} \tag{16.28}$$

Problem 16.2.4. *Check some or all of the coefficients in the following formulas which are the planetary equations of the Kepler problem solved for the time derivatives of the orbit elements.*

$$\begin{aligned}
\dot{a} &= -\frac{2a^2}{K/m}\overline{\frac{\partial R}{\partial \tau}},\\
\dot{\epsilon} &= -\frac{am(1-\epsilon^2)}{\epsilon K}\overline{\frac{\partial R}{\partial \tau}} - \frac{1}{\epsilon}\sqrt{\frac{m(1-\epsilon^2)}{aK}}\overline{\frac{\partial R}{\partial \beta_2}},\\
\dot{\tau} &= \frac{2a^2 m}{K}\overline{\frac{\partial R}{\partial a}} + \frac{am(1-\epsilon^2)}{\epsilon K}\overline{\frac{\partial R}{\partial \epsilon}},\\
\dot{i} &= \sqrt{\frac{m}{(1-\epsilon^2)aK}}\frac{1}{\sin i}\left(\cos i\,\overline{\frac{\partial R}{\partial \beta_2}} - \overline{\frac{\partial R}{\partial \beta_3}}\right),\\
\dot{\beta}_3 &= \sqrt{\frac{m}{(1-\epsilon^2)aK}}\frac{1}{\sin i}\overline{\frac{\partial R}{\partial i}},\\
\dot{\beta}_2 &= \sqrt{\frac{m(1-\epsilon^2)}{aK}}\frac{1}{\epsilon}\left(\overline{\frac{\partial R}{\partial \epsilon}} - \frac{\epsilon\cot i}{1-\epsilon^2}\overline{\frac{\partial R}{\partial i}}\right).
\end{aligned} \tag{16.29}$$

Problem 16.2.5. *Complete the calculation of the precession of Kepler orbits caused by the second term of Eq. (16.18).*

16.3
Iterative Analysis of Anharmonic Oscillations

Consider a system that executes simple harmonic motion for sufficiently small amplitudes but to include large amplitudes needs to be described by an equation

$$\left(\frac{d^2}{dt^2} + \omega_0^2\right) x = R(x) = \alpha x^2 + \beta x^3. \tag{16.30}$$

For now we assume the system is *autonomous* which means that R does not depend explicitly on t. But R is here allowed us to depend on "nonlinear" powers of x higher than first order. Such terms could have been derived from a potential energy function $V = -\alpha x^3/3 - \beta x^4/4$. (Note that function R is not the same as in the previous section). Like all one-dimensional problems this one could therefore be studied using methods explained in Chapter 1. The motion oscillates between the (readily calculable) turning points closest to the origin. Such motion is trivially periodic but the presence of nonlinear terms causes the time dependence to be not quite sinusoidal and the system is therefore called "anharmonic."

We now wish to apply a natural iterative method of solution to this problem. This may seem to be an entirely academic undertaking since the solution described in the previous paragraph has to be regarded as already highly satisfactory. Worse yet, on a first pass the proposed method will yield an obviously wrong result. We are then led to a procedure that overcomes the problem, thereby repairing the iterative "tool" for use in multidimensional or nonconservative situations where no exact method is available. The previously mentioned unphysical behavior is ascribed to so-called "secular terms" and the procedure for eliminating them is known as "Linstedt's method."

By choosing the initial position x_0 and velocity v_0 small enough, it is possible to make the terms on the right-hand side of Eq. (16.30) negligibly small. In this approximation the solution to Eq. (16.30) takes the form

$$x = a \cos \omega_0 t, \tag{16.31}$$

where we have simplified to the maximum extent possible (with no loss of generality) by choosing the initial time to be such that the motion is described by a pure cosine term. This solution will be known as *the zeroth-order solution.*

We are primarily interested in larger amplitudes where the anharmonic effects have become noticeably large. This region can be investigated mathematically by keeping only the leading terms in power series in the amplitude a. In fact one usually works only to the lowest occurring power unless there is a good reason (and there often is) to keep at least one more term. An intuitively natural procedure then is to approximate the right-hand side of Eq. (16.30) by substituting the zeroth-order solution to obtain

$$\left(\frac{d^2}{dt^2} + \omega_0^2\right) x = \alpha a^2 \cos^2 \omega_0 t + \beta a^3 \cos^3 \omega_0 t$$
$$= \frac{\alpha a^2}{2}(1 + \cos 2\omega_0 t) + \frac{\beta a^3}{4}(3 \cos \omega_0 t + \cos 3\omega_0 t). \qquad (16.32)$$

The terms on the right-hand side have been expanded into Fourier series with period $2\pi/\omega_0$. Note that $R(x)$ could have been any function of x whatsoever and the right-hand side would still have been expressible as a Fourier series with the same period – any function of a periodic function is periodic. In general the Fourier series would be infinite, but for our simple perturbation the series terminates.

Though Eq. (16.30) was autonomous, Eq. (16.32) is nonautonomous. In fact the terms on the right-hand side are not different from the terms that would describe external sinusoidal drive at the four frequencies 0, ω_0, $2\omega_0$, and $3\omega_0$. Furthermore the equations have magically become "linear" – that was the purpose of the Fourier expansion. Methods for solving equations like these have been illustrated in the problems of the Chapter 1, such as Problem 1.4.5.

The drive term $(3\beta a^3/4) \cos \omega_0 t$ is troublesome. Solving, for example, by the Laplace transform technique, one finds its response to be proportional to $t \sin \omega_0 t$ which becomes arbitrarily large with increasing time. This occurs because the "drive" frequency is equal to the natural frequency of the unperturbed system (which is an ideal lossless simple harmonic oscillator). The infinite buildup occurs because the drive enhances the response synchronously on every cycle, causing the amplitude to grow inexorably. This is known as "resonance." This infinite buildup is clearly unphysical and a perturbing term like this is known as a "secular term." The rate of growth is proportional to β but the motion will eventually blow up no matter how small the parameter β.

Having identified this problem, it is pretty obvious what is its source. It is only because of the parabolic shape of the potential well that the frequency of a simple harmonic oscillator is independent of amplitude. Since the extra terms that have been added distort this shape they can be expected to cause the frequency to depend on amplitude. This is known as "detuning with amplitude." This detuning will disrupt the above-mentioned synchronism and this is presumably what prevents the unphysical behavior.

Having identified the source of the problem, it is not hard to repair the solution. We need to include a term $2\omega_0 \delta\omega\, x$ on the left-hand side of Eq. (16.32) to account for amplitude-dependent shift of the "natural frequency of oscillation;" $\omega_0 \to \omega = \omega_0 + \delta\omega$. (A term $\delta\omega^2\, x$ that might also have been expected will be dropped because it is quadratically small.) The result is

$$\left(\frac{d^2}{dt^2} + \omega_0^2 + 2\omega_0\delta\omega\right) x \tag{16.33}$$
$$= \frac{\alpha a^2}{2}(1 + \cos 2\omega_0 t) + \left(\frac{3\beta a^3}{4} + 2a\omega_0\delta\omega\right)\cos\omega_0 t + \frac{\beta a^3}{4}\cos 3\omega_0 t.$$

With a term having been added to the left-hand side of the equation, it has been necessary to add the same term to the right-hand side in order to maintain the equality. But (consistent with the iterative scheme) this term has been evaluated using the zeroth approximation to the motion. The only way for this equation to yield a steady periodic solution is for the coefficient of $\cos\omega_0 t$ to vanish; this yields a formula for ω;

$$\omega \overset{q}{=} \omega_0 - \frac{3\beta}{8\omega_0} a^2. \tag{16.34}$$

For want of a better term we will call this procedure "the Linstedt trick." We have made this only a "qualified" equality since it will be shortly seen to be not quite right unless $\alpha = 0$. Making this substitution, the equation of motion becomes

$$\left(\frac{d^2}{dt^2} + \omega^2\right) x \overset{q}{=} \frac{\alpha a^2}{2} + \frac{\alpha a^2}{2}\cos 2\omega t + \frac{\beta a^3}{4}\cos 3\omega t. \tag{16.35}$$

Because none of the frequencies on the right-hand side of the equation are close to ω, they have been approximated by $\omega_0 \to \omega$. At this particular amplitude the frequency ω_0 has lost its significance and all functions of x have become periodic with frequency ω. A particular integral of this equation can be obtained by inspection

$$x(t) \overset{q}{=} \frac{\alpha a^2}{2}\frac{1}{\omega^2} - \frac{\alpha a^2}{2}\cos 2\omega t\,\frac{1}{3\omega^2} - \frac{\beta a^3}{4}\cos 3\omega t\,\frac{1}{8\omega^2} + \cdots, \tag{16.36}$$

where ... is a reminder that the solution of an inhomogeneous equation like Eq. (16.35) remains a solution when it is augmented by any solution of the "homogeneous equation" (obtained by dropping the terms of the right-hand side). Augmenting Eq. (16.36) by the zeroth-order solution yields

$$x(t) \overset{q}{=} a\cos\omega t + \frac{\alpha a^2}{2}\frac{1}{\omega^2} - \frac{\alpha a^2}{2}\frac{1}{3\omega^2}\cos 2\omega t - \frac{\beta a^3}{4}\frac{1}{8\omega^2}\cos 3\omega t + \cdots. \tag{16.37}$$

Each of these terms comes from the general formula for the response to drive term $\cos r\omega t$, where r is any integer, which is

$$\frac{1}{(-r^2+1)\omega^2}\cos r\omega t. \tag{16.38}$$

It was the vanishing of the denominator factor $r^2 - 1$ for $r = 1$ that made it necessary to suppress the secular term before proceeding. This problem is ubiquitous in mechanics; it goes by the name "the problem of small denominators." Had we solved the equation by the Laplace transform technique the vanishing denominator problem would have manifested itself by the structure of the formula for $\overline{x}(s)$;

$$\frac{1}{s+i\omega}\frac{1}{s-i\omega}\frac{s}{s^2+r^2\omega^2}. \tag{16.39}$$

For $r = 1$ the poles become double.

We have to leave it ambiguous whether Linstedt's trick constitutes a *theorem of mathematics* for solving the equation or a *principle of physics* stating that "nature" shifts the frequency of oscillation to avoid the infinity. As it happens, nature has another way of handling the problem, namely by creating *chaos*. Speaking very loosely, for small amplitudes nature chooses to maintain *regular* motion with shifted frequency, but for large amplitudes has to resort to *irregular*, chaotic motion. The way nature proceeds from the regular to the irregular regime is tortuous and not easily subject to description in closed form. Chaotic motion is much studied, especially numerically, but we will stay in the regular regime.

Returning to Eq. (16.37), even given that we are suppressing terms proportional to a^4 and above, the ... and the $\stackrel{q}{=}$ are still present for two reasons; the last term is not quite correct (which we rectify below) and, more fundamental, but also more easily taken care of, the solution is not unique. To make the solution unique we should make it match given initial conditions $x(0) = x_0$, $\dot{x}(0) = v_0$. Since the choice of time origin has been left arbitrary we can make the replacement $t \to t - t_0$ in Eq. (16.37) and then adjust a and t_0 to provide the required match.

One may find it surprising that only the anharmonic term proportional to x^3 in Eq. (16.30) has led to an amplitude-dependent frequency shift in Eq. (16.34). Figure 16.3 should make this result at least plausible however. If one considers the restoring force in the problem as being due to a spring then pure simple harmonic motion requires perfect adherence to "Hooke's law" by the spring. An actual spring may violate Hooke's law either by being "soft" and giving too little force at large extension or "hard" and giving too much. These behaviors are illustrated in Fig. 16.3. If a spring is "hard" the natural frequency increases with increasing amplitude. But if the spring is soft on the left and hard on the right the frequency shifts on the left and right will tend to cancel.

There is one more thing we should be concerned about though. We have been a bit cavalier in dropping small terms, and we have in fact made a mistake in the treatment so far. In our haste at accepting the absence of frequency shift coming from the quadratic force term αx^2 we should have registered on

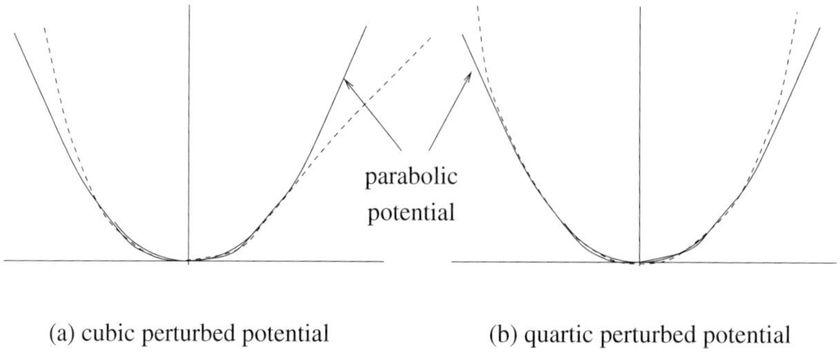

Fig. 16.3 Perturbed potential energy functions leading to anharmonic oscillations. (a) "Cubic" deformation makes the spring "hard" on the left, "soft" on the right. (b) "Quadratic" deformation make the string symmetrically hard on the left and on the right.

dimensional grounds, by comparison with Eq. (16.34), that the frequency shift it would have caused (but didn't because of a vanishing coefficient) would have been proportional to $\alpha a/\omega_0$. Since formula (16.34) can be said to be the a^2 correction in a formula for the natural frequency as a power series in a, we have included only the effect of the βx^3 term in the original equation, but not yet the αx^2 term in the next order of approximation.

We must therefore perform another iteration stage, this time substituting for $x(t)$ from Eq. (16.37) into the right-hand side of Eq. (16.30). Exhibiting only the secular term from Eq. (16.33) and the only new term that may contain a part oscillating at the same frequency, the right-hand side of Eq. (16.32) becomes

$$\left(\frac{3\beta a^3}{4} + 2a\omega\delta\omega\right)\cos\omega_0 t + \alpha\left(a\cos\omega_0 t + \frac{\alpha a^2}{2}\frac{1}{\omega_0^2} - \frac{\alpha a^2}{2}\frac{1}{3\omega_0^2}\cos 2\omega_0 t\right)^2 + \cdots. \quad (16.40)$$

In making this step we have been justified in dropping some terms because they lead only to terms with higher powers of a than the terms being kept. To apply the Linstedt trick to this expression we need only isolate the term on the right-hand side that varies like $\cos\omega_0 t$ and set its coefficient to zero. In a later section a formula will be given that performs this extraction more neatly, but here it is simple enough to do it using easy trigonometric formulas. Completing this work, and setting the coefficient of the secular term to zero, we obtain

$$\omega = \omega_0 + \left(-\frac{3\beta}{8\omega_0} - \frac{5\alpha^2}{12\omega_0^2}\right)a^2. \quad (16.41)$$

One sees that "in a second approximation" the quadratic force term gives a term of the same order that the cubic force term gave in the first approximation. Here "the same order" means the same dependence on a. Of course, one or the other of α^2/ω_0 and β may dominate in a particular case.

Already at this stage the solution is perhaps adequate for most purposes. One has determined the frequency shift at amplitude a and, in Eq. (16.36) has obtained the leading "overtone" amplitude of the motion at "second harmonic frequencies" 2ω as as well as the "DC offset" at zero frequency. Since the third harmonic amplitude is proportional to a^3 it is likely to be negligible. But if it is not, the final term in Eq. (16.37) needs to be corrected. This is left as an exercise.

Problem 16.3.1. *Complete the second interaction step begun in the text in order to calculate to order a^3 the third harmonic response (at frequency 3ω) for the system described by Eq. (16.30).*

Let us consider the degree to which Eq. (16.36) (with its last term dropped or corrected) or any similar "solution" obtained in the form of a truncated Fourier series, "solves" the problem. By construction such a solution is perfectly periodic. But as the amplitude a increases the convergence of the Fourier series, which has only been hoped for, not proved, becomes worse. From our knowledge of motion in a one-dimensional potential we know that the true behavior at large a depends on the values of α and β. If $\beta = 0$ the potential energy becomes negative either on the left or on the right, depending on the sign of α. In this case our periodic solution will eventually become flagrantly wrong. On the other hand, if $\alpha = 0$ and $\beta < 0$ (in which case we have what is known as "Duffing's equation") the restoring force becomes arbitrarily large both on the left and on the right and the motion remains perfectly periodic. Even in this case, direct calculation of the total energy would show that it is only approximately conserved according to our solution. If we declare that periodicity is the essential "symplectic" feature in this case, then we might say that our solution satisfies symplecticity (by construction) but not energy conservation. This example illustrates the difficulty in characterizing the strong and weak points of any particular method of approximation.

Most methods of solution that derive power series solutions one term at a time do not converge. Like most valid statements in this area, there is a theorem by Poincaré to this effect. (There is however a method due to Kolmogorov, called "superconvergent perturbation theory," to be discussed later, that can yield convergent series.) It is not so much mathematical ineptitude that causes these procedures to not yield faithful solutions as it is the nature of the systems – most systems exhibit chaotic motion when the amplitude is great enough to cause the Fourier series convergence to deteriorate. (This last is a phenomenological observation, not a mathematical theorem.) In spite of all these reservations, solutions like Eq. (16.36) can describe the essential behavior of anharmonic systems.

16.4
The Method of Krylov and Bogoliubov

The method of Krylov and Bogoliubov (to be abbreviated here as "the K–B method") is probably the closest thing there is to a universal method for analyzing oscillatory systems, be they single- or multidimensional, harmonic or anharmonic, free or driven. The book by Bogoliubov and Mitropolsky listed at the end of the chapter is perhaps the best (and quite elementary) reference but, unfortunately, it is not easily available. The method starts by an exact change of variables resembling that of Section 14.3.3 and then continues by combining the variation of constants, averaging, and Linstedt methods described in two previous sections of this chapter. Perhaps its greatest defect (at least as the method is described here) is that it is not explicitly Hamiltonian. It is however based on action-angle variables, or rather on amplitude and phase variables, that are very much like action-angle variables.

It cannot be said that the K–B method is particularly illustrative of the geometric ideas that this text has chosen to emphasize. But the method does lend itself to qualitative description that is well motivated. In any case, every mechanics course should include study of this method. Furthermore it will be appropriate in a later section to compare a symplectic perturbation technique with the K–B method.

There is little agreement as to where credit is due. Even Bogoliubov and Mitropolsky credit Van der Pol for the method that is now commonly ascribed to Krylov and Bogoliubov. What is certain is that this school of Russians validated, expanded the range of applicability, and otherwise refined the procedures.

We will derive this method only in one dimension, but the method is easily extended to multiple dimensions. Since the method is so well-motivated and "physical" this extension is neither difficult nor dubious.

16.4.1
First Approximation

We continue to analyze oscillatory systems and assume that the motion is simple harmonic for sufficiently small amplitudes, so the equation of motion has the form

$$\frac{d^2x}{dt^2} + \omega_0^2 x = \epsilon f(x, dx/dt), \tag{16.42}$$

where $f(x, dx/dt)$ is an arbitrary perturbing function of position and velocity, and ϵ is a small parameter. The unperturbed motion can be expressed as

$$x = a\cos\Phi, \quad \dot{x} = -a\omega_0 \sin\Phi, \qquad \text{where} \qquad \Phi = \omega_0 t + \phi, \tag{16.43}$$

where ϕ is a constant, and perturbed motion will be later expressed in the same form. These equations can be regarded as a transformation $x, \dot{x} \to a, \Phi$ for which the inverse transformation is given by

$$a = \sqrt{x^2 + \frac{\dot{x}^2}{\omega_0^2}}, \quad \Phi = -\tan^{-1}\frac{\dot{x}}{\omega_0 x}. \tag{16.44}$$

The variables a and Φ are not essentially different from action and angle variable, but it will not be assumed that a is an adiabatic invariant. Since this will be a "variation of constants" method, the "constants" being a and ϕ, the motion in configuration and phase space will be as illustrated in Fig. 16.4. Since the parameter ω_0 will remain fixed it is not wrong to think of ϕ as an angle measured in a phase space that is rotating at constant angular velocity ω_0. Viewed in such a frame the system point moves slowly, both in radial position and in angle. We will continue to retain both ϕ and Φ. It is important to remember that they are redundant, always satisfying $\Phi = \omega_0 t + \phi$. But ϕ will be used to express the argument of "slowly varying" functions while Φ will be the argument of "rapidly varying" functions.

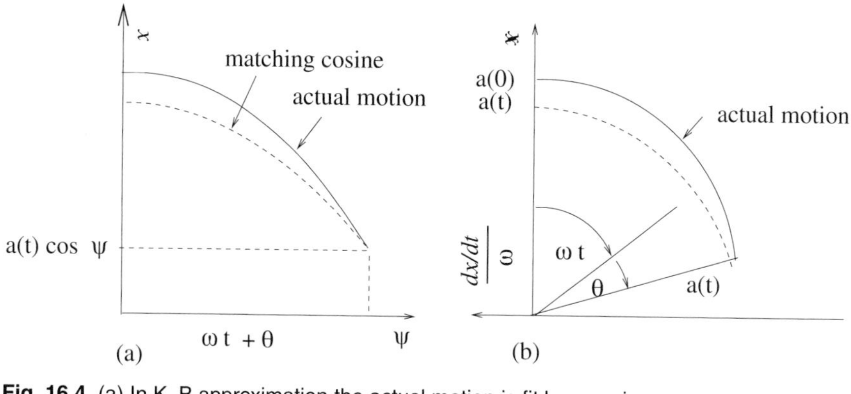

Fig. 16.4 (a) In K–B approximation the actual motion is fit by a cosine function modulated by amplitude $a(t)$. (b) The angle in (normalized) phase space advances as $\omega_0 t + \phi(t)$ where ω_0 is constant and $\phi(t)$ varies slowly.

Equation (16.42) can be transformed into two first-order equations for a and Φ. Differentiating the first of Eqs. (16.44) and re-substituting from Eq. (16.42) yields

$$\dot{a} = \frac{1}{a}\left(x\dot{x} + \frac{\dot{x}\ddot{x}}{\omega_0^2}\right) = -\frac{\epsilon}{\omega_0}\sin\Phi\, f(a\cos\Phi, -a\omega_0\sin\Phi). \tag{16.45}$$

The arguments of the function f have also been re-expressed in terms of a and Φ. Since this function will appear so frequently in this form, we abbreviate it as $F(a,\Phi) \equiv f(a\cos\Phi, -a\omega_0\sin\Phi)$. An expression like (16.45) can also be found for $\dot{\Phi}$. Together we have[2]

$$\begin{aligned}\dot{a} &= -\frac{\epsilon}{\omega_0}\sin\Phi\, F(a,\Phi) \equiv \epsilon G(a,\phi),\\ \dot{\Phi} &= \omega_0 - \frac{\epsilon}{a\omega_0}\cos\Phi\, F(a,\Phi) \equiv \omega_0 + \epsilon H(a,\phi).\end{aligned} \tag{16.46}$$

These are exact equations. They are said to be "in standard form." They have much the same character as Eqs. (14.57), but here we are dealing with an autonomous equation with the perturbation expressed as a direct drive. There we were dealing with a nonautonomous system with the perturbation expressed as a parametric drive. It is nevertheless natural to contemplate approximating the equations by averaging the right hand sides, for Φ ranging from 0 to 2π. This yields

$$\begin{aligned}&\dot{a} \approx <\dot{a}> = \epsilon G_{\text{av}}(a),\\ &\qquad\text{where } G_{\text{av}}(a) = -\frac{1}{2\pi\omega_0}\int_0^{2\pi} F(a,\Phi)\sin\Phi\, d\Phi,\\ &\dot{\Phi} \approx <\dot{\Phi}> = \omega_0 + \epsilon H_{\text{av}}(a),\\ &\qquad\text{where } H_{\text{av}}(a) = -\frac{1}{2\pi a\omega_0}\int_0^{2\pi} F(a,\Phi)\cos\Phi\, d\Phi.\end{aligned} \tag{16.47}$$

These equations constitute "the first K–B approximation." They are ordinary differential equations of especially simple form. Since the first depends only on a it can be solved by quadrature. Then the second can be solved by integration.

Example 16.4.1. Conservative forces. *If the force is derivable from a potential, then $f(x, dx/dt)$ is, in fact, independent of dx/dt. In this case we have*

$$G_{\text{av}}(a) = -\frac{1}{2\pi\omega_0}\int_{-\pi}^{\pi} f(a\cos\Phi)\sin\Phi\, d\Phi = 0, \tag{16.48}$$

because the integrand is an odd function of Φ. The first of Eqs. (16.47) then implies that a is constant – a gratifying result. The second of Eqs. (16.47) then yields

$$\omega_1(a) \equiv \dot{\Phi} = \omega_0 - \frac{1}{2\pi a\omega_0}\int_0^{2\pi} f(a\cos\Phi)\cos\Phi\, d\Phi. \tag{16.49}$$

Here the frequency at amplitude a has been expressed by $\omega_1(a)$ where the subscript indicates "first K–B approximation." For a gravity pendulum with natural frequency

2) The functions $G(a,\phi)$ and $H(a,\phi)$ are introduced primarily for later convenience. Since Hamiltonian methods are not being employed there should be no danger that $H(a,\phi)$ will be interpreted as a Hamiltonian.

$\omega_0 = \sqrt{g/l}$ the equation of motion for the angle x is

$$\ddot{x} = -\omega_0^2 \sin x \approx -\omega_0^2 \left(x - \frac{x^3}{6} \right). \tag{16.50}$$

We have $F(a, \Phi) = \omega_0^2 a^3 \cos^3 \Phi / 6$ and the equations in standard form are

$$\dot{a} = -\epsilon \frac{\omega_0\, a^3}{6} \sin \Phi \cos^3 \Phi, \qquad \dot{\Phi} = \omega_0 - \epsilon \frac{\omega_0\, a^2}{6} \cos^4 \Phi. \tag{16.51}$$

Averaging the second equation and setting $\epsilon = 1$,

$$\omega_1(a) = \omega_0 \left(1 - \frac{a^2}{16} \right). \tag{16.52}$$

This dependence on amplitude makes it important that pendulum clocks run at constant amplitude if they are to keep accurate time. The importance of this consideration and the quality of the approximation can be judged from the following table.

Radians	Degrees	$\omega_1(a)/\omega_0$	$\omega_{\text{exact}}(a)/\omega_0$
0.0	0.0	1.0	1.0
1.0	57.3	0.938	0.938
2.0	114.6	0.75	0.765
3.0	171.9	0.438	0.5023

Example 16.4.2. Van der Pol oscillator. *Consider the equation*

$$L \frac{d^2Q}{dt^2} + (-|R| + cQ^2) \frac{dQ}{dt} + \frac{Q}{C} = 0. \tag{16.53}$$

The parameters in this equation (for charge Q on capacitor C) have obviously been chosen to suggest an electrical LRC circuit. For small Q we can neglect the term $cQ^2 dQ/dt$. But the term $-|R| dQ/dt$ has the wrong sign to represent the effect of a resistor in the circuit. Stated differently, the resistance in the circuit is negative. Normally the effect of a resistor is to damp the oscillations which would otherwise be simple harmonic. With negative resistance one expects (and observes) growth. In fact the circuit should spring into oscillation, even starting from $Q = 0$ (because of inevitable tiny noise terms not shown in the equation) followed by steady growth. But with growth it will no longer be valid to neglect the $cQ^2 dQ/dt$ term. This term has the "correct" sign for a resistance and at sufficiently large amplitude it "wins." We anticipate some compromise therefore between the growth due to one term and the damping due to the other.

This system, known as the "Van der Pol" oscillator, is readily analyzed using the K–B method (and was done so by Van der Pol, early in the twentieth century). Eliminating superfluous constants the perturbing term becomes

$$f\left(x, \frac{dx}{dt} \right) = (1 - x^2) \frac{dx}{dt}, \quad \text{or} \quad F(a, \Phi) = -a\omega_0 (1 - a^2 \cos^2 \Phi) \sin \Phi, \tag{16.54}$$

and the equations in standard form are

$$\dot{a} = \epsilon\, a\,(1 - a^2 \cos^2\Phi)\, \sin^2\Phi,$$
$$\dot{\Phi} = \omega_0 + \epsilon(1 - a^2\cos^2\Phi)\, \sin\Phi\cos\Phi. \qquad (16.55)$$

After averaging these become

$$\dot{a} = \epsilon\frac{a}{2}\left(1 - \frac{a^2}{4}\right),$$
$$\dot{\Phi} = \omega_0. \qquad (16.56)$$

Problem 16.4.1. *By solving Eqs. (16.56) show that, for a Van der Pol oscillator starting with amplitude a_0 in the range $0 < a_0 < 2$, the motion is given by*

$$x = \frac{2}{\sqrt{1 + \frac{4-a_0^2}{a_0^2}e^{-\epsilon t}}}\cos(\omega_0 t - \phi_0), \qquad (16.57)$$

which inexorably settles to pure harmonic oscillation at $a = 2$ after a time long compared to $1/\epsilon$. For $2 < a_0$ the solution settles to the same amplitude with the same time constant.

According to the previous problem the motion settles to a "limit cycle" at $a = 2$ independent of its starting amplitude. The following graph of "growth rate" da/dt makes it clear that this result could have been expected. Only at $a = 0$ and at $a = 2$ is $da/dt = 0$ and only at $a = 2$ is the sign "restoring." For amplitudes in the vicinity of $a = 2$ it is sound to approximate da/dt by a straight line. Then one obtains

$$\frac{da}{dt} = \epsilon(2 - a), \quad \text{and} \quad a = 2 - (2 - a_0)e^{-\epsilon t}. \qquad (16.58)$$

The growth rate is plotted in Fig. 16.5, which also shows the linear approximation.

16.4.2
Equivalent Linearization

We have seen that the first K–B approximation accounts fairly accurately for some of the most important nonlinear aspects of oscillators, such as amplitude dependence of frequency and limit cycles. Since autonomous linear equations do not exhibit oscillation it can be said that *autonomous oscillators are inherently nonlinear*. Unfortunately this takes away from us our best tool – the ability to solve linear equations. For multidimensional systems this problem is especially acute. In this section we study the method of "equivalent linearization" that is based on the K–B approximation (or similar methods) and imports

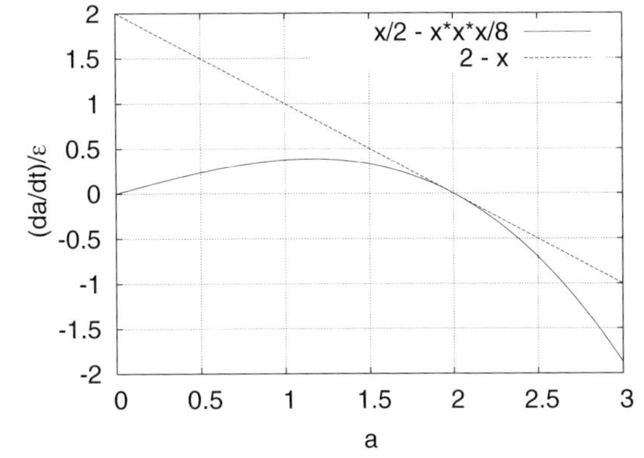

Fig. 16.5 Graph of $(1/\epsilon)da/dt$ for Van der Pol oscillator in the lowest K–B approximation and its approximation near $a = 2$.

much of the effect of nonlinearity into a description using linear equations. Nowadays such approaches find their most frequent application in the design of electrical circuits. Such circuits can have many independent variables and it is attractive to be able to apply linear circuit theory even when some of the branches of the circuit are weakly nonlinear.

Consider again Eq. (16.42) which we rewrite with slightly modified coefficients, intended to suggest "mass and spring;"

$$m\frac{d^2x}{dt^2} + k\,x = \epsilon f(x, dx/dt). \tag{16.59}$$

The small amplitude frequency is $\omega_0 = \sqrt{k/m}$ and the nonlinear forces are contained in $f(x, dx/dt)$. We define an "equivalent system" to be one for which the equation of motion is

$$m\frac{d^2x}{dt^2} + \lambda_e(a)\frac{dx}{dt} + k_e(a)\,x = 0, \qquad \text{and} \qquad \omega_e^2(a) = \frac{k_e(a)}{m}. \tag{16.60}$$

It is not quite accurate to say that this is a "linear equation" since the parameters depend on amplitude a. But if a is approximately constant (and known) this may provide an acceptable level of accuracy. By applying the K–B approximation we find that the two equations "match," as regards their formulas for $\dot{a}$ and $\dot{\Phi}$, if we define the "equivalent damping coefficient" $\lambda_e(a)$ and the "equivalent spring constant" $k_e(a)$ by copying from Eqs. (16.47)

$$\begin{aligned} \lambda_e(a) &= \frac{\epsilon}{\pi a \omega_0}\int_0^{2\pi} F(a,\Phi)\sin\Phi\, d\Phi, \\ k_e(a) &= k - \frac{\epsilon}{\pi a}\int_0^{2\pi} F(a,\Phi)\cos\Phi\, d\Phi \equiv k + k_1(a). \end{aligned} \tag{16.61}$$

These formulas are equivalent to making in Eq. (16.59) the replacement

$$\epsilon f(x, dx/dt) \rightarrow -k_1(a)\, x - \lambda_e(a)\frac{dx}{dt}, \tag{16.62}$$

and the averaged equations are

$$\frac{da}{dt} = -\frac{\lambda_e(a)}{2m}\,a \quad \text{and} \quad \frac{d\Phi}{dt} = \omega_e(a). \tag{16.63}$$

The fractional change in amplitude after one period, $(\lambda_e/\omega_e)(\pi/m)$ is sometimes known as the "damping decrement."

16.4.3 Power Balance, Harmonic Balance

If we wish we can interpret Eq. (16.59) as describing the interplay of an "agent" providing force $\epsilon\, f(x, dx/dt)$ and acting on a linear system described by the terms on the left-hand side of the equation. The work done by the agent during one period of duration $T = 2\pi/\omega_0$ is given by

$$\int_0^T \epsilon f(x, dx/dt)\,\frac{dx}{dt}\,dt = -\epsilon\omega_0 a \int_0^{2\pi} F(a,\Phi)\,\sin\Phi\, d\Phi. \tag{16.64}$$

Our "equivalent agent" provides force $-k_1(a)x - \lambda_e(a)dx/dt$, and hence does an amount of work per cycle given by

$$-k_1(a)\int_0^T x\frac{dx}{dt}\,dt - \lambda_e(a)\int_0^T \left(\frac{dx}{dt}\right)^2 dt. \tag{16.65}$$

The first term here gives zero and the second gives $-\pi a^2\omega_0\lambda_e(a)$. Equating the results of these two calculations we recover the first of Eqs. (16.61). The expression $\epsilon f(x, dx/dt)(dx/dt)$ is the *instantaneous power* dissipated in lossy elements and we have matched the *average power* dissipated in the equivalent agent to that of the actual agent. To obtain $k_e(a)$ by a similar argument it is necessary to define average *reactive power* by $\epsilon f(x, dx/dt)x/T$. The equivalent parameters can then be said to have been determined by the "principle of power balance."

Another (and equivalent) approach to establishing an "equivalent" linear model is to express the function $F(a,\Phi)$ as a Fourier series;

$$F(a,\Phi) = \frac{1}{2}g_0(a) + \sum_{n=1}^{\infty} g_n(a)\cos n\Phi + \sum_{n=1}^{\infty} h_n(a)\sin n\Phi. \tag{16.66}$$

The coefficients in this expansion are given by

$$g_n = \frac{1}{\pi}\int_0^{2\pi} F(a,\Phi)\,\cos n\Phi\, d\Phi, \qquad h_n = \frac{1}{\pi}\int_0^{2\pi} F(a,\Phi)\,\sin n\Phi\, d\Phi. \tag{16.67}$$

The "in phase," "fundamental component" of force is therefore given by

$$\frac{\epsilon}{\pi}\cos\Phi \int_0^{2\pi} F(a,\Phi')\,\cos n\Phi'\,d\Phi' = -k_1(a)\,a\cos\Phi, \tag{16.68}$$

where the defining Eq. (16.61) for $k_1(a)$ has been employed. This is equal to the in-phase portion of the "equivalent" force. The out-of-phase term can be similarly confirmed. This is known as the "principle of harmonic balance."

16.4.4 Qualitative Analysis of Autonomous Oscillators

From the analysis of the Van der Pol oscillator, and especially from Fig. 16.5, it is clear that much can be inferred about the qualitative behavior of an oscillator from the equation:

$$\frac{da}{dt} = \epsilon G(a). \tag{16.69}$$

The function $G(a)$ may be approximated by G_{av} obtained using the first or higher K–B approximation or even phenomenologically. Points a_e at which $G(a_e) = 0$ are especially important because $da/dt = 0$ there, but it is not *a priori* known whether this "equilibrium" is stable or unstable. The linearized dependence on deviation from equilibrium δa is given by

$$G(a_e + \delta a) = G'(a_e)\,\delta a. \tag{16.70}$$

As in Eq. (16.58), it is clear that an initial deviation $\delta a|_0$ evolves according to

$$\delta a = \delta a|_0\, e^{\epsilon G'(a_e)t}. \tag{16.71}$$

Stability is therefore governed by the sign of $G'(a_e)$.

Some possible oscillator profiles are illustrated in Fig. 16.6. In every case $G(a)$ becomes negative for sufficiently large a since, otherwise, infinite amplitude oscillation would be possible. Points where the curve crosses the horizontal axis are possible equilibrium points, but only those with negative slope are stable and this is indicated by arrows that indicate the direction of system evolution. Stability at the origin is a bit special in that it depends on the sign of $G(0)$ rather than the sign of $G'(0)$. In case (d) the system springs into oscillation spontaneously and evolves to the first zero crossing. In case (c), stable oscillation is possible at the second zero crossing but the system cannot proceed there spontaneously from the origin because the slope at the origin is negative. In case (a) the origin is stable and in case (b), like the Van der Pol oscillator, the system moves spontaneously to the first zero crossing (after the origin).

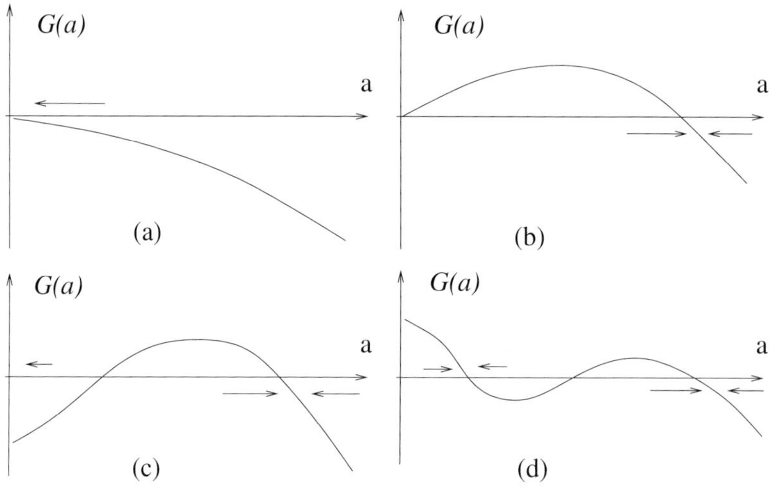

Fig. 16.6 Growth-rate profiles $G(a)$ for various autonomous oscillators. Arrows indicate directions of progress toward stable points or stable limit cycles.

The sorts of behavior that are possible can be discussed in connection with a slightly generalized version of the Van der Pol oscillator: pictorially in Fig. 16.7 and analytically. Let its equation of motion be

$$\frac{d^2x}{dt^2} + (\lambda_1 + \lambda_3\, x^2 + \lambda_5\, x^3)\frac{dx}{dt} + \omega_0^2 x = 0. \tag{16.72}$$

The coefficient of dx/dt could also have even powers, but it is only the odd powers that contribute to da/dt in first K–B approximation. The first of Eqs. (16.47) yields

$$\frac{da}{dt} = \epsilon G_{\text{av}}(a) = -\frac{\lambda_1\, a}{2} - \frac{\lambda_3\, a^3}{8} - \frac{\lambda_5\, a^5}{16}. \tag{16.73}$$

Let us assume that $\lambda_1 > 0$ so that self-excitation is absent. Other than the root at the origin, the zeros of $G_{\text{av}}(a)$ are given by

$$a^2 = -\frac{\lambda_3}{\lambda_5} \pm \sqrt{\left(\frac{\lambda_3}{\lambda_5}\right)^2 - 8\,\frac{\lambda_1}{\lambda_5}}. \tag{16.74}$$

Points at which a qualitative feature of the motion undergoes discontinuous change are known as points of "bifurcation." Assuming the first term is positive, the condition establishing a bifurcation point, as any one of the parameters is varied, is that the square root term vanishes;

$$\lambda_3^2 = 8\lambda_1\lambda_5. \tag{16.75}$$

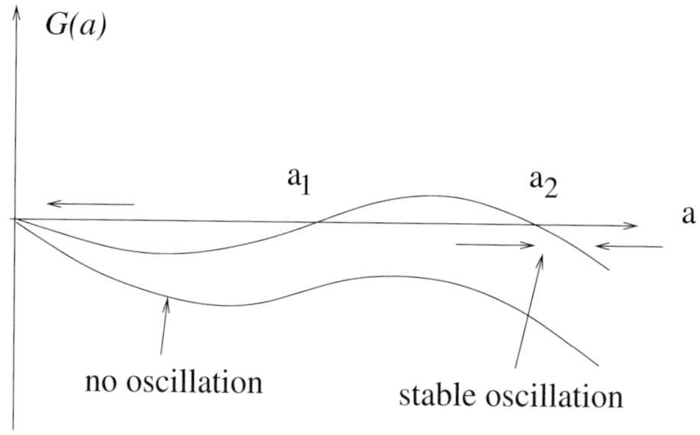

Fig. 16.7 Small change of a parameter can move the system curve from the lower, nonoscillatory case, to the upper curve that indicates the possibility of stable oscillation at amplitude a_1. "Bifurcation" between these states occurs when $a_1 = a_2$.

Rather than having multiple parameters the qualitative behavior of the oscillator can be more clearly understood if one dominant "control parameter" or "stabilizing parameter," call it μ, is singled out. Suppose $G(a)$ is given by

$$G(a) = -\frac{a}{\mu} + G_r(a), \tag{16.76}$$

where the relative strength of a leading term is regarded as externally controllable via the parameter μ. Small μ corresponds to very negative $G(a)$ and no possibility of oscillation. In Fig. 16.8, the separate terms of Eq. (16.76) are plotted (with the sign of the control term reversed). Different control parameter values are expressed by different straight lines from the origin and, because of the negative sign in Eq. (16.76), stability is governed by whether and where the straight line intersects the curve of $G_r(a)$.

If the curve of $G_r(a)$ at the origin is concave downward, as in part (a) of the figure, then, as μ is increased, when the initial slope of the control line

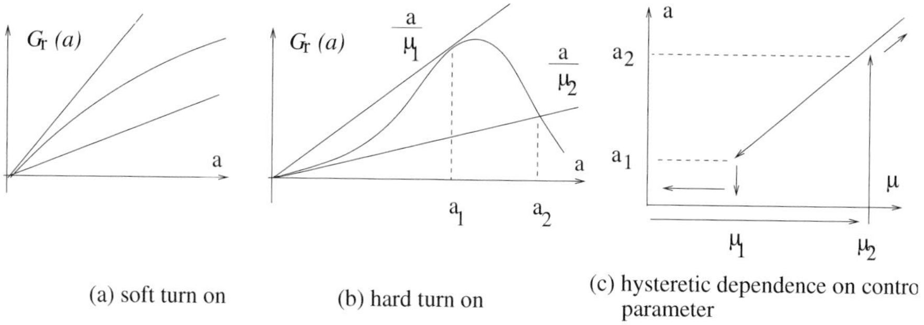

Fig. 16.8 Characteristics of autonomous oscillator exhibiting hysteretic turn on and extinction.

matches that of $G_r(a)$, the oscillator is self-excited and settles to the first intersection point. This is known as "soft turn-on." But if the curve of $G_r(a)$ at the origin is concave upward, as in the (b) figure, as μ is increased from very small values, a point is reached at which self-sustaining oscillation would be possible but does not if fact occur because the origin remains stable. This point is indicated by μ_1 in the (b) figure. As μ is increased further a point μ_2 is reached where the origin is unstable and the system undergoes "hard turn-on" and continues to oscillate at the large amplitude a_2. From this point, if μ is increased the amplitude increases. Furthermore, if μ is reduced only modestly, the amplitude will follow down below a_2 without extinguishing. But when μ is dropped below μ_1 the oscillator turns off suddenly. The overall "hysteresis cycle" is illustrated in the (c) figure. It is beyond the capability of this model to describe the turn-on and turn-off in greater detail, but the gross qualitative behavior is given.

Problem 16.4.2. *A grandfather clock keeps fairly regular time because it oscillates at constant amplitude but, as lossless as its mechanism can be, it still has to be kept running by external intervention and this can affect its rate. For high precision, its amplitude has to be kept constant. A "ratchet and pawl" or "escapement" mechanism by which gravitational energy is imparted to the pendulum to make up for dissipation is illustrated schematically in Fig. 16.9. This mechanism administers a small impulse I, once per cycle, at an approximately optimal phase in the cycle. An equation of motion for a system with these properties is*

$$m\frac{d^2x}{dt^2} + \lambda\frac{dx}{dt} - I\frac{\frac{dx}{dt} + \left|\frac{dx}{dt}\right|}{2}\delta(x - x_0) + kx = 0, \tag{16.77}$$

where the δ-function controls the phase of the impulse and the other factor in the term proportional to I assures that the impulse occurs on only one or the other of the "back" and "forth" trips. For "too small" amplitude, the K–B approximation yields $da/dt = -(\lambda/2m)a$ and the clock stops. Find the amplitude x_0 such that the clock continues to run if $a > x_0$. Find the condition on x_0, I, and λ which must be satisfied for the clock to keep running if the pendulum is started with initial amplitude exceeding x_0. In the same approximation find the dependence on a of the frequency of oscillation.

16.4.5
Higher K–B Approximation

This section follows closely R.L. Stratonovich, *Topics in the theory of random noise*, Vol. II, p. 97. However the concentration on random processes (though lucidly explained in this remarkable book) would probably be disconcerting to someone interested only in mechanics. According to Stratonovich the procedure for proceeding to higher approximation is due to Bogoliubov.

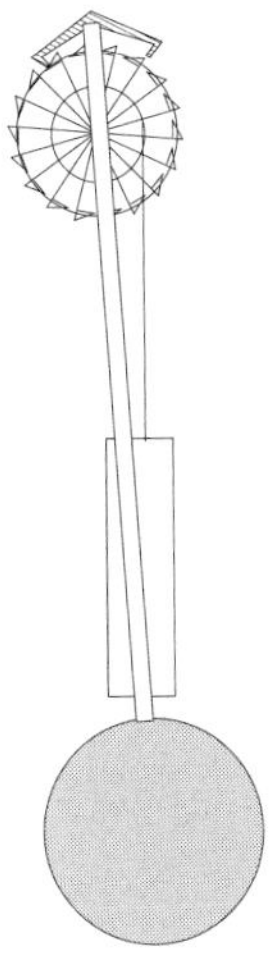

Fig. 16.9 Grandfather clock with "escapement mechanism" exhibited.

Proceeding to an improved approximation in solving Eq. (16.42) may be necessary, especially if higher harmonics are to be accurately evaluated. Since this discussion is somewhat complicated, to make this section self-contained, we will rewrite some equations rather than referring to their earlier versions. The solution is sought in the form

$$x(t) = a\cos\Phi. \tag{16.78}$$

and the phase is separated into "fast" and "slow" parts;

$$\Phi = \omega_0\, t + \phi. \tag{16.79}$$

The equations satisfied by a and ϕ are

$$\frac{da}{dt} = \epsilon\, G(a,\phi), \qquad \frac{d\phi}{dt} = \epsilon\, H(a,\phi), \tag{16.80}$$

where $G(a,\phi)$ and $H(a,\phi)$ are known functions, appearing in the "equations in standard form," Eqs. (16.46). If the system were nonautonomous these functions would also depend explicitly upon t. The following development would still proceed largely unchanged but we will simplify by restricting discussion to autonomous systems.

To solve these equations we anticipate transforming the variables $(a,\Phi) \to (a^*,\Phi^*)$ according to

$$a = a^* + \epsilon\, u(a^*,\Phi^*), \qquad \Phi = \Phi^* + \epsilon\, v(a^*,\Phi^*). \tag{16.81}$$

For the time being the functions u and v are arbitrary. Later they will be chosen to simplify the equations. The "small parameter" ϵ will be used to keep track of the "order" of terms. Corresponding to Φ^* we define also ϕ^* related as in Eq. (16.79);

$$\Phi^* = \omega_0 t + \phi^* \qquad \text{and hence} \qquad \dot{\Phi}^* = \omega_0 + \dot{\phi}^*. \tag{16.82}$$

(From here on time derivatives will be indicated with dots, as here.) The equations of motion will be assumed to have the same form as in Eq. (16.80)

$$\dot{a}^* = \epsilon\, G^*(a^*, \phi^*), \qquad \dot{\phi}^* = \epsilon\, H^*(a^*, \phi^*), \tag{16.83}$$

so the new functions G^* and H^* will also have to be found. Since Eqs. (16.80) are to be satisfied by values of a and Φ given by Eqs. (16.81) we must have

$$\begin{aligned} \dot{a} &= \epsilon\, G\big(a^* + \epsilon\, u(a^*, \Phi^*), \phi^* + \epsilon\, v(a^*, \Phi^*)\big), \\ \dot{\phi} &= \epsilon\, H\big(a^* + \epsilon\, u(a^*, \Phi^*), \phi^* + \epsilon\, v(a^*, \Phi^*)\big). \end{aligned} \tag{16.84}$$

These are the same as Eqs. (16.80) except the arguments are expressed in terms of the new variables. They are exact. From here on it will be unnecessary to exhibit arguments explicitly since the arguments of G^* and H^* will always be (a^*, ϕ^*) and the arguments of u and v will always be (a^*, Φ^*). (Since Φ^* and ϕ^* are equivalent variables, the distinction in arguments here is essentially cosmetic; the rationale behind the distinction should gradually become clear.)

There is an alternate way of determining the quantities appearing on the left-hand side of Eqs. (16.84). It is by time-differentiating equations (16.81) and using Eqs. (16.83);

$$\begin{aligned} \dot{a} &= \dot{a}^* + \epsilon\, \dot{u} = \epsilon\, G^* + \epsilon\, \frac{\partial u}{\partial a^*}\, G^* + \epsilon\, \frac{\partial u}{\partial \Phi^*}(\omega_0 + \epsilon\, H^*), \\ \dot{\phi} &= \dot{\phi}^* + \epsilon\, \dot{v} = \epsilon\, H^* + \epsilon\, \frac{\partial v}{\partial a^*}\, G^* + \epsilon\, \frac{\partial v}{\partial \Phi^*}(\omega_0 + \epsilon\, H^*). \end{aligned} \tag{16.85}$$

Equating Eqs. (16.84) and (16.85) we obtain

$$\begin{aligned} G^* + \omega_0 \frac{\partial u}{\partial \Phi^*} &= G(a^* + \epsilon\, u, \phi^* + \epsilon\, v) - \epsilon\, \frac{\partial u}{\partial a^*}\, G^* - \epsilon\, \frac{\partial u}{\partial \Phi^*}\, H^*, \\ H^* + \omega_0 \frac{\partial v}{\partial \Phi^*} &= H(a^* + \epsilon\, u, \phi^* + \epsilon\, v) - \epsilon\, \frac{\partial v}{\partial a^*}\, G^* - \epsilon\, \frac{\partial v}{\partial \Phi^*}\, H^*. \end{aligned} \tag{16.86}$$

These are exact functional identities; that is, they are true for arbitrary functions u and v. But terms have been grouped with the intention of eventually exploiting the smallness of ϵ. This is a "high frequency approximation" in that terms proportional to ω_0 are not multiplied by ϵ. u and v will be determined next.

We assume that all functions are expanded in powers of ϵ;

$$\begin{aligned} G^* &= G_1^* + \epsilon\, G_2^* + \cdots, \qquad & H^* &= H_1^* + \epsilon\, H_2^* + \cdots, \\ u &= u_1 + \epsilon\, u_2 + \cdots, \qquad & v &= v_1 + \epsilon\, v_2 + \cdots. \end{aligned} \tag{16.87}$$

Since all the functions that have been introduced have to be periodic in Φ^* one is to imagine that all have also been expanded into Fourier series. Then averaging over one period amounts to extracting the term in the Fourier series that is independent of Φ^*. The guidance in determining the functions u_i and v_i is that they are to contain all the terms that depend on Φ^* and only those terms. According to Eqs. (16.81), the quantities a^* and ϕ^* will then contain no oscillatory factors. Then, because of Eq. (16.83), the terms G_i^* and H_i^* will also be independent of Φ^*. That this separation is possible will be demonstrated by construction. The formalism has been constructed so that, at each stage, Φ^*-dependent terms enter with an extra power of ϵ because u and v entered with a multiplicative factor ϵ. This is also legitimized constructively, but the overall convergence of the process is only conjectural.

Since all functions are Fourier series it is too complicated to make these procedures completely explicit, but all functions can be determined sequentially using Eq. (16.85). Since these equations contain only derivatives of u and v only those derivatives will be determined directly. But the antiderivatives of terms in a Fourier series are easy – the antiderivatives of $\sin r\Phi$ and $\cos r\Phi$ are $-\cos r\Phi\,/r$ and $\sin r\Phi\,/r$. Since all coefficients will be functions of a^* it will be necessary to evaluate the antiderivatives of the terms $\partial u_i/\partial a^*$ and $\partial v_i/\partial a^*$ to obtain the u_i and v_i functions themselves. All this is fairly hard to describe but fairly easy to accomplish. It is easiest to understand by example.

Substituting Eq. (16.87) into Eq. (16.86) and setting $\epsilon = 0$ we obtain the first approximation;

$$
\begin{aligned}
G_1^* + \omega_0 \frac{\partial u_1}{\partial \Phi^*} &= G(a^*, \phi^*),\\
H_1^* + \omega_0 \frac{\partial v_1}{\partial \Phi^*} &= H(a^*, \phi^*).
\end{aligned} \tag{16.88}
$$

The functions on the right-hand side are unambiguous since G and H are the functions we started with, but with "old" arguments replaced by "new" arguments. We separate these equations into Φ^*-independent terms

$$
\begin{aligned}
G_1^* &= \langle G(a^*, \phi^*)\rangle,\\
H_1^* &= \langle H(a^*, \phi^*)\rangle,
\end{aligned} \tag{16.89}
$$

and Φ^*-dependent terms

$$
\begin{aligned}
\omega_0 \frac{\partial u_1}{\partial \Phi^*} &= \langle\!\langle G(a^*, \phi^*)\rangle\!\rangle,\\
\omega_0 \frac{\partial v_1}{\partial \Phi^*} &= \langle\!\langle H(a^*, \phi^*)\rangle\!\rangle,
\end{aligned} \tag{16.90}
$$

where the *ad hoc* notation $\langle\!\langle\,\rangle\!\rangle$ stands for the bracketed quantity after constant terms have been removed.

Before continuing we illustrate using the Van der Pol oscillator as example. From the equations in standard form, Eq. (16.55), after using some trigonometric identities to convert them to Fourier series, we have

$$\begin{aligned} G(a,\phi) &= \frac{a}{2} - \frac{a^3}{8} - \frac{a}{2}\cos 2\Phi + \frac{a^3}{8}\cos 4\Phi, \\ H(a,\phi) &= \left(\frac{1}{2} - \frac{a^2}{4}\right)\sin 2\Phi - \frac{a^2}{8}\sin 4\Phi. \end{aligned} \tag{16.91}$$

Applying Eq. (16.89) we obtain

$$G_1^* = \frac{a}{2}\left(1 - \frac{a^2}{4}\right), \qquad H_1^* = 0, \tag{16.92}$$

which recovers the result Eq. (16.56) obtained in the first K–B approximation. Applying Eq. (16.90) and integrating we also obtain

$$\begin{aligned} u_1 &= -\frac{a}{4\omega_0}\sin 2\Phi + \frac{a^3}{32\omega_0}\sin 4\Phi, \\ v_1 &= -\frac{1}{4\omega_0}\left(1 - \frac{a^2}{2}\right)\cos 2\Phi + \frac{a^2}{32\omega_0}\cos 4\Phi. \end{aligned} \tag{16.93}$$

All that remains in this order of approximation is to substitute these into Eqs. (16.81) and from there into Eq. (16.78) to obtain the harmonic content of the self-sustaining oscillations.

We will show just one more step, namely the equations corresponding to Eq. (16.88) in the second approximation.

$$\begin{aligned} G_2^* + \omega_0\frac{\partial u_2}{\partial \Phi^*} &= \frac{\partial G}{\partial a}u_1 + \frac{\partial G}{\partial \phi}v_1 - \frac{\partial u_1}{\partial a^*}G_1^* - \frac{\partial u_1}{\partial \Phi^*}H_1^*, \\ H_2^* + \omega_0\frac{\partial v_2}{\partial \Phi^*} &= \frac{\partial H}{\partial a}u_1 + \frac{\partial H}{\partial \phi}v_1 - \frac{\partial v_1}{\partial a^*}G_1^* - \frac{\partial v_1}{\partial \Phi^*}H_1^*. \end{aligned} \tag{16.94}$$

All functions required are available from the previous step and the separation is performed the same way.

Problem 16.4.3. *Complete the next iteration step in the Krylov–Bogoliubov analysis of the Van der Pol oscillator. That is to say, complete Eqs. (16.94) and perform the separation into constant and varying terms. Show that* $G_2^* = 0$ *and evaluate* H_2^**. Write Eqs. (16.56) with the newly-calculated term included.*

Problem 16.4.4. *Find the term proportional to* a^4 *in the amplitude dependence of the ordinary gravity pendulum. In other words, extend Eq. (16.52) to one more term.*

16.5 Superconvergent Perturbation Theory

Because transformations based on generating functions, such as have been discussed in Section 14.1.1, are automatically symplectic, there has been a strong historical tendency to base perturbation schemes on this type of transformation. G.D. Birkoff was the leader of the successive canonical transformation approach. His book, *Dynamical Systems*, reprinted in 1991 by the American Mathematical Society, is both important and readable. The down side of this approach, as has been noted previously, is that it mixes old and new variables, giving implicit, rather than explicit transformation formulas. The only systematic way to obtain explicit formulas is by the use of series expansion. When one truncates such series (as one always must) one loses the symplecticity that provided the original motivation for the method. It is my opinion therefore, that the more "direct" methods described to this point are more valuable than this so-called canonical perturbation theory.

There is, however, a theoretically influential development, due to Kolmogorov, and known as "superconvergent perturbation theory," based on this approach. This is the basis for Kolmogorov's name being attached to the important "KAM" or "Kolmogorov, Arnold, Moser" theorem.

16.5.1 Canonical Perturbation Theory

For this discussion, we return to a one dimensional, oscillatory system, described by Hamiltonian

$$H(q,p) = H_0 + H_1, \tag{16.95}$$

where the term H_1 is the perturbation. We assume that the unperturbed system for Hamiltonian H_0 has been solved using the action/angle approach described in Section 14.3.3. When described in terms of action variable I_0 and angle variable φ_0 the unperturbed system is described by the relations

$$H = H_0(I_0), \qquad \omega_0 = \frac{\partial H_0}{\partial I_0}, \qquad \varphi_0 = \omega_0 t + \text{constant}. \tag{16.96}$$

When q is expressed in terms of I_0 and φ_0 and substituted into the function H_1 the result $H_1(\varphi_0, I_0)$ is periodic in φ_0 with period 2π; it can therefore be expanded in a Fourier series, much as was done on the right-hand side of Eq. (16.32);

$$H_1(\varphi_0, I_0) = \sum_{k=-\infty}^{\infty} h_k^{(0)}(I_0) e^{ik\varphi_0}. \tag{16.97}$$

To "simplify" the perturbed system we now seek a generating function $S(\varphi_0, I_1)$ to be used in a transformation from "old" variables I_0 and φ_0 to "new" variables I_1 and φ_1 that are action/angle variables of the perturbed system. The generating function for this transformation has the form

$$S(\varphi_0, I_1) = \varphi_0 I_1 + \Phi(\varphi_0, I_1). \tag{16.98}$$

According to Eqs. (14.13), the generated transformation formulas are then given by

$$\begin{aligned} I_0 &= I_1 + \frac{\partial \Phi(\varphi_0, I_1)}{\partial \varphi_0}, \\ \varphi_1 &= \varphi_0 + \frac{\partial \Phi(\varphi_0, I_1)}{\partial I_1}; \end{aligned} \tag{16.99}$$

the second terms are of lower order than the first terms. Substituting into Eq. (16.95), the new Hamiltonian is

$$\begin{aligned} H &= H_0\left(I_1 + \frac{\partial \Phi(\varphi_0, I_1)}{\partial \varphi_0}\right) + H_1(\varphi_0, I_0) \\ &= H_0(I_1) + \left(\frac{\partial H_0}{\partial I_0}\frac{\partial \Phi}{\partial \varphi_0}(\varphi_0, I_1) + H_1(\varphi_0, I_1)\right) + \cdots \\ &= H_0(I_1) + \langle H_1(\varphi_0, I_1)\rangle + \omega_0 \frac{\partial \Phi}{\partial \varphi_0}(\varphi_0, I_1) + \langle\!\langle H_1(\varphi_0, I_1)\rangle + \cdots . \end{aligned} \tag{16.100}$$

Here we have used the same notation for averaging that was used in Eqs. (16.90); operating on a periodic function $\langle\ \rangle$ yields the average and $\langle\!\langle\ \rangle$ yields what is left over. It has been unnecessary to distinguish between I_0 and I_1 where they appear as arguments in terms of reduced order since the ensuing errors are of lower order yet.

By choosing $\Phi(\varphi_0, I_1)$ appropriately the angle-dependent part of Eq. (16.100) (the last two terms) can be eliminated. This determines Φ according to

$$\omega_0 \frac{\partial \Phi}{\partial \varphi_0}(\varphi_0, I_1) = -\langle\!\langle H_1(\varphi_0, I_1)\rangle. \tag{16.101}$$

This is known as "killing" these angle-dependent terms. The task of obtaining Φ is straightforwardly faced, just as in Eqs. (16.90). Doing this makes the transformation equations (16.99) explicit. Since the Hamiltonian is then, once again, independent of the angle coordinate φ_1, the variable I_1 is the action variable to this order. After this procedure the newly "unperturbed" Hamiltonian is

$$H = H_0(I_1) + \langle H_1(\varphi_0, I_1)\rangle \tag{16.102}$$

and its frequency is given by

$$\omega_1 = \frac{\partial H}{\partial I_1} = \omega_0 + \frac{\partial}{\partial I_1}\langle H_1(\varphi_0, I_1)\rangle. \tag{16.103}$$

By choosing (16.102) as another "unperturbed Hamiltonian" the whole procedure can (in principle) be iterated. In practice the formulas rapidly become very complicated. It is easiest to follow an explicit example such as the following.

16.5.2 Application to Gravity Pendulum

To illustrate the preceding formulas and to see how they can be extended to higher order let us consider the gravity pendulum, closely following the treatment of Chirikov listed at the end of the chapter. The Hamiltonian is

$$H = \frac{p^2}{2} + (\cos\theta - 1) = \frac{p^2}{2} + \frac{\theta^2}{2!} - \frac{\theta^4}{4!} + \frac{\theta^6}{6!} - \frac{\theta^8}{8!} + \cdots. \tag{16.104}$$

The constants have been chosen to simplify this as much as possible. In particular, mass $= 1$ and $\omega_0 = 1$. Taking the quadratic terms as the unperturbed part of this Hamiltonian, we have

$$H_0 = I_0, \quad \theta = \sqrt{2I_0}\cos\varphi_0. \tag{16.105}$$

Expressing H in the form (16.97),

$$H(I_0,\varphi_0) = I_0 - \frac{4I_0^2}{4!}\cos^4\varphi_0 + \frac{8I_0^3}{6!}\cos^6\varphi_0 - \frac{16I_0^4}{8!}\cos^8\varphi_0. \tag{16.106}$$

These series have been truncated arbitrarily after four terms. For the first approximation only the first two terms have any effect, but to later illustrate the Kolmogorov superconvergence idea it is appropriate to complete some calculations to a higher order than might initially seem to be justified.

Define Fourier expansions based on the identities

$$\langle\cos^n\theta\rangle = \begin{cases} \frac{1}{2^n}\binom{n}{n/2} & \text{for } n \text{ even}, \\ 0 & \text{for } n \text{ odd}, \end{cases} \tag{16.107}$$

and the definitions

$$f_n = \cos^n\varphi - \langle\cos^n\varphi\rangle, \qquad f_n' = \frac{df^n}{d\varphi}, \qquad F_n = \langle\!\langle\overline{F}_n\rangle,$$

$$\text{where} \quad \frac{d\overline{F}_n}{d\varphi} = f_n. \tag{16.108}$$

For example,

$$f_4 = \cos^4\varphi - \frac{3}{8} = \frac{1}{2}\cos 2\varphi + \frac{1}{8}\cos 4\varphi,$$
$$f_4' = -\sin 2\varphi - \frac{1}{2}\sin 4\varphi, \tag{16.109}$$
$$F_4 = \left\langle\!\!\left\langle \int \left(\frac{1}{2}\cos 2\varphi' + \frac{1}{8}\cos 4\varphi'\right) d\varphi' \right\rangle\!\!\right\rangle = \frac{\sin 2\varphi}{4} + \frac{\sin 4\varphi}{32}.$$

Rearranging the terms of Eq. (16.106) yields

$$H(\phi_0, I_0) = I_0 - \frac{I_0^2}{16} + \frac{I_0^3}{288} - \frac{I_0^4}{9216} - \frac{I_0^2}{6} f_4(\varphi_0) + \frac{I_0^3}{90} f_6(\varphi_0) - \frac{I_0^4}{2520} f_8(\varphi_0). \tag{16.110}$$

The angle-dependent part of this Hamiltonian is of order I^2, and the averaged Hamiltonian is

$$\langle H_1(I_0)\rangle = I_0 - \frac{I_0^2}{16} + \frac{I_0^3}{288} - \frac{I_0^4}{9216}. \tag{16.111}$$

It is *a priori* unclear how many of these terms are valid, so the same is true of the perturbed frequency derived from this formula.

Choosing to "kill" only the $f_4(\varphi_0)$ term, the leading term of Eq. (16.101) yields

$$\frac{\partial\Phi}{\partial\varphi_0} = \frac{I_1^2}{6} f_4, \qquad \text{and hence} \qquad \Phi = \frac{I_1^2}{6} F_4(\varphi_0). \tag{16.112}$$

Substituting this into Eqs. (16.99) yields

$$I_0 = I_1 + \frac{I_1^2}{6} f_4(\varphi_0),$$
$$\varphi_1 = \varphi_0 + \frac{I_1}{3} F_4(\varphi_0). \tag{16.113}$$

As mentioned previously, because of the generating function formalism, the new and old coordinates are still inconveniently coupled at this point. The cost of uncoupling them is further truncated Taylor expansion;

$$I_0 = I_1 + \frac{I_1^2}{6} f_4(\varphi_1) - \frac{I_1^3}{18} f_4'(\varphi_1) F_4(\varphi_1),$$
$$\varphi_0 = \varphi_1 - \frac{I_1}{3} F_4(\varphi_1) + \frac{I_1^2}{9} f_4(\varphi_1) F_4(\varphi_1). \tag{16.114}$$

The result of re-expressing Hamiltonian (16.110) in terms of the new variables (keeping only terms up to I_1^4) is

$$H(\phi_1, I_1) = I_1 - \frac{I_1^2}{16} + \frac{I_1^3}{288} - \frac{I_1^4}{9216} - \frac{I_1^3}{6}\left(\frac{f_4}{48} - \frac{f_6}{90} + \frac{f_4^2}{18}\right)$$
$$+ I_1^4\left(\frac{f_4}{576} - \frac{f_8}{2520} - \frac{f_4^2}{576} - \frac{f_4^3}{216} + \frac{f_4 f_6}{180}\right)$$
$$+ \frac{I_1^4}{18}\left(\frac{f_4'}{8} - \frac{f_6'}{15} + \frac{f_4'^2}{3}\right)F_4. \qquad (16.115)$$

At this point the angle-dependent terms in the Hamiltonian are of order I^3. The reason for this is that the order increased from I^2 (the previous order) by the order of $\partial\Phi/\partial I_1$ which was I^1.

16.5.3 Superconvergence

For the particular problem (pendulum) being discussed, an analytic solution in the form of elliptic functions is known, so it is possible to check the formulas that have been obtained. One finds that Eq. (16.110) is correct only up to the I^2 term and Eq. (16.115) is correct only up to the I^3 term. This is the same "rate of improvement" as has been obtained with the methods described previously in this chapter. What is remarkable is that, when we have completed the next iteration step using the current method, the next result will be correct up to the I^5 term. The step after that will be correct up to the I^9 term. In general the nth iteration yields $2^n + 1$ correct powers of I. This is Kolmogorov's superconvergence.

To see how this comes about let us determine the generating function $\Phi(\varphi_1, I_2)$ that "kills" the leading angle-dependent term of Eq. (16.115). By Eq. (16.101) we have

$$\frac{\partial\Phi}{\partial\varphi_1}(\varphi_1, I_2) = \frac{I_1^3}{6}\left\langle\!\!\left\langle\frac{f_4}{48} - \frac{f_6}{90} + \frac{f_4^2}{18}\right\rangle\right. \qquad (16.116)$$

which is of order I^3. The order of $\partial\Phi/\partial I_2$ is therefore I^2. After this iteration the angle-dependent part of the Hamiltonian will be of order I^5. The other statements in the previous paragraph are confirmed similarly.

This is superconvergence. The key to its success is the appropriate segregation of time-dependent and time-independent terms at each stage, since this prevents the pollution of lower order terms in higher order calculations. The accelerated number of valid terms in each order is inherent to the scheme of iteration.

Bibliography

References for Further Study

Section 16.1

1 F.T. Geyling and H.R. Westerman, *Introduction to Orbital Mechanics,* Addison-Wesley, Reading, MA, 1971.

Section 16.3

2 L.D. Landau and E.M. Lifshitz, *Mechanics,* Pergamon, 1976.

Section 16.4

3 N.N. Bogoliubov and Y.A. Mitropolsky, *Asymptotic Methods in the Theory of Oscillations,* Gordon and Breach, New York, 1961.

Section 16.4.5

4 R.L. Stratonovich, *Topics in the Theory of Random Noise,* Vol. 2, Gordon and Breach, New York, 1973, p. 97.

Section 16.5

5 G.D. Birkhoff, *Dynamical Systems,* American Mathematical Society, Providence, RI, 1991.

6 B.V. Chirikov, A universal instability of many-dimensional oscillators, Physical Reports, **52**, (1979).

17 Symplectic Mechanics

"Symplectic mechanics" is the study of mechanics using "symplectic geometry," a subject that can be pursued with no reference whatsoever to mechanics. However, we will regard "symplectic mechanics" and "Hamiltonian mechanics" as essentially equivalent. We have seen that Newtonian and Lagrangian mechanics is naturally pictured in configuration space while Hamiltonian mechanics is based naturally in phase space. This distinction is illustrated in Fig. 17.1. In configuration space one deals with spatial trajectories (they would be rays in optics) and "wavefront-like" surfaces that are transverse to the trajectories. A useful concept is that of a "congruence" or bundle of space-filling, nonintersecting curves. A point in phase space fixes both position and slope of the trajectory passing through that point and as a result there is only one trajectory through any point and the valid trajectories of the mechanical system naturally form a congruence of space-filling, nonintersecting curves. This is in contrast to configuration space, where a rule relating initial velocities with initial positions must be given to define a congruence of trajectories.

In Newtonian mechanics it is natural to work on finding trajectories starting from the n second order, ordinary differential equations of the system. In Hamilton–Jacobi theory one first seeks the wavefronts, starting from a partial differential equation. As stated already, both descriptions are based on configuration space. If the coordinates in this space are the $3n$ Euclidean spatial components, the usual Pythagorean metric of distances and angles applies and, for example, it is meaningful for the wavefronts to be orthogonal to the trajectories. Also the distance along a trajectory or the distance between two trajectories can be well defined.

Even in Hamiltonian mechanics one usually starts from a Lagrangian $L(\mathbf{q}, \dot{\mathbf{q}}, t)$. But, after introducing canonical momenta, the natural geometry of Hamiltonian mechanics is phase space and one seeks the trajectories as solutions of $2n$ first order, ordinary differential equations. In this space the geometry is much more restrictive since there is a single trajectory through each point. Also there is no natural metric by which distances and angles can be defined. "Symplectic geometry" is the geometry of phase space. It is frequently convenient, especially in phase space, to refer a bundle of system

Geometric Mechanics: Toward a Unification of Classical Physics. 2nd Edition. Richard Talman

ISBN: 978-3-527-40683-8

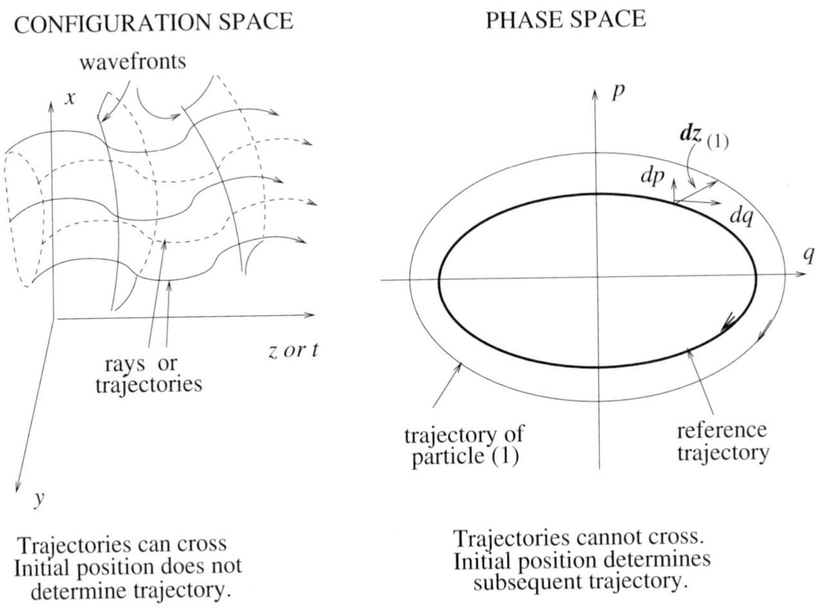

Fig. 17.1 Schematic representation of the essential distinctions between configuration space and phase space. Especially in phase space it is convenient to define a "reference trajectory" as shown and to relate nearby trajectories to it.

trajectories to a single nearby "reference trajectory" as shown in Fig. 17.1. But because there is no metric in phase space the "length" of the deviation vector is not defined.

17.1 The Symplectic Properties of Phase Space

17.1.1 The Canonical Momentum 1-Form

Why are momentum components indicated by subscripts, when position components are indicated by superscripts? Obviously it is because momentum components are covariant whereas position components are contravariant. How do we know this? Most simply it has to do with behavior under coordinate transformations. Consider a transformation from coordinates q^i to $Q^i = Q^i(\mathbf{q})$. Increments to these coordinates are related by

$$dQ^i = \frac{\partial Q^i}{\partial q^j} dq^j \equiv \Lambda^i\, j(\mathbf{q})\, dq^j, \tag{17.1}$$

which is the defining equation for the Jacobean matrix $\Lambda^i j(\mathbf{q})$. This is a linear transformation in the *tangent space* belonging to the *manifold* M whose coordinates are $\mathbf{q}$. The momentum components $\mathbf{P}$ corresponding to new coordinates $\mathbf{Q}$ are given by

$$P_i = \frac{\partial}{\partial \dot{Q}^i} L\Big(\mathbf{q}(\mathbf{Q}), \dot{\mathbf{q}}(\mathbf{Q}, \dot{\mathbf{Q}}, t), t\Big) = \frac{\partial L}{\partial \dot{q}^j} \frac{\partial q^j}{\partial Q^i} = \Big((\mathbf{\Lambda}^{-1})^T\Big)_i^{\ j} p_j, \tag{17.2}$$

where $(\Lambda^{-1})^j{}_l = \partial q^j / \partial Q^l$.[1] This uses the fact that the matrix of derivatives $\partial q^j / \partial Q^i$ is the inverse of the matrix of derivatives $\partial Q^j / \partial q^i$ and, from Eq. (5.12), $\partial \dot{q}^j / \partial \dot{Q}^i = \partial q^j / \partial Q^i$. It is the appearance of the *transposed inverse* Jacobean matrix in this transformation that validates calling $\mathbf{p}$ a covariant vector. With velocity $\dot{\mathbf{q}}$ (or displacement $\mathbf{dq}$) residing in the *tangent* space, one says that $\mathbf{p}$ resides in the *cotangent* space. From Eq. (2.3) we know that these transformation properties assure the existence of a certain *invariant* inner product. In the interest of making contact with notation used there, we therefore introduce, temporarily at least, the symbol $\widetilde{\mathbf{p}}$ for momentum. Then the technical meaning of the statement that $\widetilde{\mathbf{p}}$ resides in the cotangent space is that the quantity $\langle \widetilde{\mathbf{p}}, \mathbf{dq} \rangle \equiv p_i dq^i$ is invariant to the coordinate transformation from coordinates $\mathbf{q}$ to coordinates $\mathbf{Q}$. As an alternate notation for given $\widetilde{\mathbf{p}}$, one can introduce a 1-form or operator $\boldsymbol{\eta}^{(1)}$ defined so that $\widetilde{\boldsymbol{\eta}}^{(1)} \equiv \langle \widetilde{\mathbf{p}}, \cdot \rangle$, which yields a real number when acting on increment $\mathbf{dq}$. (The $\cdot$ in $\langle \widetilde{\mathbf{p}}, \cdot \rangle$ is just a place-holder for $\mathbf{dq}$.)

It is necessary to distinguish mathematically between $\mathbf{p} \cdot \mathbf{dq}$ and $p_i dq^i$, two expressions that a physicist is likely to equate mentally. Mathematically the expression $p_i dq^i$ is a 1-form definable on any manifold, whether possessed of a metric or not, while $\mathbf{p} \cdot \mathbf{dq}$ is a more specialized quantity that only is definable if it makes sense for $\mathbf{p}$ and $\mathbf{dq}$ to be subject to scalar multiplication because they reside in the same metric space.

The operator $\widetilde{\boldsymbol{\eta}}^{(1)}$ is known as a "1-form" with the tilde indicating that it is a form and the superscript (1) meaning that it takes one argument. Let the configuration space, the elements of which are labeled by the generalized coordinates $\mathbf{q}$, be called a "manifold" M. At a particular point $\mathbf{q}$ in M, the possible velocities $\dot{\mathbf{q}}$ are said to belong to the "tangent space" at $\mathbf{q}$, denoted by $TM_{\mathbf{q}}$. The operator $\widetilde{\boldsymbol{\eta}}^{(1)}$ "maps" elements of $TM_{\mathbf{q}}$ to the space R of real numbers;

$$\widetilde{\boldsymbol{\eta}}^{(1)} : \mathrm{TM}_{\mathbf{q}} \rightarrow R. \tag{17.3}$$

Consider a real-valued function $f(\mathbf{q})$ defined on M,

$$f : M \rightarrow R. \tag{17.4}$$

1) Our convention is that matrix elements such as $\Lambda^j{}_l$ do not depend on whether the indices are up or down, but their order matters; in this case, the fact that l is the column index is indicated by its slight displacement to the right.

As introduced in Section 2.3.5, the prototypical example of a one form is the "differential" of a function such as f; it is symbolized by $\widetilde{\boldsymbol{\eta}}^{(1)} = \widetilde{\mathbf{df}}_{\mathbf{q}}$. An incremental deviation $\mathbf{dq}$ from point $\mathbf{q}$ is necessarily a local tangent vector. The corresponding (linearized) change in value of the function, call it $df_{\mathbf{q}}$ (not bold face and with no tilde) depends on the "direction" of $\mathbf{dq}$. Consider the lowest order Taylor approximation,

$$f(\mathbf{q} + \mathbf{dq}) - f(\mathbf{q}) \approx \frac{\partial f}{\partial q^i} dq^i. \tag{17.5}$$

By the "linearized" value of df we mean this approximation to be taken as exact so that

$$df_{\mathbf{q}} \equiv \widetilde{\mathbf{dq}}(f) = \frac{\partial f}{\partial q^i} dq^i; \tag{17.6}$$

this is "proportional" to $\mathbf{dq}$ in the sense that doubling $\mathbf{dq}$ doubles $df_{\mathbf{q}}$. If $\mathbf{dq}$ is tangent to a curve γ passing through the point $\mathbf{q}$ then, except for a scale factor proportional to rate of progress along the curve, $df_{\mathbf{q}}$ can be regarded as the rate of change of f along the curve. Except for the same scale factor, $df_{\mathbf{q}}$ is the same for any two curves that are parallel as they pass through $\mathbf{q}$. Though it may seem convoluted at first, for the particular function f, $\widetilde{\mathbf{df}}_{\mathbf{q}}$ therefore maps tangent vector $\mathbf{dq}$ to real number $df_{\mathbf{q}}$;

$$\widetilde{\mathbf{df}}_{\mathbf{q}} : \mathrm{TM}_{\mathbf{q}} \rightarrow R. \tag{17.7}$$

Recapitulating, the quantity $\langle \mathbf{p}, \cdot \rangle$, abbreviated as $\widetilde{\mathbf{p}}$ or later even just as $\mathbf{p}$, is said to be a "1-form," a linear, real-valued function of one vector argument. The components p_i of $\widetilde{\mathbf{p}}$ in a particular coordinate system, which in "classical" terminology are called *covariant* components, in "modern" terminology are the coefficients of a 1-form. We are to some extent defeating the purpose of introducing 1-forms by insisting on correlating their coefficients with covariant components. It is done because components are to a physicist what insulin is to a diabetic. A physicist says "$p_i q^i$ is *manifestly covariant*, (meaning invariant under coordinate transformation), because q^i is *contravariant* and p_i is *covariant*." A mathematician says the same thing in coordinate-free fashion as "cotangent space 1-form $\widetilde{\mathbf{p}}$ maps tangent space vector $\mathbf{q}$ to a real number."

What about the physicist's quantity $\mathbf{p} \cdot \mathbf{q}$? Here physicists (Gibbs initially I believe) have also recognized the virtue of *intrinsic* coordinate-free notation and adopted it universally. So $\mathbf{p} \cdot \mathbf{q}$ is the well-known coordinate-independent product of three factors, the magnitudes of the two vectors and the cosine of their included angle. But this notation implicitly assumes a Euclidean coordinate system, whereas the "1-form" notation does not. This may be the source of the main difficulty a physicist is likely to have in assimilating the language of modern differential geometry: traditional vector calculus, with its obvious power, already contains the major benefits of *intrinsic* description without being burdened by unwelcome abstraction. *But traditional vector analysis contains an implicit specialization to Euclidean geometry.* This makes it all the more difficult to grasp the more abstract analysis required when Euclidean geometry is inappropriate. Similar comments apply with even greater force to cross products $\mathbf{p} \times \mathbf{q}$ and even more yet to curls and divergences.

For a particular coordinate q^i, the coordinate 1-form $\widetilde{\mathbf{dq}}^i$ picks out the corresponding component V^i from arbitrary vector $\mathbf{V}$ as $V^i = \langle \widetilde{\mathbf{dq}}^i, \mathbf{V} \rangle$. Since the components p_i are customarily called "canonically conjugate" to the coordinates q^i, the 1-form

$$\widetilde{\boldsymbol{\eta}}^{(1)} \equiv \widetilde{\mathbf{p}} = p_i \widetilde{\mathbf{dq}}^i \tag{17.8}$$

is said to be the "canonical momentum 1-form." Incidentally, when one uses $\widetilde{\mathbf{p}}$ expanded in terms of its components as $p_i \widetilde{\mathbf{dq}}^i$, the differential form $\widetilde{\mathbf{dq}}^i$ will eventually be replaced by an ordinary differential dq^i and manipulations of the form will not be particularly distinguishable from the manipulations that would be performed on the ordinary differential. Nevertheless it seems somewhat clearer, when describing a possible multiplicity of mechanical systems, to retain the form $\widetilde{\mathbf{dq}}^i$ which is a property of the coordinate system, than to replace it with dq^i which is a property of a particular mechanical system.

17.1.2
The Symplectic 2-Form $\widetilde{\omega}$

In spite of having just gone to such pains to explain the appropriateness of using the symbol $\widetilde{\mathbf{p}}$ for momentum in order to make the notation expressive, we now drop the tilde. The reason for doing this is that we plan to work in *phase space* where $\mathbf{q}$ and $\mathbf{p}$ are to be treated on a nearly equal footing. Though logically possible it would be simply too confusing, especially when introducing forms on phase space, to continue to exhibit the intrinsic distinction between displacements and momenta explicitly, other than by continuing to use subscripts for the momentum components and superscripts for generalized coordinates. By lumping $\mathbf{q}$ and $\mathbf{p}$ together we get a vector space with dimension $2n$, double the dimensionality of the configuration space. (As explained previously, there is no absolute distinction between covariant and contravariant vectors *per se*.) Since we previously identified the $\mathbf{p}$'s with forms in configuration space and will now proceed to introduce forms that act on $\mathbf{p}$ in phase space, we will have to tolerate the confusing circumstance that $\mathbf{p}$ is a form in configuration space and a portion of a vector in phase space.

Since "phase space" has been newly introduced it is worth mentioning a notational limitation it inherits from configuration space. A symbol such as x can mean either where a particle actually *is* or where, in principle, it *could be*. It is necessary to tell by context which is intended. Also, when the symbol $\dot{x}$ appears, it usually refers to an *actual* system velocity, but it can also serve as a formal argument of a Lagrangian function. The same conventions have to be accepted in phase space. But the q's and the p's are not quite equivalent, since the q's are defined independent of any particular Lagrangian while the p's depend on the Lagrangian. Still they can refer either to a particular evolving system or to a possible configuration of the system. Mainly then, in phase space, the combined sets $\mathbf{q}, \mathbf{p}$ play the same role as $\mathbf{q}$ plays in configuration space.

In Problem 7.1.1 it was found that the quantity $x_1(z)p_2(z) - x_2(z)p_1(z)$ calculated from two rays in the same optical system is constant, independent of longitudinal coordinate z. This seemingly special result can be generalized to play a central role in Lagrangian (and hence Hamiltonian) mechanics. That is the immediate task. The simultaneous analysis of more than one trajectory at a time characterizes this newer-than-Newtonian approach.

We start by reviewing some topics from Chapter 2. Recall Eq. (2.81) by which *tensor product* $\mathbf{f} = \mathbf{x} \otimes \mathbf{y}$ is defined as a function of 1-forms $\widetilde{\mathbf{u}}$ and $\widetilde{\mathbf{v}}$;

$$\mathbf{f}(\widetilde{\mathbf{u}}, \widetilde{\mathbf{v}}) = \langle \widetilde{\mathbf{u}}, \mathbf{x} \rangle \langle \widetilde{\mathbf{v}}, \mathbf{y} \rangle. \tag{17.9}$$

Furthermore, a (mixed) tensor product $\mathbf{f} = \widetilde{\mathbf{u}} \otimes \mathbf{y}$ can be similarly defined by

$$\mathbf{f}(\mathbf{x}, \widetilde{\mathbf{v}}) = \langle \widetilde{\mathbf{u}}, \mathbf{x} \rangle \langle \widetilde{\mathbf{v}}, \mathbf{y} \rangle. \tag{17.10}$$

"Wedge products" or "exterior products" are defined by

$$\begin{aligned} \mathbf{x} \wedge \mathbf{y}(\widetilde{\mathbf{u}}, \widetilde{\mathbf{v}}) &= \langle \mathbf{x}, \widetilde{\mathbf{u}} \rangle \langle \mathbf{y}, \widetilde{\mathbf{v}} \rangle - \langle \mathbf{x}, \widetilde{\mathbf{v}} \rangle \langle \mathbf{y}, \widetilde{\mathbf{u}} \rangle, \\ \widetilde{\mathbf{u}} \wedge \widetilde{\mathbf{v}}(\mathbf{x}, \mathbf{y}) &= \langle \widetilde{\mathbf{u}}, \mathbf{x} \rangle \langle \widetilde{\mathbf{v}}, \mathbf{y} \rangle - \langle \widetilde{\mathbf{u}}, \mathbf{y} \rangle \langle \widetilde{\mathbf{v}}, \mathbf{x} \rangle. \end{aligned} \tag{17.11}$$

Another result from Chapter 4 was the construction of a multicomponent *bivector* from two vectors, $\mathbf{x}$ and $\mathbf{y}$, with the components being the 2×2 determinants constructed from the components of the two vectors, as illustrated in Fig. 4.4. The bivector components can be interpreted as the areas of the projections onto the coordinate axes of the parallelogram formed from the two vectors. This figure is repeated as Fig. 17.2, but with axes labeled by q^1, p_1 and q^2. The projected areas are (except for a possible combinatorial factor) the components of an antisymmetric two component tensor, x^{ij}, with $x^{12} = x^1 y^2 - x^2 y^1$ *etc.*

We now intend to utilize these quantities in phase space. As in geometric optics, we will consider not just a solitary orbit, but rather a congruence of orbits or, much of the time, two orbits. As stressed already, in phase space

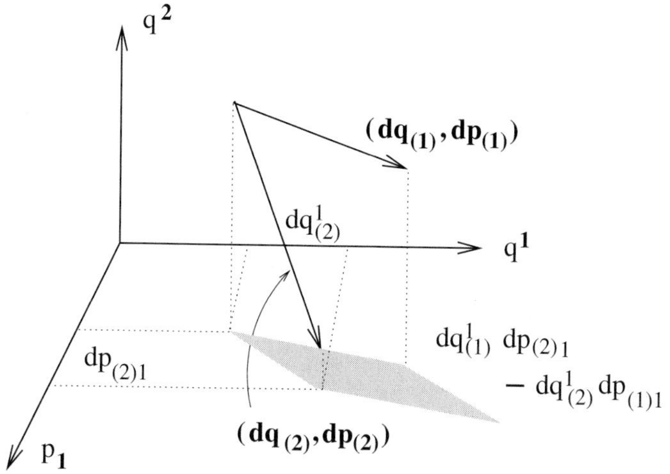

Fig. 17.2 The "projected area" on the first coordinate plane (q^1, p_1) defined by tangent vectors $\mathbf{dz}_{(1)} = (\mathbf{dq}_{(1)}, \mathbf{dp}_{(1)})^T$ and $\mathbf{dz}_{(2)} = (\mathbf{dq}_{(2)}, \mathbf{dp}_{(2)})^T$.

there can be only one valid orbit through each point, which is the major formal advantage of working in phase space. To discuss two particular close orbits without giving preference to either, it is useful to refer them both to a reference path as in Fig. 17.1. Though it would not be necessary, this reference path may as well be thought of as a valid orbit as well. A point on one nearby orbit can be expressed by $\mathbf{dz}_{(1)} = (\mathbf{dq}_{(1)}, \mathbf{dp}_{(1)})^T$ and on the other one by $\mathbf{dz}_{(2)} = (\mathbf{dq}_{(2)}, \mathbf{dp}_{(2)})^T$.

Consider a particular coordinate q, say the first one, and its conjugate momentum p. Since these can be regarded as functions in phase space, the differential forms $\widetilde{\mathbf{dq}}$ and $\widetilde{\mathbf{dp}}$ are everywhere defined.[2] As in Eq. (2.2), when "coordinate 1-form" $\widetilde{\mathbf{dq}}$ operates on the vector $\mathbf{dz}_{(1)}$ the result is

$$\widetilde{\mathbf{dq}}(\mathbf{dz}_{(1)}) = dq_{(1)}, \qquad \text{and similarly} \qquad \widetilde{\mathbf{dp}}(\mathbf{dz}_{(1)}) = dp_{(1)}. \tag{17.12}$$

Notice that it has been necessary to distinguish $\widetilde{\mathbf{dq}}$, say, which is a form specific to the coordinate system, from $dq_{(1)}$, which is specific to particular mechanical system (1). As usual, the placing of the (1) in parenthesis, as here,

2) Recall that, since we are working in phase space, the symbol $\widetilde{\mathbf{dp}}$ has a meaning different from what it would have in configuration space. Here it expects as argument a phase-space tangent vector $\mathbf{dz}$. A notational ambiguity we will have is that it is not obvious whether the quantity $\widetilde{\mathbf{dq}}$ is a 1-form associated with one particular coordinate q or the set of 1-forms $\widetilde{\mathbf{dq}}^i$ corresponding to all the coordinates q^i. We shall state which is the case every time the symbol is used. Here it is the former.

"protects" it from being interpreted as a vector index. Consider then the wedge product[3]

$$\widetilde{\omega} = \widetilde{\mathbf{dq}} \wedge \widetilde{\mathbf{dp}}. \tag{17.13}$$

Copying from Eq. (17.11), when $\widetilde{\boldsymbol{\omega}}$ operates on the two system vectors, the result is

$$\widetilde{\boldsymbol{\omega}}(\mathbf{dz}_{(1)}, \mathbf{dz}_{(2)}) = dq_{(1)}dp_{(2)} - dq_{(2)}dp_{(1)}. \tag{17.14}$$

This quantity vanishes when the components are proportional, but not otherwise in general.

So far q and p have either referred to a one-dimensional system or are one pair of coordinates in a multidimensional system. To generalize to more than one configuration space coordinate we define

$$\widetilde{\omega} = \sum_{i=1}^{n} \widetilde{\mathbf{dq}}^{i} \wedge \widetilde{\mathbf{dp}}_{i}. \tag{17.15}$$

This is known as the "symplectic 2-form" or, because conjugate coordinates are singled out, "the canonical 2-form". (To avoid addressing the question of the geometric character of the individual terms the sum is expressed explicitly rather than by the repeated index convention.) Acting on vectors **u** and **v** this expands to

$$\widetilde{\boldsymbol{\omega}}(\mathbf{u}, \mathbf{v}) = \sum_{i=1}^{n} \Big(\langle \widetilde{\mathbf{dq}}^{i}, \mathbf{u} \rangle \langle \widetilde{\mathbf{dp}}_{i}, \mathbf{v} \rangle - \langle \widetilde{\mathbf{dq}}^{i}, \mathbf{v} \rangle \langle \widetilde{\mathbf{dp}}_{i}, \mathbf{u} \rangle \Big). \tag{17.16}$$

When $\widetilde{\boldsymbol{\omega}}$ acts on $\mathbf{dz}_{(1)}$ and $\mathbf{dz}_{(2)}$, the result is

$$\widetilde{\boldsymbol{\omega}}(\mathbf{dz}_{(1)}, \mathbf{dz}_{(2)}) = \sum_{i=1}^{n} (dq^{i}_{(1)} dp_{(2)i} - dq^{i}_{(2)} dp_{(1)i}). \tag{17.17}$$

If the two terms are summed individually they are both scalar invariants but it is more instructive to keep them paired as shown. Each paired difference, when evaluated on two vectors, produces the directed area of a projection onto one of the (q^i, p_i) coordinate planes; see Fig. 17.2. For example, $dq^1_{(1)} dp_{(2)1} - dq^1_{(2)} dp_{(1)1}$, is the area of a projection onto the q^1, p_1 plane. For one-dimensional motion there is no summation and no projection needed, and $\widetilde{\boldsymbol{\omega}}(\mathbf{dz}_{(1)}, \mathbf{dz}_{(2)})$ is simply the area defined by $(dq_{(1)}, dp_{(1)})$ and $(dq_{(2)}, dp_{(2)})$.

3) To be consistent we should use $\widetilde{\boldsymbol{\omega}}^{(2)}$ to indicate that it is a 2-form but the symbol will be used so frequently that we leave off the superscript (2).

It can be noted in passing that, as in Section 2.2, the 2-form $\widetilde{\omega}^{(2)}$ can be obtained by exterior differentiation of $\widetilde{\omega}^{(1)}$. Applying Eq. (2.40)

$$\widetilde{\mathbf{d}}\,\widetilde{\mathbf{p}} = \widetilde{\mathbf{d}}\,(p_i\widetilde{\mathbf{dq}}^i) = -\widetilde{\mathbf{dq}}^i \wedge \widetilde{\mathbf{dp}}_i. \tag{17.18}$$

17.1.3
Invariance of the Symplectic 2-Form

Now consider a coordinate transformation such as was discussed in Section 17.1.1, from q^i to $Q^i = Q^i(\mathbf{q})$. Under this transformation

$$\widetilde{\mathbf{dQ}}^j = \frac{\partial Q^j}{\partial q^i}\widetilde{\mathbf{dq}}^i, \qquad \text{and} \qquad \widetilde{\mathbf{dP}}_j = \frac{\partial q^k}{\partial Q^j}\widetilde{\mathbf{dp}}_k + p_k\widetilde{\mathbf{dq}}^l \frac{\partial}{\partial q^l}\left(\frac{\partial q^k}{\partial Q^j}\right). \tag{17.19}$$

(The expression for the differential of P_j is more complicated than the expression for the differential of Q^i because the coefficients $\partial Q^j/\partial q^i$ are themselves functions of position.) The Jacobean matrix elements satisfy

$$\frac{\partial q^k}{\partial Q^j}\frac{\partial Q^j}{\partial q^i} = \frac{\partial q^k}{\partial q^i} = \delta_{ki}. \tag{17.20}$$

After differentiation this yields

$$0 = \frac{\partial}{\partial q^l}\left(\frac{\partial q^k}{\partial Q^j}\frac{\partial Q^j}{\partial q^i}\right) = \frac{\partial Q^j}{\partial q^i}\frac{\partial}{\partial q^l}\left(\frac{\partial q^k}{\partial Q^j}\right) + \frac{\partial q^k}{\partial Q^j}\frac{\partial^2 Q^j}{\partial q^l \partial q^i}. \tag{17.21}$$

The factor $\frac{\partial}{\partial q^l}\left(\frac{\partial q^k}{\partial Q^j}\right)$ in the final term in Eq. (17.19) can be evaluated using these two results. In the new coordinates the wedge product is

$$\widetilde{\mathbf{dQ}}^j \wedge \widetilde{\mathbf{dP}}_j = \frac{\partial Q^j}{\partial q^i}\widetilde{\mathbf{dq}}^i \wedge \left(\frac{\partial q^k}{\partial Q^j}\widetilde{\mathbf{dp}}_k + p_k\frac{\partial}{\partial q^l}\left(\frac{\partial q^k}{\partial Q^j}\right)\widetilde{\mathbf{dq}}^l\right) = \widetilde{\mathbf{dq}}^j \wedge \widetilde{\mathbf{dp}}_j. \tag{17.22}$$

Here the terms proportional to $\widetilde{\mathbf{dq}}^i \wedge \widetilde{\mathbf{dq}}^l$ with equal index values have vanished individually and those with unequal indices have canceled in pairs since they are odd under the interchange of i and l, whereas the coefficient $\frac{\partial^2 Q^j}{\partial q^l \partial q^i}$, entering by virtue of Eq. (17.21), is even under the same interchange.

To obtain the canonical 2-form $\widetilde{\omega}$ and demonstrate its invariance under coordinate transformation all that has been assumed is the existence of generalized coordinates q^i and some particular Lagrangian $L(\mathbf{q}, \dot{\mathbf{q}}, t)$, since momenta p_i were derived from them. One can therefore say that the phase space of a Lagrangian system is sure to be "equipped" with the form $\widetilde{\omega}$. It is this form that will permit the identification of 1-forms and vectors in much the same way that a metric permits the identification of covariant and contravariant vectors (as was discussed in Section 4.2.5.) This is what will make up for the absence of the concept of orthogonality in developing within mechanics the analog of rays and wavefronts in optics. One describes these results as "symplectic geometry," but the results derived so far, in particular Eq. (17.22), can be regarded simply as differential calculus. The term "symplectic calculus" might therefore be as justified.[1]

Another conclusion that will follow from Eq. (17.22) is that the 2-form $\widetilde{\mathbf{dq}}^i \wedge \widetilde{\mathbf{dp}}_i$ evaluated for any two phase-space trajectories is "conserved" as time advances. We will put off deriving this result (which amounts to being a generalized Liouville theorem) for the time being. It is mentioned at this point to emphasize that it follows purely from the structure of the equations – in particular from the definition in Eq. (1.11) of the momenta p_j as a derivative of the Lagrangian with respect to velocity $\dot{q}^j$. Since the derivation could have been completed before a Hamiltonian has even been introduced, it cannot be said to be an essentially Hamiltonian result, or of any property of a system other than that of being characterized by a Lagrangian.

For paraxial optics in a single transverse plane, a result derived in Problem 7.1.1, was the invariance of the combination $x_{(1)}p_{(2)} - x_{(2)}p_{(1)}$ for any two rays. This is an example of Eq. (17.22). Because that theory had already been linearized, the conservation law applied to the full amplitudes and not just to their increments. In general however the formula applies to small deviations around a reference orbit, even if the amplitude of that reference orbit is great enough for the equations of motion to be arbitrarily nonlinear.

17.1.4
Use of $\widetilde{\omega}$ to Associate Vectors and 1-Forms

To motivate this discussion recall, for example from Eq. (2.111) (which read $x_i = g_{ij}x^j$), that a metric tensor can be used to obtain covariant components x_i from contravariant components x^k. This is "lowering the index." This amounts to defining a "dot product" operation and allows the orthogonality of two vectors x^i and y^i to be expressed in the form $\mathbf{x}\cdot\mathbf{y} \equiv x_i y^i = 0$.

The symplectic 2-form $\widetilde{\boldsymbol{\omega}}$ discussed in the previous section, can be written in the form $\widetilde{\boldsymbol{\omega}}(\cdot,\cdot)$ to express the fact that it is waiting for two *vector* argu-

1) Explanation of the source of the name *symplectic* actually legitimizes the topic as *geometry* since it relates to the vanishing of an antisymmetric form constructed from the coordinates of, say, a triplex of three points. The name "symplectic group" (from a Greek word with his intended meaning) was coined by Hermann Weyl as a replacement for the term "complex group" that he had introduced even earlier, with "complex" used in the sense "is a triplex of points on the same line?". He intended "complex" to mean more nearly "simple" than "complicated" and certainly *not* to mean $\sqrt{-1}$. But the collision of meanings become an embarrassment to him. Might one not therefore call a modern movie complex a "cineplectic group?"

ments, from which it will *linearly* produce a real number. It is important also to remember that, as a tensor, $\widetilde{\boldsymbol{\omega}}$ is *antisymmetric*. This means, for example, that $\widetilde{\boldsymbol{\omega}}(\vec{\mathbf{u}}, \vec{\mathbf{u}}) = 0$ where $\vec{\mathbf{u}}$ is any vector belonging to the tangent space $TM_{\mathbf{x}}$ at system configuration $\mathbf{x}$. For the time being here we are taking a "belt and suspenders" approach of indicating a vector $\vec{\mathbf{u}}$ with both bold face and overhead arrow. It is done only to stress the point and this notation will be dropped when convenient.

Taking $\vec{\mathbf{u}}$ as one of the two vector arguments of $\widetilde{\boldsymbol{\omega}}$ we can define a new quantity (a 1-form) $\widetilde{\mathbf{u}}(\cdot)$ by the formula

$$\widetilde{\mathbf{u}}(\cdot) = \widetilde{\boldsymbol{\omega}}(\vec{\mathbf{u}}, \cdot). \tag{17.23}$$

This formula "associates" a 1-form $\widetilde{\mathbf{u}}$ with the vector $\vec{\mathbf{u}}$. Since the choice of whether to treat $\vec{\mathbf{u}}$ as the first or second argument in Eq. (17.23) was arbitrary, the sign of the association can only be conventional.

The association just introduced provides a one-to-one linear mapping from the tangent space $TM_{\mathbf{x}}$ to the cotangent space $TM^*_{\mathbf{x}}$. These spaces have the same dimensionality. For any particular choices of bases in these spaces the association could be represented by matrix multiplication $u_i = A_{ij}u^j$ where A_{ij} is an antisymmetric, square matrix with nonvanishing determinant, and which would therefore be invertible. Hence the association is one-to-one in both directions and can be said to be an *isomorphism*. The *inverse* map can be symbolized by

$$I : TM^*_{\mathbf{x}} \rightarrow TM_{\mathbf{x}}. \tag{17.24}$$

As a result, for any 1-form $\widetilde{\boldsymbol{\eta}}$ there is sure to be a vector

$$\vec{\boldsymbol{\eta}} = I\,\widetilde{\boldsymbol{\eta}} \qquad \text{such that} \qquad \widetilde{\boldsymbol{\eta}} = \widetilde{\boldsymbol{\omega}}(\vec{\boldsymbol{\eta}}, \cdot). \tag{17.25}$$

An immediate (and important) example of this association is its application to $\widetilde{\mathbf{df}}$ which is the standard 1-form that can be constructed from any function f defined over phase space; Eq. (17.25) can be used to generate a vector $\vec{\mathbf{df}} = I\,\widetilde{\mathbf{df}}$ from the 1-form $\widetilde{\mathbf{df}}$ so that

$$\vec{\mathbf{df}} = I\,\widetilde{\mathbf{df}} \qquad \text{satisfies} \qquad \widetilde{\mathbf{df}} = \widetilde{\boldsymbol{\omega}}(\vec{\mathbf{df}}, \cdot). \tag{17.26}$$

17.1.5 Explicit Evaluation of Some Inner Products

Let q be a specific coordinate, say the first one, and p be its conjugate momentum and let $f(q, \ldots, p, \ldots)$ be a function defined on phase space. Again we use $\widetilde{\mathbf{dq}}$ and $\widetilde{\mathbf{dp}}$ temporarily as the 1-forms corresponding to these particular coordinates. The 1-form $\widetilde{\mathbf{df}}$ can be expressed two different way, one according to its original definition, the other using the association (17.26) with $\widetilde{\omega}$ spelled out as in Eq. (17.16);

$$\frac{\partial f}{\partial q}\widetilde{\mathbf{dq}} + \frac{\partial f}{\partial p}\widetilde{\mathbf{dp}} + \cdots = \widetilde{\mathbf{df}} = \widetilde{\omega}(\vec{\mathbf{df}},\cdot)$$
$$= \langle\widetilde{\mathbf{dq}},\vec{\mathbf{df}}\rangle\,\langle\widetilde{\mathbf{dp}},\cdot\rangle - \langle\widetilde{\mathbf{dq}},\cdot\rangle\,\langle\widetilde{\mathbf{dp}},\vec{\mathbf{df}}\rangle + \cdots \tag{17.27}$$
$$= \langle\widetilde{\mathbf{dq}},\vec{\mathbf{df}}\rangle\,\widetilde{\mathbf{dp}} - \langle\widetilde{\mathbf{dp}},\vec{\mathbf{df}}\rangle\,\widetilde{\mathbf{dq}} + \cdots.$$

It follows that

$$\langle\widetilde{\mathbf{dq}},\vec{\mathbf{df}}\rangle = \frac{\partial f}{\partial p} \quad \text{and} \quad \langle\widetilde{\mathbf{dp}},\vec{\mathbf{df}}\rangle = -\frac{\partial f}{\partial q}. \tag{17.28}$$

These equations will be used shortly to evaluate $\widetilde{\omega}(\vec{\mathbf{dz}},\cdot)$.

17.1.6
The Vector Field Associated with $\widetilde{\mathbf{dH}}$

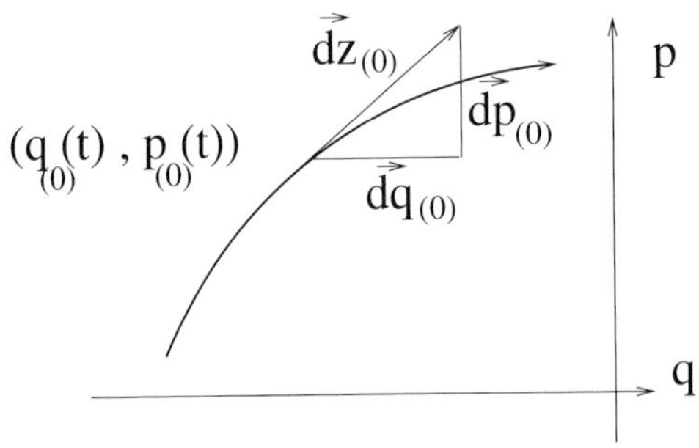

Fig. 17.3 The vector $\vec{\mathbf{dz}}_{(0)} = (\mathbf{dq}_{(0)}, \mathbf{dp}_{(0)})^T$ is tangent to a phase-space trajectory given by $\mathbf{z}_{(0)}(t) = (\mathbf{q}_{(0)}(t), \mathbf{p}_{(0)}(t))^T$. The trajectory is assumed to satisfy Hamilton's equations.

Since the Hamiltonian is a function on phase space, its differential 1-form $\widetilde{\mathbf{dH}}$ is well defined;

$$\widetilde{\mathbf{dH}} = \sum\left(\frac{\partial H}{\partial \mathbf{p}}\widetilde{\mathbf{dp}} + \frac{\partial H}{\partial \mathbf{q}}\widetilde{\mathbf{dq}}\right) \equiv \frac{\partial H}{\partial p_i}\widetilde{\mathbf{dp}}^i + \frac{\partial H}{\partial q^i}\widetilde{\mathbf{dq}}_i. \tag{17.29}$$

What is the associated vector $\vec{\mathbf{dH}} = I\,\widetilde{\mathbf{dH}}$? Figure 17.3 shows the unique trajectory $(\mathbf{q}(t),\mathbf{p}(t))$ passing through some particular point $(\mathbf{q}_{(0)},\mathbf{p}_{(0)})$, and an incremental tangential displacement at that point is represented by a $2n$ component column vector

$$\vec{\mathbf{dz}} = \begin{pmatrix}\mathbf{dq}\\ \mathbf{dp}\end{pmatrix} = \begin{pmatrix}\vec{\dot{\mathbf{q}}}\\ \vec{\dot{\mathbf{p}}}\end{pmatrix} dt. \tag{17.30}$$

(Our notation is inconsistent since, this time, **dq** and **dp** *do* stand for a full array of components. Also to reduce clutter we have suppressed the subscripts (0) which was only introduced to make the point that what follows refers to one particular point.) Hamilton's equations state

$$\begin{pmatrix} \dot{q}^i \\ \dot{p}_i \end{pmatrix} = \begin{pmatrix} \partial H/\partial p_i \\ -\partial H/\partial q^i \end{pmatrix}, \tag{17.31}$$

and these equations can be used to evaluate the partial derivatives appearing in Eq. (17.29). The result is

$$\widetilde{\mathbf{dH}} = \dot{q}^i\, \widetilde{\mathbf{dp}}_i - \dot{p}_i\, \widetilde{\mathbf{dq}}^i. \tag{17.32}$$

On the other hand, evaluating the symplectic 2-form on $\vec{\mathbf{dz}}$ yields

$$\begin{aligned} \widetilde{\omega}(\vec{\mathbf{dz}},\, \cdot) &= \langle \widetilde{\mathbf{dq}}^i, \vec{\mathbf{dz}} \rangle\, \widetilde{\mathbf{dp}}_i - \langle \widetilde{\mathbf{dp}}_i, \vec{\mathbf{dz}} \rangle\, \widetilde{\mathbf{dq}}^i \\ &= dq^i\, \widetilde{\mathbf{dp}}_i - dp_i\, \widetilde{\mathbf{dq}}^i, \\ &= \left(\dot{q}^i\, \widetilde{\mathbf{dp}}_i - \dot{p}_i\, \widetilde{\mathbf{dq}}^i\right) dt, \end{aligned} \tag{17.33}$$

where the inner products have been evaluated using Eq. (17.28). Dividing by dt, the equation implied by Eqs. (17.32) and (17.33) can therefore be expressed using the isomorphism introduced in Eq. (17.25);

$$\vec{\dot{\mathbf{z}}} = \mathbf{I}\, \widetilde{\mathbf{dH}}. \tag{17.34}$$

Though particular coordinates were used in deriving this equation, in the final form the relationship is coordinate-independent, which is to say that the relation is *intrinsic*. This is in contrast with the *coordinate-dependent* geometry describing the Hamilton–Jacobi equation in the earlier chapters.

In configuration space one is accustomed to visualizing the *force* as being directed parallel to the gradient of a quantity with dimensions of energy, namely the potential energy. Here in phase space we find the system *velocity* related to (though not parallel to) the "gradient" of the Hamiltonian, also an energy. In each case the motivation is to represent vectorial quantities in terms of (possibly) simpler-to-obtain scalar quantities. This motivation should be reminiscent of electrostatics, where one finds the *scalar* potential and from it the *vector* electric field.

17.1.7
Hamilton's Equations in Matrix Form

This section builds especially on the geometry of covariant and contravariant vectors developed in Section 2.1. It is customary to represent first order differential equations such as Eqs. (17.34) in matrix form. Since there are $2n$

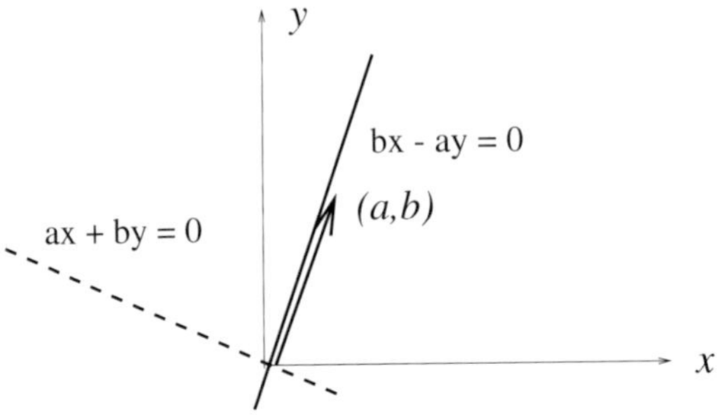

Fig. 17.4 The line $bx + ay = 0$ is perpendicular to the line $ax - by = 0$, to which the vector (a, b) is parallel.

equations, one wishes to represent the operator I, which has so far been entirely formal, by a $2n \times 2n$ matrix. According to Eq. (17.26), Eq. (17.34) can be written even more compactly as $\vec{\mathbf{z}} = \overrightarrow{\mathbf{dH}}$, but the right-hand side remains to be made explicit. When expressed in terms of canonical coordinates, except for sign changes and a coordinate-momentum interchange, the components of $\vec{\mathbf{dH}}$ are the same as the components of the ordinary gradient of H.

In metric geometry a vector can be associated with the hyperplane to which it is orthogonal. If the dimensionality is even, and the coordinates are arranged in (q^i, p_i) pairs, the equation of a hyperplane through the origin takes the form $a_i q^i + b^i p_i = 0$. The vector with contravariant components $(q^i, p_i) = (a_i, b^i)$ is normal to this plane (see Fig. 17.4). In this way a *contravariant* vector $\vec{\mathbf{dz}}$ is associated with a *covariant* vector, or 1-form. If one insists on defining a dot product operation, in the usual way, as $\vec{\mathbf{u}} \cdot \vec{\mathbf{v}} = u_i v^i$, then the dot product of a vector with itself is zero. The isomorphism (17.25) can be specified, for an arbitrary vector $\vec{\mathbf{w}}$, as relating its contravariant and covariant components by

$$\begin{pmatrix} w^1 \\ w^2 \\ w^3 \\ w^4 \end{pmatrix} = -\mathbf{S} \begin{pmatrix} w_1 \\ w_2 \\ w_3 \\ w_4 \end{pmatrix}, \quad \begin{pmatrix} w_1 \\ w_2 \\ w_3 \\ w_4 \end{pmatrix} = \mathbf{S} \begin{pmatrix} w^1 \\ w^2 \\ w^3 \\ w^4 \end{pmatrix}, \quad \text{where} \quad \mathbf{S} = \begin{pmatrix} \mathbf{0} & -\mathbf{1} \\ \mathbf{1} & \mathbf{0} \end{pmatrix}, \tag{17.35}$$

where $\mathbf{0}$ is an $n \times n$ matrix of 0's and $\mathbf{1}$ is an $n \times n$ unit matrix. With this definition,

$$\vec{\mathbf{dH}} = -\mathbf{S} \begin{pmatrix} \partial H/\mathbf{dq} \\ \partial H/\mathbf{dp} \end{pmatrix}. \tag{17.36}$$

Notice that $\mathbf{S}$ is a rotation matrix yielding rotation through 90° in the q, p plane when $n = 1$, while for $n > 1$ it yields rotation through 90° in each of the q^i, p_i

planes separately. Using $\mathbf{S}$, the $2n$ Hamilton's equations take the form

$$\dot{\mathbf{z}} = -\mathbf{S}\frac{\partial H}{\partial \mathbf{z}}. \tag{17.37}$$

At this point it might have seemed more natural to have defined $\mathbf{S}$ with the opposite sign, but the choice of sign is conventional. When the alternate symbol $\mathbf{J} \equiv -\mathbf{S}$ is used, Hamilton's equations become $\dot{\mathbf{z}} = \mathbf{J}(\partial H/\partial \mathbf{z})$.

It should be emphasized that, though a geometric interpretation has been given to the contravariant/covariant association, it is coordinate-dependent and hence artificial. Even changing the units of, say, momenta, but not displacements, changes the meaning of, say, orthogonality. It does not, however, change the solutions of the equations of motion.

17.2 Symplectic Geometry

In the previous sections the evolution of a mechanical system in phase space was codified in terms of the antisymmetric bilinear form $\widetilde{\omega}$ and it was stated that this form plays a role in phase space analogous to the metric form in Euclidean space. The geometry of a space endowed with such a form is called "symplectic geometry." The study of this geometry can be formulated along the same lines that ordinary geometry was studied in the early chapters of this text. In Chapter 3 one started with rectangular axes for which the coefficients of the metric tensor were those of the identity matrix. When skew axes were introduced the metric tensor, though no longer diagonal, remained symmetric. Conversely it was found that, given a symmetric metric tensor, axes could be found such that it became a diagonal matrix – the metric form became a sum of squares (possibly with some signs negative). It was also shown that orthogonal matrices play a special role describing transformations that preserve the Pythagorean form, and the product of two such transformations has the same property. Because of this and some other well known properties, these transformations were said to form a group, the *orthogonal group*. Here we will derive the analogous "linearized" properties and will sketch the "curvilinear" properties heuristically.

17.2.1 Symplectic Products and Symplectic Bases

For symplectic geometry the step analogous to introducing a metric tensor was the step of introducing the "canonical 2-form"

$$\widetilde{\omega} = \widetilde{\mathbf{q}}^1 \wedge \widetilde{\mathbf{p}}_1 + \widetilde{\mathbf{q}}^2 \wedge \widetilde{\mathbf{p}}_2 + \cdots + \widetilde{\mathbf{q}}^n \wedge \widetilde{\mathbf{p}}_n = \widetilde{\mathbf{q}}^i \wedge \widetilde{\mathbf{p}}_i. \tag{17.38}$$

Here, analogous to neglecting curvilinear effects in ordinary geometry, we have removed the differential "d" symbols since we now assume purely linear geometry for all amplitudes. Later, when considering "variational" equations that relate solutions in the vicinity of a given solution, it will be appropriate to put back the "d" symbols. (Recall that for any vector $\mathbf{z} = (q^1, p_1, q^2, p_2, \cdots)^T$ one has $\widetilde{\mathbf{q}}^1(\mathbf{z}) \equiv \langle\widetilde{\mathbf{q}}^1, \mathbf{z}\rangle = q^1$ and so on.)

The form $\widetilde{\boldsymbol{\omega}}$ accepts two vectors, say $\mathbf{w}$ and $\mathbf{z}$, as arguments and generates a scalar. One can therefore introduce an abbreviated notation

$$[\mathbf{w}, \mathbf{z}] = \widetilde{\boldsymbol{\omega}}(\mathbf{w}, \mathbf{z}), \tag{17.39}$$

and this "skew-scalar" or "symplectic" product is the analog of the dot product of ordinary vectors. If this product vanishes the vectors $\mathbf{w}$ and $\mathbf{z}$ are said to be "in involution." Clearly one has

$$[\mathbf{w}, \mathbf{z}] = -[\mathbf{z}, \mathbf{w}] \qquad \text{and} \qquad [\mathbf{z}, \mathbf{z}] = 0, \tag{17.40}$$

so every vector is in involution with itself. The concept of vectors being in involution will be most significant when the vectors are solutions of the equations of motion. A set of n independent solutions in involution is said to form a "Lagrangian set."

The skew-scalar products of pairs drawn from the $2n$ basis vectors

$$\mathbf{e}_{q1}, \mathbf{e}_{q2}, \ldots \qquad \text{and} \qquad \mathbf{e}^{p1}, \mathbf{e}^{p2}, \ldots \tag{17.41}$$

are especially simple; (with no summation implied)

$$[\mathbf{e}_{q(i)}, \mathbf{e}^{p(i)}] = 1, \qquad \text{and all other basis vector products vanish.} \tag{17.42}$$

Expressed in words, as well as being skew-orthogonal to itself, each basis vector is also skew-orthogonal to all other basis vectors except that of its conjugate mate, and for that one the product is ± 1. Any basis satisfying these special product relations is known as a "symplectic basis."

Though the only skew-symmetric form that has been introduced to this point was that given in Eq. (17.38), in general a similar skew-product can be defined for any skew-symmetric form $\widetilde{\boldsymbol{\omega}}$ whatsoever. Other than linearity, the main requirements on $\widetilde{\boldsymbol{\omega}}$ are those given in Eq. (17.40), but to avoid "degenerate" cases it is also necessary to require that there be no nonzero vector orthogonal to all other vectors.

With these properties satisfied, the space together with $\widetilde{\boldsymbol{\omega}}$ is said to be symplectic. Let N stand for its dimensionality. A symplectic basis like (17.41) can be found for the space. To show this one can start by picking any arbitrary vector $\mathbf{u}_1$ as the first basis vector. Then, because of the nondegeneracy requirement, there has to be another vector, call it $\mathbf{v}_1$, that has a nonvanishing skew-scalar product with $\mathbf{u}_1$, and the product can be made exactly 1 by appropriate choice of a scale factor multiplying $\mathbf{v}_1$. If $N = 2$ then $n = N/2 = 1$ and the basis is complete.

For $N > 2$, by subtracting an appropriate multiple of $\mathbf{u}_1$ from a vector in the space the resulting vector either vanishes or has vanishing skew-scalar product with $\mathbf{u}_1$. Perform this operation on all vectors. The resulting vectors form a space of dimensionality $N - 1$ that is said to be "skew complementary" to $\mathbf{u}_1$; call it U_1. It has to contain $\mathbf{v}_1$. Similarly one can find a space V_1 of dimensionality $N - 1$ skew complementary to $\mathbf{v}_1$. Since V_1 does not contain $\mathbf{v}_1$ it follows that U_1 and V_1 do not coincide, and hence their intersection, call it W, has dimension $N - 2$.

On W we must and can use the same rule $[\cdot, \cdot]$ for calculating skew-scalar products, and we now check that this product is nondegenerate. If there were a vector skew-orthogonal to all elements of W, because it is also skew-orthogonal to $\mathbf{u}_1$ and $\mathbf{v}_1$ it would have been skew-orthogonal to the whole space which is a contradiction.

By induction on n we conclude that the dimensionality of the symplectic space is even, $N = 2n$. Also, since a symplectic basis can always be found (as in Eq. (17.42)), all symplectic spaces of the same dimensionality are isomorphic, and the skew-scalar product can always be expressed as in Eq. (17.38).

The arguments of this section have assumed linearity but they can be generalized to arbitrary curvilinear geometry and, when that is done, the result is known as Darboux's theorem. From a physicist's point of view the generalization is obvious since, looking on a fine enough scale, even nonlinear transformations appear linear. A variant of this "argument" is that, just as an ordinary metric tensor can be transformed to be Euclidean over small regions, the analogous property should be true for a symplectic "metric." This reasoning is only heuristic however (see Arnold, p. 230, for further discussion).

17.2.2 Symplectic Transformations

For symplectic spaces the analog of orthogonal transformation matrices (which preserve scalar products) are symplectic matrices $\mathbf{M}$ (that preserve skew-scalar products.) The "transform" $\mathbf{Z}$ of vector $\mathbf{z}$ by $\mathbf{M}$ is given by

$$\mathbf{Z} = \mathbf{M}\mathbf{z}. \tag{17.43}$$

The transforms of two vectors $\mathbf{u}$ and $\mathbf{v}$ are $\mathbf{M}\mathbf{u}$ and $\mathbf{M}\mathbf{v}$ and the condition for $\mathbf{M}$ to be symplectic is, for all $\mathbf{u}$ and $\mathbf{v}$,

$$[\mathbf{M}\mathbf{u}, \mathbf{M}\mathbf{v}] = [\mathbf{u}, \mathbf{v}]. \tag{17.44}$$

If $\mathbf{M}_1$ and $\mathbf{M}_2$ are applied consecutively, their product $\mathbf{M}_2\mathbf{M}_1$ is necessarily also symplectic. Since the following problem shows that the determinant of a symplectic matrix is 1, it follows that the matrix is invertible, and from this it follows that the symmetric transformations form a group.

Problem 17.2.1. *In a symplectic basis the skew-scalar product can be reexpressed as an ordinary dot product by using the isomorphism I defined in Eq. (17.25), and I can be represented by the matrix* **S** *defined in Eq. (17.36). Using the fact that* $\det|\mathbf{S}| = 1$, *adapt the argument of Section 4.1 to show that* $\det|\mathbf{M}| = 1$ *if* **M** *is a symplectic matrix.*

17.2.3
Properties of Symplectic Matrices

Vectors in phase space have dimensionality $2n$ and, when expressed in a symplectic basis, have the form $(q^1, p_1, q^2, p_2, \dots)^T$ or $(q^1, q^2, \cdots, p_1, p_2, \dots)^T$, whichever one prefers. Because it permits a more compact partitioning, the second ordering is more convenient for writing compact, general matrix equations. But when motion in one phase-space plane, say (q^1, p_1), is independent of, or approximately independent of, motion in another plane, say (q^2, p_2), the first ordering is more convenient. In Eq. (17.36) the isomorphism from covariant to contravariant components was expressed in coordinates for a particular form $\widetilde{\mathbf{dH}}$. The inverse isomorphism can be applied to arbitrary vector $\vec{\mathbf{z}}$ to yield a form $\widetilde{\mathbf{z}}$;

$$\widetilde{\mathbf{z}} \overset{q}{=} \mathbf{S}\begin{pmatrix}\mathbf{q}\\ \mathbf{p}\end{pmatrix} \qquad \text{where} \qquad \mathbf{S} = \begin{pmatrix}\mathbf{0} & -\mathbf{1}\\ \mathbf{1} & \mathbf{0}\end{pmatrix}, \tag{17.45}$$

(The qualified equality symbol $\overset{q}{=}$ acknowledges that the notation is a bit garbled, with the left-hand side appearing to be intrinsic and the right-hand side expressed in components; as it appears in this equation, $\widetilde{\mathbf{z}}$ has to be regarded as a column array of the covariant coefficients of the form $\widetilde{\mathbf{z}}$.) Using Eq. (17.45) it is possible to express the skew-scalar product $[\mathbf{w}, \mathbf{z}]$ of vectors $\mathbf{w}$ and $\mathbf{z}$ (defined in Eq. (17.39)) in terms of ordinary scalar products and from those, as a quadratic form;

$$[\vec{\mathbf{w}}, \vec{\mathbf{z}}] \equiv \langle \widetilde{\mathbf{w}}, \vec{\mathbf{z}}\rangle \overset{q}{=} \langle \mathbf{S}\vec{\mathbf{w}}, \vec{\mathbf{z}}\rangle = \mathbf{S}\vec{\mathbf{w}} \cdot \vec{\mathbf{z}} = \vec{\mathbf{z}} \cdot \mathbf{S}\vec{\mathbf{w}} = z^i S_{ij} w^j = -w^i S_{ij} z^j. \tag{17.46}$$

Since displacements and momenta are being treated homogeneously here it is impossible to retain the traditional placement of the indices for both displacements and momenta. Equation (17.46) shows that the elements $-S_{ij}$ are the coefficients of a quadratic form giving the skew-scalar product of vectors $\vec{\mathbf{z}}_a$ and $\vec{\mathbf{z}}_b$ in terms of their components;

$$[\vec{\mathbf{z}}_b, \vec{\mathbf{z}}_a] = \begin{pmatrix} q_b^1 & p_{b1} & q_b^2 & p_{b2}\end{pmatrix} \begin{pmatrix} 0 & 1 & 0 & 0\\ -1 & 0 & 0 & 0\\ 0 & 0 & 0 & 1\\ 0 & 0 & -1 & 0\end{pmatrix} \begin{pmatrix} q_a^1\\ p_{a1}\\ q_a^2\\ p_{a2}\end{pmatrix}. \tag{17.47}$$

This combination, that we have called a "symplectic product," is sometimes called "the Poisson bracket" of the vectors $\vec{\mathbf{z}}_b$ and $\vec{\mathbf{z}}_a$ but it must be distinguished from the Poisson bracket of scalar functions to be defined shortly.

When the condition Eq. (17.44) for a linear transformation $\mathbf{M}$ to be symplectic is expressed with dot products, as in Eq. (17.46), it becomes

$$\mathbf{Su} \cdot \mathbf{v} = \mathbf{SMu} \cdot \mathbf{Mv} = \mathbf{M}^T\mathbf{SM}\,\mathbf{u} \cdot \mathbf{v}. \tag{17.48}$$

This can be true for all $\mathbf{u}$ and $\mathbf{v}$ only if

$$\mathbf{M}^T\mathbf{SM} = \mathbf{S}. \tag{17.49}$$

This is an algebraic test that can be applied to a matrix $\mathbf{M}$ whose elements are known explicitly, to determine whether or not it is symplectic. Equivalently $\mathbf{MSM}^T = \mathbf{S}$.

Problem 17.2.2. *Hamilton's equations in matrix form are*

$$\dot{\mathbf{z}} = -\mathbf{S}\frac{\partial H}{\partial \mathbf{z}}, \tag{17.50}$$

and a change of variables with symplectic matrix $\mathbf{M}$

$$\mathbf{z} = \mathbf{MZ}, \tag{17.51}$$

is performed. Show that the form of Hamilton's equations is left invariant. Such transformations are said to be "canonical."

A result equivalent to Eq. (17.49) is obtained by multiplying it on the right by $\mathbf{M}^{-1}$ and on the left by $\mathbf{S}$;

$$\mathbf{M}^{-1} = -\mathbf{SM}^T\mathbf{S}. \tag{17.52}$$

This provides a handy numerical shortcut for determining the inverse of a matrix that is known to be symplectic; the right-hand side requires only matrix transposition and multiplication by a matrix whose elements are mainly zero, and the others ± 1. Subsequent formulas will be abbreviated by introducing $\overline{\mathbf{A}}$, to be called the "symplectic conjugate" of arbitrary matrix $\mathbf{A}$ by

$$\overline{\mathbf{A}} = -\mathbf{SA}^T\mathbf{S}. \tag{17.53}$$

A necessary and sufficient condition for matrix $\mathbf{M}$ to be symplectic is then

$$\mathbf{M}^{-1} = \overline{\mathbf{M}}. \tag{17.54}$$

From here on, until further notice, when a matrix is symbolized by $\mathbf{M}$, it will implicitly be assumed to be symplectic and hence to satisfy this equation.

For any 2×2 matrix $\mathbf{A}$, with $\mathbf{S}$ given by Eq. (17.45), substituting into Eq. (17.53) yields

$$\overline{\mathbf{A}} \equiv \overline{\begin{pmatrix} a & b \\ c & d \end{pmatrix}} = \begin{pmatrix} d & -b \\ -c & a \end{pmatrix} = \mathbf{A}^{-1} \det |\mathbf{A}|, \tag{17.55}$$

assuming the inverse exists. Hence, using Eq. (17.54), for $n = 1$ a necessary and sufficient condition for symplecticity is that $\det |\mathbf{M}| = 1$. For $n > 1$ this condition will shortly be shown to be necessary. But it can obviously not be sufficient, since Eq. (17.54) implies more than one independent algebraic condition.

For most practical calculations it is advantageous to list the components of phase-space vectors in the order $\mathbf{z} = (q^1, p_1, q^2, p_2)^T$ and then to streamline the notation further by replacing this by $\mathbf{z} = (x, p, y, q)^T$. (Here, and when the generalization to arbitrary n is obvious, we exhibit only this $n = 2$ case explicitly.) With this ordering the matrix $\mathbf{S}$ takes the form

$$\mathbf{S} = \begin{pmatrix} 0 & -1 & 0 & 0 \\ 1 & 0 & 0 & 0 \\ 0 & 0 & 0 & -1 \\ 0 & 0 & 1 & 0 \end{pmatrix}. \tag{17.56}$$

Partitioning a 4×4 matrix $\mathbf{M}$ into 2×2 blocks, it and its symplectic conjugate are

$$\mathbf{M} = \begin{pmatrix} \mathbf{A} & \mathbf{B} \\ \mathbf{C} & \mathbf{D} \end{pmatrix}, \qquad \overline{\mathbf{M}} = \begin{pmatrix} \overline{\mathbf{A}} & \overline{\mathbf{C}} \\ \overline{\mathbf{B}} & \overline{\mathbf{D}} \end{pmatrix}. \tag{17.57}$$

The eigenvalues of a symplectic matrix $\mathbf{M}$ will play an important role in the sequel. The "generic" situation is for all eigenvalues to be unequal, and that is much the easiest case for the following discussion. The degeneracy of equal eigenvalues causes the occurrence of indeterminant ratios which require special treatment in the algebra. Unfortunately there are two cases where equality of eigenvalues is unavoidable. (i) Systems often exhibit symmetries which, if exactly satisfied, force equality among certain eigenvalues or sets of eigenvalues. This case is more a nuisance than anything else since the symmetry can be removed either realistically (as it would be in nature) or artificially; in the latter case the perturbation can later be reduced to insignificance. It is very common for perturbing forces of one kind or another, in spite of being extremely small, to remove degeneracy in this way. (ii) It is often appropriate to idealize systems by one or more variable "control parameters" that characterize the way the system is adjusted externally. Since the eigenvalues depend continuously on these control parameters the eigenvalues may have to become exactly equal as a control parameter is varied. It may happen that the system refuses to allow this (see Problem 1.11.2). In other cases the eigenvalues can pass gracefully through each other. Typically the possibility of such "collisions" of the eigenvalues contributes to the "essence" of the system under study and following the eigenvalues through the collision or avoidance of collision is essential to the understanding of the device. For example a "bifurcation" can occur at the point where the eigenvalues become equal and in that case the crossing point marks the boundary of regions of qualitatively different behavior.

In spite of this inescapability of degeneracy, in the interest of simplifying the discussion, for the time being we will assume all eigenvalues of **M** are distinct. When discussing approximate methods in Chapter 16 the problem of equal eigenvalues was mainly ignored.

The eigenvalues λ and eigenvectors $\boldsymbol{\psi}_\lambda$ of any matrix **A** satisfy the "eigenvalue" and the "eigenvector" equations

$$\det|\mathbf{A} - \lambda\mathbf{1}| = 0, \qquad \text{and} \qquad \mathbf{A}\boldsymbol{\psi}_\lambda = \lambda\boldsymbol{\psi}_\lambda. \tag{17.58}$$

Since the determinant is unchanged when **A** is replaced by $\mathbf{A}^T$, a matrix and its transpose share the same set of eigenvalues. From Eq. (17.53) it follows that the symplectic conjugate $\overline{\mathbf{A}}$ also has the same set of eigenvalues. Then, from Eq. (17.54), it follows that the eigenvalue spectrum of a symplectic matrix **M** and its inverse $\mathbf{M}^{-1}$ are identical. For any matrix, if λ is an eigenvalue, then $1/\lambda$ is an eigenvalue of the inverse. It follows that if λ is an eigenvalue of a symplectic matrix, then so also is $1/\lambda$.

Even if all the elements of **M** are real (as we assume) the eigenvectors can be complex and so can the eigenvalues. But here is where symplectic matrices shine. Multiplying the second of Eqs. (17.58) by $\mathbf{M}^{-1}$ and using Eq. (17.54), one concludes both that

$$\mathbf{M}\boldsymbol{\psi}_\lambda = \lambda\boldsymbol{\psi}_\lambda, \qquad \text{and} \qquad \overline{\mathbf{M}}\boldsymbol{\psi}_\lambda = \frac{1}{\lambda}\boldsymbol{\psi}_\lambda. \tag{17.59}$$

Writing $\lambda = re^{i\theta}$ then $1/\lambda = (1/r)e^{-i\theta}$ is also an eigenvalue and these two eigenvalues are located in the complex λ-plane as shown in Fig. 17.5(a). But it also follows from the normal properties of the roots of a polynomial equation that if an eigenvalue $\lambda = re^{i\theta}$ is complex then its complex conjugate $\lambda^* = re^{-i\theta}$ is also an eigenvalue. This is illustrated in Fig. 17.5(b). It then follows, as shown in figures (c) and (d), that the eigenvalues can only come in real reciprocal pairs, or in complex conjugate pairs lying on the unit circle, or in quartets as in (c). For the cases illustrated in Fig. 17.5(d), these requirements can be exploited algebraically by adding the equations (17.59) to give

$$(\mathbf{M} + \overline{\mathbf{M}})\boldsymbol{\psi}_\lambda = \Lambda\boldsymbol{\psi}_\lambda \qquad \text{where} \qquad \Lambda = \lambda + \lambda^{-1}, \tag{17.60}$$

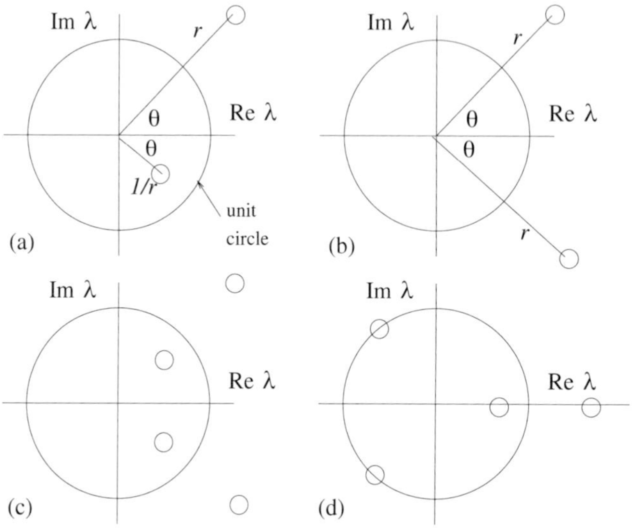

Fig. 17.5 (a) If $\lambda = re^{i\theta}$ is an eigenvalue of a symplectic matrix, then so also is $1/\lambda = (1/r)e^{-i\theta}$. (b) If an eigenvalue $\lambda = re^{i\theta}$ is complex then its complex conjugate $\lambda^* = re^{-i\theta}$ is also an eigenvalue. (c) If any eigenvalue is complex with absolute value other than 1, the three complementary points shown are also eigenvalues. (d) Eigenvalues can come in pairs only if they are real (and reciprocal) or lie on the unit circle (symmetrically above and below the real axis).

which shows that the eigenvalues Λ of $\mathbf{M}+\overline{\mathbf{M}}$ are real. Performing the algebra explicitly in the 4×4 case yields

$$\mathbf{M}+\overline{\mathbf{M}} = \begin{pmatrix} \mathbf{A}+\overline{\mathbf{A}} & \mathbf{B}+\overline{\mathbf{C}} \\ \mathbf{C}+\overline{\mathbf{B}} & \mathbf{D}+\overline{\mathbf{D}} \end{pmatrix} = \begin{pmatrix} (\mathrm{tr}\,\mathbf{A})\mathbf{1} & \overline{\mathbf{E}} \\ \mathbf{E} & (\mathrm{tr}\,\mathbf{D})\mathbf{1} \end{pmatrix}, \tag{17.61}$$

where the off-diagonal combination $\mathbf{E}$ and its determinant $\mathcal{E}$ are defined by

$$\mathbf{E} = \mathbf{C}+\overline{\mathbf{B}} \equiv \begin{pmatrix} e & f \\ g & h \end{pmatrix}, \qquad \text{and} \qquad \mathcal{E} \equiv \det|\mathbf{E}| = eh - fg. \tag{17.62}$$

The eigenvalue equation is[5]

$$\det\begin{vmatrix} (\mathrm{tr}\,\mathbf{A}-\Lambda)\mathbf{1} & \overline{\mathbf{E}} \\ \mathbf{E} & (\mathrm{tr}\,\mathbf{D}-\Lambda)\mathbf{1} \end{vmatrix} = \Lambda^2 - (\mathrm{tr}\,\mathbf{A} + \mathrm{tr}\,\mathbf{D})\Lambda + \mathrm{tr}\,\mathbf{A}\ \mathrm{tr}\,\mathbf{D} - \mathcal{E} = 0, \tag{17.63}$$

5) It is not in general valid to evaluate the determinant of a partitioned matrix treating the blocks as if they were ordinary numbers, but it is valid if the diagonal blocks are individually proportional to the identity matrix as is the case here.

whose solutions are

$$\Lambda_{\mathbf{A,D}} = (\operatorname{tr}\mathbf{A} + \operatorname{tr}\mathbf{D})/2 \pm \sqrt{(\operatorname{tr}\mathbf{A} - \operatorname{tr}\mathbf{D})^2/4 + \mathcal{E}}. \tag{17.64}$$

The eigenvalues have been given subscripts A and D to facilitate discussion in the common case that the off-diagonal elements are small so the eigenvalues can be associated with the upper left and lower right blocks of $\mathbf{M}$ respectively. Note that the eigenvalues satisfy simple equations:

$$\Lambda_A + \Lambda_D = \operatorname{tr}\mathbf{A} + \operatorname{tr}\mathbf{D}, \qquad \Lambda_A\Lambda_D = \operatorname{tr}\mathbf{A}\ \operatorname{tr}\mathbf{D} - \mathcal{E}. \tag{17.65}$$

Though we have been proceeding in complete generality and this result is valid for any $n = 2$ symplectic matrix, the structure of these equations all but forces one to contemplate the possibility that $\mathcal{E}$ be "small," which would be true if the off-diagonal blocks of $\mathbf{M}$ are small. This would be the case if the x and y motions were independent or almost independent. Calling x "horizontal" and y "vertical" one says that the off-diagonal blocks $\mathbf{B}$ and $\mathbf{C}$ "couple" the horizontal and vertical motion. If $\mathbf{B} = \mathbf{C} = 0$ the horizontal and vertical motions proceed independently. The remarkable feature of Eqs. (17.65) is that, though $\mathbf{B}$ and $\mathbf{C}$ together have eight elements each capable of not vanishing, they shift the eigenvalues only through the combination $\mathcal{E}$.

In Eq. (17.64) we should insist that $A(D)$ go with the $+(-)$ sign respectively when $\operatorname{tr}\mathbf{A} - \operatorname{tr}\mathbf{D}$ is positive and vice versa. This choice assures, if $\mathcal{E}$ is in fact small, that the perturbed eigenvalue $\Lambda_{\mathbf{A}}$ will correspond to approximately horizontal motion and $\Lambda_{\mathbf{D}}$ to approximately vertical.

Starting from a 4×4 matrix one expects the characteristic polynomial to be quartic in λ but here we have found a characteristic polynomial quadratic in Λ. The reason for this is that the combination $\mathbf{M} + \overline{\mathbf{M}}$ has nothing but pairs of degenerate roots so the quartic characteristic equation factorizes exactly as the square of a quadratic equation. We have shown this explicitly only for $n = 2$ (and for $n = 3$ in an example below) but the result holds for arbitrary n.

Anticipating results to appear later on, multiplying $\mathbf{M}$ by itself repeatedly will be of crucial importance for the behavior of Hamiltonian systems over long times. Such powers of $\mathbf{M}$ are most easily calculated if the variables have been transformed to make $\mathbf{M}$ diagonal, in which case the diagonal elements are equal to the eigenvalues. Then, evaluating $\mathbf{M}^l$ for large (integer) l, the diagonal elements are λ^l and their magnitudes are $|\lambda|^l$ which approach 0 if $|\lambda| < 1$ or ∞ if $|\lambda| > 1$. Both of these behaviors can be said to be "trivial." This leaves just one possibility as the case of greatest interest. It is one of the two cases illustrated in Fig. 17.5(d) – the one in which each of a pair of eigenvalues lies on the unit circle. In this case there are real angles μ_A and μ_D satisfying

$$\begin{aligned}\Lambda_A &= e^{i\mu_A} + e^{-i\mu_A} = 2\cos\mu_A,\\ \Lambda_D &= e^{i\mu_D} + e^{-i\mu_D} = 2\cos\mu_D.\end{aligned} \tag{17.66}$$

In the special uncoupled case, for which **B** and **C** vanish, these angles degenerate into μ_x and μ_y, the values appropriate for pure horizontal and vertical motion, and we have

$$\Lambda_{A,D} = \text{tr}\, A, D = 2\cos\mu_{x,y} = 2\cos\mu_{A,D}. \tag{17.67}$$

The *sign* of determinant $\mathcal{E}$ has special significance if the uncoupled eigenvalues are close to each other. This can be seen most easily by rearranging Eqs. (17.64) and (17.66) into the form

$$(\cos\mu_A - \cos\mu_D)^2 = \tfrac{1}{4}(\text{tr}\,\mathbf{A} - \text{tr}\,\mathbf{D})^2 + \mathcal{E}. \tag{17.68}$$

If the unperturbed eigenvalues are close the first term on the right-hand side is small. Then for $\mathcal{E} < 0$ the perturbed eigenvalues Λ can become complex (which pushes the eigenvalues λ off the unit circle, leading to instability.) But if $\mathcal{E} > 0$ the eigenvalues remain real and the motion remains stable, at least for sufficiently small values of $\mathcal{E} > 0$.

An even more important inference can be drawn from Eqs. (17.64) and (17.66). If the parameters are such that both $\cos\mu_A$ and $\cos\mu_D$ lie in the (open) range $-1 < \cos\mu_A < \cos\mu_A < 1$, then both angles μ_A and μ_D are real and the motion is "stable." What is more, for sufficiently small variations of the parameters the eigenvalues, because they must move smoothly, cannot leave the unit circle and these angles necessarily remain real. This means the stability has a kind of "robustness" against small changes in the parameters. Pictorially, the eigenvalues in Fig. 17.5(d) have to stay on the unit circle as the parameters are varied continuously. Only when an eigenvalue "collides" with another eigenvalue can the absolute value of either eigenvalue deviate from 1. Furthermore, if the collision is with the complex conjugate mate it can only occur at at either ± 1.

The reader who is not impressed that it has been possible to find closed form algebraic formulas for the eigenvalues of a 4×4 matrix should attempt to do it for a general matrix $((a, b, c, d), (e, f, g, h), \ldots)$. It is symplecticity that has made it possible. To exploit our good fortune we should also find closed form expressions for the eigenvectors. One can write a 4-component vector in the form

$$\mathbf{z} = \begin{pmatrix} \chi \\ \xi \end{pmatrix} \quad \text{where} \quad \chi = \begin{pmatrix} x \\ p \end{pmatrix} \quad \text{and} \quad \xi = \begin{pmatrix} y \\ q \end{pmatrix}. \tag{17.69}$$

One can then check that the vectors

$$\mathbf{X} = \begin{pmatrix} \chi \\ \frac{\mathbf{E}}{\Lambda - \text{tr}\,\mathbf{D}}\chi \end{pmatrix} \quad \text{and} \quad \mathbf{Y} = \begin{pmatrix} \frac{\overline{\mathbf{E}}}{\Lambda - \text{tr}\,\mathbf{A}}\xi \\ \xi \end{pmatrix} \tag{17.70}$$

satisfy the (same) equations $(\mathbf{M} + \mathbf{M}^{-1})\mathbf{X} = \Lambda\mathbf{X}$ and $(\mathbf{M} + \mathbf{M}^{-1})\mathbf{Y} = \Lambda\mathbf{Y}$ for either eigenvalue and arbitrary χ or ξ. If we think of $\mathcal{E}$ as being small, so that the eigenvectors are close to the uncoupled solution, than we should select the Λ factors so that Eqs. (17.70) become

$$\mathbf{X} = \begin{pmatrix} \chi \\ \frac{\mathbf{E}}{\Lambda_A - \mathrm{tr}\,\mathbf{D}}\chi \end{pmatrix} \quad \text{and} \quad \mathbf{Y} = \begin{pmatrix} \frac{\overline{\mathbf{E}}}{\Lambda_D - \mathrm{tr}\,\mathbf{A}}\xi \\ \xi \end{pmatrix}. \tag{17.71}$$

In each case the denominator factor has been chosen to have a "large" absolute value so as to make the factor multiplying its 2-component vector "small." In this way, the lower components of $\mathbf{X}$ and the upper components of $\mathbf{Y}$ are "small." In the limit of vanishing $\mathcal{E}$ only the upper components survive for x-motion and only the lower for y. This formalism may be mildly reminiscent of the 4-component wavefunctions describing relativistic electrons and positrons.

There is another remarkable formula that a 4×4 symplectic matrix must satisfy. A result from matrix theory is that a matrix satisfies its own eigenvalue equation. Applying this to $\mathbf{M} + \overline{\mathbf{M}}$ one has

$$(\mathbf{M} + \overline{\mathbf{M}})^2 - (\Lambda_A + \Lambda_D)(\mathbf{M} + \overline{\mathbf{M}}) + \Lambda_A\Lambda_D = 0. \tag{17.72}$$

Rearranging this yields

$$\mathbf{M}^2 + \overline{\mathbf{M}}^2 - (\Lambda_A + \Lambda_D)(\mathbf{M} + \overline{\mathbf{M}}) + 2 + \Lambda_A\Lambda_D = 0. \tag{17.73}$$

By using Eq. (17.65) this equation can be expressed entirely in terms of the coefficients of $\mathbf{M}$

$$\mathbf{M}^2 + \overline{\mathbf{M}}^2 - (\mathrm{tr}\,\mathbf{A} + \mathrm{tr}\,\mathbf{D})(\mathbf{M} + \overline{\mathbf{M}}) + 2 + \mathrm{tr}\,\mathbf{A}\,\mathrm{tr}\,\mathbf{D} - \mathcal{E} = 0. \tag{17.74}$$

Problem 17.2.3. *Starting with* $\mathbf{M} + \overline{\mathbf{M}}$ *expressed as in Eq. (17.61), verify Eq. (17.72) explicitly.*

Problem 17.2.4. *Find the equation analogous to Eq. (17.73) that is satisfied by a* 6×6 *symplectic matrix.*

$$\mathbf{M} = \begin{pmatrix} \mathbf{A} & \mathbf{B} & \mathbf{E} \\ \mathbf{C} & \mathbf{D} & \mathbf{F} \\ \mathbf{G} & \mathbf{H} & \mathbf{J} \end{pmatrix}. \tag{17.75}$$

It is useful to introduce off-diagonal combinations

$$\mathbf{B} + \overline{\mathbf{C}} = \begin{pmatrix} h & -f \\ -g & e \end{pmatrix}, \quad \mathbf{E} + \overline{\mathbf{G}} = \begin{pmatrix} n & -l \\ -m & k \end{pmatrix}, \quad \mathbf{F} + \overline{\mathbf{H}} = \begin{pmatrix} s & -q \\ -r & p \end{pmatrix}. \tag{17.76}$$

The eigenvalue equation for $\Lambda = \lambda + 1/\lambda$ *is cubic in this case,*

$$\Lambda^3 - p_1\Lambda^2 - p_2\Lambda - p_3 = 0, \tag{17.77}$$

but it can be written explicitly and there is a procedure for solving a cubic equation. The roots can be written in term the combinations

$$Q = \frac{p_1^2 + 3p_2}{9}, \qquad R = \frac{-2p_1^2 - 9p_1p_2 - 27p_3}{54}, \qquad \theta = \cos^{-1}\frac{R}{Q^{3/2}}. \quad (17.78)$$

This is of more than academic interest since the Hamiltonian motion of a single particle in three-dimensional space is described by such a matrix.

17.3
Poisson Brackets of Scalar Functions

Many of the relations of Hamiltonian mechanics can be expressed compactly in terms of the Poisson brackets that we now define.

17.3.1
The Poisson Bracket of Two Scalar Functions

Consider two functions $f(\mathbf{z}) \equiv f(\mathbf{q}, \mathbf{p})$ and $g(\mathbf{z}) \equiv g(\mathbf{q}, \mathbf{p})$ defined on phase space. From them can be formed $\widetilde{\mathbf{df}}$ and $\widetilde{\mathbf{dg}}$ and from them (using the symplectic 2-form $\widetilde{\boldsymbol{\omega}}$ and the standard association) the vectors $\vec{\mathbf{df}}$ and $\vec{\mathbf{dg}}$. The "Poisson bracket" of functions f and g is then defined by

$$\{f, g\} = \widetilde{\boldsymbol{\omega}}(\vec{\mathbf{df}}, \vec{\mathbf{dg}}). \quad (17.79)$$

Spelled out more explicitly, as in Eq. (17.16), this becomes

$$\{f, g\} = \langle \widetilde{\mathbf{dq}}^i, \vec{\mathbf{df}} \rangle \langle \widetilde{\mathbf{dp}}_i, \vec{\mathbf{dg}} \rangle - \langle \widetilde{\mathbf{dq}}^i, \vec{\mathbf{dg}} \rangle \langle \widetilde{\mathbf{dp}}_i, \vec{\mathbf{df}} \rangle = \frac{\partial f}{\partial q^i}\frac{\partial g}{\partial p_i} - \frac{\partial f}{\partial p_i}\frac{\partial g}{\partial q^i}, \quad (17.80)$$

where the scalar products have been obtained using Eqs. (17.28). Though the terms in this sum are individually coordinate-dependent, by its construction, the Poisson bracket is itself coordinate-independent.

One application of the Poisson bracket is to express time evolution of the system. Consider the evolution of a general function $f(\mathbf{q}(t), \mathbf{p}(t), t)$, as its arguments follow a phase space system trajectory. Its time derivative is given by

$$\dot{f} = \sum\left(\frac{\partial f}{\partial q}\dot{q} + \frac{\partial f}{\partial p}\dot{p}\right) + \frac{\partial f}{\partial t} = \sum\left(\frac{\partial f}{\partial q}\frac{\partial H}{\partial p} - \frac{\partial f}{\partial p}\frac{\partial H}{\partial q}\right) + \frac{\partial f}{\partial t} = \{f, H\} + \frac{\partial f}{\partial t}. \quad (17.81)$$

In the special case that the function f has no explicit time dependence, its time derivative $\dot{f}$ is therefore given directly by $\{f, H\}$.

17.3.2
Properties of Poisson Brackets

The following properties are easily derived:

$$\text{Jacobi identity :} \quad \Big\{f,\{g,h\}\Big\} + \Big\{g,\{h,f\}\Big\} + \Big\{h,\{f,g\}\Big\} = 0. \tag{17.82}$$

$$\text{Leibnitz property :} \quad \{f_1 f_2, g\} = f_1\{f_2, g\} + f_2\{f_1, g\}. \tag{17.83}$$

$$\text{Explicit time dependence :} \quad \frac{\partial}{\partial t}\{f_1, f_2\} = \Big\{\frac{\partial f_1}{\partial t}, f_2\Big\} + \Big\{f_1, \frac{\partial f_2}{\partial t}\Big\}. \tag{17.84}$$

Theorem 17.3.1 (Jacobi's theorem). *If* $\{H, f_1\} = 0$ *and* $\{H, f_2\} = 0$, *then* $\{H, \{f_1, f_2\}\} = 0$.

Proof.

$$\begin{aligned}
\frac{d}{dt}\{f_1, f_2\} &= \frac{\partial}{\partial t}\{f_1, f_2\} + \{H, \{f_1, f_2\}\} \\
&= \Big\{\frac{\partial f_1}{\partial t}, f_2\Big\} + \Big\{f_1, \frac{\partial f_2}{\partial t}\Big\} - \Big\{f_1, \{f_2, H\}\Big\} - \Big\{f_2, \{f_1, H\}\Big\} \\
&= \Big\{\frac{\partial f_1}{\partial t} + \{H, f_1\}, f_2\Big\} + \Big\{f_1, \frac{\partial f_2}{\partial t} + \{f_2, H\}\Big\} \\
&= \Big\{\frac{d f_1}{dt}, f_2\Big\} + \Big\{f_1, \frac{d f_2}{dt}\Big\} = 0.
\end{aligned} \tag{17.85}$$

Corollary: If f_1 and f_2 are "integrals of the motion," then so also is $\{f_1, f_2\}$. This is the form in which Jacobi's theorem is usually remembered. □

Perturbation theory: Poisson brackets are of particular importance in perturbation theory when motion close to integrable motion is studied. Using the term "orbit element," frequently used in celestial mechanics to describe an integral of the unperturbed motion, the coefficients in a "variation of constants" perturbative procedure are expressible in terms of Poisson brackets of orbit elements, which are therefore themselves also orbit elements whose constancy throughout the motion leads to important simplification as in Section 16.1.

17.3.3
The Poisson Bracket and Quantum Mechanics

17.3.3.1 Commutation Relations

In Dirac's formulation of quantum mechanics there is a close correspondence between the Poisson brackets of classical mechanics and the commutation relations of quantum mechanics. In particular, if u and v are dynamical variables their quantum mechanical "commutator" $[u, v]_{QM} \equiv uv - vu$ is given by

$$[u, v]_{QM} = i\hbar\{u, v\}, \tag{17.86}$$

where $\hbar$ is Planck's constant (divided by 2π) and $\{u, v\}$ is the classical Poisson bracket. Hence, for example,

$$[q, p]_{QM} = qp - pq = i\hbar\{q, p\} = i\hbar. \tag{17.87}$$

In the Schrödinger representation of quantum mechanics one has $q \to q$ and $p \to -i\hbar\partial/\partial q$, where q and p are to be regarded as operators that operate on functions $f(q)$. One can then check that

$$[q, p]_{QM} = -i\hbar\left(q\frac{\partial}{\partial q} - \frac{\partial}{\partial q}q\right) = i\hbar, \tag{17.88}$$

in agreement with Eq. (17.86).

17.3.3.2 Time Evolution of Expectation Values

There needs to be "correspondence" between certain quantum mechanical and classical mechanical quantities in order to permit the "seamless" metamorphosis of a system as the conditions it satisfies are varied from being purely quantum mechanical to being classical. One such result is that the expectation values of quantum mechanical quantities should evolve according to classical laws. A quantum mechanical system is characterized by a Hamiltonian H, a wavefunction Ψ, and the wave equation relating them;

$$i\hbar\frac{\partial\Psi}{\partial t} = H\Psi. \tag{17.89}$$

The expectation value of a function of position $f(q)$ is given by

$$\overline{f} = \int \Psi^* f(q)\Psi dq. \tag{17.90}$$

Its time rate of change is then given by

$$\begin{aligned}
\dot{\overline{f}} &= \int\left(\frac{\partial\Psi^*}{\partial t}f\Psi + \Psi^*\frac{\partial f}{\partial t}\Psi + \Psi^* f\frac{\partial\Psi}{\partial t}\right)dq \\
&= \int\left(\left(\frac{H\Psi}{i\hbar}\right)^* f\Psi + \Psi^*\frac{\partial f}{\partial t}\Psi + \Psi^* f\frac{H\Psi}{i\hbar}\right)dq \\
&= \int \Psi^*\left(\frac{\partial f}{\partial t} + \frac{1}{i\hbar}(-Hf + fH)\right)\Psi dq.
\end{aligned} \tag{17.91}$$

In the final step the relation $H^* = H$ required for H to be a "Hermitian" operator has been used. To assure that $\dot{\overline{f}} = \overline{\dot{f}}$, we must then have

$$\dot{f} = \frac{\partial f}{\partial t} + \frac{i}{\hbar}[H, f]. \tag{17.92}$$

When the commutator quantum mechanical commutator $[H, f]$ is related to the classical Poisson bracket $\{H, f\}$, as in Eq. (17.86), this result corresponds with the classical formula for $\dot{f}$ given in Eq. (17.81).

17.4
Integral Invariants

17.4.1
Integral Invariants in Electricity and Magnetism

In anticipation of some complications that will arise in studying integral invariants it would be appropriate at this time to digress into the distinction between local and global topological properties in differential geometry. Unfortunately, discussions of this subject, known as "cohomology" in mathematics texts, is formidably abstract. Fortunately, physicists have already encountered some of the important notions in concrete instances. For this reason we digress to develop some analogies with vector integral calculus. Since it is assumed the reader has already encountered these results in the context of electromagnetic theory, we employ that terminology here, but with inessential constant factors set equal to 1; this includes not distinguishing between the magnetic vectors **B** and **H**. In the end the subject of electricity and magnetism will have played no role other than heuristic.

We have already encountered the sort of analysis to be performed in geometric optics. Because of the "eikonal equation" Eq. (7.12), $n(\mathbf{dr}/ds) = \nabla\phi$ was the gradient of the single-valued eikonal function ϕ. The invariance of the line integral of $n(\mathbf{dr}/ds)$ for different paths connecting the same end points then followed, which was the basis of the "principle of least time." There was potential for fallacy in this line of reasoning however, as Problem 17.4.1 is intended to illustrate.

Problem 17.4.1. *The magnetic field* **H** *of a constant current flowing along the z-axis has only x and y components and depends only on x and y. Recalling (or looking up) the formula for* **H** *in this case, and ignoring constant factors, show that* **H** *is equal to the gradient of a "magnetic potential"* Φ_M*,*

$$\mathbf{H} = \nabla\Phi_M \qquad \textit{where} \qquad \Phi_M = \tan^{-1}\frac{y}{x}. \tag{17.93}$$

In terms of polar coordinates r and θ, one has $x = r\cos\theta$ *and* $y = r\sin\theta$*. After expressing* **H** *in polar coordinates evaluate* $\oint_\gamma \mathbf{H}\cdot\mathbf{ds}$ *where* γ *is a complete circle of radius* r_0 *centered on the origin. Comment on the vanishing or otherwise of this integral. Also evaluate the same integral for a path that does not enclose the origin.*

After doing this problem you are expected to be convinced that having **H** derivable from a potential does not guarantee the γ-independence of the integral $\oint_\gamma \mathbf{H}\cdot d\mathbf{s}$. However, the integral *is* invariant to changes in path that avoid going "on the other side of" field sources.

This form of line integral is more commonly encountered in electrostatics where, based on the existence of a *single-valued* potential $\mathbf{\Phi}_E$ such that $\mathbf{E} = -\boldsymbol{\nabla}\mathbf{\Phi}_E$, the field is said to be "conservative," meaning that $\int_{P_1}^{P_2} \mathbf{E}\cdot d\mathbf{s}$ is independent of the path from P_1 to P_2. Poincaré introduced the terminology of calling such a path-independent integral an "integral invariant" or an "absolute integral invariant." But the *single-valued* requirement for $\mathbf{\Phi}_E$ is not easy to apply in practice.

To prove the electric field is conservative it is more concise in electrostatics to start from $\boldsymbol{\nabla}\times\mathbf{E} = 0$. This assures $\oint_\gamma \mathbf{E}\cdot d\mathbf{s} = 0$ (Problem 17.4.2). Though $\boldsymbol{\nabla}\times\mathbf{E} = 0$ implies the existence of Φ_E such that $\mathbf{E} = -\boldsymbol{\nabla}\mathbf{\Phi}_E$, the converse does not follow.

Problem 17.4.2. *Use Stokes' theorem to show that the condition* $\boldsymbol{\nabla}\times\mathbf{E} = 0$ *is sufficient to guarantee that the integral* $\int_\gamma \mathbf{E}\cdot d\mathbf{s}$ *is independent of path* γ.

Before investigating similar issues in mechanics, we will review the situation in magnetostatics. For symplectic mechanics it is the mathematical equivalent of Ampère's law that we will need to employ.

"Physical" argument: Ampère's law follows from the equation

$$\boldsymbol{\nabla}\times\mathbf{H} = \mathbf{J}. \tag{17.94}$$

Integrating this relation over a surface Γ_1 bounded by closed curve γ_1 and using Stokes' theorem the result is

$$\int_{\gamma_1} \mathbf{H}\cdot d\mathbf{s} = \int_{\Gamma_1} (\boldsymbol{\nabla}\times\mathbf{H})\cdot d\mathbf{a} = \int_{\Gamma_1} \mathbf{J}\cdot d\mathbf{a}, \tag{17.95}$$

giving the "flux" of $\mathbf{J}$ through surface Γ_1. As shown in Fig. 17.6, since $\mathbf{J}$ is "current density" it is natural to visualize the flow lines of $\mathbf{J}$ as being the paths of steady current. The flow lines through γ_1 form a "tube of current." The flux of $\mathbf{J}$ through γ_1 can also be said to be the "total current" flowing through γ_1. If another closed loop γ_2 is drawn around the same tube of current then it would be linked by the same total current. From this "physical" discussion the constancy of this flux seems to be "coming from" the conservation of charge, but the next section will show that this may be a misleading interpretation.

"Mathematical" Argument: Much the same argument can be made with no reference whatsoever to the vector $\mathbf{J}$. Rather, referring again to Fig. 17.6, let $\mathbf{H}$ be any vector whatsoever, and consider the vector $\boldsymbol{\nabla}\times\mathbf{H}$ obtained from it. The flow lines of $\boldsymbol{\nabla}\times\mathbf{H}$ passing through closed curve γ_1 define a "tube." Further along this tube is another closed curve γ_2 linked by the same tube. Let the part of the tube's surface between γ_1 and γ_2 be called Σ. The tube can be visualized as being "capped" at one end by a surface Γ_1 bounded by γ_1 and at the

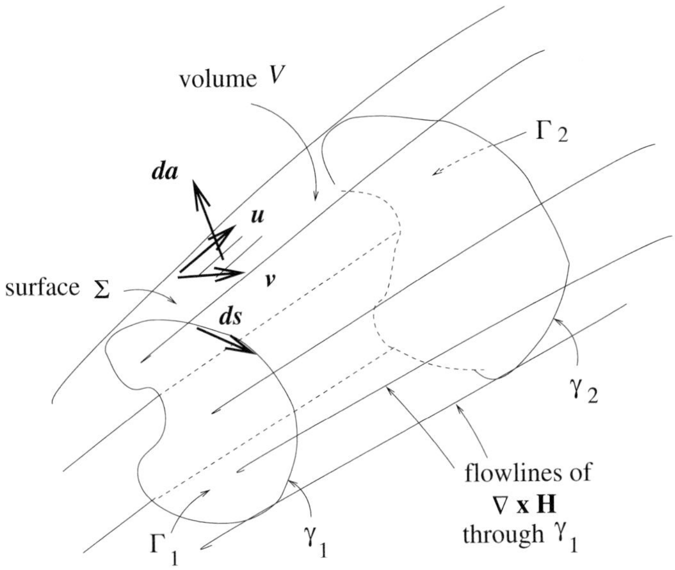

Fig. 17.6 A "tube" formed by flowlines of $\nabla \times \mathbf{H}$ passing through closed curve γ_1. The part Σ between γ_1 and another closed curve γ_2 around the same tube forms a closed volume when it is "capped" by surfaces Γ_1 bounded by γ_1 and Γ_2 bounded by γ_2.

other end by a surface Γ_2 bounded by γ_2 to form a closed volume V. Because it is a curl, the vector $\nabla \times \mathbf{H}$ satisfies

$$\nabla \cdot (\nabla \times \mathbf{H}) = 0 \tag{17.96}$$

throughout the volume, and it then follows from Gauss's theorem that

$$\Big(\int_{S_1} + \int_{S_2} + \int_{\Sigma} \Big) (\nabla \times \mathbf{H}) \cdot \mathbf{da} = \int_V \nabla \cdot (\nabla \times \mathbf{H})\, dV = 0, \tag{17.97}$$

where dV is a volume differential and $\mathbf{da}$ is a normal, outward-directed, surface area differential. By construction the integrand vanishes everywhere on the surface Σ.[6] Then, applying Stokes' theorem again yields

$$\oint_{\gamma_1} \mathbf{H} \cdot \mathbf{ds} = \oint_{\gamma_2} \mathbf{H} \cdot \mathbf{ds}. \tag{17.98}$$

Arnold refers to this as "Stokes' lemma." Poincaré introduced the terminology "relative integral invariant" for such quantities. Since $\mathbf{H}$ can be any (smooth) vector, the result is purely mathematical and does not necessarily have anything to do with the "source" of $\mathbf{H}$.

6) Later in the chapter there will be an analogous "surface" integral, whose vanishing will be similarly essential.

This same mathematics is important in hydrodynamics where **H** is the velocity of fluid flow and the vector $\nabla \times \mathbf{H}$ is known as the "vorticity;" its flow lines are known as "vorticity lines" and the tube formed from these lines is known as a "vorticity tube." This terminology has been carried over into symplectic mechanics. One reason this is being mentioned is to point out the potential for this terminology to be misinterpreted. The terminology is in one way apt and in another way misleading. What would be misleading would be to think of **H** as in any way representing particle velocity even though **H** stands for velocity in hydrodynamics. What *is* apt though is to think of **H** as being like a static "magnetic field," or rather to think of $\mathbf{J} = \nabla \times \mathbf{H}$ as the static "current density" that would cause **H**. It is the flow lines of **J** that are to be thought as the analog of the configuration space flow lines of a mechanical system. These are the lines that will be called vortex lines and will form vortex tubes. **H** tends to wrap around the flow lines and Ampére's law relates its "circulation" $\oint_\gamma \mathbf{H} \cdot \mathbf{ds}$ for various curves γ linked by the vortex tube.

17.4.2
The Poincaré–Cartan Integral Invariant

Having identified these potential hazards, we boldly apply the same reasoning to mechanics as we applied in deriving the principle of least time in optics.

In the space of **q**, **p** and t – known as the *time-extended* or simply *extended* phase space – we continue to analyze the set of system trajectories describable by function $S(\mathbf{q}, t)$ satisfying the Hamilton–Jacobi equation. The "gradient" relations of Eq. (8.12) were $\partial S/\partial t = -H$ and $\partial S/\partial q^i = p_i$. If we assume that S is single-valued, it follows that the integral from $P_1 : (\mathbf{q}_{(1)}, t_1)$ to $P : (\mathbf{q}, t)$,

$$\text{I.I.} = \int_{P_1}^{P} (p_i \widetilde{\mathbf{dq}}^i - H\widetilde{\mathbf{dt}}), \tag{17.99}$$

which measures the change in S in going from P_1 to P, is independent of path. This is called the "Poincaré–Cartan integral invariant" which, for brevity we designate by I.I. The integration path is a curve in "extended configuration space" which can also be regarded as the projection onto the extended coordinate space of a curve in extended phase space; it need not be a physically realizable orbit, but the functions p_i and H must correspond to a particular function S such as in Eq. (8.12). Unfortunately it will turn out that the requirement that S be nonsingular and single-valued throughout space is too restrictive in practice and a more careful statement of the invariance of I.I. is

$$\int_{\gamma_1} (p_i \widetilde{\mathbf{dq}}^i - H\widetilde{\mathbf{dt}}) = \int_{\gamma_2} (p_i \widetilde{\mathbf{dq}}^i - H\widetilde{\mathbf{dt}}) \tag{17.100}$$

where the integration paths γ_1 and γ_2 are closed (in phase space, though not necessarily in time-extended phase space) and encircle the same tube of system trajectories.

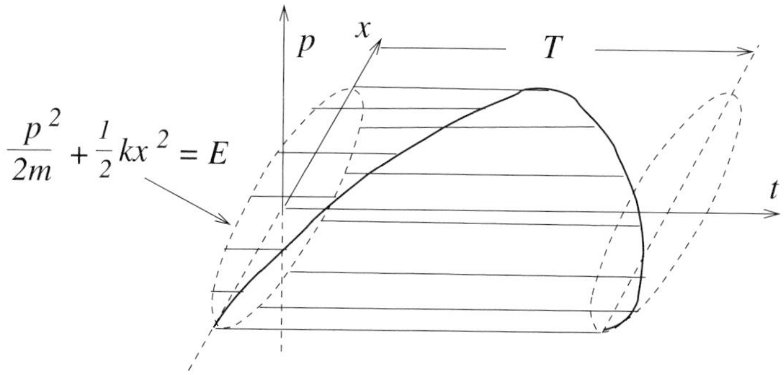

Fig. 17.7 Extended phase space for a one-dimensional simple harmonic oscillator. The heavy curve is a valid system trajectory and also a possible path of integration for the evaluation the Poincaré–Cartan integral invariant.

The evaluation of I.I. for a one-dimensional harmonic oscillator is illustrated in Fig. 17.7 – in this case the solid curve *is* a valid system path in extended phase space. Because the form in the integrand is expanded in terms of coordinates, the differential form $\widetilde{\mathbf{dq}}$ can be replaced by ordinary differential dx. Energy conservation in simple harmonic motion is expressed by

$$\frac{p^2}{2m} + \frac{1}{2}kx^2 = E, \tag{17.101}$$

as the figure illustrates. This is the equation of the ellipse which is the projection of the trajectory onto a plane of constant t. Its major and minor axes are $\sqrt{2mE}$ and $\sqrt{2E/k}$. Integration of the first term of Eq. (17.99) yields

$$\oint p(x)\,dx = \iint dp\,dx = \pi\sqrt{2mE}\,\sqrt{2E/k} = 2\pi E\sqrt{m/k} = ET, \tag{17.102}$$

since the period of oscillation is $T = 2\pi\sqrt{m/k}$. The second term of Eq. (17.99) is especially simple because $H = E$ and it yields $-ET$. Altogether I.I. $= 0$.

If the path defining I.I. is restricted to a hyperplane of fixed time t, like curve γ_1 in Fig. 17.8, then the second term of (17.99) vanishes. If the integral is performed over a closed path γ, the integral is called the "Poincaré *relative* integral invariant" R.I.I.

$$\text{R.I.I.}(t) = \oint_\gamma p_i \widetilde{\mathbf{dq}}^i. \tag{17.103}$$

This provides an invariant measure of the tube of trajectories bounded by curve γ_1 and illustrated in Fig. 17.8. Using the differential form terminology of Section 4.3.4, this quantity is written as

$$\text{R.I.I.}(t) = \oint_\gamma \widetilde{\mathbf{p}}(t), \tag{17.104}$$

and is called the circulation of $\mathbf{p}(t)$ about γ.

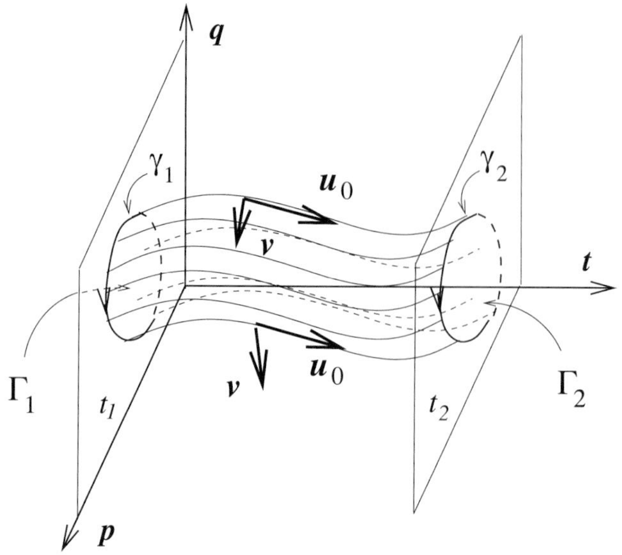

Fig. 17.8 A bundle of trajectories in extended phase space, bounded at time t_1 by curve γ_1. The constancy of R.I.I., the Poincaré relative integral invariant expresses the equality of line integrals over γ_1 and γ_2. This provides an invariant measure of the tube of trajectories bounded by curve γ_1.

Since this integral is performed over a closed path its value would seem to be zero under the conditions hypothesized just before Eq. (17.99). But we have found its value to be $2\pi E\sqrt{m/k}$, which seems to be a contradiction. Clearly the R.I.I. acquires a nonvanishing contribution because S is not single-valued in a region containing the integration path. Looking at Eq. (8.50), obtained as the Hamilton–Jacobi equation was being solved for this system, one can see that the quantity $\partial S_0/\partial q$ is doubly defined for each value of q. This invalidates any inference about the R.I.I. integral that can be drawn from Eqs. (8.12). This shows that, though the Hamilton–Jacobi gradient relations for $\mathbf{p}$ and H provide an excellent mnemonic for the integrands in the I.I. integral, it is not valid to infer integral invariance properties from them.

17.5 Invariance of the Poincaré–Cartan Integral Invariant I.I.

This section depends on essentially all the geometric concepts that have been introduced in the text. Since this makes it particularly difficult it should perhaps only be skimmed initially. But, because the proof of Liouville's theorem and its generalizations, probably the most fundamental results in classical mechanics, and the method of canonical transformation depend on the proof, the section cannot be said to be unimportant. Arnold has shown that the more elementary treatment of this topic by Landau and Lifshitz is incorrect. Other texts, such as Goldstein, do not go beyond proving a special case of Liouville's theorem even though it just scratches the surface of the rigorous demands that being symplectic places on mechanical systems.

17.5.1 The Extended Phase Space 2-Form and its Special Eigenvector

Equation (17.104) shows that the integral appearing in I.I. is the circulation of a 1-form $\widetilde{\boldsymbol{\omega}}^{(1)}$. To analyze it using the analog of Stokes' lemma, it is necessary to define the vortex tube of a 1-form. This requires first the definition of a vortex line of a 1-form. We start by finding the exterior derivative $\widetilde{\mathbf{d}}\widetilde{\boldsymbol{\omega}}^{(1)}$ as defined in Eq. (2.42). To make this definite let us analyze the "extended momentum 1-form"

$$\widetilde{\boldsymbol{\omega}}_E^{(1)} = p_i\,\widetilde{\mathbf{dq}}^i - H\,\widetilde{\mathbf{dt}}, \tag{17.105}$$

which is summed on i and is, in fact, the 1-form appearing in I.I. This is the canonical coordinate version of the standard momentum 1-form with the 1-form $H\,\widetilde{\mathbf{dt}}$ subtracted. In this case the integral is to be evaluated along an $n+1$-dimensional curve in the $2n+1$-dimensional, time-extended phase space. For all the apparent similarities between Figs. 17.6 and 17.8, there are important differences, with the most important one being that the abscissa axis in the latter is the time t. Since all the other axes come in canonical conjugate pairs the dimensionality of the extended phase space is necessarily odd. A 2-form $\widetilde{\boldsymbol{\omega}}_E^{(2)}$ can be obtained by exterior differentiation of $-\widetilde{\boldsymbol{\omega}}_E^{(1)}$ as in Eq. (2.42);

$$\widetilde{\boldsymbol{\omega}}_E^{(2)} = -\widetilde{\mathbf{d}}\,\widetilde{\boldsymbol{\omega}}_E^{(1)} = \widetilde{\mathbf{dq}}^i \wedge \widetilde{\mathbf{dp}}_i + \frac{\partial H}{\partial q^i}\,\widetilde{\mathbf{dq}}^i \wedge \widetilde{\mathbf{dt}} + \frac{\partial H}{\partial p_i}\,\widetilde{\mathbf{dp}}_i \wedge \widetilde{\mathbf{dt}}. \tag{17.106}$$

As in Eq. (17.46), this 2-form can be converted into a 1-form $\widetilde{\mathbf{z}}_E = \widetilde{\boldsymbol{\omega}}_E^{(2)}(\vec{\mathbf{z}}_E,\cdot)$ by applying it to an arbitrary extended phase-space displacement vector

$$\vec{\mathbf{z}}_E = \begin{pmatrix} dq^1 & dp_1 & dq^2 & dp_2 & dt \end{pmatrix}^T. \tag{17.107}$$

Then one can define an extended skew-scalar product of two vectors;

$$[\vec{\mathbf{z}}_{Eb}, \vec{\mathbf{z}}_{Ea}]_E = \widetilde{\boldsymbol{\omega}}_E^{(2)}(\vec{\mathbf{z}}_{Eb}, \vec{\mathbf{z}}_{Ea}). \tag{17.108}$$

This can in turn be expressed as a quadratic form as in Eq. (17.47).

$$[\vec{\mathbf{z}}_{Eb}, \vec{\mathbf{z}}_{Ea}]_E = \begin{pmatrix} dq_b^1 \\ dp_{b1} \\ dq_b^2 \\ dp_{b2} \\ dt_b \end{pmatrix}^T \begin{pmatrix} 0 & 1 & 0 & 0 & -\partial H/\partial q^1 \\ -1 & 0 & 0 & 0 & -\partial H/\partial p_1 \\ 0 & 0 & 0 & 1 & -\partial H/\partial q^2 \\ 0 & 0 & -1 & 0 & -\partial H/\partial p_2 \\ \frac{\partial H}{\partial q^1} & \frac{\partial H}{\partial p_1} & \frac{\partial H}{\partial q^2} & \frac{\partial H}{\partial p_2} & 0 \end{pmatrix} \begin{pmatrix} dq_a^1 \\ dp_{a1} \\ dq_a^2 \\ dp_{a2} \\ dt_a \end{pmatrix}. \tag{17.109}$$

The partial derivatives occurring as matrix elements are evaluated at the particular point in phase space that serves as origin from which the components in the vectors are reckoned.

Problem 17.5.1. *Show that the determinant of the matrix in Eq. (17.109) vanishes but that the rank of the matrix is 4. Generalizing to arbitrary dimensionality n, show that the corresponding determinant vanishes and that the rank of the corresponding matrix is $2n$.*

Accepting the result of the previous problem the determinant vanishes and, as a result, it is clear that zero is one of the eigenvalues. One confirms this immediately by observing that the vector

$$\mathbf{u}_E^{(H)} = \begin{pmatrix} \frac{\partial H}{\partial p_1} & -\frac{\partial H}{\partial q^1} & \frac{\partial H}{\partial p_2} & -\frac{\partial H}{\partial q^2} & 1 \end{pmatrix}^T \tag{17.110}$$

(or any constant multiple of this vector) is an eigenvector of the matrix in Eq. (17.109) with eigenvalue 0. Furthermore, one notes from Hamilton's equations that this vector is directed along the unique curve through the point under study.

It has been established then, that the vector $\mathbf{u}_E^{(H)}$, because it is an eigenvector with eigenvalue 0, has the property that

$$[\mathbf{u}_E^{(H)}, \mathbf{w}_E] = 0 \tag{17.111}$$

for arbitrary vector $\mathbf{w}_E$.

Recapitulating, it has been shown that the Hamiltonian system evolves in the direction given by the eigenvector of the $(2n+1) \times (2n+1)$ matrix derived from the 2-form $\widetilde{d}\,\widetilde{p}_E$. This has been demonstrated explicitly only for the case $n = 2$, but it is not difficult to extend the arguments to spaces of arbitrary dimension. Also, though specific coordinates were used in the derivation, they no longer appear in the statement of the result.

17.5.2
Proof of Invariance of the Poincaré Relative Integral Invariant

Though we have worked only on a particular 2-form we may apply the same reasoning to derive the following result known as Stokes' lemma. Suppose that $\widetilde{\boldsymbol{\omega}}^{(2)}$ is an arbitrary 2-form in a $2n+1$ odd-dimensional space. For reasons hinted at already, we start by seeking a vector $\mathbf{u}_0$ having the property that $\widetilde{\boldsymbol{\omega}}^{(2)}(\mathbf{u}_0, \mathbf{v}) = 0$ for arbitrary vector $\mathbf{v}$. As before, working with specific coordinates, we can introduce a matrix $\mathbf{A}$ such that the skew scalar product of vectors $\mathbf{u}$ and $\mathbf{v}$ is given by

$$\widetilde{\boldsymbol{\omega}}^{(2)}(\mathbf{u}, \mathbf{v}) = \mathbf{A}\mathbf{u} \cdot \mathbf{v}. \tag{17.112}$$

Problem 17.5.2. *Following Eqs. (17.46), show that the matrix* $\mathbf{A}$ *is antisymmetric. Show also that the determinant of an arbitrary matrix and its transpose are equal, and also that, if it is odd-dimensional, changing the signs of every element has the effect of changing the sign of its determinant. Conclude therefore that* $\mathbf{A}$ *has zero as one eigenvalue.*

Accepting the result of the previous problem, we conclude (if the stated conditions are met) that a vector $\mathbf{u}_0$ can be found, for arbitrary $\mathbf{v}$, such that

$$\widetilde{\boldsymbol{\omega}}^{(2)}(\mathbf{u}_0, \mathbf{v}) = 0. \tag{17.113}$$

This relation will be especially important when $\widetilde{\boldsymbol{\omega}}^{(2)}$ serves as the integrand of an area integral as in Eq. (4.77) and the vector $\mathbf{u}_0$ lies in the surface over which the integration is being performed, since this will cause the integral to vanish.

Vortex Lines of a 1-Form: If the 2-form $\widetilde{\boldsymbol{\omega}}^{(2)}$, for which the vector $\mathbf{u}_0$ was just found, was itself derived from an arbitrary 1-form $\widetilde{\boldsymbol{\omega}}^{(1)}$ according to $\widetilde{\boldsymbol{\omega}}^{(2)} = \widetilde{\mathbf{d}}\,\widetilde{\boldsymbol{\omega}}^{(1)}$, then the flow lines of $\mathbf{u}_0$ are said to be the "vortex lines" of $\widetilde{\boldsymbol{\omega}}^{(1)}$.

We now wish to employ Stokes' theorem for forms, (4.76), to a vortex tube such as is shown in Fig. 17.8. The curve γ_1 can be regarded, on the one hand, as bounding the surface Γ_1 and, on the other hand, as bounding the surface consisting of both Σ (formed from the vortex lines) and the surface Γ_2 bounded by γ_2. Applying Stokes' theorem to curve γ_1, the area integrals for these two surfaces are equal. But we can see from the definition of the vortex lines, that there is no contribution to the area integral coming from the area Σ. (The vortex lines belong to $\widetilde{\boldsymbol{\omega}}^{(1)}$), and the integrand is $\widetilde{\mathbf{d}}\,\widetilde{\boldsymbol{\omega}}^{(1)}$ and the grid by which the integral is calculated can be formed from differential areas each having one side aligned with a vortex line. Employing Eq. (17.113), the contribution to the integral from every such area vanishes.) We conclude therefore that

$$\oint_{\gamma_1} \widetilde{\boldsymbol{\omega}}^{(1)} = \oint_{\gamma_2} \widetilde{\boldsymbol{\omega}}^{(1)}. \tag{17.114}$$

This is known as "Stokes' lemma for forms."

It is the vanishing of the integral over Σ that has been essential to this argument and this is supposed to be reminiscent of the discussion given earlier of Ampère's law in electromagnetism. All that was required to prove that law was the vanishing of a surface integral and the argument has been repeated here.

We again specialize to phase space and consider a vortex tube belonging to the extended momentum 1-form $p_i\widetilde{\mathbf{dq}}^i - H\widetilde{dt}$. The vortex lines for this form are shown in Fig. 17.8 and we have seen that these same curves are valid trajectories of the Hamiltonian system. This puts us in a position to prove

$$\text{R.I.I.} = \oint_\gamma p_i\widetilde{\mathbf{dq}}^i = \text{independent of time.} \tag{17.115}$$

In fact the proof has already been given, because Eq. (17.114) implies Eq. (17.115). This completes the proof of the constancy in time of the Poincaré relative integral invariant R.I.I.

The constancy of R.I.I. is closely related to the invariance of $\sum\widetilde{dp_i} \wedge \widetilde{dq}^i$ under coordinate transformations, which was shown earlier. The new result is that the system evolution in time preserves the invariance of this phase-space area. This result is most readily applicable to the case in which many noninteracting systems are represented on the same figure, and the curve γ encloses all of them. Since points initially within the tube will remain inside, and the tube area is preserved, the density of particles is preserved.

The dimensionality of R.I.I. is

$$\begin{aligned}[\text{R.I.I.}] &= [\text{arbitrary}] \times \frac{[\text{energy}]}{[\text{arbitrary}/\text{time}]} \\ &= [\text{energy} \times \text{time}] = [\text{action}].\end{aligned} \tag{17.116}$$

Knowing that Planck's constant h is called "the quantum of action" one anticipates connections between this invariant and quantum mechanics. Pursuit of this connection led historically to the definition of *adiabatic invariants* as physical quantities subject to quantization. It has been shown in Chapter 14 that R.I.I. is an adiabatic invariant.

17.6 Symplectic System Evolution

According to Stokes' theorem for forms, an integral over surface Γ is related to the integral over its bounding curve γ by

$$\oint_\gamma \widetilde{\omega} = \int_\Gamma \widetilde{\mathbf{d}\omega}. \tag{17.117}$$

Also, as in Eq. (17.5.2), we have

$$\widetilde{\mathbf{d}}\,(p_i\widetilde{\mathbf{dq}}^i) = -\widetilde{\mathbf{dq}}^i \wedge \widetilde{\mathbf{dp}}_i. \tag{17.118}$$

With $\widetilde{\boldsymbol{\omega}} = p_i\widetilde{\mathbf{q}}^i$ these relations yields

$$\text{R.I.I.} = \oint_\gamma p_i\widetilde{\mathbf{dq}}^i = -\int_\Gamma \widetilde{\mathbf{dq}}^i \wedge \widetilde{\mathbf{dp}}_i. \tag{17.119}$$

Since the left-hand side is an integral invariant, so also is the right-hand side. Because it is an integral over an *open* region, the latter integral is said to be an *absolute* integral invariant, unlike R.I.I., which is a *relative* integral invariant because its range is closed. It is not useful to allow the curve γ of R.I.I. to become infinitesimal, but it *is* useful to extract the integrand of the absolute integral invariant in that limit, noting that it is the same quantity that has previously been called the canonical 2-form

$$\widetilde{\mathbf{dq}}^i \wedge \widetilde{\mathbf{dp}}_i \equiv \text{canonical 2-form } \widetilde{\boldsymbol{\omega}} = \text{ invariant.} \tag{17.120}$$

The "relative/absolute" terminology distinction does not seem particularly helpful to me, but the invariance of the canonical 2-form *does* lead immediately to the conclusion that the evolution of a Hamiltonian system can be represented by a symplectic transformation.

For the simple harmonic oscillator the R.I.I. was derived in Eq. (17.102) using

$$\oint p(x)\,dx = \iint dp\,dx. \tag{17.121}$$

Two important comments can be based on this formula. One is that, for area integrals in a plane, the relation (17.119) here reduces to the formula familiar from elementary calculus by which areas (two dimensional) are routinely evaluated by one-dimensional integrals. The other result is that the phase-space area enclosed is independent of time. Because this system is simple enough to be analytically solvable, the constancy of this area is no surprise, but for more general systems this is an important result.

One visualizes any particular mechanical system as one of a cloud of noninteracting systems each one represented by one point on the surface Γ of Eq. (17.119). Such a distribution of particles can be represented by a surface number density, which we may as well regard as uniform, since Γ can be taken arbitrarily small. (For a relative integral invariant there would be no useful similar limit.) As time increases the systems move, always staying in the region $\Gamma(t)$ internal to the curve $\gamma(t)$ formed by the systems that were originally on the curve γ. (It might be thought that points in the interior could in time change places with points originally on γ but that would require phase-space trajectories to cross, which is not allowed.)

Consider systems close to a reference system that is initially in configuration $\mathbf{z}(0)$ and later at $\mathbf{z}(t)$. Then let $\mathbf{\Delta z}(t)$ be the time-varying displacement of a general system relative to the reference system. By analogy with Eq. (17.43), the evolution can be represented by

$$\mathbf{\Delta z}(t) = \mathbf{M}(t)\,\mathbf{\Delta z}(0). \tag{17.122}$$

We can now use the result derived in Section 17.2.2. As defined in Eq. (17.39), the skew-scalar product $[\mathbf{z}_a(t), \mathbf{z}_b(t)]$ formed from two systems evolving according to Eq. (17.122), is the quantity R.I.I. discussed in the previous section. To be consistent with this invariance, the matrix $\mathbf{M}(t)$ has to be symplectic.

17.6.1 Liouville's Theorem and Generalizations

In Sections 4.2.2 and 4.2.3 the geometry of bivectors and multivectors was discussed. This discussion can be carried over, including Fig. 4.4, to the geometry of the canonical 2-form. Consider a two-rowed matrix

$$\begin{pmatrix} q_a^1 & p_a^1 & q_a^2 & p_a^2 & \cdots & q_a^n & p_a^n \\ q_b^1 & p_b^1 & q_b^2 & p_b^2 & \cdots & q_b^n & p_b^n \end{pmatrix}, \tag{17.123}$$

whose elements are the elements of phase-space vectors $\mathbf{z}_a$ and $\mathbf{z}_b$. By picking two columns at a time from this matrix and evaluating the determinants one forms the elements x^{ij} of a bivector $\mathbf{z}_a \wedge \mathbf{z}_b$;

$$\begin{aligned} x^{12} &= -x^{21} = \begin{vmatrix} q_a^1 & p_a^1 \\ q_b^1 & p_b^1 \end{vmatrix}, \\ x^{13} &= -x^{31} = \begin{vmatrix} q_a^1 & q_a^2 \\ q_b^1 & q_b^2 \end{vmatrix}, \qquad x^{23} = -x^{32} = \begin{vmatrix} p_a^2 & q_a^2 \\ p_a^2 & q_a^2 \end{vmatrix}, \end{aligned} \tag{17.124}$$

etc. By introducing p vectors and arraying their elements in rows one can form p-index multivectors similarly. As in Eq. (4.19), after introducing a metric tensor g^{ij} and using it to produce covariant components $x_{ij\ldots k}$ one can define an area" or "volume" (as the case may be) V by

$$V_{(p)}^2 = \frac{1}{p!} x_{ij\ldots k} x^{ij\ldots k} = \det \begin{vmatrix} [\mathbf{z}_1, \mathbf{z}_1] & [\mathbf{z}_1, \mathbf{z}_2] & \cdots & [\mathbf{z}_1, \mathbf{z}_p] \\ [\mathbf{z}_2, \mathbf{z}_1] & [\mathbf{z}_2, \mathbf{z}_2] & \cdots & [\mathbf{z}_2, \mathbf{z}_p] \\ \cdots & \cdots & \cdots & \cdots \\ [\mathbf{z}_p, \mathbf{z}_1] & [\mathbf{z}_p, \mathbf{z}_2] & \cdots & [\mathbf{z}_p, \mathbf{z}_p] \end{vmatrix}. \tag{17.125}$$

We have used the notation of Eq. (17.39) to represent the skew-invariant products of phase-space vectors. For $p = 1$ we obtain

$$V_{(1)}^2 = z_i z^i = \det \big| [\mathbf{z}, \mathbf{z}] \big| = 0; \tag{17.126}$$

like the skew-invariant product of any vector with itself, it vanishes. For $p = 2$ we obtain

$$V_{(2)}^2 = \frac{1}{p!} x_{ij\ldots k} x^{ij\ldots k} = \det \begin{vmatrix} 0 & [\mathbf{z}_1, \mathbf{z}_2] \\ [\mathbf{z}_2, \mathbf{z}_1] & 0 \end{vmatrix} = [\mathbf{z}_1, \mathbf{z}_2]^2. \tag{17.127}$$

If the vectors $\mathbf{z}_i$ represent (time-varying) system configurations we have seen previously that the elements of the matrix in Eq. (17.125), such as

$$[\mathbf{z}_a, \mathbf{z}_b] = (q_a^1 p_b^1 - q_b^1 p_a^1) + (q_a^2 p_b^2 - q_b^2 p_a^2) + \cdots \tag{17.128}$$

are invariant. (As shown in Fig. 17.2, the first term in this series can be interpreted as the area defined by the two vectors after projection onto the q^1, p_1 plane, and similarly for the other terms.) Since its elements are all invariant it follows that $V_{(p)}$ is also invariant. In Section 4.2.3. this result was called "the Pythagorean relation for areas." One should not overlook the fact that, though the original invariant given by Eq. (17.128) is a *linear* sum, of areas, the new invariants given by Eq. (17.125), are *quadratic* sums. The former result is a specifically symplectic feature while the new invariants result from metric (actually skew-metric in our case) properties. A device to avoid forgetting the distinction is always to attach the adjective Pythagorean to the quadratic sums.

By varying p we obtain a sequence of invariants. For $p = 2$ we obtain the original invariant, which we now call $V_{(2)} = [\mathbf{z}_1, \mathbf{z}_2]$. Its (physical) dimensionality is [action] and the dimensionality of $V_{(2p)}$ is $[\text{action}]^p$. The sequence terminates at $p = 2n$ since, beyond there, all multivector components vanish, and for $p = n$, except for sign, all multivector components have the same value. Considering $n = 2$ as an example, the phase space is 4-dimensional and the invariant is

$$V_{(4)}^2 = \frac{1}{4!} x_{ijkl} x^{ijkl} = \det \begin{vmatrix} 0 & [\mathbf{z}_1, \mathbf{z}_2] & [\mathbf{z}_1, \mathbf{z}_3] & [\mathbf{z}_1, \mathbf{z}_4] \\ [\mathbf{z}_2, \mathbf{z}_1] & 0 & [\mathbf{z}_2, \mathbf{z}_3] & [\mathbf{z}_2, \mathbf{z}_4] \\ [\mathbf{z}_3, \mathbf{z}_1] & [\mathbf{z}_3, \mathbf{z}_2] & 0 & [\mathbf{z}_3, \mathbf{z}_4] \\ [\mathbf{z}_4, \mathbf{z}_1] & [\mathbf{z}_4, \mathbf{z}_2] & [\mathbf{z}_4, \mathbf{z}_3] & 0 \end{vmatrix}. \tag{17.129}$$

If the vectors have been chosen so the first two lie in the q^1, p^1 plane and the last two lie in the q^2, p^2 plane, the matrix elements in the upper right and lower left quadrants vanish and $V_{(4)}$ is equal to the product of areas defined by the first, second and third, fourth pairs. This is then the "volume" defined by the four vectors.

It is the invariance of this volume that is known as Liouville's theorem. If noninteracting systems, distributed uniformly over a small volume of phase space are followed as time advances, the volume they populate remains constant. Since their number is constant, their number density is also constant.

Hence one also states Liouville's theorem in the form *the density of particles in phase space is invariant if their evolution is Hamiltonian.* Liouville's theorem itself could have been derived more simply, since it follows from the fact that the determinant of a symplectic matrix is 1. But obtaining the other invariants requires the multivector algebra.

Bibliography

General References

1 V.I. Arnold, *Mathematical Methods of Classical Mechanics,* 2nd ed., Springer, Berlin, 1989.

2 C. Lanczos, *The Variational Principles of Mechanics,* University of Toronto Press, Toronto, 1949.

Index

Geometric Mechanics: Toward a Unification of Classical Physics. 2nd Edition. Richard Talman

ISBN: 978-3-527-40683-8

a

m

n

q

r

t

u

Related Titles

Bayin, S.

Mathematical Methods in Science and Engineering

2006. Hardcover

ISBN: 978-0-470-04142-0

Kusse, B., Westwig, E. A.

Mathematical Physics

Applied Mathematics for Scientists and Engineers

2005. Softcover

ISBN: 978-3-527-40672-2

Eckert, M.

The Dawn of Fluid Dynamics

A Discipline between Science and Technology

2005. Hardcover

ISBN: 978-3-527-40513-8

Heard, W.B.

Rigid Body Mechanics

Mathematics, Physics and Applications

2006. Softcover

ISBN: 978-3-527-40620-3

McCall, M.W.

Classical Mechanics – A Modern Introduction

2000. Hardcover

ISBN: 978-0-471-49711-0

Moon, F.C.

Applied Dynamics

With Applications to Multibody and Mechatronic Systems

1998. Hardcover

ISBN: 978-0-471-13828-0